Metric Equivalents

Linear Units

25.4 millimeters (mm) = 1 inch
1 millimeter = .03937 inches
1 meter = 3.281 feet
1 meter = 1.094 yards
1 centimeter (cm) = .03281 feet
1 cm = .3937 inches
1 cm = .01 meter (m)
1 cm = 10 millimeters
1 kilometer (km) = 3281 feet
1 km = .6214 miles
1 km = 1094 yards

Area

1 square centimeter (cm^2) = .001076 square ft ($ft.^2$)
1 square centimeter (cm^2) = .1550 square inches ($in.^2$)
1 square meter (m^2) = 10.76 $ft.^2$
1 square meter = 1.196 square yards ($yd.^2$)

Angular

1 radian = 57.30 degrees
1 radian = 3438 minutes
1 degree (angular) = .01745 radians

Weight

1 pound (lb) = 453.59237 grams (g)
1 kilogram (kg) = 2.205 pounds
1 kilogram = .001102 tons

Volume

1 liter (l) = 1000 cubic centimeters (cm^3)
    = .03531 cubic feet ($ft.^3$)
    = 61.02 cubic inches ($in.^3$)
    = .001 cubic meters (m^3)
    = .2642 gallons
    = 1.057 quarts
1 cubic centimeter (cm^3) = .001 liter (milliliter or ml)
1 cubic foot ($ft.^3$) = 28.32 liters
1 cm^3 = .06102 $in.^3$
1 cubic inch ($in.^3$) = .01639 liters
1 gallon = 3785 cm^3
1 gallon = 3.785 liters

Metric Prefixes and Equivalents

Metric System Prefixes

Tera (T)	1,000,000,000,000 (trillion)
Giga (G)	1,000,000,000 (billion)
Mega (M)	1,000,000 (million)
Kilo (K)	1,000 (thousand)
Hecto (H)	100 (hundred)
Deka (DA)	10 (ten)
Unit	1
Deci (d)	.1 (tenth)
Centi (c)	.01 (hundredth)
Milli (m)	.001 (thousandth)
Micro (mu)	.000001 (millionth)
Nano (n)	.000000001 (billionth)
Pico (p)	.000000000001 (trillionth)

Machine Tool Practices

ELEVENTH EDITION

RICHARD KIBBE

ROLAND MEYER

JON STENERSON

KELLY CURRAN

Vice President, Portfolio Management: Andrew Gilfillan
Portfolio Management Assistant: Lara Dimmick
Senior Vice President, Marketing: David Gesell
Marketing Coordinator: Elizabeth MacKenzie-Lamb
Director, Digital Studio and Content Production:
Brian Hyland
Managing Producer: Jennifer Sargunar
Content Producer (Team Lead): Faraz Sharique Ali
Content Producer: Deepali Malhotra
Manager, Rights Management: Johanna Burke

Operations Specialist: Deidra Smith
Cover Design: Pearson CSC
Cover Credit: Nordroden/Shutterstock
Full-Service Management and Composition:
Integra Software Service Pvt. Ltd.
Printer/Binder: LSC Communications,
Inc.
Cover Printer: LSC Communications, Inc.
Text Font: Times LT Pro

Credits and acknowledgments for materials borrowed from other sources and reproduced, with permission, in this textbook appear on the appropriate page within text.

Library of Congress Cataloging-in-Publication Data

Names: Kibbe, Richard R., author.
Title: Machine tool practices / Richard Kibbe, Roland Meyer, Warren White,
John Neely, Jon Stenerson, Kelly Curran.
Description: [Eleventh Edition] | Boston : PEARSON Prentice Hall, [2020] |
Includes index.
Identifiers: LCCN 2018028023| ISBN 9780134893501 | ISBN 0134893506
Subjects: LCSH: Machine-tools. | Machine-shop practice.
Classification: LCC TJ1185 .K458 2020 | DDC 621.9/02—dc23
LC record available at https://lccn.loc.gov/2018028023

5 2022

ISBN 10:   0-13-489350-6
ISBN 13: 978-0-13-489350-1

Contents

Preface

As a definitive text in the field for more than 30 years, *Machine Tool Practices*, 11th edition, is geared toward successfully training machinists and tool & die makers about quality, blueprint reading, traditional machining processes, and CNC operation and programming. It is ideal for those enrolled in apprenticeship training, technical college programs, community college programs, and university courses. Presented in a student-friendly manner, the book lends itself well to classes that take a combined lecture/laboratory approach, as well as to a self-paced instructional environment.

STRENGTHS AND UNIQUE SELLING POINTS

With hundreds of color illustrations and well over a thousand color pictures, *Machine Tool Practices* is the best-illustrated book in this field. The text emphasizes practical shop knowledge and machine tool technology throughout and superbly illustrates the tools, equipment, and techniques that students will encounter in an industrial machine shop.

CLASSICAL PRACTICE/CURRENT TRENDS

Machine tools and machining practices have changed dramatically. This text has been aligned with standards that were developed by the National Institute of Metalworking Skills (NIMS). The National Institute for Metalworking Skills (NIMS) was formed in 1995 to develop and maintain a globally competitive American workforce. NIMS developed skills standards for industry, NIMS certifies individual skills against the standards and also accredits training programs that meet NIMS requirements. This text was developed for students studying machining who need to acquire the knowledge and skills required by industry and to obtain NIMS certifications, if they so desire.

Machine Tool Practices has the information that is essential for the reader to be effective in all areas of machining. With the solid background this text provides, readers will confidently understand and operate manual and CNC machines as well as other manufacturing processes.

ORGANIZATION OF THE BOOK: TOTAL FLEXIBILITY TO SUIT YOUR TEACHING STYLE

The book is divided into 13 major sections and provides total flexibility to suit your teaching style. Appendix 1 contains Answers to Self-Tests, Appendix 2 offers practical General Tables, and Appendix 3 showcases Precision Vise Project Drawings. For the student, this project embodies many setups and techniques used in general precision machine shop work. The text also contains a Glossary and an Index. Many units are designed around specific projects that provide performance experience for students. The book structure makes it easy for instructors to include additional projects that are applicable to their specific needs.

NEW TO THIS EDITION

This edition has been dramatically updated and improved to reflect changes in the machining field. The eleventh edition includes the following improvements:

- Vast improvements in the readability of the text to make it easier to read and understand.

- We have added coverage of several topics that graduates are sure to encounter in the workplace. Coverage of process plans and job packets (routings) has been added to help the student understand what they will encounter in the workplace. Practical coverage of ISO systems, calibration and the machinist's role in them has also been added.

- Blueprint reading coverage has been completely rewritten and expanded with additional coverage and extensive student questions and exercises. Geometric Dimensioning and Tolerancing (GD&T) has been added along with extensive questions and student exercises. We believe that there is enough coverage of blueprint reading/GD&T that it could be the basis of a course if the instructor augments the material with additional worksheets, blueprints, and lecture to meet their individual needs. The supplements for this text, including the PowerPoints and the test bank, are additional materials that can be used for a BPR/GD&T class.

- CNC coverage has been completely rewritten and expanded. Coverage begins with the basics of CNC, then machining center programming followed by turning center programming. Coverage of canned cycles has been added for machining centers and turning centers. We believe that there is enough coverage in the CNC section that it could be the basis of a CNC course if the instructor augments the material with additional worksheets, blueprints, and lecture to meet their specific needs. The supplements for this text, including the PowerPoints and the test bank, provide additional materials that can be used for a CNC class.

- Speeds and feeds, carbide tooling, inserts and tool holders as well as their specification and selection have been dramatically expanded and improved.

- Extensive improvements in color photos and figures.

- Hundreds of color illustrations that ease comprehension and visually reinforce learning.

- Expanded self-test questions and exercises in each unit.

- A list of useful websites at the end of appropriate units that refer the reader to state-of-the-art information on cutting tools and machine shop equipment.

Guided Tour

Machine Tool Practices is divided into sections comprised of several units. We invite you to take the Guided Tour.

HALLMARK FEATURES

Introductory Overview

Introductions summarize and provide an overview of the main themes in each major section and help reinforce topics.

Objectives

Clearly stated objectives enable you to focus on what you should achieve by the end of each unit.

> **OBJECTIVES**
>
> After completing this unit, you should be able to:
>
> - Install and remove a bronze bushing using an arbor press.
> - Press on and remove a ball bearing from a shaft on an arbor press using the correct tools.
> - Press on and remove a ball bearing from a housing using an arbor press and correct tooling.
> - Install and remove a mandrel using an arbor press.
> - Install and remove a shaft with key in a hub using the arbor press.

Photographs

Extensive use of color photographs provides you with views of actual machining operations.

Figure B-19 Pressure being applied to straighten shaft.

Graphic Explanations

These detailed explanations highlight important concepts, common errors, and difficulties that machinists encounter.

Shop Tip

Shop Tip and **Shop Tip–Craftsmanship** boxes offer helpful tips and techniques to make the student a better and more intuitive machinist.

> ### SHOP TIP
>
> When reading any drawings in the machine shop, be sure to
>
> 1. Read any and all notes on the drawing.
> 2. Be sure that you are using the latest revision of a drawing. Errors can occur by machining parts to older drawing revisions. If you are unsure about the revision, check with your supervisor or instructor.

New Technology

Directs students to the latest technology in the field.

> ### NEW TECHNOLOGY
>
> Twist drills ordinarily do not produce a smooth hole. Often, reamers are used to produce a more dimensionally accurate hole with a smoother bore. However, this process will not produce a finish smooth enough for all purposes. Burnishing drills are available, designed not only to drill the hole, but also to finish it to size and increase the hole quality. They can be used for production drilling of cast iron, die-cast aluminum, and other nonferrous materials. Burnishing drills are presently used in the automotive, aerospace, compressor, and computer parts industries.

Professional Practice

Professional Practice provides tips from professional work environments.

> ### PROFESSIONAL PRACTICE
>
> The way a worker maintains his or her hand tools reveals the kind of machinist he or she is. Dirty, greasy, or misused tools carelessly thrown into a drawer are difficult to find or use the next time around. After a hand tool is used, it should be wiped clean with a shop towel and stored neatly in the proper place. If the tool was drawn from a tool room, the attendant may not accept a dirty tool.

Safety First

Safety First boxes provide safety warnings related to handling and working with various pieces of equipment.

> ### SAFETY FIRST
>
> The primary danger in operating the vertical band machine is accidental contact with the cutting blade. Workpieces are often hand guided. One advantage in sawing machines is that the pressure of the cut tends to hold the workpiece against the saw table. However, hands are often in close proximity to the blade. If you should contact the blade accidentally, an injury is almost sure to occur. You will not have time even to think about withdrawing your fingers before they are cut. Keep this in mind at all times when operating a band saw.
>
> Always use a pusher against the workpiece whenever possible. This will keep your fingers away from the blade. Be careful when you are about to complete the cut: As the blade clears through the work, the pressure that you are applying is suddenly released, and your hand or finger can be carried into the blade. As you approach the end of the cut, reduce the feeding pressure as the blade cuts through.

Operating Tip

Advice on how to operate machinery students may come across in their studies or careers.

Self-Test

End-of-unit self-tests gauge how well you mastered the material.

> ### SELF-TEST
>
> 1. What is the kerf?
> 2. What is the set on a saw blade?
> 3. What is the pitch of the hacksaw blade?
> 4. What determines the selection of a saw blade for a job?
> 5. Hand hacksaw blades fall into two basic categories. What are they?
> 6. What speed should be used in hand hacksawing?
> 7. Give four causes that make saw blades dull.
> 8. Give two reasons why hacksaw blades break.
> 9. A new hacksaw blade should not be used in a cut started with a blade that has been used. Why?
> 10. What dangers exist when a hacksaw blade breaks while it is being used?

Internet References

The end of each unit lists pertinent Internet sites.

> ### INTERNET REFERENCES
>
> http://en.wikipedia.org/wki/Engineering_drawing
> http://design-technology.info

COMPREHENSIVE TEACHING AND LEARNING PACKAGE

FOR THE INSTRUCTOR

Instructor's Guide with Lesson Plans

This handy manual contains suggestions on how to use the text for both conventional and competency-based education. The manual has additional student exercises for the instructor to utilize as class assignments. It also includes unit post-tests and answer keys (ISBN–10: 0-13-498588-5).

Blueprint and GD&T Exercises

Worksheets and exercises are provided that can be used by an instructor to supplement a blueprint/GD&T course. Worksheets and exercises are provided that cover sketching views, sketching parts from views, identification of views, and blueprint reading and GD&T reading prints.

CNC

Worksheets, exercises and CNC projects are provided for machining centers and turning centers that can be used by an instructor to supplement a CNC course.

PowerPoint Presentations

PowerPoint presentations are designed to aide the instructor's lecture on topics covered in the book (ISBN–10: 0-13-498580-X).

TestGen (Computerized Test Bank)

TestGen contains text-based questions in a format that enables instructors to select questions and create their own quizzes and tests (ISBN–10: 0-13-498587-7).

To access supplementary materials online, instructors need to request an instructor access code. Go to www.pearsonhighered.com/irc, where you can register for an instructor access code. Within 48 hours of registering, you will receive a confirming e-mail, including an instructor access code. Once you have received your code, log on to the site for full instructions on downloading the materials you wish to use.

Acknowledgments

The authors wholeheartedly thank the reviewers for the eleventh edition for their insight and feedback:

Larry Crain
West Kentucky Community & Technical College

Jason Dixon
Bakersfield College

Billy Graham
Northwest Technical Institute

Raymond A. Miller
University of Cincinnati

Samuel Obi
San Jose State University

John Shepherd
Mt. San Antonio College

Ed VanAvery
Mid Michigan Community College

The authors would also like to thank Chris Banyai-Riepl of OMAX Corporation for his major contribution to the content on waterjet technology.

The authors would also like to thank Sandra McLain of OMAX Corporation for her assistance.

The authors would like to thank the following companies and schools for their contributions:

The 600 Group Plc.

Aloris Tool Technology Co. Inc.

American Iron and Steel Institute

American Society of Mechanical Engineers

ArcelorMittal

ASM International®

ATTCO, Inc.

Barnes International, Inc.

Bazell Technologies

Besly Cutting Tools, Inc.

Bryant Grinder Corporation

Buck Chuck Company

California Community Colleges

Cinetic Landis Ltd.

Clausing Industrial, Inc.

Climax Portable Machine Tools, Inc.

CMPC Surface Finishes, Inc.

Cogsdill Tool Products, Inc.

Command Tooling Systems/EWS

Confederation College

Criterion Machine Works

Dake Corporation

Desmond-Stephan Manufacturing Company

DoALL Company

Enco Manufacturing Company

Engis Corporation

ERIX TOOL AB

Fadal Engineering

Fine Tools

Fox Valley Technical College

Haas Automation, Inc.

Hardinge Inc.

Harig Mfg. Corp.

HE&M Saw

Illinois Tool Works, Inc.

IMI Machine Tools Pvt. Ltd.

Ingersoll Cutting Tool

Kalamazoo Machine Tool

Kennametal, Inc.

K&M Industrial Machinery Co.

Lane Community College

Louis Levin & Son, Inc.

Lovejoy Tool Company, Inc.

MAG Giddings & Lewis, Inc.

MAG Industrial Automation Systems LLC

Magna-Lock USA, Inc.

Mahr Federal Inc.

Maximum Advantage–Carolinas

Mazak Corp.

Metal Web News

Micro-Mark

Mitsubishi Laser

Mitutoyo America Corp.

Monarch Lathes

Monarch Machine Tool Company

North American Tool Corporation

Okamoto Corporation

Okuma America Corporation

Olson Saw Company

Pacific Machinery & Tool Steel Co.

Peerless Chain Company

Rank Scherr-Tumico, Inc.

Regal Cutting Tools

Reishauer Corp.

Reko Automation & Machine Tool

Renishaw, Inc.

Renishaw Plc

SCHUNK Inc.

Sii Megadiamond, Inc.

Sipco Machine Company

Southwestern Industries, Inc.

Supfina Machine Company, Inc.

TE-CO INC.

The duMONT Company, LLC

The L.S. Starrett Co.

TRUARC Company LLC

Ultramatic Equipment Co.

United Grinding Walter-EWAG

Vannattabros.com

Vermont Gage

Walker Magnetics

Weldon Tool–A Dauphin Precision Tool Brand

Wilson® Instruments

Wilton Corporation

About the Authors

Richard Kibbe (Late) served his apprenticeship in the shipbuilding industry and graduated as a journeyman marine machinist. He holds an associate in arts degree in applied arts from Yuba Community College with an emphasis in machine tool technology. He also holds bachelor's and master's degrees from the California State University with an emphasis in machine tool manufacturing technology.

Mr. Kibbe has considerable machine shop experience as well as community college and industrial teaching experience and is the author and co-author of several publications in the machine tool manufacturing field.

Roland Meyer spent the first 20 years of his career in the metal-working industry as a tool and die maker and machinist in machine design and manufacturing. He completed his apprenticeship as a tool and die maker at Siemens in Germany and then worked in die shops in Toronto and Windsor, Canada, before moving to Chicago, where he worked as a gage maker at Ford Motor Company. He was in charge of the U.S. Army machine shops in Korea and Italy for five years. When he returned to the United States, he worked in a manufacturing company designing and building experimental machines used in the timber and plywood industry.

He next entered academia and became the lead instructor at Lane Community College's manufacturing technology program in Eugene, Oregon, where he taught for 25 years. As CNC became the new method in machining, he developed a CNC curriculum and program. When CAM became available, he also developed a state-of-the-art CAM program with the assistance of a local software company.

Jon Stenerson served an apprenticeship in Tool Making with Mercury Marine. He has a BBA from the University of Wisconsin-Oshkosh and a Masters Degree from the University of Wisconsin-Stout. He held certifications for Certified Quality Engineer, Certified Quality Auditor, and Certified Lead Auditor.

Jon is the author and co-author of several books in the machining and automation field. Jon spent many years teaching and developing self-paced machine tool and automation curriculum for Fox Valley Technical College.

Kelly Curran grew up in Michigan's Upper Peninsula where he started working in machine shops at a very young age. He holds an Associate of Applied Science degree from Ferris State College with an emphasis in machine tool technology. He also holds an Associates of Applied Science degree from the Northern Michigan University with an emphasis in Business and a Bachelor of Science Degree in Career, Technical Education from University of Wisconsin Stout.

Mr. Curran has considerable machine shop experience as well as industrial teaching experience and is the author and co-author of several publications in the machine tool manufacturing field. Mr. Curran has spent many years developing self-paced machine tool curriculum for the State of Wisconsin and Fox Valley Technical College.

SECTION A
Introduction

INTRODUCTION

Machining processes are among the most important of the manufacturing processes. The fundamental cutting processes in machining such as milling and turning are still the most prominent. What has changed is the way in which these processes are applied. Changes include the cutting tool materials, the materials the parts are made from, and the methods of material removal. These new methods include the use of lasers, electrical energy, electrochemical processes, and high-pressure water jets.

Computer numerically controlled (CNC) machines have provided an exceptional degree of accuracy, reliability, and repeatability. This application of computer-driven automated equipment is creating many employment opportunities for skilled machinists and machine operators.

The computer has found its way into almost every other phase of manufacturing as well. One important area is computer-aided design (CAD). Product design is done on computers and CNC programs are generated directly from those designs.

In order to be competitive, industry will continue to automate. There will be increased use of manufacturing cells of machines, and increased use of robotics for loading and unloading parts to/from CNC machines. The automation will continue to improve the working conditions in manufacturing, reduce the manual labor required, and increase the need for skilled and knowledgeable workers.

Units in This Section

The units in this section deal with careers, competitive manufacturing, safety, threads and fasteners, blueprint reading and GD&T.

Careers and the Machinist's Role in Process Plans

The increased use of computer technology and processes in industry has had a significant effect on the types and numbers of jobs available in manufacturing. Many exciting career opportunities are available in manufacturing and in machining in particular.

Manufacturing creates wealth. Manufacturing is crucial to economic success. There is a large demand for skilled manufacturing people. Wages and benefits for skilled manufacturing people are excellent. Wages and benefits for manufacturing are usually higher than other types of jobs. Working conditions have also dramatically changed. Most modern manufacturing facilities are clean, bright, and well organized. Modern computer-controlled equipment has dramatically changed the role of the machinist from manual dexterity and manual skills to more knowledge-based skills.

Helpers and Limited Machinists

A person's first job in manufacturing might be as a helper or limited machinist. They learn the trade from the floor up. They might be sweeping the floor or cutting stock to rough lengths for further machining. They are given additional responsibilities as they prove themselves in work. Remember that the supervisor is watching. The supervisor is looking to see if the new employee has work ethic and a good attitude. This determines whether they keep the job and how quickly they advance to becoming machinists.

Machine Operator/Production Machinist

A machine operator's responsibilities are to operate computer-driven (CNC) machine tools such as turning or machining centers. The operator observes machine functions and tool performance, inspects parts, and has limited duties in setting up and making minor changes to programs. Figure A-1 shows a machinist programming a part on a CNC turning center.

Machine operators should be familiar with basic machining processes, tooling selection and application, speeds and feeds, basic blueprint reading, basic math calculations, basic machine setup, and basic use of measurement tools. Machine operators are generally taught to use one or more CNC machine controls. Machine operators may receive training from a technical college, industrial training programs, or they may learn on the job.

Figure A-1 Machinist programming a part on a CNC turning center (*Courtesy of Fox Valley Technical College*).

Apprentice Machinist

The apprentice machinist learns the trade by entering a formal training program sponsored by a private company, a trade union, or a government entity. The period of training is typically four years and is a combination of on-the-job experiences and formal classroom education. Serving an apprenticeship represents one of the best methods of learning a skilled trade.

Journeyman Machinist

The journeyman machinist will have the capability to set up and operate most CNC equipment. A person becomes a journeyman by serving an apprenticeship. A machinist apprenticeship is generally four years long. States have apprenticeship divisions that facilitate apprenticeships. An apprenticeship is a contract between the apprentice, the company, and the state. The contract is called an indenture. The role of the state apprenticeship division is to make sure that the apprentice learns a broad range of knowledge and skills during the apprenticeship. The apprenticeship agreement will specify the number of hours to complete the apprenticeship. A four-year apprenticeship would be approximately 8,320 hours. The pay for the apprentice generally increases every six months until the apprentice finishes and gets full pay. The agreement also specifies the hours of required education. The apprentice must attend the required classes during the apprenticeship. Classes are generally one day a week during the normal school year. Classes are generally offered at a technical or community college. Employers might give a beginning apprentice credit for prior experience or classes already taken. This boosts their pay and shortens the length of time required to complete the apprenticeship. When an apprentice finishes, he/she is awarded a journeyman's card by the state. The card is proof that he or she has completed an apprenticeship. A journeyman's card is held in high regard by potential employers if the machinist wants to change employers.

Tool and Die Maker

A tool and die maker is usually an experienced machinist who served a tool and die apprenticeship. The tool and die apprenticeship is usually five years and requires more classes than a machinist apprenticeship. Tool and die makers may receive training through industrial apprenticeships and/or college and trade school programs. Although tool and die makers are often chosen only after several years of on-the-job experience, it is possible to start out in tool and die work through an apprenticeship program. Tool and die makers often receive premium pay for their work and are involved with many high-precision machining applications, tool design, material selection, metallurgy, and general manufacturing processes. Tool and die makers may be involved with making tooling and fixtures for automated processes and production.

Quality Control Technician

Quality control technicians are involved with quality control and inspection. They utilize common measurement equipment as well as computerized measurement machines to inspect parts to make sure they meet the specifications (see Figure A-2).

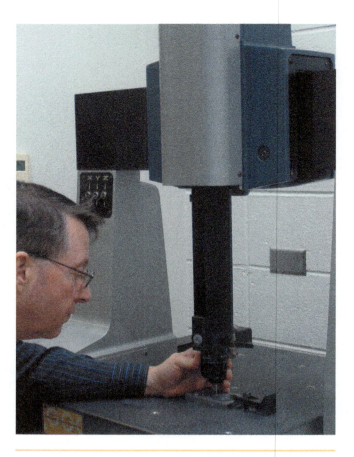

Figure A-2 Quality control technician using a CMM to inspect a part (*Courtesy of Fox Valley Technical College*).

When parts are complex and or costly, the quality control technician often checks the first part coming out of a machine to be sure they meet requirements before more parts are run. Quality control technicians often have the responsibility for measurement tool calibration systems as well. They make sure that measurement equipment is accurate throughout the plant. Many quality control technicians began as machinists and moved into a quality control position. The quality control technician may receive training through college and trade school programs or on the job.

Supervisor

Many shop supervisors begin their careers as machinists. A machinist who proves his or her ability might be chosen to be a supervisor. A supervisor is responsible for a group of people and a part of the production process. They may further their education by taking supervisory management classes at a local technical college, community college, or university.

CNC Programmer

Some shops centralize the programming of CNC machines. Instead of having machinists write their own programs, they have specialized people doing the CNC programming with specialized software (Figure A-3). A machinist is very well prepared to be a CNC programmer. A machinist has knowledge of

Figure A-3 A CNC programmer utilizing a CAD/CAM system to program a CNC machine (*Courtesy of Fox Valley Technical College*).

tooling, machining operations, feeds and speeds, and materials. To become a CNC programmer, a machinist would take a class on the software or be trained by another employee on the software that the company uses for programming.

Manufacturing Engineer

The manufacturing engineer and/or industrial engineer is involved with the application of manufacturing technology. These individuals may be involved with the design of manufacturing tooling, setting up manufacturing systems, applying computers to manufacturing requirements, new product development, and improving efficiency.

Estimator/Bidder

Most machine shops do not make complete products. They produce parts for larger companies that do produce products. Shops that produce parts for other companies are called job shops. Job shops must bid on work. An estimator/bidder looks at potential parts that need to be made and develops an estimate of how much it will cost to make. They consider costs such as material, tooling, machine, special processes, labor, shipping, and so on. They then develop a quote that is sent to the company that is requesting quotes on the parts. If the quote is accepted, the shop will get an order for the parts. The estimator/bidder must be skilled at developing the quotes. The quote price must be high enough to make a profit, but low enough to beat bids from other shops.

Shop Owner

Most machine shops were started by machinists. Many shop owners were machinists who learned the trade, gained experience, and then bought one or more CNC machines and started their own company. Many manufacturing companies started this way and grew very rapidly to become large, successful businesses.

SPECIAL OCCUPATIONS

Automotive Machinist

The automotive machinist will work in an engine rebuilding shop where engines are overhauled. This person's responsibilities will be somewhat like those of the general machinist, with specialization in engine work, including boring, milling, and some types of grinding applications. Training for this job may be obtained on the job or through college or trade school programs.

Maintenance Machinist

The maintenance machinist has broad responsibilities. The person may be involved in plant equipment maintenance, machine tool rebuilding, or general mechanical repairs, including welding and electrical repairs. The maintenance machinist is often involved with general machining as well as with general industrial mechanical work. The maintenance machinist normally just uses simple manual machines to make or repair parts for machines.

MACHINING AND YOU

There are many exciting occupations available in manufacturing. These occupations provide a good working environment with wonderful wages and benefits as well as opportunities for advancement. Industry really needs people with good attitude and work ethic. There is a real demand for, and shortage of, individuals with good attitude and good work ethic. Industry is more than willing to invest in training and advancement for those individuals. Professional behavior will also serve you well. Appropriate clothing and language help promote a professional personal image. Dealing well with people is also an important attribute. A working knowledge of the machining processes and the related subjects described in this book will provide an excellent basis on which to build a successful career in manufacturing.

FURTHER INFORMATION ABOUT OCCUPATIONS IN MANUFACTURING

If you have further interest in machine shop career opportunities, discuss opportunities with an employment counselor at your school or a program instructor. Study the job advertisements in local newspapers to see what types of jobs are available. Talk to an employment agency about jobs that are in demand and the skills required to get them. Contact local manufacturing companies that employ machinists. Ask them for an opportunity to talk about employment opportunities and a tour of their facility. You may well find an employer who is willing to hire you as an intern while you go to school.

Job Packets and Process Planning

Most manufacturing companies produce parts for other companies. They must bid on jobs to get work for their shop. A larger manufacturer typically solicits bids on work they need done. For example, assume a manufacturer (ACME Manufacturing) that makes and sells snow blowers. They do much of the work themselves, especially the assembly of the parts into the completed machine. But they have other companies produce many of the component parts that go into their final machine. These smaller machine shops are typically called job shops. Job shops do not have their own products. They make parts for other companies. Imagine ACME needs small axle-shafts made for their snow blowers. ACME puts out bid requests to their list of suppliers. The bid request would normally be transmitted electronically to all suppliers. The bid request would include a blueprint, material specifications, number of parts needed, due dates, any special requirements for packing shipping, etc. Each job shop can then evaluate the bid request to determine whether or not to bid on the work.

Imagine how one job shop (XYZ Machining) might evaluate the bid. XYZ receives the bid request in an email. XYZ has a customer service representative (CSR) who works with customers. The CSR looks at the bid request to see how many parts and what type of machining is required to quickly decide if XYZ has the machining capability to make the parts. The CSR must also determine if XYZ can produce the parts by the due date while considering all of the other work XYZ already has to do. Assuming XYZ has the machining capability and the time to make the parts, a price would need to be calculated. The CSR might confer with their CNC turning center supervisor to determine

feasibility and cost per part. XYZ would also need to get a price on the steel that is required and make sure it will be available in the needed timeframe. Tooling costs, machine times, packaging, and shipping would all need to be considered. XYZ would also evaluate whether or not this is a customer they want to work for. XYZ would evaluate based on past history, if they have worked for them before. Did they pay their bills on time? Were they difficult to work with, and so on? If everything looks good at this point, the CSR would put together a bid price for the parts. Note that the price would need to be low enough to get the job, but high enough to make a profit. When there is a lot of work, companies tend to bid higher. When work is scarce bids get very competitive. In fact, in slow times a company might take work they normally would not take, just to keep their workers working. Note that ACME might accept bids from more than one company for parts. Their real need might be 10,000 shafts and they might split it between two job shops to reduce ACME's risk.

Assume that ACME Manufacturing accepts the bid from XYZ Machining. The order calls for 1,000 parts a month from January through May. The CSR orders steel once the bid is awarded by ACME. The CSR begins to prepare a job packet and schedules the job. Assume the parts must be shipped three days before the month end and it will take 38 hours of setup and machine time to make them. The CSR must evaluate the shop schedule and determine when the parts should be scheduled so these parts and all others can be made and shipped on time. The CSR might have to consult with the supervisor of that area in the shop. The CSR then puts a job packet together (Figure A-4). The job packet will include a current revision of the blueprint with all of the part specifications. The packet may include

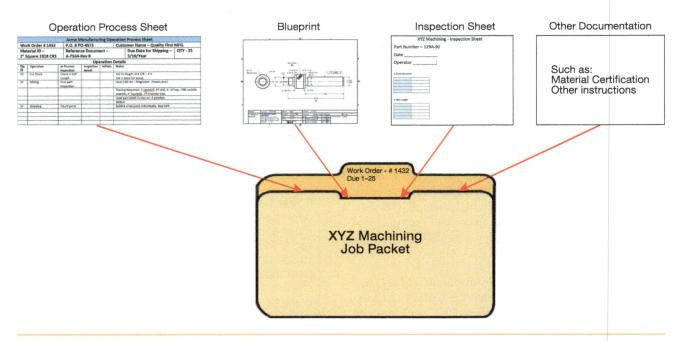

Figure A-4 Job Packet.

inspection sheets if special inspections are required. XYZ has a person who does process plans. The process planner develops an operation process sheet for the job. The process planner might also develop the CNC program for the parts, although the machinists program their own parts in many shops. If XYZ had run these parts before for ACME, the program would have already been written and the program number and machine would be specified on the operation process sheet.

Figure A-4 shows an example of the types of things that might go into a job packet. Job packets may also be called shop packets.

Process Plans

A process plan details all of the things that need to be done to manufacture a part (Figure A-5). They are step-by-step instructions that are done by the machinist. The steps detail each machining operation, inspections that need to be done, deburring operations, and so on. The process plan also references the blueprint and correct revision level that needs to be used. The process plan may list the CNC program to be used, the machine, and any special setup instructions. The process plan would also detail and special handling, packaging, and or shipping requirements.

Process plans are also called routings or routers.

Reference Document

The reference document is typically the customer-supplied blueprint. This is very important information. The machinist must always check to make sure the blueprint is the current revision. In this example, the blueprint number is A-7564-Rev B. The blueprint number must be the same exact number and revision level or the parts will be wrong.

Work Order Number

The work order number is an internal number. The company the machinist works for assigns a work order number to each order it accepts from customers. This number is used for internal tracking of all information about this part run.

P.O. Number

This is the purchase number. This number ties the order to a particular customer and order. It is the number the company will use to get paid by the customer after delivery of the parts.

Material Specification

This is where the material for this part is specified. This specification must match the material specified by the blueprint.

Due Date for Shipping

A machine shop must meet their customers' requirements. One of the most important is on-time delivery. A customer must have parts when they are needed. If a machine shop is not consistently on time with deliveries, the customer will find a new machine shop to work with.

Figure A-5 An operation process plan.

XYZ Manufacturing Operation Process Plan					
Work Order # 1432	P.O. # PO-4573		Customer Name – ACME		
Material Spec. – 2″ Square 1018 CRS	Reference Document – A-7564-Rev B		Due Date for Shipping – 5/25/Year		QTY - 1000
Operation Details					
Op. ID	Operation	In-Process Inspection	Inspection Result	initials	Notes
10	Cut Stock	Check 4 1/8″ Length			Cut to length of 4 1/8 – 4 ¼ Cut 1 extra for setup.
20	Milling	First part Inspection			Haas CNC #3 – Program# - PHAAS-3412
					Tooling Required: .5 spotdrill, #7 drill, ¼ -20 tap, .500 carbide endmill, 4″ facemill, .75 chamfer tool.
					Load part blank in vise on .5 parallels
					Deburr
30	Shipping	Count parts			Bubble wrap parts individually. Ship UPS

Operation

Operation is used to describe the operation that will be performed in this step. Note that there may be several steps and operations to be performed. In this example, there are only three operations. The first is to cut 25 blanks of 2″ square 1018 CRS (cold roll steel) to a length of 4 1/8″ to $4\frac{1}{4}$″. The note also calls for an extra blank to be cut for setting up a first piece.

SELF-TEST

1. Develop a personal career plan. What job would you like to start your career in? What would you like to be doing in five years? Fifteen years?
2. List five things you can do to achieve your 15-year goal.
3. Companies are looking for people with good attitude and work ethic. List at least five things that describe the attributes of good attitude. List at least five things that describe the attributes of work ethic.
4. How does a company get jobs to do?
5. Name some considerations that go into a bid.
6. What is a process control plan?
7. What is a router?
8. What typically is included in a job packet?

Manufacturing Competitiveness and Improvement

MANUFACTURING COMPETITIVENESS

Manufacturing is very competitive. A large percentage of manufacturing is done in job shops. They must bid on jobs to get work. The shop with the lowest bid that meets specifications will generally get the work. One of the techniques that manufacturers have been adopting to increase their efficiency is *lean manufacturing*.

Lean manufacturing is a technique that views the investment of a company's resources for any reason other than the creation of value for the customer to be wasted. The goal is to reduce or eliminate costs that do not create value for the customer. Customer value could be defined as any process that the customer would be willing to pay for. Many of the concepts and techniques of lean manufacturing have been around for a very long time.

Lean manufacturing is concerned with producing the product with fewer resources. Many of the concepts are based on principles from the Toyota Production System (TPS). The TPS was based on reducing waste and it identified seven principal wastes.

Overproduction

Overproduction is defined as making a product before it is actually required. Overproduction is very costly because it prevents the efficient flow of materials and adversely affects productivity and quality. Overproduction often hides the real problems and opportunities for improvement in manufacturing. The goal is to make only what can be sold or shipped immediately. This is often called just-in-time.

Wait Time

If a product is not moving or is being worked on, waste occurs. In a typical company, more than 99 percent of a product's time in the plant might be spent in waiting to be processed. This is often because distances between work centers are excessive, production runs are too long, and material flow is often very inefficient. Waiting can be reduced by having each process feed directly into the next process.

Unnecessary transportation

Transporting products between processes is a cost that does not add any value to the product. Excessive transportation and product handling can cause damage. When people have to excessively transport parts in a plant, it takes time from their real role of adding value to the product. This is another cost that adds no value.

Excess Inventory

Excess inventory or work in progress (WIP) is a direct result of overproduction and waiting. Excess inventory hides problems on the plant floor. Excess inventory wastes floor space, increases lead times, and dramatically increases costs.

Unnecessary Motion

Unnecessary motion by people or equipment is a waste. This waste is related to ergonomics and is seen in all unnecessary movement of people performing an operation or task.

Over Processing

Over processing often involves the use of expensive, high-cost equipment where simpler tools would do the job. The use of high-cost technology often results in higher quantity runs to try and recover the high cost of the equipment. Toyota is famous for using low-cost automation and well-maintained, older machines. If a machine is needed to produce X number of parts, Toyota would typically by a low-cost machine to produce that many parts. If production demands increase, they would buy another low-cost machine to meet the increased demand. American manufacturers would typically buy a more expensive machine that could produce many more than the required parts. There is less risk involved with the Toyota method. When possible, companies should invest in smaller and more flexible equipment.

Defective Product

Quality defects that cause scrap or rework are a huge cost to manufacturers. When scrap is produced, many additional costs occur. These additional costs include quarantining the defective product, reinspecting it, and rescheduling the line to reproduce the scrapped product. This results in the loss of capacity as the time and resources used to reproduce the product cannot be used to create additional product. The total cost of defects is often a very significant percentage of the total cost of manufacturing.

An eighth waste has recently been added: underutilization of employees. Enterprises often hire employees to physically produce products but do not utilize their brain. Manufacturers must utilize their employees' creativity to eliminate the seven wastes and increase productivity.

LEAN MANUFACTURING TOOLS FOR IMPROVEMENT

Lean manufacturing is a set of tools that assist in the identification and elimination of waste. When waste is eliminated, quality improves, and production time and cost are reduced. Some of the tools that are utilized in lean manufacturing include: Five S, Value Stream Mapping, Kanban (pull systems), and Poka-Yoke (error-proofing).

The Five S Improvement Tool

The term *Five S* comes from the Japanese. The five Japanese words used to describe the method all begin with the letter s.

Sorting (Seiri) All unnecessary parts, tools, and instructions should be eliminated. All tools, materials, instructions, and so on in the work area and overall plant should be thoroughly examined. Only essential items should be kept. Essential tools should be easily accessible in the work station. Everything that is nonessential should be stored or discarded.

Stabilizing or Straightening Out (Seiton) This technique stresses that there should be a place for everything and everything should be in its place. There should be a place for each item in a station and it should be clearly labeled. Parts, tools, supplies, and equipment should be located close to where they are needed. Everything should be arranged to promote efficient work flow.

Cleaning or Shining (Seiso) Clean the workspace and equipment, and then keep it clean and organized. At the end of each shift, clean the work area and be sure everything is restored to its place. Spills, leaks, and other messes also then become a visual signal for equipment or process steps that need attention. Maintaining cleanliness must be an integral part of the daily work.

Standardizing (Seiketsu) All work practices should be consistent and standardized. Work stations for a particular operation should be exactly the same. Employees with the same role should be able to work in any station. The same tools should be in the same location in every station.

Sustaining the Practice (Shitsuke) The new standards that have been implemented as a result of the previous tools must be reviewed, maintained, and improved. After the first four S's have been established, they must continue to be the new way to operate. This new way must be sustained. The plant must not be allowed to slide back to the old ways of operating. Everyone must also commit to continue looking for ways to improve.

Value Stream Mapping

Value stream mapping is a lean manufacturing tool that can be used to analyze and design the flow of materials and information required to bring a product to a customer. It originated at Toyota.

Implementation Steps

1. Identify the product to be studied.
2. Draw a state value stream map based on the shop floor that reflects the current steps, waits, and information flows required to deliver the product. This can be done by drawing it on paper or with special value stream mapping software.
3. Assess the efficiency of the current value stream map in terms of improving flow by eliminating waste.
4. Develop the redesigned value stream map.
5. Implement the redesign.

Shigeo Shingo was a Japanese industrial engineer who is one of the world's leading experts on manufacturing and the TPS. Shigeo Shingo suggested that the value-adding steps be drawn horizontally across the center of the value stream map and the non–value-adding steps be drawn vertically at right angles to the value stream.

This makes it easy to distinguish the value stream from the wasted steps. Shingo calls the value stream the process and the non–value streams the operations. Shingo viewed each vertical line as the story of a person or workstation. He viewed the horizontal line as the story of the product being created. Value stream mapping is a very visual and easy to understand tool to study and improve a process.

Kanban

Kanban is not an inventory control system. It is a scheduling system that helps determine what to produce, when to produce it, and how many to produce.

In the 1940s, Toyota studied supermarkets to try and apply store and shelf-stocking techniques to the factory floor. Toyota observed that in a supermarket, customers get what they need at the needed time, and in the needed quantity. Toyota saw that a supermarket only stocks what it thinks will sell, and customers only buy as many as they need because they are not worried about getting additional products in the future. Toyota saw a process as being a customer of the processes that preceded it. The customer (worker) goes to the store to get the parts they need, and the store restocks with additional product like a store would.

Kanban uses the rate of demand to control the rate of production. Demand is passed from the end customer through the chain of customer (worker) processes. Toyota first applied this technique in 1953 in their main machine shop.

Imagine that one of the components needed to make the product is a ¼−20 bolt and it arrives in a box. There are 100 of the bolts in a box. The worker might have two boxes of bolts available. When the first box is empty, the assembler installing the bolts would take the card that was attached to the box and send it to the bolt making department. The assembler would then begin to use the bolts from the second box of bolts. The card would serve as an order for another box of 100 bolts. The box of bolts would be made and sent to the assembler. A new box of bolts would not be produced until a card is received. In this way the demand drives the rate of production.

Kanban is a pull-type production system. The number of bolts that are produced depends on the actual demand. In this example, the demand is the number of cards received by the bolt manufacturing area from the customer (assembler).

Poka-Yoke   Poka-Yoke is a Japanese word that means mistake-proofing or fail-safing. A Poka-Yoke is any mechanism in a manufacturing process that helps an operator avoid mistakes. For example, the addition of a pin in a fixture to prevent the part from being mislocated would be a Poka-Yoke. Poka-Yoke attempts to eliminate product defects by preventing, correcting, or warning the operator an error is about to occur. Shigeo Shingo first used the term Poka-Yoke as part of the TPS.

A Poka-Yoke technique alerts the operator when a mistake is about to be made, or the Poka-Yoke device actually prevents the mistake from being made.

Summary of Lean Manufacturing

Toyota's belief is that *lean* is not the tools, but the actual reduction of three types of waste: non–value-adding work, overburden, and unevenness of production. The tools are used to expose problems and improve processes.

Lean manufacturing attempts to get the correct things to the correct place at the correct time in the correct quantity to achieve an efficient work flow. Lean manufacturing also attempts to minimize waste and make processes flexible and able to react to change. For lean manufacturing techniques to be effective, they must be understood and be actively supported by the employees who build the product and add the value.

SELF-TEST

1. How is lean manufacturing supposed to make a company more efficient and more profitable?

2. What was the basis of improvement in the Toyota Production System?

3. The Toyota Production System originally listed seven wastes. An eighth was added. What is it and what does it mean for you in your career?

4. What is the basis of the Five S tool?

5. How can Value Stream Mapping improve a process?

6. What is Kanban?

7. What is Poka-Yoke?

Shop Safety

After completing this unit, you should be able to:

- Identify common shop hazards.
- Identify and use common shop safety equipment.
- Explain the classes of fires that are applicable to a machine shop.
- Given a fire extinguisher, explain what types of fires it can be used on.
- Demonstrate safe working practices in the shop.
- Explain Lockout/Tagout.

SAFETY FIRST

We generally do not think about safety until it is too late. Safety is often not thought about as you do your daily tasks. Often you expose yourself to needless risk because you have experienced no harmful effects in the past. Unsafe habits become almost automatic. You may drive your automobile without wearing a seat belt. You know this to be unsafe, but you have done it before and so far, no harm has resulted. None of us really likes to think about the possible consequences of an unsafe act. However, safety can and does have an important effect on anyone who makes his or her living in a potentially dangerous environment. An accident can end your career as a machinist. Accidents are always unexpected! You may spend several years learning the trade and more years gaining experience. Years spent in training and gaining experience can be wasted in an instant if you have an accident, not to mention a possible permanent physical handicap for you and hardship on your family. Safety is an attitude that should extend far beyond the machine shop and into every facet of your life. You must constantly think about safety in everything you do. Safe habits must be developed and utilized. Safety glasses are just one example. You must develop the habit to wear them at all times in the shop to the point that you feel naked without them.

PERSONAL SAFETY

Grinding Dust, Hazardous Fumes, and Chemicals

Grinding dust is produced by abrasive wheels and consists of extremely fine metal particles and abrasive wheel particles. These should not be inhaled. Many grinding machines have a vacuum dust collector (Figure A-6). Grinding may be done with coolants that aid in dust control. You should wear an approved respirator if you are exposed to grinding dust. Change the respirator filter at regular intervals. Grinding dust can present a great danger to health.

Some metals, such as zinc, give off toxic fumes when heated above their boiling point. When inhaled, some of these fumes cause only temporary sickness, but other fumes can be severe or even fatal. The fumes of mercury and lead are especially dangerous, as their effect is cumulative in the body and can cause irreversible damage. Cadmium and beryllium compounds are also very poisonous. Therefore, when welding, burning, or heat-treating metals, adequate ventilation is an absolute necessity. This is also true when parts are being carburized with compounds containing potassium cyanide. These cyanogen compounds are deadly poisons, and every precaution should be taken when using them. Kasenit, a trade name for a nontoxic carburizing compound, is often found in school shops and in machine shops.

There may also be chemical hazards in the machine shop; lubricating oils, cutting oils, coolants, solvents, and some types of degreasing agents may be used. Any of these chemical agents can cause both short- and long-term health problems. Cutting oils may smoke when heated and give off noxious fumes. Inhaling any type of smoke can have short- and long-term health risks. Coolants may cause contact dermatitis, a skin irritation problem, and prolonged exposure can cause other long-term health problems. You should seek chemical safety data regarding these products and determine what health problems that short- and long-term exposure can cause. Chemical hazard awareness programs label chemicals to make employees aware of particular fire, health, and reactivity issues.

Figure A-6 Dust collector installed on a grinder.

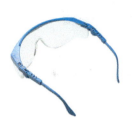

Figure A-7 Safety glasses, face shield, and safety goggles.

Material Safety Data Sheets (MSDS)

A material safety data sheet or MSDS contains information describing the properties of particular chemicals, materials, and other substances. Technical data are provided defining the chemical and physical properties of materials, such as melting point, boiling point, flash point, and any toxic elements that are likely to be present during processing or handling. Other items that may be included on an MSDS are proper disposal techniques, first-aid issues from exposure to hazardous materials, and protective equipment required to safely handle the material. Material safety data sheets are available from many different sources including the Occupational Safety and Health Administration (OSHA) and manufacturers' published information. MSDS sheets should be readily available in the shop. Modern industrial operations often use many hazardous materials. Safety in using, handling, and disposing of these materials has become extremely important.

Eye Protection

Eye protection is a primary safety consideration in the machine shop. Machine tools produce hot, sharp metal chips that may be ejected from a machine at high velocity. Sometimes they can fly many feet. Furthermore, most

cutting tools are made from hard, brittle materials. They can occasionally break or shatter from the stress applied to them during a cut. The result can be flying metal particles.

Eye protection must be worn at all times in the machine shop. Several types of eye protection are available. Plain safety glasses are all that are required in most shops (see Figure A-7). These have shatterproof lenses that may be replaced if they become scratched. The lenses have a high resistance to impact.

Side shield safety glasses should be worn around any grinding operation. The side shield protects the side of the eye from flying particles. Side shield safety glasses may be of the solid or perforated type. The perforated side shield type fits closer to the eye.

Prescription glasses may be covered with safety goggles. The full-face shield may also be used (Figure A-7). Prescription safety glasses are available. In industry, prescription safety glasses are sometimes provided free to employees.

Foot Protection

Generally, the machine shop presents a modest, but real hazard to the feet. There is always a possibility that you could drop something on your foot. Safety shoes with a steel toe shield to protect the foot are available. Some safety shoes also have an instep guard. Shoes must be worn at all times in the machine shop. A solid leather shoe is recommended. Sandals should not be worn as they expose the foot to hot chips produced in machining processes. Many shops provide rubber mats by machines that can dramatically reduce the fatigue on the feet caused by standing on a hard-concrete floor.

Ear Protection

Machine shops are noisy environments. Hearing protection should be worn. Excess noise can cause permanent hearing loss. Usually, this occurs over a period of time, depending on the intensity of the exposure. Noise is considered an industrial hazard if it is continuously above 85 **decibels (dB)**. Decibel is the unit used for measuring the relative intensity of sounds. If the noise level is over 115 dB, even for short periods of time, ear protection must be worn (Figure A-8). Several types of sound suppressors and noise-reducing earplugs may be worn. Earmuffs or earplugs should be used wherever high-intensity noise occurs. Table A-1 shows the decibel level of various sounds; sudden sharp or high-intensity noises are the most harmful to the ears.

Figure A-8 Ear plugs.

Table A-1 Decibel Level of Various Sounds

R	Sound
130	Painful, Jet engine on the ground
120	Airplane on the ground
110	Power saw at 3 feet
100	Motorcycle or snowmobile
90	Truck traffic
80	Telephone dial tone
60–70	Normal conversation at 3–5 feet
30	Whisper or a quiet library
0	Faintest audible sound

Clothing, Hair, and Jewelry

Shirts should be short-sleeved or long sleeves should be rolled up above the elbow. Shirts should be tucked in. If a shop apron is worn, it should be kept tied behind the back. If apron strings were to get entangled in the machine, you may be reeled in as well. Loose, hanging clothing is dangerous. A shop coat may be worn as long as long sleeves are rolled up above the elbow. Fuzzy sweaters should not be worn around machines. Polyester clothing is very susceptible to hot chips and should be avoided. Cotton clothing is a better choice.

If you have long hair, keep it secured properly. In industry, you may be required to keep your hair tied back so that your hair cannot become tangled in a moving machine. Long hair is easily caught up in moving tools such as drills and chucks. The result of this can be disastrous.

Remove your wristwatch and rings before operating any machine tool. These can cause serious injury if they are caught in a moving machine part.

Hand Protection

No device will totally protect your hands from injury. Next to your eyes, your hands are the most important tools you have. It is up to you to keep them out of danger. Use a brush to remove chips from a machine (Figure A-9). Do not use your hands. Chips are razor sharp and can be extremely hot. Never grab chips as they come from a cut. Long chips are extremely dangerous. Long chips can be eliminated by sharpening your cutting tools properly. Chips should not be removed with a rag. The metal particles become embedded in the cloth and they may cut you. Furthermore, the rag may be caught in a moving machine. Gloves should never be worn around

Figure A-9 Use of a brush to remove chips safely.

moving machinery. If a glove is caught in a moving part, it will be pulled in, along with the hand inside it.

Do not be careless around machines. Do not put your hands near moving machinery. Do not lean on a machine that is running. Do not lean over the back of a machine and talk to the employee who is running the machine. Accidents happen very quickly and they are never expected.

Various cutting oils, coolants, and solvents may affect your skin. The result may be a rash or an infection. Avoid direct contact with these products as much as possible, and wash your hands as soon as possible after contact.

Lifting

Improper lifting (Figure A-10) can result in a permanent back injury that can limit or end your ability to earn a living and reduce your quality of life by restricting what you are able to do. Back injury can be avoided if you lift properly at all times. If you must lift a large or heavy object, get some help or use a hoist or forklift. Don't try to be a "superman" and lift something that you know is too heavy. It is not worth the risk. Most shops have lifting devices available for heavy objects. Objects within your lifting capability can be lifted safely by using the following procedure:

1. Keep your back straight.
2. Squat down, bending your knees.
3. Lift smoothly, using the muscles in your legs to do the work. Keep your back straight. Bending over the load puts an excessive stress on your spine.
4. Position the load so that it is comfortable to carry. Watch where you are walking when carrying a load.
5. If you replace the load back at floor level, lower it in the same manner in which you picked it up.

Scuffling and Horseplay

The machine shop is no place for scuffling and horseplay. This activity can result in a serious injury to you or those around you. Practical joking is also hazardous. What might appear comical to you could result in a disastrous accident to

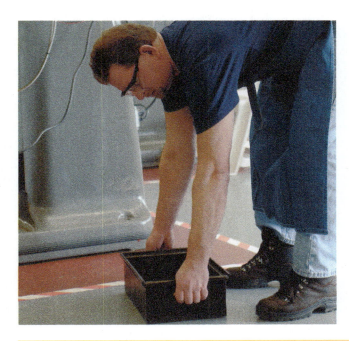

Figure A-10 The picture on the left shows the incorrect method to lift object. The picture on the right shows the correct method for lifting objects.

someone else. In industry, horseplay and practical joking are often grounds for firing an employee. Loud noises should also be avoided in the plant. Yelling, whistling, and making loud noises can startle an individual and may cause them to injure themselves.

Injuries

If you should be injured, report it immediately to your instructor. In many schools and companies, there are incident report forms that must be filled out to report injuries to help fix the problem and help prevent it in the future.

IDENTIFYING SHOP HAZARDS

A machine shop is not so much a dangerous place as it is a potentially dangerous place. One of the best ways to be safe is to be able to identify shop hazards before they result in an accident. By being aware of potential dangers, you can be very safe as you do your work in a machine shop.

Compressed Air

Machine shops use compressed air to operate certain machine tools. Often, flexible air hoses hang throughout the shop. There is a large amount of energy stored in a compressed gas such as air. Releasing this energy presents an extreme danger. You may be tempted to blow chips from a machine tool using compressed air. This is not good practice. The air will propel metal particles at high velocity. They can injure you or someone on the other side of the shop. Use a brush to clean chips from the machine. Do not blow compressed air on your clothing or skin. The air may be dirty, and the force can implant dirt and germs into your skin. Air can be a hazard to ears as well. An eardrum can be ruptured.

Should an air hose break or the nozzle on the end come unscrewed, the hose will whip about wildly. This can result in an injury if you happen to be standing nearby. When an air hose is not in use, it is good practice to shut off the supply valve. The air trapped in the hose should be vented. When removing an air hose from its supply valve, be sure that the supply is turned off and the hose has been vented. Removing a charged air hose will result in a sudden venting of air. This can surprise you, and an accident might result.

Housekeeping

Keep the floor and aisles clear of stock and tools. This will ensure that all exits are clear if the building should have to be evacuated. Material on the floor, especially round bars, can cause falls. Clean up oils or coolants that spill on the floor. Several products designed to absorb oil are available. Keep a broom handy when you are machining if chips are flying onto the floor. Keep the floor clean around the machine. Keep oily rags in an approved safety can (Figure A-11). This will prevent possible fire from spontaneous combustion.

Fire Extinguishers

Fire extinguishers work by cooling the fuel, stopping the reaction, or by removing the oxygen from the fire. Fire extinguishers are used to spray a pressurized media out of a nozzle. The spray should not be aimed at the flames. The spray should be aimed at the fuel of the fire, which is at the bottom of the flames.

OSHA has identified five basic types of fires and specified types of extinguishers for each type of fire. Four of the five types of fire may be found in a machine shop.

Class A   Class A fires involve common combustibles such as paper, cloth, wood, rubber, many types of plastics, and so on.

Figure A-11 A safety can for oily rags to prevent fires.

An extinguisher rated for a Class A fire uses pressurized water as the media. A Class A fire extinguisher should only be used on a Class A fire, never on any other type of fire.

Class B Class B fires involve oils, gasoline, lacquers, some types of paints, solvents, grease, and most other flammable liquids. Class B fire extinguishers use carbon dioxide or dry chemicals as the media.

Class C Class C fires are electrical fires. These can occur in wiring, energized electrical equipment, fuse boxes, computers, and so on. Class C fire extinguishers utilize carbon dioxide or a dry chemical as the extinguishing media.

There are fire extinguishers available that can be used for a Class A, B, or C fires. They are labeled for Class A, B, and C.

Class D Class D fires involve metals such as powders, flakes, or shavings of combustible metals such as magnesium, titanium, potassium, and sodium. These can be very dangerous fires. Class D fire extinguishers can contain a sodium chloride–based dry powder extinguishing agent, but most of the fire extinguishers labeled Class D have components that are geared to a specific metal.

Class K Class K fires involve combustible cooking fluids such as oils and fats. These are not applicable to a machine shop.

Know where all fire extinguishers are located in the plant. Know what type they are and what types of fires they can be used on. Know how they are operated. You should be fully aware of this information before they are needed. Precious minutes can be saved by knowing this ahead of time.

Carrying Objects

If material is over 6 feet long, it should be carried in the horizontal position. If it must be carried in the vertical position, be careful of electrical buses, light fixtures, and ceilings. If the material is both long and over 40 lb in weight, it should be carried by two people, one at each end.

Machine Hazards

There are many types of machine hazards. Each section of this book will discuss the specific dangers applicable to that type of machine tool. Remember that a machine cannot distinguish between cutting metal and cutting fingers. Do not think that you are strong enough to stop a machine should you become tangled in moving parts. You are not. When operating a machine, think about what you are going to do before you do it. Go over a **safety checklist**:

1. Do I know how to operate this machine?
2. What are the potential hazards involved?
3. Are all guards in place?
4. Are my procedures safe?
5. Am I doing something that I probably should not do?
6. Have I made all the proper adjustments and tightened all locking bolts and clamps?
7. Is the workpiece secured properly?
8. Do I have proper safety equipment?
9. Do I know where the stop switch is?
10. Do I think about safety in everything that I do?

INDUSTRIAL SAFETY AND FEDERAL LAW

In 1970, Congress passed the Williams-Steiger Occupational Safety and Health Act. This act took effect on April 28, 1971. The purpose and policy of the act is "to assure so far as possible every working man and woman in the Nation safe and healthful working conditions and to preserve our human resources."

The Occupational Safety and Health Act is commonly known as OSHA. Prior to its passage, industrial safety was the individual responsibility of each state. The establishment of OSHA added a degree of standardization to industrial safety throughout the nation. OSHA encourages states to assume full responsibility in administration and enforcement of federal occupational safety and health regulations.

Duties of Employers and Employees

Every employer under OSHA has the general duty to furnish places of employment free from recognized hazards causing or likely to cause death or serious physical harm. The employer has the specific duty of complying with safety and health standards as defined under OSHA. Each employee has the duty to comply with safety and health standards and all rules and regulations established by OSHA.

Occupational Safety and Health Standards

Job safety and health standards consist of rules for avoiding hazards that have been proven by research and experience to be harmful to personal safety and health. These rules may apply to all employees, as in the case of fire protection standards. Many standards apply only to workers engaged in specific types of work. A typical rule might state that aisles and passageways shall be kept clear and in good repair, with no obstruction across or in aisles that could create a hazard.

Complaints of Violations

Any employee who believes that a violation of job safety or health standards exists may request an inspection by sending a signed written notice to OSHA. This includes anything that threatens physical harm or represents an imminent danger. A copy must also be provided to the employer; however, the name of the person complaining need not be revealed to the employer.

Enforcement of OSHA Standards

OSHA inspectors may enter a business or school at any reasonable time and conduct an inspection. They are not permitted to give prior notice of this inspection. They may question any employer, owner, operator, agent, or employee in regard to any safety violation. The employer and a representative of the employees have the right to accompany the inspector during the inspection.

If a violation is discovered, a written citation is issued to the employer. A reasonable time is permitted to correct the condition. The citation must be posted at or near the place of the violation. If after a reasonable time the condition has not been corrected, a fine may be imposed on the employer. If the employer is making an attempt to correct the unsafe condition but has exceeded the time limit, a hearing may be held to determine progress.

Penalties

Willful or repeated violations may incur monetary penalties. Citations issued for serious violations incur mandatory penalties. A serious violation may be penalized for each day of the violation.

OSHA Education and Training Programs

The Occupational Safety and Health Act provides for programs to be conducted by the Department of Labor. These programs provide for education and training of employers and employees in recognizing, avoiding, and preventing unsafe and unhealthful working conditions. The act also provides for training an adequate supply of qualified personnel to carry out OSHA's purpose.

Lockout and Tagout Procedures

On October 30, 1989, the Department of Labor released "The control of hazardous energy sources (Lockout/Tagout)" standard. It is the Lockout/Tagout standard numbered 29 CFR 1910.147. The standard was intended to reduce the number of deaths and injuries related to servicing and maintaining machines and equipment.

The Lockout/Tagout standard covers the servicing and maintenance of machines and equipment in which the unexpected startup or energization of the machines or equipment or the release of stored energy could cause injury to employees. The standard covers energy sources such as electrical, mechanical, hydraulic, chemical, nuclear, and thermal. The standard establishes minimum standards for the control of these hazards.

Normal production operations are excluded from the lockout/tagout restrictions. Normal production operation is the use of a machine or equipment to perform its intended production function. Any work performed to prepare a machine or equipment to perform its normal production operation is called setup.

If an employee is working on cord and plug-connected electrical equipment for which exposure to unexpected energization or start up of the equipment is controlled by the unplugging of the equipment from the energy source, and the plug is under the exclusive control of the employee performing the servicing or maintenance, this activity is also excluded from requirements of the standard.

The Lockout/Tagout standard defines an energy source as any source of electrical, mechanical, hydraulic, pneumatic, chemical, thermal, or other energy. Machinery or equipment is considered to be energized if it is connected to an energy source or contains residual or stored energy. Stored energy can be found in pneumatic and hydraulic systems, capacitors, springs, and even gravity. Heavy objects have stored energy. If they fall they can cause injury.

Servicing and/or maintenance includes activities such as installing, constructing, setting up, adjusting, inspecting, modifying, and maintaining and/or servicing equipment. These activities include lubrication, cleaning or unjamming of machines or equipment, and making adjustments or tool changes, where personnel may be exposed to the unexpected energization or start up of the equipment or a release of hazardous energy.

Employers are required to establish (lockout/tagout) procedures and employee training to ensure that before any employee performs any servicing or maintenance on a machine or equipment where the unexpected energizing, startup, or release of stored energy could occur and cause injury, the machine or equipment shall be isolated, and rendered inoperative. Most companies impose strong sanctions on employees who do not follow the procedures to the letter. Many companies terminate employees who repeatedly violate the procedures. The procedures are designed to keep personnel safe. Make sure you follow procedures. You should also be aware of the penalties of not following them. The greatest penalty could be severe injury or death.

Employers are also required to conduct periodic inspections of the procedures at least annually to ensure that the procedures and the requirements of the standard are being followed.

Only authorized employees may lock out machines or equipment. An authorized employee is defined as being one who has been trained and has been given the authority to lock or tag out machines or equipment to perform servicing or maintenance on that machine or equipment.

An energy-isolating device is defined as being a mechanical device that can physically prevent the transmission or release of energy. A disconnect on an electrical enclosure is a good example of an energy-isolating device (see Figure A-12). If the disconnect is off, it physically prevents the transmission of electrical energy. Energy-isolating devices include the following: disconnects, manual electrical circuit breakers, manually operated switches by which the conductors of a circuit can be disconnected from all ungrounded supply conductors and, in addition, no pole can be operated independently; line valves (see Figure A-12); locks and any similar device used to block or isolate energy. Push buttons, selector switches and other control circuit type devices are not considered to be energy-isolating devices.

Due to the invisible and potentially fatal hazard of electrical energy, manufacturing companies developed **lockout/tagout** procedures for safely working on electrical equipment. When working on electrical equipment, it is of critical importance to absolutely prevent the accidental energizing of an electrical circuit. In lockout/tagout procedures, the source of the power is turned off and the control switches, circuit breakers, or main switches are physically locked out, often using a keyed lock. The circuit is also tagged and signed off by the electrician or other maintenance workers and can be unlocked and reenergized only by the person directly responsible for the lockout/tagout procedure. Make sure that you follow lockout/tagout procedures. Failure to follow them may cause serious injury or death.

Figure A-12 Electrical disconnect on the left and a pneumatic disconnect on the right (*Courtesy of Fox Valley Technical College*).

Most companies also have severe penalties up to and including, being fired for not following lockout/tagout procedures.

SELF-TEST

1. What is the primary piece of safety equipment in the machine shop?
2. What are side shields and what are they designed to do?
3. Describe proper dress for the machine shop.
4. What can be done to control grinding dust?
5. What hazards exist from coolants, oils, and solvents?
6. Describe proper lifting procedure.
7. Describe at least two compressed-air hazards.
8. Describe good housekeeping procedures.
9. How should long pieces of material be carried?
10. List at least five points from the safety checklist for a machine tool.
11. Describe the purpose of lockout/tagout procedures.

INTERNET REFERENCES

Information on industrial safety, lockout/tagout and safety equipment:

http://www.osha.gov

http://www.seton.com

Threads and Fasteners

Many precision-machined products are useless until assembled into a machine, tool, or other mechanism. This assembly requires many types of fasteners and other mechanical hardware. In this unit you will be introduced to many of these important hardware items.

OBJECTIVES

After completing this unit, you should be able to:

- Identify threads and threaded fasteners.
- Identify thread nomenclature on drawings.
- Discuss standard series of threads.
- Identify and describe applications of common mechanical hardware found in the machine shop.

THREADS

This unit will introduce types of fasteners and their proper use. Threads are mainly used as fastening devices. Threads are also used for adjustment, measurement, and the transmission of power. The thread is an extremely important mechanical device. It derives its usefulness from the principle of the inclined plane, one of the six simple machines. Almost every mechanical device is assembled with threaded fasteners. One of the fundamental tasks of a machinist is to produce external and internal threads using machine tools and hand tools.

THREAD FORMS

There are a number of thread forms. In later units you will examine these in detail, and you will have the opportunity to make several of them on a machine tool. The most common is the unified thread form (Figure A-13). The unified thread form is an outgrowth of the American National Standard form. It was developed to help standardize manufacturing in the United States, Canada, and Great Britain.

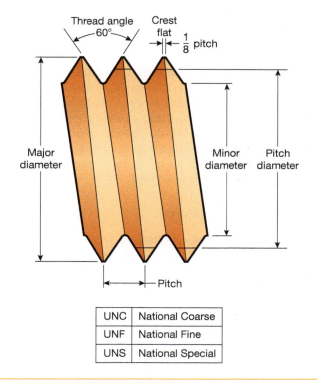

UNC	National Coarse
UNF	National Fine
UNS	National Special

Figure A-13 Unified thread form.

In 1948, representatives from America, Canada, and Great Britain agreed on a common standard for threads. It utilized the best of both standards and was called the Unified Screw Thread System.

America has not adopted the metric system. Great Britain adopted the metric system in 1965. The International Standards Organization (ISO) adopted an international screw thread standard in 1969. The ISO metric standard is the most widely used in the world. Unified threads, a combination of the American National and the British Standard Whitworth forms, are divided into the following series: Unified National Course (UNC); Unified National Fine (UNF); and Unified National Special (UNS).

IDENTIFYING THREADED FASTENERS

Unified coarse and *unified fine* refer to the number of threads per inch of length on standard threaded fasteners. A specific diameter of bolt or nut will have a specific number of threads per inch of length. For example, a $\frac{1}{2}$ − inch − 13 Unified National Coarse bolt will have 13 threads per inch of length. This bolt is identified by the following marking:

$$\frac{1}{2} \text{ in.} - 13 \text{ UNC}$$

The $\frac{1}{2}$ inch is the **major diameter**, and 13 is the number of threads per inch of length. A $\frac{1}{2}$ inch diameter Unified National Fine bolt will be identified by the following marking:

$$\frac{1}{2} \text{ in.} - 20 \text{ UNF}$$

The $\frac{1}{2}$ inch is the major diameter and 20 is the number of threads per inch.

The Unified National Special threads are identified in the same manner. A $\frac{1}{2}$ inch diameter UNS bolt may have 12, 14, or 18 threads per inch. These are less common than the standard UNC and UNF. There are many other series of threads used for different applications. Information and data on these can be found in machinists' handbooks. You might wonder why there needs to be a UNC and a UNF series. A fine thread is stronger in harder materials like steels and a coarse thread is stronger in softer materials like aluminum. The choice of coarse versus fine series also has to do with the application. For example, an adjusting screw may require a fine thread, while a common bolt may require only a coarse thread.

Thread Terminology

Some of the more commonly used thread terms are:

Thread Angle—The angle of the thread is the included angle between the sides of the thread (see Figure A-13). For example, the thread angle for Unified Screw Thread form is 60 degrees.

Major Diameter—Commonly known as the outside diameter (see Figure A-13). On a straight screw thread, the major diameter is the largest diameter of the thread on the screw or nut.

Minor Diameter—Commonly known as the root diameter (see Figure A-13). On a straight screw thread, the minor diameter is the smallest diameter of the thread on the screw or nut.

Number of Threads—Refers to the number of threads per inch of length.

Pitch—Is the distance from a given point on one thread to a corresponding point on the next thread.

Lead—Lead and pitch are closely related concepts. Lead is how far a thread advances by one complete rotation of the screw (360 degrees). For example, if we turn a bolt one revolution clockwise, the distance it would move into the nut would be the lead of the thread. If a bolt has 10 threads per inch, the lead would be equal to 1 inch/10 or .100 inch. One complete revolution of the bolt would advance the bolt into the nut .100 inch. Lead and pitch are normally the same for a thread. Lead and pitch would be different if the threads are not a single thread. For example, the lead on a double thread would be twice as much as the pitch.

CLASSES OF THREAD FITS

Some applications can tolerate loose threads, while other applications require tight threads. For example, the head of your car's engine is held down by a threaded fastener called a stud bolt, or simply a stud. A stud is threaded on both ends. One end is threaded into the engine block. The other end receives a nut that bears against the cylinder head. When the head is removed, it is desirable to have the stud remain screwed into the engine block. This end requires a tighter thread fit than the end of the stud accepting the nut. If the fit on the nut end is too tight, the stud may unscrew as the nut is removed.

Unified thread fits are classified as 1A, 2A, 3A, or 1B, 2B, 3B. The A symbol indicates an external thread. The B symbol indicates an internal thread. This notation is added to the thread size and number of threads per inch. Let us consider the $\frac{1}{2}$-inch diameter bolt discussed previously. The complete notation reads:

$$\frac{1}{2} - 13 \text{ UNC 2A}$$

On this particular bolt, the class of fit is 2. The symbol A indicates an external thread. If the notation had read:

$$\frac{1}{2} - 13 \text{ UNC 3B}$$

it would indicate an internal thread with a class 3 fit. This could be a nut or a hole threaded with a tap. Taps are a common tool for producing an internal thread.

Classes 1A and 1B have the greatest manufacturing tolerance. They are used where ease of assembly is desired and a loose thread is acceptable. Class 2 fits are used on the largest percentage of threaded fasteners. Class 3 fits will be tight when assembled. Each class of fit has a specific tolerance on major diameter and pitch diameter. These specifications can be found in machinists' handbooks and are required for the manufacture of threaded fasteners.

STANDARD SERIES OF THREADED FASTENERS

Threaded fasteners range from very small machine screws through very large bolts. Fasteners under $\frac{1}{4}$-inch diameter are given a number. Common UNC and UNF series threaded fasteners are listed in Table A-2. Above size 12, the major diameter is expressed in fractional form. Both series continue up to about 4 inches.

Metric Threads

With all of the importation of foreign products, metric threads have become much more common.

Table A-2 UNC and UNF Threaded Fasteners

	UNC			UNF	
Size	Major Diameter (inches)	Threads/ Inch	Size	Major Diameter (inches)	Threads/ Inch
			0	.059	80
1	.072	64	1	.072	72
2	.085	56	2	.085	64
3	.098	48	3	.098	56
4	.111	40	4	.111	48
5	.124	40	5	.124	44
6	.137	32	6	.137	40
8	.163	32	8	.163	36
10	.189	24	10	.189	32
12	.215	24	12	.215	28
$\frac{1}{4}$ inch	.248	20	$\frac{1}{4}$ inch	.249	28
$\frac{5}{16}$ inch	.311	18	$\frac{5}{16}$ inch	.311	24
$\frac{3}{8}$ inch	.373	16	$\frac{3}{8}$ inch	.373	24
$\frac{7}{16}$ inch	.436	14	$\frac{7}{16}$ inch	.436	20
$\frac{1}{2}$ inch	.498	13	$\frac{1}{2}$ inch	.498	20
$\frac{9}{16}$ inch	.560	12	$\frac{9}{16}$ inch	.561	18
$\frac{5}{8}$ inch	.623	11	$\frac{5}{8}$ inch	.623	18
$\frac{3}{4}$ inch	.748	10	$\frac{3}{4}$ inch	.748	16
$\frac{7}{8}$ inch	.873	9	$\frac{7}{8}$ inch	.873	14
1 inch	.998	8	1 inch	.998	12

Table A-3 ISO Metric Threads

Diameter (mm)	Pitch (mm)	Diameter (mm)	Pitch (mm)
1.6	.35	20.0	2.5
2.0	.40	24.0	3.0
2.5	.45	30.0	3.5
3.0	.50	36.0	4.0
3.5	.60	42.0	4.5
4.0	.70	48.0	5.0
5.0	.80	56.0	5.5
6.3	1.00	64.0	6.0
8.0	1.25	72.0	6.0
10.0	1.50	80.0	6.0
12.0	1.75	90.0	6.0
14.0	2.00	100.0	6.0
16.0	2.00		

The metric thread form is similar to the unified and based on an equilateral triangle. The root may be rounded and the depth somewhat greater. An attempt has been made through international efforts (ISO) to standardize metric threads. The ISO metric thread series now has 25 thread sizes with major diameters ranging from 1.6 to 100 mm.

Metric thread notation takes the following form:

$$M\ 10 \times 1.5$$

M is the major diameter and 1.5 is the thread pitch in millimeters. This thread would have a major diameter of 10 mm and a pitch (or lead) of 1.5 mm. ISO metric thread major diameters and respective pitches are shown in Table A-3.

ISO METRIC THREADS CLASSES OF FIT

Numbers and letters after the pitch number in the notation specify the tolerance for both the pitch and the major diameter of external threads and the pitch and the minor diameter of internal threads. The numbers indicate the amount of tolerance allowed. The smaller the number, the smaller the amount of tolerance allowed. The letters indicate the position of the thread tolerance in relation to its basic diameter. Lowercase letters are used for external threads, with the letters e indicating a large allowance, g a small allowance, and

h no allowance. Conversely, uppercase letters are used for internal threads, with a G used to indicate a small allowance and an H used to indicate no allowance.

The tolerance classes 6H/6g are used for general-purpose applications. They are comparable to the Unified National 2A/2B classes of fit. The designation 4H5H/4h5h is approximately equal to the Unified National class 3A/3B.

Taper Pipe Threads

Standard taper pipe threads (NPT) are unlike standard threads in the way they are cut and in their use. Taper pipe threads are one of the more challenging thread-cutting operations. Cutting taper pipe threads requires greater accuracy than standard threads. Pipe threads are designed to mechanically seal a threaded joint for pressure and to prevent leakage. To create the mechanical seal, 100 percent of the thread height must be cut by the tap to maintain the standard thread profile. The NPT standard limits for both external and internal are identical, so that hand-tight engagement can result in thread flank contact only or crest and root contact only. Wrench tightening normally produces an assembly with crest and root clearances. Excessive tightening with a wrench without the use of some kind of tape or chemical sealer is not advisable.

Taper threads are also unlike standard threads in their designations. A machinist looking at a $\frac{1}{4}$–18 NPT would immediately notice that the outside diameter of the tap is much bigger than $\frac{1}{4}$ of an inch. This is due to the fact that the tap size designation refers to the inside diameter of the standard iron pipe it is designed to fit. Another major difference between taper pipe tapping and standard thread tapping is the way in which the machinist must control the thread diameter. Since the tap is tapered, the machinist can control the thread diameter by adjusting the depth that the tap threads into the hole. To get the proper thread depth, a machinist drives the tap into the work 12 full turns. More turns may result in the tapped hole being too large. Less turns may result in the tapped hole being too small. There is no one particular class of fit for NPT threads, such as 3A or 3B. Proper taper thread gauging is the only acceptable method for determining proper depth for taper pipe tapping. One method of monitoring tap depth is to wrap a piece of wire or tape around the tap as a line to serve as a depth indicator. This practice works well to get you "close." This trick works well for hand tapping when cutting fluids are used.

COMMON EXTERNALLY THREADED FASTENERS

Common mechanical hardware includes threaded fasteners such as bolts, screws, nuts, and thread inserts. All these are used in a variety of ways to hold parts and assemblies together.

Bolts and Screws

A general definition of a *bolt* is an externally threaded fastener that is inserted through holes in an assembly. A bolt is tightened with a nut (Figure A-14, right). A *screw* is an externally threaded fastener that is inserted into a threaded hole and tightened or released by turning the head (Figure A-14, left). From these definitions, it is apparent that a bolt can become a screw or vice versa. This depends on how it is used. Bolts and screws are the most common of the threaded fasteners. Threaded fasteners are used to assemble parts quickly, and they also make disassembly possible.

The strength of an assembly of parts depends to a large extent on the diameter of the screws or bolts used. In the case of screws, strength depends on the amount of thread engagement. *Thread engagement* is the distance that a screw extends into a threaded hole. The minimum thread engagement should be a distance equal to the diameter of the screw used; preferably, it should be a minimum of $1\frac{1}{2}$ times the screw diameter. Should an assembly fail, it is better to have the screw break

than to have the internal thread stripped from the hole. It is generally easier to remove a broken screw than to drill and tap for a larger screw size. With a screw engagement of $1\frac{1}{2}$ times its diameter, the screw will usually break rather than strip the threads in the hole.

Machine bolts are made with hexagonal or square heads. These bolts are often used in the assembly of parts that do not require a precision bolt. The body diameter of machine bolts is usually slightly larger than the nominal or standard size of the bolt. Body diameter is the diameter of the unthreaded portion of a bolt below the head. A hole that is to accept a common bolt must be slightly larger than the body diameter. Common bolts are made with a class 2A unified thread and come in both UNC and UNF series. Hexagonal head machine bolt sizes range from $\frac{1}{4}$-inch diameter to 4 inches. The size of the head on a square-head machine bolt is typically $1\frac{1}{2}$ times the thread diameter.

Stud bolts (Figure A-15) have threads on both ends. Stud bolts are used where one end is semipermanently screwed into a threaded hole. A good example of the use of stud bolts is in an automobile engine. The stud bolts are tightly held in the cylinder block, and easily changed nuts hold the cylinder heads in place. The end of the stud bolt screwed into the tapped hole has a class 3A thread, while the nut end is a class 2A thread. Studs are also widely used to hold vises and or parts down on machine tables.

Machine screws are made with either coarse or fine thread and are used for general assembly work. Machine screws are available in many sizes and lengths (Figure A-16). Several head styles are also available (Figure A-17). Machine screw sizes fall into two categories. Fraction sizes range from diameters of $\frac{1}{5}$ to $\frac{3}{4}$ inch. Screws that are under $\frac{1}{4}$-inch diameter are identified by numbers from 0 to 12. A No. 0 machine screw has a diameter of .060 inch (60 thousandths of an inch). For each number above zero, add .013 in. to the diameter.

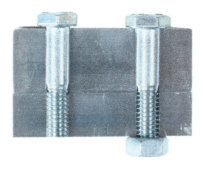

Figure A-14 On the left a fastener is used as a screw and on the right the screw is used as a bolt because it is used with a nut.

Figure A-15 Stud bolts, T-nuts, and a clamping nut (*Courtesy of Fox Valley Technical College*).

Figure A-16 Machine screw head styles.

Figure A-17 Various cap screws. From the left: a stainless-steel socket-head cap screw, a regular socket-head cap screw, a button-head cap screw, and a flat-head cap screw (*Courtesy of Fox Valley Technical College*).

EXAMPLE

Find the diameter of a No. 6 machine screw:

No. 0 diameter = .060 in.

No. 6 diameter = .060 in. + (6 × .013 in.)

= .060 in. + .078 in.

= .138 in.

Machine screws 2 inches long or less typically have threads extending all the way to the head. Longer machine screws have a $1\frac{3}{4}$-inch thread length.

Machine screws (Figure A-16) are made with a variety of different head shapes and are used where precision bolts or screws are needed. Machine screws are made with coarse, fine, or special threads. Machine screws with a 1-inch diameter have a class 3A thread. Those with greater than a 1-inch diameter have a class 2A thread. The strength of screws depends mainly on the kind of material used to make the screw. Materials that screws are made from include aluminum, brass, bronze, low-carbon steel, medium-carbon steel, alloy steel, stainless steel, titanium, and various plastics.

Socket Head Cap Screws

Socket head screws are made with a variety of different head shapes and are used where precision bolts or screws are needed. Socket head screws are manufactured to close tolerances. Cap screws can have round heads, low heads, flat heads, button head, or can be in a shoulder screw configuration (see Figure A-17).

Socket head cap screws are the strongest of the head styles. The height of the head is equal to the shank diameter.

Low Head Cap Screw

These are designed for applications in which head height is limited. These are not as strong as the regular head. The head height is approximately half the shank diameter.

Flat Head Cap Screw

These are designed for flush mount applications. Note that the inch and metric screws have different countersink angles. You must use the correct countersink or the screw may fail.

Button Head Cap Screw

These are well suited to holding thin materials because of the large head. They are good for tamper proofing because of the hex drive style makes it less susceptible to tampering with a screwdriver. Also good for applications where a regular screw (slotted or Phillips) is prone to stripping.

Socket Shoulder Screw

These are typically used as pivot points or axles for other parts to rotate around. The shoulders are ground to tight tolerances.

Bolt Grades and Torque Factors

In some instances, bolts need to be fastened with the correct amount of pressure (torque). In these instances, the manufacturer of the product will specify a certain torque be applied to a particular fastener. Insufficient torque will usually result in parts working loose. Overtightening, on the other hand, can cause stress or warping which also might disturb alignment on assemblies. The "armstrong" (overtightening) method of tightening fasteners can also cause broken castings, broken bolts, or stretching of the fastener.

Steel has excellent elasticity. Elasticity is the ability to stretch slightly, like a spring, and then return to its original shape. Any fastener must reach its limits of stretching to exert clamping force. But also, like a spring, an overstretched fastener takes on a set, loses its elasticity, and cannot snap back to its original shape. Proper torqueing will prevent this condition.

A popping or snapping sound is sometimes heard during the final tightening of a fastener. This popping sound indicates that the fastener is undergoing set. When a new fastener is being used and the popping occurs, the remedy is to back it off and retighten to the proper torque specifications. When an old fastener is being used and you hear popping, take the fastener out, and clean the bolt and the internal threads out completely. The safer, more economical thing to do is replace the old fastener with a new one.

SCREW STRENGTH GRADES

The selection of the right grade of fastener is just as important as proper torqueing. The strength grades of bolts are identified by the markings on the heads. The grade indicates the strength of the fastener. Figure A-18 shows a few examples of grade markings. Use a manufacturer's chart as a guide for the proper torque of fasteners.

Metric Strength Property Classes

The correct terminology for the grade of a metric fastener is property class. The metric system designates strength capabilities via property classes rather than grades. The numbering system is simple. Figure A-19 shows an example of the strength class marking on a metric bolt. The number that appears before the decimal, when multiplied by 100, will provide the approximate minimum tensile strength of the bolt. In this example, the 8 in 8.8 multiplied by 100 tells the user that this bolt has an approximate minimum tensile strength of 800 MPa (Mega (million) Pascals). The

Bolt head marking	SAE – Society of Automotive Engineers ASTM – American Society for Testing and Materials ASTM Definitions	Material	Minimum tensile strength in pounds per square inch (PSI)
 No marks	SAE grade 1 SAE grade 2 Indeterminate quality	Low carbon steel Low carbon steel	65,000 PSI
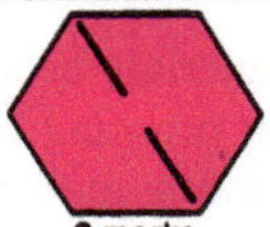 2 marks	SAE grade 3	Medium carbon steel, cold worked	110,000 PSI
 3 marks	SAE grade 5 ASTM – A 325 Common commercial quality	Medium carbon steel, quenched and tempered	120,000 PSI
 Letters BB	ASTM – A 354	Low alloy steel or medium carbon steel, quenched and tempered	105,000 PSI
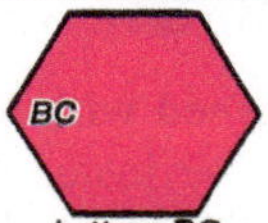 Letters BC	ASTM – A 354	Low alloy steel or medium carbon steel, quenched and tempered	125,000 PSI
 4 marks	SAE grade 6 Better commercial quality	Medium carbon steel, quenched and tempered	140,000 PSI
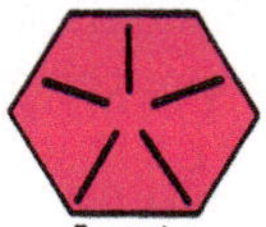 5 marks	SAE grade 7	Medium carbon alloy steel, quenched and tempered, roll threaded after heat treatment	133,000 PSI
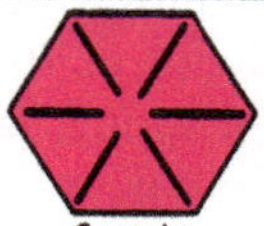 6 marks	SAE grade 8 ASTM – A 345 Best commercial quality	Medium carbon alloy steel, quenched and tempered	150,000 PSI

Figure A-18 Fastener strength grade chart.

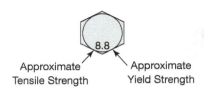

Figure A-19 Strength class marking for metric bolts (*Courtesy of Fox Valley Technical College*).

Pascal (symbol: **Pa**) is the unit of pressure, stress, and tensile strength, named after the French physicist, mathematician, inventor, and philosopher Blaise Pascal. It is a measure of force per unit area, defined as one newton per square meter.

The number which appears after the decimal, when multiplied by 10, will provide the approximate yield strength percentage in relation to the minimum tensile strength. For the 8.8 bolt, the 8 after the decimal point tells the user that the yield strength of the bolt is approximately 80 percent of the first number: 800 MPa. Thus, the 8.8 bolt has a yield strength of approximately 640 MPa.

There are nine common metric property classes: 4.6, 4.8, 5.8, 8.8, 9.8, 10.9, and 12.9. In the United States, property classes 8.8.3 and 10.9.3 also defined. The 3 after the second decimal point indicates the fastener is made of weather resistant steel. Hex head bolts M5 and larger have the property class marked on the head of the bolt. Fasteners smaller than M5, and those with slotted or recessed heads do not have to be marked.

Setscrews are used to lock pulleys or collars on shafts. Setscrews can have square heads with the head extending above the surface; more often, the setscrews are slotted or have socket heads. The heads of slotted or socket head setscrews are usually set below the surface of the part to be fastened. A pulley or collar with the setscrews below the surface is much safer for persons working around it. Socket head setscrews may have hex socket heads or spline socket heads. Setscrews are manufactured in number sizes from 0 to 10 and in fractional sizes from $\frac{1}{4}$ to 2 inches. Setscrews are usually made from carbon or alloy steel and hardened.

Setscrews can have various point configurations (Figure A-20). The flat point setscrew will make the least amount of indentation on a shaft and is used where frequent adjustments are made. A flat point setscrew is also used to provide a jam screw action when a second setscrew is tightened on another setscrew to prevent its release through vibration. The oval point setscrew will make a slight indentation as compared with the cone point. With a half-dog or full-dog point setscrew holding a collar to a shaft, alignment between shaft and collar will be maintained even when the parts are disassembled and reassembled. This is because the shaft is drilled with a hole of the same diameter as the dog point. Cup-pointed setscrews will make a ring-shaped depression in the shaft and will give a slip-resistant connection. Square head setscrews have a class 2A thread and are usually supplied with a coarse thread. Slotted and socket head setscrews have a class 3A UNC or UNF thread.

Thread-forming screws form their own threads and eliminate the need for tapping. These screws are used in the assembly of sheet metal parts, plastics, and nonferrous material. Thread-forming screws form threads by displacing material with no cutting action. These screws require an existing hole of the correct size.

Thread-cutting screws make threads by cutting and producing chips. Because of the cutting action these screws need less driving torque than thread-forming screws. Applications are similar to those for thread-forming screws. These include fastening sheet metal, aluminum, brass, die castings, and plastics.

COMMON INTERNALLY THREADED FASTENERS

Nuts

Common nuts (Figure A-21) are manufactured in as many sizes as bolts are. Most nuts are either hex (hexagonal) in shape. Nuts are identified by the size of the bolt they fit, not by their outside size. Common hex nuts are made in different thicknesses. A thin hex nut is called a *jam nut*. It is used where space is limited or where the strength of a regular nut is not required. Jam nuts are often used to lock other nuts. Regular hex nuts are slightly thinner than their size designation. A $\frac{1}{2}$-inch regular hex nut is $\frac{7}{16}$-inch thick. A $\frac{1}{2}$-inch heavy hex nut is $\frac{31}{64}$-inch thick. A $\frac{1}{2}$-inch-high hex nut measures $\frac{11}{16}$ inch in thickness. Other common nuts include various stop nuts or locknuts. Two common types are the elastic stop nut

Figure A-20 Setscrew points.

Figure A-21 Common nuts.

and the compression stop nut. They are used in applications where the nut might vibrate off the bolt. Wing nuts and thumb nuts are used where quick assembly or disassembly by hand is desired. Other hex nuts are slotted or castle nuts. These nuts have slots cut into them. When the slots are aligned with holes in a bolt, a cotter pin may be used to prevent the nut from turning. Axles and spindles on vehicles have slotted nuts with cotter pins to prevent wheel bearing adjustments from slipping.

Cap or acorn nuts are often used where decorative nuts are needed. These nuts also protect projecting threads from accidental damage. Nuts are made from many different materials, depending on their application and strength requirements.

INTERNAL THREAD INSERTS

Thread inserts can be used when an internal thread is damaged or stripped and it is not possible to drill and tap for a larger size. A thread insert retains the original thread size; however, it is necessary to drill and tap a somewhat larger hole to accept the thread insert.

One common type of internal thread insert is the key–lock type. The thread insert has both external and internal threads. This type of thread insert is screwed into a hole tapped to the same size as the thread on the outside of the insert. The four keys are driven in using a special driver (Figure A-22). This holds the insert in place. The internal thread in the insert is the same as the original hole.

A second type of thread insert is also used in repair applications as well as in new installations. Threaded holes are often required in products made from soft metals such as aluminum. If bolts, screws, or studs were to be screwed directly into the softer material, excessive wear could result, especially if the bolt was taken in and out a number of times. To overcome this problem, a thread insert made from a more durable material may be used. Stainless steel inserts are frequently used in aluminum (Figure A-23). This type of thread insert requires an insert tap, an insert driver, and a thread insert (Figure A-24). After the hole for the thread insert is tapped, the insert driver is used to screw the insert into the hole (Figure A-25). The end of

Figure A-23 Stainless steel thread insert used in an aluminum valve housing.

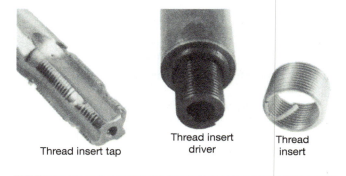

Thread insert tap

Thread insert driver

Thread insert

Figure A-24 Thread insert tap, driver, and repair insert for spark plug hole repair.

Figure A-25 Thread insert driver.

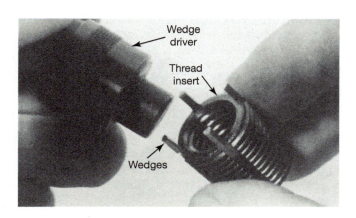

Wedge driver

Thread insert

Wedges

Figure A-22 Key–lock internal thread insert.

the insert coil must be broken off and removed after the insert is screwed into place. The insert in the illustration is used to repair spark plug threads in engine blocks.

WASHERS, PINS, RETAINING RINGS, AND KEYS

Washers

Flat washers (Figure A-26) are used under nuts and bolt heads to distribute the pressure over a larger area. Washers also prevent the marring of a finished surface when nuts or screws are tightened. Washers can be manufactured from many different materials. The nominal size of a washer is intended to be used with the same nominal-size bolt or screw. Standard series of washers are narrow, regular, and wide. For example, the outside diameter of a $\frac{1}{4}$-inch narrow washer is $\frac{1}{2}$ inch the outside diameter of a $\frac{1}{4}$-inch regular washer is almost $\frac{3}{4}$ inch and the diameter of a wide $\frac{1}{4}$-inch washer measures 1 inch.

Lock washers (Figure A-27) are manufactured in many styles. The helical spring lock washer provides hardened bearing surfaces between a nut or bolt head and the components of an assembly. The spring-type construction of this lock washer will hold the tension between a nut and bolt assembly even if a small amount of looseness should develop. Helical spring lock washers are manufactured in series: light, regular, heavy, extra duty, and hi-collar. The hi-collar lock washer has an outside diameter equal to the same nominal-size socket head cap screw. This makes the use of these lock washers in a counterbored bolt hole possible. Counterbored holes have the end enlarged to accept the bolt head. A variety of standard tooth lock washers are produced, the external type providing the greatest amount of friction or locking effect between fastener and assembly. For use with small head screws and where a smooth appearance is desired, an internal tooth lock washer is used. When a large bearing area is desired or where the assembly holes are oversized, an internal–external tooth lock washer is available. A countersunk tooth lock washer is used for a locking action with flat head screws.

Pins

Pins (Figure A-28) find many applications in the assembly of parts. **Dowel pins** are heat-treated and precision ground. Their diameter varies from the nominal dimension by only plus or minus .0001 inch (1/10,000th of an inch). Dowel pins are used where accurate alignments must be maintained between two or more parts. Holes for dowel pins are reamed to provide a slight press fit. Reaming is a machining process during which a drilled hole is slightly enlarged to provide a smooth finish and accurate diameter. Dowel pins only

Figure A-26 A few examples of flat washers (*Courtesy of Fox Valley Technical College*).

CONE SPRING TYPE The cone provides a spring compression locking force when the fastener is tightened, compressing the cone feature.	**CONE SPRING TYPE SERRATED PERIPHERY** Same general usage as the cone type with plain periphery but with the added locking action of a serrated periphery. Takes high tightening torque.	**COUNTERSUNK TYPE** Countersunk washers are used with either flat or oval head screws in recessed countersunk applications. Available for 82° and 100° heads and also internal or external teeth.
DISHED TYPE PLAIN PERIPHERY Recommended for the same general applications as the dome type washers but should be used where more flexibility rather than rigidity is desired. Plain periphery for reduced marring action on surfaces.	**DISHED TYPE TOOTHED PERIPHERY** Recommended for the same general applications as the dome type washers but should be used where more flexibility rather than rigidity is desired. Toothed periphery offers additional protection against shifting.	**DOME TYPE PLAIN PERIPHERY** For use with soft or thin materials to distribute holding force over larger area. Used also for oversize or elongated holes. Plain periphery is recommended to prevent surface marring.
DOME TYPE PLAIN PERIPHERY For use with soft or thin materials to distribute holding force over larger area. Used also for oversize or elongated holes. Toothed periphery should be used where additional protection against shifting is required.	**DOUBLE SEMS** Two washers securely held from slipping off, yet free to spin and lock. Prevents gouging of soft metals.	**EXTERNAL-INTERNAL TYPE** For use where a larger bearing surface is needed such as extra large screw heads or between two large surfaces. More biting teeth for greater locking power. Excellent for oversize or elongated screw holes.
EXTERNAL TYPE External type lock washers provide greater torsional resistance due to teeth being on the largest radius. Screw heads should be large enough to cover washer teeth. Available with left hand or alternate twisted teeth.	**FIBER AND ASBESTOS** In cases where insulation or corrosion resistance is more important than strength, fiber or asbestos washers are available.	**FINISH TYPE** Recommended where marring or tearing of surface material by turning screw head must be prevented and for decorative use.
FLAT TYPE For use with oversize and elongated screw holes. Spreading holding force over a larger area. Used also as a spacer. Available in all metals.	**HEAVY DUTY INTERNAL TYPE** Recommended for use with larger screws and bolts on heavy machinery and equipment.	**HELICAL SPRING LOCK TYPE** Spring lock washers may be used to eliminate annoying rattles and provide tension at fastening points.
INTERNAL TYPE For use with small screw heads or in applications where it is necessary to hide washer teeth for appearance or snag prevention.	**PYRAMIDAL TYPE** Specially designed for situations requiring very high tightening torque. The pyramid washer offers bolt locking teeth and rigidity yet is flexible under heavy loads. Available in both square and hexagonal design.	**SPECIAL TYPES** Special washers with irregular holes, cup types, plate types with multiple holes, or tab types may be supplied upon request.

Figure A-27 Lock washers.

Figure A-28 Pins.

locate. Clamping pressure is supplied by the screws. Dowel pins may be driven into a blind hole. A blind hole is closed at one end. When this kind of hole is used, provision must be made to let the air that is displaced by the pin escape. This can be done by drilling a small through hole or by grinding a narrow flat the full length of the pin.

One disadvantage of dowel pins is that they tend to enlarge the hole in an unhardened workpiece if they are driven in and out several times. When parts are intended to be disassembled frequently, **taper pins** can provide accurate alignment. Taper pins have a taper (diminishing diameter) of $\frac{1}{4}$ inch per foot of length and are fitted into reamed taper holes. If a taper pin hole wears larger because of frequent disassembly, the hole can be reamed larger to receive the next larger size of taper pin. Diameters of taper pins range in size from $\frac{1}{16}$ inch to $\frac{11}{16}$ inch measured at the large end. Taper pins are identified by a number from $\frac{7}{0}$ (small diameter) to number 10 (large diameter) as well as by their length. The large end diameter is constant for a given size pin, but the small diameter changes with the length of the pin. Some taper pins have a threaded portion on the large end. A nut can be threaded on the pin and used to pull the pin from the hole much like a screw jack. This facilitates removal of the pin.

A **grooved pin** is either a cylindrical or a tapered pin with longitudinal grooves pressed into the pin body. This causes the pin to deform. A grooved pin will hold securely in a drilled hole even after repeated removal.

Roll pins can also be used in drilled holes with no reaming required. These pins are manufactured from flat steel bands and rolled into cylindrical shape. Roll pins, because of their spring action, will stay tight in a hole even after repeated disassembly.

Cotter pins are used to retain parts on a shaft or to lock a nut or bolt as a safety precaution. Cotter pins make quick assembly and disassembly possible.

Retaining Rings

Retaining rings are fasteners that are used in many assemblies. Retaining rings can be easily installed in machined grooves, internally in housings, or externally on shafts or pins (Figure A-29). Some types of retaining rings do not require grooves but have a self-locking spring-type action. The most common application of a retaining ring is to provide a

shoulder to hold and retain a bearing or any other part on an otherwise smooth shaft. They may also be used in a bearing housing (Figure A-30). Special pliers are used to install and remove retaining rings.

Keys

Keys (Figure A-31) are used to prevent the rotation of gears or pulleys on a shaft. Keys are fitted into key seats in both the shaft and the external part. Keys should fit the key seats

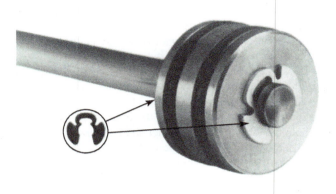

Figure A-29 External retaining ring used on a shaft (Rotor Clip Company, Inc.).

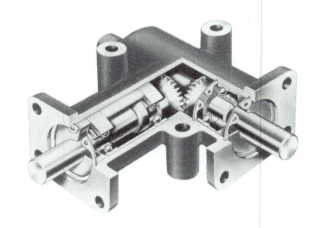

Figure A-30 Internal retaining rings used to retain bearings (Rotor Clip Company, Inc.).

Figure A-31 Keys. A square key, a Woodruff key, a rectangular key, and a gib-head taper key.

rather snugly. **Square keys**, where the width and the height are equal, are preferred on shaft sizes up to a $6\frac{1}{2}$-inch diameter. Above a $6\frac{1}{2}$-inch diameter, rectangular keys are recommended. Woodruff keys, which almost form a half circle, are used where relatively light loads are transmitted. One advantage of **Woodruff keys** is that they cannot change their axial location on a shaft because they are retained in a pocket. A key fitted into an end-milled pocket will also retain its axial position on the shaft. Most of these keys are held under tension with one or more setscrews threaded through the hub of the pulley or gear. Where extremely heavy shock loads or high torques are encountered, a taper key is used. **Taper keys** have a taper (diminishing diameter) of $\frac{1}{8}$ inch per foot. When a tapered key is used, the key seat in the shaft is parallel to the shaft axis, and a taper to match the key is in the hub. Where only one side of an assembly is accessible, a gib-head taper key is used instead of a plain taper key. When a gib-head taper key is driven into the key seat as far as possible, a gap remains between the gib and the hub of the pulley or gear. The key is removed for disassembly by driving a wedge into the gap to push the key out.

SELF-TEST

1. Define the term pitch diameter.
2. Name two ways to measure a thread.
3. What is the rule of thumb for the length of internal thread to maximize the strength of the bolt? (In other words, that the bolt will break before the threads pull out.)
4. Describe when class two fits are used.
5. Describe UNC and UNF.
6. What is the formula for calculating the OD of a machine screw? What are setscrews used for?
7. When are stud bolts used?
8. Describe the strength classification system and marking for bolts.
9. Explain two reasons why flat washers are used.
10. What is the purpose of a helical spring lock washer?
11. When is an internal–external tooth lock washer used?
12. When are dowel pins used?
13. When are taper pins used?
14. When are roll pins used?
15. What are retaining rings?
16. What is the purpose of a key?
17. When is a Woodruff key used?
18. When is a gib-head key used?

Blueprint Reading Fundamentals

From earliest times, people have communicated their thoughts through drawings. The pictorial representation of an idea is vital communication between the designer and the people who produce the final product. The drawing can also provide an important testing phase for an idea. Many times, an idea may be rejected at the drawing board or CAD stage before a large investment is made.

OBJECTIVES

After completing this unit, you should be able to:

- Find information in a title block.
- Explain terms such as surface finish, isometric, oblique, first angle projection, and third angle projection.
- Make isometric and oblique sketches of objects.
- Find information on blue prints.

Almost anything can be represented by a drawing. It is important that the designer be aware machining methods and limitations. On the other hand, a machinist must fully understand all the symbols and terminology on the designer's drawing. The machinist must be able to understand the drawings to transform the ideas of the designer into useful products.

The name "blueprint" originated from one of the first processes used to make copies of drawings. One of the first processes developed to duplicate drawings produced white lines on a blue background, hence the name "blueprint."

TITLE BLOCK

The title block is normally located in the lower-right corner (Figure A-32). The title block contains the drawing number, the name of the part or assembly, and the material to be used. It also includes the scale of the drawing, drafting record, authentication, date, and may include more information depending on the complexity of the part.

Drawing Number

Each blueprint has a drawing number in the title block (Figure A-32). On blueprints with more than one sheet, the information in the number block shows the sheet number and the number of sheets in the series. For example, note that the title blocks shown in Figure A-32 shows sheet 1 of A complex part or assembly might have several sheets.

Revision Block

If a revision has been made, there will be a revision block on the drawing, as shown in Figure A-32. Note that each revision is numbered and the revisions are shown with the number on the actual drawing where the feature was changed. The revision block may be attached to the title block or be in a separate location on the print. All revisions in this block are identified by a number and a brief description of the revision. A revised drawing is shown by the addition of a letter to the original number as shown in the figure. When the print is revised, the revision will use the next available letter. For example, if the first revision was labelled A, the next revision will be labelled B. The letter is used to identify the current revision of the drawing.

Reference Number

Reference numbers that appear in the title block refer to numbers of other blueprints. On some prints you may see a notation above the title block such as "34512 RH shown; 34512-1 LH opposite." RH means right hand, and LH means left hand. Note that both the right-hand and left-hand parts carry the same number, but the left-hand part number has a "−1."

Zone Number

Zone numbers serve the same purpose as the letters and numbers on the borders of maps. They are used to locate a particular feature on a drawing. It makes it easier to convey information when you are communicating with

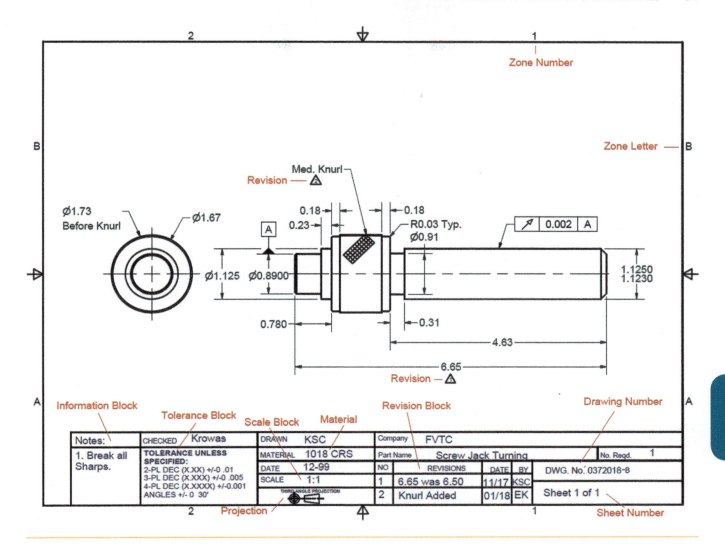

Figure A-32 An example of a typical blueprint.

someone on the phone or through email. You can give them the letter and number that helps them find the feature you are interested in. To find the feature, they can mentally draw a horizontal and vertical line from the letters and numbers. The lines will intersect at the feature you want to find.

Scale Block

Parts may be too large to draw them full-size on the print. They may also be small and it might be advantageous to draw them larger for clarity. The scale block in the title block of the blueprint shows the size of the drawing compared with the actual size of the part. For example, the part may be drawn $\frac{1}{2}$ scale. That would mean that the drawing is $\frac{1}{2}$ the size of the actual part. If it were $\frac{2}{1}$, the drawing would twice the size of the part. The scale is chosen to fit the object being drawn and space available on the print.

Never measure a drawing to find a dimension; use the dimensions that are shown on the print. The print may have been reduced in size from the original drawing.

Bill of Material Block

The bill of material block specifies the material to be used for the part (Figure A-32). If there are multiple parts that go into an assembly of parts, they are also listed in the material block, as well as the number required for the assembly. Parts and materials may be specified by their stock number or other appropriate number.

Information Blocks

Information blocks are used to give the machinist information about materials, specifications, that may not be on the blueprint or that may need additional explanation. Heat treat specification would be one example of information that might be in an information block.

Standard Tolerance Information Blocks

Not all features will have specific tolerances shown on a blueprint. For those features that do not have a specific tolerance shown, shop tolerances are shown in an information block. For example, if a feature on a print does not have a specified tolerance, the general tolerance is used. For example, if the feature's size has two decimal places, the general tolerance might be plus or minus .015. A three-place decimal feature might have a general tolerance of plus or minus .005. These tolerances vary from shop to shop and print to print. Always look for general tolerances in an information block (Figure A-32).

Finish Marks

A machinist often works on parts that are already partially shaped. An example of this might be a casting or forging. Finish marks (Figure A-33) are used to indicate which surfaces are to be machined. Finish marks may also indicate a required degree of surface finish. For example, a finish mark notation of 4, 32, or 64 refers to a specific surface finish. The number with the surface finish symbol indicates the average height of the surface deviations measured in micro inches. A micro inch equals one millionth of an inch (.000001 inch). The smaller the number, the better the finish. Higher cost is usually associated with finer finishes. Figure A-34 shows some manufacturing processes and what finishes are possible with each. Machining provides a better surface appearance and a better fit with closely mated parts, but is also adds cost. Cast parts, for example, often do not have every surface machined. Figure A-35 shows various surface finish symbols.

Figures A-36 and A-37 show explanations of surface finish specifications. Surface finish, also known as surface

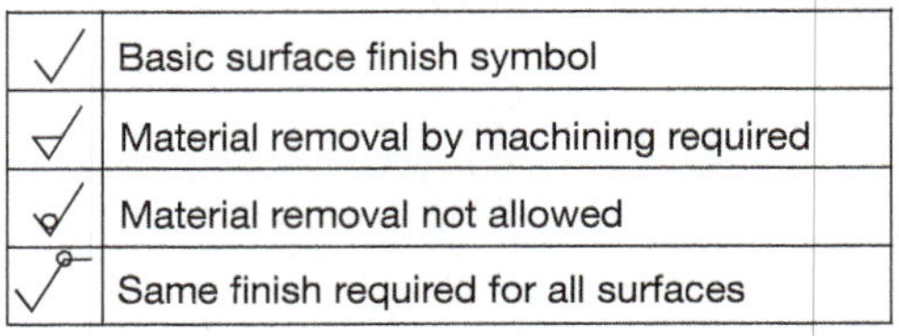
Basic surface finish symbol
Material removal by machining required
Material removal not allowed
Same finish required for all surfaces

Figure A-35 Surface finish symbols.

Figure A-33 Drawing showing finish marks.

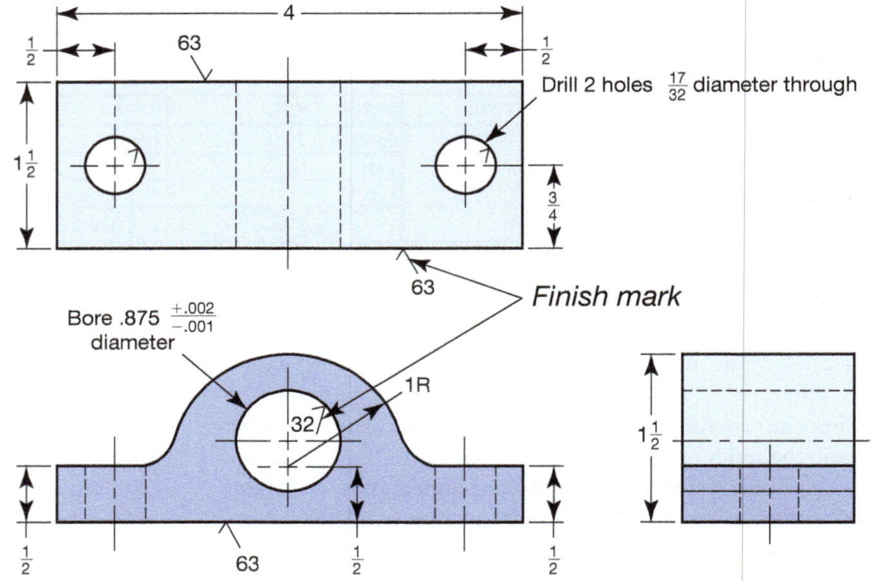
4
1/2
63
1/2
Drill 2 holes 17/32 diameter through
1 1/2
3/4
63
Finish mark
Bore .875 +.002 −.001 diameter
1R
32
1/2
63
1/2
1/2
1 1/2

Figure A-34 Roughness for various machining methods.

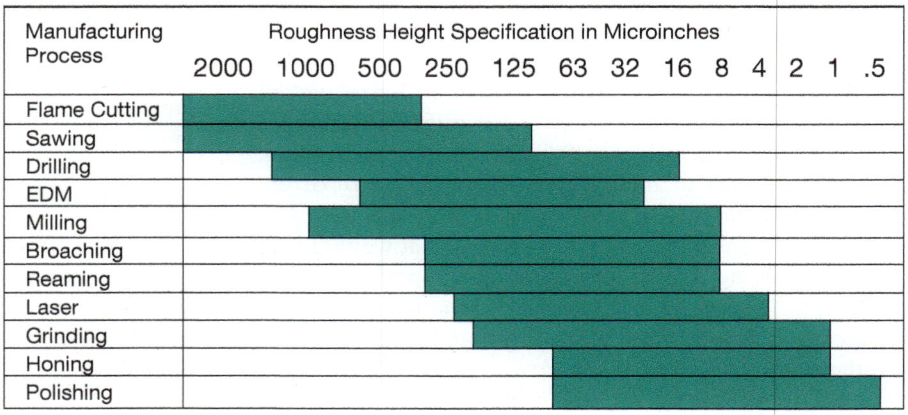
Manufacturing Process
Roughness Height Specification in Microinches
2000 1000 500 250 125 63 32 16 8 4 2 1 .5
Flame Cutting
Sawing
Drilling
EDM
Milling
Broaching
Reaming
Laser
Grinding
Honing
Polishing

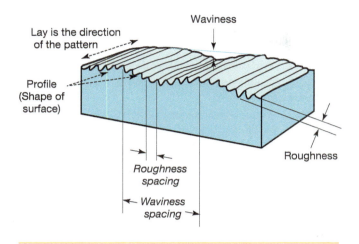

Figure A-36 Surface finish characteristics.

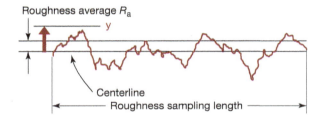

Figure A-37 Roughness (Ra) illustration.

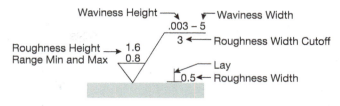

Figure A-39 How surface finish is represented on a blueprint.

Isometric and Oblique (Perspective) Drawings

An isometric drawing (Figure A-40) is one method used to represent an object in three dimensions. In the isometric format, the lines of the object remain parallel and the object is drawn about three isometric axes that are 120 degrees apart. In an isometric drawing the baselines of the part are drawn at a 30-degree angle to horizontal.

Figure A-41 shows an oblique drawing of a cube. Note that the front view is drawn horizontally. Object lines in the oblique drawing are parallel. The oblique differs from the isometric in that one axis of the object is parallel to the plane of the drawing. Isometric and oblique are not used as working drawings for the machinist. You may occasionally see them in the machine shop.

texture, is the characteristic of a surface. Surface finish has three components: lay, surface roughness, and waviness.

On a machined part, form is the result of fluctuations when the machine produces the part. These could be called straightness errors. Waviness is a result of vibrations in the machine tool and from the surroundings. Roughness results from the chosen tool geometry, the condition of the tool, the feed-rate, and variations in material and its hardness.

The most common measure of surface finish is average roughness (Ra). Ra is calculated by calculating the average difference between the peaks and valleys and the deviation from the mean line within the sampling length (see Figure A-37). Ra averages the peaks and valleys of the part surface.

Figure A-38 shows the symbols that are used when direction of lay is specified.

Figure A-39 shows several surface finish characteristics that can be specified with a finish symbol.

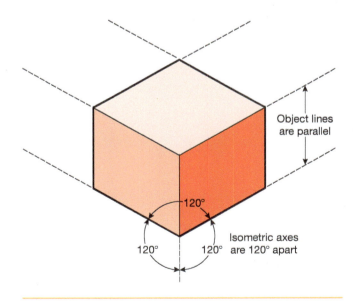

Figure A-40 An isometric drawing.

Figure A-38 Illustrations of lay and their symbols.

Parallel	Across	Crossed	Different	Concentric	Radial	Non-Directional
=			X	M	C	R	P

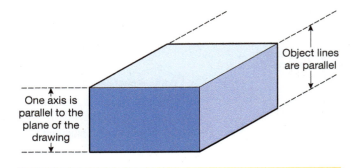

Object lines are parallel

One axis is parallel to the plane of the drawing

Figure A-41 Oblique drawing of a cube.

SELF-TEST

1. Draw the symbol for third-angle projection.
2. What is the difference between first and third angle projection?
3. Draw an oblique sketch of a shoe box.
4. Draw an isometric sketch of a shoe box.
5. List at least five things that are found in a title block.
6. Explain where revisions are shown on a blueprint and how they are indicated on the feature that was changed.

Views and Line Types

The basis of blueprints is orthographic projection. This unit will examine how views of objects are chosen and projected to create blueprints. This unit will also examine line types. Different types and weights of lines are used to convey information on blueprints.

VIEWS

Imagine you are a mechanical designer. You have a part that you need to draw. You can pick up the part, turn it around, and look at the features of the part. You can look at the object so that certain features are visible and also true size and shape. You must determine how many views of the object you need to draw to make every detail of the part understandable. Determining which views to provide on the drawing is a critical decision.

Too many views can make a drawing difficult to read. Omitting views that provide true size views of features will make it impossible to correctly dimension the drawing. The designer should provide views that will provide the part information needed in production and a clear representation of the geometry of the part. The machinist must have the ability to look at the views provided on a blueprint and visualize what the part will actually look like.

Orthographic Projection

In almost every case, the working drawing will be a three-view or orthographic drawing. The typical orthographic format always shows an object in the three-view combination of front, side, and top (Figure A-42). In many cases, an object can be completely shown by a combination of only two orthographic views. However, any orthographic drawing must have a minimum of two views to show an object completely. The top, front, and right-side views are the most common on typical drawings. Left-side, rear, and bottom views are also seen occasionally, depending on the complexity of the part being illustrated.

Third-Angle Projection versus First-Angle Projection

There are two main methods to project the part into views. Third-angle projection is the one used in America. A blueprint will have a symbol on it to declare which method was used. Figure A-43 shows the symbols for third-angle and first-angle projection. First angle is common in Europe and

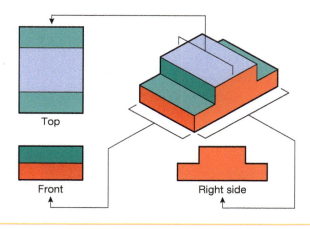

Figure A-42 An orthographic drawing.

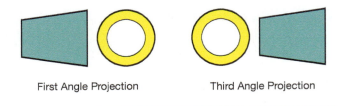

First Angle Projection   Third Angle Projection

Figure A-43 Symbols for first-angle and third-angle projection.

many other countries. It is always wise to look for the projection symbol on a blueprint to be sure which system is being used. If you think the print was drawn in third angle projection but it was really drawn in first angle projection you will make parts that are essentially left-handed instead of right-handed and the parts will be scrap.

There are six possible standard views. Two or three are usually enough to show all the required information for a part. Figure A-44 shows how the three most common views are projected to create the front, top, and right-side views for a blueprint.

Figure A-44 also shows examples of third-angle and first-angle projection. In these examples the views were projected onto the walls of an imaginary glass box. In third-angle projection (normally used in America) the front of the glass box shows the front view. The right side of the glass box shows the right-side view. The top of the glass box shows the top view. Note also the arrangement of the views. If there was a hinge on the edge of the box between the front view and the right-side view, we could open the side of the box out to the front and it would be in the correct position for the drawing. If there was a hinge on the edge of the glass box between the front view and top view, we could open the top of the glass box and the top view would be in the correct position for the drawing.

LINE TYPES AND MEANINGS

There are several types of lines used on blueprints. The type and weight of the line are used to convey information. Figure A-45 shows the use of standard lines in a simple drawing. Line characteristics such as width, breaks in the line, and zigzags have meaning.

Object Lines

Object lines (Figure A-45) are thick solid lines that outline all surfaces of a part that are visible to the eye.

Hidden Lines

Hidden or invisible lines are lines that have short, evenly spaced dashes. Hidden lines are used to outline invisible or hidden surfaces (Figure A-45). They are thin lines, about half the weight of visible lines. Hidden lines begin with a dash, except if the dash would create a continuation of a solid line.

Extension Lines

Extension lines are short, solid, thin lines used to show the limits of dimensions (Figure A-45). They may be placed inside or outside the outline of an object. They extend from the outline or surface of a part. Extension lines do not touch the part.

Dimension Lines

A dimension line is a thin solid line that shows the extent and direction of a dimension. Arrows are used at the ends of dimension lines to show the limits of the dimension. Dimension lines are broken to insert the dimension (Figure A-45).

Figure A-44 The example on the top shows a third-angle projection. Note that the projections on the glass box are between the viewer and the object. The example on the bottom shows a first-angle projection. Note that the object is between the viewer and the projections on the glass box. First-angle projects different views. Make sure you understand which system was used before making parts.

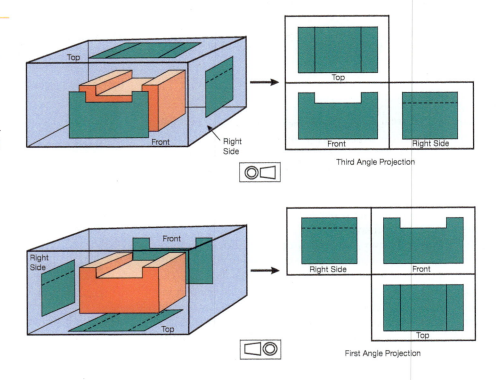

Third Angle Projection

First Angle Projection

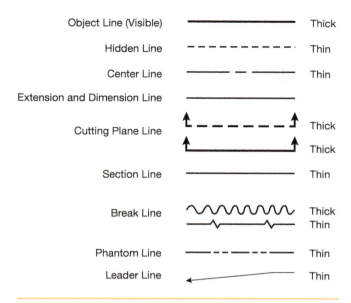

Figure A-45 Line types.

Leader Lines

A Leader line is a thin solid line that may have an arrow that is used to indicate a feature on a part (Figure A-45). Leader lines indicate the part or feature to which a number, note, or other reference applies.

Centerlines

Centerlines have alternating long and short evenly spaced dashes. Centerlines have the same weight as invisible lines. They have a long dash at each end and short dashes at points of intersection (Figure A-45). Centerlines are used to show the central axis of an object or part. If a circle, such as a bolt circle or hole is shown, horizontal and vertical centerlines are used to show the center point.

Break Lines

There are two types of lines for long and short breaks (see Figure A-45). Short breaks are shown by solid, thick, freehand lines. Long breaks are shown by solid, thin, lines broken by freehand zigzags. These can be used to indicate that a part is broken out or removed. They also are used to reduce the size of the drawing of a long part having a uniform cross section. For example, a shaft that is 10 feet long with no details between the ends would be drawn with a break in the middle, rather than having to show 10 feet of the shaft. Breaks on shafts, rods, tubes, and pipes are curved (Figure A-45).

Phantom Lines

Phantom lines (Figure A-45) are thin lines that can be used to indicate the locations of absent parts, alternate positions of the parts of an object, or repeated details.

Cutting Plane Lines

Cutting plane lines bisect a part and provide a view of the part's interior features (Figures A-45 and A-46). A cutting plane is a plane that cuts through a part drawing to create a sectional view. The sectional view is used to show internal features that cannot be seen from the outside. Imagine cutting a candy bar in half to see what is inside. The portion of the part that is in front of the cutting plane surface is removed to expose the internal details on the section view. Figure A-46 shows an example of a full- and half-section view. In the section view, features that would be cut to produce the section view are indicated by section lines.

Section Lines

Section lines, also called crosshatch lines, are used to show the portion of a part that would be cut to create a section view (Figure A-46). Section lines can also be used to distinguish between two separate parts that meet at a given point. Each part is lined or hatched in opposite directions with thin parallel lines at 30, 45, or 60 degrees on the exposed cut surface.

When internal features are complex and showing them on a drawing as hidden lines would be confusing, a section drawing may be used. Two common styles of sections are used. In the full section (Figure A-46, Section AA), the object has been cut completely through. In the half section (Figure A-46, Section BB), one quarter of the object is removed. The section indicator line shows the plane at which the section is taken. For example (Figure A-46, Section AA), the end view of the object shows the section line marked AA. Section line BB (Figure A-46, Section BB) indicates the portion removed in the half section. An object may be sectioned at any plane as long as the section plane is indicated on the drawing.

Figure A-47 shows an example of a drawing with various types of lines and their use identified.

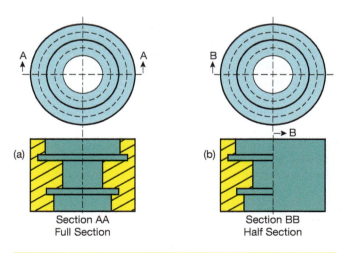

Figure A-46 Full- and half-section views.

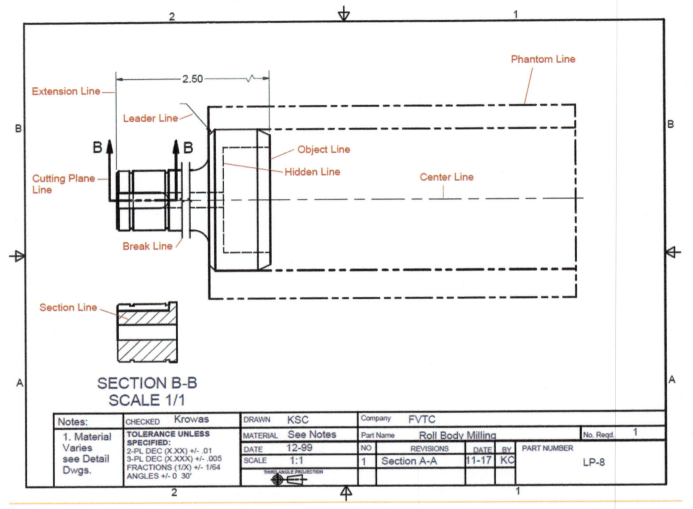

Figure A-47 Drawing showing various types of lines.

Auxiliary Views

When creating drawings, features should be shown in a view where they appear to be true to their size so that they can be dimensioned. The object is normally positioned such that the major surfaces and features are either parallel or perpendicular to the X, Y, and Z planes. For example, normally the front, top, and right-side views are shown on a part drawing. Views of the part are normally chosen so that most of the features will be visible in the front, top, and right-side views.

Study the angular surface in Figure A-48. To get a true view of the angled surface and its size, the view had to be projected at an angle. If lines had been projected from the part vertically or horizontally, the view and size of the surface would be distorted. Lines had to be projected at the angle perpendicular to the surface to provide a true view of the surface.

If a part is complex, three views may not be enough to show all of the part's features. Some part features may not appear to be true size and shape in those views, or may be hidden. In examples like this, it may be necessary to draw one or more auxiliary views. Primary auxiliary views are projected off one of the main part views. If a secondary auxiliary view is required, it is projected off a primary auxiliary view.

Exploded Views

An exploded drawing (Figure A-49) is a type of pictorial drawing designed to show several parts in their proper location prior to assembly. Exploded views appear extensively in manuals used for repair and assembly of machines and other mechanisms.

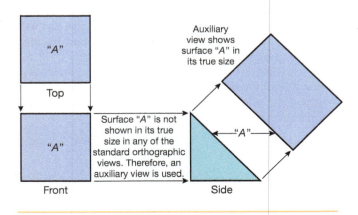

Figure A-48 Drawing with an auxiliary view.

7. Draw the missing lines in the views shown in Figure A-56.
8. Draw the missing lines in the views shown in Figure A-57.
9. Draw the missing lines in the views shown in Figure A-58.
10. Draw the missing lines in the views shown in Figure A-59.

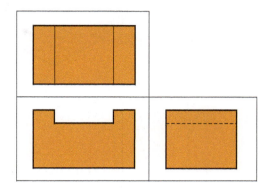

Figure A-51 Drawing for question 2.

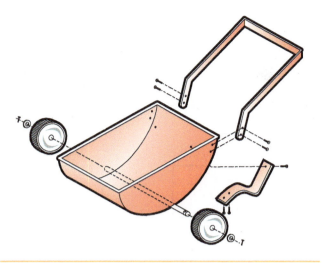

Figure A-49 Example of an exploded view.

SELF-TEST

1. Identify the types of lines in the drawing in Figure A-50 (Hint: object, invisible, phantom, extension, dimension, broken, hidden, center, cutting plane line, part, heavy, break, section, parting)

2. Assume third-angle projection. Label the views in Figure A-51.

3. Draw the three third-angle projection views for the part shown in Figure A-52.

4. Draw the missing lines in the three views shown in Figure A-53.

5. Fill in the numbers that represent the surface on the part shown in Figure A-54 in each view.

6. Fill in the numbers that represent the surface on the part shown in Figure A-55 in each view.

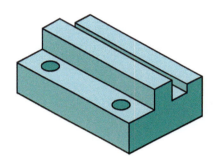

Figure A-52 Drawing for question 3.

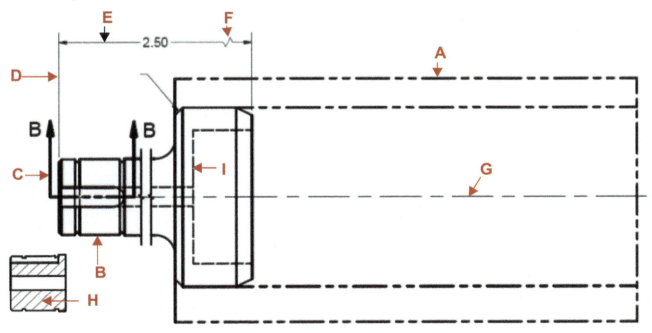

Figure A-50 Part drawing for the self check.

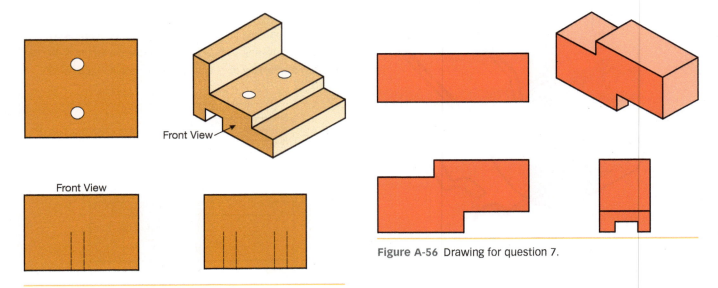

Front View

Front View

Figure A-53 Drawing for question 4.

Figure A-54 Drawing for question 5.

Figure A-56 Drawing for question 7.

Figure A-55 Drawing for question 6.

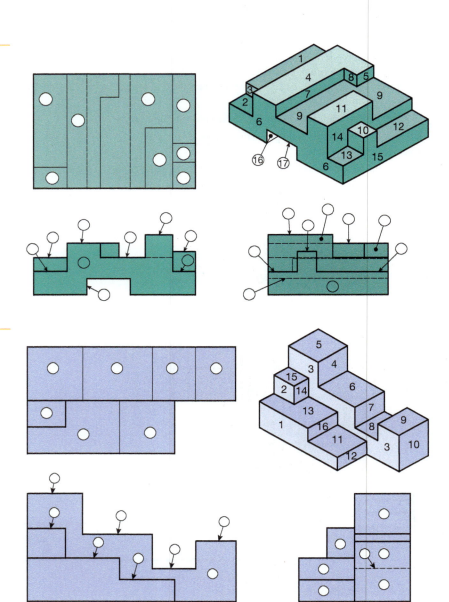

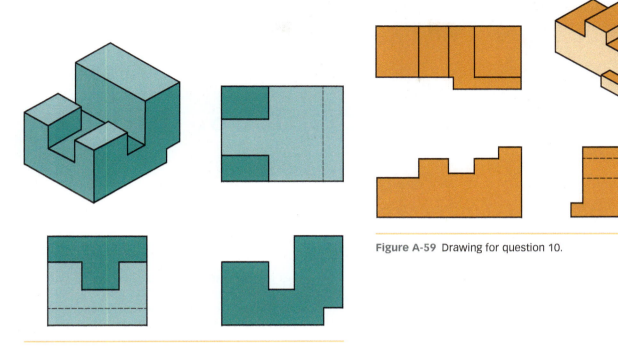

Figure A-59 Drawing for question 10.

Figure A-57 Drawing for question 8.

Figure A-58 Drawing for question 9.

Dimensions, Tolerances, and Fits

A key to producing functional parts is the development of dimensions for key features and their tolerances. Prints can be dimensioned using absolute dimensioning or incremental dimensions. Both are used in some cases. Mating parts also need to be dimensioned and toleranced. Parts have various functions. In one application one part, like a shaft, needs to be turned by a gear. In this example, a drive fit would be needed so that the gear is tight on the shaft and would turn the shaft. There are a variety of types of fits. This unit will cover the system of fits and their specification.

OBJECTIVES

After completing this unit, you should be able to:

- Explain terms such as absolute, incremental, detail drawing, bilateral, unilateral, limits, counterbore, spotface, and countersink.
- Develop unilateral and bilateral tolerances for part features.
- Given an application that involves fit, specify the tolerances for the clearance and the mating parts.

DIMENSIONS

A machinist typically works with a drawing type known as detail drawing. A detail drawing is a drawing of an individual part. A detail drawing contains all the essential information a machinist needs to make a particular part. Most important are the dimensions. Dimensioning refers generally to the sizes specified for the part and the locations of its features. Furthermore, dimensions reflect many design considerations, such as the fit of mating parts so they will function correctly.

A dimension defines the size or specification of a feature on a part. A basic dimension is the numerical value that would be the theoretically exact size (true size) of a feature. A reference dimension is a numerical value provided for information only. Reference dimensions are enclosed in parentheses on blueprints. They are for reference only and are not used in the manufacturing of the part.

The coordinate or absolute system of dimensioning (Figure A-60) is used for most part drawings. In this system, all dimensions are specified from the same zero point. This is very important. If some part features were specified from the left side of the part and some from the right, that part might not be functional because the length of the part will vary. If the part is longer, the features specified from the right side will be further away from the part features specified from the left side. If all of the part feature dimensions are taken from one side, the part should be functional because all of the part features will be located in relation to each other. Note that this method of dimensioning also makes it easier to program the part on a CNC machine. The other method of dimensioning is called incremental dimensioning. In incremental dimensioning, one feature is specified as the distance and direction from another feature, not from a zero point.

The figure shows the dimensions expressed in decimal form. General tolerances that would be shown in the title block apply to the print unless specific tolerances are shown.

With the increasing use of the metric system in recent years, some companies put dual dimensions on drawings with both metric and inch notation. In Figure A-61, metric dimensions appear above the line and inch dimensions appear below the line.

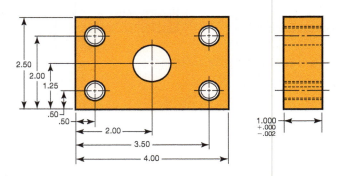

Figure A-60 Example of absolute dimensioning.

Figure A-61 Dual dimensioning in metric and English.

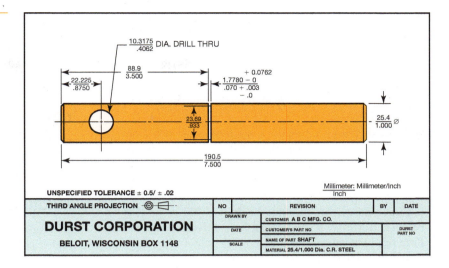

Limit and Tolerance

Since it is impossible to machine a part to an exact size, a designer must specify an acceptable range of sizes that will still permit the part to function as designed. Tolerances also make it possible for different manufacturers to make parts and have them work together. The difference between the maximum and minimum limits is the tolerance, or the total amount by which a part dimension may vary. Tolerances on drawings are often indicated by specifying a limit, or by plus and minus notations (Figure A-62). A limit dimension shows the upper and lower limit for the allowable size. With plus/minus tolerance, when the tolerance is both above and below the nominal (true theoretical) size, it is said to be bilateral (two-sided). When the tolerance is indicated all on one side of nominal, it is said to be unilateral (one-sided).

Plus and minus dimensioning is the allowable positive and negative variance from the dimension specified. Note that the tolerance does not have to be the same in the positive and negative specification for a feature. For example, a bore might be specified as 1.000 +.002 −.001.

Standard Tolerances

On many drawings, tolerances will be specified at the specific feature's dimension. If no specific tolerances are specified at the feature's dimension, general tolerances are used. These are often listed in the title block.

Note that in Figure A-63, no tolerance is specified for the overall dimensions of the part or the locations of the various features. In this case the general tolerances are applied. In this example the general tolerances are shown at the top of the figure. The general tolerances apply when no specific tolerance is shown for a particular part feature. Consider the overall length of the part (4.00 in). Since there is no specific tolerance shown for that dimension on the print, the general tolerance would apply. In this example from Figure A-63 the tolerance for 4.00 (a 2-place decimal) would be ±.01.

How Tolerance Affects Mating Parts

When parts mate in an assembly, tolerance becomes crucial. Consider the following example shown in Figure A-64. The shaft must fit the bore in the bearing and be able to turn freely. The diameter of the shaft is specified as 1.000 + or − .001. This means that the maximum limit of the shaft is 1.001 and the minimum is .999. The total tolerance is .002 and bilateral. The maximum limit of the bearing bore is 1.002 and the minimum limit is 1.000. The total tolerance is .002. Will the shaft made by one machine shop fit the bearing made by another machine shop using the tolerances specified? If the shaft is turned to the maximum limit of 1.001 and the bearing is bored to its minimum limit of 1.000, both parts will be within acceptable

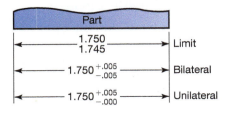

Figure A-62 Types of tolerance.

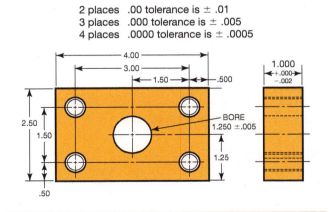

Figure A-63 Illustration of tolerancing a part.

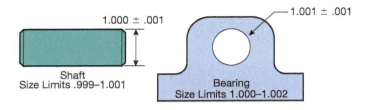

Figure A-64 The shaft diameter is specified to be 1.000 ± .001. The maximum diameter would be 1.001 and the minimum diameter would be .999. The bore could be 1.000 and still be in tolerance. The shaft could be larger than the bore, thus preventing proper fit.

tolerance but will not fit, since the shaft is .001 larger than the bearing. However, if the bearing bore was specified in limit form or unilateral tolerance of 1.002 + .002 − .000, the parts would fit as intended. Even if the shaft was turned to the high limit of 1.001, it would still fit the bearing even though the bore was machined to the low limit of 1.002.

FITS

Fit refers to the amount of clearance or lack of clearance between two mating parts. Fits can range from free running or sliding, where a certain amount of clearance exists between mating parts, to press or interference fits, where parts are forced together under pressure. Clearance fits can range from a few millionths of an inch, such as in the parts of a ball or roller bearing, to a clearance of several thousandths of an inch, for a low-speed drive bushing. There are charts in the Machinery's Handbook and other references that have classes of fits and the allowances for each.

There are three basic types of fit: clearance; transition, and interference. A clearance fit is a fit that has clearance between the hole and shaft. A transition fit is a fit where, depending on the size of the shaft and hole, clearance or interference may occur in the coupling. An interference fit is a fit that always has some interference between the hole and shaft in the coupling.

Running and Sliding Fits (RC)

Running and sliding fit ranges are designated with the classifications of RC1 to RC9. The lower RC numbers are, the tighter the fit. The higher the number, the looser the fit.

Close Sliding Fit (RC1)

A close sliding fit is intended for the accurate location of parts which must assemble without noticeable play.

Sliding Fit (RC2)

RC2 fits are designed for accurate location but with greater maximum clearance than class RC1. Parts made to this fit, turn and move easily. RC2 fits are still relatively close and in larger sizes may seize with temperature changes.

Precision Running Fit (RC3)

An RC3 fit is about the closest fit that will run freely. Precision running fits are intended for precision applications at low speed with low bearing pressures. RC3 fits should not be used where noticeable temperature differences can occur.

Close Running Fit (RC4)

RC4 fits are used for running fits on accurate machinery with moderate speed and moderate bearing pressures where accurate location and minimal play is required.

Medium Running Fit (RC5 and R6)

RC5 and RC6 fits are designed for machines running at higher speeds with large bearing pressures.

Free Running Fit (RC7)

RC7 fits are used where accuracy is not crucial. An RC7 fit allows large temperature variations.

Loose Running Fit (RC8 and RC9)

RC8 and RC9 fits are used where wide tolerances are acceptable for the shaft. These loose running fits may be used where the parts are exposed to the effects of corrosion and contamination.

Locational Fits (LC, LT, LN)

Locational fits are divided into three groups; locational clearance fits (LC), locational transition fits (LT), and interference fits (LN). These fits are used to accurately locate mating parts. They provide accurate location with an interference fit or a minor amount of location variation with a clearance fit. Locational fits are not used for moving parts.

Locational Clearance Fit (LC)

Locational fits are used for non-moving parts that need to be easily assembled and disassembled. Locational clearance fits can be snug fits for parts that require accurate location or looser fits. The tolerance of mating parts and clearance increases with increasing class of the fit. The lower classes such as LC1 and LC2 are for precise location of parts. Higher classes such as LC10 and LC11 are for free fits with greater clearances and maximum tolerances. LC fits are tighter than RC fits.

Locational Transition Fit (LT)

Locational transition fits lie between clearance and interference. Locational interference fits are for applications where accuracy of location is important. A small amount of clearance or interference is permissible with a locational transition fit. These fits are a compromise between LC and LN (interference/force) fits. LT1 and LT2 fits can be used for bushings, hubs of gears or pulleys, etc. LT3 and LT4 fits can be used for pulleys, wheels, etc. The parts can be assembled or disassembled without any great force using a rubber hammer. LT5 and LT6 can be used for motor armatures, driven bushings, shafts and so on. LT fit parts can be assembled using low force.

Locational Interference Fit (LN)

Locational Interference fits are used where accuracy of location is very important. Locational interference fits are used for parts requiring rigidity and alignment. The parts can be assembled or disassembled using a press. These fits are not designed for applications where one part is used to drive another part. For example, a shaft and a pulley. The fit would be insufficient to drive the pulley without additional parts such as a key or setscrew.

Force or Shrink Fit (FN)

Force or shrink fits are used to move parts using the friction between holes and shafts. Force and shrink fits are a type of interference fit used to maintain constant hole pressure for all sizes of components. The amount of interference varies with the diameter of the parts. The amount of interference of the fit increases with increasing class of the fit. During assembly, steel parts are usually greased to prevent seizing or galling of the components. High pressures are normally required to assemble parts of this class of fit.

Light Drive Fit (FN1)

Light drive fits are used for parts that require light assembly pressure. Light drive fits are used for more permanent assemblies.

Medium Drive Fit (FN2)

Medium drive fits are used for steel parts. They are also used for shrink fits on light, fragile sections of material. Medium drive fits are about the tightest fits that can be used with cast-iron parts. Cast iron is quite brittle.

Heavy Drive Fit (FN3)

Heavy drive fits are used for heavy steel parts or can be used for shrink fits in medium sections of material.

Force Drive Fits (FN4, FN5)

Force drive fits are for parts which can be heavily stressed. They can also be used for shrink fits where heavy pressing forces are not practical.

HOW TO DETERMINE SIZES FOR A GIVEN FIT

Table A-4 shows a partial chart showing sizes for a few of the running and sliding (RC) fits. Note that only RC1–RC3 are shown and only a few diameters are included. Comprehensive charts that show all classes and types of fits for all diameters are readily available in reference texts. The first step for a machinist or designer is to choose the correct type of fit for the application. Let's assume an RC3 fit is chosen of our example. An RC3 fit is a close fit that will run freely. Precision running fits are used for precision applications at low speed with low bearing pressures and light journal pressures. Assume we have a 1.5-inch shaft that turns slowly in a bronze bearing on a conveyor. We find that a 1.5-inch part would be in the bottom row of the chart in Table A-4 in the basic size ranges column of the chart. Then we go to the RC3 columns on the right side of the chart. We see that the clearance between the shaft and bore should be between .001 and .0026 of an inch. The values shown in the chart are in thousandths of an inch. From the chart we see that tolerance of the bore should be 0 to .001. For our 1.5-inch application, the bore should be toleranced as 1.500 to 1.501. The shaft tolerance from the table is −.001 to −.0016. The shaft specification should be 1.4984 to 1.499.

A machinist often deals with press or interference fits. For a press or interference fit, two parts are forced together, usually by mechanical or hydraulic pressing. The frictional forces hold the parts together without any additional hardware such as keys or setscrews. Tolerances for press fits are critical because parts can easily be damaged if there is an excessive difference in their mating dimensions. In addition, press fitting physically deforms the parts to some extent. This can result in damage, mechanical binding, or the need for a secondary resizing operation such as hand reaming or honing after the parts are pressed together.

Table A-4 Partial Chart of RC Fits for Various Diameters of Shafts and Bores

| Basic Size Ranges | | RC1 | | | RC2 | | | RC3 | | |
| | | Clearance Limits | Standard Tolerance Limits | | Clearance Limits | | | Clearance Limits | | |
Over	To		Hole	Shaft		Hole	Shaft		Hole	Shaft
0	0.12	.1 to .45	0 to +.2	−.1 to−.25	.1 to .55	.25 to 0	−.1 to −.3	.3 to .95	.4 to 0	−.3 to −.55
0.12	0.24	.15 to .5	.2 to 0	−.15 to −.3	.15 to .65	.3 to 0	−.15 to −.35	.4 to 1.12	.5 to 0	−.4 to −.7
0.24	0.40	.2 to .6	.25 to 0	−.2 to −.35	.2 to .85	.35 to 0	−.2 to −.45	.5 to 1.5	.6 to 0	−.5 to −.9
0.40	0.71	.25 to .75	.3 to 0	−.2 to −.45	.25 to .95	.4 to 0	−.25 to −.55	.6 to 1.7	.7 to 0	−.6 to −1
0.71	1.19	.3 to .95	.4 to 0	−.3 to −.55	.3 to 1.2	.5 to 0	−.3 to −.7	.8 to 2.1	.8 to 0	−.8 to −1.3
1.19	1.97	.4 to 1.1	.4 to 0	−.4 to −.7	.4 to 1.4	.6 to 0	−.4 to −.8	1 to 2.6	1 to 0	−1 to −1.6

Values Below Are in Thousandths of Inches

A typical example of press fit is the pressing of a ball bearing inner race onto a shaft, or an outer race into a bore. Thus, the bearing is retained by friction, and the free-running feature is obtained within the bearing itself. Ball-to-race clearance is only a few millionths of an inch in precision bearings. If a bearing is pressed into a bore or onto a shaft with excessive force, the bearing may be physically deformed and cause mechanical binding. This would cause excess heat and friction and result in early failure of the bearing. On the other hand, insufficient frictional retention of the part resulting from a poor press fit can cause the wrong part to turn under load or cause the mechanism to come loose during operation.

PRESS FIT ALLOWANCES

Press fit allowances depend on factors such as the diameter, the length of engagement, the material, and the need for later disassembly of parts. Soft materials, such as aluminum, may deform, and may not stand up to repeated pressings. Parts made of the same material pressed without lubrication may gall, making them difficult if not impossible to press apart. Thin parts such as thin-walled tubing may bend or deform to such a degree that the press fit fails to hold the parts together under load.

Generally, pressing tolerances range from a few ten thousandths to a few thousandths of an inch depending on the diameter of the parts and other factors. Proper measurement tools and techniques must be used to make accurate measurements. For additional dimensions on pressing allowances, consult a Machinery's Handbook or other reference.

PRESS FITS AND SURFACE FINISHES

The surface finish (texture) of parts being pressed is an important factor. Smooth-finished parts will press together more easily than rough-finished parts. If the roughness height of the surface texture is large, more frictional forces will be generated, and the chances for misalignment, galling, and seizing will be increased, especially if no lubricant is used. Lubrication will improve this situation. Lubrication, however, can be detrimental to press fit retention in some cases. Lubricant between fit surfaces, especially if they are quite smooth, can cause the parts to slip when subjected to force.

SHRINK AND EXPANSION FITS

Parts can be fitted by using the characteristics of metals to expand or contract when heated or cooled. Heating parts expand them, and they can then be slipped onto a mating part. When the part cools, the heated part will contract and grip the mating part with tremendous force. Parts may also be mated by cooling one so that it shrinks. When the part warms to ambient temperature, it will expand to meet the mating part. Shrink and expansion fits can have superior holding power over press fits. As with press fits, allowances are extremely important. Consult a machining handbook for proper allowance specifications.

A clearance fit exists when two mating parts have a clearance between them when assembled. An interference fit exists when two mating parts interfere when assembled. A transition fit exists when two mating parts may have an interference fit or a clearance fit when assembled.

ABBREVIATIONS FOR MACHINE OPERATIONS

Drawings contain symbols and abbreviations that convey important information to the machinist (Figure A-65). Many machining operations can be abbreviated. Countersinking is a machining operation in which the end of a hole is shaped to accept a flat head screw. On a drawing, countersinking may be abbreviated as CS or CSK. The desired angle will also be specified. A counterbore is a hole that is enlarged in diameter so that a bolt head may be recessed. Counterboring may be abbreviated as C'BORE. Spot facing is usually spelled out. This operation is similar to counterboring except that the spot facing depth is only enough to provide a smooth and flat surface around a hole. Figure A-65 shows the symbols for counterbore, spotface, and countersink. Note that counterbore and spotface use the same symbol. The depth would be different for spotface and counterbore.

OTHER COMMON SYMBOLS AND ABBREVIATIONS

External and internal radii are generally indicated by the symbol R and the specified size. Chamfers may be indicated by size and angle (Figure A-66). Threads on drawings are generally represented by symbols (Figure A-66). Thread notations on drawings include form, size, number of threads per inch or per millimeter, and the class of fit.

Bolt circles are indicated by the abbreviation B.C. or D.B.C (diameter bolt circle), meaning bolt circle (see Figure A-66). The size of the diameter of a bolt circle is often indicated with an abbreviation such as 2 1/2 B.C.

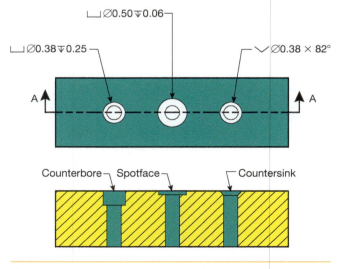

Figure A-65 Symbols for various machining operations.

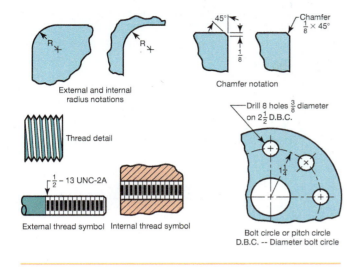

External and internal radius notations

Thread detail

$\frac{1}{2}$ – 13 UNC-2A

External thread symbol   Internal thread symbol

45°

Chamfer $\frac{1}{8}$ × 45°

$\frac{1}{8}$

Chamfer notation

Drill 8 holes $\frac{3}{8}$ diameter on $2\frac{1}{2}$ D.B.C.

$1\frac{1}{4}$

Bolt circle or pitch circle
D.B.C. -- Diameter bolt circle

Figure A-66 Common machining symbols and abbreviations.

SELF-TEST

1. What is absolute dimensioning?
2. What is incremental dimensioning?
3. What is a general tolerance and where are they found?
4. Write down an example of a bilateral tolerance.
5. Write down an example of a unilateral tolerance.
6. What are the three basic types of fits?
7. Describe an RC3 fit.
8. Describe an LT fit.
9. Describe a LN fit.
10. An RC3 fit is desired for a one inch shaft and bore. Give the tolerance specification for the clearance between the shaft and bore, the tolerance for the shaft, and the tolerance for the bore in thousandths of an inch.

Fundamentals of GD&T

Geometric dimensioning and tolerancing (GD&T) has had profound effects on manufacturing. GD&T actually enables a part to have more tolerance and still assure that it will function as designed. One interesting fact about GD&T is that a tolerance can actually change based on the size of a part. For example, if a bored hole is at the upper end of its size the location tolerance might be larger, and yet assure that it will function.

OBJECTIVES

After completing this unit, you should be able to:

- Explain the difference between traditional rectangular tolerances and GD&T tolerances.
- Explain the use and function of datums.
- Explain maximum material condition, least material condition, and regardless of feature size as they affect tolerances.
- Calculate tolerances based on MMC and LMC.
- Explain what a feature control frame is.

GEOMETRIC DIMENSIONING AND TOLERANCING

It is often more important to control the form, orientation, location, profile, or runout of a part than the basic tolerance. In other words, the function of the parts is the most important. Geometric tolerancing helps establish specifications so that parts will function as designed.

Geometric Dimensioning and Tolerancing (GD&T)

Tolerances allow parts to vary from their perfect sizes, but only within defined limits. The difference with GD&T tolerancing is that it is based on the part's function. Tolerance limits allow a part's size to vary, but ensure that the part will function as intended. A designer must always balance between high cost, narrow tolerances and lower cost, wide tolerances.

GD&T can actually allow larger tolerances while ensuring the parts will function properly. Figure A-67 shows an example. A hole position is specified using normal rectangular tolerances on the left of the drawing. The hole position is 1.000 plus or minus .005 in the X and the Y axes. This tolerance zone is illustrated in gray on the right-hand side of the figure. The hole center's location could be anywhere in the gray area and it would be in tolerance for location. A hole could be located at the extreme corner positions and still be within the + or − .005 tolerance. But as the figure shows, the hole position would actually be located .007 from the center on the diagonal. That means that the hole location (if properly engineered) would actually be functional if it were .007 out of location. The problem is that with rectangular tolerancing there is no way to show that, so the stated tolerance must be + or − .005. Geometric tolerancing utilizes a circular tolerance zone. Note that any hole location in the blue would also be a good part. In effect the tolerance has been increased with the use of GD&T while still ensuring the proper function of the part.

Geometric tolerances are used to control the form and shape of parts. They are used when the shape or form has a particular function and inaccuracies would result in poor performance. In these cases, GD&T tolerances are applied in addition to dimensional tolerances.

Datums and Basic Dimensions

Datums are reference points, lines, areas, and planes that are theoretically exact for the purpose of calculations and measurements. The first machined surface on a casting, for example, might be selected as a datum surface and used as a reference from which to locate other part features. Datums are usually not changed by subsequent machining operations and are identified by letters inside a rectangular frame (see Figure A-68). In this figure, the bottom of the plate is Datum B and the left side of the part is Datum A. In this example, it would be important to take dimensions from Datum A and Datum B.

The term basic dimension (or true position) represents a theoretically exact dimension describing the location or

Figure A-67 This figure compares rectangular tolerances versus GD&T tolerances.

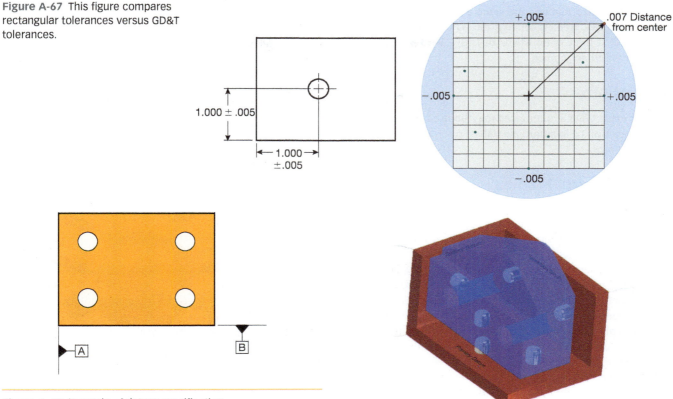

Figure A-68 Example of datum specification.

Figure A-69 Example of how part datums are established.

shape of a part feature. Basic dimensions have no tolerance. The tolerance comes from the associated geometric tolerances used with the basic dimension.

Primary, Secondary, and Tertiary Datum Planes

Figure A-69 shows how datums are established. Datum A is the primary, Datum B is the secondary, and Datum C is the tertiary datum. The primary datum (Datum A) is selected to provide functional relationships, standardizations, and repeatability between surfaces. Datum B is a secondary datum and is perpendicular to the primary datum so measurements can be referenced from them. Datum C, when needed, is perpendicular to both the primary and secondary datums ensuring the part's fixed position from three related datums. A datum does not have to be a surface. A datum can be a feature, such as the center of a bore.

After the primary datum is located using three points (see Figure A-69), the secondary datum will typically use two points to define a line. The third (tertiary) datum is defined by a single point.

Maximum Material Condition, Least Material Condition, and Regardless of Feature Size

Three material condition modifier symbols are shown in Figure A-70. These symbols identify at which size the geometric tolerance applies. MMC refers to the maximum

amount of material remaining on a feature or size. On an external cylindrical feature this would be the high limit of the feature tolerance. For example, a shaft with a diameter of .750+ or −0.010 would have an MMC diameter of .760, since this would leave maximum material remaining on the part (largest diameter allowed for the shaft). An internal cylindrical feature such as a hole with a diameter of .750 Plus or minus 0.010 would have an MMC diameter of .740, since this would leave maximum material remaining on the part (smallest hole, most remaining material).

Least material condition (LMC) is the opposite. The LMC of the shaft would be .740 diameter, and the LMC of the hole would be .760 diameter. These sizes would have the least amount of material remaining on the part. RFS (regardless of feature size) means that the geometric tolerance applies no matter what the feature size is. The tolerance zone size remains the same, unlike MMC or LMC, which allows a tolerance zone size growth as the feature size changes. RFS is the default condition for all geometric tolerances. Unless

MMC or Ⓜ for Maximum Material Condition

LMC or Ⓛ for Least Material Condition

RFS or Ⓢ for Regardless of Feature Size

Figure A-70 Symbols for maximum material condition, least material condition, and regardless of feature size.

MMC or LMC is specified on the drawing by M or L, the tolerance zone size remains the same even though the feature size changes. Remember that the RFS symbol is very seldom used. If there is no MMC or LMC, RFS is implied. LMC is not used very much.

Limits of Size

Many different GD&T tolerances may be used to control a feature's position or location. No datum reference is needed for limits of size.

Feature Control Frames

Figure A-71 shows a GD&T feature control frame. The leftmost box would hold a geometric symbol that describes what type of tolerance this is. For example, it might be the symbol for perpendicularity. The next box would hold the tolerance zone size. For example, if the tolerance is .005, the tolerance zone would be .005 and the feature controlled by this specification would have to be perpendicular within the .005 tolerance zone. The third box would have a datum reference. This will all be clarified later in this section.

Figure A-72 shows an example of a feature control frame with three datums specified. The location control symbol is used in this example. The location of this feature would have to be within .005 at MMC in relation to Datum A, Datum B, and Datum C.

A sample drawing that includes a feature control frame (FCF) is shown in Figure A-73. In this example the flatness symbol is shown in the left rectangle of the FCF. The right box of the FCF has the tolerance for flatness for this example. The FCF denotes a flatness tolerance of .002s. Tolerance values in FCFs are a total tolerance, not a plus or minus value. In this example, you could think of it as a ±.001 tolerance of flatness. This flatness tolerance controls how flat this surface is.

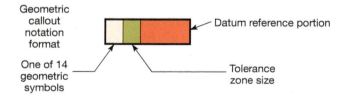

Figure A-71 A feature control frame.

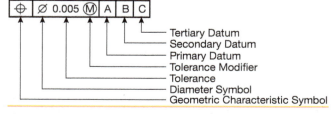

Figure A-72 Feature control frame with three datums specified.

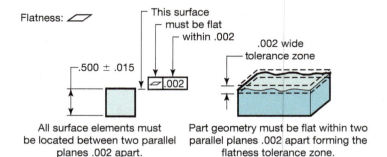

Figure A-73 Example of flatness specification.

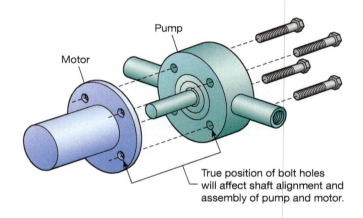

Figure A-74 Pump and motor assembly drawing illustrating the importance of true position.

True Position

Study the pump and motor shown in Figure A-74. For these to fit together, the holes in the pump and the threaded holes in the motor need to be accurately located.

SELF-TEST

1. How is it possible that GD&T can have larger tolerances than traditional rectangular tolerancing and still assure that the part will function as designed?
2. What is a datum?
3. How do you decide where a datum should be?
4. Describe how a part would be located based on primary, secondary and tertiary datums.
5. Describe the term MMC as it would apply to a shaft.
6. Describe the term MMC as it would apply to a bore.
7. Describe the term LMC as it would apply to a shaft.
8. Describe the term LMC as it would apply to a bore.
9. Write down the symbol for MMC and LMC.
10. What is RFS and why isn't it used much?
11. Draw a typical feature control frame for location with a primary datum, a secondary datum, a tolerance, and a MMC condition.

Geometric Tolerancing

The real power of GD&T is in its ability to describe the required relationships between features. For example, maybe the top and bottom of a part need to be parallel within .001. GD&T enables relationships like these to be specified. This unit will cover the ways in which GD&T can specify these relationships between features.

OBJECTIVES

After completing this unit, you should be able to:

- Explain terms such as form, profile, orientation, location, and runout.
- Evaluate feature control frames that specify control features.
- Apply MMC and LMC to feature control frame tolerances.
- Read blueprints that utilize GD&T feature control frames.

GEOMETRIC CONTROLS

Form tolerances are for individual features of a part. Figure A-75 shows the four forms of tolerance types and symbols.

Tolerance Type	Characteristic	Symbol
Form	Straightness	—
	Flatness	▱
	Circularity (Roundness)	○
	Cylindricity	⌀

Figure A-75 Forms of tolerance symbols.

Straightness

A straightness tolerance specifies a tolerance zone within which an axis or element must lie (Figure A-76). Straightness is a condition in which a feature is a straight line. Straightness is a two-dimensional geometric tolerance. Straightness is a condition where one-line element of a surface or an axis must lie in a straight line. It controls how much one line on a surface can deviate from a straight line. The feature control frame is attached to the surface with an extension line or a leader. Tolerance zone for both cylindrical and flat surfaces are applied along the entire surface. When straightness is inspected each line checked on the surface is independent.

Figure A-77 shows how the straightness tolerance is applied. In this example the tolerance for straightness is .005.

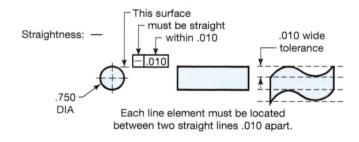

Figure A-76 Straightness specification example.

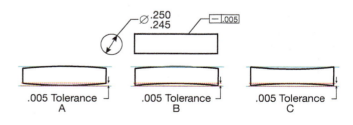

Figure A-77 Illustration of straightness tolerance.

Figure A-78 Straightness tolerance with MMC.

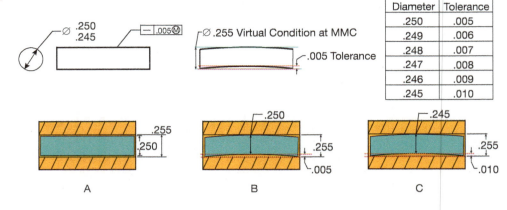

Diameter	Tolerance
.250	.005
.249	.006
.248	.007
.247	.008
.246	.009
.245	.010

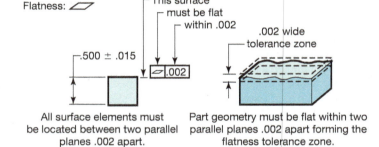

There is no modifier (MMC or LMC) to the tolerance so the tolerance is regardless of feature size (RFS). The diameter of the part must be .245 to .250. Note examples A, B, and C in the figure. Note that the tolerance zone applies to individual lines on the part. Each line would be independent. Note also that the part can have different shapes as long as each line falls within the .005 straightness tolerance zone.

Figure A-78 shows how the straightness tolerance is applied. In this example the tolerance for straightness is .005. There is a maximum material condition (MMC) modifier to the tolerance. The diameter of the part must be .245 to .250. Note that the tolerance zone applies to individual lines on the part. Each line would be independent. Note also that the part can have different shapes as long as each line falls within the .005 straightness tolerance zone. In this example there is a MMC modifier so the tolerance zone changes depending on the diameter of the part. The table shows the size of the tolerance zone for each possible diameter. Note that if the part is at the maximum diameter of .250, the tolerance is the specified .005 (MMC). If the part diameter is at the smallest allowable size of .245, the tolerance zone is .010 (larger).

A gage could be used to inspect this part. The gage is illustrated in gray in A, B, and C in Figure A-78. Note that the bore though the gage is .255 in all three examples. In illustration A, the part diameter is .250 and the part appears to be straight. It fits in the gage with .005 clearance. In illustration B, the part diameter is .250 and we can see that the straightness tolerance allows the part to be bent (not straight) up to the .005 tolerance (part at MMC of .250). In illustration C, the part is at its smallest allowable size. In illustration C, the part can be bent (not straight) up to .010. This should make sense. If the part is smaller, it can be less straight and still fit into the gage.

Flatness

A flatness tolerance specifies a tolerance zone defined by two parallel planes within which all points on the surface must lie (Figure A-79). The tolerance is specified by a zone formed by two parallel planes.

A flatness tolerance does not have to be related to a datum. Flatness is applied to an individual surface. A feature control frame points to the surface with a leader line. When a feature

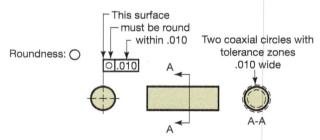

Figure A-79 Flatness specification example.

control frame with a flatness tolerance is applied with a size dimension, the flatness tolerance applies to the median plane.

Roundness or Circularity

A roundness (circularity) tolerance specifies a tolerance zone bounded by two concentric circles within which each circular element of the surface must lie. Circularity tolerance is used to control the roundness of circular features. Circularity tolerance is applied to an individual surface. Circular features can be defined by cylinders, cones, or spheres.

Figure A-80 shows an example of a circularity tolerance. Note that the circularity (or roundness) symbol is

Figure A-80 Circularity specification example.

Figure A-81 Cylindricity specification example.

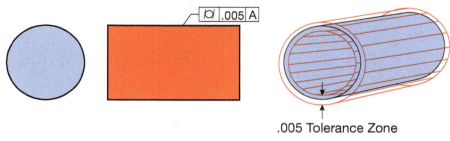

.005 Tolerance Zone

used in the feature control frame and a tolerance of .010 is specified. Section A-A shows the two concentric circles spaced .010 apart. All points on that circle on the part must fall within the zone shown by the two concentric circles. Note that more than one circle might be checked on a diameter, but all checks are independent for circularity.

Cylindricity

Figure A-81 shows the concept of cylindricity. The tolerance zone for cylindricity is the red inner and outer cylinder. All points on the surface of the part (gray) must fall within the inner and outer tolerance zones.

QUICK CHECK 1

1. Explain the form tolerance shown in Figure A-82.
2. Explain the form tolerance shown in Figure A-83.
3. Explain the difference between a flatness tolerance and a straightness form tolerance.
4. Explain the form tolerance shown and complete the tolerance table shown in Figure A-84.
5. Explain the form tolerance shown in Figure A-85.
6. Explain the form tolerance shown in Figure A-86.

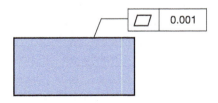

Figure A-82 Drawing for question 1.

Figure A-84 Drawing for question 4.

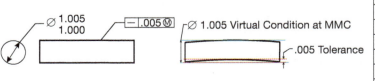

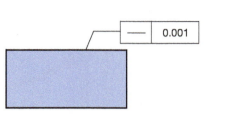

Figure A-83 Drawing for question 2.

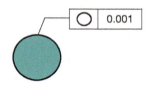

Figure A-85 Drawing for question 5.

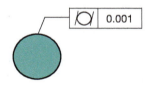

Figure A-86 Drawing for question 6.

PROFILE TOLERANCES

A profile tolerance specifies a tolerance zone along the true profile within which all elements of the surface must lie. Figure A-87 shows the profile symbols.

Tolerance Type	Characteristic	Symbol
Profile	Profile of a Line	⌒
	Profile of a Surface	⌓

Figure A-87 Symbols for profile of a line and surface.

Diameter	Tolerance

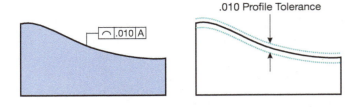

Figure A-88 Line profile specification example.

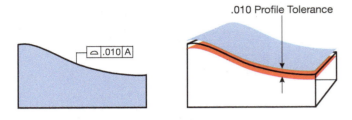

Figure A-89 Surface profile specification example.

Profile of a Line

The profile of a line describes the tolerance zone around a line, usually of a curved shape. Profile of a line is a tolerance zone that can be applied to any linear tolerance. A profile of a line takes a cross section at any point along the surface and sets a tolerance zone on either side of the specified profile (Figure A-88). This would be inspected by checking lines on the surface to make sure all points on the line were within the tolerance zone. With a line profile, the tolerance applies to each checked line independently. The points on one line are not compared with the points on a different line.

Profile of a Surface

Figure A-89 shows the tolerance zone for a surface. The tolerance is .010. The tolerance zone is illustrated on the right of the figure. In a surface profile all points on the surface must fall within the tolerance zone.

QUICK CHECK 2

1. Explain the profile tolerance shown in Figure A-90.
2. Explain the profile tolerance shown in Figure A-91.

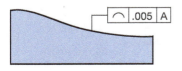

Figure A-90 Drawing for question 1.

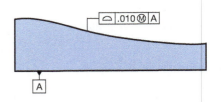

Figure A-91 Drawing for question 2.

ORIENTATION, LOCATION, AND RUNOUT TOLERANCING

Orientation, location, and runout tolerances are for related features of a part. Figure A-92 shows the symbols for orientation, location, and runout.

Angularity

The angularity tolerance symbol is shown as "∠." An angularity tolerance specifies a tolerance zone defined by two parallel planes at the specified angle. In this example, the top surface must be at a 15-degree angle in relation to the bottom (Datum A) within a tolerance zone of .010. LMC or MMC can apply to features of size (Figure A-93).

Perpendicularity

The perpendicularity tolerance symbol is "⊥." A perpendicularity tolerance is a three-dimensional geometric tolerance that controls how much a surface, axis, or plane can

Tolerance Type	Characteristic	Symbol
Orientation	Angularity	∠
	Perpendicularity	⊥
	Parallelism	//
Location	Position	⊕
	Concentricity	◎
	Symmetry	=
Runout	Circular Runout	↗
	Total Runout	↗↗

Figure A-92 Orientation, location, and runout symbols.

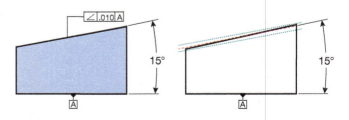

Figure A-93 Angularity specification example.

deviate from a 90-degree angle. A perpendicularity tolerance zone can be defined by two parallel planes or a cylindrical zone perpendicular to a datum. LMC or MMC can apply to features of size. Figure A-94 shows an example of a pin that must be perpendicular within .005 to a surface (Datum A). Note the cylindrical tolerance zone.

Figure A-95 shows another example of perpendicularity. The left side of the figure shows the perpendicularity specification for the surface. The right side of the figure shows the .010 MMC tolerance. The tolerance at MMC is .010. If the width is 1.105, the tolerance zone will be .010. The table in the figure shows the tolerance zone for different size widths. Note that the smaller the width is the larger the tolerance zone becomes.

Parallelism

The parallelism symbol is used in the feature control frame in Figure A-96. A parallelism tolerance zone is the condition such that a surface is equidistant at all points from a datum plane, or an axis. The distance between the parallel lines, or surfaces, is specified by the geometric tolerance zone. In this example all the points in the top surface must be within a .002 tolerance zone that is parallel to the bottom surface (Datum B). LMC or MMC can apply to features of size.

Figure A-97 shows an example of a parallelism specification for a bored hole. In this example the hole must be parallel to surface Datum A within .002. Note the tolerance zone in the figure.

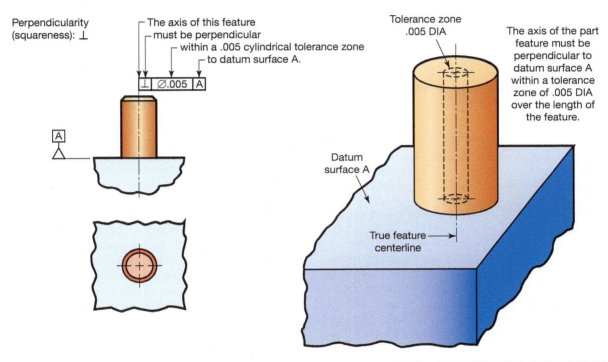

Figure A-94 Perpendicularity example.

Figure A-95 Example of perpendicularity of a surface.

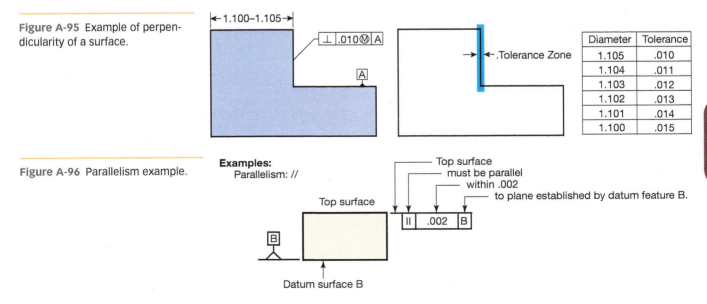

Diameter	Tolerance
1.105	.010
1.104	.011
1.103	.012
1.102	.013
1.101	.014
1.100	.015

Figure A-96 Parallelism example.

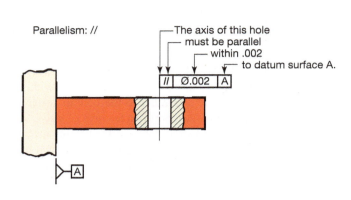

Parallelism: //

The axis of this hole
must be parallel
within .002
to datum surface A.

| // | Ø.002 | A |

A

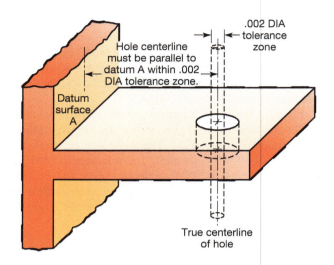

.002 DIA
tolerance
zone

Hole centerline
must be parallel to
datum A within .002
DIA tolerance zone.

Datum
surface
A

True centerline
of hole

Figure A-97 Parallelism example for a bored hole.

QUICK CHECK 3

1. List the three types of orientation tolerances.
2. Explain the orientation tolerance and complete the tolerance table in Figure A-98.

CONTROLLING THE LOCATION OF FEATURES

Figure A-77 shows an exploded view of a motor and pump that are bolted together. The pattern of bolt holes in the pump must match the pattern of bolt holes in the motor for them to fit together. Also, the bore in the pump housing must match the boss on the motor so that the shaft and the motor will align. If the position of either bolt pattern deviates far from the specified dimensions, the parts will not fit together. This is an example of a situation in which the location or the true position of the holes could be more critical than the size of the holes themselves.

Location Tolerances

Location tolerances can be stated by three tolerance zones. These are position, concentricity, and symmetry tolerances (Figure A-99). Concentricity and symmetry are used to control the center distance of feature elements. Position is used to control coaxial features, the center distance between

features, and the location of features as a group. Position, concentricity, and symmetry tolerances are associated with datums. LMC or MMC can apply to features of size.

Position Tolerance Position tolerances control how much the location of a feature can vary from its true position. Positional tolerances are used to locate features of size from datum planes. A position tolerance is the total permissible variation in the location of a feature about its exact true position. For every part feature there is an exact position (true position) where the feature would be if it was perfect. A tolerance zone allows deviation from perfection. The exact, or true, position of the feature is indicated by its basic dimensions.

For cylindrical features, the position tolerance zone is typically a cylinder within which the axis of the feature must lie (Figure A-100). The tolerance defines a zone that the axis or center plane of a feature of size may vary within. LMC or MMC can apply to features of size.

Tolerance Type	Characteristic	Symbol
	Position	⊕
Location	Concentricity	◎
	Symmetry	≡

Figure A-99 Location tolerance symbols.

Figure A-98 Drawing for question 2.

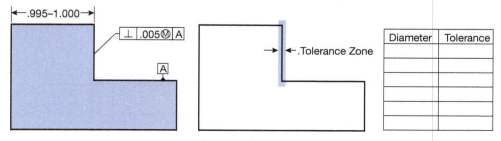

.995–1.000

| ⊥ | .005Ⓜ | A |

A

.Tolerance Zone

Diameter	Tolerance

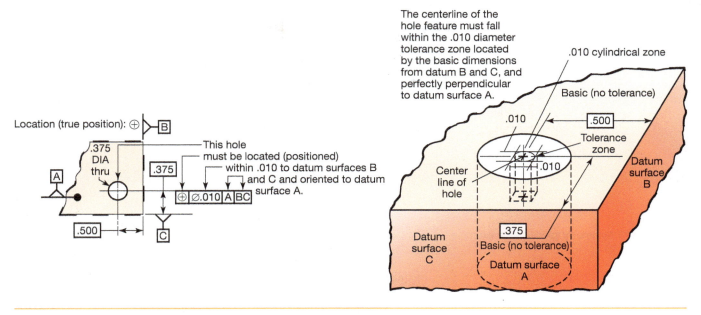

Figure A-100 Position (location) tolerance.

True Position Using MMC

True position controls where the datum locations need to be. MMC specifies a minimum size and positional location of the hole. It does this by allowing a bonus tolerance to be added to the feature location tolerance. As a hole size gets closer to the MMC, the hole must be closer to its specified location. If the hole is larger, it can vary from its true position more and still function as designed.

Consider the bored hole in Figure A-101. The diameter of the hole must be between .625 and .630. The location of the bore is 1.000 from the left edge and 1.000 from the bottom edge of the part. There is also a GD&T specification for the hole. The feature control symbol is a location symbol. The feature control frame specifies a location tolerance of .005 at MMC. MMC for a hole is the smallest hole size allowed (most material left). The MMC specification adds a bonus tolerance to location based on the hole size. If the hole is larger than minimum size, the location tolerance increases. Study the table in Figure A-101. If the hole size is .625 (smallest allowed) the location tolerance is .005. If the hole is .001 larger (.626), the location tolerance increases to .006. If the hole is at its largest allowed size the location tolerance is .010. Basically, this means that if the hole size is

larger (but within specifications) the location can vary more and the part will still be functional.

This adds some complexity to the inspection of parts. In industry a company may require first piece inspection on a coordinate measuring machine or measurement arm. If the parts are made in high quantities, the company might have a gage made to inspect parts.

Concentricity

A concentricity tolerance specifies a cylindrical tolerance zone whose axis coincides with the datum axis. A concentricity tolerance indicates that a cylinder, cone, hex, square, or surface of revolution shares a common axis with a datum feature. It controls the location for the axis of the indicated feature within a cylindrical tolerance zone whose axis coincides with the datum axis.

An example of a concentricity tolerance is shown in Figure A-102. The larger diameter of the part is specified to be Datum A. The smaller diameter must be concentric with Datum A within .005. Note that there was no modifier so the tolerance is regardless of feature size (RFS). Note that the tolerance zone is cylindrical around the centerline of Datum A. Runout is easier to inspect and is used more often than concentricity.

Figure A-101 Use of MMC to control a hole location.

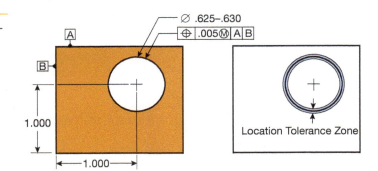

Diameter	Tolerance
.625	.005
.626	.006
.627	.007
.628	.008
.629	.009
.630	.010

Figure A-102 Concentricity example.

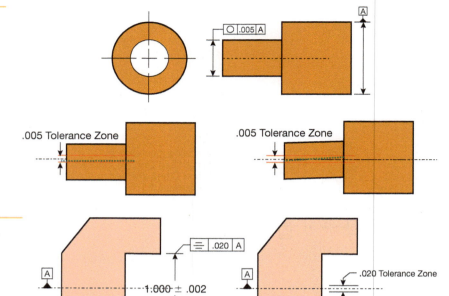

Figure A-103 Symmetry specification example.

Symmetry A symmetry tolerance controls the median points of a feature of size. Symmetry tolerances are applied to non-circular features. A symmetry tolerance is a three-dimensional geometric tolerance that controls how much the median points between two features may deviate from a specified center plane or axis. It is generally recommended to use position, parallelism, or straightness instead of a symmetry tolerance, when possible.

An example of a symmetry tolerance is shown in Figure A-103. The left half of the figure shows the symmetry symbol applied to a slot in the part. The symmetry symbol in this example specifies that the median points of the opposing surfaces of the slot must be symmetric about Datum A within a tolerance zone of .020. In this example, the slot width must be 1.000+ or −.002 and it must be centered around Datum A within a tolerance zone of .020.

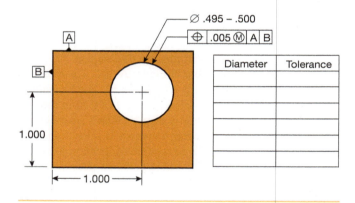

Figure A-104 Drawing for question 1.

Diameter	Tolerance

QUICK CHECK 4

1. Explain the location tolerance and complete the tolerance table in Figure A-104.

2. Complete the feature control frame for a location tolerance. The tolerance should be related to datum A and B and should be at maximum material condition. Complete the tolerance table shown in Figure A-105.

3. Explain what a concentricity tolerance is used for.

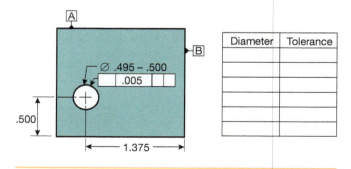

Figure A-105 Drawing for question 2.

Diameter	Tolerance

Runout

Runout is how much one feature varies with respect to a datum when the part is rotated 360° around the datum axis. It is essentially a control of a circular feature, and how much variation it has with the rotational axis. Runout can be called out on any feature that is rotated about an axis. There are two symbols used for runout. They are shown in Figure A-106.

Circular Runout   Circular runout is used to provide control of the circular elements of a surface (Figure A-107). The tolerance is applied independently at any circular measuring position as the part is rotated 360 degrees. One way this part could be inspected would be to put the large diameter in a V-block and an indicator at one point on the small diameter (Figure A-108). The part would then be rotated and the smallest and largest readings on the indicator would be used to determine the runout. More than one circle can be checked, but the checks are independent. When applied to surfaces at right angles to the datum axis, it controls the circular elements of the plane surface (Figure A-109).

Total Runout   Total runout is very similar to circular runout except that total runout applies to the whole controlled feature, not just one circle. See Figure A-110. The GD&T symbol for total runout is used. This means that the whole length of the small diameter would need to be within .005 when checked. To inspect this part the operator would need to check several circles along the whole length of the small diameter (see Figure A-111). All points of the checks would need to have .005 or less runout.

Runout and total runout can also be on a surface of a cylindrical part. In the example shown in Figure A-111 the surface must have runout of less than .005 in relation to Datum A.

Tolerance Type	Characteristic	Symbol
Runout	Circular Runout	↗
	Total Runout	↗↗

Figure A-106 Runout symbols.

SELF-TEST

1. Explain a circular runout tolerance.
2. How is circular runout checked?
3. How is total runout checked?
4. Thoroughly explain the numbered features and tolerances (red 1-5) shown on the print in Figure A-112.
5. Thoroughly explain the numbered features and tolerances (red 1-5) shown on the print in Figure A-113. Complete the tolerance table.

Size	Tolerance
.203	
.202	
.200	
.201	
.200	
.199	
.198	
.197	

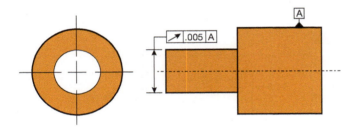

Figure A-107 Circular runout specification.

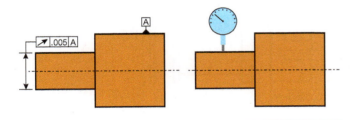

Figure A-108 Method to check runout. Note that the diameter specified by Datum A would be put in a precision V-block and the part would be rotated.

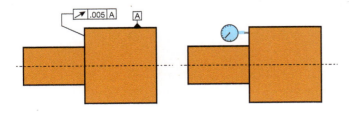

Figure A-109 Inspecting for runout on a surface.

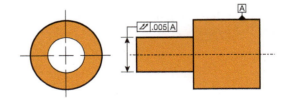

Figure A-110 Total runout specification.

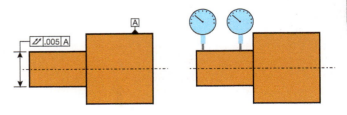

Figure A-111 Inspecting total runout.

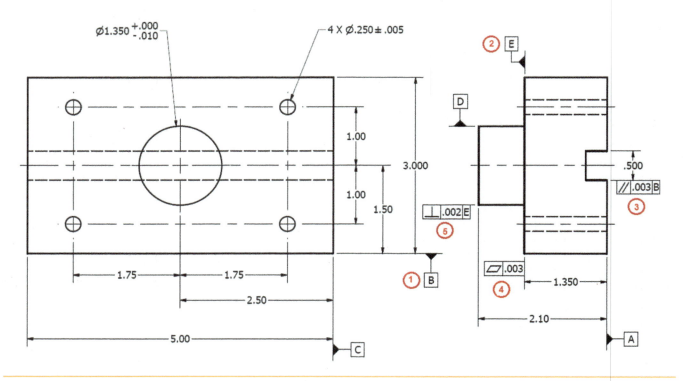

Figure A-112 Mill part.

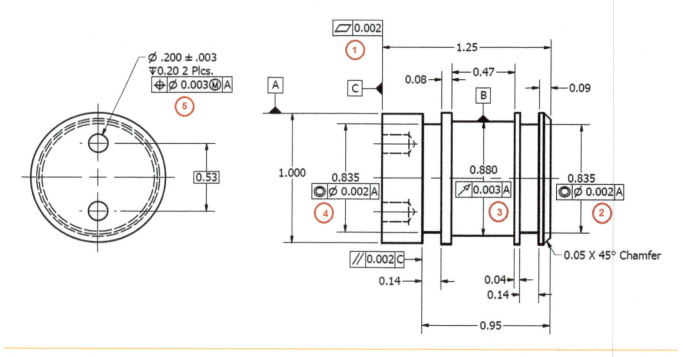

Figure A-113 Lathe part.

6. Explain the following based on the blueprint in Figure A-114.

1.

2.

3.

4.

5.

6.

7.

8.

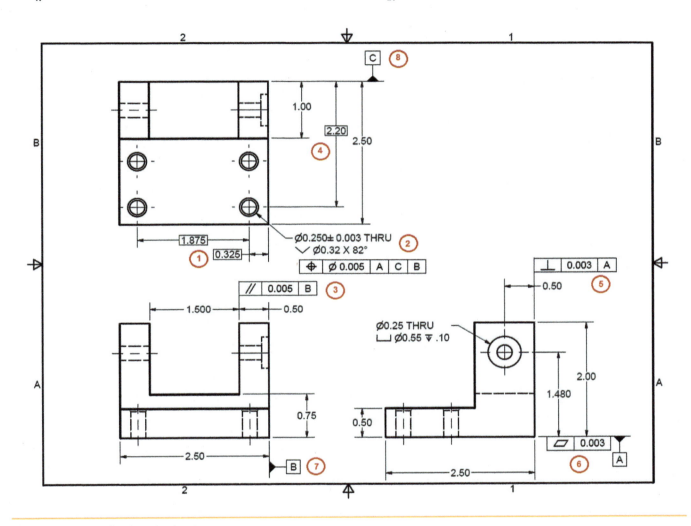

Figure A-114 Drawing for question 6.

SECTION B
Hand Tools

The ability to make and use tools has been directly responsible for much of human progress. Prior to the use of metals, natural materials such as stone, flint, and wood were the only tool materials. When metals and metalworking techniques became more widely used, rapid advances occurred. In this section you will be introduced to the basic hand tools used in manufacturing. The units that follow in this section will instruct you in the identification, selection, use, and safety of these important hand tools and hand-operated machines.

SAFETY FIRST

Hand Tool Safety Tools are quite safe if they are used as they were designed to be used. For example, a screwdriver is not meant to be a chisel, and a file is not meant to be a pry bar. Wrenches should be the correct size for the nut or bolt head so they will not slip. Inch measure wrenches should not be used on metric fasteners. When a wrench slips, skinned knuckles or severe cuts are often the result. Files should never be used without a handle because the tang can damage the hand or wrist. Safety precautions are noted throughout this section.

Arbor and Shop Presses

The arbor press is very common in machine shops. Arbor presses can be dangerous to you and destructive to the workpiece if not used properly.

OBJECTIVES

After completing this unit, you should be able to:

- Install and remove a bushing using an arbor press.
- Press on and remove a ball bearing from a shaft on an arbor press using the correct methods and tools.
- Press on and remove a ball bearing from a housing using an arbor press and correct methods and tooling.
- Install and remove a mandrel using an arbor press.
- Install and remove a shaft with a key in a hub using the arbor press.

TYPES OF PRESSES

The arbor press is an essential piece of equipment in the machine shop. Two basic types of hand-powered arbor presses are used: the hydraulic (Figure B-1) and the mechanical (Figure B-2). The lever gives a "feel" or a sense of pressure applied, which is not possible with the power-driven presses. This pressure sensitivity is needed when small delicate parts are being pressed so that a worker will know when to stop before collapsing the piece. The bolster plate supports the workpiece and allows the press to push shafts through or out of the workpiece. The bolster plate has slots of different widths for different application needs.

USES OF PRESSES

The major uses of the arbor press are installing and removing bushings and ball bearings (Figure B-3), pressing shafts into hubs (Figure B-4) and mandrels into workpieces,

Figure B-1 Small hydraulic shop press.

straightening and bending, and broaching keyways. A **keyway** (also known as a keyseat) is a rectangular groove in a shaft or hub. Keyways in shafts are cut in milling machines. Keyways in hubs of gears, sprockets, or other driven members must be either broached or cut in a keyseater, a special type of vertical shaper.

Bolster Plate
with Various Slot
Widths

Figure B-2 Simple ratchet floor-type arbor press.

Bolster
Plate

Figure B-3 Roller bearing being removed from a shaft.

PROCEDURES

Installing Bushings

A **bushing** is a short metal tube, machined inside and out to precision dimensions, and usually made to fit into an accurately machined hole. Many types of bushings are available for various purposes and are usually installed with an **interference fit** (or press fit). This means that the bushing is slightly larger than the hole into which it is pressed. The amount of interference will be covered in a later unit. Bushings are available in various materials, including bronze and hardened steel. Bushings should be lubricated with high-pressure lube before they are pressed into a bore. Oil is not used, as it will simply wipe off and allow the bushing to seize in the bore. Seizing is the condition in which two unlubricated metals tend to weld together under pressure. Seizing may cause the bushing, or bore, to be damaged beyond repair.

Figure B-4 Shaft being pressed into hub.

The bore should always have a chamfer—that is, an angled or beveled edge, since a sharp edge would cut into the bushing and damage it (Figure B-5). The bushing should also have a chamfer, so it will not dig in and be misaligned. Bushings can go in crooked if there is a sharp edge. Care should be taken to see that the bushing is straight when entering the bore and that it continues into the bore in proper alignment. This should not be a problem if the tooling is right—that is, if the end of the press ram is square and if it is not loose or worn. The proper bolster plate should also be used under the part so that it cannot tilt out of alignment. A bolster plate is a thick solid piece of steel on the bottom of the press that the workpiece sits on. There are often slots of various sizes in the bolster plate so that the part can sit on the plate and the bushing or shaft can be pressed out using the slot under the part. Sometimes, special tooling is used to guide the bushing (Figure B-6). When the workpiece is resting on a solid bolster plate, the minimum pressure needed to force the bushing into place should be used, especially if the bushing is longer than the bore length. Excessive pressure will distort and collapse the bushing (Figure B-7). Some bolster plates provide various size holes so that a bushing or part can extend through. In that case, the press ram can be brought to the point where it contacts the top of the workpiece, and there is no danger of upsetting the bushing.

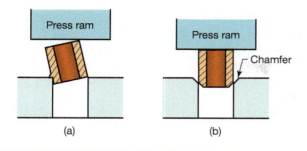

Figure B-5 (*a*) Bushing being pressed where bore is not chamfered and bushing is misaligned; (*b*) bushing being pressed into correctly chamfered hole in correct alignment.

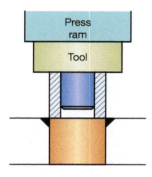

Figure B-6 Special tool to keep bushing square to press ram.

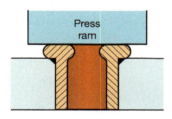

Figure B-7 Effect of excessive pressure on bushing that exceeds bore length.

Ball and Roller Bearings

Ball bearings pose special problems when they are installed and removed. The pressure must be applied directly against the race and not through the ball bearings or rollers, since this could destroy the bearing. When removing ball bearings from a shaft, the inner race is frequently hidden by a shoulder and cannot be supported in the normal way. In this case, a special tool called a bearing puller is used (Figure B-8). Bearings may be installed by pressing on the inner (over a shaft) or outer race (into a bore) with a steel tube of the proper diameter. As with bushings, high-pressure lubricant should be used.

Sometimes there is no other way to remove an old ball bearing except by exerting pressure through the ball bearings. When this is done, there is a real danger that the race may violently shatter. In this case, a scatter shield must be used. A scatter shield is a heavy steel tube to cover the work. The shield is placed around the bearing during pressing to keep shattered parts from flying out and injuring the operator. It

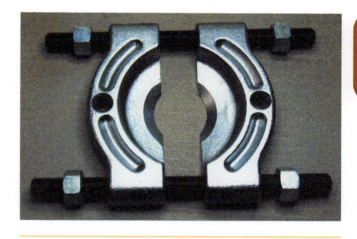

Figure B-8 Bearing puller, a special tool for supporting inner bearing race.

is a good safety practice to use a scatter shield whenever ball bearings are removed from a shaft by pressing. Safety glasses should be worn during all press operations.

Bores and Shafts

Holes in gears, sprockets, and other machine parts are frequently designed for a force fit. There is usually a keyway (also called keyseat) that needs to be aligned. A keyway is a groove in which a key is placed. The key also fits into a slot in the hub of the gear or pulley and secures the part against the shaft, keeping it from rotating. When pressing shafts with keys into hubs with keyway (Figure B-9), it is helpful to chamfer the leading edge of the key so that it will align itself properly. Seizing may occur in this operation if high-pressure lubricant is not used (Figure B-10).

Mandrels

Mandrels are cylindrical pieces of steel with a slight taper. Mandrels are pressed into bores in much the same way as shafts are pressed into hubs. There is one important difference, however; since the mandrel is tapered about .006 inch per foot, it can only be installed with the small end in first. The small end should start into the hole, but the large end should not. Apply lubricant and press the mandrel in until definite resistance is felt (Figure B-11). Mandrels are typically used between centers on a lathe. A round part that needs to be turned is pressed onto the mandrel and held for machining. The part can then be removed from the mandrel.

Keyway (Keyseat) Broaching

Broaching is the process of cutting shapes on a metal part. Broaching can be done on both internal and external surfaces. In keyway broaching, a groove is cut through the inside of a hub or pulley so that a key can be used.

Keyway broaching is often done on arbor presses (see Figure B-12). Keyways are only one type of cutting that can be done by the push-type procedure. Internal shapes such as splines, squares, hexagonal can also be cut by this method. All

Figure B-9 Chamfer on key helps in alignment of parts being pressed together.

Figure B-11 Mandrel being lubricated and pressed into part for further machining.

Figure B-10 This shaft was forced into an interference-fit bore. No lubrication was used, and it seized and welded to the bore, which was also ruined.

Figure B-12 Broaching a keyway (*Courtesy of Fox Valley Technical College*).

that is needed for broaching is an arbor press, a set of keyway broaches, and a guide (Figure B-13). Keyway broaches are hardened cutters with stepped teeth so that each tooth cuts only a small amount when pushed or pulled through a part. Broaches for keyways are available in inch and metric dimensions.

Figure B-13 Keyway (keyseat) broach.

Figure B-14 Broach with guide bushing inserted into a hub.

The procedure for broaching keyways (multiple-pass method) is as follows:

Step 1 Choose the bushing that fits the bore and the broach, and put it in place in the bore.

Step 2 Insert the correct-size broach into the bushing slot (Figure B-14).

Step 3 Place this assembly in the arbor press (Figure B-15).

Step 4 Lubricate.

Step 5 Push the broach through.

Step 6 Clean the broach.

Step 7 Place the second-pass shim in place.

Step 8 Insert the broach.

Step 9 Lubricate.

Step 10 Push the broach through. Measure to check for proper depth.

Step 11 If more than one shim is needed to obtain the correct depth, repeat the procedure (Figure B-16).

Figure B-15 Broach with guide bushing placed in arbor press that is ready to be lubricated and have the first pass performed.

Figure B-16 Shims in place behind broach that is ready to be lubricated and have the final cut made on the part.

Shim goes behind broach after first pass

Shim

Clean the tools and return them to their box. Then deburr and clean the finished keyway.

Production or single-pass broaching requires no shims or second-pass cuts, and with some types no bushings need to be used.

Two important things in broaching are alignment and lubrication. Misalignment, caused by a worn or loose ram, can cause the broach to hog (dig in) or break. Sometimes this can be avoided by facing the teeth of the broach toward the back of the press and permitting the bushing to protrude above the work to provide more support for the broach. After starting the cut, relieve the pressure to allow the broach to center itself. Repeat this procedure during each cut.

At least two or three teeth should be in contact with the work. If needed, stack two or more workpieces to lengthen the cut. The cut should never exceed the length of the standard bushing used with the broach. Never use a broach on material harder than Rockwell C35 or you will damage the broach. If it is suspected that a part is harder than mild steel, its hardness should be determined before any broaching is attempted.

Use good high-pressure lubricant. Also apply cutting oil to the teeth of the broach. Lubricate the back of the keyway broach to reduce friction, regardless of the material to be cut. Brass is usually broached dry, but bronzes cut better with oil or soluble oil. Cast iron is broached dry, and kerosene (solvent) or cutting oil is recommended for aluminum.

Bending and Straightening

A shaft can be straightened by placing it between two nonprecision V blocks (Figure B-17). The shaft is rotated to detect runout, or the amount of bend in the shaft. The runout is measured on a dial indicator. The high point is found and marked on the shaft. After the indicator is removed, a soft metal pad such as brass is placed between the shaft and the ram and pressure is applied (Figure B-18). The shaft should

Figure B-18 Pressure being applied to straighten shaft.

be bent to a straight position and then slightly beyond that point. The pressure is then removed and the dial indicator is again put in position. The shaft is rotated as before, to find the high point, as well as the amount of runout. If improvement has been found, continue the process; but if the first mark is opposite the new high point, too much pressure has been applied. Repeat the same steps until the shaft is straight.

SAFETY FIRST

Bending and straightening are frequently done on hydraulic shop presses. Mechanical arbor presses are only used for smaller diameter shafts. There is a definite safety hazard in this type of operation, as a poor setup can allow pieces under pressure to fly out of the press. Brittle materials such as cast iron or hardened steel bearing races can break under pressure and explode into fragments.

Other straightening jobs on flat stock and other shapes are done in a similar fashion. Frequently, two or more bends will be found that may be opposite or not in the same direction. This condition is best corrected by straightening one bend at a time and checking with a straightedge and feeler gage.

Figure B-17 Part being indicated for runout prior to straightening.

SELF-TEST

1. List several uses of the arbor press.

2. A newly machined steel shaft with an interference fit is pressed into the bore of a steel gear. The result is a shaft ruined beyond repair; the bore of the gear is also badly damaged. What are some of the things that could have caused this failure?

3. The ram of an arbor press is loose in its guide and the pushing end is rounded off. What kind of problems could this cause?

4. When a bushing is pushed into a bore that is located over a hole in the bolster plate of a press, how much pressure should you apply to install the bushing: 30 tons, 10 tons, or just enough to seat the bushing into the bore?

5. Where should the bearing be supported on the bolster plate of the press when pressing a shaft from the inner race of a ball bearing?

6. What difference is there in the way a press fit is obtained between mandrels and ordinary shafts?

7. Prior to installing a bushing with the arbor press, what two important steps must be taken?

8. Name five ways to avoid tool breakage and other problems when using push broaches for making keyways in the arbor press.

INTERNET REFERENCES

Information on shop presses:

www.dakecorp.com

www.buffalohydraulic.com

Work-Holding and Hand Tools

Hand tools are essential in all of the mechanical trades. This unit will cover the names and uses of hand tools commonly used by machinists.

OBJECTIVES

After completing this unit, you should be able to:

- Identify various types of vises, their uses, and their maintenance.
- Identify the proper tool for a given job.
- Determine the correct use of a selected tool.

TYPES OF VISES

Vises of various types are used by machinists when doing hand or bench work. Some bench vises have a solid base, and others have a swivel base (Figure B-19). Bench vises are measured by the width of the jaws. So if the jaws measure 6 inches in width, the vise would be a 6-inch vise.

Toolmakers often use small vises that pivot on a ball and socket for holding delicate work. Handheld vises, called *pin vises*, are made for holding very small or delicate parts.

Most bench vises have hardened insert jaws that are serrated for greater gripping power. These crisscross serrations are sharp and will dig into finished workpieces enough to mar them. Soft jaws (Figure B-20) made of copper, brass, rubber, or other soft metals can be used to protect a finished surface on a workpiece. Soft jaws are made to slip over vise jaws.

USES OF VISES

Vises are used to hold work for filing, hacksawing, chiseling, and bending light metal. They are also used for holding work when assembling and disassembling parts.

Vises should be mounted at the correct working height. The top of the vise jaws should be at elbow height. Poor quality work is produced when the vise is mounted too high or too low.

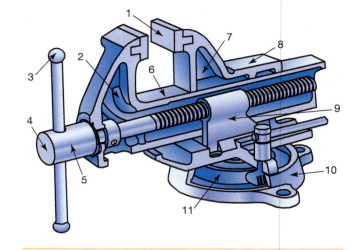

Figure B-19 Cutaway view of a vise: (1) replaceable hardened tool steel faces pinned to jaw; (2) malleable iron front jaw; (3) steel handle with ball ends; (4) cold-rolled steel screw; (5) bronze thrust bearing; (6) front jaw beam; (7) malleable iron back jaw body; (8) anvil; (9) nut, mounted in back jaw keyway for precise alignment; (10) malleable iron swivel base; (11) steel tapered gear and lock bolt.

CARE OF VISES

Like any other tool, vises have limitations. "Cheater" bars or pipes should not be used on the handle to tighten the vise. Heat from a torch should not be applied to work held in the jaws, as the hardened insert jaws will then become softened. There is usually a vise in a shop reserved for heating and bending.

Heavy hammering should not be done on a bench vise. The force of bending should be against the fixed jaw rather than the movable jaw. Bending light flat stock or small round stock in the jaws is permissible if a light hammer is used. The movable jaw slide bar should never be hammered on, as it is usually made of thin cast iron and can be cracked quite easily. An anvil is often provided behind the solid jaw for the purpose of light hammering.

Figure B-20 View of the soft jaws placed on the vise.

Bench vises should occasionally be taken apart so that the screw, nut, and thrust collars can be cleaned and lubricated. The screw and nut should be cleaned in solvent. Heavy grease should be applied to the screw and thrust collars before reassembly.

CLAMPS

C-clamps are used to clamp parts to tables, angle plates, and so on. The size of the clamp is determined by the largest opening of its jaws. Heavy-duty C-clamps (Figure B-21) are used by machinists to hold heavy parts such as steel plates together for drilling or other machining operations. Parallel clamps (Figure B-22) are used to hold small parts. Parallel clamps are usually used for more delicate work. Precision measuring setups are usually held in place with parallel clamps.

Figure B-21 Heavy-duty C-clamp (*Courtesy of Fox Valley Technical College*).

Figure B-22 Parallel clamps.

PLIERS

Pliers are available in various shapes and with several types of jaws. Simple combination, or slip joint pliers (Figure B-23) are adequate for many needs. The slip joint allows the jaws to expand to grasp a larger size workpiece. They are measured by overall length and are made in 5-, 6-, 8-, and 10-inch sizes.

Interlocking joint pliers (Figure B-24), or water pump pliers, are useful for a variety of jobs. Pliers should never be used as a substitute for a wrench, as the nut or bolt head can be damaged by the serrations in the jaws. Needle nose pliers are used for holding small delicate workpieces in tight spots. They are available in both straight and bent nose (Figure B-25) types. Linemen's pliers (Figure B-26) can be used for wire cutting and bending. Some types have wire stripping grooves and insulated handles. Diagonal cutters (Figure B-27) are used to cut wire.

Vise grip wrenches have high gripping power. The screw in the handle adjusts the lever action to the work size (Figure B-28). It is made with special jaws for various uses such as the C-clamp types used in welding (Figure B-29).

Figure B-23 Slip joint or combination pliers.

Figure B-24 Interlocking joint or water pump pliers.

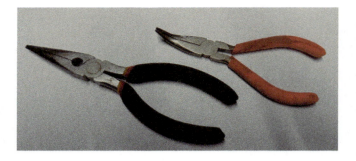

Figure B-25 Needle nose pliers, straight and bent.

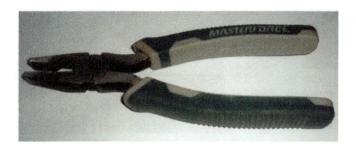

Figure B-26 Linesmen's pliers (*Courtesy of Fox Valley Technical College*).

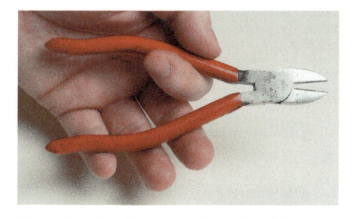

Figure B-27 Diagonal cutters.

Figure B-28 Vise grip wrench.

Figure B-29 Vise grip C-clamp (*Courtesy of Fox Valley Technical College*).

HAMMERS

Hammers are classified as either hard or soft. The ball peen hammer (Figure B-30) is the one most frequently used by machinists. It has a rounded surface on one end of the head and a hardened striking surface on the other. Two hammers should never be struck together on the face, as pieces could break off. Hammers are specified according to the weight of the head. Ball peen hammers range from 2 oz to 3 lb. A 10 oz ball peen hammer is commonly used for layout.

Soft hammers are made of plastic (Figure B-31), brass, copper, lead alloy (Figure B-32), or rawhide and are used to position workpieces that have finishes that would be damaged by a hard hammer. A dead blow hammer is often used instead of a lead-alloy hammer because, like the lead-alloy hammer, the dead blow hammer does not have a tendency to rebound. Dead blow hammers are often made of **polyurethane**. The head of a dead blow hammer is loaded with steel shot.

In precision work the part is often put on parallels to be sure it is held flat in the vise. When a vise is tightened, the movable jaw on the vise tends to move slightly upward when tightened against the workpiece. Thus, the workpiece is moved upward and out of position. The part is then tapped on the top with a dead blow hammer to seat the part against the parallels. The machinist then checks to see if the parallels are tight between the part and bottom of the vise.

Figure B-30 Ball peen hammer.

Figure B-31 Plastic hammers.

Figure B-32 Lead-alloy hammer. Commonly called a lead hammer.

WRENCHES

A large variety of wrenches are available for various uses. The adjustable wrench, commonly called a *Crescent* wrench (Figure B-33), is a general-purpose tool and will not suit every job. An adjustable wrench should be rotated toward the movable jaw and should fit the nut or bolt tightly. The size of the wrench is determined by its overall length in inches.

Open-end wrenches (Figure B-34) usually fit two sizes, one on each end of the wrench. The ends of this type of wrench are often angled so they can be used in close quarters. Box wrenches (Figure B-35) are also double ended and offset to clear the user's hand. The box completely surrounds the nut or bolt and usually has 12 points so that the wrench can be reset after rotating only a partial turn. Box wrenches have the advantage of precise fit. Combination wrenches are made with a box at one end and an open end at the other (Figure B-36).

Socket wrenches are similar to box wrenches in that they also surround the bolt or nut and usually are made with 12 points contacting the six-sided nut (Figure B-37). Sockets are made to be detached from various types of drive handles.

Pipe wrenches (Figure B-38) are used for holding and turning pipe. These wrenches have sharp serrated teeth and

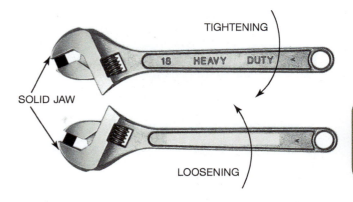

Figure B-33 Adjustable wrench showing the correct direction of pull. The movable jaw should always face the direction of rotation.

Figure B-34 Open-end wrench.

Figure B-35 Box wrench.

Figure B-36 Combination wrench.

will damage any finished surface on which they are used. Strap wrenches (Figure B-39) are used for extremely large parts or to avoid marring the surface of tubular parts.

Spanner wrenches come in several basic types, including face and hook (Figure B-40). Face types are sometimes called pin spanners. Spanners are made in fixed sizes or adjustable types.

Socket head wrenches, commonly called Allen wrenches, are six-sided bars having a 90-degree bend near one end. They are used with socket head capscrews and socket setscrews (Figure B-41).

Torque wrenches (Figure B-42) are used to provide the correct amount of tightening torque on a screw or nut. The dial on a torque wrench reads in English measure (inch-pounds and foot-pounds) or in metric measure (kilogram-centimeters and newton-meters).

Figure B-37 Socket wrench set.

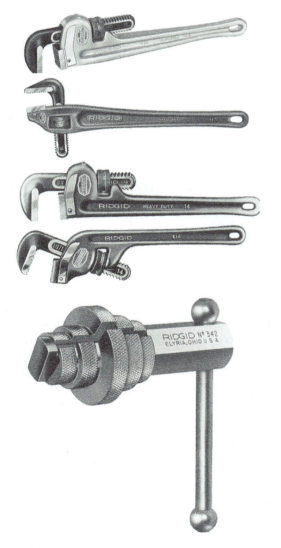

Figure B-38 Pipe wrenches, external and internal.

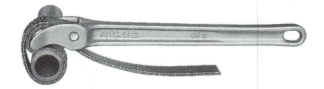

Figure B-39 Strap wrench.

| Fixed-Face Spanner | Adjustable Face Spanner | Hook Spanner | Adjustable Hook Spanner |

Figure B-40 Spanner wrenches.

BALL TIP

Figure B-41 Socket head wrench.

Figure B-42 Dial and click–type torque wrench.

The hand-tap wrench (Figure B-43) is used for medium- and large-sized taps. The T-handle tap wrench (Figure B-44) is used for small taps $\frac{1}{4}$ inch and under, as its more sensitive "feel" results in less tap breakage.

Figure B-43 Hand tap wrench.

Figure B-44 T-handle tap wrench.

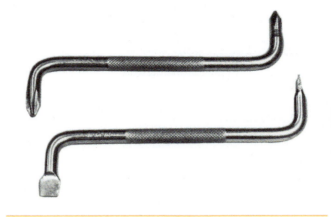

Figure B-46 Standard and Phillips offset screwdrivers.

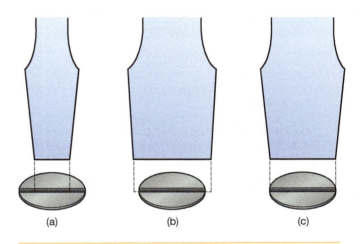

Figure B-47 Width of a screwdriver blade: (*a*) too narrow; (*b*) too wide; (*c*) correct width.

SCREWDRIVERS

The most common types of screwdrivers are the standard and Phillips (Figure B-45). Both types are made in various sizes and in various styles: straight, shank, and offset (Figure B-46). It is important to use the right width and thickness of blade when installing or removing screws (Figure B-47). The shape of the tip is also important. If the tip is badly worn or incorrectly ground, it will tend to jump out of the slot. Never use a screwdriver as a chisel or pry bar. Keep a screwdriver in proper shape by using it only on the screws for which it was meant.

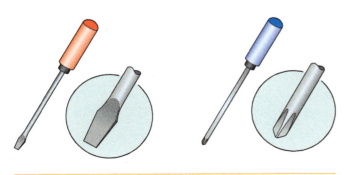

Figure B-45 Screwdriver, standard and Phillips.

CHISELS AND PUNCHES

Chisels and punches (Figure B-48) are very useful tools. The tool at the top of the illustration is a pin punch, used to drive out straight, taper, and roll pins. The drift punch below it is used as a starting punch for driving out pins. In the middle is a center punch that makes a starting point for drilling. The two bottom tools are cold chisels. Cold chisels are made in many shapes and are useful for cutting off rivet heads and welds.

PROFESSIONAL PRACTICE

The way a worker maintains their tools reveals a great deal about them and their work habits. Dirty, greasy, or misused tools carelessly thrown into a drawer are difficult to find the next time around. After a hand tool is used, it should be wiped clean with a shop towel and stored neatly in the proper place.

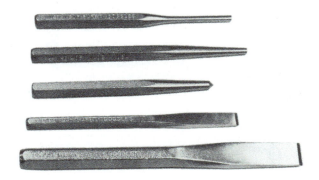

Figure B-48 Common chisels and punches used by machinists. Top to bottom: pin punch, drift punch, center punch, small flat chisel, and large flat chisel.

SAFETY FIRST

Here are safety hints for using wrenches:

1. Make sure that the wrench you select fits properly. If it is a loose fit, it may round off the corners of the nut or bolt head.
2. Pull on a wrench instead of pushing to avoid injury.
3. Never use a wrench on moving machinery.
4. Do not hammer on a wrench or extend the handle for additional leverage. Use a larger wrench.

SELF-TEST

1. Name two types of bench vises.
2. How is vise size specified?
3. How can the finished surface of a part be protected in a vise?
4. Name three things that should never be done to a vise.
5. How should a vise be lubricated?
6. Parallel clamps are used for heavy-duty clamping work, and C-clamps are used for holding precision setups. True or false?
7. What advantage does the lever-jawed wrench offer over other similar tools such as pliers?
8. Some objects should never be struck with a hard hammer—a finished machine surface or the end of a shaft, for instance. What could you use to avoid damage?
9. Why should pipe wrenches never be used on bolts, nuts, or shafts?
10. What are the two important things to remember about standard screwdrivers that will help you avoid problems in their use?

INTERNET REFERENCES

Information on hand tools:

http://wiltontool.com

http://www.armstrongtoolsupply.com

Hacksaws

The hand hacksaw is a relatively simple tool to use. The information presented in this unit will help you to correctly use a hacksaw.

OBJECTIVE

After completing this unit, you should be able to:

- Identify, select, and use hand hacksaws.

HACKSAW DESIGN

The hacksaw consists of three parts: the frame, handle, and saw blade (Figure B-49). Frames are either solid or adjustable. The solid frame can be used with only one length of saw blade. The adjustable frame can be used with hacksaw blades from 8 to 12 inches in length.

The blade can be mounted to cut in line with the frame or at a right angle to the frame (Figure B-50). By turning the

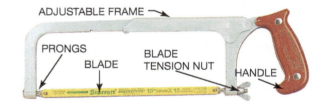

Figure B-49 Parts of a hacksaw.

blade at right angles to the frame, you can continue a cut that is deeper than the capacity of the frame.

Most hacksaw blades are made from high-speed steel and in standard lengths of 8, 10, and 12 inches. Blade length is the distance between the centers of the holes at each end. Hand hacksaw blades are generally $\frac{1}{2}$-inch wide and .025-inch thick. The *kerf*, or cut, produced by the hacksaw is wider than the .025-inch thickness of the blade because of the set of the teeth.

The term *set* refers to the bending of teeth outward from the blade itself. Two kinds of sets are found on hand

Figure B-50 Left—Straight sawing with a hacksaw. Right—Sawing with the blade set at 90 degrees to the frame.

Blade Vertical with Frame

Blade at 90° to Frame

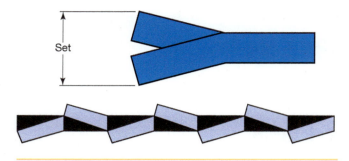

Figure B-51 Straight (alternate) set.

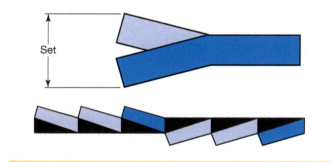

Figure B-52 Wavy set.

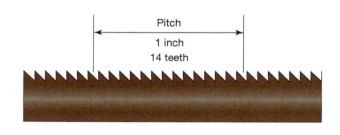

Figure B-53 The pitch of the blade is expressed as the number of teeth per inch.

hacksaw blades. The first is the straight (also called alternate set) (Figure B-51), in which one tooth is bent to the right and the next tooth to the left for the length of the blade. The second kind of set is the wavy set, in which a number of teeth are gradually bent to the right and then to the left (Figure B-52). A wavy set is found on most fine-tooth hacksaw blades.

The spacing of the teeth on a hand hacksaw blade is called the *pitch* and is expressed in teeth per inch of length (Figure B-53). Standard pitches are 14, 18, 24, and 32 teeth per inch, with the 18-pitch blade used as a general-purpose blade.

HACKSAW USE

The hardness and thickness of a workpiece determines which blade pitch to use. As a rule, you should use a coarse-tooth blade on soft materials, to have sufficient clearance for the chips, and a fine-tooth blade on harder materials. But you should also have at least three teeth cutting at any time, which may require a fine-tooth blade on soft materials with thin cross sections.

Hand hacksaw blades fall into two categories: soft-backed or flexible blades and all-hard blades. On the flexible blades only the teeth are hardened, the back being tough and flexible. The flexible blade is less likely to break when used in places that are difficult to get at, such as in cutting off bolts on machinery. The all-hard blade is, as the name implies, hard and very brittle, and should be used only where the workpiece can be rigidly supported, as in a vise. On an all-hard blade, even a slight twisting motion may break the blade. All-hard blades, in the hands of a skilled person, will cut true, straight lines and give long service.

The blades are mounted in the frame with the teeth pointing away from the handle so that the blade cuts on the forward stroke. No cutting pressure should be applied to the blade on the return stroke as this tends to dull the teeth. The sawing speed with the hacksaw should be from 40 to 60 strokes per minute. To get the best performance from a blade, make long, slow, steady strokes using the full length of the blade. Sufficient pressure should be maintained on the forward stroke to keep the teeth cutting. Teeth on a saw blade will dull rapidly if too little or too much pressure is used. The teeth will also dull if the cutting stroke is too fast; a speed in excess of 60 strokes a minute will dull the blade because friction will overheat the teeth.

The saw blade may break if it is too loose in the frame or if the workpiece slips in the vise while sawing. Too much pressure may also cause the blade to break. A badly worn blade, one which the set has been worn down, will cut a narrow kerf, which will cause binding and perhaps breakage of the blade. When this happens and a new blade is used to finish the cut, turn the workpiece over and start with the new blade from the opposite side and make a cut to meet the first one (Figure B-54). The set on the new blade is wider than the old kerf. Forcing the new blade into an old cut will immediately ruin it by wearing the set down.

Cuts should be started with light cutting pressure, with a thumb or finger acting as a guide for the blade. Sometimes it helps to start a blade in a small v-notch filed into the workpiece. When a workpiece is supported in a vise, make sure that the cutting is done close to the vise jaws for a rigid setup free of chatter (Figure B-55). Work should be positioned in

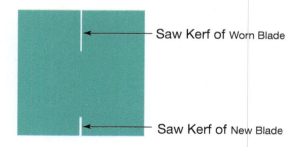

Figure B-54 A new blade must be started on the opposite side of the work, not in the same kerf made by the old blade.

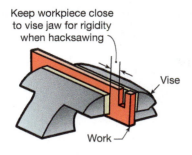

Figure B-55 The workpiece is being sawed close to the vise to present vibration and chatter.

a vise so that the saw cut is vertical. This makes it easier for the saw to follow a straight line. At the end of a saw cut, reduce the cutting pressure or you may be caught off balance when the pieces come apart and cut your hands on the sharp edges of the workpiece. To saw thin material, sandwich it between two pieces of wood for a straight cut. Avoid bending the saw blades, because they are likely to break, and when they do, they may shatter in all directions and could injure you or others nearby.

SELF-TEST

1. What is the kerf?
2. What is the set on a saw blade?
3. What is the pitch of the hacksaw blade?
4. What determines the selection of a saw blade for a job?
5. Hand hacksaw blades fall into two basic categories. What are they?
6. Give four causes that make saw blades dull.
7. Give two reasons why hacksaw blades break.
8. A new hacksaw blade should not be used in a cut started with a blade that has been used. Why?

Files

Files are often used to put the finishing touches on a machined workpiece, either to remove burrs or sharp edges or as a final fitting operation. In this unit you are introduced to the types and uses of files in metalworking.

OBJECTIVE

After completing this unit, you should be able to:

- Identify eight common files and some of their uses.

TYPES OF FILES

Machinists will use files for a variety of tasks. Files are made in many lengths ranging from 4 to 18 inches (Figure B-56). Files are manufactured in many different shapes and are used for various purposes. Figure B-57 shows the parts of a file. When a file is measured, the length is taken from the heel to the point, with the tang excluded. Most files are made from high-carbon steel and are heat-treated to the correct hardness. They are manufactured in four different cuts: single, double, curved tooth, and rasp. The single cut, double cut, and curved tooth are the most common in machine shops. Rasps are generally used with wood. Curved tooth files will give excellent results with soft materials such as aluminum, brass, plastic, or lead.

Figure B-56 Files are made in several different lengths.

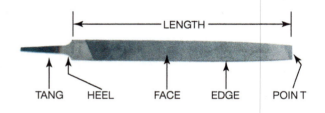

Figure B-57 Parts of a file.

Files are available in various coarseness; rough, coarse, bastard, second cut, smooth, and dead smooth. The files most often used are the bastard, second-cut, and smooth grades. Different sizes of files within the same coarseness designation will have varying sizes of teeth (Figure B-58): the longer the file, the coarser the teeth. For maximum metal removal a double-cut file is used. A single-cut file is recommended for finer finishes.

The face of most files is slightly convex because they are made thicker in the middle than on the ends. Because of this curvature only some of the teeth are cutting at any one time, which makes them penetrate better. If the face were flat, it would be difficult to obtain an even surface because of the tendency to rock a file while filing. Some of this curvature is also offset by the pressure applied to make the file cut.

Figure B-58 These two files are both bastard cut, but since they are of different lengths, they have different coarsenesses.

Files are either straight (blunt) or tapered (Figure B-59). A straight file has the same cross-sectional area from heel to point, whereas a tapered file narrows toward the point. Files fall into five basic categories: mill and saw files, machinists' files, Swiss pattern files, curved tooth files, and rasps. Machinists', mill, and saw files are classified as American pattern files. Mill files (Figure B-60) were originally designed to sharpen large saws in lumber mills, but now they are used for draw filing. The lathe file was designed for filing on a lathe (Figure B-61), or filing a finish on a workpiece. Mill files are single cut and work well on brass and bronze. The flat file (Figure B-62) is usually a double-cut file. Double-cut files are used when fast cutting is needed. The finish produced by a double-cut file is relatively rough.

Pillar files (Figure B-63) have a narrower but thicker cross section than flat files. Pillar files are parallel in width and taper slightly in thickness. They also have one or two safe edges that allow filing into a corner without damaging the shoulder. Square files (Figure B-64) usually are double cut and are used to file in keyways, slots, or holes.

Figure B-59 Blunt and tapered file shapes.

Figure B-60 Mill file.

Figure B-61 The lathe file has a longer angle on the teeth to clear the chips when filing on the lathe.

Figure B-62 The flat file is usually a double-cut file.

Figure B-63 Two pillar files.

Figure B-64 Square file.

If a thin file is needed with a rectangular cross section, a warding file (Figure B-65) is used. This file is often used by locksmiths when filing notches into locks and keys. Another file that will fit into narrow slots is a knife file (Figure B-66). The included angle between the two faces of this file is approximately 10 degrees.

Three-square files (Figure B-67), also called three-cornered files, are triangular in shape with the faces at 60-degree angles to each other. These files are used for filing internal angles between 60 and 90 degrees as well as to make sharp corners in square holes. Half-round files (Figure B-68) are available to file large internal curves. Half-round files, because of their tapered construction, can be used to file many different radii. Round files (Figure B-69) are used to file small radii or to enlarge holes. These files are available in many diameter sizes.

Figure B-65 Warding file.

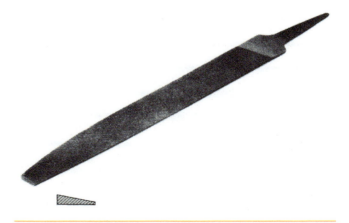

Figure B-66 Knife file.

Figure B-67 Three-square files are used for filing angles between 60 and 90 degrees.

Figure B-68 Half-round files are used for internal curves.

Figure B-69 Round files are used to file a small radius or to enlarge a hole.

Swiss pattern files (Figure B-70) are manufactured to close tolerances in many shapes. Swiss pattern files range in length from 3 to 10 inches, and their coarseness is indicated by numbers from 00 (coarse) to 6 (fine). Swiss pattern files are made with tangs to be used with file handles or as needle files with round or square handles that are a part of the files. Another type of Swiss pattern file is the die sinkers' riffler (Figure B-71). These files are double-ended with cutting

Figure B-70 Set of Swiss pattern files. Since these small files are very delicate and can be broken quite easily, great care must be exercised in their use.

Figure B-71 Die sinker's rifflers.

surfaces on either end. Swiss pattern files are used primarily by tool and die makers, mold makers, and other workers engaged in precision filing on delicate parts.

Curved tooth files (Figure B-72) cut very freely and remove material rapidly. The teeth on curved tooth files are all of equal height, and the gullets or valleys between teeth are deep and provide sufficient room for the filings to curl and drop free. Curved tooth files are manufactured in three grades of cut—standard, fine, and smooth—and in lengths from 8 to 14 inches. Curved tooth file shapes are flat, half round, pillar, and square.

The bastard-cut file (Figure B-73) has a safe edge that is smooth. Flat filing may be done up to the shoulders of the workpiece without fear of damage. Files of other cuts and coarseness are also available with safe edges on one or both sides.

Thread files (Figure B-74) are used to clean up and reshape damaged threads. They are square in cross section and have eight different thread pitches on each file. The thread file of the correct pitch is typically held or stroked against the thread while it is rotating in a lathe.

Figure B-72 Curved tooth files are used on soft metals.

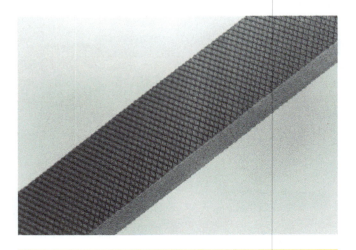

Figure B-73 A file with a safe edge will not cut into shoulders or corners when filing is being done.

Figure B-74 Thread files.

CARE AND USE OF FILES

Files only cut efficiently when they are sharp. Files and their teeth are hard and brittle. Do not use a file as a hammer or as a pry bar. When a file breaks, particles will fly at high speed and may cause an injury. Files should be stored so that they are not in contact with other files. The same applies to files on a workbench. Do not let files lie on top of one another because one file will damage teeth on the other file. Teeth on files will also break if too much pressure is put on them while filing. On the other hand, if not enough pressure is applied while filing, the file only rubs the workpiece and dulls the teeth. A dull file can be identified by its shiny, smooth teeth and by the way it slides over the work without cutting. Dulling of teeth is also caused by filing hard materials or by filing too fast. A good filing speed is 40 to 50 strokes per minute, but remember that the harder the material, the slower the strokes should be; the softer the material, the coarser the file should be.

Too much pressure on a new file may cause **pinning**. Pinning is the term used to describe the teeth of a file becoming loaded with filings. Pinning can cause deep scratches on a work surface. If the pinning cannot be removed with a file card (Figure B-75), try a piece of brass, copper, or mild steel, and push it through the teeth. Do not use a scriber or other hard object for this operation. A file will not pin as much if some blackboard chalk is applied to the face (Figure B-76). *Never* use a file without a file handle, or the pointed tang may cause serious hand or wrist injury (Figure B-77).

Many filing operations are performed with the workpiece held in a vise. Clamp the workpiece securely, but

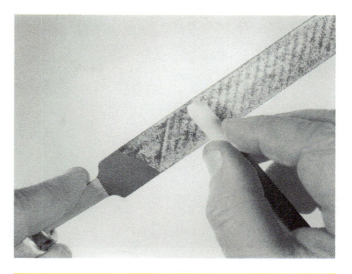

Figure B-76 Using chalk on the file to help reduce pinning.

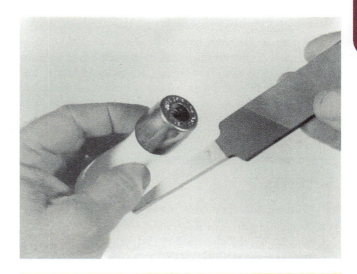

Figure B-77 A file should never be used without a file handle. This style of handle is designed to screw on rather than be driven on the tang.

remember to protect it from the serrated vise jaws with a soft piece of material such as copper, brass, wood, or paper. The workpiece should extend out of the vise so that the file clears the vise jaws by $\frac{1}{8}$ to $\frac{1}{4}$ inch. Since a file cuts only on the forward stroke, no pressure should be applied on the return stroke. Letting the file drag over the workpiece on the return stroke helps release the small chips so that they can fall from the file. However, this can also dull the file and scratch the part, so do it cautiously.

Strokes should be as long as possible; this will make the file wear out evenly instead of just in the middle. To file a flat surface, change the direction of the strokes frequently to produce a crosshatch pattern (Figure B-78). You can determine where the high spots are that have to be filed away by using a straight edge.

Figure B-75 Using a file card to clean a file.

Figure B-78 The crosshatch pattern shows that this piece has been filed from two directions, resulting in a flatter surface (*Courtesy of Fox Valley Technical College*).

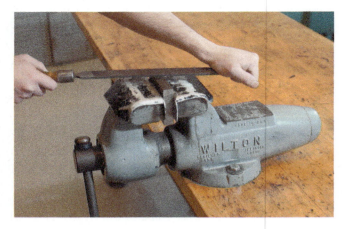

Figure B-79 Proper filing position.

Figure B-80 Draw filing.

Figure B-79 shows how a file should be held to file a flat surface. A smooth finish can be accomplished by draw-filing (Figure B-80). To drawfile, a single-cut file is held with both hands and drawn back and forth on a workpiece. The file should not be pushed beyond the ends of the work-piece, as this would cause rounded edges.

SELF-TEST

1. What are the four different cuts found on files?
2. Name four coarseness designations for files.
3. Which of the two kinds of files—single cut or double cut—is designed to remove more material?
4. What are the coarseness designations for needle files?
5. What happens if too much pressure is applied when filing?
6. What causes a file to get dull?
7. Why should a handle be used on a file?
8. How does the hardness of a workpiece affect the selection of a file?
9. Should pressure be applied to a file on the return stroke?
10. Why is a round file rotated while it is being used?

UNIT FIVE

Hand Reamers

Holes produced by drilling are not accurate in size and often have rough surfaces. Hand reamers are used to finish a previously drilled hole to an exact dimension with a smooth surface. When parts of machine tools are aligned and fastened with capscrews or bolts, the final operation is often hand reaming a hole in which a dowel pin is used to maintain the alignment. Hand reamers are designed to remove a small amount of material from a hole, usually from .001 to .005 inch. These tools are made from high-speed steel. Hand reamers are not used in production work. They are better suited to maintenance type applications. This unit describes commonly used hand reamers and how they are used.

OBJECTIVES

After completing this unit, you should be able to:

- Identify at least three types of hand reamers.
- Hand ream a hole to a specified size.

FEATURES OF HAND REAMERS

Figure B-81 shows the major parts of a typical hand reamer. The square on the end of the shank permits the clamping of a tap wrench or T-handle wrench to provide the driving torque for reaming. The diameter of this square is between .004 and .008 inch smaller than the reamer size, and the shank of the reamer is between .001 and .006 inch smaller, to guide the reamer and permit it to pass through a reamed hole without marring it. It is important that these tools not be put into a drill chuck, because a burred shank can ruin a reamed hole as the shank is passed through it.

Hand reamers have a long starting taper that is usually as long as the diameter of the reamer. This starting taper is usually slight and may not be apparent at a glance. Hand reamers do their cutting on this tapered portion. The gentle taper and length of the taper help start the reamer straight and keep it aligned in the hole.

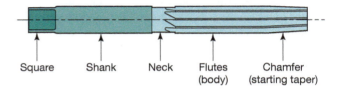

Square Shank Neck Flutes (body) Chamfer (starting taper)

Figure B-81 Major features of the hand reamer.

Details of the cutting end of the hand reamer are shown in Figure B-82. The full diameter or actual size of the hand reamer is measured where the starting taper ends and the margin of the land appears. The diameter of the reamer should be measured only at this junction, as the hand reamer is generally back tapered or reduced in outside diameter by about .0005 to .001 inch per inch of length toward the shank. This back tapering is done to reduce tool contact with the workpiece. When hand reamers become dull, they are resharpened at the starting taper, using a tool and cutter grinder.

The hand reamer cuts much like a scraper rather than an aggressive cutting tool like most drills and machine reamers. For this reason, hand reamers typically have zero or negative

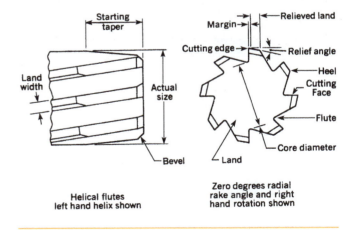

Figure B-82 Functional details of the hand reamer (*Besly Cutting Tools, Inc.*).

radial rake on the cutting face rather than the positive radial rake characteristic of most machine reamers. The right-hand cut with a left-hand helix is considered standard for hand reamers. The left-hand helix contributes to a smooth cutting action.

Most reamers, hand or machine types, have staggered spacing on teeth, which means that the flutes or body channels are not uniformly spaced. The difference is very small, only a degree or two, but it tends to reduce chatter by reducing harmonic effects between cutting edges. Harmonic chatter is especially a problem with adjustable hand reamers, which often leave a tooth pattern in the work.

Hand reamers are made with straight flutes (Figure B-83) or with helical flutes (Figure B-84). Most hand reamers are manufactured with a right-hand cut, which means they will cut when rotated in a clockwise direction. Helical or spiral fluted reamers are available with a right-hand helix or a left-hand helix. Helical flute reamers are especially useful when reaming a hole having keyways or grooves cut into it, as the helical flutes tend to bridge the gaps and reduce binding or chattering.

Hand reamers are made as solid (Figures B-83 and B-84) or expansion types (Figures B-85 and B-86). Expansion reamers are designed for use where it is necessary to enlarge a hole slightly for proper fit, such as in maintenance applications. These reamers have an adjusting screw that allows limited expansion to an exact size. The maximum expansion of these reamers is approximately .006 inch for diameters up to $\frac{1}{2}$ inch, .010 inch for diameters between $\frac{1}{2}$ and 1 inch, and .012 inch for diameters between 1 and $1\frac{1}{2}$ inches. These tools can be broken if they are expanded beyond these limits.

Figure B-83 Straight flute hand reamer.

Figure B-84 Helical flute hand reamer.

Figure B-85 Straight flute expansion hand reamer.

Figure B-86 Adjustable hand reamers. The lower reamer is equipped with a pilot and tapered guide bushing for reaming in alignment with a second hole.

Helical flute expansion reamers are especially useful for reaming holes with keyways or straight grooves. Expansion reamers have a slightly undersized pilot on the end that guides the reamer and helps keep it in alignment.

The adjustable hand reamer (Figure B-86) is different than the expansion reamer in that it has inserted blades. These cutting blades fit into tapered slots in the body of the reamer and are held in place by two locking nuts. The blades have a taper corresponding to the taper of the slots that keeps them parallel at any setting. Adjustments in reamer size are made by loosening one nut while tightening the other. Adjustable hand reamers are available in diameters from $\frac{1}{4}$ to 3 inches. The adjustment range varies from $\frac{1}{32}$ inch on the smaller-diameter reamers to $\frac{5}{16}$ inch on the larger size reamers. Small amounts of material should be removed at a time, as large cuts may cause chatter.

Taper pin reamers (Figures B-87 and B-88) are used for reaming holes for standard taper pins used in the assembly of machine tools and other parts. Taper pin reamers have a taper of $\frac{1}{4}$ inch per foot of length and are manufactured in 18 different sizes numbered from 8/0 to 0 and on up to size 10. The smallest size, number 8/0, has a large-end diameter of .0514 inch, and the largest reamer, a number 10, has a large-end diameter of .7216 inch. The sizes of these reamers are designed to allow the small end of each reamer to enter a hole reamed by the next smaller size reamer. A helical flute reamer will cut with less chatter, especially on interrupted cuts.

Morse taper socket reamers are designed to produce holes for American Standard Morse taper shank tools. These reamers are available as roughing reamers (Figure B-89) and as finishing reamers (Figure B-90). The roughing reamer has notches ground at intervals along the cutting edges. These

Figure B-87 Straight flute taper pin hand reamer.

Figure B-88 Straight flute taper pin hand reamer.

Figure B-89 Morse taper socket roughing reamer.

Figure B-90 Morse taper socket finishing reamer.

notches act as chip breakers and make the tool more efficient at the expense of a fine finish. The finishing reamer is used to impart the final size and finish to the socket. Morse taper socket reamers are made in sizes from No. 0, with a large-end diameter of .356 inch, to No. 5, with a large-end diameter of 1.8005 inches. There are two larger Morse tapers, but they are typically sized by boring rather than reaming.

USING HAND REAMERS

A hand reamer should be turned with a tap wrench or T-handle wrench rather than with an adjustable wrench. The use of a single-end wrench makes it almost impossible to apply torque without disturbing the alignment of the reamer with the hole. A hand reamer should be rotated slowly and evenly, allowing the reamer to align itself with the hole to be reamed. Use a tap wrench large enough to give a steady torque and to prevent vibration and chatter. Use a steady and large feed; feeds up to one-quarter of the reamer diameter per revolution can be used. Small and lightweight workpieces can be reamed by fastening the reamer vertically in a bench vise and rotating the work over the reamer by hand (Figure B-91).

Never rotate a reamer backward to remove it from the hole, as this will dull the reamer rapidly. If possible, pass the reamer through the hole and remove it from the far side without stopping the forward rotation. If this is not possible, it should be withdrawn while maintaining the forward rotation.

The preferred stock allowance for hand reaming is between .001 and .005 inch. Reaming more material than this makes it very difficult to force the reamer through the workpiece. Reaming too little, on the other hand, results in excessive tool wear because it forces the reamer to work in the zone of material work-hardened during the drilling operation. This stock allowance does not apply to taper reamers, for which a hole has to be drilled at least as large as the small diameter of the reamer. The hole size for a taper pin is determined by the taper pin number and its length. These data can be found in machinist handbooks.

Since hand reaming is restricted to removing small amounts of material, you must be able to drill a hole of predictable size and of a surface finish that will assure a finished cleanup cut by the reamer. It is a good idea to drill a test hole in a piece of scrap of similar composition and carefully measure both for size and for an enlarged or bell-mouth entrance. You may find it necessary to drill a slightly smaller hole before drilling the correct reaming size to assure a more accurate size. Carefully spot drill the location before drilling the hole in your part. The hole should then be lightly chamfered with a countersinking tool to remove burrs and to promote better reamer alignment.

The use of cutting fluid improves the cutting action and the surface finish when reaming most metals. Exceptions are cast iron and brass, which should be reamed dry.

When a hand reamer is started it should be checked for squareness on two sides of the reamer, 90 degrees apart. Another way to ensure alignment of the reamer with the drilled holes is to use the drill press as a reaming fixture. Put a piece of cylindrical stock with a 60-degree center in the drill chuck (Figure B-92) and use it to guide and follow the squared end of the reamer as you turn the tool with the tap wrench. Be sure to plan ahead so that you can drill, countersink, and ream the hole without moving the table or head of the drill press between operations.

On deep holes, or especially on holes reamed with taper reamers, remove the chips frequently from the reamer flutes to prevent clogging. Remove chips with a brush to avoid cutting your hands.

Reamers should be stored so they do not contact other reamers and damage the cutting edges. They should

Figure B-91 Hand reaming a small workpiece with the reamer held in a vise.

Figure B-92 Using a drill press as a reaming fixture.

be kept in their original shipping tubes or set up in a tool stand. Always check reamers for burrs before you use them. Otherwise, the reamed hole can be oversized or marred with a rough finish.

SELF-TEST

1. How is a hand reamer identified?
2. What is the purpose of a starting taper on a reamer?
3. What is the advantage of a spiral flute reamer over a straight flute reamer?
4. How does the shank diameter of a hand reamer compare with the diameter measured over the margins?
5. When are expansion reamers used?
6. What is the difference between an expansion and an adjustable reamer?
7. What is the purpose of cutting fluid in reaming?
8. Why should reamers not be rotated backward?
9. How much reaming allowance is left for hand reaming?
10. If you were repairing the lathe tailstock taper, you would use a _____ reamer.

INTERNET REFERENCE

Information on reamers:

http://www.icscuttingtools.com

Identification and Use of Taps

Most internal threads are made with taps. Taps are available in a variety of styles, each one designed to perform a specific type of tapping operation efficiently. This unit will help you identify and select taps for threading operations.

IDENTIFYING COMMON TAP FEATURES

Taps are used to cut internal threads in holes. This process is called tapping. Tap parts and terminology are illustrated in Figure B-93. The active cutting part of the tap is the *chamfer*, which is produced by grinding away the tooth form at an angle, with relief behind the cutting edge, so that the cutting action is distributed over a number of teeth. The fluted portion of the tap provides space for chips to accumulate and for the passage of cutting fluids. Two-, three-, and four-flute taps are common.

The *major diameter* (Figure B-93) is the outside diameter of the tap as measured over the thread crests at the first full thread behind the chamfer. This is the largest diameter of the cutting portion of the tap, as most taps are back tapered or reduced slightly in thread diameter toward the shank. This back taper reduces the amount of tool contact with the thread during the tapping process, making the tap easier to turn.

Taps are made from high-speed steel and have a hardness of about Rockwell C63. High-speed steel taps typically are ground after heat treatment to ensure accurate thread geometry.

Another identifying characteristic of taps is the amount of chamfer at the cutting end of the tap (Figure B-94). A set consists of three taps—taper, plug, and bottoming taps—which are identical except for the number of chamfered threads. The taper

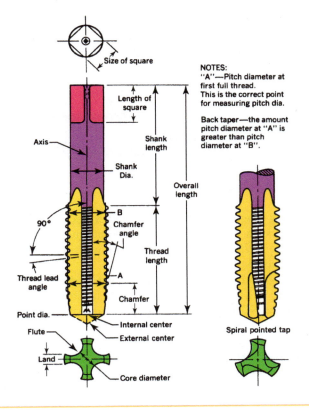

Figure B-93 General tap terms (*Besly Cutting Tools, Inc.*).

tap is useful in starting a tapped thread square with the part. The most commonly used tap, both in hand and machine tapping, is a plug tap. Bottoming taps are used to produce threads that extend almost to the bottom of a blind hole. A *blind hole* is one that is not drilled entirely through a part.

Serial taps are made in sets of three taps for any given size of tap. Each of these taps has one, two, or three rings cut on the shank near the square. The No. 1 tap has smaller major and pitch diameters and is used for rough cutting the thread. The No. 2 tap cuts the thread slightly deeper, and the No. 3 tap finishes it to size. Serial taps are used when

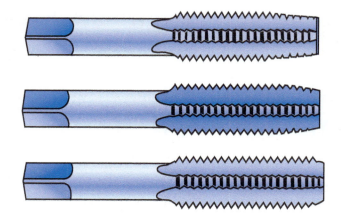

Figure B-97 Set of spiral pointed (or gun) taps.

Figure B-94 Chamfer designations for cutting taps. Top to bottom: starting tap, plug tap, and bottoming tap.

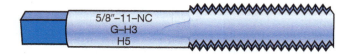

Figure B-95 Interrupted thread tap.

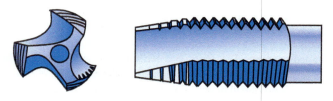

Figure B-98 Cutting action of spiral pointed taps.

5/8"–11–NC
G–H3
H5

Figure B-96 Identifying marking on a tap.

Spiral pointed taps can be operated at higher speeds and require less torque to drive than ordinary hand taps. Figure B-99 shows the design of the cutting edges. The cutting edges (*a*) at the point of the tap are ground at an angle (*b*) to the axis. Fluteless spiral pointed taps (Figure B-100) are recommended for production tapping of through holes in sections no thicker than the tap diameter. This type of tap is strong and rigid, which reduces tap breakage caused by misalignment. Fluteless spiral point taps give excellent results when tapping soft materials or sheet metal.

Spiral fluted taps are made with helical flutes instead of straight flutes (Figure B-101). Helical flutes draw the chips

tapping tough metals by hand. Another tap used for tough metal, such as stainless steel is the interrupted thread tap (Figure B-95). This tap has alternate teeth removed to reduce tapping friction.

Figure B-96 shows the identifying markings on a tap, where $\frac{5}{8}$ inch is the nominal size, 11 is the number of threads per inch, and NC refers to the standardized National Coarse thread series. G is the symbol used for ground taps. H3 identifies the tolerance range of the tap. Left-handed taps will also be identified by an LH or left-hand marking on the shank. More information on taps may be found in the *Machinery's Handbook*.

OTHER KINDS AND USES OF TAPS

Spiral pointed taps (Figure B-97), often called *gun taps*, are especially useful for machine tapping of through holes or blind holes with sufficient chip room below the threads. When the spiral point is turned, the chips are forced ahead of the tap (Figure B-98). Since the chips are pushed ahead of the tap, the problems caused by clogged flutes, especially breakage of taps, are eliminated if it is a through hole. If a spiral pointed tap is used to tap a blind hole, sufficient hole depth is necessary to accommodate the chips that are pushed ahead of the tap. Also, since they are not needed for chip passage, the flutes of spiral pointed taps can be made shallower, thus increasing the strength of the tap.

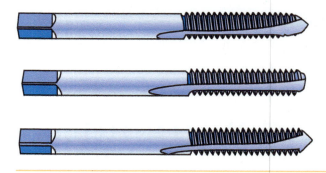

Figure B-99 Detail of spiral pointed tap.

Figure B-100 Fluteless spiral pointed tap for thin materials.

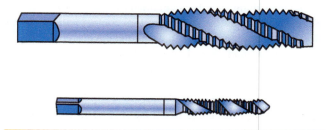

Figure B-101 Spiral fluted taps—regular spiral.

because of the flow of the displaced metal. Threads produced in this manner have improved surface finish and increased strength because of the cold working of the metal. The size of the hole to be tapped must be closely controlled, since too large a hole will result in a poor thread form, and too small a hole will result in the breaking of the tap.

A tapered pipe tap (Figure B-105) is used to tap holes with a taper of $\frac{3}{4}$ inch per foot for pipes with a matching thread and to produce a leak proof fit. The nominal size of a pipe tap is that of the pipe fitting and not the actual size of the tap. When taper pipe threads are tapped, every tooth of the tap engaged with the work is cutting until the rotation is stopped. This takes much more torque than does the tapping of a straight thread in which only the chamfered end and the first full thread are actually cutting. Straight pipe taps (Figure B-106) are used for tapping holes or couplings to fit taper-threaded pipe and to secure a tight joint when a sealant is used.

A pulley tap (Figure B-107) is used to tap setscrew and oil cup holes in the hubs of pulleys. The long shank also permits tapping in places that might be inaccessible for regular

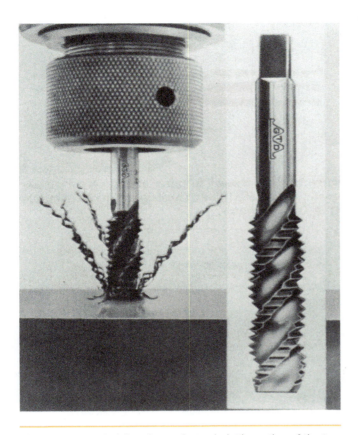

Figure B-102 Spiral fluted tap—fast spiral. The action of the tap lifts the chips out of the hole to prevent binding.

Figure B-103 Fluteless thread-forming tap.

out of the hole. This kind of tap is also used when tapping a hole that has a keyway or spline, as the helical lands of the tap will bridge the interruptions. Spiral fluted taps are recommended for tapping deep blind holes in ductile materials such as aluminum, magnesium, brass, copper, and die-cast metals. Fast-spiral fluted taps (Figure B-102) are similar to regular spiral fluted taps, but the faster spiral flutes increase the chip lifting action and permit the spanning of comparably wider spaces.

Thread-forming taps (Figure B-103) are fluteless and do not cut threads in the same manner as conventional taps. They are forming tools, and their action can be compared with external thread rolling. On ductile materials such as aluminum, brass, copper, die castings, lead, and leaded steels, these taps give excellent results. Thread-forming taps are held and driven just like conventional taps, but because they do not cut the threads, no chips are produced. Problems of chip congestion and removal often associated with the tapping of blind holes are eliminated. Figure B-104 shows how the thread-forming tap displaces metal. The crests of the thread at the minor diameter may not be flat but will be slightly concave

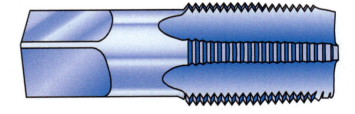

Figure B-104 The thread-forming action of a fluteless thread-forming tap.

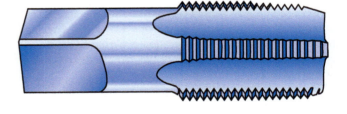

Figure B-105 Taper pipe tap.

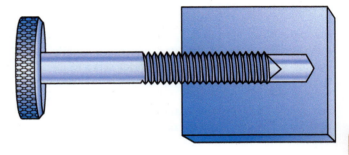

Figure B-106 Straight pipe tap.

Figure B-107 Pulley tap.

Figure B-108 Nut tap.

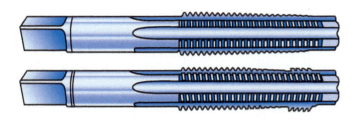

Figure B-109 Set of Acme thread taps. The upper tap is used for roughing, the lower tap for finishing.

Figure B-110 Tandem Acme tap designed to rough and finish cut the thread in one pass.

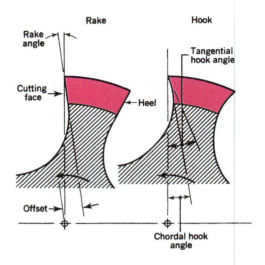

Figure B-111 Rake and hook angles on cutting taps (*Besly Cutting Tools, Inc.*).

Table B-1 Recommended Tap Rake Angles

0–5 Degrees	8–12 Degrees	16–20 Degrees
Bakelite	Bronze	Aluminum and alloys
Plastics	Hard rubber	Zinc die castings
Cast iron	Cast steel	Copper
Brass	Carbon steel	Magnesium
Hard rubber	Alloy steel	
	Stainless steel	

hand taps. When used for tapping pulleys, these taps are inserted through holes in the rims, which are slightly larger than the shanks of the taps. These holes serve to guide the taps and assure proper alignment with the holes to be tapped.

Nut taps (Figure B-108) differ from pulley taps in that their shank diameters are smaller than the root diameter of the thread. The smaller shank diameter makes the tapping of deep holes possible. Nut taps are used when small quantities of nuts are made or when nuts have to be made from tough materials such as some stainless steels or similar alloys.

Figure B-109 shows Acme taps for roughing and finishing. Acme threads are used to provide accurate movement—for example, in lead screws on machine tools—and for applying pressure in various mechanisms. On some Acme taps the roughing and finishing operation is performed with one tap (Figure B-110). The length of this tap usually requires a through hole.

RAKE AND HOOK ANGLES ON CUTTING EDGES

Cutting face geometry is an important factor in efficient cutting. Cutting face geometry should vary depending on the material to be tapped. Cutting face geometry is expressed in terms of rake and hook (Figure B-111). The *rake* of a tap is the angle between a line through the flat cutting face and a radial line from the center of the tool to the tooth tip. The rake can be negative, neutral, or positive. *Hook angle*, on the other hand, relates to the concavity of the cutting face. Unlike the rake angle, the hook angle cannot be negative. Table B-1 gives the rake angle recommendations for workpiece materials. In general, the softer or more ductile the material, the greater the rake angle should be. Harder and more brittle materials call for reduced rake angles.

REDUCING FRICTION IN TAPPING

A tap is usually back tapered along the thread to relieve the friction between the tool and the workpiece. Another form of relief is often applied to taps. When the fully threaded portion of the tap is cylindrical (other than back taper), it is called a *concentric thread* (Figure B-112). If the pitch diameter on each flute decreases from face to back (heel), it has *eccentric relief* (Figure B-112). This means less tool contact with the workpiece and less friction. A third form of friction relief combines the concentric thread and the eccentric thread relief and is termed *con-eccentric*. The concentric margin helps guide the tap, and the relief following the margin reduces friction. Relief is also provided behind the chamfer of the tap to provide radial clearance for the cutting edge. Relief may also be provided in the form of a channel that runs lengthwise down the center of the land (Figure B-113).

Other steps may also be taken to reduce friction and to increase tap life. Surface treatment of taps is often an answer if poor thread forming or tap breakage is caused by chips adhering to the flutes or welding to the cutting faces. These surface treatments improve the wear life of taps by increasing their abrasion resistance.

Various kinds of surface treatments are available. Liquid nitride produces a hard, shallow surface on high-speed steel

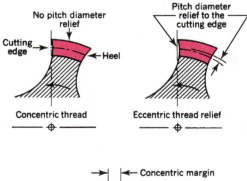

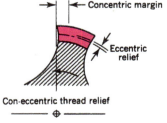

Figure B-112 Pitch diameter relief forms on taps (*Besly Cutting Tools, Inc.*).

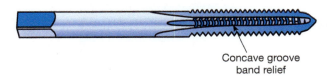

Figure B-113 Tap with concave groove land relief.

tools. Oxide finishes are usually applied in steam tempering furnaces and can be identified by their bluish black color. The oxide acts as a solid lubricant. It also holds liquid lubricant at the cutting edges during a tapping operation. Oxide treatments prevent chips from welding to the tool and reduce friction between the tool and the work. Chrome plating is an effective treatment for taps used on nonferrous metals and some soft steels. The chromium deposit is shallow and often referred to as *flash chrome plating*. Titanium nitride is another effective coating for extending the working life of taps.

SELF-TEST

1. What type of tap is used to produce threads that extend almost to the bottom of a blind hole.
2. When should a gun tap be used?
3. What are fluteless spiral pointed taps typically used for?
4. When is a spiral fluted tap used?
5. How are thread-forming taps different from conventional taps?
6. How are taper pipe taps identified?
7. Why are finishing and roughing Acme taps used?
8. Why are rake angles varied on taps for different materials?

INTERNET REFERENCES

Information on taps:

www.e-taps.com

www.osgtool.com

Tapping Procedures

A machinist must have a good understanding of the factors that affect the tapping of a hole. Factors such as the work material and its cutting speed, the proper cutting fluid, and the size and condition of the hole all affect the quality of the tapped hole. A good machinist can analyze a tapping operation, determine whether it is satisfactory, and find a solution if it is not. You will learn about tapping procedures in this unit.

OBJECTIVES

After completing this unit, you should be able to:

- Select the correct tap drill for a specific percentage of thread.
- Determine the cutting speed for a given work material–tool combination.
- Select the correct cutting fluid for tapping.
- Tap holes by hand or with a drill press.
- Identify and correct common tapping problems.

TAP USE

Taps are used to cut internal threads in holes. The cutting process is called *tapping* and can be performed by hand or with a machine (Figure B-114). A tap wrench or a T-handle tap wrench (Figure B-115) is used to provide driving torque while hand tapping. To obtain a greater accuracy in hand tapping, a hand tapper (Figure B-116) can be used. This fixture acts as a guide for the tap to ensure that it stays in alignment and cuts concentric threads.

Holes can also be tapped in a drill press that has a spindle reverse switch, which is often foot operated for convenience. Drill presses without reversing switches can be used for tapping with a tapping attachment (Figure B-117). Some of these tapping attachments have an internal friction clutch, so that downward pressure on the tap turns the tap forward and feeds it into the work. Releasing downward pressure will automatically reverse the tap and back it out of the workpiece. Some tapping attachments have lead screws that provide tap feed rates equal

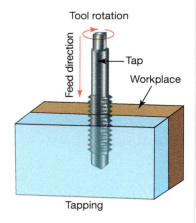

Figure B-114 Tapping a hole.

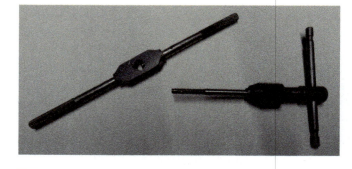

Figure B-115 Tap wrench and a T-handle tap wrench.

to the lead of the tap. Most tapping attachments have an adjustment to limit the torque to the size of the tap, which eliminates most tap breakage. Figure B-118 shows a tapping machine. Tapping machines can be used to quickly tap holes.

THREAD PERCENTAGE AND STRENGTH

The strength of the thread in a tapped hole depends largely on the workpiece material, the percentage of full thread depth, and the length of the thread. The workpiece material is

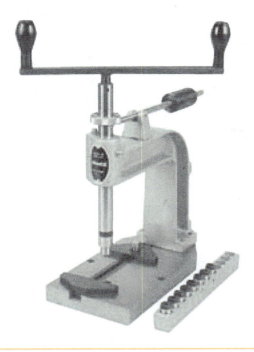

Figure B-116 Hand tapper (*Courtesy of Enco*).

Figure B-117 Drill press tapping attachment.

usually selected by the designer, but the machinist can often control the percentage of thread produced and the depth of the thread. The percentage of thread produced is dependent on the diameter of the drilled hole. Tap drill charts generally give tap drill sizes to produce 75 percent thread. (See Appendix Tables 3 and 4 for tap drill charts.)

An example will illustrate the relationships between the percentage of thread, torque required to drive the tap, and the resulting thread strength. An increase in thread depth from 60 to 72 percent in AISI 1020 steel requires twice the torque to drive the tap, but it increases the strength of the thread by only 5 percent. The practical limit is approximately 75 percent full

Figure B-118 Tapping machine (*Courtesy of Fox Valley Technical College*).

thread, since a greater percentage of thread does not increase the strength of the threaded hole in most materials.

In some difficult-to-machine materials such as titanium alloys, high-tensile steels, and some stainless steels, 50 to 60 percent thread depth will give sufficient strength to the tapped hole. Threaded assemblies are usually designed so that the bolt breaks before the thread strips. The rule of thumb is to have a bolt engage a tapped hole by 1 to $1\frac{1}{2}$ times its diameter. This assures that the bolt will break before the threads are stripped.

DRILLING THE CORRECT HOLE SIZE

The condition of the drilled hole affects the quality of the thread produced. An out-of-round hole leads to an out-of-round thread. Bell-mouthed holes will produce bell-mouthed threads. The size of the hole to be drilled is usually obtained from tap drill charts, which are based on 75 percent thread depth. If a thread depth other than 75 percent is needed, the following formula can be used to determine the proper hole size. For the American National Unified form:

$$\text{Outside diameter of thread} - \frac{.01266 \times \% \text{ of thread depth}}{\text{number of threads per inch (TPI)}}$$

$$= \text{hole diameter}$$

For example, to calculate the hole size for a 1-inch 12-thread fastener with a 70 percent thread depth:

$$1.0 - \frac{.01266 \times 70\%}{12} = .926$$

This formula will work for any thread system.

SPEEDS FOR TAPPING

When a tapping machine or tapping attachment is used, speed (RPM) is very important. The quality of the thread is dependent on the speed at which the tap is operated. The selection of the best speed for tapping is limited, because the feed per revolution is fixed by the lead of the thread. Excessive speed

Table B-2 Recommended Cutting Speeds and Lubricants for Machine Tapping

Material	Speeds (ft/min)	Lubricant
Aluminum	90–100	Kerosene and light base oil
Brass	90–100	Soluble oil or light base oil
Cast iron	70–80	Dry or soluble oil
Magnesium	20–50	Light base oil diluted with kerosene
Phosphor bronze	30–60	Mineral oil or light base oil
Plastics	50–70	Dry or air jet
Steels		
Low carbon	40–60	Sulfur-base oil
High carbon	25–35	Sulfur-base oil
Free machining	60–80	Soluble oil
Molybdenum	10–35	Sulfur-base oil
Stainless	10–35	Sulfur-base oil

develops high temperatures that cause rapid wear of the tap's cutting edge. Dull taps produce rough, torn, or off-size threads. High cutting speeds prevent adequate lubrication at the cutting edges and often create a problem of chip disposal.

When selecting the best speed for tapping, you should consider the material being tapped, the size of the hole, and the lubricant being used. Table B-2 gives guidelines in selecting a speed and a lubricant for materials when using high-speed steel taps.

These cutting speeds in feet per minute have to be translated into rpm to be useful. For example, calculate the rpm when tapping a $\frac{3}{8}-24$ UNF hole in free-machining steel. The cutting speed chart gives a cutting speed between 60 and 80 feet per minute. Use the lower figure; you can increase the speed once you see how the material taps. The formula for calculating rpm is

$$\frac{\text{cutting speed (CS)} \times 4}{\text{diameter }(D)} \text{ or } \frac{60 \times 4}{3/8} = 640 \text{ rpm}$$

Lubrication is one of the most important factors in a tapping operation. Cutting fluids used when tapping serve as coolants but are more important as lubricants. It is important to select the correct lubricant. For lubricants to be effective, they should be applied in sufficient quantity to the actual cutting area in the hole.

SOLVING TAPPING PROBLEMS

In Table B-3, common tapping problems are listed with some possible solutions. Occasionally, it becomes necessary to remove a broken tap from a hole. If part of the broken tap extends out of the workpiece and you are very lucky, removal is relatively easy with a pair of pliers. If the tap breaks flush with or below the surface of the workpiece, a tap extractor can be used (Figure B-119). Before trying to remove a broken tap, remove the chips in the flutes. A jet of compressed air or cutting fluid can be helpful in removing chips.

Figure B-119 Tap extractor.

SAFETY FIRST

Be careful when cleaning holes with compressed air, as chips and particles fly out at high velocity.

When the chips are packed so tightly in the flutes or the tap is jammed in the work so that a tap extractor cannot be used, the tap may be broken up with a pin punch and removed piece by piece.

It may be necessary to use an electrical discharge machine (EDM), sometimes called a *tap disintegrator*, to remove the broken tap. These machines electrically erode material while they are immersed in a fluid.

TAPPING PROCEDURE, HAND TAPPING

Step 1 Determine the size of the thread to be tapped and select the tap.

Step 2 Select the proper tap drill from a tap drill chart. Choose a taper tap for hand tapping; or if a drill press or tapping machine is to be used for alignment, use a plug tap.

Step 3 Fasten the workpiece securely in a drill press vise. Calculate the correct rpm for the drill used:

$$\text{rpm} = \frac{\text{CS} \times 4}{D}$$

Drill the hole using the recommended coolant. Check the hole size.

Step 4 Countersink the top of the hole to a diameter slightly larger than the major diameter of the threads (Figure B-120). This allows the tap to be started more easily, and it protects the start of the threads from damage.

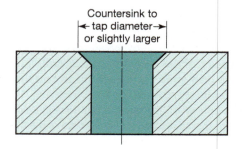

Figure B-120 Preparing the workpiece.

Table B-3 Common Tapping Problems and Possible Solutions

Causes of Tap Breakage	Solution
Tap hitting bottom of hole or bottoming on packed chips	Drill the hole deeper. Eject chips with air pressure. (*Caution*: Stand aside when you do this and always wear safety glasses.) Use spiral fluted taps to pull chips out of hole. Use a thread-forming tap.
Chips are packing in flutes	Use a tap style with more flute space. Tap to a lesser depth or use a smaller percentage of threads. Select a tap that will eject chips forward (spiral point) or backward (spiral fluted).
Hard materials or hard spots	Anneal the workpiece. Reduce cutting speed. Use longer chamfers on tap. Use taps with more flutes.
Inadequate lubricant	Use the correct lubricant and apply sufficient amount of it under pressure at the cutting zone.
Tapping too fast	Reduce cutting speed.
Excessive wear	
Abrasive materials	Improve lubrication. Use surface-treated taps. Check the alignment of tap and hole to be tapped.
Chips clogging flutes	
Insufficient lubrication	Use a better lubricant and apply it with pressure at the cutting zone.
Excessive speed	Reduce cutting speed.
Wrong-style tap	Use a more-free-cutting tap such as a spiral pointed tap, spiral fluted tap, interrupted thread tap, or a surface-treated tap.
Torn or rough threads	
Dull tap	Use a new tap.
Chip congestion	Use a tap with more chip room. Use lesser percentage of thread. Drill a deeper hole. Use a tap that will eject chips.
Inadequate lubrication and chips clogging flutes	Correct as suggested previously.
Hole improperly prepared	Torn areas on the surface of the drilled, bored, or cast hole will be shown in the minor diameter of the tapped thread.
Undersize threads	
Pitch diameter of tap too small	Use a tap with a large pitch diameter.
Excessive speed	Reduce tapping speed.
Thin-wall material	Use a tap that cuts as freely as possible. Improve lubrication. Hold the workpiece so that it cannot expand while it is being tapped. Use an oversize tap.
Dull tap	Use a new tap.
Oversize or bell-mouth threads	
Loose spindle or worn holder	Replace or repair spindle or holder.
Misalignment	Align spindle, fixture, and work.
Tap oversized	Use a smaller-pitch-diameter tap.
Dull tap	Use a new tap.
Chips packed in flutes	Use a tap with deeper flutes, spiral flutes, or spiral points.
Buildup on cutting edges of tap	Use correct lubricant and tapping speed.

Step 5 Mount the workpiece in a bench vise so that the hole is in a vertical position.

Step 6 Tighten the tap in the tap wrench.

Step 7 Cup your hand over the center of the wrench (Figure B-121) and place the tap in the hole in a vertical position. Start the tap by turning two or three turns in a clockwise direction for a right-hand thread. At the same time, keep a steady pressure downward on the tap. When the tap is started, it may be turned as shown in Figure B-122.

Step 8 After the tap is started for a few turns, remove the tap wrench without disturbing the tap. Place the blade of a square against the solid shank of the tap to check for squareness (Figure B-123). Check from two positions 90 degrees apart. If the tap is not square with the work, it will ruin the thread and possibly break in the hole if you continue tapping. Back the tap out of the hole and restart.

Step 9 Use the correct cutting oil on the tap when cutting threads.

Figure B-121 Starting the tap (*Courtesy of Fox Valley Technical College*).

Figure B-122 Tapping a thread by hand (*Courtesy of Fox Valley Technical College*).

Figure B-123 Checking the tap for squareness.

Figure B-124 Using the drill press as a tapping fixture.

Step 10 Turn the tap clockwise one-quarter to one-half turn and then turn it back three-quarters of a turn to break the chip. Do this with a steady motion to avoid breaking the tap.

Step 11 When tapping a blind hole, use the taps in the order: starting, plug, and then bottoming. Remove the chips from the hole before using the bottoming tap, and be careful not to hit the bottom of the hole with the tap.

Step 12 Figure B-124 shows a 60-degree point center chucked in a drill press to align a tap squarely with the previously drilled hole. Only very slight follow-up pressure should be applied to the tap. Too much downward pressure will cut a loose, oversize thread.

SELF-TEST

1. What kind of tools are used to drive taps when hand tapping?
2. What is a hand tapper?
3. What is a tapping attachment?
4. Which three factors affect the strength of a tapped hole?
5. How deep should the usable threads be in a tapped hole?
6. What causes taps to break while tapping?
7. What causes rough and torn threads?
8. Give three methods of removing broken taps from holes.

Thread-Cutting Dies and Their Uses

A die is used to manually cut external threads on a bolt or rod. In this unit you will be introduced to some thread-cutting dies and their uses.

Dies are used to cut external threads on round materials. Industrial dies are made from high-speed steel. Dies are identified by markings on the face as to the size of thread, number of threads per inch, and form of thread, such as NC, UNF, or other standard designations.

COMMON TYPES OF HAND THREADING DIES

The die shown in Figure B-125 is an example of a round split adjustable die, also called a *button die*. These dies are made in all standardized thread sizes up to 1 ½-inch thread diameters and 1.2-inch pipe threads. The outside diameters of these dies vary from ⅝ to 3 inches.

Adjustments on these dies are made by turning a fine-pitch screw that forces the sides of the die apart or allows them to spring together. The range of adjustment of round split adjustable dies is very small, allowing only for a loose or tight fit on a threaded part. Excessive expansion may cause the die to break.

Some round split adjustable dies do not have an adjusting screw. Adjustments are then made with the three screws in the diestock (Figure B-126). Two of these screws on opposite sides of the diestock hold the die in the diestock and also provide closing pressure. The third screw engages the

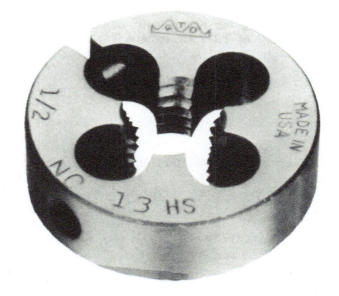

Figure B-125 Markings on a die. (Example shown is a round split adjustable die.)

split in the die and provides opening pressure. These dies are used in a diestock for hand threading or in a machine holder for machine threading.

Another type of threading die is the two-piece die, whose halves (Figure B-127) are called *blanks*. The blanks are assembled in a collet consisting of a cap and the guide (Figure B-128).

Figure B-126 Diestock for round split adjustable dies.

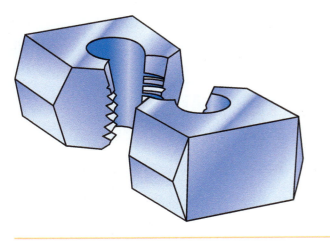

Figure B-127 Die halves for two-piece die.

Figure B-128 Components of a split adjustable die collet.

The normal position of the blanks in the collet is indicated by witness marks (Figure B-129). The adjusting screws allow for precise control of the cut thread size. The blanks are inserted in the cap with the tapered threads toward the guide. Each of the two die halves is stamped with a serial number. Make sure that the halves you select have the same numbers. The guide used in the collet serves as an aid in starting and holding the dies square with the work being threaded. Each thread size uses a guide of the same nominal or indicated size. Collets are held securely in diestocks (Figure B-130) by a knurled screw that seats in a dimple in the cap.

Hexagon rethreading dies (Figure B-131) are used to recut slightly damaged or rusty threads. Rethreading dies are driven with a wrench large enough to fit the die. All the die types discussed previously are also available in pipe thread sizes.

HAND THREADING PROCEDURES

Threading throat side of the die should be used to start a thread. This side is identified by the chamfer on the first two or three threads and also by the size markings. The chamfer distributes the cutting load over a number of threads, which produces better threads and less chance of chipping the cutting edges of the die. Cutting oil or other threading fluids are important in obtaining quality threads and maintaining long die life. Once a cut is started with a die, it will tend to follow its own lead, but uneven pressure on the diestock will make the die cut variable helix angles or "drunken" threads.

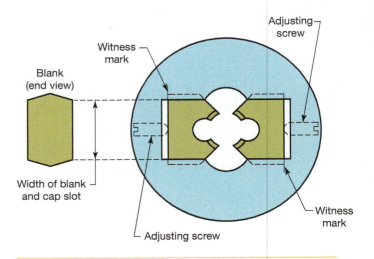

Figure B-129 Setting the die position to the witness marks on the die and collet assembly.

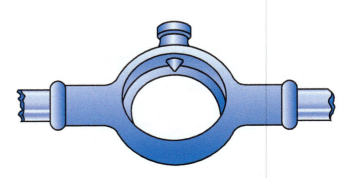

Figure B-130 Diestock for adjustable die and collet assembly.

Figure B-131 Hexagon rethreading die.

Threads cut by hand often show considerable accumulated lead error. The *lead* of a screw thread is the distance a nut will move on the screw if it is turned one full revolution. This problem results because the dies are relatively thin compared with the diameter of thread they cut. Only a few threads in the die can act as a guide on the already cut threads. This error usually does not cause problems when standard or thin nuts are used on the threaded part. However, when an item with a long internal thread is assembled with a threaded rod, it can tighten and lock, not because the thread depth is insufficient, but because there is a lead error. This lead error can be as much as one-fourth of a thread in 1 inch of length.

The outside diameter of the material to be threaded should not be more than the nominal size of the thread and preferably a few thousandths of an inch (.002 to .005 inch) smaller. After a few full threads are cut, the die should be removed so that the thread can be tested with a nut or thread ring gage. A thread ring gage set usually consists of two gages, a *go* and a *no go* gage. As the names imply, a go gage should screw on the thread, whereas the no go gage will not go more than $1\frac{1}{2}$ turns on a thread of the correct size. Do not assume that the die will cut the correct size thread; always check by gaging or assembling. Adjustable dies should be spread open for the first cut and set progressively smaller for each pass after checking the thread size.

It is important that a die be started squarely on the rod to be threaded. A lathe can be used as a fixture for cutting threads with a die (Figure B-132). The rod is held in a lathe chuck for rotation, while the die is held square because it is supported by the face of the tailstock spindle. The carriage or the compound rest prevents the diestock from turning while the chuck is rotated by hand. As the die advances, the tailstock spindle is also advanced to stay in contact with the die. Do not force the die with the tailstock spindle, or a loose thread may result. A die may be used to finish a long thread that has been rough threaded on the lathe.

It is good practice to chamfer the end of a workpiece before starting a die (Figure B-133). The chamfer on the end

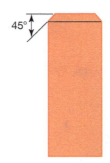

Figure B-133 Chamfer workpiece before using die.

of a rod can be made by grinding on a pedestal grinder, by filing, or with a lathe. This will help in starting the cut and will also leave a finished thread end. When cutting threads with a hand die, reverse the rotation of the die after each full turn to break the chips into short pieces that will fall out of the die. Chips jammed in the clearance holes will tear the thread.

THREADING PROCEDURE, THREADING DIES

Step 1 Select the part to be threaded and measure its diameter. Then, chamfer the end, either on a grinder or with a file. The chamfer should be at least as deep as the thread to be cut.

Step 2 Select the correct die and mount it in a diestock. Do not over tighten the screws in the diestock.

Step 3 Mount the workpiece in a bench vise. Short workpieces are mounted vertically and long pieces usually are held horizontally.

Step 4 To start the thread, place the die over the workpiece. Holding the diestock with one hand (Figure B-134), apply downward pressure and turn the die.

Step 5 Apply cutting fluid to the workpiece and die, and start turning the diestock with both hands (Figure B-135). After each complete revolution forward, reverse the die one-half turn to break the chips.

Figure B-132 Threading a rod with a hand die in a lathe.

Figure B-134 Start the die with one hand.

Figure B-135 Use both hands to turn the threading die.

Step 6 Check to see that the thread is started square, using a machinist's square. Make any necessary corrections by applying slight downward pressure on the high side while turning.

Step 7 When several turns of the thread have been completed, check the fit of the thread with a nut, thread ring gage, thread micrometer, or the mating part. If the thread fit is incorrect, adjust the die with the adjustment screws and take another cut with the adjusted die. Continue making adjustments until the proper fit is achieved.

Step 8 Continue threading to the required thread length. To cut threads close to a shoulder, invert the die after the normal threading operation and cut the last two or three threads with the side of the die that has less chamfer.

SELF-TEST

1. What is a die?
2. What tool is used to drive a die?
3. How much adjustment is possible with a round split adjustable die?
4. What are important points to watch when assembling two-piece dies in a collet?
5. Why do dies have a chamfer on the cutting end?
6. Why are cutting fluids used?
7. What diameter should a rod be before being threaded?
8. Why should a rod be chamfered before being threaded?

Off-Hand Grinding

The pedestal grinder is used for many hand-grinding operations, especially sharpening and shaping drills and tool bits. In this unit you will study the setup, use, and safety aspects of a pedestal grinder.

OBJECTIVE

After completing this unit, you should be able to:

■ Describe the setup, use, and safety of the pedestal grinder.

OFF-HAND GRINDING ON PEDESTAL GRINDERS

The pedestal grinder is a machine. However, since the workpiece is handheld, we will cover it as a hand tool. The pedestal grinder gets its name from the floor stand or pedestal that supports the motor and abrasive wheels. The pedestal grinder is a common machine tool that you will use almost daily in the machine shop. This grinding machine is used for general-purpose, off-hand grinding in which the workpiece is handheld and applied to the rotating abrasive wheel. One of the primary functions of the pedestal grinder is shaping and sharpening tool bits and drills in machine shop work. Pedestal grinders can also have rotary wire brushes or buffing wheels installed.

Large, heavy-duty pedestal grinders are sometimes found in machine shops. These grinders are used for rough grinding (snagging) welds, castings, and general rough work. These machines are generally set up in a separate location from the tool grinders. Rough grinding of metal parts should never be done on tool grinders because it causes the wheels to become grooved, uneven, and out-of-round. In that condition they are useless for tool grinding.

Setup of the Pedestal Grinder

The pedestal grinder in your shop is ready for use most of the time. If you do need to do maintenance on the grinder, you must disconnect power (lockout) before the maintenance is performed. If it becomes necessary to replace a worn wheel, the side of the guard must be removed and the tool rest moved out of the way. A piece of wood may be used to prevent the wheel from rotating so that the spindle nut can be turned and removed (Figure B-136). Remember that the left side of the spindle has left-handed threads, and the right side has right-handed threads.

A new wheel should be ring tested to determine whether there are any cracks or imperfections (Figure B-137). Gently tap the wheel near its rim with a screwdriver handle or a piece of wood and listen for a clear ringing sound like that of a bell. A clear ring indicates that the wheel is safe to use; if a dull thud is heard, the wheel may be cracked and should not be used. Check that the flanges and the spindle are clean before mounting the wheel. Be sure that the center hole in

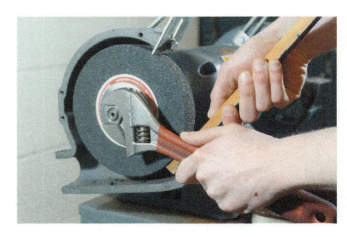

Figure B-136 Using a piece of wood to hold the wheel while removing the spindle nut.

Figure B-137 The ring test is made before mounting the wheel.

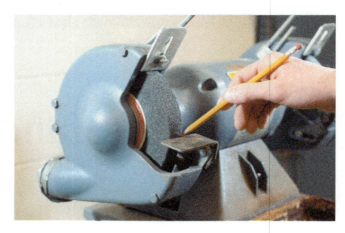

Figure B-139 The tool rest is adjusted.

Figure B-138 The wheel is mounted with the proper bushing in place.

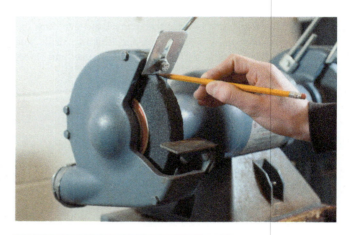

Figure B-140 The spark guard is adjusted.

the wheel is the correct size for the grinder spindle. If you must use a bushing, be sure that it is the correct size and installed properly (Figure B-138). Place a clean, undamaged blotter on each side between wheel and flanges. Tighten the spindle nut just enough to hold the wheel firmly. Excessive tightening will break the wheel.

After you have replaced the guard and cover plate, bring the tool rest up to the wheel so that there is $\frac{1}{16}$ to $\frac{1}{8}$-inch clearance between the rest and wheel (Figure B-139). If there is excessive space between the tool rest and the wheel, a small workpiece may flip up and catch between the wheel and tool rest.

Dressing the Grinding Wheel

Stand aside, out of the line of the rotation of the grinding wheel and turn on the grinder. Let the wheel run for a full minute. A new wheel does not always run exactly true and therefore must be dressed. A Desmond dresser (Figure B-141) may be used to

SAFETY FIRST

Some of the machines you would assume are the safest in the machine shop are often the most dangerous. If your fingers slip on the workpiece they may contact the wheel and be cut severely. If the tool rest is not adjusted properly, your finger may be caught between the tool rest and the grinding wheel, resulting in a serious injury. Use caution when using grinders. Injuries are very painful.

The clearance between the tool rest and wheel should never exceed $\frac{1}{8}$ inch. The spark guard, located on the upper side of the wheel guard (Figure B-140), should be adjusted to within $\frac{1}{16}$ inch of the wheel. This protects the operator if the wheel should shatter.

sharpen and true the face of the wheel. Pedestal grinder wheels often become grooved, out-of-round, glazed, or misshapen and therefore must be frequently dressed to obtain proper grinding results. The dresser is moved back and forth across the front of the wheel.

Figure B-141 The wheel is being dressed.

Figure B-143 Flat chisel being ground to produce a 60-degree angle.

Using the Pedestal Grinder

Bring the workpiece into contact with the wheel gently, without bumping. Grind only on the face of the wheel. The workpiece will heat from friction during the grinding operation. It may become too hot to hold in just a few seconds. To prevent this, frequently cool the workpiece in the water pot attached to the grinder. Be especially careful when grinding drills and tool bits so that they do not become overheated. Excessive heat may permanently affect the tool's hardness.

Center punches can be sharpened on a grinder. Center punches should be hollow ground (Figure B-142) when they are sharpened. They should be evenly rotated while the point is ground to the correct angle. Flat cold chisels should also be hollow ground (Figure B-143) when they are sharpened.

Layout tools, such as spring dividers and scribers, should be kept sharp by honing on a fine, flat stone in the same way a knife is sharpened. However, when it becomes necessary to

reshape a layout tool, exercise extreme care to avoid overheating the thin point. A fine-grit wheel should be used for sharpening most cutting tools, and they should be cooled frequently in water.

Safety Checkpoints on the Pedestal Grinder

Nonferrous metals such as aluminum and brass should never be ground on the aluminum oxide wheels found on most pedestal grinders. These metals fill the voids or spaces between the abrasive particles in the grinding wheel so that more pressure is needed to accomplish the desired grinding. This additional pressure sometimes causes the wheel to break or shatter. Pieces of grinding wheel may be thrown out of the machine at extreme velocities. Always use silicon carbide abrasive wheels for grinding nonferrous metals. Excessive pressure should never be used in any grinding operation. If this seems to be necessary, it means that the improper grit or grade of abrasive is being used, or the wheel is glazed and needs to be dressed. Always use the correct abrasive grit and grade for the particular grinding that you are doing. (For grinding wheel selection see Section L, Unit 2.)

Figure B-142 A punch being correctly sharpened to produce a 90-degree angle for use as a center punch. It must be rotated while being ground. A sharper angle of 60 degrees may be ground to produce a prick punch.

SAFETY FIRST

Always wear appropriate eye protection when dressing wheels or grinding on the pedestal grinder. Be sure that grinding wheels are rated at the proper speed for the grinder you are using. The safety shields, wheel guards, and spark guard must be kept in place at all times while grinding. The tool rest must be adjusted and the setting corrected as the diameter of the wheel decreases from use. Grinding wheels and rotary wire brushes may catch loose clothing or long hair. Long hair should be tied back. Wire wheels often throw out small pieces of wire at high velocities.

SELF-TEST

1. What are pedestal grinders primarily used for in a machine shop?

2. Why should a tool grinder never be used for rough grinding metal?

3. When a wheel on a pedestal grinder needs to be reshaped, sharpened, or trued, what tool is usually used?

4. When sharpening layout tools and other tools, what is the most important consideration?

5. Name at least three safety factors to remember when using the pedestal grinder.

6. How far from the wheel should the work (tool) rest be adjusted?

7. Why do you allow the grinder to run for a short time after changing a wheel?

8. What is the purpose of the wheel blotter?

9. What is the primary safety consideration when selecting a grinding wheel?

10. What does the wheel ring test do?

SECTION C
Dimensional Measurement

INTRODUCTION

*M*easurement can be defined as the assignment of a value to time, length, weight, and so on. We cannot escape measurements. Our daily lives are greatly influenced by the clock, a device that measures time. The weight of almost every product we buy is measured, and the measure of length is incorporated into every product, ranging from the microscopic components of an integrated circuit to the many thousands of highway miles extending across the country.

Measurement has developed into an exact science known as **metrology**. Mass production of goods has increased the importance of metrology to check and control critical dimensions. Components of an automobile, for example, may be manufactured at many different locations and then brought to a central assembly point, with the assurance that all parts will fit and function as designed. The units in this section cover the common measuring tools used in machining.

Much of your work as a machinist will be dedicated to meeting dimensional specifications of the parts that you make. Meeting tolerance specifications ensures the correct fit between parts and that they will function properly.

Measurement and Common Measuring Tools

MEASUREMENT AND THE MACHINIST

A machinist is mainly concerned with the measurement of **length**—that is, the distance along a line between two points (Figure C-1).

Length defines the **size** of most objects. **Width** and **depth** are simply other names for length. A machinist typically measures length in the basic units of measure such as **inches** or **millimeters**. In addition, a machinist sometimes needs to measure the relationship of one surface to another, commonly called **angularity** (Figure C-2).

Squareness is the measure of deviation from true perpendicularity. A machinist will measure angularity in the basic units of angular measure—**degrees**, **minutes**, and **seconds of arc**.

A machinist also needs to measure such things as **surface finish**, **concentricity**, **straightness**, and **flatness**. A machinist also occasionally sees measurements that involve circularity, sphericity, and alignment (Figure C-3). Many of these more specialized measurement techniques are performed by a quality control inspector on computerized measurement equipment.

Figure C-1 The measurement of length may appear under several different names.

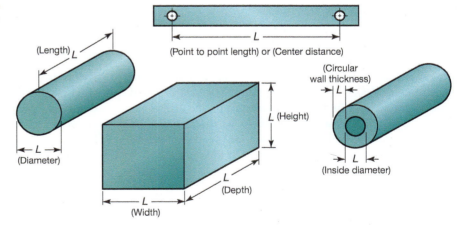

Figure C-2 Measurement of surface relationships, or angularity.

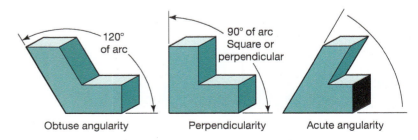

Measurement of surface relationships or angularity

111

Figure C-3 Other measurements encountered by the machinist.

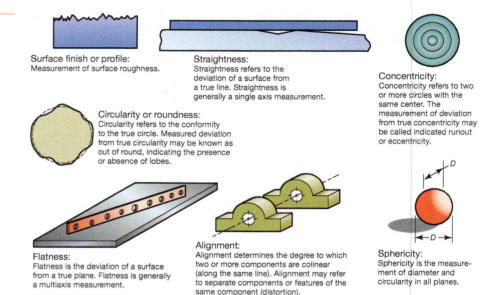

Surface finish or profile: Measurement of surface roughness.

Straightness: Straightness refers to the deviation of a surface from a true line. Straightness is generally a single axis measurement.

Concentricity: Concentricity refers to two or more circles with the same center. The measurement of deviation from true concentricity may be called indicated runout or eccentricity.

Circularity or roundness: Circularity refers to the conformity to the true circle. Measured deviation from true circularity may be known as out of round, indicating the presence or absence of lobes.

Flatness: Flatness is the deviation of a surface from a true plane. Flatness is generally a multiaxis measurement.

Alignment: Alignment determines the degree to which two or more components are colinear (along the same line). Alignment may refer to separate components or features of the same component (distortion).

Sphericity: Sphericity is the measurement of diameter and circularity in all planes.

GENERAL PRINCIPLES OF METROLOGY

There are many types of measuring tools available that were designed for use in various applications. Not every tool is equally suited for a specific measurement. Selecting the proper measuring tool for a specific application is a must for a machinist. A machinist must be familiar with the important terms and principles of dimensional metrology.

Accuracy

Accuracy can mean a couple of different things. First, accuracy can refer to whether or not a specific measurement actually is its stated size. For example, a certain drill has its size stamped on its shank. A doubtful machinist decides to verify the drill size using a micrometer. The size is found to be correct. Therefore, the size stamped on the drill is accurate. Second, accuracy can refer to whether or not the chosen measuring tool is appropriate for the measurement. If the machinist had used a steel rule to measure the drill to see if it was a number 7 drill (.201) or a number 6 drill (.204), it would be an inappropriate tool for the job. In this example, the act of measurement is not accurate because the wrong measuring tool was selected. **User accuracy** is also an important consideration. If the machinist is not very good at using or reading a micrometer, the measurement will be inaccurate. Or if the machinist used a micrometer that was not calibrated for accuracy, the reading may not be accurate.

Precision

The term **precision** refers to the specific measurement being made, with regard to the degree of exactness required. For example, the distance from the earth to the moon, measured to within one mile, would be a precise measurement.

Likewise, a clearance of five thousandths of an inch between a certain bearing and journal might be precise for that specific application. However, five thousandths of an inch clearance between ball and race on a ball bearing would not be considered precise, as this clearance would, in fact, be only a few millionths of an inch. There are many degrees of precision dependent on the application and design requirements. For a machinist, any measurement made to a degree finer than one sixty-fourth of an inch or one-half millimeter can be considered a precision measurement and must be made with the appropriate measuring instrument.

Reliability

Reliability in measurement is the ability to obtain the desired result to the degree of precision required. The selection of the proper measuring tool is crucial for reliability. A certain tool may be reliable for a certain measurement but totally unreliable in another application. For example, if it were desired to measure the distance to the next town, the odometer on an automobile speedometer would yield a reliable result, provided a degree of precision of less than one-tenth mile was not required. On the other hand, to measure the length of a city lot with an odometer would not yield a reliable result.

Discrimination

Discrimination refers to how finely the basic unit of length of a measuring instrument is divided. The mile on an automobile odometer is divided into 10 parts; therefore, it discriminates to the nearest tenth of a mile. An inch on a micrometer is divided into 1,000 or, in some cases, 10,000 parts. Therefore, the micrometer discriminates to .001 or .0001 of an inch. If a measuring instrument is used beyond its discrimination, it will not be reliable.

10:1 Ratio for Discrimination

A rule of thumb for choosing which measurement device to use is that the measuring instrument should **discriminate 10 times finer** than the smallest unit that it will be used to measure. If the total tolerance for the length of a part is .010, we would choose a measuring instrument that could at least measure to .001 (.010/10 = .001) Note that this is not always possible.

Position of a Measuring Instrument in Relation to the Axis of Measurement

Most measurements are linear. Linear measurements measure the shortest distance between two points. In order to make an accurate, reliable linear measurement, the measuring instrument must be exactly in line with the axis of that measurement. If this condition is not met, the reading will not be reliable. The alignment of the measuring instrument with the axis of measurement applies to all linear measurements (Figure C-4). The figure illustrates the alignment of the instrument with the axis of measurement using a rule. Misalignment of the instrument, as illustrated in the unreliable situation, will result in inaccurate measurements.

Responsibility of the Machinist in Measurement

It is the responsibility of a machinist to select the proper measuring instrument for the job. When choosing a measurement device, the following should be considered:

1. What degree of accuracy and precision must this measurement meet?

2. What degree of measuring discrimination does this measurement require?

3. What is the most reliable tool for this application?

Calibration   In a manufacturing facility all measuring tools have their accuracy checked by the quality department. The term calibration means that the tools are calibrated against very accurate standards. If the tools are found to be out of calibration, they are adjusted or replaced. The tools are also logged into a logbook and the results of the calibration are recorded along with the next required calibration date. Stickers are applied to each tool to indicate when the tools are required to be calibrated next. Calibration helps assure that a company's measurement equipment is accurate so that they can produce accurate measurements and parts that meet specifications. Workers cannot use any measurement equipment that is beyond the calibration date. Machinists are also responsible for taking good care of measuring equipment and reporting any possible damage to a device so that its accuracy can be checked.

Measurement Error   Measurements are never perfect. There are always measurement errors. The machinist must be aware of potential sources of measurement errors and try to reduce them. Sources of measurement error include the measuring tool selected, the procedure used, the temperature of the part, the cleanliness of the part, and many others. One example of a possible method error is the incorrect use of a micrometer. A user must develop consistent "feel" to accurately use a micrometer. A machinist, who uses a micrometer like a C-clamp, will get a much different measurement than a machinist with the correct "feel."

Part designers are aware of measurement error. They are also aware that no part can be made perfectly, so they develop part tolerances. Every measurement has a tolerance, meaning that the measurement is acceptable within a specific range. A machinist must understand the basic principles of measurement error and assume the proper responsibility in the selection, calibration, and application of measuring instruments.

TOOLS FOR DIMENSIONAL MEASUREMENT

There are measuring instruments that can be applied to any conceivable measurement. A machinist must be concerned with the use, care, and application of common measuring instruments. There are also a large variety of instruments designed for specialized uses. Some of these are not common in a school or machine shop. Others are intended for use in the tool room or quality department, where they are used in the calibration process.

A machinist must be skilled in the use of all common measuring instruments. A machinist should also be familiar with the instruments used in production machining, inspection, and calibration. In the following pages, many of these tools will be briefly described so that you may become familiar with the wide selection of measuring instruments that are available.

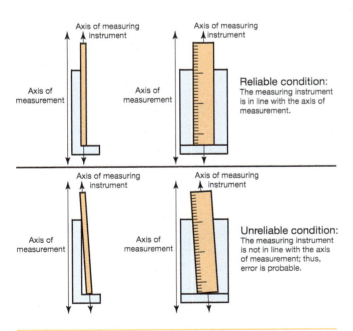

Figure C-4 The axis of a linear measuring instrument must be in line with the axis of measurement.

Fixed Gages and Air Gages

Fixed Gages   When large numbers of duplicate parts are produced, it may not be necessary to know the exact size of the part. The machinist may only have to determine if the part is good or bad (within acceptable tolerance). A go/no-go gage is often used for this type of inspection. The **adjustable limit snap gage** (Figure C-5) can be used to check outside diameter. One anvil is set to the minimum limit of the tolerance to be measured. The other anvil is set to the maximum limit of the tolerance. If both anvils slip over the part, an undersized condition is indicated. If neither anvil slips over the part, an oversized condition is indicated. The gage is set initially to a known standard by using gage blocks. This is a go/no-go type gage. The gage only tells us if the parts pass or fail. They do not tell us the exact size of the part.

Threaded parts are often checked with fixed gages. A **thread plug gage** (Figure C-6) is used to check internal threads. A **thread ring gage** (Figure C-6) is used to check external threads. These are frequently called **go/no-go** gages. One end of the plug gage is at the low limit of the tolerance, and the other end is at the high limit of the tolerance. The thread gage functions in the same manner. When using a thread plug gage, the blueprint may give a specification about how many revolutions the go thread gage must be able to turn into the hole to be a good part.

Fixed gages can also be used to check internal and external tapers (Figure C-7). Go/no-go plug gages are used for internal holes (Figure C-8). In this example the green end (go) should fit

Figure C-5 Adjustable limit snap gage.

Figure C-6 External and internal thread gages (*Courtesy of Fox Valley Technical College*).

Figure C-7 Cylindrical external and internal taper gages (*Courtesy of Fox Valley Technical College*).

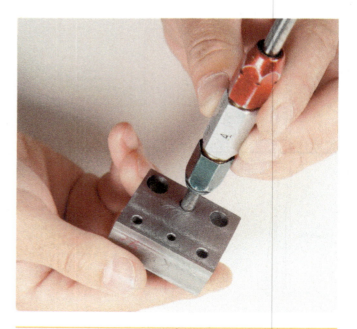

Figure C-8 Go/no-go plug gage being used to inspect a hole.

into the hole, but the red end (no go) should not fit into the hole. A ring gage is used for external diameters (Figure C-9). These are go/no-go type gages. The gages only tell us if the parts pass or fail. They do not tell us the exact size.

Air Gages   Air gages are also known as **pneumatic comparators**. Two types of air gages are used in comparison measuring applications. Air gage heads may be ring, snap (Figure C-10), or plug types (Figure C-11). Air flowing through the reference and the measuring channel is adjusted until the pressure meter reads zero with the setting master in place. A difference in workpiece size above or below the master size will cause more or less air to escape from the gage head. This, in turn, will change the pressure on the reference channel. The pressure change will be indicated on the pressure meter. The meter scale is graduated in suitable units. Workpiece size above or below the master can be directly read.

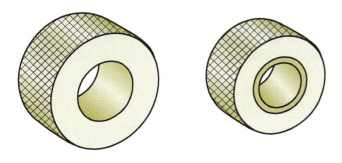

Figure C-9 Cylindrical ring gages.

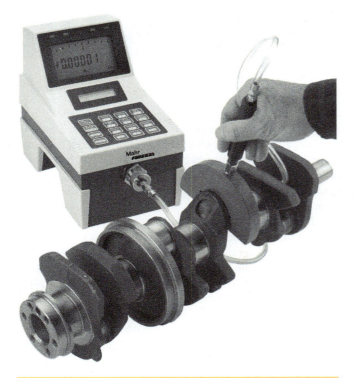

Figure C-10 Pressure-type air snap gage (*Courtesy of Mahr Federal Inc.*).

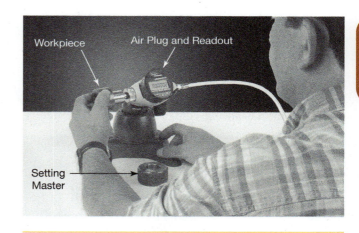

Figure C-11 Air plug gage being set with the use of a setting master (*Courtesy of Mahr Federal Inc.*).

Air gages have several advantages. The gage head does not make much contact with the workpiece because of the air pressure between the gage and the part. Consequently, there is very little wear on the gage head and no damage to the finish of the workpiece.

Mechanical Dial Measuring Instruments

Dial instruments have an advantage over their vernier instruments in that they are easier to read. Dial measuring equipment is frequently found in the inspection department, where many types of measurements must be made quickly and accurately.

Dial Thickness Gage   A **dial thickness gage** (Figure C-12) can be used to measure thickness. Discrimination for this one is .001 inch.

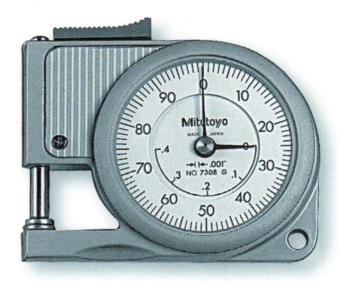

Figure C-12 Dial thickness gage (*Mitutoyo America Corp.*).

Dial Indicating Snap Gages

Dial indicating snap gages (Figure C-13) are used for determining whether workpieces are within acceptable limits. They are first set with a master gage. Part size deviation is shown on the dial indicator.

Dial Bore Gage

A **dial bore gage** (Figure C-14) uses a three-point measuring contact. This more accurately

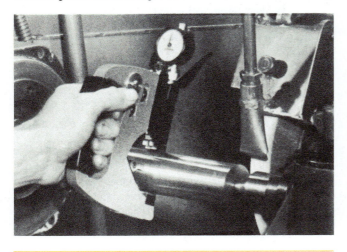

Figure C-13 Using the indicating snap gage (*Courtesy of Mahr Federal Inc.*).

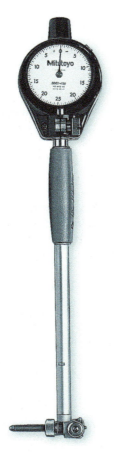

Figure C-14 Dial Bore Gage.

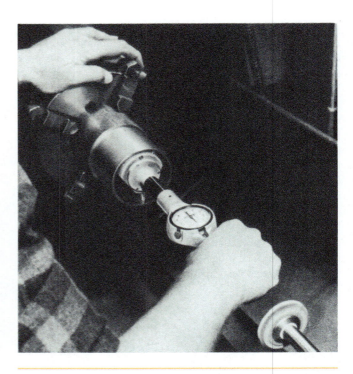

Figure C-15 Using a dial bore gage to measure a bore (*Courtesy of Mahr Federal Inc.*)

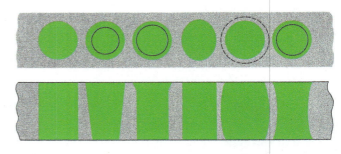

Figure C-16 Hole geometry detectable with the dial bore gage (*Mahr Federal, Inc.*).

measures the true shape of a bore (Figure C-15). The dial bore gage is useful for checking machined bores for size, taper, bellmouth, ovality, barrel shape, and hourglass shape (Figure C-16). Dial bore gages are set to a master ring and then compared to a bore diameter. Discrimination ranges from .001 to .0001 inch.

The **indicating expansion plug gage** is used to measure the inside diameter of a hole or bore. This type of gage is built to check a range of dimensions. The expanding plug is retracted and the instrument inserted into the hole to be measured (Figure C-17). As shown in the figure this gage can detect ovality, bellmouth, barrel shape, and taper.

Dial Indicating Screw Thread Snap Gage

The dial **indicating screw thread snap gage** (Figure C-18) is used to measure an external thread. The instrument may be fitted with suitable anvils for measuring the major, minor, or pitch diameter of screw threads. They are also available for internal thread measurement.

Figure C-17 Indicating expansion plug gage being used to measure a bore.

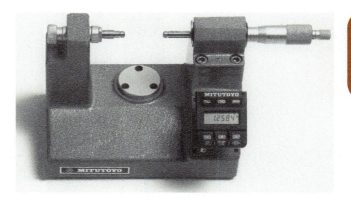

Figure C-19 Digital electronic bench micrometer (*Mitutoyo America Corp*.).

Figure C-20 Surface roughness comparison gage.

Figure C-18 Dial-indicating thread snap gage (*Courtesy of Mahr Federal Inc*.).

Inspection and Calibration through Mechanical Measurement

Parts produced on a machine tool must be inspected to see if they meet design requirements. Parts that are out of tolerance can greatly increase the cost of production. Scrap must be kept to a minimum. This is the purpose of part inspection and calibration of measuring instruments.

Indicating Bench Micrometer

The **indicating bench micrometer**, commonly called a **supermicrometer** (Figure C-19), is used to inspect tools, parts, and gages. It is much more accurate than a regular micrometer.

Surface Finish Visual Comparator

Surface finish may be approximated by visual inspection using a **surface roughness gage** (Figure C-20). Samples of finishes produced by various machining operations are indicated on the gage. These can be visually compared to a machined surface to determine the approximate surface finish.

Coordinate Measuring Machine

A **coordinate measuring machine (CMM)** (Figure C-21) is an extremely accurate instrument that can measure the workpiece in three dimensions. Coordinate measuring machines are useful for determining the location of a part feature relative to a reference plane, line, or point. They are almost a necessity for inspecting complex tolerances.

The coordinate measuring machine is an indispensable tool for the inspector. Printouts indicating measurements, as well as graphics illustrating the parts being measured, are easily produced. The CMM integrates electronics and computers into a precision mechanical system for high-precision measurement.

Portable Measurement Arm

A portable measurement absolute arm is a three-dimensional, portable, coordinate measuring device (Figure C-22). The device duplicates and enhances the movement and reach of the human arm. The device can duplicate several degrees of freedom. This means that a portable measurement arm has six axes.

Measurement with Electronic Measurement Devices

Surface Finish Indicators

Surface finish is critical on parts such as bearings, gears, and hydraulic cylinders. Surface finish is a measure of **surface roughness** or **profile**. A surface

Figure C-21 Three-axis coordinate measuring machine (*Mitutoyo America Corp.*).

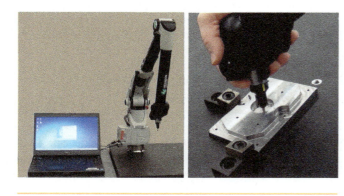

Figure C-22 Portable measurement arm.

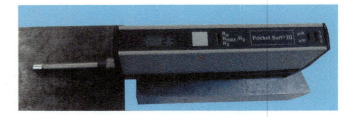

Figure C-23 Surface roughness instrument used to check a machined surface.

Figure C-24 Digital readout system installed on a vertical mill (*Courtesy of Fox Valley Technical College*).

finish measurement is in **microinches**. A **microinch** is **one millionth of an inch**. A surface finish indicator (Figure C-23) has a diamond-tipped stylus that measures the roughness of a surface. The surface deviations are calculated and displayed in various standard measures, such as **roughness average** (R_a).

Digital Readouts **Digital Readouts** use linear sensors attached to the machine tool axes. These systems typically discriminate to .0001 inch. The travel of the axis is displayed on a digital display (Figure C-24). Digital readouts are useful

for accurate positioning of machine axes on many types of machines. Digital readouts dramatically reduce the amount of error due to backlash in machine screws. Digital readouts can be switched between English and metric units of measurement.

Electronic Comparators **Electronic comparators** (Figure C-25) take advantage of the sensitivity of electronic equipment. They are used to make comparison measurements of parts versus a gage. For example, a part height might be compared to a set of gage blocks set to the desired size.

Measurement with Light

Toolmaker's Microscope A toolmaker's microscope (Figure C-26) is used to inspect parts, cutting tools, and measurement tools. The microscope has a stage that can be

Figure C-25 Electronic comparator.

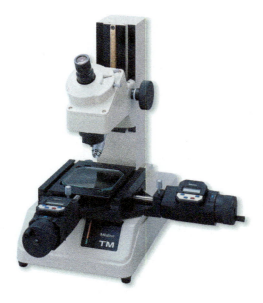

Figure C-26 Toolmaker's microscope with digital measuring (*Mitutoyo America Corp.*).

precisely rotated and moved in two perpendicular axes. The instrument may be equipped with a measuring system that can discriminate to .0001 inch. Thus, stage movement can be recorded, permitting very accurate part measurements to be made.

The **optical comparator** (Figure C-27) is used to inspect parts, cutting tools, etc. Optical comparators project a greatly magnified shadow of the object on a screen. The surface of the workpiece may also be illuminated. Shape patterns or graduated patterns can be placed on the screen and used to make measurements on the workpiece projection.

Optical flats are used to measure flatness. They can be used, for example, to reveal the surface geometry of the measuring faces of a micrometer or a gage block. Optical flats take advantage of the principles of light interferometry

to make measurements in millionths of an inch. Optical flats are optical-grade pieces of glass that have been lapped and polished to be extremely flat on one or both sides to within a few millionths of an inch. When an optical flat is placed on another surface and illuminated, the light waves reflect off both the bottom surface of the flat and the surface it is resting on. The reflected waves interfere and create a pattern of interference fringes, which are visible as light and dark bands. The distance between the fringes is smaller where the gap is changing more rapidly, which indicates a difference in flatness in one of the two surfaces. This is similar to the contour lines on a map. A flat surface is indicated by a pattern of equally spaced straight parallel fringes, while other patterns would indicate an uneven surface. Two adjacent fringes indicate a difference in elevation of one-half of the wavelength of the light that is used. If the fringes are counted, differences in elevation of the surface can be measured to millionths of an inch.

Laser Interferometer   A laser light beam is a coherent beam. This means that each ray of light follows the same path. Thus, it does not disperse over long distances. This property makes the laser beam useful in many measurement and alignment applications. For example, the laser beam

Figure C-27 Horizontal optical comparator (*Mitutoyo America Corp.*).

Figure C-28 Laser interferometer being used for straightness determination (*Courtesy of Renishaw PLC*).

may be used to determine how straight a machine tool table travels (Figure C-28). Other uses include checking the accuracy of machine tool measuring systems.

SELF-TEST

1. Describe some of the factors that contribute to accuracy in measurement.
2. Define discrimination as it applies to measurements.
3. What is the rule of thumb for discrimination in choosing the correct measurement tool?
4. What is calibration?
5. Name a tool that can be used to check an internal thread.
6. Describe an application that a go/no-go gage would be used for and how it would be used.
7. How can a bore be measured?
8. Name two ways that surface finish can be determined.
9. What is a measurement arm?
10. What is a CMM?

INTERNET REFERENCES

Information on coordinate measuring machines and inspections systems:

http://www.starrett.com

http://www.prattandwhitney.com

Systems of Measurement

Throughout history, there have been many systems of measurement. Prior to the era of national and international industrial operations, an individual was often responsible for making a complete product. Since the same person made all the necessary parts and did the required assembly, he or she needed to conform only to his or her particular system of measurement. However, as machines replaced people and diversified mass production was established on a national and international basis, the need for standardization of measurement became crucial. Measurements use either the English (inch-pound-second) or the metric (meter-kilogram-second) system. Metric measurement is now predominant in most of the industrialized nations of the world. The English system is the primary system in U.S. manufacturing. The import of products built to metric specifications has made the use of metric tools and measurements more common in the United States.

Machinists must be able to understand and use metric measurement. This unit will cover the basic length standards of both systems, examine mathematical and other methods of converting from system to system, and look at techniques by which a machine tool can be converted to work in the metric system.

OBJECTIVES

After completing this unit, you should be able to:

- Identify common methods of measurement conversion.
- Convert inch dimensions to metric equivalents, and convert metric dimensions to inch equivalents.

ENGLISH SYSTEM OF MEASUREMENT

The English system of measurement uses units of inches, pounds, and seconds to represent the measurement of length, mass, and time, respectively. Since we are primarily concerned with the measurement of length in machining, we will simply refer to the English system as the inch system.

Subdivisions and Multiples of the Inch

The following table shows the common subdivisions and multiples of the inch that are used by the machinist.

Common Subdivisions

.000001	millionth
.00001	hundred-thousandth
.0001	ten-thousandth
.001	thousandth
.01	Ten thousandths
.1	One hundred thousandths
1.00	Unit inch

Common Multiples and Other Subdivisions

12.00	1 foot
36.00	1 yard
$\frac{1}{128}$	.007810 (decimal equivalent)
$\frac{1}{64}$	.015625
$\frac{1}{32}$	.031250
$\frac{1}{20}$	.050000
$\frac{1}{16}$	.062500
$\frac{1}{8}$	.125000
$\frac{1}{4}$	.250000
$\frac{1}{4}$ ½	.500000

Multiples of Feet

3 feet = 1 yard

5,280 feet = 1 mile

Multiples of Yards

$$1,760 \text{ yards} = 1 \text{ mile}$$

METRIC SYSTEM AND INTERNATIONAL SYSTEM OF UNITS

The basic unit of length in the metric system is the *meter*. Originally the length of the meter was defined by a natural standard: specifically, a portion of the earth's circumference. Later, more convenient metal standards were constructed. In 1886, the metric system was legalized in the United States, but its use was not made mandatory. Since 1893, the yard has been defined in terms of the metric meter by the ratio

$$1 \text{ yard} = \frac{3,600}{3,937} \text{ meter}$$

Although the metric system had been in use for many years in many different countries, it still was not completely standardized among its users. Therefore, an attempt was made to modernize and standardize the metric system. The result was the **Systéme International d'Unités**, known as **SI** or the **International Metric System**.

The basic unit of length in SI is the meter, or metre (in the common international spelling). The SI meter is defined by a physical standard that can be reproduced anywhere with unvarying accuracy:

1 meter = the length of the path traveled by light in a vacuum during a time interval of r = 1/299,792,458 of a second

Probably the primary advantage of the metric system is that of convenience in computation. All subdivisions and multiples use 10 as a divisor or multiplier, as can be seen in the following table:

.000001	(one-millionth meter or micrometer)
.001	(one-thousandth meter or millimeter)
.01	(one-hundredth meter or centimeter)
.1	(one-tenth meter or decimeter)
1.00	*Unit meter*
10	(10 meters or 1 dekameter)
100	(100 meters or 1 hectometer)
1,000	(1,000 meters or 1 kilometer)
1,000,000	(1 million meters or 1 megameter)

METRIC SYSTEM EXAMPLES

1. One meter (m) = _____ millimeter (mm).
   Since a millimeter is 1/1,000 part of a meter, there are 1,000 mm in a meter.
2. 50 mm = _____ centimeters (cm).
   Since 1 cm = 10 mm, 50/10 = 5 cm in 50 mm.
3. Four kilometers (km) = _____ m.
   Since 1 km = 1,000 m, then 4 km = 4,000 m.
4. 582 mm = _____ cm.
   Since 10 mm = 1 cm, 582/10 = 58.2 cm.

CONVERSION BETWEEN SYSTEMS

Much of the difficulty with working in two-systems is converting from one system to the other. This can be a matter of concern to the machinist, as he or she must exercise caution in making conversions. Errors can be easily made. Therefore, the use of a calculator is recommended.

Conversion Factors and Mathematical Conversion

Since the historical evolution of the inch and metric systems was quite different, there are no obvious relationships between length units of the two systems. You simply have to memorize the basic conversion factors. We know from the preceding discussion that the yard has been defined in terms of the meter. Knowing this relationship, you can mathematically derive any length unit in either system. However, the conversion factor

$$1 \text{ yard} = \frac{3,600}{3,937} \text{ meter}$$

is a less common factor for the machinist. A more common factor can be determined by the following:

$$1 \text{ yard} = \frac{3,600}{3,937} \text{ meter}$$

Therefore,

$$1 \text{ yard} = .91440 \text{ meter}$$

$$\left(\frac{3,600}{3,937} \text{ expressed in decimal form} \right)$$

Then

$$1 \text{ inch} = \frac{1}{36} \text{ of } .91440 \text{ meter}$$

So

$$\frac{.91440}{36} = .025400$$

We know that

$$1 \text{ m} = 1,000 \text{ mm}$$

Therefore,

$$1 \text{ inch} = .025400 \times 1,000$$

or

$$\textbf{1 inch} = \textbf{25.4000 mm}$$

The conversion factor 1 inch = 25.4 mm is common and should be memorized. To find inches knowing millimeters, you must divide inches by 25.4.

$$1,000 \text{ mm} = _____ \text{ inches}$$

$$\frac{1,000}{25.4} = 39.37 \text{ inches}$$

To simplify the arithmetic, any conversion can always take the form of a multiplication problem.

EXAMPLE

Instead of 1,000/25.4, multiply by the reciprocal of 25.4, which is 1/25.4 or .03937. Therefore,

$$1,000 \times .03937 = 39.37 \text{ inches}$$

EXAMPLES OF CONVERSIONS (INCH TO METRIC)

1. 17 inches = _____ cm. Knowing inches, to find centimeters multiply inches by 2.54:
2.54 × 17 inches = 43.18 cm.

2. .807 inch = _____ mm. Knowing inches, to find millimeters multiply inches by 25.4:
25.4 × .807 inch = 20.49 mm.

EXAMPLES OF CONVERSIONS (METRIC TO INCH)

1. .05 mm = _____ inch. Knowing millimeters, to find inches multiply millimeters by .03937:
.05 × .03937 = .00196

2. 1.63 m = _____ inch. Knowing meters, to find inches multiply meters by 39.37: 1.63 × 39.37 = 64.173 inches

Conversion Factors to Memorize

1 inch = 25.4 mm or 2.54 cm

1 mm = .03937 inch

Other Methods of Conversion

Many calculators can convert directly from one system to another. Of course, any calculator can be used to do a conversion problem. The direct converting calculator does not require that any conversion constant be remembered.

CNC machines can typically work in the inch or metric system. Digital readouts can also be switched to display measurements in the inch or metric system.

SELF-TEST

Perform the following conversions:

1. 35 mm = _____ inch.
2. 125 inches = _____ mm
3. 6.273 inches = _____ mm
4. Express the tolerance ±.050 in metric terms to the nearest mm.
5. To find centimeters knowing millimeters, multiply/divide by 10.
6. Express the tolerance ±.02mm in terms of inches to the nearest $\frac{1}{10,000}$ inch.
7. What is meant by *SI*?
8. Describe methods to convert between metric and inch measurement systems.
9. How large is a yard compared to a meter?
10. Can an inch machine tool be converted to work in metric units?

INTERNET REFERENCES

Information on systems of measurement:

http://www.webmath.com

http://ts.nist.gov/WeightsAndMeasures

Using Steel Rules

One of the most used measurement tools is the steel rule. It is important to be able to select and use steel rules.

SCALES AND RULES

The terms **scale** and **rule** are often used incorrectly. A rule is a linear measuring instrument whose graduations represent real units of lengths and their subdivisions. In contrast, a **scale** is graduated into imaginary units either smaller or larger than the real units they represent. This is done for convenience where proportional measurements are needed. For example, an architect uses a scale that has graduations representing feet and inches. However, the actual length of the graduations on the architect's scale is quite different from full-size dimensions.

DISCRIMINATION OF STEEL RULES

Discrimination refers to how finely a unit of length has been divided. If the smallest graduation on a specific rule is $\frac{1}{32}$ inch, then the rule has a discrimination of, or discriminates to, $\frac{1}{32}$ inch. Likewise, if the smallest graduation of the rule is $\frac{1}{64}$ inch, then this rule discriminates to $\frac{1}{64}$ inch.

The maximum discrimination of a steel rule is generally $\frac{1}{64}$ inch, or in the case of the decimal inch rule, $\frac{1}{100}$ inch. The metric rule has a discrimination of .05 mm. Remember that a measuring tool should never be used beyond its discrimination. This means that a steel rule will not be reliable in trying to ascertain a measurement increment smaller than $\frac{1}{64}$ or $\frac{1}{100}$ inch. If a measurement falls between the markings on the rule, only this can be said of this reading: It is more or less than the amount of the nearest mark. No further data as to how much more or less can be reliably determined. You should not read between the graduations on a steel rule as you will not get reliable readings.

RELIABILITY AND EXPECTATION OF ACCURACY IN STEEL RULES

Be very careful when using the steel rule at its maximum discrimination. The markings on the rule have a width. A good-quality steel rule has engraved graduations. This means that the markings are actually cuts in the metal. Of all types of graduations, engraved ones occupy the least width along the rule. Other rules, graduated by other processes, may have markings of greater width. These rules are not necessarily any less accurate, but they may require more care in reading. Generally, the reliability of the rule will diminish as its maximum discrimination is approached. The smaller graduations are more difficult to see without the aid of a magnifier. Of particular importance is the point from which the measurement is taken. This reference point must be carefully aligned at the point where the length being measured begins.

The use of a steel rule is most appropriate for measurements no smaller than $\frac{1}{32}$ inch on a fractional rule or $\frac{1}{50}$ inch on a decimal rule. This does not mean that the rule cannot measure to its maximum discrimination because under the proper conditions it certainly can. However, at or near maximum discrimination, the time consumed to ensure reliable measurement is really not justified. You will be more productive if you use a measuring instrument with finer discrimination for measurements under the nearest $\frac{1}{32}$ or $\frac{1}{50}$ inch. It is good practice to take more than one reading when using a steel rule. After determining the desired measurement, apply the rule once again to see if the same result is obtained. This procedure increases the reliability of measurements.

TYPES OF RULES

Rules may be selected in many different shapes and sizes, depending on the need. The common **rigid steel rule** is 6-inch-long, $\frac{3}{4}$-inch wide, and $\frac{3}{64}$-inch thick. It is engraved with No. 4 standard rule graduations. A No. 4 graduation consists of $\frac{1}{8}$- and $\frac{1}{16}$-inch divisions on one side (Figure C-29) and $\frac{1}{32}$ and $\frac{1}{64}$-inch divisions on the reverse side (Figure C-30).

Other common graduations are summarized in the following table:

Graduation Number	Front Side	Back Side
3	32nds	10ths
	64ths	50ths
16	50ths	32nds
	100ths	64ths

The No. 16 graduated rule is often found in the aircraft industry, where dimensions are specified in decimal fraction notations. Decimal fractions are based on 10 or a multiple of 10 divisions of an inch rather than 32 or 64 divisions.

Figure C-29 Six-inch rigid steel rule (front side).

Figure C-30 Six-inch rigid steel rule (back side).

Another common rule is the **flexible type** (Figure C-31). This rule is 6-inch-long, $\frac{1}{2}$-inch wide, and $\frac{1}{64}$-inch thick. Flexible rules are made from hardened and tempered spring steel. One advantage of a flexible rule is that it will bend, permitting measurements to be made where a rigid rule would not work. Most flexible rules are 6 or 12-inch long.

The **narrow rule** (Figure C-32) is convenient when measuring in small openings, slots, or holes. Most narrow rules have only one set of graduations on each side. These can be No. 10, which is thirty-seconds and sixty-fourths, or No. 11, which is sixty-fourths and hundredths.

The **standard hook rule** (Figure C-33) makes it possible to reach through an opening; the rule is hooked on the far side to measure a thickness or the position of a slot (Figure C-34). When a workpiece has a chamfered edge, a hook rule will be advantageous over a common rule. If the hook is not loose or excessively worn, it will provide an easy-to-locate reference point.

The **short rule set** (Figure C-35) consists of a set of rules with a holder. Short rule sets have a range of $\frac{1}{4}$ or 1 inch. They can be used to measure shoulders in holes or steps in slots, where space is extremely limited. The holder will attach to the rules at any angle, making these versatile tools.

The **slide caliper rule** (Figure C-36) is a versatile tool used to measure round bars, tubing, and other objects when it is difficult to measure at the ends and difficult to estimate the diameter with a rigid steel rule. The small slide caliper rule can also be used to measure internal dimensions from $\frac{1}{4}$ inch up to the capacity of the tool.

The **rule depth gage** (Figure C-37) consists of a slotted steel head in which a narrow rule slides. For depth measurements the head is held securely against the surface with the rule extended into the cavity or hole to be measured (Figure C-38). The locking nut is tightened and the rule depth gage can then be removed and the dimension determined.

Figure C-31 Flexible steel rule (metric).

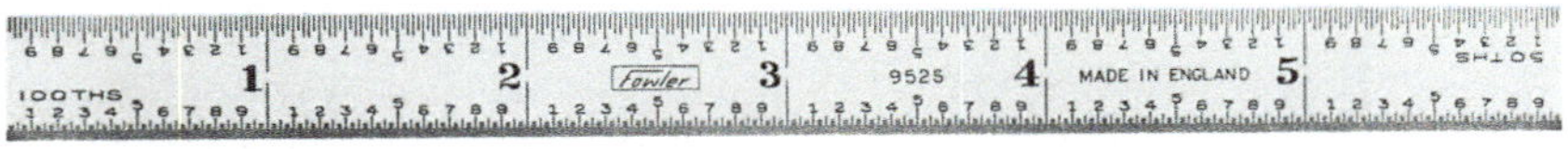

Figure C-32 Narrow rule (decimal inch).

Figure C-33 Hook rule.

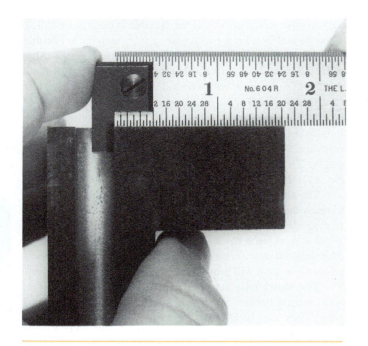

Figure C-34 Use of a hook rule to measure the width of a part.

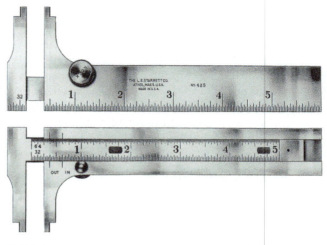

Figure C-36 Slide caliper rule (*Courtesy of The L.S. Starrett Co.*).

Figure C-35 Short rule set with holder (*Courtesy of The L.S. Starrett Co.*).

Figure C-37 Rule depth gage.

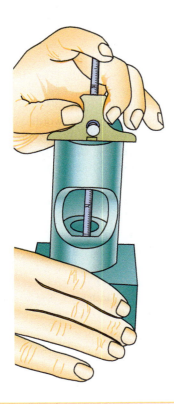

Figure C-38 Rule depth gage in use.

CARE OF RULES

Rules are measurement tools, and they should be properly cared for. A rule should not be used as a screwdriver. Rules should be kept separate from hammers, wrenches, files, and other hand tools to protect them from possible damage. Wiping a steel rule occasionally with a light coat of oil will keep it clean and free from rust.

APPLYING STEEL RULES

When using a steel rule close to the moving parts of a machine, always keep safety in mind. Stop the machine before attempting to make any measurements. Attempting to measure with the machine running may cause the rule to be caught by a moving part. This may damage the rule, and worse, may result in serious injury to the operator.

One of the problems associated with the use of rules is that of parallax error. Parallax error results when the observer making the measurement is not in line with the workpiece and the rule. You may see the graduation either too far left or too far right of its real position (Figure C-39).

Parallax error occurs when the rule is read from a point other than one directly above the point of measurement. The point of measurement is the point at which the measurement is read. It may or may not be the true reading of the size, depending on what location was used as the reference point on the rule. Parallax can be controlled by always observing the point of measurement from directly above. Furthermore, the graduations on a rule should be placed as close as possible to the surface being measured. A thin rule is preferred to a thick rule.

Using a rule causes wear, usually on the ends. The outside inch markings on a worn rule are less than 1 inch from the end. This has to be considered when measurements are made. A reliable way to measure (Figure C-40) is to use the 1-inch mark on the rule as the reference point. In the figure, the measured point is at $2\frac{1}{32}$ inches. Subtracting 1-inch results in a size of $1\frac{1}{32}$ inches.

Round bars and tubing should be measured with the rule placed on the end of the tube or bar (Figure C-41). Select a reference point and set it carefully at a point on the circumference of the round part to be measured. Using the reference point as a pivot, move the rule back and forth slightly to find the largest distance across the diameter. When the largest distance is determined, read the measurement at that point.

Figure C-39 Parallax error.

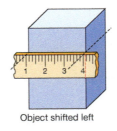

Object shifted left

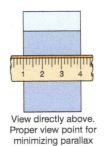

View directly above. Proper view point for minimizing parallax

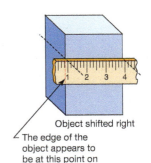

Object shifted right

The edge of the object appears to be at this point on the rule.

When viewed from directly above, the rule gradations are exactly in line with the edge of the object being measured. However, when the object is shifted right or left of a point directly above the point of measurement, the alignment of the object edge and the rule graduations appears to no longer coincide.

Figure C-40 Using the 1-inch mark as the reference point.

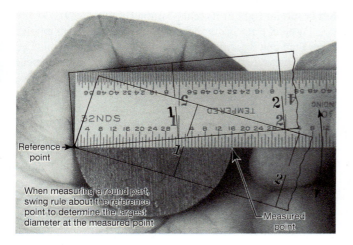

When measuring a round part, swing rule about the reference point to determine the largest diameter at the measured point

Reference point

Measured point

Figure C-41 Measuring round objects.

READING FRACTIONAL INCH RULES

Most dimensions are expressed in inches and fractions of inches. These dimensions are measured with fractional inch rules. The typical machinist's rule is broken down into 1, $\frac{1}{2}$, $\frac{1}{4}$, $\frac{1}{8}$, $\frac{1}{16}$, $\frac{1}{32}$, and $\frac{1}{64}$-inch graduations. To facilitate reading, the 1, $\frac{1}{2}$, $\frac{1}{4}$, $\frac{1}{8}$, and $\frac{1}{16}$-inch graduations appear on one side of the rule (Figure C-42). The reverse side of the rule has one edge graduated in $\frac{1}{32}$-inch increments and the other edge graduated in $\frac{1}{64}$-inch increments. On the $\frac{1}{32}$-inch side, every fourth mark is numbered, and on the $\frac{1}{64}$-inch side, every eighth mark is numbered (Figure C-43). This

eliminates the need to count graduations from the nearest whole inch mark. On these rules, the length of the graduation line varies, with the 1-inch line being the longest, the $\frac{1}{2}$ inch line being next in length, and the $\frac{1}{4}$, $\frac{1}{8}$ and $\frac{1}{16}$-inch lines each being consecutively shorter. The difference in line lengths is an important aid in reading a rule. The smallest graduation on any edge of a rule is marked by small numbers on the end. Note that the words *8THS* and *16THS* appear at the ends of the rule. The numbers 32NDS and 64THS appear on the reverse side of the rule, thus indicating thirty-seconds and sixty-fourths of an inch.

EXAMPLES OF FRACTIONAL INCH READINGS

Figure C-44 Distance *A* falls on the third $\frac{1}{8}$-inch graduation. This reading would be $\frac{3}{8}$ inch.

Distance *B* falls on the longest graduation between the end of the rule and the first full inch mark. The reading is $\frac{1}{2}$ inch.

Distance *C* falls on the sixth $\frac{1}{8}$-inch graduation, making it $\frac{6}{8}$ or $\frac{3}{4}$ inch.

Distance *D* falls at the fifth $\frac{1}{8}$-inch mark beyond the 2-inch graduation. The reading is $2\frac{5}{8}$ inches.

Figure C-45 Distance *A* falls at the thirteenth $\frac{1}{16}$-inch mark, making the reading $\frac{13}{16}$ inch.

Distance *B* falls at the first $\frac{1}{16}$-inch mark past the 1-inch graduation. The reading is $1\frac{1}{16}$ inches.

Distance *C* falls at the seventh $\frac{1}{16}$-inch mark past the 1-inch graduation. The reading is $1\frac{7}{16}$ inches.

Distance *D* falls at the third $\frac{1}{16}$-inch mark past the 2-inch graduation. The reading is $2\frac{3}{16}$ inches.

Figure C-46 Distance *A* falls at the third $\frac{1}{32}$-inch mark. The reading is $\frac{3}{32}$ inch.

Distance *B* falls at the ninth $\frac{1}{32}$-inch mark. The reading is $\frac{9}{32}$ inch.

Distance *C* falls at the eleventh $\frac{1}{32}$-inch mark past the 1-inch graduation. The reading is $1\frac{11}{32}$ inches.

Distance *D* falls at the fourth $\frac{1}{32}$-inch mark past the 2-inch graduation. The reading is $2\frac{4}{32}$ inches, which reduced to lowest terms becomes $2\frac{1}{8}$ inches.

Figure C-47 Distance *A* falls at the ninth $\frac{1}{64}$-inch mark, making the reading $\frac{9}{64}$ inch.

Distance *B* falls at the fifty-seventh $\frac{1}{64}$-inch mark, making the reading $\frac{57}{64}$ inch.

Figure C-42 Front-side graduations of the typical machinist's rule.

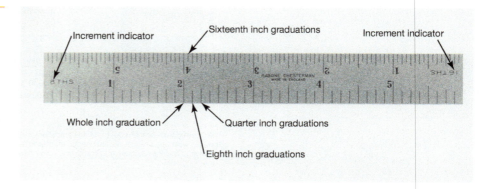

Increment indicator

Sixteenth inch graduations

Increment indicator

Whole inch graduation

Quarter inch graduations

Eighth inch graduations

Figure C-43 Back-side graduations of the typical machinist's rule.

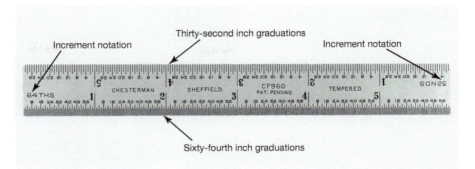

Figure C-44 Examples of readings on the $\frac{1}{8}$-inch discrimination edge.

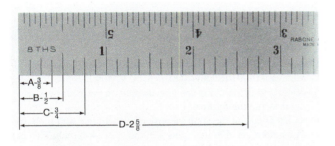

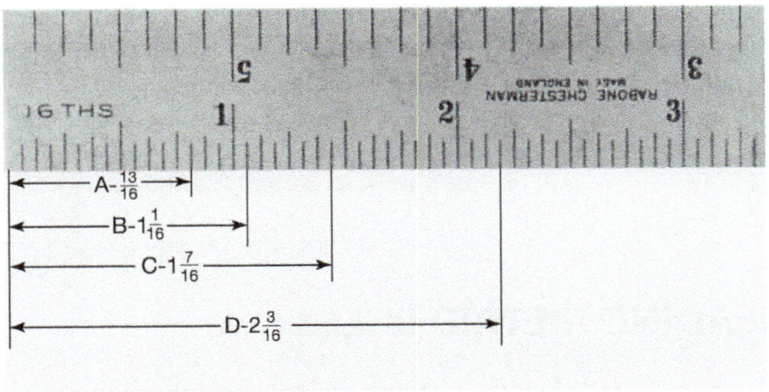

Figure C-45 Examples of readings on the $\frac{1}{16}$-inch discrimination edge.

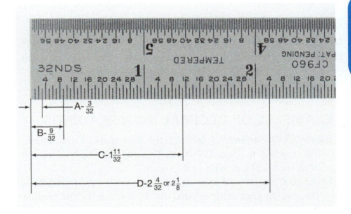

Figure C-46 Examples of readings on the $\frac{1}{32}$-inch discrimination edge.

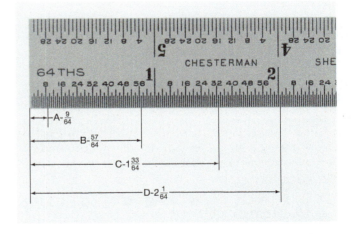

Distance C falls at the thirty-third $\frac{1}{64}$-inch mark past the 1-inch graduation. The reading is $1\frac{33}{64}$ inches.

Distance D falls at the first $\frac{1}{64}$-inch mark past the 2-inch graduation, making the reading $2\frac{1}{64}$ inches.

Figure C-47 Examples of readings on the $\frac{1}{64}$-inch discrimination edge.

READING DECIMAL INCH RULES

Most dimensions in industry are specified in **decimal notation**, which refers to the division of the inch into 10 parts or a multiple of 10 parts, such as 50 or 100 parts. In this case, a **decimal rule** is used. Decimal inch dimensions are specified and read as thousandths of an inch. Decimal rules, however, do not discriminate to the individual thousandth because the width of an engraved or etched division on the rule is approximately .003 inch (3 thousandths of an inch). Decimal rules are commonly graduated in increments of $\frac{1}{10}$, $\frac{1}{50}$, or $\frac{1}{100}$ inch.

A typical decimal rule may have $\frac{1}{50}$-inch divisions on the top edge and $\frac{1}{100}$-inch divisions on the bottom edge (Figure C-48). The inch is divided into 10 equal parts, making each numbered division $\frac{1}{10}$ inch or .100 inch (100 thousandths of an

Figure C-48 Six-inch decimal rule.

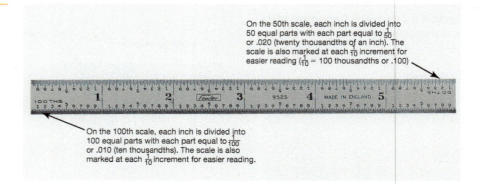

On the 50th scale, each inch is divided into 50 equal parts with each part equal to $\frac{1}{50}$ or .020 (twenty thousandths of an inch). The scale is also marked at each $\frac{1}{10}$ increment for easier reading ($\frac{1}{10}$ = 100 thousandths or .100)

On the 100th scale, each inch is divided into 100 equal parts with each part equal to $\frac{1}{100}$ or .010 (ten thousandths). The scale is also marked at each $\frac{1}{10}$ increment for easier reading.

inch). On the top scale each $\frac{1}{10}$-inch increment is further subdivided into five equal parts, which makes the value of each of these divisions .020 inch (20 thousandths of an inch).

EXAMPLES OF DECIMAL INCH READINGS

Figure C-49 Distance *A* falls on the first marked graduation. The reading is $\frac{1}{10}$ or .100 inch. This can also be read on the 50th-inch scale, as seen in the figure.

Distance *B* can be read only on the 100th-inch scale, as it falls at the seventh graduation beyond the .10-inch mark. The reading is .100 inch plus .070 inch, or .170 inch. This distance cannot

be read on the 50th-inch scale because discrimination of the 50th-inch scale is not sufficient.

Distance *C* falls at the second mark beyond the .400-inch line. This reading is .400 inch plus .020 inch, or .420 inch. Since .020 inch is equal to $\frac{1}{50}$ inch, this can also be read on the 50th-inch scale, as shown in the figure.

Distance *D* falls at the sixth increment beyond the .400-inch line. The reading is .400 inch plus .060 inch, or .460 inch. This can also be read on the 50th-inch scale, as seen in the figure.

Distance *E* falls at the sixth division beyond the .700-inch mark. The reading is .700 inch plus .040 inch, or .740 inch. This can also be read on the 50th-inch scale.

Distance *F* falls at the eighth mark beyond the .700-inch line. The reading is .700 inch plus .080 inch, or .780 inch. This cannot be read on the 50th-inch scale.

Distance *G* falls at the .100 graduation on top and at the .900 graduation on the bottom. The reading is .900, or $\frac{9}{10}$ inch.

Distance *H* falls two marks past the first full inch mark. The reading is 1.00 inch plus .020 inch, or 1.020 inches.

READING METRIC RULES

Many products are made in metric dimensions requiring a machinist to use a **metric rule**. The typical metric rule has millimeter (mm) and half-millimeter graduations (Figure C-50).

EXAMPLES OF READING METRIC RULES

Figure C-51 Distance *A* falls at the fifty-third graduation on the millimeter scale. The reading is 53 mm.

Distance *B* falls at the twenty-second graduation on the millimeter scale. The reading is 22 mm.

Distance *C* falls at the sixth graduation on the millimeter scale. The reading is 6 mm.

Distance *D* falls at the seventeenth $\frac{1}{2}$-mm mark. The reading is 8 mm plus an additional $\frac{1}{2}$ mm giving a total of 8.5 mm.

Distance *E* falls $\frac{1}{2}$ mm beyond the 3-cm graduation. Since 3 cm is equal to 30 mm, the reading is 30.5 mm.

Distance *F* falls $\frac{1}{2}$ mm beyond the 51-mm graduation. The reading is 51.5 mm. In machine design, all dimensions are specified in mm. Hence 1.5 meters (m) would be 1,500 mm.

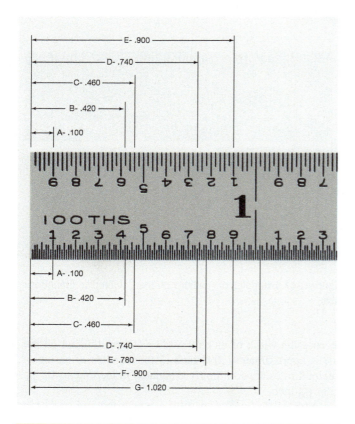

Figure C-49 Examples of decimal rule readings.

Figure C-50 150-mm metric rule.

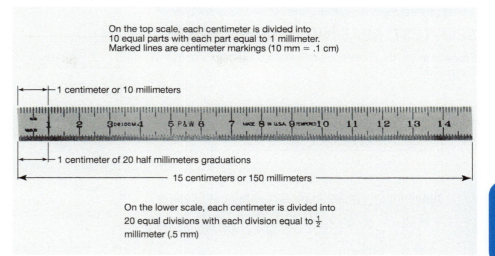

On the top scale, each centimeter is divided into
10 equal parts with each part equal to 1 millimeter.
Marked lines are centimeter markings (10 mm = .1 cm)

1 centimeter or 10 millimeters

1 centimeter of 20 half millimeters graduations

15 centimeters or 150 millimeters

On the lower scale, each centimeter is divided into
20 equal divisions with each division equal to $\frac{1}{2}$
millimeter (.5 mm)

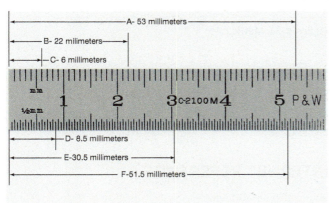

Figure C-51 Examples of metric rule readings.

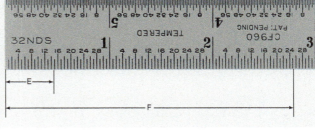

Figure C-52b Readings for C-52b.

SELF-TEST: READING INCH RULES

Read and record the dimensions indicated by the letters *A* to *H* in
Figures C-52*a* to C-52*d*.

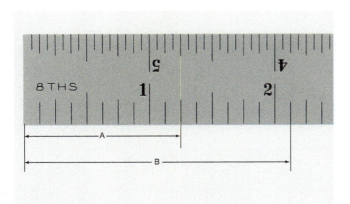

Figure C-52a Readings for C-52a.

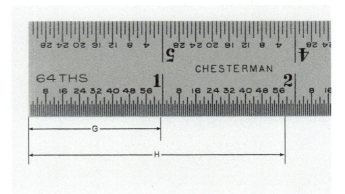

Figure C-52c Readings for C-52c.

Figure C-52d Readings for C-52d.

SELF-TEST: READING DECIMAL RULES

Read and record the dimensions indicated by the letters *A* to *E* in Figure C-53.

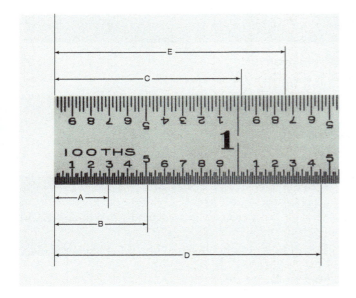

Figure C-53 Decimal inch rule.

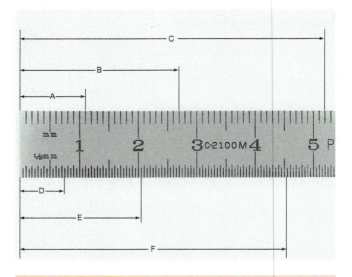

Figure C-54 Metric rule.

SELF-TEST: READING METRIC RULES

Read and record the dimensions indicated by the letters *A* to *F* in Figure C-54.

INTERNET REFERENCE

Information on steel rules:

http://www.fine-tools.com/mass.htm

Using Vernier, Dial, and Digital Instruments for Direct Measurements

The inspection and measurement of machined parts require various kinds of measuring tools. Often, the discrimination of a rule is sufficient, but in many cases the discrimination of a rule with a vernier scale is required. This unit explains the types, use, and applications of common vernier, dial, and digital instruments.

OBJECTIVES

After completing this unit, you should be able to:

- Measure dimensions to an accuracy of plus or minus .001 inch with a vernier caliper.
- Measure dimensions to an accuracy of plus or minus .02 mm using a metric vernier caliper.
- Measure dimensions using a vernier depth gage.

USE OF THE VERNIER

A vernier is used to divide the units in which a measuring tool measures into smaller increments, permitting high discrimination measurement. Although the vernier is highly reliable and accurate, it is somewhat difficult to read, thus requiring more time and skill on the part of the machinist. Modern precision manufacturing techniques and the wide application of digital tools have produced measurement tools that are easy to use and are very reliable. Thus, the use of the mechanical vernier has declined. As digital electronics replace more mechanical measurement, use of the vernier may in time disappear completely. However, vernier measuring equipment is still used in industry, and it is likely that vernier applications will be around for some time to come.

PRINCIPLE OF THE VERNIER

The principle of the vernier may be used to increase the discrimination of all graduated scale measuring tools. A vernier system consists of a main scale and a vernier scale.

The vernier scale is placed adjacent to the main scale so that graduations on both scales can be observed together. The spacing of the vernier scale graduations is shorter than the spacing of the main scale graduations. For example, consider a main scale divided as shown (Figure C-55a). It is desired to further subdivide each main scale division into 10 parts with the use of a vernier. The spacing of each vernier scale division is made $\frac{1}{10}$ of a main scale division shorter than the spacing of a main scale division. This may sound confusing, but think of it as 10 vernier scale divisions corresponding to 9 main scale divisions (Figure C-55a). The vernier now permits the main scale to discriminate to $\frac{1}{10}$ of its major divisions. Therefore, $\frac{1}{10}$ is known as the **least count** of the vernier.

The vernier functions in the following manner. Assume that the zero line on the vernier scale is placed as shown (Figure C-55b). The reading on the main scale is 2 plus a fraction of a division. You want to know the amount of the fraction over 2 to the nearest tenth, or least count, of the vernier. As you inspect the alignment of the vernier scale and the main scale lines, you note that they move closer together until one line on the vernier scale **coincides** with a line on the main scale. This is the coincident line of the vernier and indicates the fraction in tenths that must be added to the main scale reading. The vernier is coincident at the sixth line. Since the least count of the vernier is $\frac{1}{10}$, the zero vernier line is $\frac{6}{10}$ past 2 on the main scale. Therefore, the main scale reading is 2.6 (Figure C-55b).

DISCRIMINATION AND APPLICATIONS OF VERNIER INSTRUMENTS

Vernier instruments used for linear measure in the inch system discriminate to .001 inch. Metric verniers generally discriminate to $\frac{1}{50}$ mm.

The most common vernier instruments include several styles of **calipers**. The common vernier caliper is used for outside and inside linear measurement. Another style of vernier caliper has the capability of depth measurement in

Figure C-55a and b Principle of the vernier.

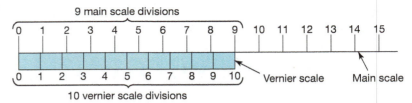

9 main scale divisions

Vernier scale   Main scale

10 vernier scale divisions

Each vernier scale division is $\frac{1}{10}$ of a main scale division shorter than the main scale division

(a)

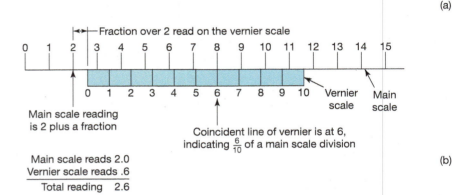

Fraction over 2 read on the vernier scale

Vernier scale   Main scale

Main scale reading is 2 plus a fraction

Coincident line of vernier is at 6, indicating $\frac{6}{10}$ of a main scale division

(b)

Main scale reads 2.0
Vernier scale reads .6
Total reading   2.6

addition to outside and inside capacity. The vernier also appears on a variety of depth gages.

Vernier scales are also used on some height gages. Vernier scales are also used on gear tooth calipers, a special vernier caliper used in gear measurement. Because the principle of the vernier can be used to subdivide a unit of angular measure as well as linear measure, it appears on various types of protractors used for angular measurement.

RELIABILITY AND EXPECTATION OF ACCURACY IN VERNIER INSTRUMENTS

Reliability in vernier calipers and depth gages is dependent on the proper use of the tool. The simple fact that the caliper or depth gage has increased discrimination over a rule does not necessarily provide increased reliability. The improved degree of discrimination in vernier instruments requires more than the mere visual alignment of a rule graduation against the edge of the object to be measured.

The vernier scale must be read carefully if a reliable measurement is to be determined. On many vernier instruments, the vernier scale should be read with the aid of a magnifier. Without this aid, the coincident line of the vernier is difficult to determine. Therefore, the reliability of the vernier readings can be in question. On the slide vernier caliper, no provision is made for the "feel" of the measuring pressure. Some calipers and the depth gage are equipped with a screw-thread fine adjustment that gives them an advantage in providing consistent pressure during the measurement.

Generally, the overall reliability of vernier instruments for measurement at maximum discrimination is .001. The vernier should never be used in an attempt to discriminate

below .001 except on instruments with that capability. With proper use and an understanding of the limitations of a vernier tool, it can be a useful tool.

VERNIER CALIPERS

A **vernier caliper** is a measuring tool that is based on a rule but with much greater discrimination. Vernier calipers have a discrimination of .001 inch. The beam or bar is engraved with the main scale. This is also called the **true scale**, as the inch markings are exactly 1 inch apart. The beam and the solid jaw are square, or at 90 degrees to each other.

The movable jaw contains the **vernier scale**. This scale is located on the sliding jaw of a vernier caliper or is part of the base on the vernier depth gage. The function of the vernier scale is to subdivide the minor divisions on the beam scale into the smallest increments that the vernier instrument is capable of measuring. For example, a 25-division vernier subdivides the minor divisions of the beam scale into 25 parts. Since the minor divisions are equal to .025 inch, the vernier divides them into increments of .001 inch. This is the finest discrimination of the instrument.

Many vernier calipers have a fine-adjustment clamp for precise adjustments of the movable jaw. Inside measurements are made over the nibs on the jaw and are read on the top scale of the vernier caliper (Figure C-56). The top scale is a duplicate of the lower scale, with the exception that it is offset to compensate for the size of the nibs.

The standard vernier caliper is a common and versatile tool because of its capacity to make outside, inside, and depth measurements (Figure C-57). Many different measuring types of measurements can be made with a vernier caliper (Figure C-58).

Figure C-56 Typical inside-outside 50-division vernier caliper.

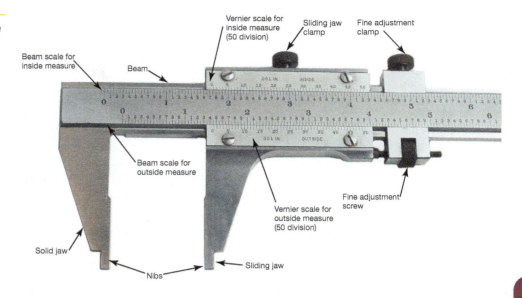

Figure C-57 Common vernier caliper.

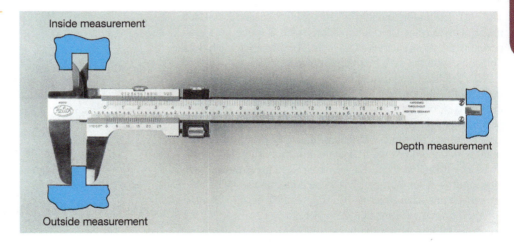

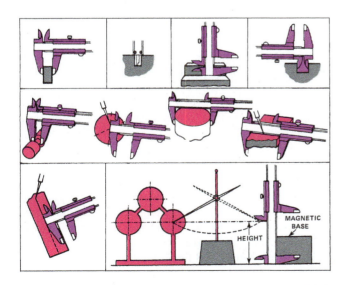

Figure C-58 The vernier caliper is a very versatile tool.

SHOP TIP

A vernier caliper is a delicate precision tool and should be treated as such. It is very important that the correct amount of pressure (feel) be developed while taking a measurement. The measuring jaws should contact the workpiece firmly. However, excessive pressure will spring the jaws and give inaccurate readings. When measuring an object, use the solid jaw as the reference point, then move the sliding jaw until contact is made. When measuring with the vernier caliper, make certain that the beam of the caliper is in line with the surfaces being measured. Whenever possible, read the vernier caliper while it is still in contact with the workpiece. Moving the instrument may change the reading. Any measurement should be taken at least twice to assure reliability.

VERNIER CALIPER PROCEDURES

To test a vernier caliper for accuracy, clean the contact surfaces of the two jaws. Bring the movable jaw with normal gaging pressure into contact with the solid jaw. Hold the caliper against a light source and examine the alignment of the solid and movable jaws. If wear exists, a line of light will be visible between the jaw faces. A gap as small as $\frac{1}{10,000}$ inch can be seen against a light. If the contact between the jaws is satisfactory, check the vernier scale alignment. The vernier scale zero mark should be in alignment with the zero on the main scale. The vernier scale can be realigned to adjust it to zero on some vernier calipers.

READING INCH VERNIER CALIPERS

Vernier scales are engraved with 25 or 50 divisions (Figures C-59 and C-60). On a 25-division vernier caliper, each inch on the main scale is divided into 10 major divisions numbered from 1 to 9. Each major division is .100 (100 thousandths). Each major division has four subdivisions with a spacing of .025 (twenty-five thousandths) of an inch. The vernier scale has 25 divisions, with the zero line being the index.

To read the vernier caliper, count all the graduations to the left of the index line. In Figure C-61 this would be 1 whole inch plus $\frac{2}{10}$ or .200, plus 1 subdivision valued at .025, plus part of one subdivision. The value of this partial subdivision is determined by the coincidence of one line on the vernier scale with one line of the true scale. For this example, the coincidence is on line 13 of the vernier scale. This is the value in thousandths of an inch that has to be added to the value read on the beam. Therefore, $1 + .100. + 100 + .025 + .013$ equals the total reading of 1.238. An aid in determining the coincidental line is that the lines adjacent to the coincidental line fall inside the lines on the true scale (Figure C-62).

The 50-division vernier caliper shown in Figure C-63 is read as follows. Each inch on the true scale is divided into

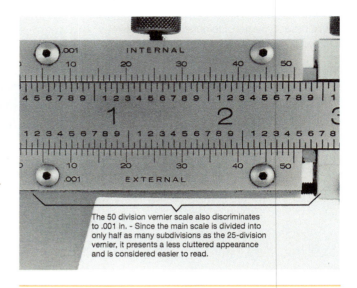

The 50 division vernier scale also discriminates to .001 in. - Since the main scale is divided into only half as many subdivisions as the 25-division vernier, it presents a less cluttered appearance and is considered easier to read.

Figure C-60 50-division vernier caliper.

10 major divisions of .100 inch each, with each major division subdivided in half, thus being .050 inch. The vernier scale has 50 divisions. The 50-division vernier caliper reading shown is read as follows:

Beam whole-inch reading	1.000
Additional major divisions	.400
Additional minor divisions	.050
Vernier scale reading	.009
Total caliper reading	1.459

READING METRIC VERNIER CALIPERS

The applications for a metric vernier caliper are exactly the same as those described for an inch system vernier caliper. The discrimination of metric vernier caliper models varies among the values .02 mm, .05 mm, or .1 mm. The most commonly used type discriminates to .02 mm. The main scale on a metric vernier caliper is divided into millimeters, with every tenth millimeter mark numbered. The 10-mm line is numbered 1, the 20-mm line is numbered 2, and so on, up to the capacity of the tool (Figure C-64). The vernier scale on the sliding jaw is divided into 50 equal spaces with every fifth space numbered. Each numbered division on the vernier represents one-tenth (0.10) of a millimeter. The five smaller divisions between the numbered lines represent two-hundredths (.02) of a millimeter each.

To determine the caliper reading, read, on the main scale, whole millimeters to the left of the zero or the index line of the sliding jaw. Figure C-64 shows 27 mm plus part of an additional millimeter. The vernier scale coincides with the main scale at the eighteenth vernier division. Each vernier scale spacing is equal to .02 mm. The reading on the vernier scale is equal to 18 times .02, or .36 mm. Therefore, .36 mm must be added to the amount showing on the main scale to obtain the final reading. The result is equal to 27 mm + .36 mm, or 27.36 mm.

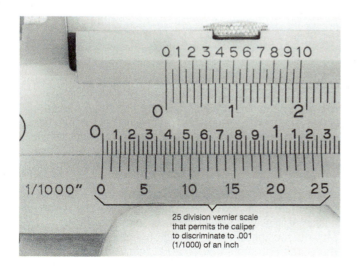

25 division vernier scale that permits the caliper to discriminate to .001 (1/1000) of an inch

Figure C-59 Lower scale is a 25-division vernier.

Figure C-61 Reading a 25-division vernier caliper.

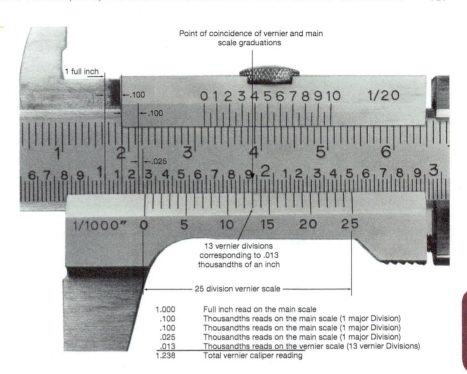

1.000	Full inch read on the main scale
.100	Thousandths reads on the main scale (1 major Division)
.100	Thousandths reads on the main scale (1 major Division)
.025	Thousandths reads on the main scale (1 major Division)
.013	Thousandths reads on the vernier scale (13 vernier Divisions)
1.238	Total vernier caliper reading

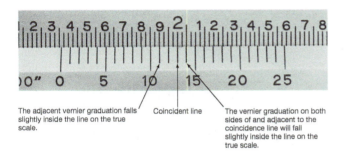

The adjacent vernier graduation falls slightly inside the line on the true scale.

Coincident line

The vernier graduation on both sides of and adjacent to the coincidence line will fall slightly inside the line on the true scale.

Figure C-62 Determining the coincident line on a vernier.

Figure C-63 Reading a 50-division vernier.

READING VERNIER DEPTH GAGES

Vernier depth gages are designed to measure the depth of holes, recesses, steps, and slots. Basic parts of a vernier depth gage include the base or anvil with the vernier scale and the fine-adjustment screw (Figure C-65). Also shown is the graduated beam or bar that contains the true scale. To make accurate measurements, the reference surface must be flat and free from nicks and burrs. The base should be held firmly against the reference surface while the beam is brought in contact with the surface being measured. The measuring pressure should approximately equal the pressure exerted when making a light dot on a piece of paper with a pencil. On a vernier depth gage, dimensions are read in the same manner as on a vernier caliper.

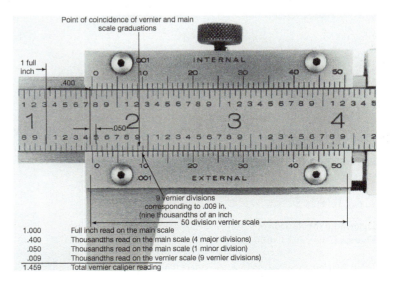

1.000	Full inch read on the main scale
.400	Thousandths read on the main scale (4 major divisions)
.050	Thousandths read on the main scale (1 minor division)
.009	Thousandths read on the vernier scale (9 vernier divisions)
1.459	Total vernier caliper reading

Figure C-64 Reading a 50-division vernier.

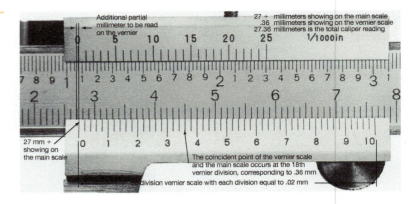

SELF-TEST: READING METRIC VERNIER CALIPERS

Determine the metric vernier caliper dimensions illustrated in Figures C-67a and C-67b.

SELF-TEST: READING VERNIER DEPTH GAGES

Determine the depth measurements illustrated in Figures C-68a and C-68b.

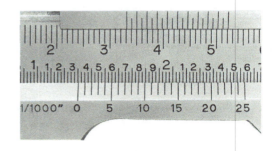

Figure C-66a Vernier reading for C-66a.

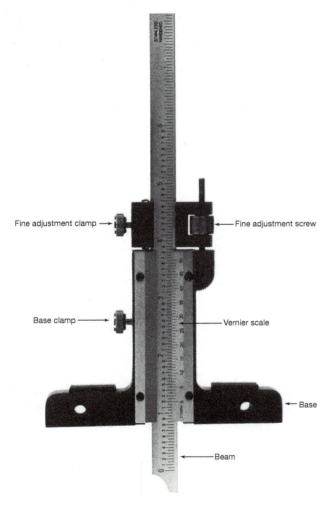

Figure C-65 Vernier depth gage with .001-inch discrimination.

SELF-TEST: READING INCH VERNIER CALIPERS

Determine the dimensions in the vernier caliper illustrations in Figures C-66a and C-66b.

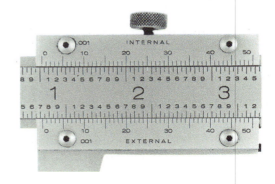

Figure C-66b Vernier reading for C-66b.

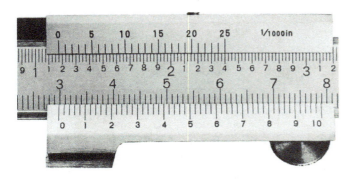

Figure C-67a Vernier reading for C-67a.

Figure C-67b Vernier reading for C-67b.

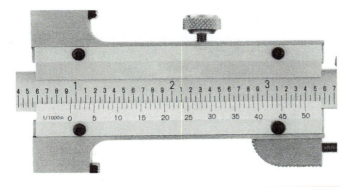

Figure C-68a Vernier reading for C-68a.

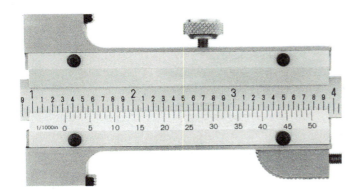

Figure C-68b Vernier reading for C-68b.

DIAL INSTRUMENTS FOR DIRECT MEASUREMENT

Dial measuring instruments are widely used. These instruments are direct reading. They are extremely reliable and easy to read. Mechanical-dial instruments and their electronic digital counterparts have become the industry standard and have replaced vernier instruments in most measurement applications. Common examples of these instruments include the dial caliper and the dial depth gage.

Dial Caliper

An outgrowth from the vernier caliper is the **dial caliper** (Figure C-69). The dial caliper does not use the principle of the vernier. The beam scale on the dial caliper is graduated into .10-inch increments. The caliper dial is graduated into either 100 or 200 divisions. The dial hand is operated by a pinion gear that engages a rack on the caliper beam. On the 100-division dial, the hand makes one complete revolution for each .10-inch movement of the sliding jaw along the beam. Therefore, each dial graduation represents $\frac{1}{100}$th of .10 inch, or .001-inch maximum discrimination. On the 200-division dial, the hand makes only half a revolution for each .10 inch of movement along the beam. Discrimination is also .001 inch.

Since the dial caliper is direct reading, there is no need to determine the coincident line of a vernier scale. This greatly facilitates reading of the measurement, and for this reason, the dial caliper has all but replaced vernier calipers.

Dial Depth Gages

The dial depth gage (Figure C-70) functions the same as the dial caliper. Readings are direct without the need to use a vernier scale. The dial depth gage has the capacity to measure over several inches of range, depending on the length of the beam. Discrimination is .001 inch.

Another type of dial depth gage uses a dial indicator (Figure C-71). The capacity and discrimination of this instrument depend on the range and discrimination of the dial indicator. The tool is used primarily in comparison measuring applications.

Figure C-69 Dial caliper.

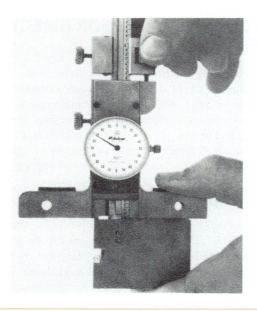

Figure C-70 Dial depth gage.

Figure C-71 Dial indicator depth gage (*Courtesy of The L.S. Starrett Co.*).

Figure C-72 Digital caliper.

Digital Calipers

Digital readout calipers are becoming more common (Figure C-72). As the sliding jaw moves along the beam, the position is shown on the digital display. The display may be set to zero at any point and may also be switched between inch and metric measurement.

SHOP TIP

Dial instruments tend to get dirt and chips in the gear rack. The teeth of the gear rack and the dial pinion gear are very small and closely spaced. Dirt and chips can cause the dial to lose its place as it moves from the zero (closed) position. This will result in errors in the measurement reading. Keep the instrument as clean as possible and check occasionally to see if the sliding jaw moves freely or seems to feel gritty as it moves. If you suspect that this is a problem, close the jaws (on calipers) fully and see if the dial reads zero. Chips or a small amount of dirt on the surfaces of the caliper jaw can also cause them not to zero out correctly. Check the instrument frequently and adjust the dial position to zero if necessary or clean any dirt and chips carefully from the gear rack. A dial caliper may also be checked with a micrometer standard to see if it reads accurately at partial or full range.

INTERNET REFERENCES

Information on electronic measuring tools:

http://www.fvfowler.com

http://www.mitutoyo.com/

http://www.starrett.com

Using Micrometers

Micrometers are the most commonly used measuring tools in industry. Correct use of them is essential to anyone making or inspecting machined parts.

OBJECTIVES

After completing this unit, you should be able to:

- Measure and record dimensions using outside micrometers to an accuracy of .001 inch.
- Measure and record diameters to an accuracy of .001 inch using an inside micrometer.
- Measure and record depth measurements using a depth micrometer to an accuracy of .001 inch.
- Measure and record dimensions using a metric micrometer to an accuracy of .01 mm.
- Measure and record dimensions using a vernier micrometer to an accuracy of .0001 inch.

TYPES OF MICROMETER INSTRUMENTS

The common types of micrometer instruments, **outside**, **inside**, and **depth**, are discussed in detail in this unit.

Blade Micrometer

The blade micrometer (Figure C-73) has a thin spindle and an anvil. Blade micrometers are used to measure narrow slots and grooves (Figure C-74) where a standard micrometer spindle and anvil will not fit.

Combination Metric/Inch or Inch/Metric Micrometer

The combination micrometer (Figure C-75) is designed for dual use in metric and inch measurement. The tool has a digital reading scale for one system, and the sleeve and thimble are used for the other system.

Figure C-73 Blade micrometer (*Courtesy of The L.S. Starrett Co.*).

Point Micrometer and Comparator Micrometer

The point micrometer (Figure C-76) is used in applications where limited space is available or where it might be desired to take a measurement at an exact location. Several point angles are available.

Disk Micrometer

The disk micrometer (Figure C-77) finds application in measuring thin materials such as paper, where a measuring face with a large area is needed. It is also useful for such measurements such as the one shown in the figure where the distance from the slot to the edge is measured.

Direct-Reading Micrometer

The direct-reading micrometer, also known as a high-precision micrometer, reads directly to 1/1000 inch (Figure C-78).

Hub Micrometer

The frame of the hub micrometer (Figure C-79) is designed so that the instrument may be put through a hole or bore to measure the hub thickness of a gear or sprocket (Figure C-80).

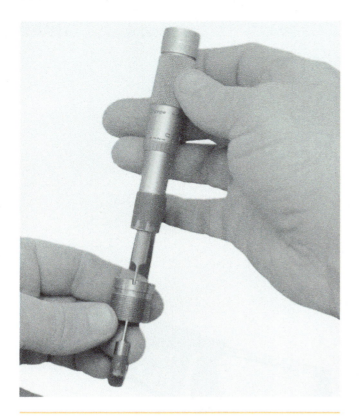

Figure C-74 Blade micrometer measuring a groove.

Figure C-75 Combination inch/metric micrometer.

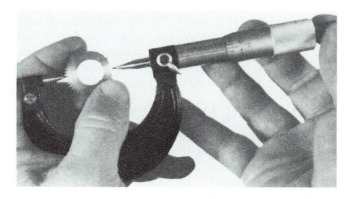

Figure C-76 Thirty-degree point comparator micrometer.

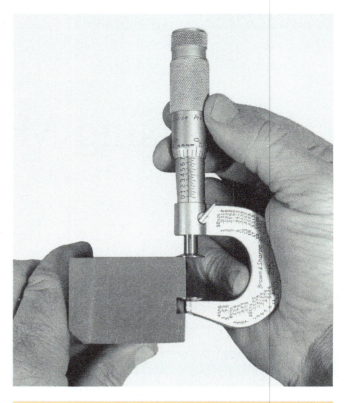

Figure C-77 Disk micrometer measuring slot-to-edge distance.

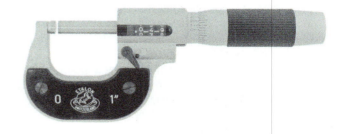

Figure C-78 Direct-reading digital micrometer.

Figure C-79 Hub micrometer.

Indicating Micrometer

The indicating micrometer (Figure C-81) is useful in determining if a part measurement is in tolerance. The instrument has an indicator that discriminates to .0005 inch. The dial can be set to the middle of the tolerance. When an object is measured, the size deviation above or below the micrometer setting will be indicated on the dial. This indicating micrometer has a range of ±.002 inch.

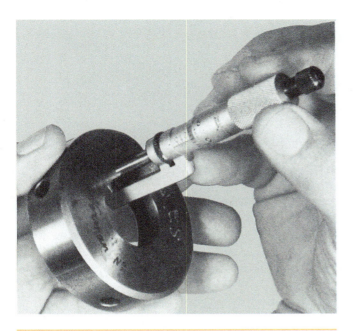

Figure C-80 Hub micrometer measuring through a bore.

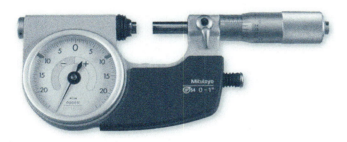

Figure C-81 Indicating micrometer (*Mitutoyo America Corp.*).

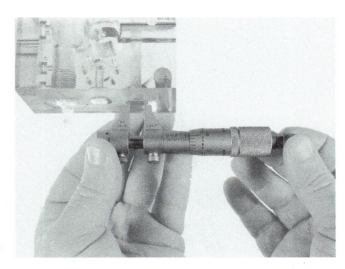

Figure C-82 Inside micrometer caliper.

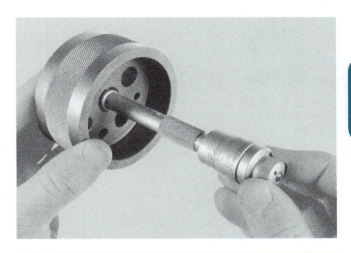

Figure C-83 Internal micrometer.

Inside Micrometer Caliper

The inside micrometer caliper (Figure C-82) has jaws like those on a vernier caliper. This instrument is designed for inside measurement. The versatility of the caliper and the reliability of the micrometer are combined in this tool.

Internal Micrometer

The internal micrometer (Figure C-83) uses a three-point measuring contact to determine the size of a bore or hole. The instrument is direct reading and is more likely to yield a reliable reading because its three-point measuring contacts make the instrument self-centering.

Interchangeable Anvil-Type Micrometer

The interchangeable anvil-type micrometer is often called a *multianvil* micrometer. It can be used in a variety of applications. A straight anvil is used to measure into a slot

(Figure C-84). A cylindrical anvil may be used for measuring from an edge to a hole (Figure C-85). Variously shaped anvils can be used to meet special measuring requirements.

Spline Micrometer

The spline micrometer has a small-diameter spindle and anvil (Figure C-86). The length of the anvil is considerably longer than that of the standard micrometer, and the frame of the instrument is also larger. This type of micrometer is well suited to measuring the minor diameter of a spline.

Screw Thread Micrometer

The screw thread micrometer (Figure C-87) is specifically designed to measure the pitch diameter of a screw thread. The anvil and spindle tips are shaped to match the form of

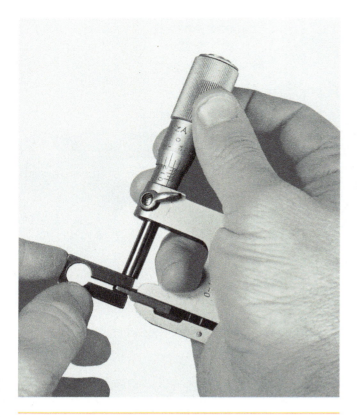

Figure C-84 Interchangeable anvil micrometer with flat anvil installed.

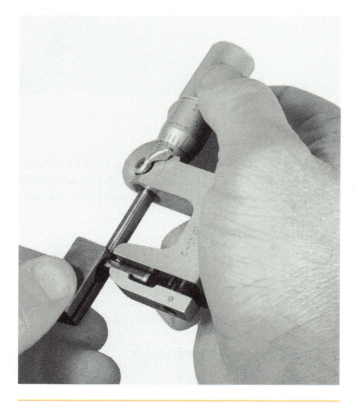

Figure C-85 Interchangeable anvil micrometer with pin anvil installed.

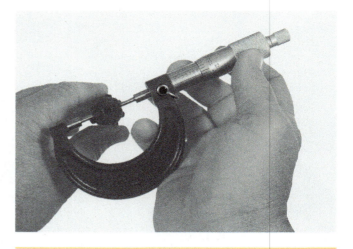

Figure C-86 Spline micrometer.

Figure C-87 Screw thread micrometer.

the thread to be measured. There is another method to measure the pitch diameters of threads. It is called the three-wire method. To use the three-wire method you would first look up the correct size of wires to use for the thread size. Then you would put two wires on one side of the thread and one wire on the opposite side of the thread and then use a regular micrometer to measure the size across the pins. You could then compare the measurement to the size specification you looked up for the pin and thread combination. There are charts to choose the wire size for a given thread size. The charts give the correct size pins to use and what the measurement across them should be for each thread size.

V-Anvil Micrometer

The V-anvil micrometer is used to measure the diameter of an object with odd-numbered, symmetrical (evenly spaced) features (Figure C-88). The type shown is for three-sided objects such as the three-fluted end mill being measured (Figure C-89). This design is also useful in checking out-of-round conditions in centerless ground parts that cannot be checked with a regular outside micrometer. The next most common V-anvil micrometer is designed for five-fluted tools such as end mills.

Tubing Micrometer

One type of tubing micrometer has a vertical anvil with a cylindrical-shaped tip (Figure C-90). Another design is like the ordinary micrometer caliper except that the anvil is a half sphere instead of a flat surface (Figure C-91). This type is designed to measure the wall thickness of tubing. The tubing micrometer can also be applied in other applications such as determining the distance of a hole from an edge.

A standard outside micrometer may also be used to determine the wall thickness of tube or pipe (Figure C-92). In this application, a ball adapter is placed on the anvil. The diameter of the ball must be subtracted from the micrometer reading to determine the actual reading.

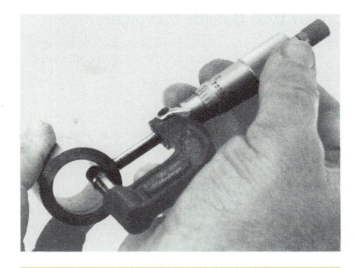

Figure C-90 Tubing micrometer measuring a tube wall.

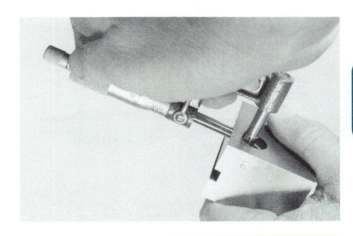

Figure C-91 Tubing micrometer measuring hole-to-edge distance.

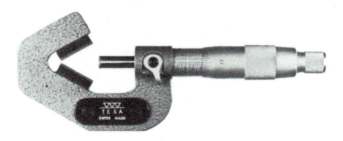

Figure C-88 V-anvil micrometer.

Figure C-89 Three-fluted end mill diameter being measured by a V-anvil micrometer.

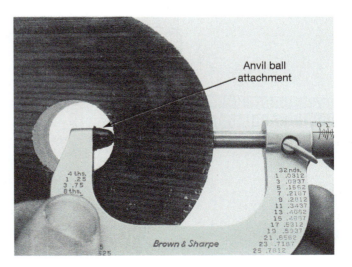

Figure C-92 Ball attachment for tubing measurement.

Caliper-Type Outside Micrometer

The caliper-type outside micrometer is used where measurements to be taken are inaccessible to a regular micrometer (Figure C-93).

Taper Micrometer

The taper micrometer can measure internal tapers (Figure C-94) or outside tapers (Figure C-95).

Groove Micrometer

The groove micrometer (Figure C-96) is well suited to measuring grooves and slots, especially in inaccessible places such as bores.

Digital Micrometers

As with calipers and many other common measuring tools, micrometers are also available in digital readout models (Figure C-97). Digital micrometers are easy to read, highly accurate, and available in many styles.

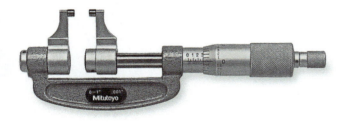

Figure C-93 Caliper-type outside micrometer (*Mitutoyo America Corp.*).

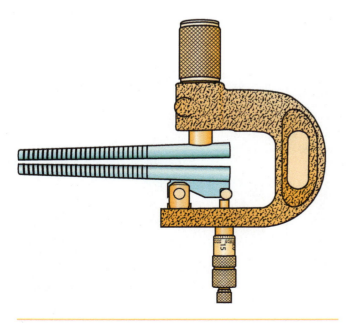

Figure C-94 Internal taper micrometer.

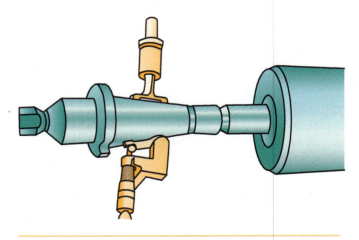

Figure C-95 Outside taper micrometer in use.

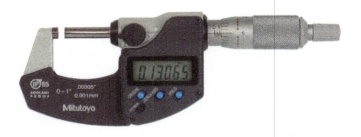

Figure C-96 Groove micrometer.

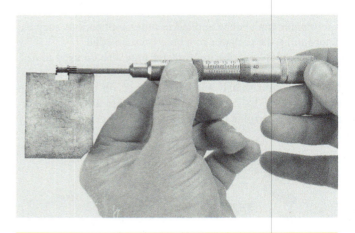

Figure C-97 Digital readout micrometers are available in a wide variety of styles (*Mitutoyo America Corp.*).

DISCRIMINATION OF MICROMETERS

The standard micrometer will discriminate to .001 inch. In the vernier form, the discrimination is increased to .0001 inch. The common metric micrometer discriminates to .01 mm. The same rules apply to micrometers as apply to all other measuring instruments. The tool should not be used beyond its discrimination. A standard micrometer with .001 discrimination should not be used in an attempt to take measurements beyond that point.

RELIABILITY AND ACCURACY IN MICROMETER INSTRUMENTS

A micrometer is more reliable than a vernier. The reason for that is that micrometers are much easier to read than verniers. The .001-inch graduations that dictate the maximum discrimination of the micrometer are placed on the circumference of the thimble. The distance between the marks is large, making them easier to see.

The micrometer will yield reliable results to .001-inch discrimination if the instrument is properly cared for, properly calibrated, and if the correct procedures are followed. Calibration is the process by which measurement tools are compared to a known standard. If the tool measurement deviates from the standard, it is adjusted to read correctly.

If the total tolerance is .001 inch, can a micrometer be used to check the dimension? The answer is no for the standard micrometer, as this violates the 10 to 1 rule for discrimination. The answer is yes for a vernier micrometer, but only under controlled conditions. What then is an acceptable expectation of accuracy that will yield maximum reliability? This is dependent to some degree on the tolerance specified and is summarized in the following table:

Tolerance Specified	Acceptability of a Standard Micrometer	Acceptability of a Vernier Micrometer
+.001 to .000 or +.000 to −.001	No	Yes
−.005 to +.005	Yes	Yes (vernier will not be required)

For a specified tolerance less than or equal to .001 inch, a vernier micrometer should be used that can discriminate to .0001 inch. Plus or minus .005 inch is a total range of .005 inch, which falls within the capability of the standard micrometer. Remember that the 10 to 1 rule for measurements is a rule of thumb and is not always possible.

The micrometer is a marvelous example of precision manufacturing. This rugged tool is produced in quantity, with each one conforming to equally high standards. Micrometer instruments, in their many forms, constitute one of the fundamental measuring instruments for the machinist.

CARE OF OUTSIDE MICROMETERS

You should be familiar with the names of the major parts of the typical outside micrometer (Figure C-98). The micrometer uses the movement of a precisely threaded rod turning in a nut for precision measurements. The accuracy of micrometer measurements depends on the quality of the tool's construction, the care it receives, and the skill of the user. A micrometer should be wiped clean of dust and oil before and after use. A micrometer should not be opened or closed by holding it by the thimble and spinning the frame around the axis of the

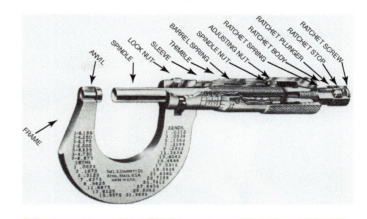

Figure C-98 Parts of the outside micrometer (*Courtesy of The L.S. Starrett Co.*).

spindle, and a micrometer should not be dropped. Even a fall of a short distance can spring the frame. This will cause misalignment between the anvil and spindle faces and destroy the accuracy of the tool. A micrometer should be kept away from dirt, oil, and chips on a machine. The instrument should be placed on a clean shop towel (Figure C-99) when not in use.

A machinist is responsible for the measurements they make. To blame an inaccurate measurement on a micrometer that was not properly adjusted or cared for would be less than professional. The machinist is responsible for the accuracy of the tool. When a micrometer is stored after use, the spindle face should not touch the anvil. Measurement tools must be kept clean. Perspiration, moisture, dirt, and other things promote corrosion between the measuring faces of a micrometer, which affect accuracy.

Clean the measuring faces of the micrometer, before using a micrometer. The measuring faces of many newer micrometers are made from tungsten carbide. Carbide-tipped micrometers have durable and long-wearing measuring faces. Screw the spindle down lightly against a piece of paper held between the spindle and the anvil (Figure C-100). Slide the paper out from between the measuring faces and blow away

Figure C-99 Micrometers should be kept on a clean shop towel when not in use.

Figure C-100 Cleaning the measuring faces.

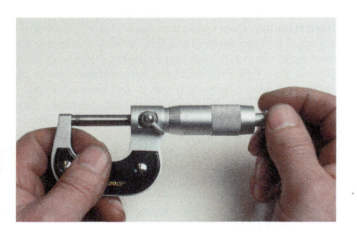

Figure C-101 Checking the zero reference.

any fuzz that clings to the spindle or anvil. Then test the zero reading of the micrometer by bringing the spindle slowly into contact with the anvil (Figure C-101). Use the ratchet stop or friction thimble to perform this operation. The ratchet stop or friction thimble found on most micrometers is designed to equalize the gaging force. When the spindle and anvil contact the workpiece, the ratchet stop or friction thimble will slip as a predetermined amount of torque is applied to the micrometer thimble. If the micrometer does not have a ratchet device, use your thumb and index finger to provide a slip clutch effect on the thimble. Never use more pressure when checking the zero reading than when making actual measurements on the workpiece. If there is a small error, it may be corrected by adjusting the index line to the zero point (Figure C-102). Follow the manufacturer's instructions when making this adjustment. Also follow the manufacturer's instructions for correcting a loose thimble-to-spindle connection or incorrect friction thimble or ratchet stop action. One drop of instrument oil applied to the micrometer thread at monthly intervals will help it provide many years of reliable service. Machinists are often judged by their coworkers by the way they handle and care for their tools. Those who care for their tools properly are more likely be held in higher professional regard.

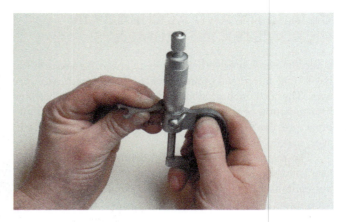

Figure C-102 Adjusting the index line to zero.

READING INCH MICROMETERS

Dimensions requiring the use of micrometers will generally be expressed in decimal form to three decimal places. In the case of an inch instrument, this is the thousandths place. You should think in terms of thousandths whenever reading decimal fractions. For example, the decimal .156 inch is read as one hundred fifty-six thousandths of an inch. Likewise, .062 inch is read as sixty-two thousandths of an inch. Machinists generally talk about all dimensions in terms of thousandths of an inch. For example, $\frac{5}{8}$ inch is referred to as six hundred and twenty-five thousandths of an inch.

On the **sleeve** of the micrometer there is a graduated scale with 10 numbered divisions, $\frac{1}{10}$ inch or .100 inch (100 thousandths) apart. Each of these major divisions is further subdivided into four equal parts, which makes the distance between these graduations $\frac{1}{4}$ of .100 inch, or .025 inch (25 thousandths) (Figure C-103). The **spindle screw** of a micrometer has 40 threads per inch. When the spindle is turned one complete revolution, it moves $\frac{1}{40}$ of an inch, or, expressed as a decimal, .025 inch (25 thousandths).

When you examine the **thimble**, you will find 25 evenly spaced divisions around its circumference (Figure C-103). Because each complete revolution of the thimble causes it to move a distance of .025 inch, each thimble graduation must be equal to $\frac{1}{25}$ of .025 inch, or .001 inch (one thousandth). On most micrometers, each thimble graduation is numbered to facilitate reading the instrument. On older micrometers, only every fifth line may be numbered.

When reading the micrometer (Figure C-104), first determine the value indicated by the lines exposed on the sleeve. The edge of the thimble exposes three major divisions. This represents .300 inch (300 thousandths). However, there are also two minor divisions showing on the sleeve. The value of these is .025, for a total of .050 inch (50 thousandths). The reading on the thimble is 9, which indicates .009 inch (9 thousandths). The final micrometer reading is determined by adding the total of the sleeve and thimble readings. In the example shown (Figure C-104), the sleeve shows a total of .350 inch. Adding this value to the thimble gives the final reading: .350 inch + .009 inch, or .359 inch.

Figure C-103 Graduations on the inch micrometer.

The sleeve is graduated into 10 equal divisions, each of which is further subdivided into 4 smaller divisions.

The length of the sleeve graduations is 1 inch, or the distance the thimble travels in 40 complete revolutions.

The thimble has 25 equal graduations on its circumference. Each graduation is equal to 1/25 of 1/40, or .001 of an inch.

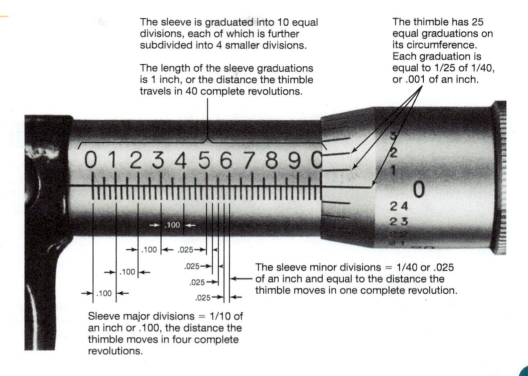

The sleeve minor divisions = 1/40 or .025 of an inch and equal to the distance the thimble moves in one complete revolution.

Sleeve major divisions = 1/10 of an inch or .100, the distance the thimble moves in four complete revolutions.

Figure C-104 Inch micrometer reading of .359, or three hundred fifty-nine thousandths.

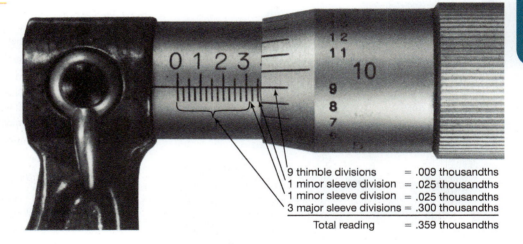

9 thimble divisions	= .009 thousandths
1 minor sleeve division	= .025 thousandths
1 minor sleeve division	= .025 thousandths
3 major sleeve divisions	= .300 thousandths
Total reading	= .359 thousandths

USING THE MICROMETER

A micrometer should be gripped by the frame (Figure C-105), leaving the thumb and forefinger free to operate the thimble. When possible, take micrometer readings while the instrument is in contact with the workpiece. Use only enough pressure on the **spindle** and **anvil** to yield a reliable result. This is what the machinist refers to as **feel**. The proper feel of a micrometer will come only from experience. Obviously, excessive pressure will not only result in an inaccurate measurement but will also distort the frame of the micrometer and possibly damage it permanently. You should also remember that not enough pressure on the part by the measuring faces can yield an unreliable result.

Figure C-105 Proper way to hold a micrometer.

The micrometer should be held in both hands whenever possible. This is especially important when measuring cylindrical workpieces (Figure C-106). Holding the instrument in one hand does not permit sufficient control for reliable readings. Cylindrical workpieces should be checked at least twice with **measurements made 90 degrees apart**. This is to check for an out-of-round condition (Figure C-107). When critical dimensions are measured, at least two consecutive measurements should be made. Both readings should indicate identical results. If two identical readings cannot be achieved, then the actual size of the part cannot be stated reliably. **All critical measurements should be made at room temperature.** A workpiece that is hot because of machining will be larger because of heat expansion.

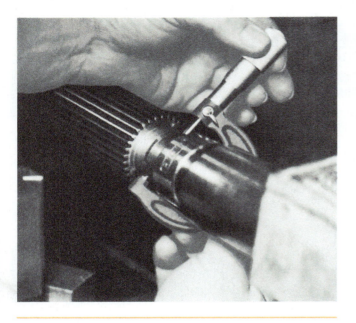

Figure C-106 Hold a micrometer in both hands when measuring a round part.

SHOP TIP

When using a 0- to 1-inch micrometer, always check the zero reference before making any measurements with the tool. When using micrometers with larger ranges, use a standard or gage blocks to verify that the tool is reading correctly.

Outside micrometers usually have a measuring range of 1 inch. They are identified by size according to the largest dimensions they measure. A 2-inch micrometer will measure from 1 to 2 inches. A 3-inch micrometer will measure from 2 to 3 inches. The capacity of the tool is increased by increasing the size of the frame. It requires a great deal more skill to get consistent measurements with large micrometers.

SELF-TEST

1. Why should a micrometer be kept clean and protected?
2. Why should a micrometer be stored with the spindle out of contact with the anvil?
3. Why are the measuring faces of the micrometer cleaned before measuring?
4. How precise is the standard micrometer?
5. What affects the accuracy of a micrometer?
6. What is the difference between the sleeve and thimble?
7. Why should a micrometer be read while it is still in contact with the object to be measured?
8. Why should a part dimension be measured more than once?
9. What effect does heat have on the size of a part?
10. What is the purpose of the friction thimble or ratchet stop on the micrometer?
11. Read and record the five outside micrometer readings in Figures C-108a to C-108e.

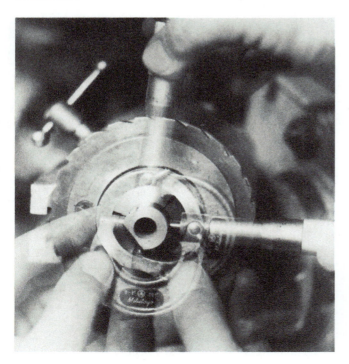

Figure C-107 When measuring round parts, take two readings 90 degrees apart.

Figure C-108a Micrometer reading for C-108a.

Figure C-108b Micrometer reading for C-108b.

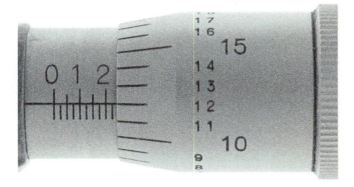

Figure C-108c Micrometer reading for C-108c.

Figure C-108d Micrometer reading for C-108d.

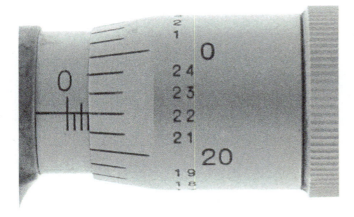

Figure C-108e Micrometer reading for C-108e.

USING INSIDE MICROMETERS

Inside micrometers have the same graduations as outside micrometers. Inside micrometers discriminate to .001 inch and have a measuring capacity ranging from 1.5 to 20.0 inches or more. A typical tubular type inside micrometer set (Figure C-109) consists of the micrometer head with detachable hardened anvils and several tubular measuring rods with hardened contact tips. The lengths of these rods differ in increments of .5 inch to match the measuring capacity of the micrometer head, which in this case is .5 inch. A handle is provided to hold the instrument in places where holding the instrument directly would be difficult. Another common type of inside micrometer comes equipped with relatively small diameter solid rods that differ in inch increments, even though the head movement is .5 inch. In this case, a .5-inch spacing collar is provided. This can be slipped over the base of the rod before it is inserted into the measuring head.

Inside micrometer heads have a range of .250, .500, 1.000, or 2.000 inches, depending on the total capacity of the set. For example, an inside micrometer set with a head range of .500 inch will be able to measure from 1.500 to 12.500 inches.

The measuring range of the inside micrometer is changed by attaching the extension rods. Extension rods may be solid or tubular. Tubular rods are lighter in weight and are often found in large-range inside micrometer sets. Tubular rods are also more rigid. There are two very important considerations when using an inside micrometer. First, accurate measurements are dependent on developing a consistent "feel" with the instrument. This means adjusting the micrometer to the correct pressure to measure accurately. Another requirement is that all parts must be extremely clean when changing extension rods (Figure C-110) and they must be installed correctly. Even dust particles can affect the accuracy of the instrument. It is always wise after installing the extension rod to check the micrometer with an outside micrometer. This will help assure that the inside micrometer is reading correctly.

When making internal measurements, set one end of the inside micrometer against one side of the hole to be measured (Figure C-111). A handle is usually provided that facilitates insertion of the micrometer into a bore or hole (Figure C-112). One end of the micrometer will become the center of the arcing movement used when finding

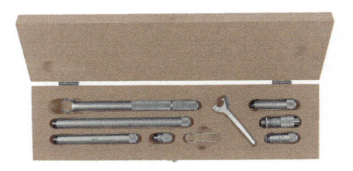

Figure C-109 Tubular inside micrometer set.

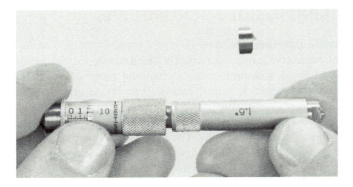

Figure C-110 Attaching a 1.5-inch extension rod to an inside micrometer head.

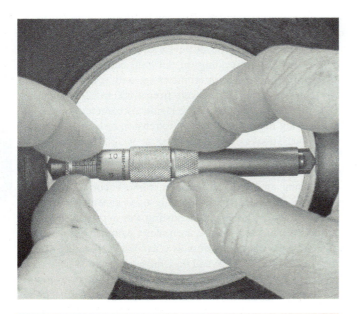

Figure C-111 Placing the inside micrometer in the bore to be measured.

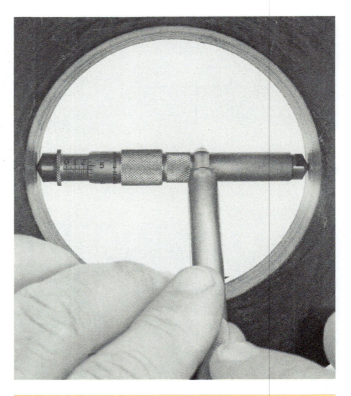

Figure C-112 Inside micrometer head used with a handle.

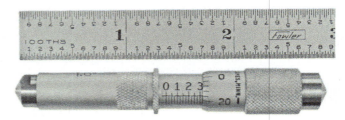

Figure C-113 Confirming the inside micrometer range using a rule.

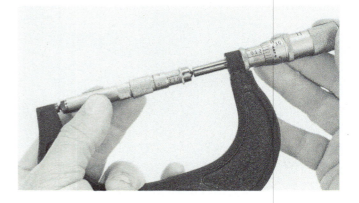

Figure C-114 Checking the inside micrometer with an outside micrometer.

the centerline of the hole to be measured. The micrometer should then be adjusted to the size of the hole. When the correct hole size is reached, there should be a light drag between the measuring tip and the work when the tip is moved through the centerline of the hole. The size of the hole is determined by adding the reading of the micrometer head, the length of the extension rod, and the length of the spacing collar, if one was used. Read the micrometer **while it is still in place if possible**. If the instrument must be removed to be read, the correct micrometer range to be used can be determined by checking with a rule (Figure C-113). A machinist will usually use an **outside micrometer to verify** a reading taken with an inside micrometer. In this case, the inside micrometer becomes an easily adjustable transfer measuring tool (Figure C-114). Take at least two readings 90 degrees apart to obtain the size of a hole or bore. The readings should be identical.

SELF-TEST

1. Read and record the five inside micrometer readings shown in Figures C-115a to C-115e. The micrometer head is 1.500 inch when zeroed.

2. Obtain an inside micrometer set from your instructor and practice using the instrument on objects around your laboratory. Measure examples such as lathe spindle holes, bushings, bores of roller bearings, hydraulic cylinders, and tubing.

Figure C-115a Micrometer reading for C-115a.

Figure C-115b Micrometer reading for C-115b.

Figure C-115c Micrometer reading for C-115c.

Figure C-115d Micrometer reading for C-115d.

Figure C-115e Micrometer reading for C-115e.

USING DEPTH MICROMETERS

A **depth micrometer** is a tool used to precisely measure the depth of holes, grooves, shoulders, and recesses. Like other micrometer instruments, it will discriminate to .001 inch. Depth micrometers usually come as a set with

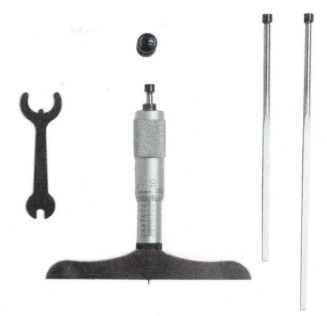

Figure C-116 Depth micrometer set.

interchangeable rods to accommodate different depth measurements (Figure C-116). The basic parts of the depth micrometer are the base, sleeve, thimble, extension rod, thimble cap, and, frequently, a rachet stop. The base of depth micrometers can be of various widths. Generally, the wider bases are more stable, but in many instances, space limitations dictate the use of narrower bases. Some depth micrometers are made with only a half base for measurements in confined spaces.

Extension rods are installed or removed by holding the thimble and unscrewing the thimble cap. Make sure that the seat between the thimble cap and rod adjusting nuts is clean before reassembling the micrometer. Do not over tighten the thimble cap. Furthermore, do not attempt to adjust the rod length by turning the adjusting nuts. These rods are calibrated and matched as a set. The measuring rods from a specific depth micrometer set should always be kept with that set. Since these rods are calibrated and matched to a specific instrument. Exchanging measuring rods from set to set will result in incorrect measurements.

When making depth measurements, it is important that the micrometer base has a smooth and flat surface on which to rest. Furthermore, sufficient pressure must be applied to keep the base in contact with the reference surface (Figure C-117).

READING DEPTH MICROMETERS

When a comparison is made between the sleeve of an outside micrometer and the sleeve of a depth micrometer, note that the graduations are numbered in the opposite direction

Figure C-117 Proper way to hold the depth micrometer.

(Figure C-118). When reading a depth micrometer, the distance to be measured is the value covered by the thimble. Consider the reading shown in Figure C-118. The thimble edge is between the numbers 5 and 6. This indicates a value of at least .500 inch on the sleeve major divisions. The thimble also covers the first minor division on the sleeve, which has a value of .025 inch. The value on the thimble circumference indicates .010 inch. Adding these three values results in a total of .535 inch, or the amount of extension of the rod from the base.

A depth micrometer should be tested for accuracy before it is used. When the 0- to 1-inch rod is used, retract the measuring rod into the base. Clean the base and contact surface of the rod. Hold the micrometer base firmly against a flat, highly finished surface, such as a surface plate, and advance the rod until it contacts the reference surface (Figure C-119). If the micrometer is properly adjusted, it should read zero. When testing for accuracy with the 1-inch extension rod, **set** the base of the micrometer on a 1-inch gage block and measure to the reference surface (Figure C-120). Other extension rods can be tested in a like manner.

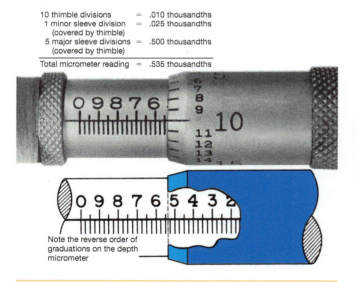

10 thimble divisions	=	.010 thousandths
1 minor sleeve division (covered by thimble)	=	.025 thousandths
5 major sleeve divisions (covered by thimble)	=	.500 thousandths
Total micrometer reading	=	.535 thousandths

Note the reverse order of graduations on the depth micrometer

Figure C-118 Sleeve graduations on the depth micrometer are numbered in the direction opposite those on the outside micrometer.

Figure C-119 Checking a depth micrometer for zero adjustment using the surface plate as a reference surface.

Figure C-120 Checking the depth micrometer calibration at the 1.000-inch position in the 0- to 1-inch rod, and a 1-inch square or Hoke gage block.

SELF-TEST

Read and record the five depth micrometer readings in Figures C-121a to C-121e.

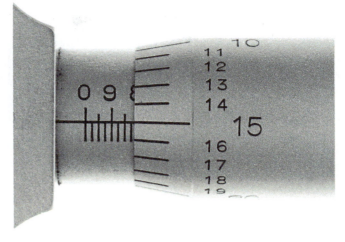

Figure C-121b Micrometer reading for C-121b.

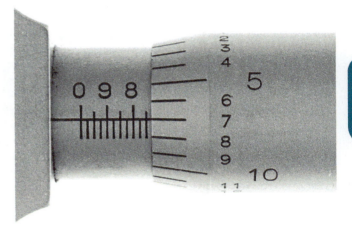

Figure C-121c Micrometer reading for C-121c.

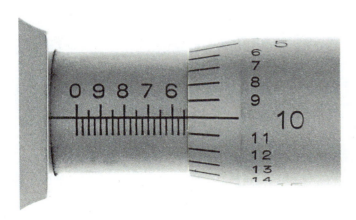

Figure C-121a Micrometer reading for C-121a.

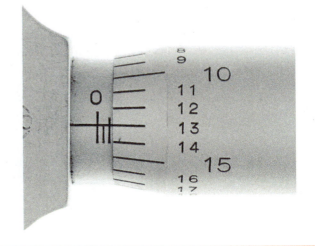

Figure C-121d Micrometer reading for C-121d.

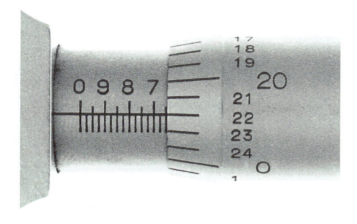

Figure **C-121e** Micrometer reading for C-121e.

the index line, with every fifth division numbered. These are whole-millimeter graduations. Below the index line are graduations that fall halfway between the divisions above the line. The lower graduations represent half, or .5-mm, values. The thimble edge in Figure C-124 leaves the 12-mm line exposed with no .5-mm line showing. The thimble reading is 32, which is .32 mm. Adding the two figures results in a total of 12.32 mm.

The 15-mm mark in Figure C-125 is exposed on the sleeve plus a .5-mm graduation below the index line. The thimble reads 20 or .20 mm. Adding these three values (15.00 + .50 + .20) results in a total of 15.70 mm.

Any metric micrometer should receive the same care as that discussed in the section on outside micrometers.

READING METRIC MICROMETERS

The metric micrometer (Figure C-122) has a spindle thread with a .5-mm lead. This means that the spindle will move .5 mm when the thimble is turned one complete revolution. Two revolutions of the thimble will advance the spindle 1 mm. In precision machining, metric dimensions are usually expressed in terms of .01 mm. On the metric micrometer the thimble is graduated into 50 equal divisions, with every fifth division numbered (Figure C-123). If one revolution of the thimble is .5 mm, then each division on the thimble is equal to .5 mm divided by 50, or .01 mm. The sleeve of the metric micrometer is divided into 25 main divisions above

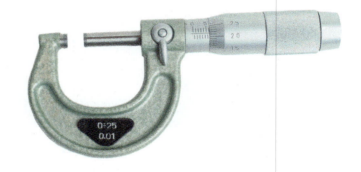

Figure **C-122** Metric micrometer.

Figure **C-123** Graduations on the metric micrometer.

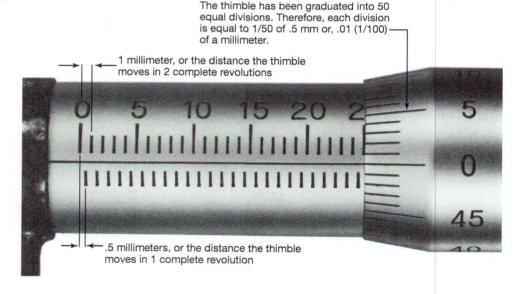

The thimble has been graduated into 50 equal divisions. Therefore, each division is equal to 1/50 of .5 mm or, .01 (1/100) of a millimeter.

1 millimeter, or the distance the thimble moves in 2 complete revolutions

.5 millimeters, or the distance the thimble moves in 1 complete revolution

Figure C-124 Metric micrometer reading of 12.32 mm.

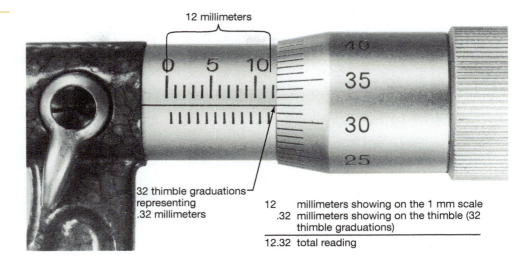

12 millimeters

32 thimble graduations representing .32 millimeters

12	millimeters showing on the 1 mm scale
.32	millimeters showing on the thimble (32 thimble graduations)
12.32	total reading

Figure C-125 Metric micrometer reading of 15.70 mm.

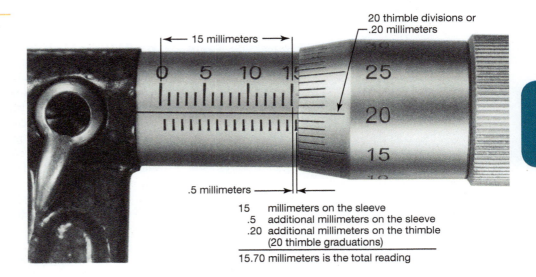

20 thimble divisions or .20 millimeters

15 millimeters

.5 millimeters

15	millimeters on the sleeve
.5	additional millimeters on the sleeve
.20	additional millimeters on the thimble (20 thimble graduations)
15.70	millimeters is the total reading

SELF-TEST

Read and record the five metric micrometer readings in Figures C-126a to C-126e.

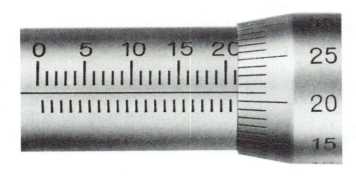

Figure C-126a Micrometer reading for C-126a.

Figure C-126b Micrometer reading for C-126b.

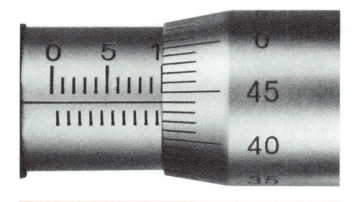

Figure C-126c Micrometer reading for C-126c.

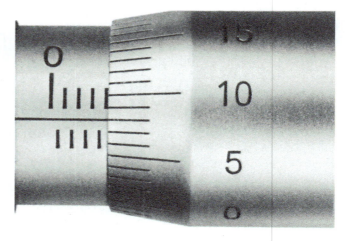

Figure C-126e Micrometer reading for C-126e.

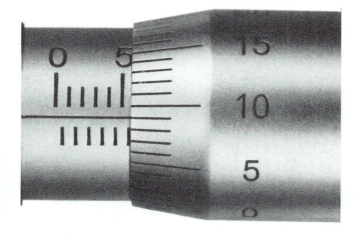

Figure C-126d Micrometer reading for C-126d.

READING VERNIER MICROMETERS

When measurements must be made finer than .001 inch, a standard micrometer is not sufficient. With a **vernier micrometer**, readings can be made to a one **ten-thousandth part of an inch** (.0001 inch). This kind of micrometer is commonly known as a "tenth mike." A vernier scale is part of the sleeve graduations. The vernier scale consists of 10 lines parallel to the index line and is located above it (Figure C-127). The word *tenth* might be a bit misleading. Used in this context, the word *tenth* refers to .0001 inch. Do not confuse this with a tenth part of an inch (.10 inch).

Figure C-127 Inch vernier micrometer reading of .2163 inch.

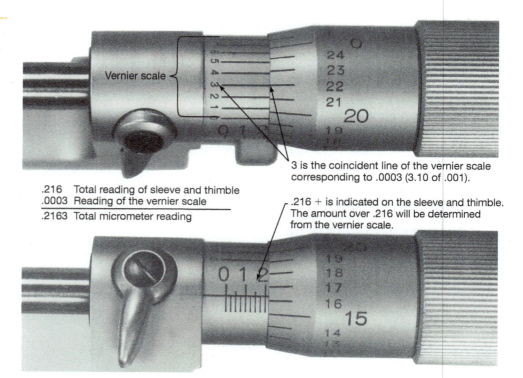

Vernier scale

3 is the coincident line of the vernier scale corresponding to .0003 (3.10 of .001).

.216 + is indicated on the sleeve and thimble. The amount over .216 will be determined from the vernier scale.

.216   Total reading of sleeve and thimble
.0003   Reading of the vernier scale
_____
.2163   Total micrometer reading

The graduations on the thimble are .001 (one thousandth) apart. Each vernier spacing must then be equal to $\frac{1}{10}$ of .001 inch, or .0001 inch (one ten-thousandth). Thus, according to the principle of the vernier, each thousandth of the thimble is subdivided into 10 parts. This permits the vernier micrometer to discriminate to .0001 inch.

To read a vernier micrometer, first read to the nearest thousandth just like you would on a standard micrometer. Then, find the line on the vernier scale that coincides with a graduation on the thimble. The value of this coincident vernier scale line is the value in ten-thousandths that must be added to the thousandths reading to make up the total reading. **Remember to add the value of the vernier scale line and not the number of the matching thimble line.**

In the lower view of Figure C-127, a micrometer reading of slightly more than .216 inch is indicated. In the top view, on the vernier scale, the line numbered 3 is in alignment with the line on the thimble. This indicates that .0003 (three ten-thousandths) must be added to the .216 inch for a total reading of .2163. This number is read "two hundred sixteen thousandths and three tenths." (Remember that the three tenths is .0003 inch)

You must be very careful when measuring to a tenth of a thousandth using a vernier micrometer. There are many conditions that can influence the reliability of such measurements. Measurements in .0001s should be carried out under **controlled conditions** for truly reliable results. The finish of the workpiece must be extremely smooth. Contact pressure of the measuring faces must be consistent. The workpiece and instrument must be temperature stabilized. Heat transferred to the micrometer by handling can cause it to deviate considerably. Furthermore, the micrometer must be carefully calibrated against a known standard. Only under these conditions can true reliability be assured.

SELF-TEST

Read and record the five vernier micrometer readings in Figures C-128a to C-128e.

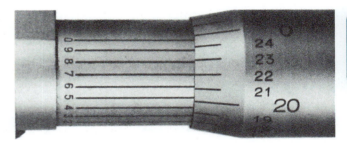

Figure C-128a Micrometer reading for C-128a.

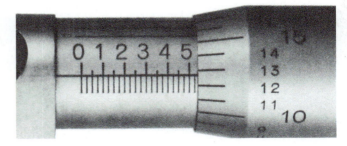

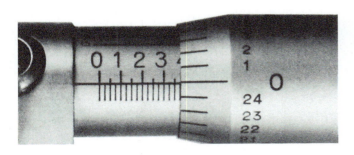

Figure C-128b Micrometer reading for C-128b.

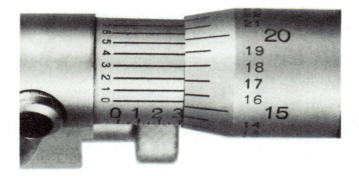

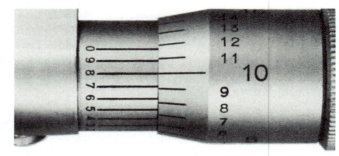

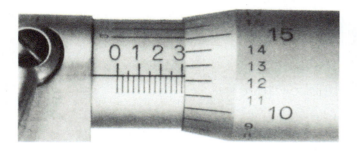

Figure C-128c Micrometer reading for C-128c.

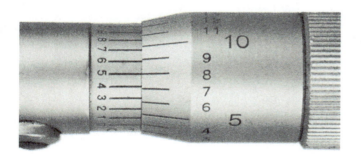

Figure C-128e Micrometer reading for C-128e.

INTERNET REFERENCES

Information on micrometers:

http://www.fvfowler.com

http://mitutoyo.com

http://www.elexp.com

http://www.auto-met.com

http://www.starrett.com

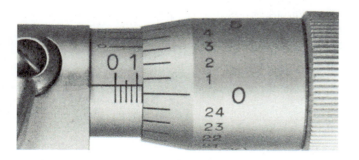

Figure C-128d Micrometer reading for C-128d.

UNIT SIX

Using Comparison Measuring Instruments

There are many measuring devices that do not show an actual measurement. These are used in comparison measurement applications, where they are compared with a known standard, or used in conjunction with an instrument that can show a measurement. This unit will introduce the principles of comparison measurement, the common tools of comparison measurement, and their applications.

OBJECTIVES

After completing this unit, you should be able to:

- Define comparison measurement.
- Identify common comparison measuring tools.
- Given a measuring situation, select the proper comparison tool for the measuring requirement.

MEASUREMENT BY COMPARISON

All of us, at some time, have probably been involved in constructing something in which we used no measuring instruments of any kind. For example, suppose that you had to build some wooden shelves. You had the required lumber available, with all boards longer than the shelf spaces. You held a board to the shelf space and marked the required length for cutting. By this procedure, you compared the length of the board (the unknown length) with the shelf space (the known length or standard). After cutting the first board to the marked length, you then used it to determine the lengths of the remaining shelves. The board has no capacity to show a measurement; however, in this case, it became a measuring instrument.

A great deal of comparison measurement often involves the following steps:

1. A device that has no capacity to show measurement is used to establish and represent an unknown dimension.
2. This representation of the unknown is then transferred to an instrument that has the capability to show a measurement.

This procedure is commonly known as transfer measurement. In the example of cutting shelf boards, the shelf space was transferred to the first board and then the length of the first board was transferred to the remaining boards.

Transfer of measurements may reduce reliability. This factor must be kept in mind when using comparison tools requiring that a transfer be made. Remember that an instrument with the capability to show measurement directly is always best. Direct-reading instruments should be used whenever possible. Measurements requiring a transfer must be accomplished with proper caution if reliability is to be maintained.

COMMON COMPARISON MEASURING TOOLS AND THEIR APPLICATIONS

Spring Calipers

The spring caliper is a common comparison measuring tool for rough measurements of inside and outside dimensions. To use a spring caliper, set one jaw on the workpiece (Figure C-129a), use this point as a pivot, and swing the other caliper leg back and forth over the largest point on the diameter. At the same time, adjust the leg spacing. When the correct feel is obtained, remove the caliper and compare it with a steel rule to determine the reading (Figure C-129b). The inside spring caliper can be used in a similar manner (Figure C-130). Spring calipers are not used much at all anymore. The spring caliper should only be used for the roughest of measurements.

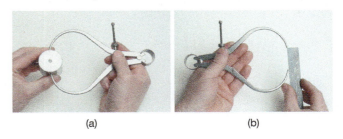

(a)   (b)

Figure C-129 Use of a caliper for an outside measurement.

Figure C-130 Using an inside spring caliper.

Telescoping Gage

The telescoping gage is also a common comparison measuring instrument. Telescoping gages are widely used in the machine shop, and they can be used for many types of measurements.

Telescoping gages generally come in a set of six gages (Figure C-131). The range of the set is usually $\frac{1}{2}$ to 6 inches (12.5 to 150 mm). The gage consists of one or two telescoping plungers with a handle and locking screw. The gage is inserted into a bore or slot, and the plungers are permitted to extend, thus conforming to the size of the feature. The gage is then removed and transferred to a micrometer, where the

Figure C-131 Set of telescoping gages.

reading is determined. The telescoping gage can be a reliable and versatile tool if proper procedure is used in its application.

Procedure for Using the Telescoping Gage

Step 1 Select the proper gage for the desired measurement range.

Step 2 Insert the gage into the bore to be measured and release the handle lock screw (Figure C-132). Rock the gage sideways to ensure that you are measuring at the full diameter (Figures C-133a and C-133b). This is especially important in large-diameter bores.

Step 3 Lightly tighten the locking screw.

Step 4 Use a downward or upward motion and roll the gage through the bore. The plungers will be pushed in, thus conforming to the bore diameter. This part of the procedure should be done only once, as rolling the gage back through the bore may cause the plungers to be pushed in farther, resulting in an inaccurate setting. If you feel that the gage is not centered properly, release the locking screw and repeat the procedure from the beginning.

Step 5 Remove the gage and measure with an outside micrometer (Figure C-134). Place the gage between the micrometer spindle and the anvil. Try to determine the same feel on the gage with the micrometer as you felt while the gage was in the bore. Excessive pressure with the micrometer will depress the gage plungers and cause an incorrect reading.

Step 6 Take at least two readings or more with the telescoping gage to verify reliability. If the readings do not agree, repeat steps 2 to 6.

Figure C-132 Inserting the telescoping gage into the bore (*Courtesy of The L.S. Starrett Co.*).

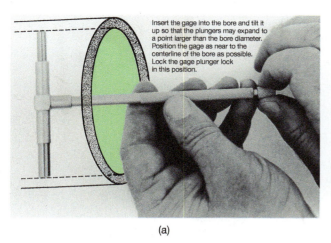

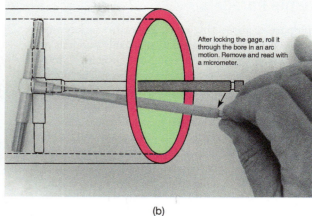

Insert the gage into the bore and tilt it up so that the plungers may expand to a point larger than the bore diameter. Position the gage as near to the centerline of the bore as possible. Lock the gage plunger lock in this position.

After locking the gage, roll it through the bore in an arc motion. Remove and read with a micrometer.

(a)

(b)

Figure C-133a and b Use of telescope gage to measure a bore.

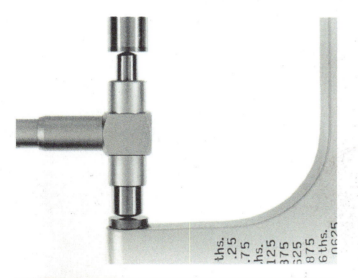

Figure C-134 Checking the telescoping gage with an outside micrometer.

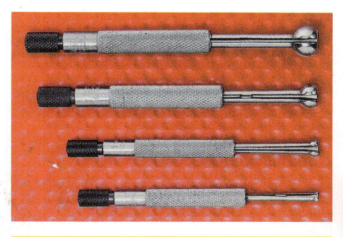

Figure C-135 Set of small hole gages (*Courtesy of The L.S. Starrett Co.*).

Small Hole Gages

Small hole gages come in sets with a range of $\frac{1}{8}$ to $\frac{1}{2}$ inch (4 to 12 mm) (see Figure C-135). One type of small hole gage consists of a split ball connected to a handle (Figure C-136a). A tapered rod is drawn between the split ball halves, causing them to expand and contact the surface to be measured. The split-ball small hole gage has a flattened end so that a shallow hole or slot may be measured. After the gage has been expanded in the feature to be measured, it should be moved back and forth to determine the proper feel. The gage is then removed and measured with an outside micrometer.

Gage Pins

Gage pins are cylindrical pins for measuring the diameter of machined holes and slots. They are normally purchased as sets of pins, although they can be purchased individually. They are available in a wide range of sizes. For example, one set for small holes might have 50 pins that range in size

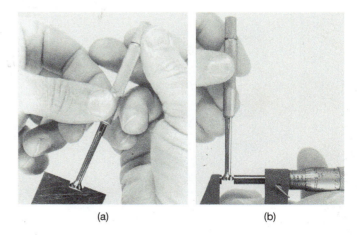

(a)

(b)

Figure C-136 Use of a small hole gage to measure a slot.

from .011 to .060. The next set might be sized from .061 to .250. Sets are available for larger size holes also. Accuracy for gage pins is often .0002 or less. They are very helpful to accurately check the sizes of small holes. They can be used as go/no-go gages with use of a handle.

Adjustable Parallels

For the purpose of measuring slots, grooves, and keyways, the adjustable parallel may be used. Adjustable parallels are available in sets ranging from about $\frac{3}{8}$ to $2\frac{1}{4}$ inch (10 to 60 mm). They are precision ground for accuracy. The typical adjustable parallel consists of two parts that slide together on an angle. Adjusting screws are provided so that clearance in the slide may be adjusted or the parallel locked after setting for a measurement. As the halves of the parallel slide, the width increases or decreases depending on the direction. The parallel is placed in the groove or slot to be measured and expanded until the parallel edges conform to the width to be measured. The parallel is then locked with a small screwdriver and measured with a micrometer (Figure C-137). If possible, an adjustable parallel should be left in place while being measured.

Radius Gages

The typical radius gage set ranges in size from $\frac{1}{32}$ to $\frac{1}{2}$ inch (.8 to 12 mm). Larger radius gages are also available. The gage can be used to measure the radii of grooves and external radii (outside rounded corners) or internal fillets (inside rounded corners) (Figure C-138). Radius gages are useful for checking the radii of rounds and fillets on castings (Figure C-139).

Thickness Gages

The thickness gage is often called a *feeler gage* (Figure C-140). It is probably best known for its various automotive applications. However, a machinist may use a thickness gage for such measurements as the thickness of a shim, setting a grinding wheel above a workpiece, or determining the height difference

Figure C-138 Using a radius gage.

Figure C-137 Using adjustable parallels.

Figure C-139 Radius gage used to check a casting fillet (*Courtesy of The L.S. Starrett Co.*).

of two parts. The thickness gage is not a true comparison measuring instrument, as each leaf is marked as to size. However, it is good practice to check a thickness gage with a micrometer, especially when leaves are stacked together.

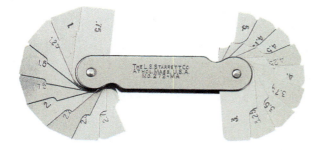

Figure C-140 A feeler or thickness gage (*Courtesy of The L.S. Starrett Co.*).

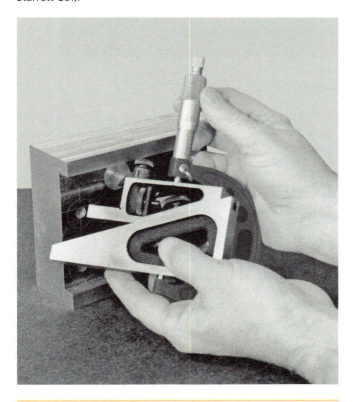

Figure C-141 Setting a planer gage with an outside micrometer.

Figures C-142a and b Use of gage blocks and indicator to set the height of a planer gage.

Planer Gage

The planer gage functions much like an adjustable parallel. Planer gages were originally used to set tool heights on shapers and planers. They can also be used as a comparison measuring tool.

The planer gage may be equipped with a scriber and used in layout. The gage may be set with a micrometer (Figure C-141) or in combination with a dial test indicator and gage blocks. Figure C-142a shows the planer gage being set using a test indicator set to zero on a gage block. This dimension is then transferred to the planer gage (Figure C-142b). After the gage has been set with the scriber attached, the instrument can be used for layout (Figure C-143).

Squares

The square is an important and useful tool for the machinist. A square is a comparison measuring instrument that is used to compare its degree of perpendicularity with an unknown degree of perpendicularity on the workpiece.

Combination Square

The combination square (Figure C-144) is part of a combination set (Figure C-145). A combination set consists of a graduated rule, square head, bevel protractor, and center head. The square head slides on the graduated rule and can be locked in any position (Figure C-146). This feature makes the tool useful

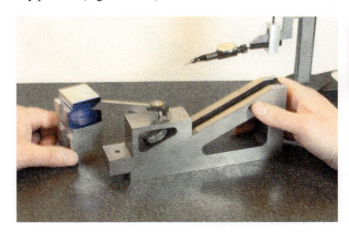

Figure C-143 Using the planer gage in layout.

(a)   (b)

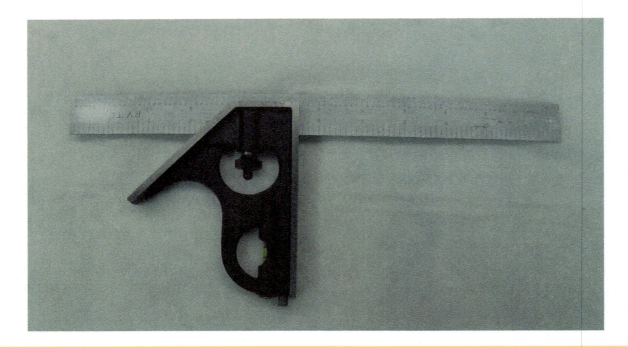

Figure C-144 A combination square (*Courtesy of Fox Valley Technical College*).

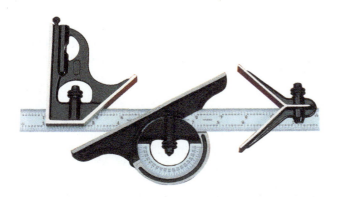

Figure C-145 Machinist's combination set (*Courtesy of The L.S. Starrett Co.*).

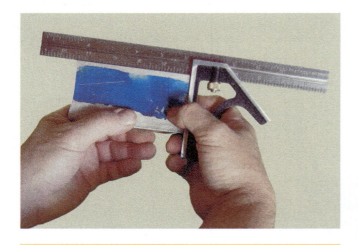

Figure C-146 Using the combination square (*Courtesy of The L.S. Starrett Co.*).

for layout, as the square head can be set according to the rule graduations. The combination square head also has a 45-degree angle along with a spirit level and layout scriber. The combination set is one of the most versatile tools of the machinist.

Solid Beam Square On the solid beam square, the beam and blade are fixed. Solid beam squares range in size from 2 to 72 inches (Figure C-147).

Precision Beveled Edge Square The precision beveled edge square is an extremely accurate square used in the toolroom and in inspection applications. The beveled edge permits a single line of contact with the part to be checked. Precision squares range in size from 2 to 14 inches (Figure C-148).

The squares discussed up to this point have no capacity to directly indicate the amount of deviation from perpendicularity. The only determination that can be made is

Figure C-147 Solid beam square.

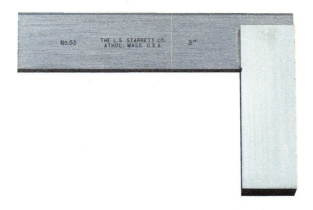

Figure C-148 Precision beveled-edge square (*Courtesy of The L.S. Starrett Co.*).

that the workpiece is perpendicular or not by comparison with the square. The actual amount of deviation from perpendicularity on the workpiece must be determined by other measurements. With the following group of squares, the deviation from perpendicularity can be measured directly. In this respect, the following instruments are not true comparison tools, since they can show a measurement directly.

Cylindrical Square

The direct-reading cylindrical square (Figure C-149) consists of an accurate cylinder with one end square to the axis of the cylinder. The other end is made slightly out of square with the cylindrical axis. When the nonsquare end is placed on a clean surface plate, the instrument is actually tilted slightly. As the square is rotated (Figure C-150), one point on the circumference of the cylinder will eventually come into perpendicularity with the surface plate. On a cylindrical square, this point is marked by a vertical line running the full length of the tool. The cylindrical square has a set of curved lines marked on the cylinder that permits deviation from squareness of the workpiece to be determined. Each curved line represents a deviation of .0002 of an inch over the length of the instrument.

Figure C-149 Cylindrical square.

Cylindrical squares are applied as follows. The square is placed on a clean surface plate and brought into contact with the part to be checked. The square is then rotated until contact is made over the entire length of the instrument. The deviation from squareness is determined by reading the amount corresponding to the line on the square that is in contact with the workpiece. Cylindrical squares are often used to check the accuracy of another square. Cylindrical squares range in size from 4 to 12 inches.

Diemaker's Square

A diemaker's square (Figure C-151) is used in applications such as checking the clearance angle on a die (Figure C-152). The instrument can be used with a straight or offset blade. A diemaker's square can be used to measure a deviation of up to 10 degrees on either side of the perpendicular.

Figure C-150 Principle of the cylindrical square.

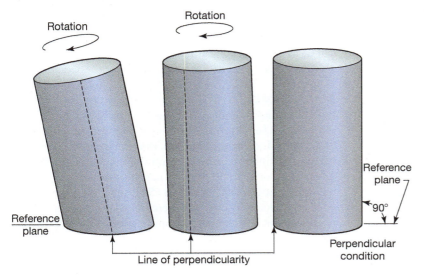

Principle of the cylindrical square

Figure C-151 Diemaker's square.

Figure C-152 Using the diemaker's square to check die clearance angle.

Micrometer Square The micrometer square (Figure C-153) is another type of adjustable square. The blade is tilted by means of a micrometer adjustment to determine the deviation of the part being checked.

The square is one of the few measurement tools that are essentially self-checking. If you have a workpiece with true parallel sides, as measured with a micrometer, one end can be observed under the beam of the square and the error observed. Then, the part can be rotated under the beam 180 degrees and rechecked. If the error is identical but reversed, the square is accurate. If there is a difference, except for simple

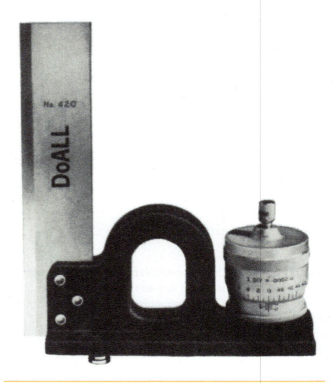

Figure C-153 Micrometer square (*Courtesy of the DoALL Company*).

reversal, the square should be considered inaccurate. It should be checked against a standard, such as a cylindrical square.

Indicators

Indicators are one of the useful tools for a machinist. There are two general types of indicators, dial indicators and dial test indicators. Both types have a spring-loaded spindle that when depressed moves the hand on the dial. The dial face is usually graduated in thousandths of an inch or subdivisions of thousandths. You might think that the indicator spindle movement corresponds exactly to the amount shown on the indicator face. This is not necessarily true. Dial test indicators should not be used to make direct linear measurements. Reasons for this will be developed in the information that follows. Dial indicators can be used to make linear measurements, but only if they are specifically designed to do so and under proper conditions.

Dial Indicators Discrimination of dial indicators typically ranges from .00005 to .001 inch. In metric dial indicators, the discrimination typically ranges from .002 to .01 mm. Indicator ranges, or the total reading capacity of the instrument, may commonly range from .003 to 2.000 inch, or .2 to 50 mm for metric instruments. On the balanced indicator (Figure C-154), the face numbering goes both clockwise and counterclockwise from zero. This is convenient for comparator applications where readings above and below zero need to be indicated.

The continuous reading indicator (Figure C-155) is numbered from zero in one direction. This indicator has a discrimination of .0005 inch and a total range of 1 inch. The small

Figure C-154 Balanced dial indicator.

Figure C-155 Dial indicator with 1 inch of travel.

center hand counts revolutions of the large hand. Note that the center dial counts each .100 inch of spindle travel. This indicator is also equipped with **tolerance hands** that can be set to mark a desired limit. Many dial indicators are designed for high discrimination and short range (Figure C-156). This indicator has .0001 discrimination and a range of .025 inch.

The **back-plunger** indicator (Figure C-157) has the spindle in the back at right angles to the face. This type of indicator usually has a range of about .200 inch with .001-inch discrimination. It is a popular type for use on a machine. The indicator usually comes with mounting accessories (Figure C-157).

Figure C-156 Dial indicator with .025-inch range and .0001-inch discrimination.

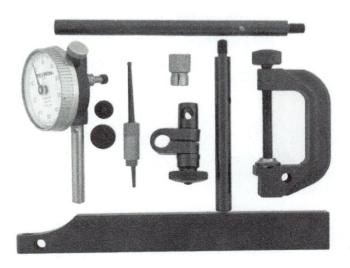

Figure C-157 Back-plunger indicator with mounting accessories.

Indicators are equipped with a **rotating face** or **bezel**. This feature permits the instrument to have a bezel lock. Dial indicators may have removable spindle tips, thus permitting use of different-shaped tips as required by the specific application (Figure C-158).

Care and Use of Indicators

Dial indicators are precision instruments and should be treated accordingly. They must not be dropped and should not be exposed to shocks. Dropping an indicator may bend the spindle and render the instrument useless. Shocks, such as hammering on a workpiece while an indicator is still in contact, may damage the delicate operating mechanism. The spindle should be kept free from dirt and grit, which can cause binding, damage, and false readings. It is important to check indicators for free travel before using. When an indicator is not in use, it should be stored carefully with a protective device around the spindle.

Indicators must be solidly mounted to be reliable. Indicators must be clamped or mounted securely when used on a machine tool. Various mounting devices are commonly used. Some have magnetic bases that permit an indicator to be attached at any convenient place on a machine tool. The permanent-magnet indicator base (Figure C-159) is a useful accessory. This type of indicator base is equipped with an adjusting screw that can be used to set the instrument to zero. Another useful magnetic base has a provision for turning off the magnet by mechanical means (Figure C-160). This feature makes for easy locating of the base prior to turning on the magnet. Bases that make use of flexible-link indicator holding arms are also used. Often, they are not adequately rigid for reliability. An indicator may also be clamped to a machine setup with suitable clamps.

Dial Test Indicators

Dial test indicators frequently have a discrimination of .0005 inch and a range of about .030 inch. The test indicator is frequently quite small (Figure C-161) so that it can be used to indicate in locations inaccessible to other indicators. The spindle or tip of the test indicator can be swiveled to any desired position. Test indicators are usually equipped with a **movement reversing lever**. This means that the indicator can be actuated by pressure from either side of the tip. The instrument need not be turned around. Test indicators, like dial indicators, have a rotating bezel for zero setting. Dial faces are generally of the balanced design.

Potential for Error in Using Dial Indicators

Indicators must be used with appropriate caution for reliable results. The spindle of a dial indicator usually consists of a gear rack that engages a pinion and other gears that drive the indicating hand. In any mechanical device, there is always clearance between the moving parts. There are also minute errors in

Figure C-158 Dial indicator tips with holder (*Courtesy of Rank Scherr–Tumico, Inc.*).

Figure C-159 Permanent-magnet indicator base (*Courtesy of the L.S. Starrett Co.*).

Figure C-160 Magnetic-base indicator holder with on/off magnet (*Courtesy of The L.S. Starrett Co.*).

Figure C-161 Dial test indicator.

the machining of the indicator parts. Consequently, small errors may creep into an indicator reading. This is especially true in long-travel indicators. For example, if a 1-inch-travel indicator with a .001-inch discrimination has plus or minus 1 percent error at full travel, the following condition could exist

if the instrument was to be used for a direct measurement: You wish to determine if a certain part is within the tolerance of .750 ± .003 inch. The 1-inch-travel indicator has the capacity for this measurement, but remember that it is accurate only to plus or minus 1 percent of full travel. Therefore, .01 × 1.000 inch is equal to plus or minus .010 or .020 total. That is why indicators are usually used for comparison and not for taking measurements (Figure C-162).

Of course, you will not know the amount of error on any specific indicator. This can be determined only by a calibration procedure. Furthermore, you would probably not use a long-travel indicator in this particular application. A moderate- to short-travel indicator would be more appropriate. Keep in mind that any indicator may contain some **travel error** and that by using a fraction of that travel, the amount of this error can be reduced considerably.

In the introduction to this section you learned that the axis of a linear measurement instrument must be in line with the axis of measurement. If a dial indicator is misaligned with the axis of measurement, the following condition will exist: In Figure C-163, line AC represents the axis of measurement, and line AB represents the axis of the dial indicator. If the distance from A to C is .100 inch, then the distance from A to B is obviously longer, since it is the hypotenuse of triangle ABC. The angle of misalignment, angle A, is equal to 20 degrees. The distance from A to B can then be calculated by the following:

$$AB = \frac{.100}{\cos A}$$
$$AB = \frac{.100}{.9396}$$
$$AB = .1064 \text{ inch}$$

This shows that a movement along the axis of measurement results in a much larger movement along the instrument axis. This **error** is known as cosine error and must be kept in mind when using dial indicators. Cosine error increases as the angle of misalignment increases.

Figure C-162 Potential for errors in indicator travel.

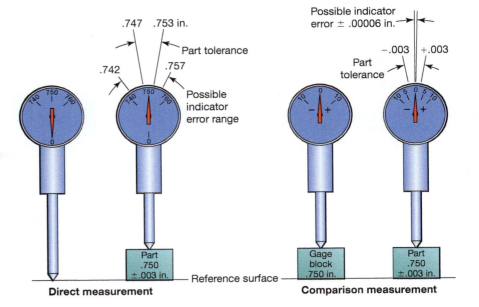

Direct measurement

Comparison measurement

Figure C-163 Cosine error.

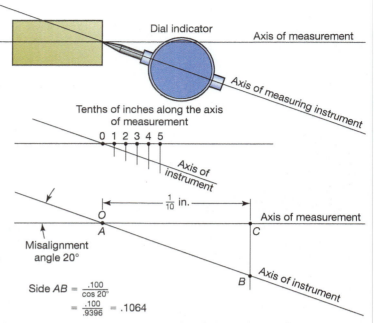

Tenths of inches along the axis
of measurement

0 1 2 3 4 5

Axis of instrument

$\frac{1}{10}$ in.

Misalignment angle 20°

$$\text{Side } AB = \frac{.100}{\cos 20°}$$
$$= \frac{.100}{.9396} = .1064$$

A movement of .100 along the axis of measurement results in a movement of .1064 along the axis of the measuring instrument.

When using dial test indicators, watch for arc versus chord length errors (Figure C-164). The tip of the test indicator moves through an arc. This distance may be considerably greater than the chord distance of the measurement axis. Dial test indicators should not be used to make direct measurements. They should only be used for comparison applications.

USING DIAL TEST INDICATORS IN COMPARISON MEASUREMENTS

The dial test indicator is very useful in making comparison measurements with the height gage and height transfer micrometer. Comparison measurement using a vernier height gage is made as follows:

Step 1 Set the height gage to zero and adjust the test indicator until it also reads zero when in contact with the surface plate (Figure C-165). It is important to use a test indicator for this procedure.

Step 2 Raise the indicator and adjust the height gage vernier until the indicator reads zero on the workpiece (Figure C-166). Read the dimension from the height scale.

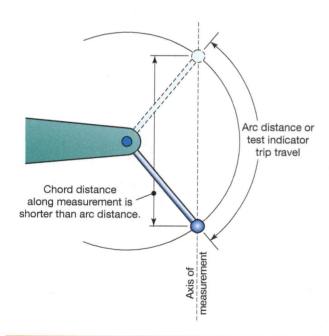

Arc distance or test indicator trip travel

Chord distance along measurement is shorter than arc distance.

Axis of measurement

Figure C-164 Potential for error in dial test indicator tip movement.

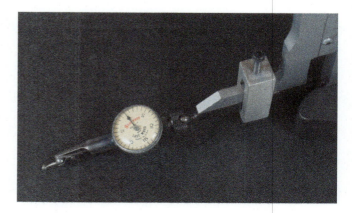

Figure C-165 Setting the dial test indicator to zero on the reference surface.

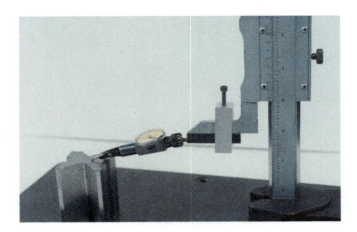

Figure C-166 Using the dial test indicator and vernier height gage to measure the workpiece.

Comparison measurement can also be made using a precision height gage (Figure C-167). The precision height gage shown consists of a series of rings moved by the micrometer spindle. The ring spacing is an accurate 1 inch, and the micrometer head has a 1-inch travel. Discrimination of the typical height micrometer is .0001 inch. Precision height gages come in various height capacities and sometimes have riser blocks as accessories. The following procedure is used:

Step 1 The height transfer micrometer is adjusted to the desired height setting. The test indicator is zeroed on the appropriate ring (Figure C-167). A height transfer gage, vernier height gage, or other suitable means can be used to hold the indicator.

Step 2 The indicator is then moved over to the planer gage, which is adjusted until the test indicator reads zero (Figure C-168).

Test indicators can also be used to make comparison measurements in conjunction with a digital electronic height gage (Figure C-169).

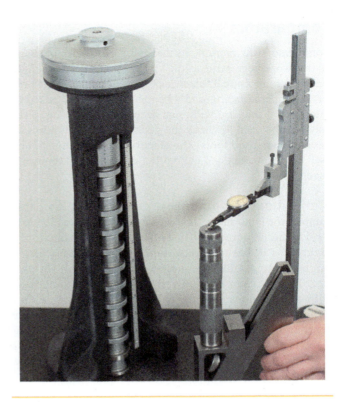

Figure C-168 Transferring the measurement to the planer gage.

Figure C-167 Setting the dial test indicator to the precision height gage.

Figure C-169 Electronic digital height gage (*Mitutoyo America Corp.*).

Comparators

Comparators are instruments that compare the size or shape of the workpiece with a known standard. Types include dial indicator, optical, electrical, and electronic comparators. Comparators are used when parts must be checked to determine if they are within an acceptable tolerance. They may also be used to check the geometry of such things as threads, gears, and tools. The electronic comparator can be used for inspection and calibration of measuring tools and gages.

Dial Indicator Comparators The dial indicator comparator is no more than a dial indicator attached to a rigid stand (Figure C-170). The comparator indicator in Figure C-170 has tolerance hands. The tolerance hands can be set to establish an upper and lower limit for part size. The indicator is set to zero at the desired dimension with gage blocks (Figure C-171). When using a dial indicator comparator, keep in mind the **potential for error** in indicator travel and instrument alignment along the axis of measurement. Once the indicator has been set to zero, parts can be checked for acceptable tolerance (Figure C-172).

Figure C-171 Setting the dial indicator to zero using gage blocks.

Figure C-170 Dial indicator comparator.

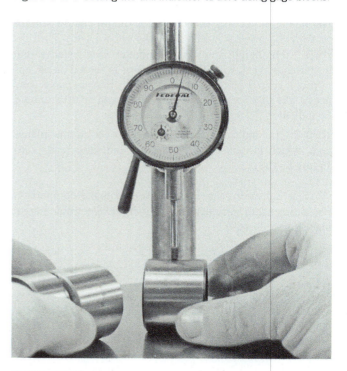

Figure C-172 Using the dial indicator comparator.

Electronic Digital Indicator Comparators Electronic technology is used for indicator comparators as well as many other instruments. With digital readouts, these instruments are easy to read and calibrate, and they demonstrate high reliability and high discrimination. One example is the digital indicator comparator (Figure C-173).

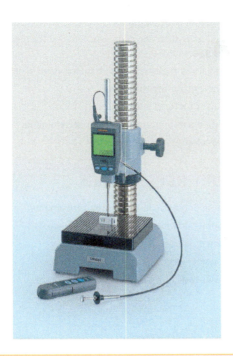

Figure C-173 Digital electronic comparator (*Mitutoyo America Corp.*).

Profile Projectors

A profile projector (Figure C-174) projects a greatly magnified profile of the object to be measured onto a screen. Various templates or patterns in addition to graduated scales can be placed on the screen and compared with the projected shadow of the part. The optical comparator is particularly useful for inspecting the geometry of screw threads, gears, and formed cutting tools.

Electronic and digital readouts are also used on the profile projector. These features increase this instrument's reliability, ease of operation, metric and inch selection, high discrimination, and high sensitivity.

Electronic Gaging

Electronic comparators convert dimensional change into changes of electric current or voltage. These changes are read on a readout. Economical mass production of high-precision parts requires that fast and reliable measurements be made so that out of tolerance parts can be sorted from those within tolerance.

The electronic comparator (Figure C-175) is a sensitive instrument. It is used in a variety of comparison measuring applications. The comparator is set to a gage block by first adjusting the coarse adjustment. This mechanically moves the measuring probe. Final adjustment to zero is accomplished electronically. This is one of the unique advantages of such instruments. The electronic comparator shown has three scales. The first scale reads $\pm.003$ inch at full range, with a discrimination of .0001 inch. The second scale reads $\pm.001$ inch at full range, with a discrimination of .00005 inch. The third scale reads $\pm.0003$ inch at full range, with a discrimination of .00001 inch.

Figure C-174 Profile projector with built-in linear scales (*Mitutoyo America Corp.*).

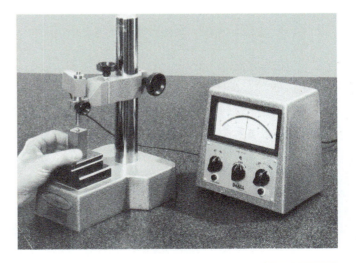

Figure C-175 Electronic comparator with maximum discrimination of .00001 inch (*Courtesy of DoALL Company*).

SHOP TIP

There can be a loss of reliability in making comparison measurements, especially when it involves the physical transfer and comparison of one tool with another, as comparison techniques often do. Control as many variables as possible, and if necessary, repeat the measurement several times. If you can obtain the same result from several repetitions of the measurement, reliability and the likelihood that the measurement is correct will be greatly enhanced.

SELF-TEST

1. What is comparison measurement?
2. Define *cosine error.*
3. How can cosine error be reduced?

Which comparison tools would you use to measure the following? You may need more than one comparison tool and a measurement device to check it with.

4. A milled slot 2-inch-wide with a tolerance of + or −.002 inch.
5. A height transfer measurement
6. The shape of a form lathe cutter
7. Checking a combination square to determine its accuracy
8. The diameter of a $1\frac{1}{2}$ inch hole

INTERNET REFERENCES

Information on comparison measuring instruments:

http://www.prattandwhitney.com

http://dialindicator.com

http://www.starrett.com

http://mitutoyo.com

Using Gage Blocks

Manufacturing can only be successful if machinists everywhere measure to the same standards. Gage blocks help ensure that measurement instruments are calibrated to recognized international standards. They are one of the most important measuring tools in the shop. Gage blocks are used in the metrology laboratory, toolroom, and machine shop for calibration of precision measuring instruments and for precise setups of angles and other operations.

OBJECTIVES

After completing this unit, you should be able to:

- Describe the care required to maintain gage block accuracy.
- Wring gage blocks together correctly.
- Disassemble gage block combinations and prepare the blocks properly for storage.
- Calculate combinations of gage block stacks with and without wear blocks.
- Describe gage block applications.

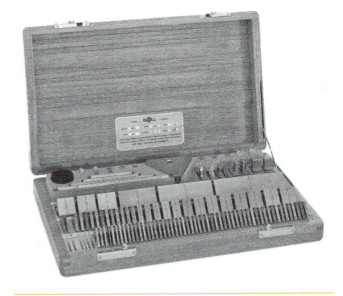

Figure C-176 Gage block set with accessories (*Courtesy of DoALL Company*).

GAGE BLOCK TYPES AND GRADES

Gage blocks are available individually or in sets. A typical gage block set will contain 81 to 88 blocks ranging in thickness from .050 to 4.000 inches. The total measuring range of the set is over 25 inches (Figure C-176). Larger and smaller sets are also available. Sets of extra-long blocks are available permitting measurements to 84 inches. Metric gage block sets contain blocks ranging from .5 to 100 mm. Angular gage blocks can measure from 0 to 30 degrees. Gage blocks for linear measurement are either rectangular or square. Gage blocks are available in various levels of accuracy.

The three grades of gage blocks are grade 1 (laboratory), grade 2 (inspection), and grade 3 (shop) (Table C-1). Grades 1 and 2 are manufactured grades. Grade 3 blocks are compromises between grades 1 and 2. Sets of grade 3 can be purchased, or they may be created by using out-of-tolerance grade 2 and 3 gage blocks that do not meet the tolerances for grade 1. Grade 3 blocks are not used by the inspection or gage laboratory but are acceptable in the shop for many typical measurement applications.

Table C-1 Gage Block Tolerances

Size	Tolerances in Microinches (.000001 inch) for Gage Block Grade		
	Grade 1 (Formerly AA)	Grade 2 (Formerly A+)	Grade 3 (between A and B)
1 inch and less	+2−2	+4−2	+8−2
2 inches	+4−4	+8−4	+16−8
3 inches	+5−5	+10−5	+20−10
4 inches	+6−6	+12−6	+24−12

THE VALUE OF GAGE BLOCKS

Gage blocks are one of the physical standards that can closely approach exact dimensions. This makes them useful as measuring instruments to check other measuring tools. From the table on gage block tolerances, you can see that the length tolerance on a grade 1 block is ±.000002 inch. This is only four millionths of an inch total tolerance. Such a small amount is hard to visualize. Consider that the thickness of a page of this book is about .003 inch. Compare this amount with total gage block tolerance and you will note that the page is 750 times thicker than the tolerance. This should indicate that a gage block would be useful for checking a measuring instrument with .001 or even .0001-inch discrimination.

As a further demonstration of gage block value, consider the following example. You want to establish a distance of 20 inches as accurately as possible. Using a typical gage block set, imagine a hypothetical situation in which each block has been made to the plus tolerance of .000002 inch over the actual size. This situation would not exist in an actual gage block set, as the tolerance of each block is most likely bilateral. If it required 30 blocks to make up a 20-inch stack, the cumulative tolerance would amount to .000060 inch (60 millionths). As you can see, the 20-inch length is still extremely close to actual size. In a real situation, because of the bilateral tolerance of the gage blocks, the 20-inch stack will actually be much closer to 20.000000 than to 20.000060 inches. Because the gage block is so close to actual size, cumulative tolerance has little effect even over a long distance.

PREPARING GAGE BLOCKS FOR USE

Gage blocks are rugged and yet delicate. During their manufacture, they are put through many heating and cooling cycles that stabilize their dimensions. For a gage block to function, its **surface** must be **extremely smooth** and **flat**.

Gage blocks are almost always used in combination, known as the *gage block stack*. The secret of gage block use lies in the ability to place two or more blocks together in such a way that most of the air between them is displaced. This is the process of **wringing**. Once this is accomplished, atmospheric pressure will hold the stack together. The space or interface between wrung gage blocks is known as the *wringing interval*. Properly wrung gage block stacks are essential if cumulative error is to be avoided. Two gage blocks simply placed against each other will have an air layer between them. The thickness of the air layer will greatly affect the accuracy of the stack.

Before gage blocks can be wrung, they must be properly prepared. Burrs, foreign material, lint, grit, and even dust from the air can prevent proper wringing and permanently damage a gage block. The main cause of gage block wear is the wringing of poorly cleaned blocks. Preparation of gage blocks should conform to the following procedure.

Step 1 Remove the desired blocks from the box and place them on a lint-free tissue. The gage blocks should be handled as little as possible so that heat from fingers will not temporarily affect size.

Step 2 A gage block must be cleaned thoroughly before wringing. This can be done with an appropriate cleaning solvent or commercial gage block cleaner (Figure C-177). Use the solvent sparingly, especially if an aerosol is applied. The evaporation of a volatile solvent can cool the block and cause it to temporarily shrink out of tolerance.

Step 3 Dry the block immediately with a lint-free tissue.

Step 4 Any burrs on a gage block can prevent a proper wring and possibly damage the highly polished surface.

Figure C-177 Applying gage block cleaner.

Figure C-178 Using the conditioning stone.

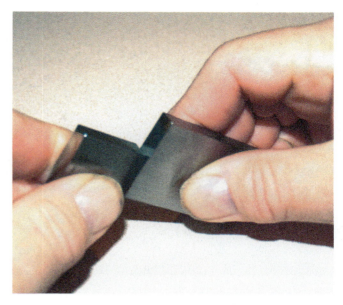

Figure C-179 Overlapping gage blocks prior to wringing.

Deburring is accomplished with a special deburring stone or dressing plate (Figure C-178). The block should be lightly moved over the stone using a single back-and-forth motion. After deburring, the block must be cleaned again.

WRINGING GAGE BLOCKS

Gage blocks should be wrung immediately after cleaning. If more than a few seconds elapse, dust from the air will settle on the wringing surface. The block may require dusting with a very soft brush. To wring rectangular gage blocks, the following procedure should be used.

Step 1 Place the freshly cleaned and deburred mating surfaces together and overlap them about $\frac{1}{8}$ inch (Figure C-179).

Step 2 Slide the blocks together while lightly pressing together. During the sliding process, you should feel an increasing resistance. This resistance should then level off.

Step 3 Position the blocks so that they are in line (Figure C-180).

Step 4 Make sure that the blocks are wrung by holding one block and releasing the other. Hold your hand under the stack in case the block should fall (Figure C-181).

Square gage blocks require the same cleaning and deburring as rectangular blocks. Square gage blocks are wrung by a slightly different technique. Since they are square, they should be placed together at a 45-degree angle. The upper block is then slid over the lower block while at the same time twisting the blocks and applying a light pressure.

Figure C-180 Wrung gage blocks in line.

During the wringing process, heat from the hands may cause the block stack to expand. Generally, gage blocks should be handled as little as possible to minimize heat problems.

If during the wring the blocks tend to slide freely, slip them apart immediately and recheck cleaning and deburring. If the blocks fail to wring after the proper preparation procedure has been followed, they may be warped or have surface imperfections.

Figure C-181 Checking for a proper wring.

USING WEAR BLOCKS

When gage blocks are used in applications that require a lot of contact, it is advisable to use **wear blocks**. For example, if you were using a gage block stack to calibrate a large number of micrometers, wear blocks would reduce the wear on the gage block. Wear blocks are usually included in typical gage blocks sets. They are made from the tungsten carbide, which is very hard and wear resistant. A wear block is placed on one or both ends of a gage block stack to protect it from possible damage by direct contact. Wear blocks are usually .050 or .100 inch thick.

PREPARING GAGE BLOCKS FOR STORAGE

Gage blocks should not be left wrung for extended periods of time. The surface finish can be damaged in a few hours if the blocks were not exceedingly clean at the time of wringing. After use, the stack should be unwrung and the blocks cleaned once again. Blocks should be handled with tissue (Figure C-182). Each block should be sprayed with a suitable gage block preservative and replaced in the box. The entire set should then be lightly sprayed with gage block preservative (Figure C-183).

CALCULATING GAGE BLOCK COMBINATIONS

In making a gage block stack, use a minimum number of blocks. Each surface or wringing interval between blocks can increase the opportunity for error, and poor wringing can make this error relatively large. To check your wringing ability, assemble a combination of blocks totaling 4.000 inches. Compare this stack with the 4.000-inch gage block using a .0001 indicator.

Table C-2 gives the specifications of a typical set of 83 gage blocks. Note that they are in four series.

Figure C-182 Handle gage blocks only with tissue.

Figure C-183 Applying gage block preservative.

In the following example, it is desired to construct a gage block stack to a dimension of 3.5752 inches. Wear blocks will be used on each end of the stack.

3.5762	First, eliminate two .050-inch wear blocks.
−.100	
3.4762	
−.1002	Then, eliminate the last figure right, by
3.3760	subtracting the .1002-inch block.
.126	Once again, eliminate the last figure right, by
3.250	subtracting the .126-inch block.
−.250	Eliminate the last figure right, using the
3.000	.250-inch block.
−3.000	Eliminate the 3.000 inches with the 3.000-inch
0.000	block.

Therefore, the blocks required to construct this stack are

Quantity	Size
2	.050-inch wear blocks
1	.1002-inch block
1	.126-inch block
1	.250-inch block
1	3.000-inch block

Table C-2 Typical 83-Piece Gage Block Set

First: .0001 Series—9 Blocks											
.1001	.1002	.1003	.1004	.1005	.1006	.1007	.1008	.1009			
Second: .001 Series—49 Blocks											
.101	.102	.103	.104	.105	.106	.107	.108	.109			
.110	.111	.112	.113	.114	.115	.116	.117	.118			
.119	.120	.121	.122	.123	.124	.125	.126	.127			
.128	.129	.130	.131	.132	.133	.134	.135	.136			
.137	.138	.139	.140	.141	.142	.143	.144	.145			
.146	.147	.148	.149								
Third: .050 Series—19 Blocks											
.050	.100	.150	.200	.250	.300	.350	.400	.450	.500	.550	
.600	.650	.700	.750	.800	.850	.900	.950				
Fourth: 1.000 Series—4 Blocks											
		1.000	2.000	3.000	4.000						
					2		.050 wear blocks				

As a second example, we shall construct a gage block stack of 4.2125 without wear blocks.

$$
\begin{array}{r}
4.2125 \\
-.1005 \\
\hline
4.1120 \\
-.112 \\
\hline
4.0000 \\
-4.0000 \\
\hline
0.0000
\end{array}
$$

Blocks for this stack are

Quantity	Size
1	.1005
1	.112
1	4.000

GAGE BLOCK APPLICATIONS

Gage blocks are used in setting sine bars for establishing precise angles. The use of the sine bar is discussed in the unit on angular measure. Gage blocks are used to set other measuring instruments such as snap gages (Figure C-184). The proper blocks are selected for the desired dimension and the stack is assembled. Since this is a direct contact application, wear blocks should be used. The stack is then used to set the gage (Figure C-185).

There are many accessories available for gage block measurement (Figure C-186). Accessories include scribers, bases, gage pins, and screw sets for holding the stack together. A torque screwdriver is used when screws are used to secure gage block stacks. A torque screwdriver assures that the correct amount of pressure will be used on the gage block stack. Gage blocks may be used with gage pins (Figure C-187), for direct gaging or for checking other measuring instruments.

Figure C-184 Gage and wear blocks for setting a snap gage.

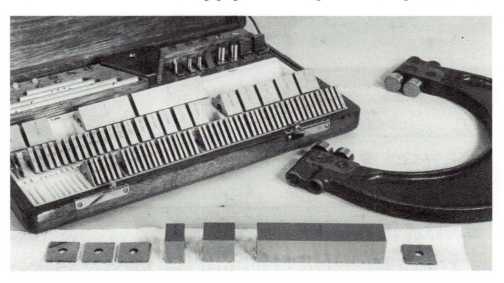

Figure C-185 Setting a snap gage using gage blocks.

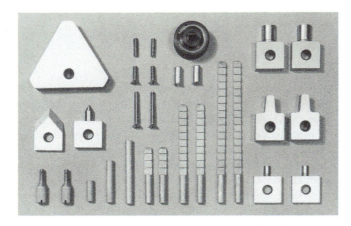

Figure C-186 Gage block accessories.

Figure C-187 Gage block stack with accessory gage pins (*Courtesy of DoALL Company*).

SHOP TIP

Gage blocks are very valuable in industry. Always handle them with care and use proper procedures for cleaning, wringing, and storing. The gage block can add a high degree of reliability and accuracy to many machine shop measurements, machine and inspection setups, and calibration procedures.

SELF-TEST

1. What is a *wringing interval*?
2. Why are wear blocks frequently used in combination with gage blocks?
3. As related to gage block use, what is meant by the term *normalize*?

4. What length tolerances are allowed for the following grades of gage blocks (under 2.000-inch sizes): grade 1; grade 2; grade 3?
5. What is a conditioning stone and how is it used?
6. What does the term *microinch* regarding surface finish of a gage block mean?
7. Describe handling precautions for the preservation of gage block accuracy.
8. What gage blocks are necessary to assemble a stack equal to 3.0213 without using wear blocks?
9. List gage blocks necessary for a stack equal to 1.9643 with wear blocks.
10. Describe at least two gage block applications.

INTERNET REFERENCES

Information on gage blocks:

http://starrett.com

http://www.mitutoyo.com/

http://www.fvfowler.com

Using Angular Measuring Instruments

Angular measurement is as important as linear measurement. The same principles of metrology apply to angular measure as to linear measure. Angular measuring instruments have various degrees of discrimination. Angular measuring instruments require the same care and handling as any other of your precision tools.

OBJECTIVES

After completing this unit, you should be able to:

- Identify common angular measuring tools.
- Read and record angular measurements using a vernier protractor.
- Calculate sine bar elevations and measure angles using a sine bar and adjustable parallels.
- Calculate sine bar elevations and establish angles using a sine bar and gage blocks.

TYPES OF ANGLES

A machinist will need to measure acute angles, right angles, and obtuse angles (Figure C-188). Acute angles are less than 90 degrees. Obtuse angles are more than 90 degrees but less than 180 degrees. Ninety-degree or right angles are generally measured with squares. However, the amount of angular deviation from perpendicularity may have to be determined. This requires that an angular measuring instrument be used.

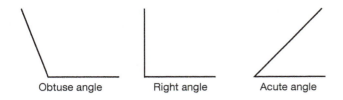

Obtuse angle   Right angle   Acute angle

Figure C-188 Acute, right, and obtuse angles.

Straight angles, or those of 180 degrees, generally fall into the category of straightness or flatness and are measured by other types of instruments.

UNITS OF ANGULAR MEASURE

In the inch system, the unit of angular measure is the degree.

Full Circle = 360 degrees

1 degree = 60 minutes of arc ($1° = 60'$)

1 minute = 60 seconds of arc ($1' = 60''$)

In the metric system, the unit of angular measure is the **radian**. A radian is a unit of angle measurement, equal to an angle at the center of a circle whose arc is equal in length to the radius (Figure C-189). Since the circumference of a circle is equal to $2\pi r$ (radius), there are 2π radians in a circle. Converting one radian to degrees gives the equivalent:

$$1 \text{ radian} = \frac{360}{2\pi r}$$

Assuming a radius of 1 unit:

$$1 \text{ radian} = \frac{360}{2\pi}$$
$$= 57°17'44'' \text{ (approximately)}$$

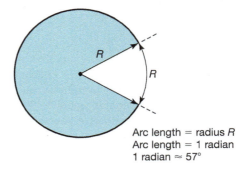

Arc length = radius R
Arc length = 1 radian
1 radian ≈ 57°

Figure C-189 Radian measure.

It is unlikely that you will come in contact with much radian measure. All the common comparison measuring tools you will use read in degrees and fractions of degrees. Metric angles expressed in radian measure can be converted to degrees by a conversion formula.

REVIEWING ANGLE ARITHMETIC

You may find it necessary to perform angle arithmetic. Many calculators have the ability to work with angles.

Adding Angles

Angles are added just like any other quantity. One degree contains 60 minutes. One minute contains 60 seconds. Any minute total of 60 or larger must be converted to degrees. Any second total of 60 or larger must be converted to minutes.

EXAMPLES

$$35° + 27° = 62°$$
$$3°15' + 7°49' = 10°64'$$

Since $64' = 1°4'$, the final result is $11°4'$.

$$265°15'52'' + 10°55'17'' = 275°70'69''$$

Since $69' = 1'9''$ and $70' = 1°10'$, the final result is $276°11'9''$.

Subtracting Angles

When subtracting angles where borrowing is necessary, degrees must be converted to minutes and minutes must be converted to seconds.

EXAMPLES

$$15° - 8° = 7°$$
$$15°3' - 6°8' \text{ becomes}$$
$$14°63' - 6°8' = 8°55'$$
$$39°18'13'' - 17°27'52'' \text{ becomes}$$
$$38°77'73'' - 17°27'52'' = 21°50'21''$$

Decimal Angles

When using digital readout measuring and dividing equipment for angular measurement, you may encounter angle fractions in decimal form rather than minutes and seconds. For example, 30 degrees 30 minutes would be $30\frac{1}{2}$ degrees or 30.5 degrees. You should familiarize yourself with angle decimal notation and the methods of converting minutes and seconds to their decimal equivalents.

Decimal fractions of angles are calculated by the following procedures. Since there are 60 minutes in each degree, a fractional portion of a degree becomes a fraction of 60.

EXAMPLES

32 minutes is $\frac{32}{60}$th of 1 degree. Converting to a decimal fraction:

$$\frac{32}{60} = .5333 \text{ degrees}$$

23 seconds is $\frac{23}{60}$th of one minute

$$\frac{23}{60} = .3833 \text{ minute}$$

To convert angle decimal fractions to minutes and seconds, multiply the decimal portion times 60.

EXAMPLES

45.5 degrees converted to minutes: The decimal portion is $.5 \times 60 = 30$ minutes, or 45.5 degrees = 45 degrees and 30 minutes

32.75 degrees converted to minutes: The decimal portion is .75 degree $\times 60 = 45$ minutes, or 32 degrees and 45 minutes

Some calculators can convert decimal fractions of angles to minutes and seconds. Other calculators require computations to perform conversions.

ANGULAR MEASURING INSTRUMENTS

Plate Protractors

Plate protractors have a discrimination of one degree and are useful in applications such as layout and checking the point angle of a drill (Figure C-190).

Bevel Protractors

The bevel protractor is part of the machinist's combination set. This protractor can be moved along the rule and locked in any position. The protractor has a flat base, permitting it to rest squarely on the workpiece (Figure C-191). The combination set protractor has a discrimination of one degree.

Universal Bevel Vernier Protractor

The universal bevel vernier protractor (Figure C-192) is equipped with a vernier that permits discrimination of $\frac{1}{12}$ of a degree, or 5 minutes. The instrument can measure obtuse angles (Figure C-193). The acute attachment facilitates the measurement of angles less than 90 degrees (Figure C-194).

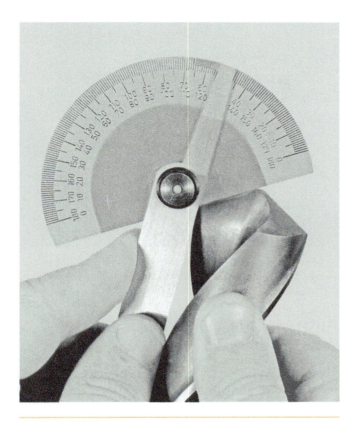

Figure C-190 Plate protractor measuring a drill point angle.

Figure C-191 Using the combination set bevel protractor.

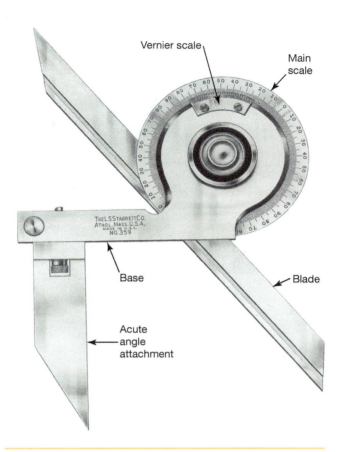

Figure C-192 Parts of the universal bevel vernier protractor (*Courtesy of The L.S. Starrett Co.*).

Figure C-193 Measuring an obtuse angle with the vernier protractor (*Courtesy of The L.S. Starrett Co.*).

When used in conjunction with a vernier height gage, the universal bevel vernier protractor allows angular measurements to be made that would be difficult by other means (Figure C-195).

Vernier protractors are read like any other vernier instrument. The main scale is divided into whole degrees. These are marked in four quarters of 0 to 90 degrees each. The vernier divides each degree into 12 parts, each equal to 5 minutes of arc.

To read the protractor, determine the nearest full degree mark between zero on the main scale and zero on the vernier scale. Always read the vernier in the same direction as you read the main scale. Determine the number of the vernier

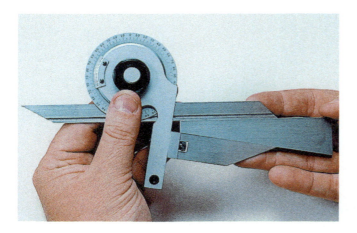

Figure C-194 Using the acute angle attachment (*Courtesy of The L.S. Starrett Co.*).

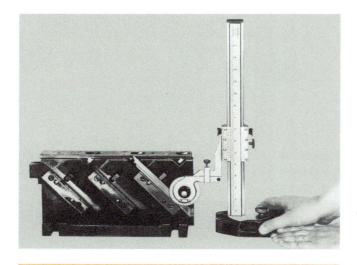

Figure C-195 Using a vernier protractor with a vernier height gage (*Courtesy of The L.S. Starrett Co.*).

coincident line. Since each vernier line is equal to 5 minutes, multiply the number of the coincident line by 5. Add this value to the main scale reading.

EXAMPLE READING

The protractor shown in Figure C-196 has a magnifier so that the vernier may be seen more easily.

> Main scale reading = 56°
> Vernier coincident at line 6
> 6 * 5 minutes = 30 minutes
> Total reading: 56°30′

For convenience, the vernier scale is marked at 0, 30, and 60, indicating minutes.

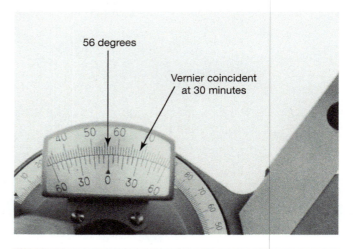

Figure C-196 Vernier protractor reading of 56 degrees and 30 minutes.

RIGHT-TRIANGLE TRIGONOMETRY: THE SINE RATIO

Trigonometry deals with triangular measurement. A working knowledge of right-triangle trigonometry is essential to being a competent machinist.

Right-triangle trigonometry deals with the relationships or ratios of the lengths of the sides of a right triangle. In Figure C-197, right triangle *ABC* is shown in the standard notation, where the angles are identified by uppercase letters (*A*, *B*, *C*) and the sides of the triangle are identified by lowercase letters (*a*, *b*, *c*). In standard triangle notation, the side opposite an angle carries the same letter as the angle. Therefore, side *a* is opposite angle *A*. Side *b* is opposite angle *B*, and side *c*, the long side, is opposite angle *C*, which is the right or 90-degree angle.

The two acute angles, *A* and *B*, are always less than 90 degrees. Angle *C* is always a right or 90-degree angle formed by the two short legs *a* and *b*. Side *c* is the longest side of the triangle and is called the **hypotenuse**. When you add the three angles in the right triangle, the sum is always 180 degrees. Since angle *C* is 90 degrees, the sum of angles *A* and *B*

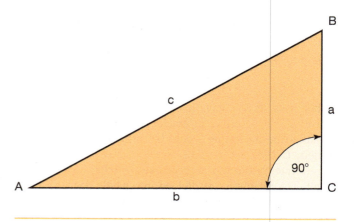

Figure C-197 Standard notation for a right triangle.

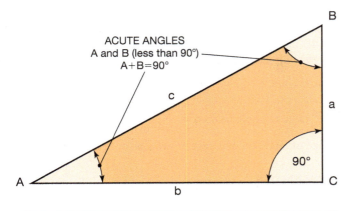

Figure C-198 Acute angles of a right triangle.

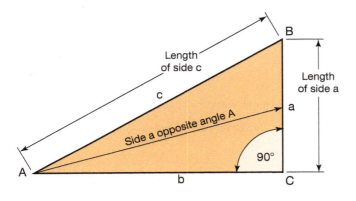

Figure C-199 The ratio of the length of side *a* to side *c* defines the sin of angle *A*.

is therefore 90 degrees (Figure C-198). This is the first piece of basic information that you need to know about right triangles.

$$\angle A + \angle B = 90°$$

The sine bar is an example of a tool used to establish and measure angles as functions of linear units. The sine bar makes use of the **sine ratio** in right-triangle trigonometry. The **sine (sin)** of an angle is defined as the ratio of the length of the side opposite an acute angle to the length of the hypotenuse, or

$$\sin = \frac{\text{side opposite}}{\text{hypotenuse}}$$

For example, the sine of angle A is defined as

$$\sin \text{ of angle } A \text{ or } \sin A = \frac{\text{side opposite}}{\text{hypotenuse}} = \frac{a}{c}$$

The size of the acute angles *A* and *B* in the right triangle controls the length of the triangle's sides. Similarly, the length of the sides *a* and *b* controls the size of the acute angles (Figure C-199). Therefore, the size of angles and the length of sides may be expressed in terms of each other. This is to say that for a given specified length of sides, as measured in linear units, the acute angles must be a corresponding size, as measured in degrees of arc.

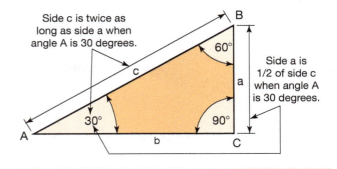

Figure C-200 30-60-90 right triangle.

Consider the following specific case. In the triangle shown in Figure C-200, the ratio of side *a* to side *c*, the hypotenuse, is .5 to 1. In other words, the side opposite angle *A*, side *a*, is exactly half as long as side *c*. For this to be true, angle *A* must then be exactly 30 degrees, which then makes angle *B* equal to 60 degrees. If the ratio of *a* to *c* were not .5 to 1, the size of angles *A* and *B* would have to be different as well. The ratio of *a* to *c* (*a/c*) defines the **sine of angle** A. When *a* is divided by *c*, the result is .5/1 = .5. You can use a trigonometric function table or a calculator to determine what angle corresponds to a sine value of .5. Look in the trig function table under the sine column and locate the .5000 entry. You will see that the corresponding angle is 30 degrees. On a calculator .5 is entered and the inverse (or second function) sin key is pressed. The displayed result is 30 degrees.

In this specific case, where angle *A* and angle *B* are 30 and 60 degrees, respectively, the length of side *c*, the hypotenuse, is always twice that of side *a*, no matter what the actual length is of sides *a* and *c*. Another way to state this relationship is simply to say that the side opposite a 30-degree angle is always half the length of the hypotenuse. This information is useful for a machinist when dealing with the 30-60-90 right triangle, which finds many applications in machine shop work. In later units, methods to calculate the length of side *b* will be developed.

Using the Sine Bar

The sine bar (Figure C-201) is a typical example of a tool used to both measure and establish precise angles by inferring their size in degrees as a function of linear measurement. The sine bar consists of a hardened and precision-ground steel bar with a cylinder attached to each end. The distance between the centerlines of the cylinders is precisely established at either 5 or 10 inches. This is done to make the sine bar calculations more convenient. The sine bar is placed on the surface plate and becomes the hypotenuse of a right triangle. The angle to be measured or established is the acute angle between the surface plate and the sine bar's top surface. This becomes angle *A* in the standard triangle. By elevating the other end of the sine bar, a specific amount, you control the ratio between the side opposite the acute angle formed between the surface plate and the sine bar's top surface (Figure C-202). Since this defines the sine ratio

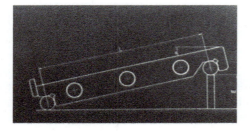

Figure C-201 Sine bar (*Mahr Gage Company*).

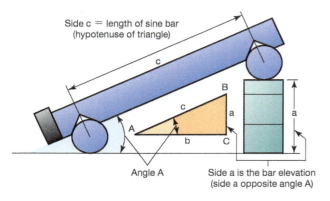

Figure C-202 Elevating the sine bar with gage blocks.

(opposite side/hypotenuse), the amount of bar elevation varies the length of the opposite side and also varies the acute angle between surface and sine bar.

Because the length of the sine bar is a fixed dimension, either 5 or 10 inches, it is easy to determine the angle from trig table entries or from a calculator. Since the sine is defined by the equation

$$\sin A = \frac{\text{side opposite}}{\text{hypotenuse}}$$

Transposing the equation gives

side opposite = (hypotenuse) × (sine of the angle desired)

On the sine bar, this becomes

bar elevation (the side opposite angle *A*)
= [bar length (the hypotenuse)]
× (sine of the angle desired)

Since standard sine bars are either 5- or 10-inch long, the bar-length multiplier in the preceding equation is simply 5 or 10 times the sine of the angle. If you are establishing an angle with the side bar, use the equation

bar elevation = (bar length) × (sine of the angle desired)

$$\text{sine of angle desired} = \frac{\text{elevation}}{\text{length of sine bar}}$$

size of angle (in degrees) = arc sine or inverse sine

EXAMPLE

Determine the elevation for 30 degrees using a 5-inch sine bar.

bar elevation = 5 inches × sine 30°
= 5 × .5
= 2.500 inch

This means that if the bar was elevated 2.500 inch, an angle of 30 degrees would be established.

EXAMPLE

Determine the elevation for 42 degrees using a 5-inch sine bar.

bar elevation = 5 inches × sine 42°
= 5 × .6691
= 3.3456 inches

Determining Workpiece Angle Using the Sine Bar and Measuring Workpiece Angle Using the Sine Bar and Adjustable Parallel

An angle may be measured using the sine bar and adjustable parallel. An adjustable parallel can be used to elevate the sine bar (Figure C-203). The workpiece is placed on the sine bar, and a

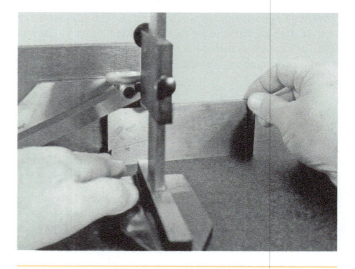

Figure C-203 Placing the adjustable parallel under the sine bar.

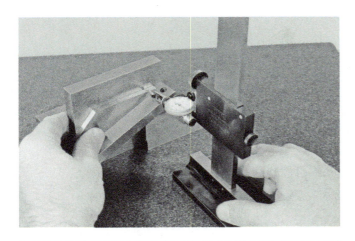

Figure C-204 Setting the test indicator to zero at the end of the workpiece.

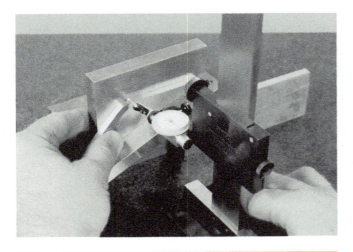

Figure C-205 Checking the zero reading at the opposite end of the workpiece.

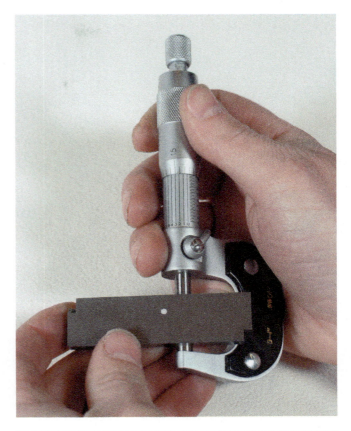

Figure C-206 Measuring the adjustable parallel with an outside micrometer.

Figure C-207 Wringing the gage block stack for the sine bar elevation.

dial test indicator is set to zero on one end of the part (Figure C-204). The parallel is adjusted until the dial indicator reads zero at each end of the workpiece (Figure C-205). The parallel is then removed and measured with a micrometer (Figure C-206). To determine the angle of the workpiece, simply transpose the sine bar elevation formula and solve for the angle.

bar elevation $=$ bar length $\times$ sine of the angle desired
sine of the angle desired $=$ elevation/bar length
sine of angle $=$ 1.9935 (micrometer reading/5)
sine $=$ .3987
angle $=$ 23°29′48″

Establishing Angles Using the Sine Bar and Gage Blocks

Extremely precise angles can be measured or established by using gage blocks to elevate the sine bar. Bar elevation is calculated in the same manner. The required gage blocks are properly prepared, and the stack is wrung (Figure C-207). The gage block stack totaling 1.9940 inch is placed under the bar (Figure C-208). This will establish an angle of 23°30′11″ using a 5-inch sine bar. The angle of the workpiece is checked using a dial test indicator (Figure C-209).

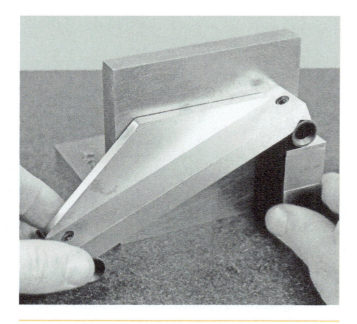

Figure C-208 Placing the gage block stack under the sine bar.

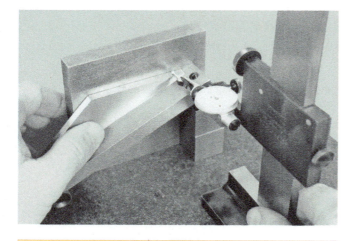

Figure C-209 Checking the workpiece using the dial test indicator.

Sine Bar Constant Tables

The elevations for angles up to about 55 degrees can be obtained directly from a table of sine bar constants. Such tables can be found in the Appendix of this text and in machinists' handbooks. The sine bar constant table eliminates the need to perform a trigonometric calculation. The sine bar table may discriminate only to minutes of arc. If discrimination to seconds of arc is required, it is better to calculate the amount of sine bar elevation required.

Figure C-210a Vernier reading for C-210a.

Figure C-210b Vernier reading for C-210b.

Figure C-210c Vernier reading for C-210c.

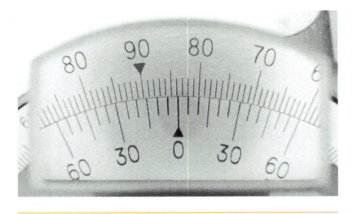

Figure C-210d Vernier reading for C-210d.

Figure C-210e Vernier reading for C-210e.

SELF-TEST

1. Name two angular measuring instruments with one degree of discrimination.
2. What is the discrimination of the universal bevel protractor?
3. Describe the use of the sine bar.
4. Read and record the vernier protractor readings in Figures C-210a to C-210e.
5. Calculate the required sine bar elevation for an angle of 37 degrees. (Assume a 5-inch sine bar.)
6. A 10-inch sine bar is elevated 2.750 inch. Calculate the angle established to the nearest minute.
7. How do 10-inch and 5-inch sine bars affect the height of gage block stacks?
8. What gage block stack would establish an angle of 35 degrees using the 5-inch sine bar?
9. What gage block stack would establish an angle of 23.5 degrees using the 5-inch sine bar?
10. A 10-inch bar is elevated 2.5 inch. What angle is established?

INTERNET REFERENCES

Information on angular measurement tools:

http://fvfowler.com

http://www.mitutoyo.com/

http://starrett.com

Quality in Manufacturing

This unit will examine the topic of quality in a typical manufacturing enterprise as well as methods to improve processes. ISO 9001, an international quality standard, will be covered first. Then the basics of statistical quality control and charting will be covered.

OBJECTIVES

After completing this unit, you should be able to:

Explain what ISO 9001 is and how it impacts a typical machine shop.

- Explain how ISO may affect a machinist.
- Explain calibration.
- Describe the general terms and purpose of SPC.
- Explain the importance of accurate, relevant data.
- Explain terms such as *attribute*, *variable*, and *histogram*.
- Explain the advantages and disadvantages of attribute and variable data.
- Organize data to make it easier to understand.
- Code data.
- Explain terms such as *assignable*, *chance*, *special*, and *common causes*.
- Explain the rules of variation.
- Define and calculate averages.
- Define and calculate measures of variation, such as range and standard deviation.
- Explain terms such as *normal distribution* and *bell curve*.
- Predict scrap rates by using statistical means.

Introduction

ISO 9001

ISO 9001 is an international standard that specifies requirements for a quality management system (QMS). A quality management system could be described as a system that details how the company operates. Companies can use the ISO standard to demonstrate the ability to consistently provide products and/or services that meet customer and regulatory requirements.

The ISO 9001 quality standards were first published in 1987 by the International Organization for Standardization (ISO). ISO consists of the national standard bodies of more than 160 countries. The current version of ISO 9001 was released in September 2015. ISO 9001:2015 can be applied to any organization, regardless of the size or industry.

More than one million enterprises from more than 160 countries have applied the ISO 9001 standard requirements to their Quality Management Systems (QMSs).

The fundamental principle of the standard is that there are some basic things any enterprise must do to be successful and meet their customer's requirements. The standard allows the individual enterprise to create their own system to meet the ISO requirements. This enables companies to develop a QMS that meets their own needs and that covers the basic requirements of ISO 9001. To be sure that their systems do meet the requirements of the standard, companies can choose to be certified. To be certified, a company hires a third-party auditor to come in and make sure their quality management system meets the requirements. If the system does meet requirements and the company is following their system, they can get certified by the third party. The certification is not permanent. They must be audited on a regular basis by the third-party auditor to retain their certification.

ISO 9001 is based on the plan-do-check-act (PDCA) methodology. ISO 9001 system is a process-oriented approach to documenting and reviewing the structure, responsibilities, and procedures required to achieve effective quality management in a company.

Changes introduced in the 2015 revision to ISO 9001 were meant to adapt to the changing environments in which organizations operate. Some of the changes in the ISO 9001:2015 revision include the introduction of some new terminology, some structural changes, an emphasis on risk- and opportunity-based thinking, increased leadership requirements, and making it more applicable for service-type enterprises.

The ISO 9001:2015 revision replaced the word *product* with the term *goods and services*. This acknowledges the growth of the service industry, and allows users in service industries to adapt the standard to their unique requirements.

The requirements for documentation have been reduced in the 2015 revision. The terms *documents* and *records* have been replaced with the term *documented information*. The term *continual improvement* has been replaced with *improvement*.

Risk management, change management, and knowledge management are given more emphasis in the ISO 9001:2015 revision. The revision allows organizations greater flexibility and recognizes the need for businesses to integrate their QMS into their overall business strategy.

The ISO 9001:2015 revision is less rigid than previous ISO 9001 versions. The ISO 9001:2015 revision incorporates more business concepts and management terminology. Companies have more control over what documentation meets their needs.

Benefits of ISO 9001:2015

ISO 9001 helps companies ensure that their customers consistently receive high quality products and services. This can help assure satisfied customers, management, and employees.

ISO 9001:2015 specifies the requirements for an effective quality management system. Companies using ISO 9001 systems have found that the standard helps:

Organize and improve processes,
Develop a Quality Management System (QMS),
Achieve satisfied customers, management, and employees,
Achieve improvement.

THE ISO 9001:2015 STANDARD HAS 10 SECTIONS

1. Scope
   What and who the standard covers.
2. Normative references
   Other standards that were used to establish a company's quality management system would be listed here.
3. Terms and definitions
   Definitions of terms referred to in the standard.
4. Context of the organization
   Requirements related to determining the purpose and the direction of the organization.
5. Leadership
   Requirements related to the commitment of the top management, quality policy, and roles and responsibilities of personnel.
6. Planning
   Requirements related to planning to address risks and opportunities and achieving quality objectives.
7. Support
   Requirements for people, infrastructure, communication, and documented information needed to achieve the purposes of the organization.
8. Operation
   Requirements for identifying customer requirements, design, delivery, and post-delivery support.
9. Performance evaluation
   Requirements related to customer satisfaction, analysis, internal audits, and management review.
10. Improvement
    Requirements related to nonconformities, corrective actions, and risk-based continual improvement.

A typical ISO 9001 system in a company would have a quality manual (not required in ISO 9001:2015) and procedures (not required) that detail how important processes are done and by whom. It should be noted that ISO does not guarantee a good quality system. Some companies have developed quality systems that meet the ISO requirements, but are way too cumbersome and paper intensive. ISO does not require complex, paper intensive systems.

A machinist interacts with the ISO system in several ways. The company's quality system will specify how the machinist gets the information to make the parts. This is typically a job folder or routing which might include blueprints, process sheets, setup sheets, inspection details, and so on. A company's ISO system probably details instructions to the operator on what happens if a piece is scrapped. There might be a form to be filled out, who to report it to, etc.

Calibration

Calibration is one of the main ways in which an operator interacts with the ISO QMS. To make parts that meet a customer's specifications, a company must have accurate gages to inspect them. In addition to the gages being accurate, the operator must know how to use them correctly. For example, if one operator uses a micrometer like a C-clamp, they will not get the same readings as another operator that uses a light "feel" when using the same micrometer.

Imagine what might happen if a micrometer was to lose its accuracy for a few months. Any operator using that micrometer to inspect parts would have been getting bad readings. They would have been accepting parts that were out of specification and rejecting parts that were in specification. Out of specification parts would have then been shipped out to the customer. This is very expensive. The customer might inspect and reject the parts. They would then have to wait for replacement parts. The customer might not discover the bad parts and have products fail because of them. The customer might decide they need a new supplier for the parts.

ISO requires that instruments used to inspect parts for quality must be calibrated on regular intervals. Companies can set their own intervals for calibration as long as they prove to be effective. For example, a company might want all micrometers calibrated once each year. Every micrometer would have a sticker on it that would show an identifying number for that micrometer and a date when it is to be calibrated next. The instrument cannot be used if it is beyond the specified calibration date.

The Calibration Log

Each instrument in the calibration system would be entered into a calibration log system. The person responsible for calibration would gather the instruments that needed calibration before their recalibration date. Calibration involves comparing a measuring instrument to standards. A micrometer, for example, might be calibrated by making measurements with very accurate gage blocks and comparing the measurement to the actual size of the gage block. The gage blocks used for calibration would be much more accurate than the micrometer. The calibration gage blocks are usually kept in a secure area and only used for calibrating instruments.

A standard used for calibration is generally at least 10 times more accurate than the instrument being calibrated. The standards are also checked for accuracy on regular intervals.

In many shops the standards would be sent out every few years to companies that check them against even more accurate standards. This helps assure that every company has standards that are checked against very accurate national standards. When the measuring instrument comes in for its regular calibration, it is compared against the standard. If it is accurate it is noted in the log. If it is out of calibration it is also noted in the log. If it is out of calibration it means that the parts it checked between this calibration and the previous calibration may not have been measured correctly.

Many out of specification parts may have been made and sent to the customer. If that is the case, the company must try to assess the situation and remedy anything it can, although that becomes difficult when the parts are gone. The measuring instrument must either be recalibrated or fixed and recalibrated. If it cannot be fixed, it is scrapped and noted in the calibration log. The company should also decide if calibration intervals should be shortened if this has been happening with other similar instruments.

Many companies have operators to do a daily or weekly quick check on their measuring instruments against a shop standard between the calibration intervals. This helps prevent the situation where an instrument may have lost its accuracy between calibration intervals.

FUNDAMENTALS OF STATISTICS IN MANUFACTURING

Statistics can be a very practical, simple tool when properly applied. The most important fundamental in the effective use of statistics is data. Data must be accurate and relevant.

INTRODUCTION TO STATISTICS IN MANUFACTURING

Manufacturing is very competitive. Enterprises have to be competitive in quality, price, delivery, and customer satisfaction. Statistical process control is one way to improve and keep machining processes under control so that they produce accurate, cost-effective parts. Machinists should understand the basics of statistics so they can apply the concepts to their work.

Cost of Scrap

Imagine a small company: Acme Manufacturing. Acme makes a widget that costs 95 cents. Acme sells the widget for $1 and makes 5 cents on every one. What types of expenses go into the 95 cents that it costs to make the widget? Certainly, the material cost and the labor cost to make it, and also some of the cost of the machine, tooling, and maintenance. Acme also has to pay the supervisor, inspectors, sales people, secretaries, president, and so on. Acme also has to pay for the building, utilities, and many other things.

What does it cost if an Acme machinist makes a defective widget? Does it cost Acme the 5-cent profit? Yes, it does. Does it cost Acme any more than 5 cents? Yes, Acme lost the 95 cents it had invested in that part, and probably more. The machinist has to fill out a scrap ticket, the supervisor has to take time to deal with it, and Acme might have to order more raw material.

How many parts does Acme have to make to make up for this loss? Acme would have to make at least 20 parts. Acme makes 5 cents on each part, so 20 * 5 will make up for the basic $1.00 loss. Acme, however, does not make any money on those 20 parts. Acme just tries to recover the loss. Worse than that, while Acme is making the 20 parts they can't be making anything else, so Acme lost potential profit there, too. As you can see, the costs of scrap are very high! Statistics and statistical thought can help reduce these costs.

Data

Data is crucial in making manufacturing successful because statistical process control cannot work without accurate data. If we have inaccurate data because of inaccurate measuring equipment (out of calibration) or because people are using it improperly, we cannot make quality parts. There are several important considerations for data collection.

Make sure that the data is accurate. Consider the gages, methods, and personnel.

Take action based on the data.

Types of Data   There are two types of data: attribute and variable.

Attribute Data

Attribute is the simplest type of data. The product either has the characteristic (attribute) or it doesn't. Attribute data is go/no-go type data. Go/no-go gages are often used to check hole sizes. If the go end of the gage fits in the hole and the no-go end doesn't, we know the hole is within tolerance.

This type of inspection gives two piles of parts: parts that are within the tolerance and parts that are not (good parts and bad parts). Attribute data does not tell us how good or how bad the parts are. We only know they are good or bad.

Variable Data

Variable data can be much more valuable. It not only tells us *whether* parts are good or bad, it tells us *how* good or how bad they are.

Variable data is generated using measuring instruments such as micrometers, verniers, and indicators. If we are measuring with these types of instruments, we generate a range of sizes, not just two (good or bad) as with attribute data. We could end up with many piles of part sizes.

Coding Data

Numbers can become difficult to work with when they have many digits. For example, look at the numbers below. Can you add them in your head and find the average?

1.7431 1.7426 1.7437 1.7438 1.7439

Even though there are only five numbers, it is cumbersome. It would be easier if we could work with one-digit numbers. Coding is the way we can do this (see Figure C-211). In this table the specified part size on the blueprint was 1.000.

Consider the following blueprint specification: 1.000 ± .005.

Set the desired size of 1.000 be equal to zero (1.000 = 0). A part whose size is 1.000 will have a coded value of zero. The coding for this part size is set to the third place to the right of the decimal point (.001). That place equals ones. If a part is 1.002, we will call it a 2 because it is two larger than

1.005 = 5	1.006 = 6	.999 = -1	.994 = -6
1.009 = 9	1.002 = 2	1.001 = 1	1.000 = 0
.999 = -1	.997 = -3	1.002 = 2	1.001 = 1

Figure C-211 Coded sample part sizes.

our desired size of 1.000 (zero). If the next part were .999, we would call it −1.

How do you decide which place to assign the value of 1? You simply look at the specification and the number of decimal places. If the specification has three decimal places (5.000, for example), you would make 1 equal one thousandth of an inch. If there were four decimal places, a 1 would equal .0001.

Which numbers would you rather work with: the actual size or a coded value? If you needed to find the average of the numbers, you could just figure the average for the coded values. Because coded values are much easier to work with, most companies use coded values when they use statistical methods.

Graphic Representation of Data

Look at the part sizes in Figure C-212.

What conclusions can you draw? It can be confusing to look at large amounts of data. Data is easier understood if we have a picture of it (graphical representation). It would also be easier if it were coded. One method of presenting data graphically is called a *histogram*.

Study the data below. These numbers represent coded values for part sizes that were produced.

1, 4, 1, −1, 1, 3, 2, 3, 3, 2, 2, 2, 4, 1,
−1, 1, 1, 0, 0, 1, −1, 0, −1, 0, 0, −2, 0, 0,
−1, −3, −2, 1, −4

Now look at Figure C-213. This histogram makes it easy to see patterns in data. You can instantly see that the most common sizes were 0 and 1. You can see that the average part size was between 0 and 1. This histogram is much easier to analyze than the list of data. The only thing we lose in the histogram is the time value. We do not know if the zeroes were the first parts, the last parts, or if they were distributed throughout the production.

BASICS OF VARIATION

Is it true that all things vary and that no two things are exactly alike? Yes. Fingerprints, snowflakes, and so on—no two are exactly alike. Variation is normal. The same is true of a stamping die that makes steel washers. If we look at the

1.023	1.021	1.025	1.021	1.021	1.019	1.023	1.023	1.019	1.022	1.017
1.021	1.020	1.024	1.023	1.012	1.021	1.014	1.022	1.020	1.023	1.023
1.019	1.018	1.022	1.022	1.019	1.021	1.021	1.018	1.019	1.021	1.018

Figure C-212 Some sample part sizes.

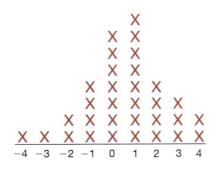

Figure C-213 Histogram of data from Figure C-212.

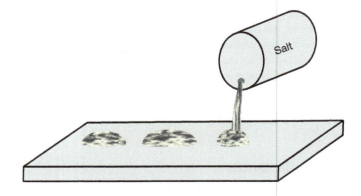

Figure C-214 A simple process.

washers, they all look the same. If we measure them with a steel rule, they all measure the same. But if we use a micrometer, we would find differences in every one. An appropriate measuring instrument will find differences between any two things.

The first rule of variation is straightforward: *No two things are exactly alike.*

Rule number two, *variation can be measured,* is also straightforward. No matter what our product or process, no two will be alike (rule 1), and we will be able to measure and find differences in every part (rule 2).

This assumes that we use an appropriate tool to measure the parts. It also assumes that we have been trained in the correct use of the measuring tool.

The third rule of variation is that *individual outcomes are not predictable.* What would happen if we flipped a coin 10,000 times? We would predict that we would get approximately 5,000 heads and 5,000 tails. We could predict that approximately 50 percent of flips would be heads and 50 percent tails. This would be a very accurate prediction.

What if we flipped the coin 10 times? Our prediction might not be as accurate. We might get seven heads and three tails (not a very good prediction). But if we flipped a larger amount of times, we could predict quite accurately that heads and tails would each occur about 50 percent of the time. Assuming we didn't cheat, 50 percent would be heads and 50 percent tails.

Can we predict that the next flip will be a tail? No! We can never predict an individual outcome.

For example, if we were running a lathe and the last piece was 1.001, can we predict that the next piece will be 1.001? No, because we cannot predict individual outcomes.

Rule number four states that *groups form patterns with definite characteristics.*

Think of a simple process involving a large salt container (see Figure C-214). The process is to dump 3 ounces of salt from a point 6 inches above the table. The product of the process is the salt piles. If we were very careful, the piles of salt would appear to be the same. The diameter and height would look identical. We could predict the size of the pile if we were to run the process again. In fact, we could make fairly accurate predictions. We could not predict where one grain of salt would fall, however (rule 3). But rule 4 tells us we can make predictions about groups.

Would all of the piles be the exact same size? No. Rule 1 tells us no two things are exactly alike.

Why wouldn't each salt pile be exactly the same? There are many reasons.

The height of drop varies. The amount of salt dropped varies. How fast the 3 ounces were poured varies. Air currents in the room change. The humidity in the room is not consistent. The shapes of salt grains vary. The surface of the table is not exactly the same all over. Some of these reasons would cause large variations, some small. Can you think of any other reasons?

There are many reasons why the piles will not be the same. Statistics tells us, however, that a small number of the reasons cause the majority of the variation. In fact, 20 percent of the reasons will cause 80 percent of the variation. This tells us that if we identify the few key causes, we can dramatically reduce variation. Note also that this applies to all manufacturing processes.

CHANCE AND ASSIGNABLE VARIATION

There are two types of causes: chance and assignable.

Assignable causes of variation are those causes that we can identify and fix. For example, the height from which the salt was dropped varied. How could anyone maintain exactly 6 inches of height every time? Now that we have identified it as an assignable cause, we could make a simple stand to maintain the correct height for the process and eliminate height as a cause of variation.

The amount of salt dropped also varied. Could we expect a person to drop exactly 3 ounces of salt every time? The amount of salt is also an assignable cause. How could we remove this cause of variation?

Statistical methods can be used to identify assignable causes of variation.

Chance causes of variation always exist. Chance causes of variation are those minor reasons which make processes vary. We cannot quantify or even identify all of the chance causes. Room humidity changes may affect the process. Normal temperature fluctuation might affect the process.

Normal fluctuations in air currents will affect the process. We really can't separate the effects of these chance causes. Some may even cancel out the effects of others.

You can never eliminate all chance causes of variation. If you are able to identify a cause and its effect, it becomes an assignable cause.

Statistical methods will help us identify assignable causes. Chance causes cannot be separated and evaluated. Chance causes, often called *common causes*, are reasons for variation that cannot be corrected unless the process itself is changed.

For example, if we moved our salt process to a temperature and humidity-controlled room, we would minimize the effects of humidity and temperature. It takes process change to correct common causes.

Assignable causes are often called *special causes*. These are things that go wrong with a process. For example, the operator starts to pour the 3 ounces but runs out of salt after 1 ounce. The operator refills the container and pours 2 ounces more. This is a special cause of variation. It is something out of the ordinary that occurred in the process. The operator could solve this problem, without a process change, by just making sure the container is full before each pour.

This is the main difference between special and common causes of variation. Special causes are things that are not normal in the process. In other words, something has changed in the process. The operator can often identify these problems and correct them without management action or process change. Common causes of variation are causes that are inherent in the process. The only way to correct the effect of the common causes is to change the process.

It should be clear that variation is normal and that all things vary. It should also be clear that predictions about individual outcomes (next flip of the coin) cannot be made. Predictions about groups of outcomes, however, can be made. Next, we look at some ways to describe the characteristics of groups of parts.

AVERAGE (MEAN)

One thing that is helpful in understanding a group of data is what the average or mean is. *Mean* is just another term for average.

If the heights of 30 students in a class were measured, the average (mean) height could be calculated. The 30 heights would be added up and then divided by 30 (the number of students in the class). They also have standard notation for data. One person's height, for example, would be called an "X." The letter X is the symbol for one data value (one height in this example). The notation for the average of the data values is $\bar{X}$ (pronounced *x-bar*). Whenever you see a letter with the bar symbol above the letter, it means average or mean. A letter with two bars above it means average of the averages.

Example: Heights of five people in inches:

$$x = 60, x = 70, x = 75, x = 80, x = 70$$
$$\text{Total} = 355/5 = 71 \text{ inches}$$

This means that the average of the individual xs was 71 inches. Remember: average and mean are the same.

MEASURES OF VARIATION

All things vary. It would be desirable to have terms that we could use to describe how much a process varies.

Range

One term used to describe variation is *range*. Range is usually symbolized with the letter "R." The range is simply the largest value in the sample minus the lowest value in the sample.

Example 1: A person shopping for a new car looked at five cars. Their prices were $8,000; $15,000; $9,500; $8,500; $12,000. The range of values would equal $15,000 minus $8,000. The range of car prices that this person looked at was $7,000.

Example 2: A food manufacturer measured five consecutive cans of a product and found the weights to be 7.1, 6.9, 7.2, 7.0, and 7.2 ounces. The range would be the highest value (7.2) minus the lowest value (6.9), or .3 ounces.

Range is a very useful, simple value and is a very useful look at variation. It gives us a good quick look at how much a process or sample varies. If we say a process range is .005, we have a good idea of how much the process varies.

STANDARD DEVIATION

The second term we can use to describe variation is *standard deviation*. Standard deviation can be thought of as the average variation of a single piece in a process. While this is a very rough definition, it may be more understandable than the technical one. If we had a sample of pieces and the mean was 10 and the standard deviation was 1, we could say that the average piece varies about 1 from the mean of 10. A statistician would define standard deviation as being the square root of the average square deviation of each variate from the arithmetic mean. (Forget this definition, by the way.) It is more useful and practical to think of standard deviation as the *average variation of a single piece*.

There is an elaborate formula to calculate standard deviation. The formula is not as complex as the definition, but it is tedious and time consuming, and mistakes are easy to make. Fortunately, many calculators will calculate standard deviation. All you have to do is input the values, and the calculator outputs the mean (average) and standard deviation.

Sample data:

$$5, 7, 9, 6, 4, 2, 1, 7, 5, 4$$

These data values were entered into a calculator, and the standard deviation for the sample was found to be approximately 2.4. The calculated mean (average) was 5.

Calculators can calculate standard deviation for a population or for a sample. Standard deviation for a sample is represented as S, SD, or $N - 1$. You should use the standard deviation for a sample, although they would both be very close. Standard deviation is a very useful term. It can help us visualize processes.

NORMAL DISTRIBUTION

If we chose 25 adult men at random and measured their height, some would be very tall and some very short, but most would be about average size. If we plotted a histogram of their heights, it would look something like the one shown in Figure C-215.

Most processes produce parts that would be normally distributed. If we measured a dimension on 25 parts from a lathe and then drew a histogram, we would expect a "bell" shape or normal distribution. A few parts would be large, a few small, but most of them would be average.

There are some very useful things about normal distributions (bell curves). Consider an example. Assume you measured the height of 25 adult men at random. Their mean height was 5 feet 10 inches (see Figure C-216). The standard deviation of their heights was also calculated and found to be 2 inches. The bell is normally drawn as being 6 standard deviations wide. In this case, the standard deviation was 2 inches, so the bell would be drawn with a width of 6 * 2 or 12. The bell is then broken into six areas.

The two areas closest to the middle of the bell each contain 34 percent of all the people's heights. The next two each contain 14 percent of the heights. The last two each

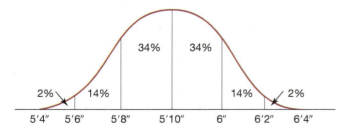

Figure C-215 A histogram of the heights of 25 men taken at random. The sizes are in inches.
This distinctive shape is called a *normal distribution or bell curve*. A normal distribution looks like a bell.

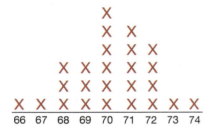

Figure C-216 Bell curve for the heights of 25 randomly chosen men. The mean height for this sample was 5 feet 10 inches tall. The standard deviation was calculated and found to be 2 inches. This means that, based on this sample, 34 percent of all men would be between 5 feet 10 inches and 6 feet tall. We can make all kinds of predictions about men's heights. We could say that approximately 14 percent of all men would be between 5 feet 6 inches and 5 feet 8 inches tall. If our sample was truly random, we could make very good predictions

contain approximately 2 percent. This is very useful because the same relationship exists for any normal process. The percentages will always be the same.

Bell curves are set up so that if we know the standard deviation, we can make predictions from the data. Thirty-four percent of all adult males' heights should be between the mean and plus 1 standard deviation. Thirty-four percent of all adult males' heights should fall between the mean and minus 1 standard deviation. Fourteen percent should fall between plus 1 standard deviation and plus 2 deviations. For this example, we know that the standard deviation equals 2 inches, so we could predict that 68 percent of all adult males will be between 5 feet 8 inches and 6 feet tall.

Statisticians have found that 99.7 percent of all part sizes, heights, or whatever we are measuring will be between ±3 standard deviations for any normal process.

Consider another example. A lathe is producing pins. The outside diameter was measured on 40 pins from the process as they were being run. The standard deviation was measured and found to be .001. The mean (average) was found to be .300.

Consider the bell curve in Figure C-217. We could now apply these actual part sizes to the bell curve. The mean is .300. This lets us predict that (assuming we don't change the process) 48 percent of all parts should be between .298 and .300.

What percent would be between .298 and .302? (96 percent); .302 and .306? (approximately 2 percent); .297 and .298? (approximately 2 percent).

This is very valuable information. We can make very accurate predictions about how this process will run in the future.

Consider the example of people's heights again. Could we make predictions about the heights of all people in Wisconsin from our data? No, because our data was based on adult males only. We could make pretty good predictions about the adult male population of Wisconsin.

The other point to remember is that the more data we have, the better our prediction can be. If we sampled five people's heights, our prediction would not be very good. The larger the sample size, the better predictions we can make.

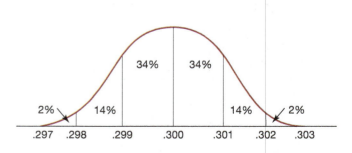

Figure C-217 A bell curve for a process producing parts of sizes .297 to .303 inches. Note that 34 percent are between .299 and .300 inches. Another 34 percent are between .300 and .301. What percentage of parts are between .302 and .303? (2 percent).

We can make very accurate predictions from relatively small samples, and it is more cost effective to use samples. Time is money, and inspection takes time. Statistical methods will help us make predictions that are almost 100 percent accurate.

SELF-TEST

1. What is the ISO 9001:2015 standard?

2. True or false? An ISO 9001 quality management system is the same in every company.

3. True or false? ISO 9001 is an American quality management standard.

4. True or false? A company must be audited on a regular basis by a third-party auditor to retain their certification.

5. True or false? Calibration is done to make sure all measuring instruments are accurate in an enterprise.

6. True or false? Out of calibration measuring instruments must be scrapped in an ISO 9001 certified company.

7. True or false? An operator cannot use a measuring instrument that is past its calibration date.

8. True or false? ISO 9001 systems are very paper intensive because everything must be documented.

9. True or false? All measuring instruments that are used to inspect for quality must be included in the calibration log.

10. True or false? Each operator must maintain a calibration log for their measuring equipment.

11. What is attribute data?

12. What is variable data?

13. What are some rules concerning the collection and use of data?

14. What is coding and why is it used?

15. What is a histogram and why is it used?

16. Code the following data: Blueprint specification = 1.126. (Hint: 1.126 = 0.)

Blueprint Specification = 1.126				
1.124 =	1.128 =	1.123 =	1.127 =	1.121 =
1.119 =	1.127 =	1.125 =	1.119 =	1.127 =
1.129 =	1.126 =	1.118 =	1.121 =	1.116 =

17. Code the following:

Blueprint Specification = 1.2755			
1.2752 =	1.2749 =	1.2752 =	1.2754 =
1.2759 =	1.2750 =	1.2761 =	1.2752 =
1.2748 =	1.2756 =	1.2752 =	1.2756 =
1.2753 =	1.2755 =	1.2747 =	1.2749 =

18. Code the following:

Blueprint Specification = 2.105			
2.109 =	2.103 =	2.108 =	2.113 =
2.102 =	2.101 =	2.096 =	2.104 =
2.101 =	2.100 =	2.100 =	2.111 =
2.098 =	2.100 =	2.099 =	2.109 =

19. Code the following:

Blueprint Specification = 5.00				
5.03 =	5.06 =	5.02 =	5.02 =	5.07 =
5.02 =	5.05 =	5.01 =	5.04 =	5.03 =
4.98 =	4.97 =	5.00 =	5.01 =	4.93 =

20. Using the coded data from question 18, construct a histogram.

21. Using the data from question 19, construct a histogram.

22. Define and explain variation.

23. How does an assignable cause of variation differ from a chance cause of variation?

24. Briefly define the following key words.
   a. mean
   b. histogram
   c. normal distribution
   d. bell curve
   e. range

25. Consider the following data. Blueprint specification = 2.250.

1	2	3	4	5	6
2.252	2.249	2.248	2.252	2.246	2.250
2.252	2.249	2.254	2.253	2.248	2.249
2.253	2.243	2.252	2.251	2.247	2.252
2.250	2.248	2.251	2.250	2.246	2.248
2.252	2.249	2.247	2.250	2.249	2.251

   a. Code the data. (Hint: 1 should equal 2.251.)
   b. Find the sample standard deviation.

26. Blueprint specification = 2.0000.

Subgroup 1	Subgroup 2	Subgroup 3
2.0005	2.0005	2.0002
1.9997	2.0004	2.0004
2.0009	2.0001	2.0000
2.0006	2.0004	1.9995
1.9999	2.0002	2.0001

   a. Code the data. (Hint: 2.0005 = 5.)
   b. Find the mean.
   c. Find the sample standard deviation.

27. Draw a bell curve and label with standard deviations and mean. Label the percentages for each deviation. Note: You don't have data concerning the actual mean and standard deviation. Draw a generic bell curve.

28. Consider the following coded data.
   5, 0, 0, 1, 1, 4, 2, −1, −1, −1, 2, −1, 0, 0, 0, −1, −2, 1, 2, 0, −2, −1, −1, −3, −2
   Use a calculator.
   a. Calculate the mean.
   b. Calculate the sample standard deviation.

29. Consider the following data.
9, 0, 1, −1, 2, 8, −1, 0, 0, 3, 7, 4, 2, −3, 8, 2, −4, 3, −5, 5, 3, −2, −2, 0, 3
   a. Calculate the mean.
   b. Calculate the sample standard deviation.

30. Consider the following data.
−3, 2, −1, −2, −4, −1, −2, 0, −2, 3, 0, −1, −1, 4, 2, 2, 0, 0, 1, 4
   a. Calculate the mean.
   b. Calculate the sample standard deviation.

31. A machining process is studied and the mean and standard deviation were calculated: x = 5, s = 2 (coded data).
   a. Draw a bell curve.
   b. Draw lines where the 6 standard deviations would be.
   c. Label them with actual sizes from this process. (Hint: 99.7 percent of all parts should lie between −1 and +11.)
   d. Label the percentages.

32. Holes are measured after a drill press operation. The mean was found to be 1 (coded data) and the standard deviation was found to be 3 (coded data).
   a. Draw a bell curve for this data.
   b. If 1,000 parts are run, how many will be between −5 and −2?
   c. How many will be between −2 and +4?
   d. If our tolerance was −2 to +7, how many scrap parts would we have?

33. Consider the following coded data.
3, 3, 3, 3, 5, 4, 2, 4, 2, 1, 5, 4, 3, 2, 4, 6, 1, 4, 3, 2
   a. Construct a histogram.
   b. Does it look like a normal distribution?

34. Consider the following coded data.
1, 4, 2, 5, 2, 5, 1, 1, 4, 1, 3, 5, 5, 1, 5, 1, 5, 3, 5, 1, 5, 4, 2, 1, 0
   a. Construct a histogram.
   b. Does it look like a normal distribution?

Statistics in Manufacturing

Statistics can be a very valuable tool in manufacturing. This unit will cover machine capability and constructing average and range charts. There are many benefits to charting. But even if a company does not use charts for processes, machinists can benefit from understanding manufacturing statistics and applying them to their work.

OBJECTIVES

After completing this unit, you should be able to:

- Define terms such as bell curve, capability, average and range, and so on.
- Calculate capability for a machine.
- Explain the benefits of thinking statistically.
- Calculate limits for an average and range chart.
- Construct an average and range chart.
- Explain the rules for machine adjustment using a chart.

STATISTICAL PROCESS CONTROL

Control charts are very valuable. If properly used, they can very accurately predict a process's performance. Charts can show when a process has changed, when it should be adjusted, when tools should be changed, when maintenance should be done, and charts can even help find out what has gone wrong with a process.

PROCESS CAPABILITY

The real reason to collect data about processes is to use the data to improve the processes. Capability gives us concrete data on how good or bad our processes or machines are.

Consider a simple example. The most common operation on a lathe would be turning an outside diameter to a specific size. A study could be performed on a lathe to establish the lathe's capability to turn diameters. The term *capability* refers to how close of a tolerance a machine can hold. If a lathe can turn diameters to within ±.001, we could say its capability for turning diameters is .002.

To actually conduct the capability study, the machinist would first make sure there was nothing wrong with the machine, tooling, or material. Then the machinist would run some pieces. Note: the machinist cannot make changes or even adjustments to the process during the study. The purpose of the study is to see how closely the machine can hold sizes, so the machine must be left alone. No changes can be made during the study, just run the parts. The machine is being studied, not the machinist's ability to adjust it. It would ruin the results if the machine was adjusted or changes were made while it was running.

For this example, assume 25 parts were run and the outside diameter was measured. The blueprint specification was 1.125. The actual sizes of the parts and their coded values are shown in Figure C-218 (1.125 = 0).

The standard deviation (sample standard deviation) was calculated and found to be 1.67. Six standard deviations are equal to 99.7 percent of all things in the process. If we multiply 6 * 1.67, we get 10.02. (Remember, this is a coded value.)

This means that if nothing changes in this process, we could run 99.7 percent of all pieces within a range of approximately 10 (coded value). In this case, 1 coded value equals .001. We could run 99.7 percent of all pieces on this

1st 5	2nd 5	3rd 5	4th 5	5th 5
1.123 = −2	1.124 = −1	1.123 = −2	1.122 = −3	1.127 = 2
1.128 = 3	1.125 = 0	1.125 = 0	1.124 = −1	1.126 = 1
1.127 = 2	1.128 = 3	1.126 = 1	1.125 = 0	1.127 = 2
1.126 = 1	1.126 = 1	1.127 = 2	1.126 = 1	1.125 = 0
1.127= 2	1.125 = 0	1.127 = 2	1.127 = 2	1.124 = −1

Figure C-218 Sizes and the coded values for the 25 diameters that were cut to study this machine.

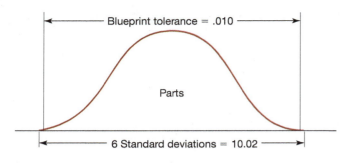

Figure C-219 Bell curve for a job that has a total tolerance of .010 inches and 6 standard deviations (SDs) that are equal to 10.02 inches.

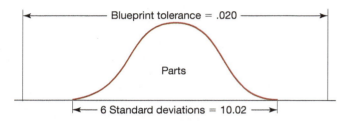

Figure C-220 Bell curve for a job where the process capability is about $\frac{1}{2}$ of the blueprint tolerance. This would be a great job, and it would be difficult to produce scrap.

lathe within approximately .010. (This lathe is obviously not very accurate, but the numbers are easy to work with for this initial example.)

If the blueprint tolerance is ±.005 (.010 total), we will get almost no scrap (see Figure C-219).

There is almost no scrap because 6 standard deviations (.010) are equal to our tolerance of .010. (This assumes that we can keep the process exactly on the mean.)

Consider a different job for the same machine. This job has a tolerance of ±.010. This means that the total tolerance is equal to .020 (see Figure C-220).

Figure C-220 makes it clear that this process should produce no scrap. The blueprint tolerance is twice as wide as the process capability.

For an additional example, assume the same process capability (6 SD = .010). This time the blueprint tolerance is equal to ±.003. Study the diagram shown in Figure C-221. (One standard deviation has been rounded from .00167 to .0017.)

The more parts run, the better our estimate of capability will be. In other words, the more we run, the more confident we could be of our results and the better prediction we could make about running these parts in the future.

Capability is usually not expressed in terms of 6 standard deviations alone.

We would like to be able to compare the process capability to the blueprint specification of the actual job we will be running. This will show how good (or bad) the job is.

One way in which capability is expressed is called CP (see Figure C-222).

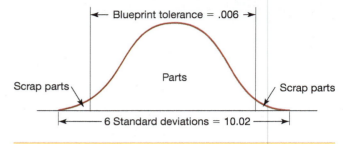

Figure C-221 Bell curve for a process that would produce approximately 4 percent scrap. In this example, the blueprint tolerance is located at about plus and minus 2 standard deviations. This means that the best this process could do would be about 96 percent good parts (if the process could be kept on the mean). Yelling at the operator will not help reduce the scrap rate. The best that the process can do is 96 percent good parts. Unless the process is changed, it will produce a minimum of 4 percent scrap.

$$\text{Capability (CP)} = \frac{\text{Total Blueprint Tolerance}}{\text{Standard Deviations}}$$

Figure C-222 The formula for simple capability.

Capability
Above 1.33—Excellent
1 to 1.33—BC
Below 1—Poor

Figure C-223 Capability chart.

You can see that we are comparing the total print tolerance to the process capability (6 standard deviations) and calling the result CP. Figure C-223 is used to grade a process for a particular job.

In the first example, the total blueprint tolerance was .010, and 6 standard deviations was .010. The CP = .010/.010 or 1 (CP = 1). If we look at our capability chart, this job would be classified as a B-C job (not really good, not really bad).

$$010/.010 * 100 = 100\%$$

In the second example, the blueprint tolerance was .020, and 6 standard deviations was .010. The CP = .020/.010 or 2 (CP = 2). This job is excellent, and it should be easy to run the job without scrap. In fact, it should be almost impossible to run scrap.

$$.010/.020 * 100 = 50\%$$

In the third example, the blueprint tolerance was .006, and 6 standard deviations equaled .010. The CP = .006/.010 = .6. This job is very, very poor. There is no way we can run this process without producing scrap.

$$.006/.010 * 100 = 60\%$$

CP is one way to express capability. Another way to express capability is easier to use because it is a percentage.

$$\text{Capability (CP)\%} = \frac{\text{Total Blueprint Tolerance}}{\text{Standard Deviations}} \times 100$$

Figure C-224 Formula used to find capability in terms of a percentage. The lower the percentage, the better the capability.

The higher the percentage, the worse the job (see Figure C-224).

Many industries have a goal that CP% should be at least 150 percent. They will try to improve processes so that they have a CP% of 150 percent or larger. There are many advantages to high CP percentages:

Better parts (parts are closer to the blueprint mean)

Processes are easier to run

Less scrap

Less frequent inspection is necessary

Higher productivity through quality and process improvement

One example of capability comes from the auto industry. One of the automakers had one model of transmission it wanted to study using statistics. The automaker made some of the transmissions and purchased some from another automaker. The transmissions were all made to the exact same specifications and tolerances.

The automaker noticed that the other manufacturer's transmissions seemed to perform more reliably (less warranty work, customer complaints about noise, and so on). They decided to choose a number of transmissions at random from the other automaker and the same number at random from their production. Inspectors were then assigned to completely disassemble the transmissions and inspect them. They checked everything possible, including torque of nuts and bolts and the sizes of all parts.

They found that all of the parts of the other automaker's transmissions were all in tolerance. It was also found that all parts of all their own transmissions were all in tolerance. If all of the transmissions met all of the specifications, why did the other automaker's transmissions perform more quietly and reliably? We would assume that if all parts met specifications, they would perform the same.

This is not true, however. When they analyzed the data, they found that the other automaker had much higher CPs than their own CPs. In other words, the other automaker's processes had better capabilities and could make parts with much less variation in size.

Note that they had exactly the same tolerances. The other automaker just had better processes because they used statistical methods and had improved them.

It should be clear that the blueprint specification (center of the blueprint tolerance) should be the ideal size (the size at which the part performs the best). The other automaker made all parts closer to the mean than they did (see Figure C-225). Thus, their transmissions were quieter and more reliable.

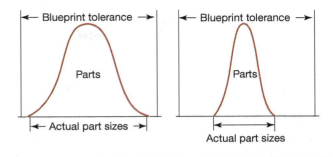

Figure C-225 Two bell curves for two processes making the same part, and the tolerance is the same. The bell curve on the left shows the process bell curve for one company making the part. The bell curve on the right shows the process bell for another company making the same part. Which company makes better parts? Which company has less variation in their parts? Which company's process runs better? Which company's parts would you want in your transmission?

Capability provides valuable information about a machine or process. For example, if a company calculated the capabilities for each of their machines, they would know how tight a tolerance could be held on each machine.

It can show us the parts of the business that we need to improve (or even drop). If the lathe department has problems and cannot hold the required tolerances of their customers, they will have to make a decision to rebuild equipment, buy new equipment, look for work that does not require close tolerances, or decide not to do lathe work anymore. It can also help discover a company's strengths.

This can help a company bid on work that is more profitable and show them what not to bid on. It can help direct their maintenance efforts and show where to invest in new equipment.

Charts can be a very beneficial tool in industry. The charts we will examine next are designed to be accurate 99.7 percent of the time.

BENEFITS OF CHARTING

Adjustment Reduction

Charts will show when a machine needs adjustment. People tend to adjust machines too much. In fact, the more conscientious a worker is, the more he or she will adjust. This is because the worker notices the small variations in a process and tries to adjust the machine; however, you cannot eliminate all variation, and by making unnecessary adjustments, you actually make the process vary more.

Dr. W. Edwards Deming was one of the most famous quality gurus in history. He is given much of the credit for Japan's amazing manufacturing success. One of Deming's studies involved a paper coating process (see Figure C-226).

The coating thickness on the paper was very important. Deming was asked to study the process because it was impossible to keep the thickness consistent.

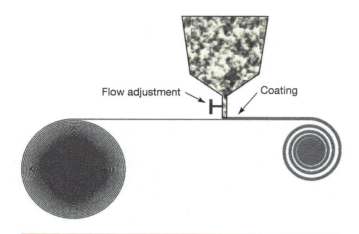

Figure C-226 Simplified paper coating process. The coating thickness is very important: The more consistent the thickness, the better the product.

Figure C-227 Simple example of a chart of part sizes as they were machined.

The worker's job was to measure the thickness and make adjustments when necessary. If the coating was too thick, he would close the valve more; if the coating was too thin, he would open it more. The worker was very diligent.

Deming insisted that the process be run with no adjustments to see how good or bad the process was. Everyone believed that the process would run terribly without adjustment. But Deming persisted, and it was run without adjustment. The process ran very well. There was much less variation in thickness. A chart was put on the machine that showed when adjustment was necessary, the job became a favorite instead of a problem.

No one knew that variation was normal. The thickness has to vary somewhat. If a machine is adjusted every time a part varies, the process produces much more variation. Charts can make the adjustment decision correctly 99.7 percent of the time.

Process Monitoring

Once a process has been studied and a chart has been constructed, we know how the process should run. The chart will immediately show us when something changes in the process. The chart is 99.7 percent accurate in these tasks.

When we chart, we also are developing historical data on the processes or machines. If we looked at charts over a machine's life, the chart would show the deterioration of the machine or process. If we project this information forward, we could predict when the machine or process will need to be rebuilt or replaced.

Process Improvement

Charts can become the basis for process improvement. Process improvement means less part variation, and less part variation means higher quality at lower cost. Continual process improvement will yield continually higher quality at continually lower cost.

Imagine the following process: coin flips (heads or tails). A coin is repeatedly flipped. If it comes up heads, an x is plotted above the centerline; tails, an x is plotted below the centerline.

You would expect that approximately half of the flips would be heads and half tails.

The odds of flipping a heads or tails would be $\frac{1}{2}$. What are the odds of flipping two heads in a row? The odds would be $\frac{1}{2} * \frac{1}{2}$, or $\frac{1}{4}$. In other words, there is only one chance in four that two consecutive flips would be heads (or tails). The odds for three in a row would be $\frac{1}{2} * \frac{1}{2} * \frac{1}{2}$, or $\frac{1}{8}$.

This should help you understand that we would expect half of our part sizes to be above the centerline (mean) and half below. We would also expect that we would not get too many in a row on one side of the centerline. The odds of getting seven in a row above (or below) the centerline (mean) is $\frac{1}{128}$.

This means that if we are plotting sizes of parts on a chart, we would expect that the sizes would occur randomly above and below the centerline (average part size). Figure C-227 shows a simple chart that a machinist used. The part specification was 1.000. The machinist put a dot on the chart for each part. The first part was 1.001. The second part size was .999. Note that these are coded values.

There is a very small chance that seven sizes in a row would be on one side of the centerline. The chart rules would say that a process can be adjusted if seven in a row fall on one side of the centerline (1.000 = 0 in this example). Assume seven-part sizes in a row fall below center. The odds are so low that this could happen that we could assume that something has changed in the process. This is the second rule for these charts: if seven fall on one side of the centerline, the process has changed and an adjustment or change is necessary. If the average of the seven sizes was calculated, it would give the exact adjustment needed. The chart not only tells when to adjust, but also how much to adjust.

In other words, if seven in a row are above or below center, it means that the mean (average) has shifted up or down.

The third rule states that a process has changed if seven in a row increase or decrease. This is called a *trend*. If each of the seven in a row gets larger (or smaller), the process has changed.

These rules are all based on making the correct decision 99.7 percent of the time.

Rule 1: Do not adjust unless a point falls outside the limits.

Rule 2: Do not adjust unless seven in a row are on one side of the centerline. (If seven are on one side, the mean has shifted.)

Rule 3: Do not adjust the process unless seven in a row trend up or down. Each one must be larger (or smaller) than the last for the trend. This kind of trend will usually indicate something more than an adjustment is needed. Something has changed in the process.

If you follow these rules, you will make the right decision the vast majority of the time.

Charting Processes

The $\overline{X}R$ (X-bar and R) chart is one of the most widely used charts in industry. An X-bar and R chart is designed to be accurate 99.7 percent of the time. X-bar and R mean average and range chart. Samples are taken of consecutive parts (usually five) and the average of the sample is plotted on the chart. The range (largest size minus the smallest size) of the sample is also plotted on the chart.

In effect, a chart is really two charts: an average chart with control limits and a range chart with limits.

$\overline{X}R$ Chart Construction

The first step in chart construction is to gather data from a process. There are often several sizes inspected on each piece. One of these must be chosen for a chart. One should try to choose the dimension (or characteristic) that seems most critical. If our process was a lathe part, we might be turning four diameters. Any of the diameters would probably be appropriate to chart. If one diameter is more important, it would be chosen. But if the operations are all very similar, one will be a good indicator of the others.

Once a particular part characteristic has been chosen, data is gathered. It is very important that the process is running well. If we know there is something wrong with the process, it should be fixed before the study is done. The process should be running as well as it can (good operator, tooling, etc.).

No adjustments can be made during the study. Accurate data is needed on how good the process is, not the operator. The more that data is gathered, the better the results will be. Twenty-five parts will give fairly good results. The data must be consecutive parts.

The data must be accurate. Gages should be appropriate, and the operator should be proficient in using the gage. The operator should also understand why the data is being gathered and the importance of accurate data. If this is not done, there is a tendency for a person to be wary and fudge the data. Remember, the process is being studied, not the person.

Rules:

1. The process should be running optimally.
2. Choose an appropriate part characteristic.
3. Do not change the process in any way during the study.

4. Make sure the operator understands the purpose.
5. Use appropriate, well-understood gages.
6. Gather the data. (It is a good idea to code the data because it is easier to use and understand. It is also more difficult to make a mistake.)

See Figure C-228 for data from a process (coded). As you can see, the data was gathered in groups of five.

The charts we are constructing work only for normal processes. To see if our process is normal, the data must be put into a histogram (see Figure C-229). A histogram is a simple, quick look that will indicate whether our process is, or is not, normal.

The data in Figure C-229 looks normally distributed (somewhat like a bell curve). The process appears to be normally distributed. Processes should be normally distributed. If a histogram shows that a process is not normally distributed, something is wrong with the process. The problem in the process must be found and corrected, and new data must be gathered.

Since the process appears to be normal, the chart can be constructed. The next step is to find the average (mean) and the range for each subgroup.

First, the total for each subgroup must be found. This is done by adding the numbers for each subgroup of 5 and writing them in the total column (see Figure C-230). Study the rest of the subgroup totals.

Next, the average of each subgroup is calculated. This is done by multiplying the total of each column by 2 and writing the result in the average column. Then put a decimal point one place from the right (see Figure C-231). The first

Subgroup 1	Subgroup 2	Subgroup 3	Subgroup 4	Subgroup 5
0	2	0	2	2
1	0	−1	1	2
1	−3	−1	−2	1
−1	0	−1	1	3
−2	−1	4	0	2

Figure C-228 Part sizes of the 25 parts that were run. Subgroup 1 contains the first five parts in order, subgroup 2 the next five in order, and so on.

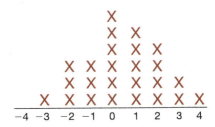

Figure C-229 Data from Figure C-228 in a histogram. Each X represents one part. For example, in this histogram there were six parts that had a coded size of 0. The histogram also gives us a rough estimate of what the average would be. It also shows the range of sizes produced. In this case, the histogram resembles a bell shape, so the process can be considered normal.

	Subgroup 1	Subgroup 2	Subgroup 3	Subgroup 4	Subgroup 5
	0	2	0	2	2
	1	0	−1	1	2
	1	−3	−1	−2	1
	−1	0	−1	1	3
	−2	−1	4	0	−2
Total	−1	−2	1	2	6
Average					
Range					

Figure C-230 The first step in finding the average for each subgroup. Add the five sizes in each subgroup and write the result in the total box.

	Subgroup 1	Subgroup 2	Subgroup 3	Subgroup 4	Subgroup 5
	0	2	0	2	2
	1	0	−1	1	2
	1	−3	−1	−2	1
	−1	0	−1	1	3
	−2	−1	4	0	−2
Total	−1	−2	1	2	6
Average	−.2	−.4	.2	.4	1.2
Range					

Figure C-231 Averages for each subgroup.

	Subgroup 1	Subgroup 2	Subgroup 3	Subgroup 4	Subgroup 5
	0	2	0	2	2
	1	0	−1	1	2
	1	−3	−1	−2	1
	−1	0	−1	1	3
	−2	−1	4	0	−2
Total	−1	−2	1	2	6
Average	−.2	−.4	.2	.4	1.2
Range	3	5	5	4	5

Figure C-232 The ranges for each subgroup were calculated by subtracting the smallest size from the largest size. For example, the largest size in the first subgroup is 1 and the smallest is −2. The range is equal to the largest minus the smallest, or 3.

column total is 1. Multiply 1 * 2, and the result is 2. Next move the decimal place one place to the left and you get .2.

That is the average for the first subgroup. Note that this method of averaging works only when there are five values in the subgroup. Study the rest of the subgroup averages.

The next step is to calculate the range for each subgroup. Remember that the range is simply the difference between the largest and smallest sizes in each subgroup. Look at the first subgroup in Figure C-232. The largest value is 1

$$UCL_{\bar{x}} = \bar{\bar{X}} + A_2(\bar{R})$$
$$LCL_{\bar{x}} = \bar{\bar{X}} - A_2(\bar{R})$$
$$UCL_{\bar{x}} = .24 + .58(4.4)$$
$$LCL_{\bar{x}} = .24 - .58(4.4)$$

Figure C-233 Formulas for upper and lower control limits.

Number in Subgroup	Value of A_2	Value of D_4
3	1.02	2.57
4	.73	2.28
5	.58	2.11
6	.48	2.0
7	.42	1.92
8	.37	1.86
9	.34	1.82
10	.31	1.78

Figure C-234 This figure shows how the values for A_2 and D_4 are found in the chart. There are five pieces in our subgroups, so the value of .58 is used.

and the smallest value is −2. The difference between these values is 3. Study the other subgroup ranges.

Next formulas will be used to calculate an upper and lower limit for the averages part of the chart. The formulas look complex, but are very simple (Figure C-233).

You should notice that the formulas are the same, except that the UCL uses a plus sign and the LCL uses a minus sign.

The formulas really just add an amount (A_2R-bar) to the process average or subtract an amount from the process average.

X-double bar is the process average (average of the averages). For our example, it is .24. You can find this by adding the five subgroup averages and dividing by 5.

R-bar is the average range for our process. We can find it by adding the five subgroup ranges and dividing by 5, or by adding the five ranges and multiplying by 2 and moving the decimal place one place to the left. For this example, the average range is 4.4.

Figure C-233 also shows the actual values substituted into the formula. Next, we need to know what A_2 is. A_2 is a constant from a table (see Figure C-234).

The number of parts in our subgroup is 5, so the value for A_2 (n) is .58.

Substitute .58 into the formula.

Make sure to perform the multiplication first, then calculate the two limits. The upper control limit (UCL X-bar) for the subgroup averages is 2.792 and the lower control limit (LCL X-bar) for the subgroup averages is −2.312.

This completes the averages portion of the subgroup averages (X-bar) chart.

The formula for the upper control limit for the range (UCL$_R$) is UCL$_R$ = D_4(R-Bar) The value of D_4 is found in the chart in Figure C-234. There are five pieces in our subgroups, so we will use a value of 2.11 for D_4.

The formula says multiply D_4 by the average range (R-Bar).

A process will always have variation. The upper control limit on the range will tell us when the variation is higher than it should be. (If a range is greater than the upper limit, there is a 99.7 percent chance that something changed in the process.)

Figure C-235 shows a completed chart for this example.

The data was transferred to the chart. Notice that the chart really has three areas. The top area contains information about the individual parts: the part name, part number, operation number, and so on, as well as the actual sizes of the parts that were made.

The actual part sizes were written down in order in the subgroup areas in groups of five. This area is also used to calculate the averages and ranges for each subgroup.

The middle portion of the chart is used to plot the average for each subgroup. This area is the chart of averages. Notice that the mean and the upper and lower control limits have dotted lines to mark the limits. The actual values for the limits are written on the left side of the chart.

If a point that we plot is between the limits, the process is acceptable. The chart will always tell us if something has changed in the process. If a point falls outside the limit, there is a 99.7 percent chance that something has changed in the process.

The subgroup averages are plotted by following the vertical line under each subgroup down to the average portion of the chart. The average for subgroup 1 is −.2. Follow the line under subgroup 1 down and you will see that a dot was drawn on the line just under 0.

Figure C-235 Completed chart.

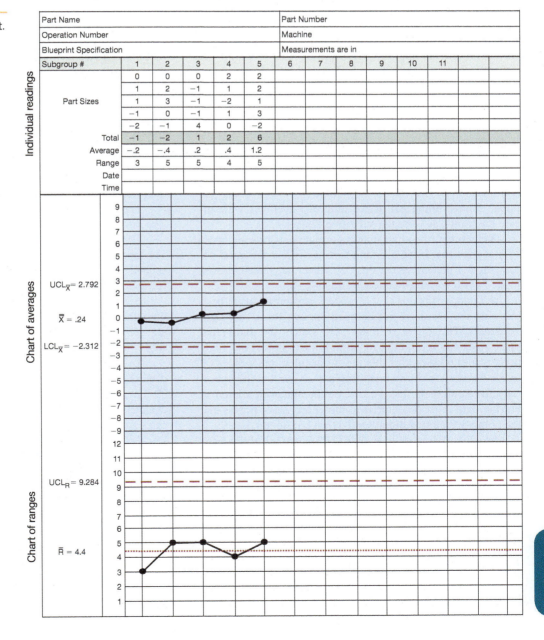

The second subgroup's average was −.4. Follow the line down from the second subgroup and you will see that a dot is drawn at −.4. The third subgroup has an average of 1.2. After all of the five averages were plotted, they were connected with lines.

The third area of the chart is the chart of ranges. The range portion of the chart is used to graphically show variation between parts in a subgroup. A small range is desirable and indicates that the variation between parts within the subgroup is small.

Note there is only an upper limit on the range chart. We want all of our subgroup ranges to be under the limit. Remember that the smaller the range the better, because it means that there is less variation between the parts in the subgroup.

The subgroup ranges are plotted by following the vertical line under each subgroup down to the range portion of the chart. The range for subgroup 1 is −3. Follow the line under subgroup 1 down and you will see that a dot was drawn at 3 in the range portion of the chart. The second subgroup's range was 5. Follow the line down from the second subgroup and you will see that a dot is drawn at 5. The third subgroup has a range of 5. After the five ranges were plotted, they were connected with lines.

ANALYZING THE CHART

Are all of the subgroup averages inside the upper and lower control limits? Are all of the subgroup ranges under the upper control limit? All of the subgroup averages were inside the limits, as were the subgroup ranges.

Remember, the chart makes correct decisions 99.7 percent of the time. If we use the chart, we will make good decisions 99.7 percent of the time. However, we have not even looked at blueprint tolerance for this job yet.

A chart assures us that a process is running at its best, but even the best may not be good enough for a very tight tolerance. The operator must still watch the individual piece sizes and scrap the bad parts. For example, how many parts would be scrap for this job if our tolerance was ±.003 inches? You should find one scrap part. Note that the operator must scrap parts that are outside the blueprint tolerance, but the data must be entered in the chart. Also, notice that the points we plotted did not tell us we had scrap.

Examine the chart in Figure C-235. Are all of the averages inside the limits? All of the ranges?

The chart would now be ready to be used on this job. Note that we only calculate limits once when we first study the job. From this time on, we would just enter the date and plot points as we run parts.

The chart assures us that the process (or machine) is running at its best. It does not assure us there is no scrap. The operator must compare the individual parts to the blueprint tolerance. If the particular job we are running has a very good CP, we can reduce our inspection. The better the job (wide tolerance compared to six or more), the less we need to inspect. We might be able to sample one subgroup an hour or one a day if we found an excellent job.

Operators should be encouraged to write notes about the job as they run it. Even hunches could prove useful. The more notes on charts the better. If we examine a chart, months or even years later, notes will improve our recall.

Now that limits have been set for this particular job on this particular machine, they never have to be recalculated again unless we change the machine or process.

More parts were run on another day. They were entered into the chart, and the averages and ranges were calculated and plotted (see Figure C-236). Note that we did not have to recalculate limits. If we run the same job on the same machine and with the same tooling, we can use the same limits and the job should run the same. This is very beneficial because we are developing some history on the job and machine. If we were to look at the same job over a long period of time, we might be able to see the variability increase as the machine wears. This might help us plan maintenance and machine replacement.

Note that in subgroup 7 there is an ellipse drawn around the −6 part. This means that there was a tool change done, which explains why the size was so far off. Note that the average for this group was all right, but the range was outside the limit. The range shows that the variability for this subgroup was high because of the tool change, which the operator noted on the chart.

The average for subgroup 8 is well within the limits. The range is fine also.

The average for subgroup 9 is above the upper limit. The operator stopped running parts when the average exceeded the limit. The chart is telling us that something changed. The operator studied the machine, the setup, and the tooling and discovered that the tool was chipped. The operator noted this on the chart and initialed the note. Notes are very important on charts. The range for the subgroup was within the limit.

You should begin to see that a chart will instantly tell us if there is a problem. The chart will make the right decision 99.7 percent of the time. The operator should not make any changes or adjustments to the machine unless the chart indicates a change has occurred in the process.

Rule 1: Do not make any changes to a machine unless the average or range is outside the limits.

Trends

The second rule for charts is that something has changed in the machine or process if there are seven points in a row either above or below the average (see Figure C-237). This partial chart shows the plots of seven points. Note that all

Figure C-236 Completed chart.

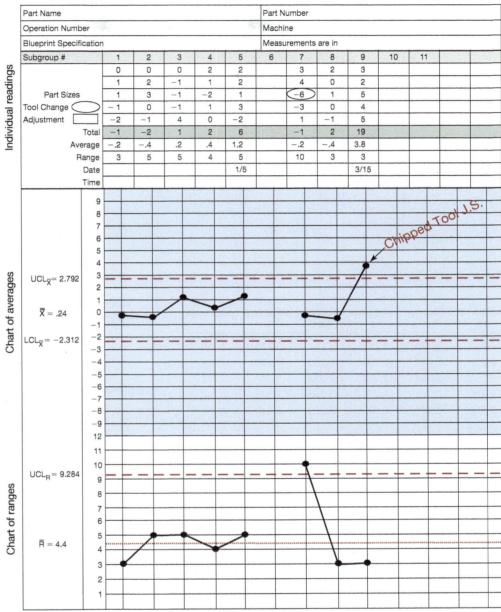

		Subgroup #	1	2	3	4	5	6	7	8	9	10	11		
Individual readings			0	0	0	2	2		3	2	3				
			1	2	−1	1	2		4	0	2				
	Part Sizes		1	3	−1	−2	1		−6	1	5				
	Tool Change ⬭		−1	0	−1	1	3		−3	0	4				
	Adjustment ☐		−2	−1	4	0	−2		1	−1	5				
		Total	−1	−2	1	2	6		−1	2	19				
		Average	−.2	−.4	.2	.4	1.2		−.2	−.4	3.8				
		Range	3	5	5	4	5		10	3	3				
		Date					1/5				3/15				
		Time													

Part Name / Part Number / Operation Number / Machine / Blueprint Specification / Measurements are in

Chart of averages: $UCL_{\overline{X}} = 2.792$ $\overline{X} = .24$ $LCL_{\overline{X}} = -2.312$

Chipped Tool, I.S.

Chart of ranges: $UCL_R = 9.284$ $\overline{R} = 4.4$

Figure C-237 Chart with seven points above the process average.

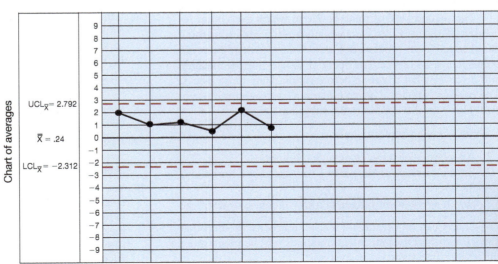

Chart of averages: $UCL_{\overline{X}} = 2.792$ $\overline{X} = .24$ $LCL_{\overline{X}} = -2.312$

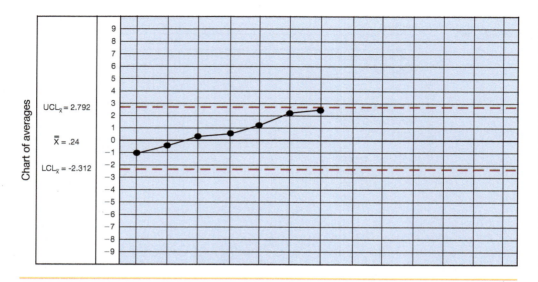

Figure C-238 Chart of averages showing 7 values increasing in a row.

Subgroup 1	Subgroup 2	Subgroup 3	Subgroup 4	Subgroup 5	Subgroup 6
2.252	2.249	2.248	2.252	2.246	2.250
2.252	2.249	2.254	2.253	2.248	2.249
2.253	2.243	2.252	2.251	2.247	2.252
2.250	2.248	2.251	2.250	2.246	2.248
2.252	2.249	2.247	2.250	2.249	2.251

Figure C-239 Use with question 10.

seven are above the average. This indicates something has changed in the process. The operator should stop the machine and see if anything obvious is wrong. If nothing is found, an adjustment should be made.

If you look at the averages and draw an imaginary line through them, you would see that their average is about 1. Because the parts are averaging about 1 over size, the operator should make an adjustment of 1. This is a major advantage of charting. The chart shows us exactly how much to adjust. (This rule also applies to subgroup ranges.)

The second kind of trend is one in which each point increases. If seven points increase in size in a row, something has changed in the process. In Figure C-239, each of the seven points increased in size. The operator should stop the machine and discover what is wrong before running any more pieces. The rule also applies if seven parts in a row each decrease in size. This rule also applies to the subgroup ranges.

Do not change anything unless:

1. An average or range is outside the limits
2. Seven averages or ranges in a row are above (or below) the average
3. Seven averages or ranges in a row increase or decrease

Charts can be invaluable if used correctly. A shop that understands and uses statistical methods will be much more successful than a shop that does not.

Note that even if you do not use charts, a fundamental understanding of statistical principles will improve quality and productivity.

SELF-TEST

1. List and explain the three rules for charts that show when a process has changed.

2. Explain at least three benefits of charts. How do they help operators, maintenance, bidding, and so forth?

3. These charts make correct decisions _____ percent of the time.

4. Why is it important that no adjustments be made to the machine or process during the initial study?

5. A machine has a standard deviation of .003. We are considering running a job on the machine that has a tolerance of ±.006.
   a. Calculate the CP.
   b. Will there be scrap? If so, how much? (Hint: Draw a bell curve to help find the answer.)

6. Using the same machine as in question 5, assume the job has a tolerance of ±.010.
   a. Calculate the CP.
   b. Will there be scrap? If so, how much?

7. Standard deviation = .0015; blueprint tolerance = ±.005.
   CP = _____

8. Standard deviation = .001; blueprint tolerance = ±.002.
   CP = _____

9. Based on the data from question 8, will this job run well? If you were asked whether to bid on the job, list at least three alternatives you could give.

10. You are asked to make a recommendation on bidding on a lathe job. The company supplied you with 30 sample pieces to run. The data for the job is shown in Figure C-239. Code the data and draw a bell curve that compares the capability (6 standard deviations) to the blueprint specifications. Blueprint Specification = 2.250 ± .004.

Figure C-240 Blank chart for question 11.

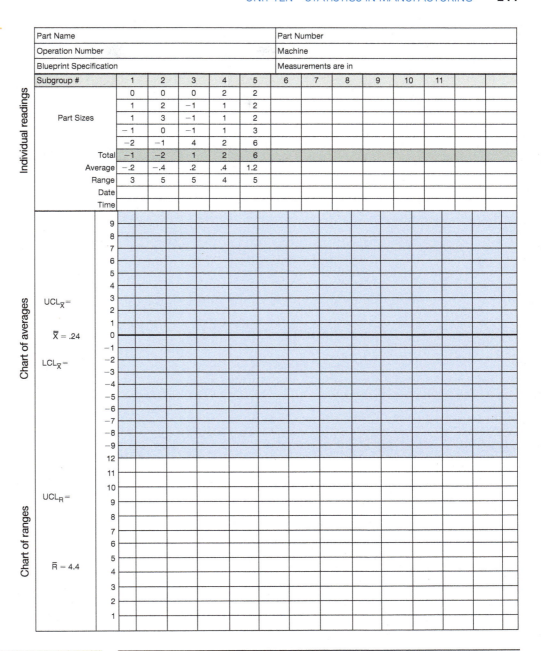

Subgroup #	1	2	3	4	5	6	7	8	9	10	11			
Part Sizes	0	0	0	2	2									
	1	2	−1	1	2									
	1	3	−1	1	2									
	−1	0	−1	1	3									
	−2	−1	4	2	6									
Total	−1	−2	1	2	6									
Average	−.2	−.4	.2	.4	1.2									
Range	3	5	5	4	5									
Date														
Time														

Part Name Part Number
Operation Number Machine
Blueprint Specification Measurements are in

Individual readings

Chart of averages

$UCL_{\overline{X}} =$

$\overline{\overline{X}} = .24$

$LCL_{\overline{X}} =$

Chart of ranges

$UCL_R =$

$\overline{R} = 4.4$

Figure C-241 Use with question 12.

Subgroup 1	Subgroup 2	Subgroup 3	Subgroup 4	Subgroup 5
1.253	1.254	1.251	1.249	1.247
1.250	1.249	1.245	1.250	1.251
1.247	1.250	1.251	1.256	1.252
1.251	1.253	1.248	1.252	1.251
1.248	1.249	1.252	1.248	1.245

11. Complete an $\overline{X}R$ chart for the process in question 10 (enter the data, calculate control limits, and plot the data).
   a. Standard deviation
   b. Process mean
   c. Average range
   d. Is the data normally distributed? (Check with a histogram.)
   e. Upper control limit averages
   f. Lower control limit averages
   g. Upper control limit range

12. The data in Figure C-241 was taken from 25 consecutive pieces. The process was a turning operation on a lathe. The blueprint specification was $1.250 \pm .003$.
   a. Code the data.
   b. Is the process normal?
   c. Mean.
   d. Average range.
   e. Complete an X-bar and R chart (enter the data, calculate control limits, and plot the data).
   f. Compute the capability.

Materials

INTRODUCTION

The tremendous progress of civilization can, in large part, be attributed to the ability to make and use tools. The discovery and use of metals is a close second, for without metal, tools would still be made of bone and stone.

Nearly everything we use in our daily lives depends on metals. Vast amounts of iron and steel are used for automobiles, ships, bridges, buildings, machines, and many other products. Almost everything that uses electricity depends on copper and many other metals. There are also hundreds of combinations of metals, called alloys.

We have come a long way since the first iron was smelted by the Hittites about 3,500 years ago. Their iron tools were not much better than tools made from the softer metals, copper and bronze, that were already in use at the time. This was because iron would bend and not hold an edge. The steelmaking process, which uses iron to make a strong and hard material by heat treatment, was still a long way from being discovered.

Metallic ores are first smelted into metals, and these metals are then formed into many different products. Many metallic ores exist in nature as oxides. Oxides are metals that are chemically combined with oxygen. Most iron is removed from ore by a process called *oxidation–reduction*. Metallic ores are also found as carbonates and silicates.

Modern metallurgy stems from the ancient desire to fully understand the behavior of metals. Long ago, the art of the metalworker was shrouded in mystery and folklore. Crude methods of making and heat-treating small amounts of steel were discovered by trial and error only to be lost and rediscovered later by others. The modern story of iron and steel begins with iron ore, coal, and limestone. Pig iron is produced from these ingredients. The steel mill refines pig iron in furnaces and then casts it into ingot molds to solidify. The ingot is then formed into various shapes to make many types of steel products.

This section will cover the metals commonly used in the machine shop. The characteristics of metals and their hardening and tempering processes will be covered. Hardening and tempering processes are used to strengthen and harden metals by heating and cooling. Different metals often have different properties (e.g., cast iron is brittle, whereas soft iron is easily bent), and the difference is in how they are made. Some steels can be made as hard as needed by heat treatment. This section will also cover methods to measure the hardness of metals.

Metals are classified for industrial use by their specific working qualities. A machinist must be able to select and identify materials using tables in a handbook or by testing processes in the shop. Some systems use numbers used to classify metals; others use color codes.

SAFETY FIRST

Always observe safety when handling material just as when using hand and machine tools.

SAFETY IN MATERIAL HANDLING

Lifting and Hoisting

Machinists were once expected to lift pieces of steel weighing a hundred pounds or more into awkward positions. This was a dangerous practice that resulted in many injuries. Hoists and cranes should be used to lift all but smaller parts. Steel weighs about 487 lb per cubic foot; water weighs 62.5 lb per cubic foot, so it is evident that steel is a heavy material for its size. You can easily be misled into thinking that a small piece of steel does not weigh much. Follow these two rules when lifting: don't lift more than you can easily handle and make sure that you bend your knees and keep your back straight. Use your leg muscles to lift. If a material is too heavy or awkward for you to position it on a machine, use a hoist. One improperly performed lift can ruin a back for life and possibly end your career and ability to earn a living.

When lifting heavy metal parts with a mechanical or electric hoist, always stand in a safe position, no matter how secure the slings and hooks seem to be. They don't often break, but it does happen, and if your foot is under the edge, a painful or crippling injury is sure to follow. Slings should not have less than a 30-degree angle with the load (Figure D-1). When hoisting long bars or shafts, a spreader bar (Figure D-2) should be used so that the slings cannot slide together and unbalance the load. When operating a crane, be careful that someone else is not standing in the way of the load or hook. If you are using a block, chain hoist, or electric winch, make sure that the lift capacity rating of the equipment and its support structure is adequate for the load.

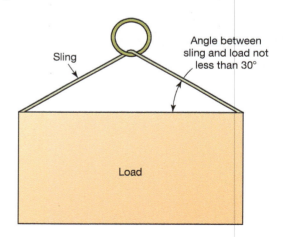

Figure D-1 Load sling.

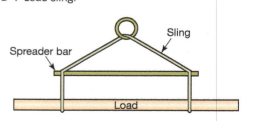

Figure D-2 Sling for lifting long bars.

Carrying Objects

Carry long stock in the horizontal position. If you must carry it in the vertical position, be careful of electrical bus bars, light fixtures, and ceilings. A better way is to have someone carry each end of a long piece of material. Do not carry sharp tools in your pockets. They can injure you or someone else.

Hot Metal Safety

Oxyacetylene torches are sometimes used for cutting shapes, circles, and plates in machine shops. Safety requires proper clothing, gloves, and eye protection. It is also important that any metal that has been heated by burning or welding be plainly marked, especially if it is left unattended. The common practice is to write the word *HOT* with soapstone on such items. Whenever arc welding is performed in a shop, the arc flash should be shielded from the other workers. *Never* look toward the arc because your eyes may be damaged.

When handling and pouring molten metals such as Babbitt, aluminum, or bronze, wear a face shield and gloves. Do not pour molten metals near a concrete floor, unless it is covered with sand. Concrete contains small amounts of water, which instantly becomes steam when the intense heat of molten metal comes in contact with it. Since concrete is a brittle material, the explosion of the steam causes pieces of the concrete floor to break off the surface and fly like shrapnel.

When heat-treating, always wear a face shield and heavy gloves (Figure D-3). There is a definite hazard to the face and eyes when cooling steel by oil quenching. The hot metal can start the oil on fire. If the metal is held with a portion of it above the oil, a fire can start. Put the metal all the way under the surface of the oil and the fire may stop. If not, most quench tanks will have a top cover that can be closed to put out the flame.

Certain metals, when finely divided as a powder or even as coarse as machining chips, can ignite with a spark or just by the heat of machining. Magnesium and zirconium are two such metals. The fire, once started, is difficult to extinguish, and if water or a water-based fire extinguisher is used, the fire will only increase in intensity. Chloride-based power fire extinguishers are commercially available. These are effective for such fires, as they prevent water absorption and form an air-excluding crust over the burning metal. Sand is also used to smother fires in magnesium.

Figure D-3 Face shield and gloves are worn for protection while heat-treating and grinding (*Courtesy of Fox Valley Technical College*).

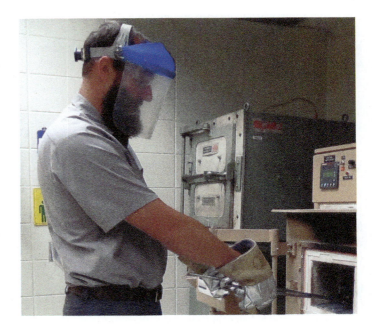

Selection and Identification of Steels

The village blacksmith only had wrought iron and carbon steel for making tools, implements, and horseshoes, so the choice of metals was relatively simple. As industry began to need more alloy steels and special metals, they were developed. There are hundreds of varieties of metals available today. This unit will introduce systems used for marking steels and some ways to choose among them.

OBJECTIVE

After completing this unit, you should be able to:

■ Identify different types of metals by testing.

WROUGHT AND CAST IRON

Iron is the basis for steel. Wrought iron is produced by refining iron ore. Iron is very soft and has very little carbon content. Wrought iron is easily shaped. It is normally used for decorative-type items such as railings. It is not commonly used in the machining industry.

Wrought iron usually contains less than 0.1 percent carbon and 1 or 2 percent slag. Iron that has high carbon content (2–4 percent) is called cast iron.

Wrought iron is better in some ways than cast iron, which is overly hard and brittle due to its high carbon content. The first iron was smelted directly by heating iron ore in a forge with charcoal. The charcoal was the fuel and also acted as a reducing agent. While still hot, the reduced iron and slag mixture was then removed and worked (wrought) with a hammer to expel most of the slag and weld the iron into a lump of mostly iron.

According to the American Iron and Steel Institute (AISI), steel can be categorized into four basic groups:

• Carbon steel
• Alloy steel
• Stainless steel
• Tool steel

There are many different grades of steel and each of them has unique properties. These unique properties can be physical, chemical, and environmental.

PLAIN CARBON STEEL

When carbon is added to molten iron, the resulting metal is called plain carbon steel. Stronger steel is made by adding more carbon, up to a point. The steel is then formed by either hot rolling or cold rolling. Hot rolling means that the steel was formed to size and shape while it was red-hot. Hot rolled steel is formed by rollers which form the steel to size and shape. Hot rolling results in a hard, black scale on the steel. The scale is abrasive and causes additional tool wear. It is good machining practice to make sure the first cut is deeper than the scale. Cold rolling forms steel to size and shape at temperatures less than red-hot. It is formed by passing through rollers which form it to shape and size. The surface of cold rolled steel is smooth and gray. Cold rolled steel is referred to as CRS or CR. Hot rolled steel is called HRS or HR.

All steel is composed of iron and carbon. It is the amount of carbon and the additional alloys that determine the properties of the steel.

Carbon steel can be divided into three categories: low carbon steel (mild steel); medium carbon steel; and high carbon steel.

Low Carbon Steel (Mild Steel)

Low carbon steels contain 0.04–0.30 percent carbon content. Low carbon steel is one of the largest groups of carbon steel. Mild steel is available in many shapes; from round, flat, tube, and various structural shapes. Low carbon steel is alloyed with other elements to modify its characteristics. Low carbon steels are not directly hardenable as they do not have high enough carbon content. They can be surface hardened by surface hardening processes.

Medium Carbon Steel

Medium carbon steel has a carbon content of 0.31–0.60 percent, and a manganese content of .060–1.65 percent. Medium carbon steels can be hardened and tempered using heat treatment. Medium carbon steel is stronger than low carbon steel. Medium carbon steel is more difficult to form, weld, and cut than low carbon steel.

High Carbon Steel

High carbon steel typically has a carbon range between 0.61 percent and 1.50 percent. It is sometimes called tool steel. High carbon steel is more difficult to cut, bend, and weld. High carbon steel can be directly hardened. High carbon steel can become extremely hard and brittle when hardened. Additional heat treatment techniques are used to reduce hardness and brittleness and to increase toughness.

ALLOY STEELS

Steel that has had one or more alloying elements added such as lead, manganese, silicon, nickel, copper, chromium is called alloy steel. These additional alloying elements add specific properties that are not found in regular carbon steel. Alloy steels are very widely used because of the additional properties that are possible.

For example, if rust and corrosion are a problem, an alloy with chromium would be used as chromium helps prevent rust and corrosion. Stainless steel is a steel alloy with increased corrosion resistance. Common alloying elements include chromium (usually at least 11 percent), nickel, or molybdenum. Stainless steel is used for food handling/processing equipment, hardware, medical equipment, etc.

Tool steel is a term used for high-hardness, abrasion-resistant steels. Tool steels can be directly heat-treated and are extremely hard. Tool steels are used in tooling to form other metal products. Specific tooling applications are cutting, stamping or extrusion dies, molds for plastics or die casting, fixtures, etc. Tool steels are also used for knife making because of their ability to be hardened and their edge holding characteristics.

STEEL IDENTIFICATION SYSTEMS

Color coding is one way to identify the type of steel. Its main disadvantage is that there is no universal color coding system. Each manufacturer has its own system for color coding. The two identification systems most used in the United States are from the Society of Automotive Engineers (SAE) and the AISI. See Table D-1.

The first two numbers denote the alloy. Carbon, for example, is denoted by the number 10. The third and fourth

Table D-1 SAE-AISI Numerical Designation of Alloy Steels (X Represents Percentage of Carbon in Hundredths)

Carbon steels	
Plain carbon	10xx
Free-cutting, resulfurized	11xx
Manganese steels	13xx
Nickel steels	
.50% nickel	20xx
1.50% nickel	21xx
3.50% nickel	23xx
5.00% nickel	25xx
Nickel–chromium steels	
1.25% nickel, .65% chromium	31xx
1.75% nickel, 1.00% chromium	32xx
3.50% nickel, 1.57% chromium	33xx
3.00% nickel, .80% chromium	34xx
Corrosion- and heat-resisting steels	303xx
Molybdenum steels	
Chromium	41xx
Chromium–nickel	43xx
Nickel	46xx and 48xx
Chromium steels	
Low-chromium	50xx
Medium-chromium	511xx
High-chromium	521xx
Chromium–vanadium steels	6xxx
Tungsten steels	7xxx and 7xxxx
Triple-alloy steels	8xxx
Silicon–manganese steels	9xxx
Leaded steels	11Lxx (example)

digits, represented by x, denote the percentage of carbon in hundredths of 1 percent. For carbon steel, it can be anywhere from .08 to 1.70 percent. For alloys the second digit designates the approximate percentage of the major alloying element. Steels having over 1 percent carbon require a five-digit number; certain corrosion- and heat-resisting alloys also use a five-digit number to identify the approximate alloy composition of the metal.

The AISI numerical system is basically the same as the SAE system with certain capital letter prefixes. These prefixes designate the process used to make the steel. The lowercase letters *a* to *i* as a suffix denote special conditions in the steel.

The AISI prefixes are

B Acid Bessemer carbon steel

C Basic open-hearth carbon steel

CB Either acid Bessemer or basic open-hearth carbon steel at the option of the manufacturer

D Acid open-hearth carbon steel

E Electric furnace alloy steel

STAINLESS STEEL

The element chromium (Cr) is what gives stainless steels their corrosion resistance. Steel must contain a minimum of about 11 percent chromium to gain resistance to atmospheric corrosion. Higher percentages of chromium make steel even more resistant to corrosion and high temperatures. Chromium produces a thin layer of oxide on the surface of steel. This layer prevents corrosion of the surface.

Stainless steel also contains carbon, silicon, and manganese. Other elements such as nickel and molybdenum may also be added to achieve other properties. For example, nickel can be added to improve ductility, corrosion resistance, and other properties.

There are three basic types of stainless steels: the martensitic and ferritic types of the 400 series, and the austenitic types of the 300 series. There also are precipitation hardening types that harden over a period of time after heat treatment.

The martensitic, hardenable type has carbon content up to 1 percent or more, so it can be hardened by heating to a high temperature and then quenching (cooling) in oil or air. The cutlery grades of stainless are to be found in this group. The ferritic type contains little or no carbon. It is essentially soft iron that has 11 percent or more chromium content. It is the least expensive of the stainless steels and is used for such things as building trim and pots and pans. Both ferritic and martensitic types are magnetic.

Austenitic stainless steel contains chromium, nickel, and little or no carbon. It cannot be hardened by quenching, but it readily work-hardens while retaining much of its ductility. For this reason, it can be work hardened until it is almost as hard as hardened martensitic steel. Austenitic stainless steel is somewhat magnetic in its work-hardened condition but nonmagnetic when annealed or soft.

Table D-2 illustrates the method of classifying the stainless steels. Only a few of the basic types are given here. Consult a manufacturer's catalog for further information.

TOOL STEELS

Special carbon and alloy steels called *tool steels* have their own classification. There are six major tool steels for which one or more letter symbols have been assigned:

1. Water-hardening tool steels

   W—High-carbon steels

2. Shock-resisting tool steels

   S—Medium carbon, low alloy

3. Cold-worked tool steels

   O—Oil-hardening types
   A—Medium-alloy air-hardening types
   D—High-carbon, high-chromium types

4. Hot-worked tool steels

   H—H1 to H19, chromium-based types
   H20 to H39, tungsten-based types
   H40 to H59, molybdenum-based types

5. High-speed tool steels

   T—Tungsten-based types
   M—Molybdenum-based types

6. Special-purpose tool steels

   L—Low-alloy types
   F—Carbon–tungsten types
   P—Mold steels P1 to P19, low-carbon types P20 to P39, other types

Several metals can be classified under each group, so that an individual type of tool steel will also have a suffix number that follows the letter symbol of its alloy group (Table D-3). The carbon content is given only in those cases in which it is considered an identifying element of that steel.

Table D-2 Classification of Stainless Steels

Alloy Content	Metallurgical Structure	Ability to Be Heat-Treated
Chromium	Martensitic	Hardenable (Types 410, 416, 420)
		Nonhardenable (Types 405, 14 SF)
	Ferritic	Nonhardenable (Types 430, 442, 446)
Chromium–nickel	Austenitic	Nonhardenable (except by cold work) (Types 301, 302, 304, 316)
		Strengthened by aging (Types 314, 17-14 CuMo, 22-4-9)
	Semi-austenitic	Precipitation hardening (PH 15-7 Mo, 17-7 PH)
	Martensitic	Precipitation hardening (17-4 PH, 15-5 PH)

Table D-3 Various Tool Steels

Type of Steel	Examples
Water hardening: straight-carbon tool steel	W1, W2, W4
Manganese, chromium, tungsten: oil-hardening tool steel	01, 02, 06
Chromium (5%): air-hardening die steel	A2, A5, A10
Silicon, manganese, molybdenum: punch steel	S1, S5
High-speed tool steel	M2, M3, M30, T1, T5, T15

Unified Numbering System

Major trade associations (Aluminum Association, the American Iron and Steel Institute, the Copper Development Association, and the Steel Founders' Society of America (SFSA)) held discussions about steel classification systems and conducted an 18-month study about the feasibility of developing a new classification system.

The study was completed in 1971. The study found that a unified numbering system (UNS) for metals and alloys was possible and would be beneficial. An advisory board was then formed by ASTM and SAE in 1972 to work on the numbering system. The new classification system was called the Unified Numbering System.

The guidelines for the system were:

- The numbers assigned must integrate numbers from existing numbering systems whenever possible for easy recognition.
- The designation for a metal or alloy must refer to a specific metal or alloy as determined by its special chemical composition, or to its mechanical properties or physical features when these are the primary defining criteria and the chemical composition is secondary or not critical.
- The numbering system must be designed to adapt current metals and alloys, and to anticipate the need to furnish numbers for new alloys in the future.
- The system must be suitable for computer use and for general indexing of metals and alloys.

The UNS Advisory Board completed "SAE/ASTM Recommended Practice for Numbering Metals and Alloys" in 1974. Though the Advisory Board worked on assigning numbers for the 18 series of numbers they developed, the specific details for each series was left to field experts. By the end of 1974, specific UNS designations for more than 1,000 metals and alloys were completed. The UNS designations were listed in a UNS Handbook in 1975.

The UNS identifies each metal by a letter followed by five numbers. The UNS system does not guarantee precise composition with impurity limits or performance specifications. Table D-4 shows the UNS lettering scheme for metals.

Table D-5 shows some examples of plain carbon steel classification in AISI-SAE and UNS systems.

Table D-4 UNS Classification System

UNS Series	Metal Type
A00001 to A99999	Aluminum and aluminum alloys
C00001 to C99999	Copper and copper alloys (brasses and bronzes)
D00001 to D99999	Specified mechanical property steels
E00001 to E99999	Rare earth and rare earth-like metals and alloys
F00001 to F99999	Cast irons
G00001 to G99999	AISI and SAE carbon and alloy steels (except tool steels)
H00001 to H99999	AISI and SAE H-steels
J00001 to J99999	Cast steels (except tool steels)
K00001 to K99999	Miscellaneous steels and ferrous alloys
L00001 to L99999	Low melting point metals and alloys
M00001 to M99999	Miscellaneous nonferrous metals and alloys
N00001 to N99999	Nickel and nickel alloys
P00001 to P99999	Precious metals and alloys
R00001 to R99999	Reactive and refractory metals and alloys
S00001 to S99999	Corrosion and heat-resistant metals
T00001 to T99999	Tool steels, wrought, and cast
W00001 to W99999	Welding filler metals
Z00001 to Z99999	Zinc and zinc alloys

Table D-5 AISI-SAE versus UNS Classifications for Mild Steel

Plain Carbon Steels

AISI-SAE Number	UNS Number	Properties
1018	G10180	AISI **1018** low carbon **steel** provides a good balance of toughness, strength and ductility. AISI **1018** hot rolled **steel** has improved machining characteristics and Brinell hardness.
1045	G10450	AISI 1045 steel is a medium carbon steel which has good tensile strength and hardness. It can also be machined easily.
1117	G11170	1117 is a low-carbon, high-manganese steel with good mechanical properties and good machinability. 1117 is best suited for machined parts that need to be case-hardened. Case hardening provides good surface hardness with a ductile core.
11L17	G11174	The addition of lead improves the machinability of 11L17 versus 1117. Manganese is added to increase hardenability. 11L17 can be used where case hardenability and good machinability are required.
1144	G11440	1144 is a medium carbon steel with good mechanical properties and free-machining characteristics.
1215	G12150	1215 steel has good machinability characteristics and good surface finish. It is generally used where mechanical properties can be sacrificed for machinability.
12L14	G12144	12L14 is a free-machining steel. This is free cutting steel is used for general machining for a wide variety of parts.

Table D-6 shows some examples stainless steel classification in AISI-SAE and UNS systems.

Table D-7 shows some examples alloy steel classification in AISI-SAE and UNS systems.

Table D-8 shows some examples of tool steel classification in AISI-SAE and UNS systems.

Table D-6 AISI-SAE versus UNS Classifications for Stainless Steels

Stainless Steels		
SAE	**UNS**	
201	S20100	Austenitic general purpose stainless steel, hardenable through cold working
303	S30300	Free machining version of 304
304	S30400	Most common grade of stainless steel
316	S31600	Second most common grade of stainless steel. Food and medical Use.
416	S41600	Good machinability
501	S50100	Heat-resisting stainless steel

Table D-7 AISI-SAE versus UNS Classifications for a Few Alloy Steels

AISI-SAE	UNS
1330	G13300
4140	G41400
4340	G43400
5117	G51170
5120	G51200
6118	G61180
8620	G86200

Table D-8 Characteristics of Tool Steel Grades

Tool Steels		
SAE	UNS	High-speed tool steels. M is molybdenum type steel. T is tungsten-type tool steel
M1	T11301	
M2	T11302	
T6	T12006	
T8	T12008	
H10	T20810	Hot-work tool steels. Oil quenching is used where maximum hardness is required. Air hardening is recommended, however, for most applications. H-type tool steels were developed for their strength and hardness during prolonged exposure to elevated temperatures
H11	T20811	
A2	T30102	Cold work tool steels, air hardening
A6	T30106	
D2	T30402	Cold work tool steels, air hardening. D-series tool steel is high-carbon chromium steel
D3	T30403	
O1	T31501	Cold work tool steels, oil hardening
O2	T31502	
S1	T41901	Shock-resisting tool steels. S-type steels are designed to resist shock at low and high temperatures
S7	T41907	
W1	T72301	Water-hardening tool steel
W2	T72302	
P4	T51604	Mold steels. They are designed to meet the requirements of zinc die casting and plastic injection molding dies
F1	T60601	Special-purpose, carbon-tungsten type steel. F-type tool steel is water hardened and substantially more wear resistant than W-type tool steel
L2	T61202	Special-purpose, low-alloy type steel tool steel is short for low alloy special purpose tool steel. L6 is extremely tough

SHOP TESTS FOR IDENTIFYING STEELS

One of the disadvantages of steel identification systems is that the marking on the steel is often lost. The end of a shaft is usually marked. If the marking is cut off and the piece is moved from its proper storage rack, it is difficult to know what it is. This shows the necessity of returning stock to its proper rack. It is also good practice to always cut off the unmarked end of the stock material so that the remaining material is still identifiable.

Unfortunately, shops always end up with short ends and otherwise useful pieces that have become unidentified. There are methods a machinist may use to identify the basic type of steel in an unknown sample. By a process of elimination, the machinist can then determine which of the several steels of that type in the shop is most comparable to the sample. The following are several methods of shop testing that you can use.

Visual Tests

Some metals can be identified by visual observation of their surface finishes. Heat scale or black mill scale is found on all hot-rolled (HR) steels. These can be low-carbon (.05–.30 percent), medium-carbon (.30–.60 percent), high-carbon (.60–1.70 percent), or alloy steels. Other surface coatings that may be detected are sherardized, plated, case-hardened, or nitrided surfaces. *Sherardizing* is a process in which zinc vapor is inoculated into the surface of iron or steel.

Cold finish (CF), also called cold rolled (CR), steel usually has a metallic luster. Ground and polished (G and P) steel has a bright, shiny finish with closer dimensional tolerances than CR. Also, cold-drawn ebonized, or black, finishes are sometimes found on alloy and resulfurized (free-machining) shafting.

Chromium–nickel stainless steel, which is austenitic and nonmagnetic, usually has a white appearance. Straight 12–13 percent chromium is ferritic and magnetic with a bluish-white color. Manganese steel is blue when polished, but copper-colored when oxidized. White cast iron fractures will appear silvery or white. Gray cast iron fractures appear dark gray and will smear a finger with a gray graphite smudge when touched.

Magnet Test

All ferrous metals such as iron and steel are magnetic—that is, they are attracted to a magnet. Nickel, which is nonferrous is also magnetic. Ferritic and martensitic (400 series) stainless steels are also magnetic and cannot be separated from other steels by this method. Austenitic (300 series) stainless steel is not magnetic unless it is work hardened.

Hardness Test

Wrought iron is quite soft, since it contains almost no carbon or any other alloying element. Generally speaking, the more carbon (up to 2 percent) and other elements that steel contains, the harder, stronger, and less ductile it becomes, even if it is in an annealed state. Thus, the hardness of a sample can help us separate low-carbon steel from an alloy steel or a high-carbon steel. The best way to test hardness is with a hardness tester. Rockwell, Brinell, and other types of hardness testing will be studied in another unit. Not all machine shops have hardness testers available, in which case the following shop methods can prove useful.

Scratch Test

Geologists and "rock hounds" scratch rocks against items of known hardness for identification purposes. The same method can be used to check metals for relative hardness. Simply scratch one sample with another and the softer sample will be marked. Be sure that all scale or other surface impurities have been removed before scratch testing. A variation of this method is to strike two similar edges of two samples together. The one receiving the deeper indentation is the softer of the two.

File Tests

Files can be used to establish the relative hardness between two samples, as in the scratch test, or they can determine an approximate hardness of a piece of steel. Table D-9 gives the Rockwell and Brinell hardness numbers for this file test when using new files. This method, however, can only be as accurate as the skill that the user has acquired through practice.

Take care not to damage the file by filing on hard materials. Testing should be done on the tip or near the edge.

Spark Testing

Spark testing is a way to test carbon content in steels. When held against a grinding wheel, the metal tested will display a particular spark pattern depending on its content. Spark testing provides a convenient means of distinguishing between tool steel (medium or high carbon) and low-carbon steel. High-carbon steel (Figure D-4) shows many more bursts than low-carbon steel (Figure D-5). It must be noted here that spark testing is by no means an exact method of identifying a particular metal. Other methods of testing should be used if there is any question of the type of metal, especially if it is vitally important that the correct metal be used.

Almost all tool steels contain alloying elements in addition to carbon, which affect the carbon burst. Chromium, molybdenum, silicon, aluminum, and tungsten suppress the carbon burst. For this reason, spark testing does little to determine the content of an unknown steel sample. It is useful, however, as a comparison test. Comparing the spark of

Figure D-4 High-carbon steel. Short, very white or light-yellow carrier lines with considerable forking, having many star-like bursts. Many of the sparks follow around the wheel.

Table D-9 File Test and Hardness Table

Steel Type	Rockwell B	Rockwell C	Brinell	Observed Action
Low-carbon steel	65		100	File will easily bite into the steel. Good machinability.
Medium-carbon steel		16	212	File will bite with pressure. Easily machined with high-speed steel tools.
High-alloy steel High-carbon steel		31	294	File will only bite into the material with difficulty. Readily machinable with carbide tools.
Tool steel		42	390	Steel can only be filed with extreme pressure. Difficult to machine even with carbide tools.
Hardened tool steel		50	481	File will mark metal, but metal is nearly as hard as the file and machining is impractical; should be ground.
Case-hardened steel and hardened tool steel		64	739	Metal is as hard as the file; should be ground.

Note: File testing is not an accurate way to test hardness, Rockwell and Brinell hardness numbers in this table are approximations.

Figure D-5 Low-carbon steel. Straight carrier lines, yellowish in color, with a small amount of branching and few carbon bursts.

Figure D-6 Cast iron. Short carrier lines with many bursts that are red near the grinder and orange-yellow farther out. Considerable pressure is required to produce sparks on cast iron.

Figure D-7 High-speed steel. Carrier lines are orange, ending in pear-shaped globules with little branching or carbon sparks. High-speed steel requires moderate pressure to produce sparks.

a known sample with that of an unknown sample can be an effective method of identification by a trained observer. Cast iron may be distinguished from steel by the characteristic spark stream (Figure D-6). High-speed steel can also be readily identified by spark testing (Figure D-7).

When spark testing, always wear safety glasses or a face shield. Adjust the wheel guard so that the spark will fly outward and downward, and away from you. Use a coarse-grit wheel that has been freshly dressed to remove contaminants.

Machinability Test

Machinability can be used in a simple comparison test to determine a specific type of steel. For example, two unknown samples identical in appearance and size can be test cut in a machine tool, using the same speed and feed for both. The ease of cutting should be compared, and chips observed for heating color and curl. See Section F for machinability ratings.

Other Tests

Metals can often be identified by their reaction to certain chemicals, usually acids or alkaline substances. Commercial spot testers, available in kits, can be used to identify some metals. Only certain metals can be identified with the chemical test, and the specific amount of an alloying element in a metal is not revealed.

Spectrographic analysis of an unknown metal or alloy can be made in a laboratory. Portable spectroscopes are used to sort metals for purposes such as scrap evaluation. This is especially useful for selecting scrap for electric furnaces that produce high-quality steels. Since each element produces characteristic emission lines, it is possible to identify the presence or absence of each element and measure its quantity. These mobile testers are calibrated to a known standard and match signal. A display on the handheld tester identifies the metal or alloy.

X-ray analyzers are among the best methods of identifying the content of metals and are available as portable units. Although X-ray and spectrographic analyzers are by far the better methods of identifying metals, they are expensive devices and are probably practical only when large amounts of material must be sorted and identified.

Material Selection

Several properties should be considered when selecting the type of steel for an application: strength, machinability, hardenability, weldability, fatigue resistance, and corrosion resistance.

Manufacturers' catalogs and reference books are available for selection of standard structural shapes, bars, and other steel products (Figure D-8). Others are available for the stainless steels, tool steels, finished carbon steels, and alloy shafting. Many of these steels are known by a trade name.

A machinist may have to select which type of steel to use for a shaft. Round stock is manufactured with three kinds of surface finish: hot rolled (HR), cold finished (CF), also called cold rolled (CR) and ground and polished (G and P). Tolerances

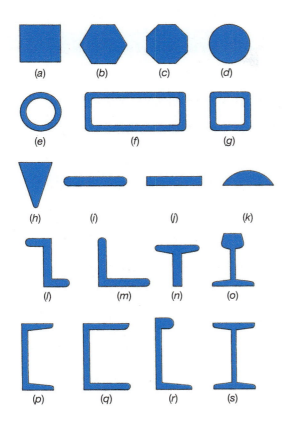

Figure D-8 Steel shapes used in manufacturing: (*a*) square HR or CR; (*b*) hexagonal; (*c*) octagon; (*d*) round; (*e*) tubing and pipe (round); (*f*) HREW (hot-rolled electric welded) rectangular steel tubing; (*g*) HREW square steel tubing; (*h*) wedge; (*i*) HR flat bar (round edge spring steel flats); (*j*) flat bar (CR and HR); (*k*) half round; (*l*) zee; (*m*) angle; (*n*) tee; (*o*) rail; (*p*) channel; (*q*) car and ship channel; (*r*) bulb angle; (*s*) beams—I, H, and wide flange (*Neely, John E; Bertone, Thomas J.*, Practical Metallurgy and Materials of Industry, *6th, ©2003. Electronically reproduced by permission of Pearson Education, Inc., Upper Saddle River, New Jersey*).

are kept much closer on ground and polished shafts. The following are some common steels used for shafting:

1. SAE 4140 is a chromium–molybdenum alloy with .40 percent carbon. It lends itself readily to heat-treating, forging, and welding. It provides a high resistance to torsional and reversing stresses, such as those to which drive shafts are exposed.

2. SAE 1140 is a resulfurized, drawn, free-machining bar stock. The term free-machining means that the steel has characteristics that make it much easier to machine. SAE 1140 steel has good resistance to bending stresses because of its fibrous qualities, and it has a high tensile strength. It is best used on shafts where the rpm (revolutions per minute) are high and the torque is low. SAE 1140 is also useful where stiffness is a requirement. It should not be heat-treated or welded.

3. Leaded steels have all the free-machining qualities and finishes of resulfurized steels. Leaded alloy steels,

such as SAE 41L40 have the superior strength of 4140 but are much easier to machine.

4. SAE 1040 is a medium-carbon steel that has a normalized tensile strength of about 85,000 psi. It can be heat-treated, but large sections will be hardened only on the surface and the core will still be in a soft condition. Its main advantage is that it is a less expensive way to obtain a higher-strength part.

5. SAE 1020 is a low-carbon steel with good machining characteristics. It normally comes as cold rolled (CR) bar stock. It is commonly used for shafting in industrial applications. It has a lower tensile strength than the alloy steels or higher-carbon steels.

COSTS OF STEEL

Steel prices, like prices of most other products, change constantly, so costs can be shown only as a rough example. Steel is usually priced by its weight. A cubic foot of mild steel weighs 489.60 lb, so a square foot 1 inch thick weighs 40.80 lb. From this, you can easily compute the weighs for flat materials such as plate. For hexagonal and rounds, it is easier to consult a table in a catalog or handbook. Given a price per pound, you should then be able to figure the rough cost of a desired steel product.

EXAMPLE

A 1-inch by 6-inch mild steel bar is 48 inches long. If current steel prices are 95 cents per pound, roughly, how much does the bar cost?

$$\frac{6 \times 48}{144} = 2 \text{ ft}^2, \text{ 1 inch thick}$$

$$2 \times 40.80 = 81.6 \text{ lb}$$

$$81.6 \times \$0.95 = \$77.52$$

SELF-TEST

1. By what universal coding system are carbon and alloy steels designated?

2. What are three basic types of stainless steels, and what is the number series assigned to them? What are their basic differences?

3. If your shop stocked the following steel shafting, how would you determine the content of an unmarked piece of each, using shop tests as given in this unit?

   a. AISI C1020 CR
   b. AISI B1140 (G and P)
   c. AISI C4140 (G and P)
   d. AISI 8620 HR
   e. AISI B1140 (Ebony)
   f. AISI C1040

4. A small part has obviously been made by a casting process. Using an inexpensive testing method, how can you determine whether it is a ferrous or a nonferrous metal, or if it is steel or white or gray cast iron?

5. What is the meaning of the symbols O1 and W1 when applied to tool steels?

6. A $2\frac{7}{16}$ inch-diameter steel shaft weighs 1.32 lb per linear inch, as taken from a table of weights of steel bars. A 40-inch length is needed for a job. At 95 cents per pound, what would the shaft cost?

7. When you check the hardness of a piece of steel with the file test, the file slides over the surface without cutting.
   a. Is the steel piece readily machinable?
   b. What type of steel is it most likely to be?

8. Steel that is nonmagnetic is called _____.

9. Which nonferrous metal is magnetic?

10. List at least four properties of steel that are important when selecting the material for a job.

INTERNET REFERENCES

Information on steels and alloy steels:

http://www.efunda.com/materials/alloys/alloy_home/alloys.cfm

http://www.autosteel.org

http://www.fordtoolsteels.com

Selection and Identification of Nonferrous Metals

Metals are designated as either ferrous or nonferrous. Iron and steel are ferrous metals, and any metal other than iron or steel is called nonferrous. Nonferrous metals such as gold, silver, copper, and tin were in use hundreds of years before the smelting of iron, and yet some nonferrous metals have only recently become available. For example, aluminum was first commercially extracted from ore in 1886 by the Hall–Heroult process, and titanium is a space age metal produced in commercial quantities only since World War II.

In general, nonferrous metals are more costly than ferrous metals. It isn't always easy to distinguish a nonferrous metal from a ferrous metal. This unit should help you to identify, select, and properly use many of these metals.

OBJECTIVES

After completing this unit, you should be able to:

- Identify and classify nonferrous metals by a numerical system.
- List the general appearance and use of various nonferrous metals.

ALUMINUM

Aluminum is white or white-gray in color and can have any surface finish from dull to shiny and polished. An anodized surface is frequently found on aluminum products. Aluminum weighs 168.5 pounds per cubic foot (lbs/ft^3) as compared with 487 lbs/ft^3 for steel, and it has a melting point of 1220°F (660°C) when pure. It is readily machinable and can be manufactured into almost any shape or form.

Magnesium is also a much lighter metal than steel, as it weighs 108.6 lb/ft^3 and looks much like aluminum. To distinguish between the two metals, it is sometimes necessary to perform a chemical test. A zinc chloride solution in water,

Table D-10 Aluminum and Aluminum Alloys

Code Number	Major Alloying Elements
1xxx	None
2xxx	Copper
3xxx	Manganese
4xxx	Silicon
5xxx	Magnesium
6xxx	Magnesium and silicon
7xxx	Zinc
8xxx	Other elements
9xxx	Unused

or a copper sulfate solution, will blacken magnesium immediately but will not change aluminum.

Several numerical systems identify types of aluminum, such as federal, military, the American Society for Testing and Materials (ASTM), and SAE specifications. The system most used by manufacturers, however, is one adopted by the Aluminum Association in 1954.

From Table D-10 you can see that the first digit of a number in the aluminum alloy series indicates the alloy type. The second digit, represented by an x in the table, indicates any modifications made to the original alloy. The last two digits indicate the numbers of similar aluminum alloys of an older marking system, except in the 1100 series, where the last two digits indicate the amount of pure aluminum above 99 percent contained in the metal.

EXAMPLES

2011 is a copper-based aluminum alloy that is commonly used for small precision gears, machine parts, pipe stems, auto fuel system components, tube (tubing) fittings, camera parts, industrial connectors, etc.

6061 is a magnesium- and silicon-based aluminum that is commonly used for structural components, frames, brackets, jigs, fixtures, machine parts, marine components, electrical connectors, couplings, valves and valves parts, gears and shafts, pistons, fasteners, bike frames, etc. 6061 is used for structural parts requiring good strength-to-weight ratio and good corrosion resistance. 6061 is easily cold worked and formed in the annealed condition. 6061 is easily machined, stamped, and bent. 6061 is also a good choice for spinning and deep drawing.

7075 aluminum has a high strength-to-weight ratio. 7075 is used for highly stressed structural parts. Applications include aircraft parts, valve parts, gears and shafts, gears, valve parts, worm gears, and various other components.

EXAMPLES

An aluminum alloy numbered 5056 is an aluminum–magnesium alloy, where the first 5 represents the alloy magnesium, the 0 represents modifications to the alloy, and 5 and 6 are the numbers of a similar aluminum of an older marking system. An aluminum numbered 1120 contains no major alloy and has .20 percent pure aluminum above 99 percent.

Aluminum and its alloys are produced as castings or as wrought (cold-worked) shapes such as sheets, bars, and tubing. Aluminum alloys are harder than pure aluminum and will scratch the softer (1100 series) aluminums. Pure aluminum and some of its alloys cannot be heat-treated, so they are tempered by other methods. The temper designations are indicated by a letter that follows the four-digit alloy series number:

—F   As fabricated. No special control over strain hardening or temper designation is noted.

—O   Annealed, recrystallized wrought products only. Softest temper.

—H   Strain-hardened, wrought products only. Strength is increased by work hardening.

The letter —H is always followed by two or more digits. The first digit, 1, 2, or 3, denotes the final degree of strain hardening:

—H1   Strain hardened only

—H2   Strain hardened and partially annealed

—H3   Strain hardened and stabilized

and the second digit denotes higher-strength tempers obtained by heat treatment:

$2\frac{1}{4}$ hard

$4\frac{1}{2}$ hard

$6\frac{3}{4}$ hard

8 Full hard

EXAMPLE

5056-H18 is an aluminum–magnesium alloy, strain hardened to a full hard temper. Some aluminum alloys can be hardened to a great extent by a process called *solution heat treatment and precipitation* or *aging*. This process involves heating the aluminum and its alloying elements until it is a solid solution. The aluminum is then quenched in water and allowed to age or is artificially aged by heating slightly. The aging produces an internal strain that hardens and strengthens the aluminum. Some other nonferrous metals are also hardened by this process. For these aluminum alloys the letter *T* follows the four-digit series number. Numbers 2 to 10 follow this letter to indicate the sequence of treatment.

—T2 Annealed (cast products only)
—T3 Solution heat-treated and cold worked
—T4 Solution heat-treated, but naturally aged
—T6 Solution heat-treated and artificially aged
—T8 Solution heat-treated, cold worked, and artificially aged
—T9 Solution heat-treated, artificially aged, and cold worked
—T10 Artificially aged and then cold worked

EXAMPLES

2024-T6 is an aluminum–copper alloy, solution heat-treated and artificially aged. Cast aluminum alloys generally have lower tensile strength than wrought alloys. Sand castings, permanent mold, and die casting alloys are of this group. They owe their mechanical properties to solution heat treatment and precipitation or to the addition of alloys. A classification system similar to that for wrought aluminum alloys is used (Table D-11).

The cast aluminum 108F, for example, has an ultimate tensile strength of 24,000 psi in the as-fabricated condition and contains no alloy. The 220.T4 copper–aluminum alloy has a tensile strength of 48,000 psi.

OTHER NONFERROUS METALS

Cadmium

Cadmium has a blue-white color and is commonly used as a protective plating on parts such as screws, bolts, and washers. It is also used as an alloying element to make metal

Table D-11 Cast Aluminum Alloy Designations

Classification	Major Alloying Element
1xx.x	None, 99 percent aluminum
2xx.x	Copper
3xx.x	Silicon with Cu and/or Mg
4xx.x	Silicon
5xx.x	Magnesium
6xx.x	Zinc
7xx.x	Tin
8xx.x	Unused series
9xx.x	Other major alloys

Source: *Neely, John E; Bertone, Thomas J., Practical Metallurgy and Materials of Industry, 6th, ©2003. Electronically reproduced by permission of Pearson Education, Inc., Upper Saddle River, New Jersey).*

alloys that melt at low temperature, such as bearing metals, solder, type casting metals, and storage batteries. Cadmium compounds such as cadmium oxide are toxic and can cause illness when breathed. Toxic fumes can be produced by welding, cutting, or machining on cadmium-plated parts. Breathing the fumes should be avoided by using adequate ventilation systems. The melting point of cadmium is 610°F (321°C) and it weighs 539.6 lb per cubic foot.

Copper and Copper Alloys

Copper is a soft, heavy metal with a reddish color. It has high electrical and thermal conductivity when pure, but loses these properties to a certain extent when alloyed. It must be strain hardened when used for electric wire. Copper is ductile and can be easily drawn into wire or tubular products. It is so soft that it is difficult to machine, and it has a tendency to adhere to tools. Copper can be work hardened or hardened by solution heat treatment when alloyed with beryllium. The melting point of copper is 1981°F (1083°C) and it weighs 554.7 lb per cubic foot.

Beryllium Copper

Beryllium copper, an alloy of copper and beryllium, can be hardened by heat-treating for making nonsparking tools and other products. This metal should be machined after solution heat treatment and aging, not when it is in the annealed state. Machining or welding beryllium copper can be hazardous if safety precautions are not followed. Machining dust or welding fumes should be removed by a heavy coolant flow or by a vacuum exhaust system. A respirator type of face mask should be worn when around these two hazards. The melting point of beryllium is 2435°F (1285°C) and it weighs 115 lb per cubic foot.

Brass

Brass is an alloy of zinc and copper. The color of brass ranges from white to yellow, and in some alloys, red to yellow. Brasses range from gilding metal used for jewelry (95 percent copper, 5 percent zinc) to Muntz metal (60 percent copper, 40 percent zinc) used for bronzing rod and sheet stock. Brasses are easily machined. Brass is usually tougher than bronze and produces a stringy chip when machined. The melting point of brasses ranges from 1616°F to 1820°F (880°C to 993°C) and their weights range from 512 to 536 lb per cubic foot.

Bronze

Bronze is found in many combinations of copper and other metals, but copper and tin are its original elements. Bronze colors range from red to yellow. Phosphor bronze contains 91.95 percent copper, 0.05 percent phosphorus, and 8 percent tin. Aluminum bronze is often used in the shop for making bushings or bearings that support heavy loads. (Brass is not normally used for making antifriction bushings.)

The melting point of bronze is approximately 1841°F (1005°C) and it weighs about 548 lb per cubic feet. Bronzes are usually harder than brasses but are easily machined with sharp tools. The chip produced is often granular. Some bronze alloys are used as brazing rods.

Chromium

Chromium, a slightly gray metal, can take a high polish. It has a high resistance to corrosion by most reagents; exceptions are diluted hydrochloric and sulfuric acids. Chromium is widely used for decorative plating on automobile parts and other products.

Chromium is not a very ductile or malleable metal, and its brittleness limits its use as an unalloyed metal. It is commonly alloyed with steel to increase hardness and corrosion resistance. Chrome–nickel and chrome–molybdenum are two common chromium alloys. Chromium is also used in electrical heating elements such as chromel or nichrome wire. The melting point of chromium is 2939°F (1615°C) and it weighs 432.4 lb per cubic foot.

Die-Cast Metals

Castings are produced with various metal alloys by the process of die casting. Die casting is a method of casting molten metal by forcing it into a mold. After the metal has solidified, the mold opens and the casting is ejected. Carburetors, door handles, and many small precision parts are manufactured using this process (Figure D-9). Die-cast alloys, often called "pot metals," are classified in six groups:

1. Tin-based alloys
2. Lead-based alloys
3. Zinc-based alloys
4. Aluminum-based alloys
5. Copper, bronze, or brass alloys
6. Magnesium-based alloys

The specific content of the alloying elements in each of the many die-cast alloys can be found in handbooks or other references on die casting.

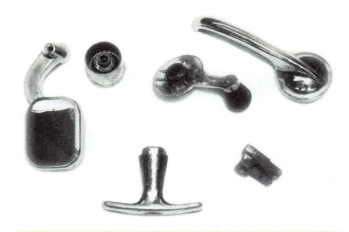

Figure D-9 Die-cast parts.

Lead and Lead Alloys

Lead is a heavy metal, silvery when newly cut and gray when oxidized. Lead has high density, low tensile strength, low ductility (cannot be easily drawn into wire), and high malleability (can be easily compressed into a thin sheet).

Lead has high corrosion resistance and is alloyed with antimony and tin for various uses. It is used as shielding material for nuclear and X-ray radiation, for cable sheathing, and for battery plates. Lead is added to steels, brasses, and bronzes to improve machinability. Lead compounds are toxic; they also accumulate in the body. Small amounts ingested over a period of time can be fatal. The melting point of lead is 621°F (327°C) and it weighs 707.7 lb per cubic foot.

A Babbitt metal is a soft, antifriction alloy metal that is sometimes used for bearings and is usually tin or lead based (Figure D-10). Tin Babbitt contains from 65 to 90 percent tin with antimony, lead, and a small percentage of copper added. These are the higher grade and generally the more expensive of the two types. Lead Babbitt contains up to 75 percent lead, with antimony, tin, and some arsenic making up the difference.

Cadmium-based Babbitt resists higher temperatures than tin- and lead-based types. These alloys contain from 1 to 15 percent nickel or a small percentage of copper and up to 2 percent silver. The melting point of Babbitt is approximately 480°F (248°C).

Magnesium

Magnesium is a soft, silver-white metal that closely resembles aluminum but weighs less. Unlike aluminum, magnesium will readily burn with a brilliant white light; thus, magnesium presents a fire hazard when machined. Magnesium, similar to aluminum in density and appearance, presents quite different machining problems. Although magnesium chips can burn in air, applying water causes the chips to burn more fiercely. Sand or special compounds should be used to extinguish these fires. Water-based coolant should never be used when machining magnesium because if the chips were to ignite, water would intensify the fire. Magnesium can be machined dry when light cuts are taken and the heat is dissipated. Compressed air is sometimes used as a coolant. Anhydrous (containing no water) oils having a high flash point and low viscosity are used in most production work. Magnesium is machined with high surface speeds and with tool angles similar to those used for aluminum.

Cast and wrought magnesium alloys are designated by SAE and ASTM numbers, which may be found in metal reference books such as the *Machinery's Handbook*. The melting point of magnesium is 1204°F (651°C) and it weighs 108.6 lb per cubic foot.

Molybdenum

Molybdenum is used for high-temperature applications; when machined, it chips like gray cast iron. It is used as an alloying element in steel to promote deep hardening and to increase tensile strength and toughness. Pure molybdenum is used for filament supports in lamps and in electron tubes. The melting point of molybdenum is 4788°F (2620°C) and it weighs 636.5 lb per cubic foot.

Nickel

Nickel is noted for its resistance to corrosion and oxidation. It is a whitish metal used for electroplating and as an alloying element in steel and other metals to increase ductility and corrosion resistance. It resembles pure iron in some ways but has greater corrosion resistance. Nickel can be used for electroplating. *Electroplating* is the coating or covering of another material with a thin layer of metal, using electricity to deposit the layer.

When spark tested, nickel throws short orange carrier lines with no sparks or sprigs (Figure D-11). Nickel is attracted to a magnet but becomes nonmagnetic near 680°F (360°C). The melting point of nickel is 2646°F (1452°C), and it weights 549.1 lb per cubic foot.

Nickel-Based Alloys

Monel is an alloy of 67 percent nickel and 28 percent copper, plus impurities such as iron, cobalt, and manganese. It is a tough but machinable, ductile, and corrosion-resistant alloy. Its *tensile strength* (resistance of a metal to a force tending to tear it apart) is 70,000–85,000 lb/in². Monel metal is used

Figure D-10 Babbitted pillow block bearings.

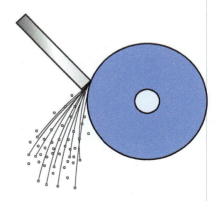

Figure D-11 Spark test for nickel.

to make marine equipment such as pumps, steam valves, and turbine blades. On a spark test, Monel shoots orange-colored, straight sparks about 10 inches long, similar to those of nickel. K-Monel contains 3–5 percent aluminum and can be hardened by heat treatment.

Chromel and nichrome are two nickel–chromium–iron alloys used as resistance wire for electric heaters and toasters. Nickel–silver contains nickel and copper in similar proportions to Monel but also contains 17 percent zinc. Other nickel alloys, such as Inconel, are used for parts exposed to high temperatures for extended periods.

Inconel, a high-temperature and corrosion-resistant metal consisting of nickel, iron, and chromium, is often used for aircraft exhaust manifolds because of its resistance to high-temperature oxidation (scaling). The nickel alloys' melting point range is 2425°F to 2950°F (1329°C to 1621°C).

Precious Metals

Gold is used in dentistry, electronics, chemical industries, and jewelry. In the past, gold was used mostly for coins. Gold coins were usually hardened by alloying with about 10 percent copper. Silver is alloyed with 8–10 percent copper for coinage and jewelry. Sterling silver is 92.5 percent silver in English coinage and has been 90 percent silver for American coinage. Silver has many commercial uses, such as an alloying element for mirrors, in photographic compounds, and electrical equipment. Silver has high electrical conductivity. Silver is used in silver solders, which are stronger and have a higher melting point than lead–tin solders.

Platinum, palladium, and iridium as well as other rare metals, are even rarer than gold. These metals are used commercially because of their special properties such as extremely high resistance to corrosion, high melting points, and high hardness. The melting points of some precious metals are as follows: gold, 1945°F (1063°C); iridium, 4430°F (2443°C); platinum, 3224°F (1773°C); and silver, 1761°F (961°C). Gold weighs 1204.3 lb per cubic foot, and silver weighs about 654 lb per cubic foot. Platinum is one of the heaviest of metals, with a weight of 1333.5 lb per cubic foot. Iridium is also a heavy metal, at 1397 lb/ft^3.

Tantalum

Tantalum, a bluish gray metal, is difficult to machine because it is quite soft and ductile, and chips cling to the cutting tool. It is immune to corrosive acids except hydrofluoric and fuming sulfuric acids. It is used for high-temperature operations above 2000°F (1093°C) Tantalum is also used for surgical implants and in electronics. Tantalum carbides are combined with tungsten carbides for cutting tools that have high abrasive resistance. The melting point of tantalum is 5162°F (2850°C) it weighs 1035.8 lb per cubic foot.

Tin

Tin has a white color with slightly bluish tinge. It is whiter than silver or zinc. Since tin has a good corrosion resistance, it is used to plate steel, especially cans for the food processing industry. Tin is used as an alloying element for solder, Babbitt, and pewter. A popular solder is an alloy of 50 percent tin and 50 percent lead. Tin is alloyed with copper to make bronze. The melting point of tin is 449°F (232°C) and it weighs 454.9 lb per cubic foot.

Titanium

The strength and light weight of this silver-gray metal make it useful in the aerospace industries for jet engine components, heat shrouds, and rocket parts. Pure finely divided titanium can ignite and burn when heated to high temperatures. Pure titanium has a tensile strength of 60,000–110,000 psi, similar to that of steel. Alloying titanium can increase its tensile strength considerably. Titanium weighs about half as much as steel and, like stainless steel, is a relatively difficult metal to machine. It can be machined with rigid setups, sharp tools, slower surface speed, and proper coolants. When spark tested, titanium throws a brilliant white spark with a single burst on the end of each carrier (Figure D-12). The melting point of titanium is 3272°F (1800°C) and it weighs 280.1 lb per cubic foot.

Tungsten

Tungsten has been used for incandescent light filaments. It has the highest known melting point 6098°F (3370°C) of any metal but is not resistant to oxidation at high temperatures. Tungsten is used for rocket engine nozzles and welding electrodes and as an alloying element with other metals. Machining pure tungsten is difficult with single-point tools, and grinding is preferred for finishing operations. Tungsten carbide compounds are used to make extremely hard and heat-resistant lathe tools and milling cutters by compressing the tungsten carbide powder into a briquette and sintering it in a furnace. Tungsten weighs about 1180 lb per cubic foot.

Zinc

The familiar galvanized steel is actually steel plated with zinc and is used mainly for its high corrosion resistance. Zinc alloys are widely used as die-casting metals. Zinc and

Figure D-12 Titanium spark testing.

zinc-based die-cast metals conduct heat much more slowly than aluminum. The rate of heat transfer on similar shapes of aluminum and zinc is a means of distinguishing between them. The melting point of zinc is 787°F (419°C) and it weighs about 440 lb per cubic foot.

Zirconium

Zirconium is similar to titanium in appearance and physical properties. It was once used as an explosive primer and as a flashlight powder for photography because, like magnesium, it readily combines with oxygen and rapidly burns. Machining zirconium, like titanium, requires rigid setups and slow surface speeds. Zirconium has an extremely high resistance to corrosion from acids and seawater. Zirconium alloys are used in nuclear reactors, flash bulbs, and surgical implants such as screws, pegs, and skull plates. When spark tested, it produces a spark similar to that of titanium. The melting point of zirconium is 3182°F (1750°C). It weighs 339 lb per cubic foot.

SELF-TEST

1. What advantages do aluminum and its alloys have over steel alloys? What are the disadvantages?
2. What is the meaning of the letter H when it follows the four-digit number that designates an aluminum alloy? The meaning of the letter T?
3. Name two ways in which magnesium differs from aluminum.
4. What is the major use of copper? How can copper be hardened?
5. What is the basic difference between brass and bronze?
6. Name two uses for nickel.
7. Lead, tin, and zinc all have one useful property in common. What is it?
8. Thoroughly describe a 2024-T6 steel.
9. Thoroughly describe a 5056-H18 steel.
10. What type of metal can be injected under pressure into a permanent mold?

INTERNET REFERENCES

Information on nonferrous metals:

http://metals.about.com

http://www.redmetals.com

http://www.engineershandbook.com/Materials/nonferrous.htm

Hardening, Case Hardening, and Tempering

Probably the most important property of carbon steels is their ability to be hardened through various types of heat treatments. Steels must be hardened if they are to be used as cutting tools. Various degrees of hardness are also desirable, depending on the application of the tool steels. Heat-treating carbon steels involves critical furnace operations. The proper steps must be carried out precisely, or a failure of the hardened steel will almost surely result. Some products require only surface hardening. This is accomplished through the process of surface or case hardening.

It is often desirable to reduce the hardness of a steel tool slightly to enhance other properties. For example, a chisel must have a hard-cutting edge. It must also have a somewhat less hard but tough shank that will withstand hammering. Toughness is acquired through the tempering process. In this unit you will study the important processes and procedures for hardening, case hardening, and tempering of plain carbon steels.

OBJECTIVES

After completing this unit, you should be able to:

- Correctly harden a piece of tool steel and evaluate your work.
- Correctly temper the hardened piece of tool steel and evaluate your work.
- Describe the proper heat-treating procedures for other tool steels.

HARDENING METALS

Most metals (except copper used for electric wire) are not used commercially in their pure states because they are too soft and ductile and have low tensile strength. When they are alloyed with other elements, such as other metals, they become harder and stronger as well as more useful. A small amount (1 percent) of carbon greatly affects pure iron when alloyed with it. The alloy metal becomes a familiar tool steel

used for cutting tools, files, and punches. Iron with 2–4 $\frac{1}{2}$ percent carbon content yields cast iron.

Iron, steel, and other metals are composed of tiny grain structures that can be seen under a microscope when the specimen is polished and etched. The grain structure, which determines the strength and hardness of a steel, can be seen with the naked eye as small crystals in the rough broken section of a piece (Figure D-13). These crystals or grain structures grow from a nucleus as the molten metal solidifies until the grain boundaries are formed. Grain structures differ according to the allotropic form of the iron or steel. An *allotropic* element is one able to exist in two or more forms with various properties without a change of chemical composition. Carbon exists in three allotropic forms: amorphous (charcoal, soot, coal), graphite, and diamond. Iron also exists in three allotropic forms (Figure D-14): ferrite (at room temperature), austenite (above 1670°F (911°C)), and

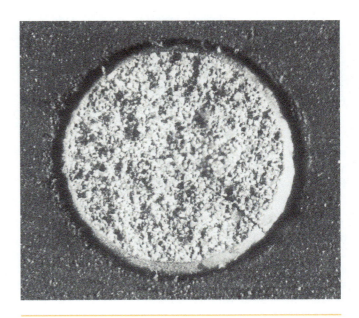

Figure D-13 Single fracture of steel.

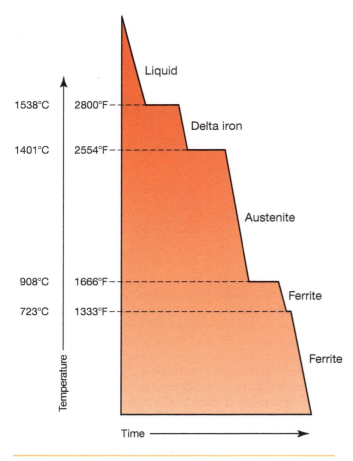

Figure D-14 Cooling curve of iron.

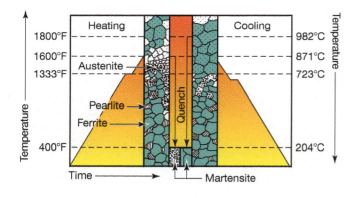

Figure D-15 Critical temperature diagram of .83 percent steel showing grain structures in heating and cooling cycles. Center section shows quenching from different temperatures and the resultant grain structure.

delta (between 2550°F and 2800°F or (1498°C and 1371°C). The points where one phase changes to another are called *critical points* by heat treaters and *transformation points* by metallurgists. Figure D-15 shows the critical temperature diagram for a steel with .83 percent carbon content. This would be slightly under 1 percent carbon content. The figure shows the grain changes as the temperature rises and falls.

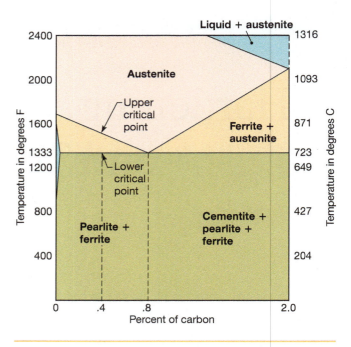

Figure D-16 Simplified phase diagram for carbon steels.

You can also see the different forms for this steel at the critical points. The critical points of water are the boiling point (212°F or 100°C) and the freezing point (32°F or 0°C).

Figure D-16 shows a critical temperature diagram for carbon steel. The lower critical point is always about 1330°F (721°C) in equilibrium or slow cooling, but the upper critical point changes as the carbon content changes.

HEAT-TREATING STEEL

When hardening tool steels, consult a manufacturer's catalog for the correct temperature, time periods, and quenching media. As steel is heated above the critical temperature of 1330°F (721°C) carbon in the form of layers of iron carbide or pearlite (Figure D-17) begins to dissolve in the iron and forms a solid solution called austenite (Figure D-18). When this solution of iron and carbon is suddenly cooled or quenched, a new microstructure is formed. This is called martensite (Figure D-19). Martensite is hard and brittle, having a much higher tensile strength than the steel with a pearlite microstructure. It is quite unstable, however, and must be tempered to relieve internal stresses to have the toughness needed to be useful.

Consider an example. AISI-C1095 has 95 pts of carbon (slightly less than 1 percent). Look at Figure D-16 and find where .95 percent of carbon would be on the horizontal line. Then look up vertically to see where the .95 percent carbon would intersect the Upper Critical Temperature line. AISI-C1095, commonly known as water-hardening (W1) steel, will begin to show hardness when quenched from a temperature just over 1330°F (721°C) but will not harden at all if quenched from a temperature lower than 1330°F (721°C). The steel will become as hard as it can get when

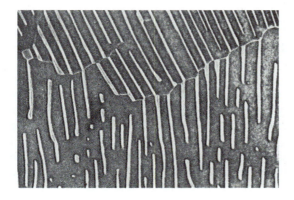

Figure D-17 Replica electron micrograph. Structure consists of lamellar pearlite (11,000×) (*Reprinted with permission of ASM International®. All rights reserved. www.asminternational.org*).

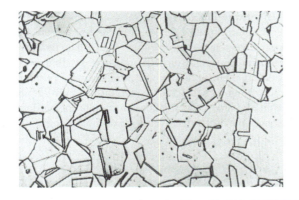

Figure D-18 Microstructure of annealed 304 stainless steel that is austenitic at ordinary temperatures (250×). (*Reprinted with permission of ASM International®. All rights reserved. www. asminternational.org*).

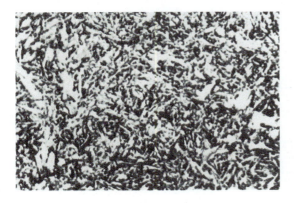

Figure D-19 1095 steel, water quenched from 1500°F (816°C) (1000×) The needle-like structure shows a pattern of fine untempered martensite (*Reprinted with permission of ASM International®. All rights reserved. www.asminternational.org*).

heated to 1450°F (788°C) and quenched in water. This quenching temperature changes as the carbon content of the steel changes. It should be 50°F (10°C) above the upper critical limit into the hardening range (Figure D-20).

Low-carbon steels such as AISI 1020 will not harden when they are heated and quenched. Oil- and air-hardening steels harden over a longer period of time and, consequently, are deeper hardening than water-hardening types, which must be cooled to 200°F (93°C) within 1 or 2 seconds. As you can see, it is quite important to know the carbon content and alloying element so that the correct temperature and quenching medium can be used. Fine-grained tool steels are much stronger than coarse-grained tool steels (Figure D-21). If a piece of tool steel is heated above the correct temperature for its specific carbon content, a phenomenon called *grain growth*

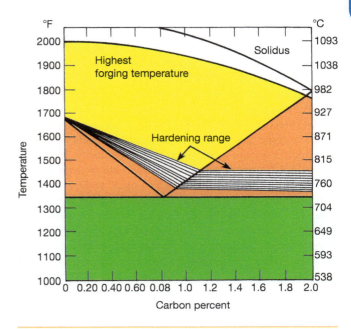

Figure D-20 Temperature ranges used for hardening carbon steel (*Neely, John E, Practical Metallurgy and Materials of Industry, 4th ©1994. Electronically reproduced by permission of Pearson Education, Inc., Upper Saddle River, New Jersey*).

Figure D-21 Fractured ends of $\frac{3}{8}$-inch-diameter 1095 water-quenched tool steel ranging from fine grain, quenched from 1475°F (802°C) to coarse grain, quenched from 1800°F (982°C).

occurs, and a coarse, weak grain structure develops. The grain growth remains when the part is quenched, and if used for a tool such as a punch or chisel, the end may break off when the first hammer blow is struck. Tempering will not remove the coarse grain structure. If the part has been overheated, simply cooling back to the quenching temperature will not help, as the coarse grain persists well down into the hardening range, as shown in Figure D-15. The part should be cooled slowly to room temperature and then reheated to the correct quenching temperature. Steels containing .83 percent carbon can get as hard (RC 67) as any carbon steel containing more carbon.

AISI-C1095, or water-hardening tool steel (W1), can be quenched in oil, depending on the size of the part. For example, for a piece of AISI-C1095 drill rod, if the section is thin or the diameter is small, oil should be used as a quenching medium. Oil is not as severe as water because it conducts heat less rapidly than water and thus avoids quench cracking. Larger sections or parts would not be fully transformed into martensite if they were oil quenched but would instead contain some softer transformation structures. Water quench should be used, but remember that W1 is shallow hardening and will harden only about $\frac{1}{8}$ inch deep.

For example, when using a furnace to harden a part made of 1095 steel, the temperature control should be set for 1450°F (788°C). If the part is small, a preheat is not necessary, but if it is thick, it should be brought up to heat slowly. If the part is left in a furnace without a controlled atmosphere for any length of time, the metal will form an oxide scale, and carbon will leave the surface. This decarburization of the surface will cause it to be soft, but the metal directly under the surface will remain hard after the part is quenched. Decarburization takes time. If the part remains at a hardening temperature for only a short time (up to 15 minutes) the loss of carbon is insignificant for most purposes, and it should not be a concern. It would take a high-temperature soaking period of several hours in an oxidizing atmosphere to remove the carbon to a depth of .010 inch. When heat-treating small parts, the oxidation scale on the surface may be more objectionable than decarburization. This oxidation can be avoided by painting the part with a solution of boric acid and water before heating or wrapping it in stainless steel foil. Place the part in the furnace with tongs, and wear gloves and a face shield for protection. When the part has become the same color as the furnace bricks, remove it by grasping one end with the tongs and *immediately* plunge it into the quenching bath. If the part is long like a chisel or punch, it should be inserted into the quench vertically (straight up and down), not at a slant. Quenching at an angle can cause unequal cooling rates and bending of the part. Also, agitate the part in an up-and-down or a figure-eight motion to remove any gases or bubbles that might cause uneven quenching.

FURNACES

Electric, gas, or oil-fueled furnaces are used for heat-treating steels (Figure D-22). They use various types of controls for temperature adjustment. These controls use a thermocouple to measure the temperature (Figure D-23). The control raises the temperature in the furnace until the set temperature is reached and then maintains that temperature. Temperatures generally range up to 2500°F (1371°C). High-temperature salt baths are also used for heating metals for hardening or annealing. One of the disadvantages of most electric furnaces

Figure D-23 Thermocouple.

Figure D-22 Electric heat-treating furnace. Part is being placed in furnace by heat treater wearing correct attire and using tongs (*Courtesy of Fox Valley Technical College*).

is that they allow the atmosphere to enter the furnace, and the oxygen causes oxides to form on the heated metal. This causes scale and decarburization of the surface of the metal. Decarburization means that carbon is burned off in a thin layer on the surface of the steel. A decarburized surface will not harden. One way to control this loss of surface carbon is to keep a slightly carbonizing atmosphere or an inert gas in the furnace.

The rate at which a part is heated up can be very important in heat-treating. When steel is first heated, it expands. If cold steel is placed in a hot furnace, the surface expands more rapidly than the still-cool core of the part. The surface will then have a tendency to pull away from the center, inducing internal stress. This can cause cracking and distortion in the part. Many furnaces can be adjusted to set the rate of heating (Figure D-24), when bringing the part up to the soaking temperature (Figure D-25). This allows a slow, gradual increase of heat that maintains an even internal and external temperature in the part. Only previously hardened

Figure D-24 Input controls on furnace (*Courtesy of Fox Valley Technical College*).

Figure D-25 Temperature control. The furnace is cold, and the temperature limit control has been set at 1550°F (1371°C) When the furnace is turned on and the controller reaches the preset limit control, the furnace will stop heating (*Courtesy of Fox Valley Technical College*).

or highly stressed parts or those having a large mass need slow heating. Small parts such as chisels and punches made of plain carbon steel can be put directly into a hot furnace.

Soaking means holding the part for a given length of time at a specified temperature. The size of the steel part helps determine the required soak time. A rule of thumb states that the steel should soak in the furnace 1 hour for each inch of thickness, but there are considerable variations to this rule. Some steels require much more soaking time than others. The correct soaking period for any specific tool steel may be found in tool steel reference books.

QUENCHING MEDIA

In general, seven media are used to quench metals. They are listed here in their order of the speed of quenching.

1. Water and salt—that is, sodium chloride or sodium hydroxide (also called *brine*)
2. Tap water
3. Fused or liquid salts
4. Molten lead
5. Soluble oil and water
6. Oil
7. Air

Most hardening is done with plain (tap) water, quenching oil, or air. The mass (size) of the part has a great influence on cooling rates in quenching and, therefore, the amount of hardening in many cases. For example, a razor blade made of plain carbon steel severely (suddenly) quenched in oil would get extremely hard; but the same metal as a 2-inch-square block would quench too slowly in oil to harden much. The hot interior of the larger mass would keep the surface too hot for the quenching medium to remove the heat rapidly. The slow rate of cooling would prevent hardening. Of course, some tool steels, unlike plain carbon steels, do not need a fast quench to harden, so large masses can be quenched in oil or air. These are called *oil-* and *air-hardening* steels.

Liquid quenching media go through three stages. The vapor-blanket stage occurs first because the metal is so hot that it vaporizes the medium, and the metal is enveloped with vapor, which insulates it from the cold liquid bath. Consequently, the cooling rate is relatively slow during this stage. The vapor transport cooling stage begins when the vapor blanket collapses, allowing the liquid medium to contact the surface of the metal. The cooling rate is much higher during this stage. The liquid cooling stage begins when the metal surface reaches the boiling point of the quenching medium. Boiling ceases at this stage, so heat must be removed by conduction and convection. This is the slowest stage of cooling. Figure D-26 shows that the heat treater should be adequately protected from the heat of the part and the quenching medium. It is important in liquid quenching baths to agitate either the quenching medium or the steel being quenched (Figure D-26). The vapor that forms around the part being quenched acts as an insulator and slows down the

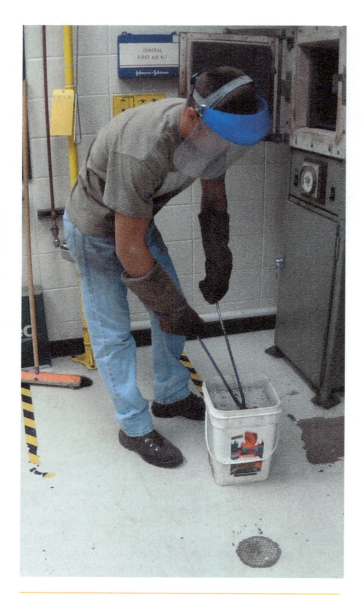

Figure D-26 Heat treater is agitating part during quench (*Courtesy of Fox Valley Technical College*).

cooling rate. This can result in incomplete or spotty hardening of the part. Agitating the part breaks up the vapor barrier. An up-and-down motion works best for long, slender parts held vertically in the quench. A figure-eight motion is sometimes used for heavier parts.

SAFETY FIRST

Gloves and face protection must be used for safety when quenching (Figure D-26). Hot oil could splash up and burn the heat treater's face if a face shield were not worn.

Molten salt is often used for quenching. This is the method of quenching used for *austempering*. Austempered parts are superior in strength and quality to those produced

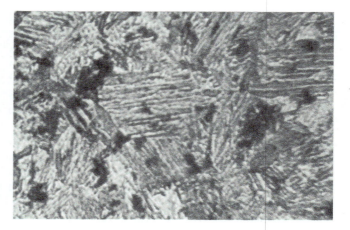

Figure D-27 Lower bainite microstructure.

by the two-stage process of quenching and tempering. The final austempered part is essentially a fine, lower bainite microstructure (Figure D-27). As a rule, only parts such as lawn mower blades and hand shovels that are thin in cross section are austempered.

Another form of isothermal quenching is called *martempering*; the part is quenched in a salt bath at about 400°F (204°C) until the outer and inner parts of the material are brought to the same uniform temperature. The part is next quenched below 200°F (93°C) to transform all the austenite to martensite. Tempering is then carried out in the conventional manner.

Steels are often classified by the type of quenching medium used to meet the requirements of the critical cooling rate. For example, water-quenched steels, which are the plain carbon steels, must have a rapid quench. Oil-hardening steels are alloy steels, and they must be quenched in oil. The air-cooled steels are alloy steels that will harden when allowed to cool from the austenitizing temperature in still air. Air is the slowest quenching medium; however, its cooling rate may be increased by movement (by use of fans, for example).

Step or multiple quenching is sometimes used when the part consists of both thick and thin sections. A severe quench will harden the thin section before the thick section has had a chance to cool. The resulting uneven contraction often results in cracking. With step quenching, the part is quenched for a few seconds in a rapid quenching medium, such as water, followed by a slower quench in oil. The surface is first hardened uniformly in the water quench, and time is provided by the slower quench to relieve stresses.

CASE HARDENING

Low-carbon steels (.08–.30 percent carbon) do not harden to any great extent even when combined with other alloying elements. Therefore, when a soft, tough core and an extremely hard outside surface are needed, case hardening is used. It should be noted that surface hardening is not necessarily case hardening. Flame hardening and induction hardening on the surfaces of gears, lathe ways, and many products depend on the carbon that is already contained in the ferrous metal. On the other hand, case hardening causes carbon from an outside source to penetrate the surface of the steel, carburizing it. Because it raises the carbon content, it also raises the hardenability of the outer surface of the steel.

Carburizing for case hardening can be accomplished in two ways. If only a shallow hardened case is needed, roll carburizing may be used. This consists of heating the part to 1650°F (899°C) and rolling it in a carburizing compound, reheating, and quenching in water. In roll carburizing, a nontoxic compound such as Kasenit should be used unless special ventilation systems are used. Roll carburizing produces a maximum case of about .003 inch. Pack carburizing (Figure D-28) can produce a case of $\frac{1}{16}$ inch. Pack carburizing is done by soaking the steel in a carbonaceous material at about 1700°F (926.6°C). The part is packed in a carburizing compound in a metal box and placed in a furnace long enough to cause the carbon to diffuse sufficiently deep into the surface of the material to produce a case of the required depth. It takes about an 8-hour soak to achieve $\frac{1}{16}$ depth of hardness. The process consist of heating the part in carburizing compound, cooling to room temperature, and reheating to the correct hardening temperature (usually about 1450°F (788°C)). The part is then removed and quenched in water. After case hardening, tempering is not usually necessary, since the core is still soft and tough. Therefore, unlike a hardened piece that is softened by tempering, the surface of a case-hardened piece remains hard, usually about RC 60 (as hard as a file).

Gas carburizing uses the carbon from a carburizing gas atmosphere in a special furnace to add carbon to the steel. A gas flame that has insufficient oxygen produces carbon in the

form of carbon monoxide (CO). This excess carbon diffuses into the heated metal surface. The same thing occurs when steel is improperly cut with an oxyacetylene torch in which the preheat flame has insufficient oxygen. This carbonizing flame allows carbon to diffuse into the molten metal, and the cut edge becomes extremely hard when it is suddenly cooled (quenched) by the mass of steel adjacent to the cut. This hardening of torch-cut steel workpieces often causes difficulties for a machinist when cutting the torch-cut edge.

Nitriding is a case hardening method in which the part is heated in a special container into which ammonia gas is released. Since the temperature used is only 950°F to 1000°F (510°C to 538°C) and the part is not quenched, warpage is kept to a minimum. The iron nitrides thus formed are even harder than the iron carbides formed by conventional carburizing methods.

TEMPERING

Tempering, or *drawing*, is the process of reheating a steel part that has been previously hardened to transform some of the hard martensite into softer structures. The higher the tempering temperature, the more martensite is transformed, and the softer and tougher (less brittle) the piece becomes. Therefore, tempering temperatures are specified according to the strength and ductility desired. The charts that show the mechanical properties, which can be found in steel manufacturers' handbooks and catalogs, give these data for each type of alloy steel. Table D-12 provides heat-treating information for plain carbon steel (AISI 1095) and other carbon and alloy steels. Expected hardnesses in Rockwell numbers refer to steel that has attained its full quenched hardness. The hardness readings after tempering would be lower if for some reason a part were not fully hard in the first place. See Unit Five in this section for Rockwell and Brinell hardness comparisons.

A part can be tempered in a furnace or oven by bringing it to the proper temperature and holding it there for a length of time, then cooling it in air or water.

Some tool steels should be cooled rapidly after tempering to avoid temper brittleness. Small parts are often tempered in liquid baths such as oil or salts. Specially prepared oils that do not ignite easily can be heated to the tempering temperature. Various salts are used for tempering because they have a low melting temperature.

When there are no facilities to harden and temper a tool by controlled temperatures, tempering by color is done. The

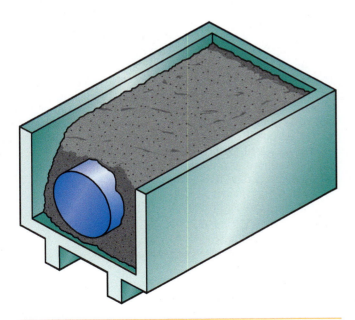

Figure D-28 Pack carburizing. The workpiece to be pack carburized should be completely covered with carburizing compound. The metal box should have a close-fitting lid.

Table D-12 Typical Heat-Treating Information for Direct-Hardening Carbon and Low-Alloy Steels[a]

Grade	Hardening Temperature	Full Hardness	Expected Hardness after Tempering 2 Hours at:								
			400	500	600	700	800	900	1000	1100	1200
8620	1650/1750	37/43	40	39	37	36	35	32	27	24	20
41310	1550/1625	49/56	47	45	43	42	38	34	32	26	22
1040	1525/1600	53/60	51	48	46	42	37	30	27	22	(14)
4140/4142	1525/1575	53/62	55	52	50	47	45	41	36	33	29
4340	1475/1550	53/60	55	52	50	48	45	42	39	34	30
1144	1475/1550	55/60	55	50	47	45	39	32	29	25	(19)
1045	1475/1575	55/62	55	52	49	45	41	34	30	26	20
4150	1500/1550	59/65	56	55	53	51	47	45	42	38	34
5160	1475/1550	60/65	58	55	53	51	48	44	40	36	31
1060/1070	1475/1550	58/63	56	55	50	43	39	38	36	35	32
1095	1450/1525	63/66	62	58	55	51	47	44	35	30	26

[a]Temperatures listed in °F and hardness in Rockwell C. Values were obtained from various recognized industrial and technical publications. All values should be considered approximations.

Source: *Courtesy of Pacific Machinery & Tool Steel Company.*

oxide color used as a guide in such tempering will form correctly on steel only if it is polished to the bare metal and is free from oil or fingerprints. An oxyacetylene torch, steel hot plate, or electric hot plate can be used. If the part is quite small, a steel plate is heated from the underside, and the part is placed on top. Larger parts such as chisels and punches can be heated on an electric hot plate (Figure D-29) until the desired color appears, then cooled in water. With this process the tempering must end when the part has come to the correct temperature, then the part must be quickly cooled in water to stop further heating of the critical areas. There is no soaking time when this method is used.

When you grind carbon steel tools, if the edge is heated enough to produce a color, you have in effect retempered the edge. If the temperature reached is above that of the original temper, the tool will be softer than before you began sharpening it. Table D-13 gives the hardness of various tools as related to their oxide colors and the temperature at which they form.

Tempering should be done as soon as possible after hardening. A part should not be allowed to cool completely, because untempered, it contains high internal stresses and tends to split or crack. Tempering will relieve the internal stresses. A hardened part left overnight without tempering may develop cracks. Furnace tempering is one of the best methods of controlling the final condition of the martensite

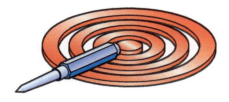

Figure D-29 Tempering a punch on a hot plate.

> ### SHOP TIP
>
> Small forged boring bars for the lathe are usually made of high-speed steel. They are used for boring small holes ($\frac{3}{4}$ inch or less). If a high-speed boring bar is not available, a substitute can be made with an O-1 drill rod by forging a 90-degree bend on one end. The bent end is then ground on a pedestal grinder to a rough shape according to the tool geometry.
>
> The tool is then heat-treated in exactly the same way as you would a punch or chisel: Heat the tool to a cherry red color and quench in oil. Wipe off the oil and then sand off the scale so that you can see the colors when it is tempered. Place the shank on a hot plate or use a torch, and make sure that the temper colors on the shank are blue or violet while the cutting edge remains a light straw. Then, cool it quickly in water.
>
> The cutting end may now be sharpened on the pedestal grinder (Figures D-31a and b). Make sure that it is not overheated in this operation or when it is used as a cutting tool. If it does become overheated, unlike the high-speed steel, it will lose its hardness and will have to be re–heat-treated. If it is used correctly, this bar makes an excellent cutting tool (Figure D-31c).
>
> When they need small boring bars, some machinists simply grind the shape they need on a high-speed steel tool bit. They grind away some of the shank behind the cutting edge to provide clearance.

to produce a tempered martensite of the correct hardness and toughness that the part requires. The still-warm part should be put into the furnace immediately. If it is left at room temperature for even a few minutes, it may develop a quench crack (Figure D-30).

Table D-13 Temper Color Chart

Degrees		Oxide Color	Suggested Uses for Carbon Tool Steels	
°C	°F			
220	425	Light straw	Steel-cutting tools, files, and paper cutters	Harder
240	462a	Dark straw	Punches and dies	
258	490	Gold	Shear blades, hammer faces, center punches, and cold chisels	
260	500	Purple	Axes, wood-cutting tools, and striking faces of tools	
282	540	Violet	Springs and screwdrivers	
304	580	Pale blue	Springs	
327	620	Steel gray	Cannot be used for cutting tools	Softer

A soaking time should be used when tempering, as it is in hardening procedures, with the length of time related to the type of tool steel used. A cold furnace should be brought up to the correct temperature for tempering. The residual heat in the bricks of a previously heated furnace may overheat the part even though the furnace has been shut off.

Double tempering is used for some alloy steels such as high-speed steels in which the austenite is incompletely transformed when they are tempered for the first time. The second time they are tempered, the austenite transforms completely into a martensitic structure.

Figure D-30 These two breech plugs were made of type L6 tool steel. Plug 1 (left) cracked in the quench because of a sharp corner and was therefore not tempered. Plug 2 (right) was redesigned to incorporate a radius in the corners of the slot, and a soft steel plug was inserted in the slot to protect it from the quenching oil. Plug 2 was oil quenched and checked for hardness (Rockwell C 62), and after tempering at 900°F (482°C) it was found to be cracked. The fact that the as-quenched hardness was measured proves that a delay between the quench and the temper was responsible for the cracking. The proper practice would be to temper immediately at a low temperature, check hardness, and retemper to the desired hardness (*ArcelorMittal*).

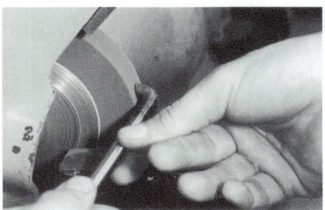

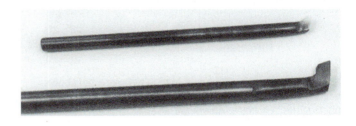

Figure D-31 Forged boring bar.

PROBLEMS IN HEAT-TREATING

Avoid overheating steels. Remember that if the furnace is set too high for the particular type of steel, a coarse grain can develop. The result is often a poor-quality part, quench cracking, or failure of the tool in use. Extreme overheating damages the grain boundaries, which cannot be repaired by heat treatment (Figure D-32); the part must be scrapped. The shape of the part itself can be a contributing factor to quench failure and quench cracking. A hole, sharp shoulder, or small extension from a larger cross section (unequal mass) can cause a crack to develop (Figure D-33). Holes can be filled with steel wool to reduce problems. The part of a tool being held by tongs may be cooled enough by the tongs that it may not harden. The tongs should therefore be heated prior to grasping the part for quenching (Figure D-34).

Figure D-32 This tool has been overheated, and the typical "chicken wire" surface markings are evident. The tool must be discarded.

Figure D-33 Drawing die made of type W1 tool steel shows characteristic cracking when water quenching is done without packing the bolt holes (*ArcelorMittal*).

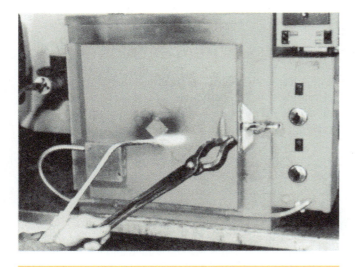

Figure D-34 Heating the tongs prior to quenching a part.

As mentioned before, decarburization is a problem in furnaces that lack a controlled atmosphere. This problem can be avoided in other ways, such as by wrapping the part in stainless steel foil, covering it with cast iron chips, or covering it with a commercial compound.

Proper selection of tool steels is necessary to prevent failures in a particular application. If there is shock load on the tool being used, shock-resisting tool steel must be selected. If heat is to be applied in use, a hot-work type of tool steel should be selected. If distortion must be kept to a minimum, air-hardening steel should be used.

Quench cracks have several easily recognized characteristics:

1. In general, the fractures run from the surface toward the center in a relatively straight line. The cracks tend to spread open.
2. Because quench cracking occurs at relatively low temperatures, the crack will not show any decarburization. That is, a black scale will form on the cracked surface if it happened at a high temperature, but no scale will appear on the surface of a quench crack. Tempering will darken a quench crack.
3. Fractured surfaces will exhibit a fine crystalline structure when tempered after quenching. The fractured surfaces may be blackened by tempering scale.

The following are some of the most common causes for quench cracks:

1. Overheating during the austenitizing cycle, which causes normally fine-grained steel to become coarse-grained.
2. Improper selection of the quenching medium; for example, the use of water or brine instead of oil for oil-hardening steel.
3. Improper selection of steel.
4. Time delays between quenching and tempering.
5. Improper design, sharp changes of section such as holes and keyways (Figure D-35).
6. Inserting the work into the quenching bath at an improper angle with respect to the shape of the part, which causes nonuniform cooling.
7. Failure to specify the correct size material to allow for clean up of the outside decarburized surface of the bar before the final part is made. Hot-rolled tool steel bars need $\frac{1}{16}$ inch removed from all surfaces before any heat-treating is done.

It is sometimes desirable to normalize the part before hardening it. This is particularly appropriate for parts and tools that have been highly stressed by heavy machining or by prior heat treatment. If they are left unrelieved, the residual stresses from such operations may add to the thermal stress produced in the heating cycle and cause the part to crack even before it has reached the quenching temperature.

There is a definite relationship between grinding and heat-treating. Development of surface temperatures ranging from 2000°F to 3000°F (1093°C to 1649°C) are generated

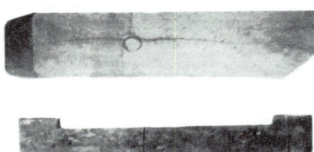

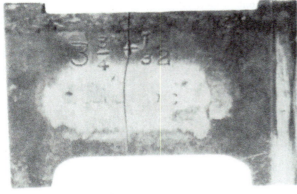

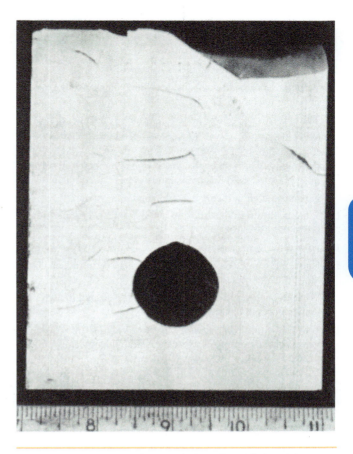

Figure D-36 Severe grinding cracks in a shear blade made of type A4 tool steel developed because the part was not tempered after quenching. Hardness was Rockwell C64, and the cracks were exaggerated by magnetic particle test. Note the geometric scorch pattern on the surface and the fracture that developed from enlargement of the grinding cracks (*ArcelorMittal*).

Figure D-35 (Top) Form tool made of type S5 tool steel, which cracked in hardening through the stamped O. The other two form tools, made of type T1 high-speed steel, cracked in heat treatment through deeply stamped + marks. Avoid stress raisers such as these deep stamp marks. Although characters with straight lines are most likely to crack, even those with rounded lines are susceptible (*ArcelorMittal*).

during grinding. This can cause two undesirable effects in hardened tool steels: development of high internal stresses, which causes surface cracks to form, and changes in the hardness and metallurgical structure of the surface area.

One of the most common effects of grinding on hardened and tempered tool steels is a reduction in the hardness of the surface by gradual tempering from the heat of grinding, where the hardness is lowest at the extreme surface but increases with distance below the surface. The depth of this tempering varies with the amount or depth of cut, the use of coolants, and the type of grinding wheel. However, if extremely high temperatures are produced locally by the grinding wheel, and the surface is immediately quenched by the coolant, a martensite with a Rockwell hardness of C65 to C70 can be formed. This hardness, being much greater than

that beneath the surface of the tempered part, can cause high stresses that can cause grinding cracks. Sometimes, grinding cracks are visible in oblique or angling light, but they can easily be detected by fluorescent magnetic particle testing.

When a part is hardened, but not tempered before it is ground, it is extremely vulnerable to stress cracking (Figure D-36). Faulty grinding procedures can also cause grinding cracks. Improper grinding operations can cause tools that have been properly hardened to fail. Sufficient stock should be allowed for a part to be heat-treated so that grinding will remove any decarburized surface on all sides to a depth of .010 to .015 inch.

SELF-TEST

1. If you heated AISI C1080 steel to 1200°F (649°C) and quenched it in water, what would happen?

2. If you heated AISI C1020 steel to 1500°F (815°C) and quenched it in water, what would happen?

3. List the advantages of using air- and oil-hardening tool steels.

4. What is the correct temperature for quenching AISI C1095 tool steel? What is the general rule for quenching temperature for any carbon steel?

5. Why is steel tempered after it is hardened?

6. What factors should you consider when you choose the tempering temperature for a tool?

7. The approximate temperature for tempering a center punch should be_____. The oxide color will be _____.

8. If a cold chisel became blue when the edge was ground on an abrasive wheel, to approximately what temperature was it raised? How would this temperature change affect the tool?

9. How soon after hardening should you temper a part?

10. What is the advantage of using low-carbon steel for parts that are to be case-hardened?

11. How can a deep case be made?

12. Name three methods by which carbon may be introduced into the surface of heated steel.

13. What can happen to carbon steel when it is heated to high temperatures in the presence of air (oxygen)?

14. Why is it absolutely necessary to allow soaking time (which varies for different kinds of steels) before quenching the piece of steel?

15. Why should the part or the quenching medium be agitated when you are hardening steel?

16. Which method of tempering gives the heat treater the most control of the final product: by the color or by the furnace control display?

17. Describe two characteristics of quench cracking that would enable you to recognize them.

18. Name four or more causes of quench cracks.

19. In what ways can you avoid decarburizing a part when heating it in a furnace?

20. When distortion must be kept to a minimum, which type of tool steel should be used?

INTERNET REFERENCE

Information on heat-treating methods and equipment, including PowerPoint slides, glossary, and definitions:

http://wisoven.com

Annealing, Normalizing, and Stress Relieving

All metals form tiny grains when cooled unless they are cooled extremely rapidly. Some grains or crystals are large enough to be seen on a section of broken metal. In general, small grains are best for tools, and larger grains for extensive cold working. Small-grain metals are stronger than large-grain ones. Slow cooling promotes the formation of large grains; rapid cooling forms small grains. Because the machinability of steel is so greatly affected by heat treatments, the processes of annealing, normalizing, and stress relieving are important to a machinist. You will learn about these processes in this unit.

OBJECTIVE

After completing this unit, you should be able to:

- Explain the principles of and differences between the various annealing processes.

ANNEALING

Annealing can be divided into several different processes: full anneal, normalizing, spheroidize anneal, stress relief (anneal), and process anneal.

Full Anneal

Full annealing is used to completely soften hardened steel, usually for easier machining of tool steels that have more than .8 percent carbon content. Lower-carbon steels are full annealed for other purposes. For instance, when welding has been done on a medium- to high-carbon steel that must be machined, a full anneal is needed. Full annealing is performed by heating the part in a furnace to 50°F (10°C) above the upper critical temperature (Figure D-37) and then cooling slowly in the furnace or in an insulating material. It is necessary to heat above the critical temperature for grains containing iron carbides (pearlite) to recrystallize them and to re-form new soft, whole grains from the old hard, distorted ones.

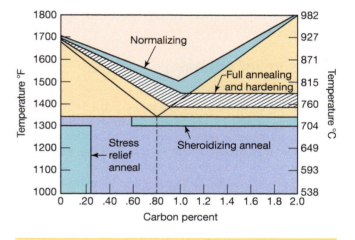

Figure D-37 Temperature ranges used for heat-treating carbon steels.

Normalizing

Normalizing is similar to annealing, but it is done for different purposes. Medium-carbon steels are often normalized to give them better machining qualities. Medium-carbon steel (.3–.6 percent) may be "gummy" when machined after a full anneal but can be made sufficiently soft for machining by normalizing. The finer, but harder, microstructure produced by normalizing gives the piece a better surface finish. The piece is heated to 100°F (38°C) above the upper critical line and cooled in still air. When the carbon content is below .3 percent or above .8 percent, higher temperatures are required (Figure D-37).

Forgings and castings that have unusually large and mixed grain structures can be corrected by using a normalizing heat treatment. Stresses are removed, but the metal is not as soft as with a full anneal. The resultant microstructure is a uniform fine-grained pearlite and ferrite including other microstructures, depending on the alloy and carbon content. Normalizing also is used to prepare steel for other forms of heat treatment such as hardening and tempering. Weldments are sometimes normalized to remove welding stresses.

Spheroidize Anneal

Spheroidizing is a form of annealing consisting of prolonged heating of iron base alloys at a temperature slightly below the critical range, usually followed by relatively slow cooling. Spheroidizing causes the graphite to assume a spheroidal shape. The spheroidal shape is where the name of the process comes from.

Spheroidizing improves the machinability of high-carbon steels (.8–1.7 percent). The cementite or iron carbide normally found in pearlite as flat plates alternating with plates of ferrite (iron) is changed into a spherical or globular form by spheroidization (Figure D-38). Low-carbon steels (.08–.3 percent) can be spheroidized, but their machinability declines because they become gummy and soft, causing tool edge buildup and poor finish. The spheroidization temperature is close to 1300°F (704°C). The steel is held at this temperature for about 4 hours. Extremely hard carbides that develop from welding on medium- to high-carbon steels can be made softer by this process.

Stress Relief Anneal

Stress relief annealing is a process of reheating low-carbon steels to 950°F (510°C). Stresses in the ferrite (mostly pure iron) grains caused by cold-working steel such as rolling, pressing, forming, or drawing are relieved by this process. The distorted grains re-form or recrystallize into new softer ones (Figures D-39 and D-40).

The pearlite grains and some other forms of iron carbide remain unaffected by this treatment, unless performed at the spheroidizing temperature and held long enough to effect spheroidization. Stress relief is often used on weldments, as the lower temperature limits the amount of distortion caused by heating. Full anneal or normalizing can cause considerable distortion in steel.

Process Annealing

Process annealing is essentially the same as stress relief annealing. It is done at the same temperatures as low- and medium-carbon steels. In the wire and sheet steel industry, the

Figure D-39 Microstructure of flattened grains of .10 percent carbon steel, cold rolled (1000×). (*Reprinted with permission of ASM International®. All rights reserved. www.asminternational.org*).

Figure D-40 The same .10 percent carbon steel as in Figure D-39, but annealed 1025°F (552°C) (1000×) at Ferrite grains (white) are mostly re-formed to their original state, but the pearlite grains (dark) are still distorted. (*Reprinted with permission of ASM International®. All rights reserved. www.asminternational.org*).

term is used for the annealing or softening necessary during cold rolling or wire drawing and for removal of final residual stresses in the material. Wire and other metal products that must be continuously formed and re-formed would become too brittle to continue the process after a certain amount of forming. A process anneal performed between a series of cold-working operations will re-form the grain to the original soft, ductile condition so that additional cold working can be done. Process annealing is sometimes referred to as *bright annealing* and is usually carried out in a closed container with inert gas to prevent oxidation of the surface.

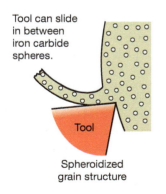
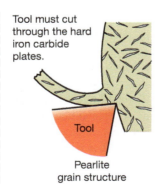

Tool can slide in between iron carbide spheres.

Tool must cut through the hard iron carbide plates.

Tool

Tool

Spheroidized grain structure

Pearlite grain structure

Figure D-38 Comparison of cutting action between spheroidized and normal carbon steels.

RECOVERY, RECRYSTALLIZATION, AND GRAIN GROWTH

When metals are heated to temperatures below the recrystallization temperature, a reduction in internal stress takes place. Elastic stresses are relieved in the lattice planes, not by re-forming the distorted grains. Recovery in annealing processes used on cold-worked metals is rarely sufficient stress relief for further extensive cold working (Figure D-41), yet it is used for some purposes and is called *stress relief anneal.* Most often, recrystallization is required to re-form the distorted grains sufficiently for further cold work.

Recovery is a low-temperature effect in which there is little or no visible change in the microstructure. Electrical conductivity is increased, and often a decrease in hardness is noted. Recrystallization releases much larger amounts of energy than recovery does.

Not only does recrystallization release much larger amounts of stored energy, but new, larger grains are formed by the nucleation of stressed grains and the joining of several grains to form larger ones. Adjacent grains are joined when grain boundaries migrate to new positions. This changes the orientation of the crystal structure. This process is called *grain growth.*

The following factors affect recrystallization:

1. A minimum amount of deformation is necessary for recrystallization to occur.
2. The larger the original grain size, the greater will be the amount of cold deformation required to yield an equal amount of recrystallization with the same temperature and time.
3. Increasing the time of anneal decreases the temperature necessary for recrystallization.
4. The recrystallized grain size depends mostly on the degree of deformation and partly on the annealing temperature.
5. Continued heating after recrystallization (re-formed grains) is complete increases the grain size.
6. The higher the cold-working temperature, the greater will be the amount of cold work required to yield equivalent deformation.

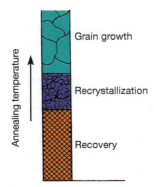

Figure D-41 Changes in metal structures that take place during the annealing process (*Neely, John E; Bertone, Thomas J.*, Practical Metallurgy and Materials of Industry, *6th, ©2003. Electronically reproduced by permission of Pearson Education, Inc., Upper Saddle River, New Jersey*).

Table D-14 Recrystallization Temperatures of Some Metals

Metal	Recrystallization Temperature (°F)
99.999% Aluminum	175
Aluminum bronze	660
Beryllium copper	900
Cartridge brass	660
99.999% Copper	250
Lead	25
99.999% Magnesium	150
Magnesium alloys	350
Monel	100
99.999% Nickel	700
Low-carbon steel	1000
Tin	25
Zinc	50

Metals subjected to cold working become hardened, and further cold working is impossible without danger of splitting or breaking the metal. Various degrees of softening can be attained by controlling the recrystallization process. The recrystallization temperatures for metals are given in Table D-14. If you need to soften work-hardened brass, for example, you need to heat it to 660°F (349°C) and cool it in water. Low-carbon steel should not be cooled in water when annealing.

SELF-TEST

1. When might normalizing be necessary?
2. At what approximate temperature should you normalize .4 percent carbon steel?
3. What is the spheroidizing temperature of .8 percent carbon steel?
4. What is the essential difference between the full anneal and stress relieving?
5. When should stress relieving be used?
6. What kind of carbon steels would need to be spheroidized to give them free-machining qualities?
7. Explain process annealing.
8. How should the piece be cooled for a normalizing heat treatment?
9. How should the piece be cooled for the full anneal?
10. What happens to machinability in low-carbon steels that are spheroidized?

INTERNET REFERENCE

http://wisoven.com

Rockwell and Brinell Hardness Testers

Rockwell and Brinell hardness testers are the most commonly used types of hardness testers in industry. Heat treaters, inspectors, and many others in industry often use these machines. Hardness testers are made in various sizes, including handheld portable units. Two other types of hardness testers are the Vickers diamond pyramid and the Knoop. All these testers use different scales, which can be compared in hardness conversion tables. This unit will cover the use of Rockwell and Brinell hardness testers.

OBJECTIVES

After completing this unit, you should be able to:

- Perform a Rockwell test using the correct penetrator, major load, and scale.
- Perform a Rockwell superficial test using the correct penetrator, major load, and scale.
- Perform a Brinell test, read the impression with a Brinell microscope, and determine the hardness number from a table.

MEASURING HARDNESS

The hardness of a metal is its ability to resist being permanently deformed. Hardness is measured three ways: resistance to penetration, elastic hardness, and resistance to abrasion. In this unit, you will study the hardness of metals by their resistance to penetration.

Hardness varies considerably among materials. This variation can be illustrated by making an indentation in a soft metal such as aluminum and in a hard metal such as alloy tool steel. The indentation could be made with an ordinary center punch and a hammer, giving a light blow of equal force to each of the two specimens (Figure D-42). Just by looking at the indentations you could tell which specimen is hardest. This is the principle of both the Rockwell

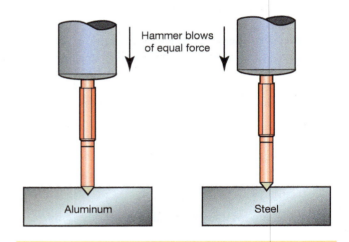

Figure D-42 Indentations made by a punch in aluminum and alloy steel.

and Brinell hardness testers: measuring the penetration of the specimen by an indenter or penetrator, such as a steel ball or diamond cone.

USING THE ROCKWELL HARDNESS TESTER

The Rockwell hardness test is performed by applying two loads to a specimen and measuring the depth of penetration in the specimen between the first, or minor, load and the major load (Figure D-43). The depth of penetration is indicated on the dial when the major load is removed. The amount of penetration decreases as the hardness of the specimen increases. Generally, the harder the material, the greater its tensile strength. Tensile strength is the ability to resist deformation and rupture when a load is applied. Table D-15 compares hardness by Brinell and Rockwell testers to tensile strength.

Two basic types of penetrators are used on the Rockwell tester (Figure D-44). One is a cone-shaped diamond called

Rockwell Test Principal

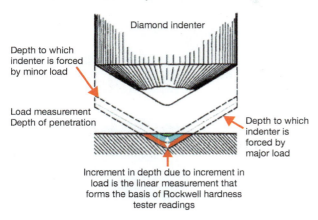

Diamond indenter

Depth to which indenter is forced by minor load

Load measurement
Depth of penetration

Depth to which indenter is forced by major load

Increment in depth due to increment in load is the linear measurement that forms the basis of Rockwell hardness tester readings

Figure D-43 Schematic showing minor and major loads being applied (*Diagram courtesy of Wilson® Instruments*).

a Brale and is only used for hard materials—that is, for materials over B-100, such as hardened steel, nitrided steel, and hard cast irons. When the "C" Brale diamond penetrator is used, the recorded readings should be preceded by the letter *C*. The major load used is 150 kgf (kilograms of force). The C scale is *not* used to test extremely hard materials, such as cemented carbides or shallow case-hardened steels and thin steel. An A Brale penetrator is used in these cases, and the A scale is used with a 60-kgf major load.

The second type of penetrator is a $\frac{1}{16}$-inch diameter ball used for testing material in the range of B-100 to B-0, including such relatively soft materials as brass, bronze, and soft steel. Using the ball penetrator on materials harder than B-100 creates a danger of flattening the ball. Ball penetrators for use on soft bearing metals are available in sizes of $\frac{1}{2}$, $\frac{1}{4}$, and $\frac{1}{8}$ inch (Table D-16).

Figure D-45 shows a digital Rockwell hardness tester.

Table D-15 Hardness and Tensile Strength Comparison Table[a]

Brinell					Brinell				
Indentation Diameter (mm)	No.[b]	Rockwell B	Rockwell C	Tensile Strength[c] (1,000 psi, approximately)	Indentation Diameter (mm)	No.[b]	Rockwell B	Rockwell C	Tensile Strength[c] (1,000 psi, approximately)
2.25	745		65.3		3.75	262	(103.0)	26.6	127
2.30	712		—		3.80	255	(102.0)	25.4	123
2.35	682		61.7		3.85	248	(101.0)	24.2	120
2.40	653		60.0		3.90	241	100.0	22.8	116
2.45	627		58.7		3.95	235	99.0	21.7	114
2.50	601		57.3		4.00	229	98.2	20.5	111
2.55	578		56.0		4.05	223	97.3	(18.8)	—
2.60	555		54.7	298	4.10	217	96.4	(17.5)	105
2.65	534		53.5	288	4.15	212	95.5	(16.0)	102
2.70	514		52.1	274	4.20	207	94.6	(15.2)	100
2.75	495		51.6	269	4.25	201	93.8	(13.8)	98
2.80	477		50.3	258	4.30	197	92.8	(12.7)	95
2.85	461		48.8	244	4.35	192	91.9	(11.5)	93
2.90	444		47.2	231	4.40	187	90.7	(10.0)	90
2.95	429		45.7	219	4.45	183	90.0	(9.0)	89
3.00	415		44.5	212	4.50	179	89.0	(8.0)	87
3.05	401		43.1	202	4.55	174	87.8	(6.4)	85
3.10	388		41.8	193	4.60	170	86.8	(5.4)	83
3.15	375		40.4	184	4.65	167	86.0	(4.4)	81
3.20	363		39.1	177	4.70	163	85.0	(3.3)	79
3.25	352	(110.0)	37.9	171	4.80	156	82.9	(0.9)	76
3.30	341	(109.0)	36.6	164	4.90	149	80.8		73
3.35	331	(108.5)	35.5	159	5.00	143	78.7		71
3.40	321	(108.0)	34.3	154	5.10	137	76.4		67
3.45	311	(107.5)	33.1	149	5.20	131	74.0		65
3.50	302	(107.0)	32.1	146	5.30	126	72.0		63
3.55	293	(106.0)	30.9	141	5.40	121	69.8		60
3.60	285	(105.5)	29.9	138	5.50	116	67.6		58
3.65	277	(104.5)	28.8	134	5.60	111	65.7		56
3.70	269	(104.0)	27.6	130					

[a]This is a condensation of Table 2, *Report J417b, SAE 1971 Handbook*. Values in parentheses are beyond the normal range and are presented for information only.
[b]Values above 500 are for tungsten carbide ball; below 500 for standard ball.
[c]The following is a formula to approximate tensile strength when the Brinell hardness is known: tensile strength = BHN × 500.

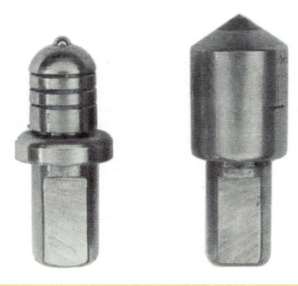

Figure D-44 Penetrators for the Rockwell hardness tester (*Photo courtesy of Wilson® Instruments*).

Figure D-45 Rockwell hardness tester (*Photo courtesy of Wilson® Instruments*).

When setting up the Rockwell hardness tester and performing a hardness test, follow these steps:

Step 1 Using Table D-16, select the proper weight (Figure D-46) and penetrator. Make sure that the crank handle is pulled completely forward.

Step 2 Place the proper anvil (Figure D-47) on the elevating screw, taking care not to bump the penetrator with the anvil. Make sure that the specimen to be tested is free from dirt, scale, or heavy oil on the underside.

Step 3 Place the specimen to be tested on the anvil (Figure D-48). Then, by turning the handwheel, gently raise the specimen until it comes in contact with the penetrator (Figure D-49). Continue turning the handwheel slowly until

Table D-16 Penetrator and Load Selection

Scale	Penetrator	Major Load (kgf)	Dial	Typical Use
B	$\frac{1}{16}$ Ball	100	Red	Soft steels, copper alloys, malleable iron, aluminum alloys, etc.
C	Brale	150	Black	Steel, pearlitic malleable iron, hard cast irons, titanium, deep case-hardened steel, and materials harder than B-100
A	Brale	60	Black	Thin steels, carbides, and shallow case-hardened steels
D	Brale	100	Black	Thin steels, medium case-hardened steel, and pearlitic malleable iron
E	$\frac{1}{8}$ Ball	100	Red	Cast iron, aluminum and magnesium alloys, and bearing metals
F	$\frac{1}{16}$ Ball	60	Red	Annealed copper alloys, thin, soft sheet metals
G	$\frac{1}{16}$ Ball	150	Red	Phosphor bronze, beryllium copper, and malleable irons. Note: Upper limit is G-92 to prevent possible flattening of ball
H	$\frac{1}{8}$ Ball	60	Red	Aluminum, zinc, and lead
K	$\frac{1}{8}$ Ball	150	Red	
L	$\frac{1}{4}$ Ball	60	Red	Bearing metals and other soft or thin materials
M	$\frac{1}{4}$ Ball	100	Red	
P	$\frac{1}{4}$ Ball	150	Red	
R	$\frac{1}{2}$ Ball	60	Red	
S	$\frac{1}{2}$ Ball	100	Red	
V	$\frac{1}{2}$ Ball	150	Red	

Source: *Courtesy of Peerless Chain Company.*

the small pointer on the dial gage is nearly vertical (near the dot). Now, watch the long pointer on the gage and continue raising the work until it is approximately vertical. It should not vary from the vertical position by more than five divisions on the dial. Set the dial to zero on the pointer by moving the bezel until the line marked zero set is in line with the pointer (Figure D-50). You have now applied the minor load. This is the actual starting point for all conditions of testing.

Figure D-48 Placing the part to be tested on the anvil.

Figure D-46 Selecting and installing the correct weight.

Figure D-47 Basic anvils used with Rockwell hardness testers. (*Photo courtesy of Wilson® Instruments*).

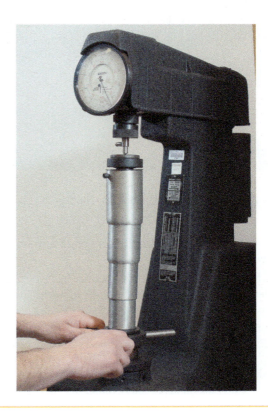

Figure D-49 Specimen being brought into contact with the penetrator. This establishes the minor load.

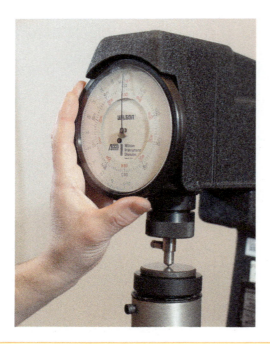

Figure D-50 Setting the bezel.

Figure D-51 Applying the major load by tripping the crank handle clockwise.

Step 4 Apply the major load by tripping the crank handle clockwise (Figure D-51).

Step 5 Wait 2 seconds after the pointer has stopped moving, then remove the major load by pulling the crank handle forward or counterclockwise.

Step 6 Read the hardness number in Rockwell units on the dial (Figure D-52). The black numbers are for the A and C scales, and the red numbers are for the B scale. The specimen should be tested in several places and an average of the test results taken, since many materials vary in hardness even on the same surface.

Superficial Testing

If you are testing a thin metal, you should examine the underside of the part after testing. If the impression of the

Figure D-52 Dial face with reading in Rockwell units after completion of the test. The reading is RC 55.

penetrator can be seen, then the reading is in error and the superficial test should be used. If the impression can still be seen after the superficial test, then a lighter load should be used. A load of 30 kgf is recommended for superficial testing. Superficial testing is also used for case-hardened and nitrided steel with a thin case.

A Brale marked N is needed for superficial testing, as A and C Brales are not suitable. Recorded readings should be preceded by the major load and the letter N when using the Brale for superficial testing—for example, 30N78. When using the $\frac{1}{16}$-inch ball penetrator, the same as that used for the B, F, and G hardness scales, the readings should always be preceded by the major load and the letter T—for example, 30T85. The $\frac{1}{16}$-inch ball penetrator, however, should not be used on material harder than 30T82. Other superficial scales, such as W, X, and Y, should also be preceded with the major load when recording hardness. See Table D-17 for superficial test penetrator selection.

The diamond spot anvil (Figure D-53) is similar to the spot anvil, but it has a diamond set into the spot. The diamond is ground and polished to a flat surface. This anvil is used only with the superficial tester, and then only in conjunction with the steel ball penetrator for testing soft metal.

Surface Preparation and Proper Use

When testing hardness, surface condition is important for accuracy. A rough or ridged surface caused by coarse grinding will not produce a reliable a result. Any rough scale caused

Table D-17 Superficial Tester Load and Penetrator Selection

Scale Symbol	Penetrator	Load (kgf)
15N	Brale	15
30N	Brale	30
45N	Brale	45
15T	$\frac{1}{16}$ inch ball	15
30T	$\frac{1}{16}$ inch ball	30
45T	$\frac{1}{16}$ inch ball	45
15W	$\frac{1}{8}$ inch ball	15
30W	$\frac{1}{8}$ inch ball	30
45W	$\frac{1}{8}$ inch ball	45
15X	$\frac{1}{8}$ inch ball	15
30X	$\frac{1}{4}$ inch ball	30
45X	$\frac{1}{4}$ inch ball	45
15Y	$\frac{1}{2}$ inch ball	15
30Y	$\frac{1}{2}$ inch ball	30
45Y	$\frac{1}{2}$ inch ball	45

Source: *Courtesy of Peerless Chain Company.*

Figure D-53 Diamond spot anvil (*Photo courtesy of Wilson® Instruments*).

by hardening must be removed before testing. Likewise, if the workpiece has been decarburized by heat treatment, the test area should have this softer "skin" ground off.

Error can also result from testing curved surfaces. This effect may be eliminated by grinding a small flat spot on the specimen. Cylindrical workpieces must always be supported in a V-type centering anvil, and the surface to be tested should not deviate from the horizontal by more than 5 degrees. Tubing is often so thin that it will deform when tested. It should be supported on the inside by a mandrel or gooseneck anvil to avoid this problem.

Several devices are made available for the Rockwell hardness tester to support overhanging or large specimens. One type, called a *jack rest* (Figure D-54), is used for supporting long, heavy parts such as shafts. It consists of a separate elevating screw and anvil support similar to that on

Figure D-54 Correct method of testing long, heavy work requires the use of a jack rest (*Photo courtesy of Wilson® Instruments*).

the tester. Without adequate support, overhanging work can damage the penetrator rod and cause inaccurate readings.

No test should be made near an edge of a specimen. Keep the penetrator at least $\frac{1}{8}$ inch away from the edge. The test block, as shown in Figure D-48, should be used every day to check the calibration of the tester, if it is in constant use.

USING THE BRINELL HARDNESS TESTER

The Brinell hardness test is made by forcing a steel ball, usually 10 mm in diameter, into the test specimen by using a known load weight and then measuring the diameter of the resulting impression. The Brinell hardness value is the load divided by the area of the impression, expressed as follows:

$$BHN = \frac{P}{(\pi D/2)\left(D - \sqrt{D^2 - d^2}\right)}$$

where

BHN = Brinell hardness number,

in kilograms per square millimeter

D = diameter of the steel ball, in millimeters

P = applied load, in kilograms

d = diameter of the impression, in millimeters

A small microscope is used to measure the diameter of the impressions (Figure D-55). Various loads are used for testing different materials: 500 kg for soft materials such as copper and aluminum, and 3,000 kg for steels and cast irons. For convenience, Table D-9 gives the Brinell hardness number and corresponding diameters of impression for a 10-mm ball and a load of 3,000 kg. The related Rockwell hardness numbers and tensile strengths are also shown. Just as for the Rockwell tests, the impression of the steel ball must not show on the underside of the specimen. Tests should not be made too near the edge of a specimen.

Figure D-56 shows an air-operated Brinell hardness tester. The testing sequence is as follows:

Step 1 Select the desired load in kilograms on the dial by adjusting the air regulator (Figure D-57).

Step 2 Place the specimen on the anvil. Make sure the specimen is clean and free from burrs. It should be smooth enough so that an accurate measurement can be taken of the impression.

Step 3 Raise the specimen to within $\frac{5}{8}$ inch of the Brinell ball by turning the handwheel.

Step 4 Apply the load by pulling out the plunger control (Figure D-58). Maintain the load for 30 seconds for nonferrous metals and 15 seconds for steel. Release the load (Figure D-59).

Step 5 Remove the specimen from the tester and measure the diameter of the impression.

Step 6 Determine the Brinell hardness number (BHN) by calculation or by using the table. Soft copper should have a BHN of about 40, soft steel from 150 to 200, and hardened tools from 500 to 600. Fully hardened high-carbon steel will have a BHN of 750. A Brinell test ball of tungsten carbide should be used for materials above 600 BHN.

Figure D-56 Air-O-Brinell air-operated metal hardness tester.

Figure D-55 The Olsen Brinell microscope provides a fast, accurate means for measuring the diameter of the impression for determining the Brinell hardness number.

Figure D-57 Select the load. An operator adjusts the air regulator as shown until the desired Brinell load in kilograms is indicated.

Figure D-58 Apply the load. An operator pulls out plunger-type control to apply load smoothly to the specimen.

Figure D-59 Release the load. As soon as the plunger is depressed, the Brinell ball retracts in readiness for the next test.

Brinell hardness testers work best for testing softer metals and medium-hard tool steels.

SELF-TEST

1. What one specific property of hardness do the Rockwell and Brinell hardness measure? How is it measured?

2. State the relationship that exists between hardness and tensile strength.

3. Explain which scale, major load, and penetrator should be used to test a block of tungsten carbide on the Rockwell tester.

4. Why can't the steel ball be used on the Rockwell tester to test the harder steels?

5. Is the same Brale used on the A, C, and D scales when testing with the Rockwell superficial tester? Explain.

6. The $\frac{1}{16}$-inch ball penetrator used for the Rockwell superficial tester is different from that used for the B, F, and G scales. True or false?

7. What is the diamond spot anvil used for?

8. How does roughness on the specimen to be tested affect the test results?

9. How does decarburization affect the test results?

10. What does a curved surface do to the test results?

11. On the Brinell tester, what load should be used for testing steel?

12. What size ball penetrator is generally used on a Brinell tester?

SECTION E
Layout

INTRODUCTION

Layout is the process of placing reference marks on the workpiece. These marks may indicate the shape and size of a part or its features. Layout marks often indicate where machining will take place. In the machine shop, you will be concerned with layout for stock cutoff, filing and off-hand grinding, drilling, milling, and occasionally in connection with lathe work.

The process of layout can generally be classified as **semiprecision** and **precision**. Semiprecision layout is usually done by rule measurement to a tolerance of plus or minus $\frac{1}{64}$ inch. Precision layout is done with tools that discriminate to .001 inch or finer, to a tolerance of plus or minus .001 inch.

Layout Tools

Surface Plates

The surface plate is an essential tool for many layout applications. A surface plate provides an accurate reference plane from which measurements for both layout and inspection may be made. These plates are often known as *layout tables*. A surface plate or layout table is a precision tool and should be treated as such. They are also used for inspection. A surface plate should be covered when not in use and kept clean when being used.

Cast Iron Surface Plates
Cast iron surface plates (Figure E-1) are made from high-quality castings that have been allowed to age, thus relieving internal stresses. Aging of the casting reduces distortion after its working surface has been finished to the desired degree of flatness. A cast iron or steel surface plate usually has several ribs on the underside to provide structural rigidity. Cast surface plates vary in size from small bench models to larger sizes 4 to 8 feet or larger. Large surface plates are generally mounted on a heavy stand with adjustments for leveling.

Granite Surface Plates
Granite surface plates are more common than cast iron (Figure E-2). Granite is superior to metal because it is harder, denser, and impervious to

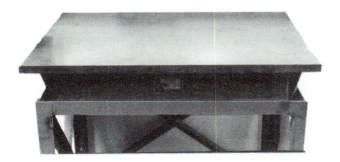

Figure E-1 Cast-iron surface plate.

Figure E-2 Granite surface plate.

water, and if it is chipped, the surrounding flat surface is not affected. Furthermore, granite, because it is a natural material, has aged in the earth for a great deal of time. Therefore, it has little internal stress. Granite surface plates also possess greater temperature stability than their metal counterparts.

Granite plates range in size from about 12 by 18 inches to 4 by 12 feet. A large granite plate may be from 10 to 20 inches thick and weigh as much as 5 to 10 tons.

Grades of Granite Surface Plates
Granite surface plates are available in three grades. Surface plate grade specifications are an indication of the plus and minus deviation of the working surface from an average plane. The tolerances are proportional to the size of the plate. As the size increases above 18 by 18 inches, the tolerance widens.

Grade	Type	Tolerance
AA	Laboratory grade	± 25 millionths inch
A	Inspection grade	± 50 millionths inch
B	Shop grade	± 100 millionths inch

Layout Dye

Layout die is used to make layout marks visible on the surface of the workpiece. Layout dyes are available in several colors. Among these are blue and red. The blue dyes are the most common. Depending on the surface color of the workpiece material, different dye colors may make layout marks more visible. Layout dye should be applied sparingly in an even coat (Figure E-3). Spray cans of layout dye provide a more uniform coat of dye.

Scribers and Dividers

Several types of scribers are in common use. The pocket scriber (Figure E-4) has a removable tip that can be stored in the handle. This permits the scriber to be safely carried in a pocket. The engineer's scriber (Figure E-5) has one straight and one hooked end. The hook permits easier access to the line to be scribed. The machinist's scriber (Figure E-6) has only one end with a fixed point.

Scribers must be kept sharp If they become dull, they must be reground or stoned to restore their sharpness. Scriber materials include hardened steel and tungsten carbide.

When scribing against a rule, hold the rule firmly. Tilt the scriber so that the tip marks as close to the rule as possible. This will ensure accuracy. An excellent scriber can be made by grinding a shallow angle on a piece of tool steel (Figure E-7). This type of scriber is particularly well suited to scribing along a rule. The flat side permits the scriber to mark close to the rule, thus permitting maximum accuracy.

Spring dividers (Figure E-8) range in size from 2 to 12 inches. The spacing of the divider legs is set by turning the adjusting screw. Dividers are usually set by using a rule. Engraved rules are best, as the divider tips can be set in the engraved rule graduations (Figure E-9). Divider tips must be kept sharp and at nearly the same length.

Figure E-3 Applying layout dye to the workpiece.

Figure E-7 Rule scriber made from a high-speed tool bit.

Figure E-4 Pocket scriber (*Courtesy of Rank Scherr-Tumico, Inc.*).

Figure E-5 Engineer's scriber (*Courtesy of Rank Scherr-Tumico, Inc.*).

Figure E-6 Machinist's scriber (*Courtesy of Rank Scherr-Tumico, Inc.*).

Figure E-8 A spring divider.

Figure E-9 Setting divider points to an engraved rule.

Hermaphrodite Caliper

The hermaphrodite caliper has one leg similar to that of a regular divider. The tip is adjustable for length. The other leg has a hooked end that can be placed against the edge of the workpiece (Figure E-10). Hermaphrodite calipers can be used to scribe a line parallel to an edge.

The hermaphrodite caliper can also be used to lay out the center of round stock (Figure E-11). The hooked leg is placed against the round stock and an arc is marked on the end of the piece. Tangent arcs can be laid out by adjusting

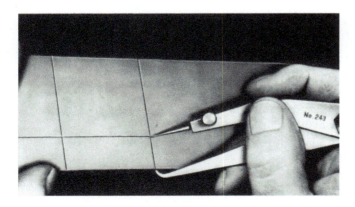

Figure E-10 Scribing a line parallel to an edge using a hermaphrodite caliper.

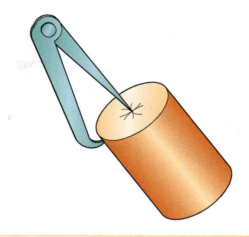

Figure E-11 Scribing the centerline of round stock with the hermaphrodite caliper.

the leg spacing. The center of the stock can be established by marking four arcs at 90 degrees.

Trammel Points

Trammel points are used for scribing circles and arcs when the distance involved exceeds the capacity of the divider. Trammel points are either attached to a bar and set to the circle's dimensions or clamped directly to a rule, where they can be set directly by rule graduations (Figure E-12).

Layout Hammers and Punches

Layout hammers are usually lightweight (4 to 10 oz.) machinist's ball-peen hammers (Figure E-13). A heavy hammer should not be used in layout, as it tends to create unnecessarily large punch marks.

The toolmaker's hammer may also be used (Figure E-14). This hammer is equipped with a magnifier that can be used to help locate a layout punch on a scribe mark.

Figure E-12 Trammel point used to scribe large arcs and circles.

Figure E-13 Layout hammer and layout prick punch.

Figure E-14 Using a toolmaker's hammer and layout punch (*Courtesy of The L.S. Starrett Co.*).

Figure E-15 Prick punch and center punch (*Courtesy of The L.S. Starrett Co.*).

Figure E-16 Automatic center punch (*Courtesy of The L.S. Starrett Co.*).

There is an important difference between a layout punch and a center punch. The layout, or prick punch, (Figure E-15, left) has an included point angle of 30 degrees. This is the only punch that should be used in layout. The slim point facilitates the locating of the punch on a scribe line. A prick punch mark is used only to preserve the location of a layout mark while doing minimum damage to the workpiece. On some workpieces, depending on the material used and the part application, layout punch marks are not acceptable, as they create a defect in the material. Layout punch marks should be of minimum depth.

The center punch (Figure E-15, right) has an included point angle of 90 degrees and is used to mark the workpiece prior to such machining operations as drilling.

A center punch should not be used in place of a layout punch. Similarly, a layout punch should not be used in place of a center punch. The center punch is used only to deepen the prick punch mark.

The automatic center punch (Figure E-16) does not require a hammer. Although it is called a center punch, its tip is shaped for layout applications (Figure E-17). Spring pressure behind the tip provides the required force for prick punching. The automatic center punch's force can be varied by changing the spring tension with an adjustment on the handle. To use the automatic center punch, the user simply pushes down on the punch and a spring activated mechanism makes the indentation.

The optical center punch (Figure E-18) consists of a locator, an optical alignment magnifier, and a punch. This type of layout punch is useful in locating punch marks precisely on a scribed line or line intersection. The locator is placed over the approximate location, and the optical alignment magnifier is inserted. The locator

Figure E-17 Using an automatic center punch for layout.

is magnetized, so it will remain in position when used on ferrous metals. The optical alignment magnifier has crossed lines etched on its lower end. By looking through the magnifier, you can move the locator about until the cross lines are matched to the scribe lines on the workpiece. The magnifier is then removed, and the punch is inserted into the locator. The punch is then lightly tapped with a layout hammer.

Centerhead

The centerhead is part of the machinist's combination set (Figure E-19). When the centerhead is clamped to the combination set rule, the edge of the rule is in line with a circle center. The other parts of the combination set are useful in layout. These include the rule, square head, and bevel protractor.

Surface Gage

A surface gage consists of a base, rocker, spindle adjusting screw, and scriber (Figure E-20). The spindle of the surface gage pivots on the base and can be moved with the adjusting screw. The scriber can be moved along the spindle and locked at any desired position. The scriber can also swivel in its clamp. A surface gage may be used as a height transfer tool. The scribe is set to a rule dimension (Figure E-21) and then transferred to the workpiece.

The hooked end of the surface gage scriber may be used to mark the centerline of a workpiece. The following procedure can be used to find the centerline of a piece. First, set the surface gage as nearly as possible to a height equal to half of the part height. Scribe the workpiece for a short distance at this position. Turn the part over and scribe it again (Figure E-22). If a deviation exists, there will be two scribe lines on the workpiece. Adjust the surface gage scriber so that it splits the difference between the two marks (Figure E-23). This ensures that the scribed line is in the center of the workpiece.

Height Gages

Height gages are among the most important instruments for precision layout. Height gages are available in mechanical, electronic, and vernier types. Use of these instruments will be covered in the unit on precision layout.

Mechanical Dial and Electronic Digital Height Gages
Mechanical dial (Figure E-24) and electronic digital (Figures E-25 and E-26) height gages eliminate the need to read a vernier scale. Once set to zero on the reference surface, the total height reading is shown on the digital display. Some electronic digital height gages can measure to .0001 inch.

Figure E-18 Using the optical center punch.

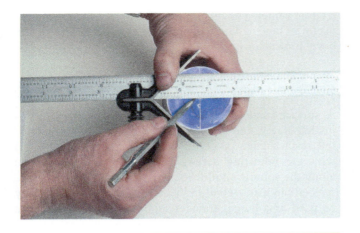

Figure E-19 Using the centerhead to lay out a centerline on round stock.

Figure E-21 Setting a surface gage to a rule.

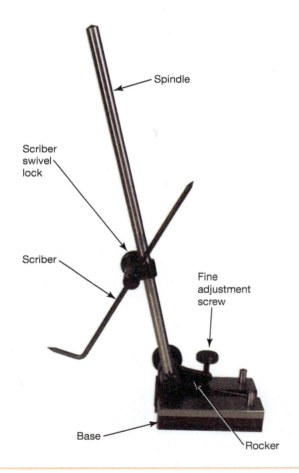

Spindle

Scriber swivel lock

Scriber

Fine adjustment screw

Base

Rocker

Figure E-20 Parts of the surface gage.

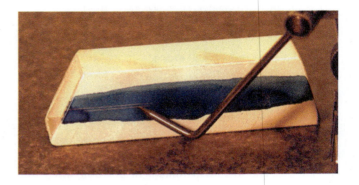

Figure E-22 Finding the centerline of the workpiece using the surface gage.

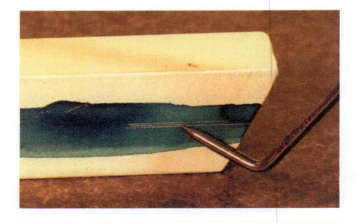

Figure E-23 Adjusting the position of the scribe line to center by inverting the workpiece and checking the existing differences in scribe marks.

Figure E-24 Mechanical dial height gage (*Mitutoyo America Corp.*).

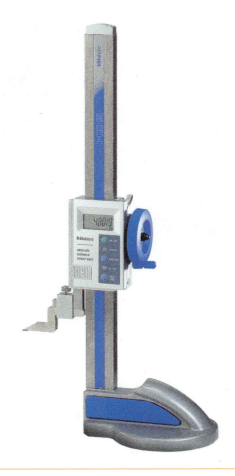

Figure E-26 The electronic digital height gage is useful for layouts and height measurements (*Mitutoyo America Corp.*).

Figure E-25 Digital height gage (*Mitutoyo America Corp.*).

Planer Gages   A planer gage may be equipped with a scriber and used as a height gage (Figure E-27). Dimensions are set by comparison with a precision height gage or height transfer micrometer.

Layout Accessories

Layout accessories are tools that aid you in accomplishing layout tasks. The layout plate or surface plate is the most common accessory, as it provides the reference surface from which to work. Other common accessories include V-blocks, parallels, angle plates, clamps, and so on that hold the workpiece during layout operations (Figure E-28).

Figure E-27 Using the planer gage as a height gage in layout.

Figure E-28 Universal right angle plate and V block used as layout accessories.

Basic Semiprecision Layout Practice

Before you begin machining, you might have to perform a layout operation. Layout for stock cutoff may involve simple chalk, pencil, or scribe marks on the material. No matter how simple the layout job may be, you should do it neatly and accurately. Accuracy is important in every layout. In this unit, you will proceed through a typical semiprecision layout that will familiarize you with basic layout practice.

OBJECTIVES

After completing this unit, you should be able to:

- Prepare the workpiece for layout.
- Measure for and scribe layout lines on the workpiece outlining the various features.
- Locate and establish hole centers using a layout prick punch and center punch.
- Lay out a workpiece to a tolerance of plus or minus $\frac{1}{64}$ inch.

Figure E-29 Applying layout dye with workpiece on a paper towel.

PREPARING THE WORKPIECE FOR LAYOUT

After the material has been cut, all sharp edges should be removed by grinding or filing before placing the stock on the layout table. Place a paper towel under the workpiece to prevent layout dye from spilling on the layout table (Figure E-29). Apply a thin, even coat of layout dye to the workpiece. Avoid breathing the vapors. You will need a drawing of the part to do the required layout (Figure E-30).

Study the drawing and determine the best way to proceed. The order of steps depends on the layout task. Before some features can be laid out, certain reference lines may have to be established. Measurements for other layouts are made from these lines.

LAYOUT OF THE DRILL AND HOLE GAGE

If possible, obtain a piece of material the same size as indicated on the drawing. Depending on the part to be made, you may be able to use material the same size as the finished job. However, certain parts may require that the edges be machined to finished dimensions. This may require material larger than the finished part to allow for machining of edges. Follow each step as described in the text. Refer to the layout drawings to determine where layout is to be done. The pictures will help you in selecting and using the required tools.

The first operation is to establish the width of the gage. Measure a distance of $1\frac{1}{8}$ inches from one edge of the material. Use the combination square and rule. Set the square at

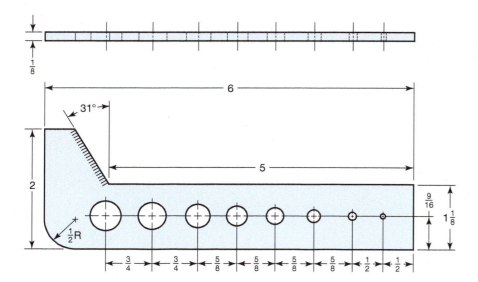

Figure E-30 Drill and hole gage layout.

the required dimension and scribe a mark at each end of the stock (Figures E-31 and E-32).

Remove the square and place the rule carefully on the scribe marks. Hold the rule firmly and scribe the line the full length of the material (Figure E-33). Be sure to use a sharp scribe and hold it so that the tip is against the rule. If the scribe is dull, regrind or stone it to restore its point. Scribe a clean visible line. Lay out the 5-inch length from the end of the piece to the start of the angle. Use the rule to determine where the corner of the angle will be (Figure E-34).

Figure E-31 Measuring and marking the width of the gage using the combination square and rule.

Figure E-33 Scribing the width line.

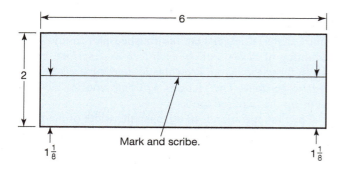

Figure E-32 Width line.

Figure E-34 Measuring the 5-inch dimension from the end to the angle vertex.

Use a plate protractor to lay out the angle. The bevel protractor from the combination set is also a suitable tool for this application. Be sure that the protractor is set to the correct angle. The edge of the protractor blade must be set exactly at the 5-inch mark (Figures E-35 and E-36). The layout of the 31-degree angle establishes its complement of 59 degrees on the drill gage. The correct included angle for general-purpose drill points will be 118 degrees, or twice 59 degrees. The corner radius is ½ inch. Establish this dimension using the square and rule.

Two measurements will be required. Measure from the side and from the end to establish the center of the circle (Figures E-37 and E-38). Prick punch the intersection of the two lines with the 30-degree included point angle layout punch. Tilt the punch so that it can be positioned exactly on the scribe marks (Figure E-39). A magnifier will be useful here. Move the punch to its upright position and tap it lightly with the layout hammer (Figure E-40).

Set the divider to a dimension of ½ inch using the rule. For maximum reliability, use the 1-inch graduation for a starting point. Adjust the divider spacing until you feel the

Figure E-37 Establishing the center point of the corner radius.

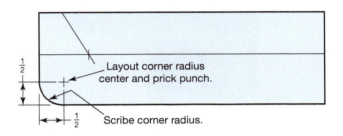

Figure E-38 Corner radius.

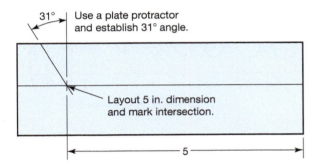

Figure E-35 Angle line.

Figure E-39 Setting the layout punch on the center point of the corner radius.

tips drop into the rule engravings (Figure E-41). Place one divider tip into the layout punch mark and scribe the corner radius (Figure E-42).

The centerline of the holes is 9/16 inch from the edge. Use the square and rule to measure this distance. Mark at each end and scribe the line to its full length.

Figure E-36 Scribing the angle line.

Figure E-40 Prick punching the center point of the corner radius.

Figure E-41 Setting the dividers to the rule engravings.

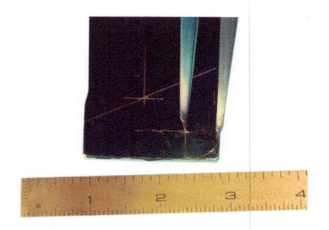

Figure E-42 Scribing the corner radius.

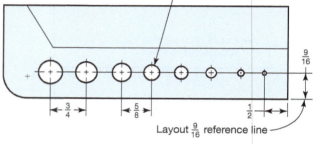

Establish hole center spacing.
Layout and prick punch hole centers.
Scribe hole diameters.

$\frac{9}{16}$

$\frac{3}{4}$ $\frac{5}{8}$ $\frac{1}{2}$

Layout $\frac{9}{16}$ reference line

Figure E-43 Punching and scribing hole centers and diameters.

Figure E-44 Center punching the hole centers.

Measure and lay out the center of each hole. Use the layout punch and mark each hole center. After prick punching each hole center, set the divider to each indicated radius and scribe all hole diameters (Figure E-43).

The last step is to center punch each hole center to deepen prick punch marks prior to drilling. Use a 90-degree included point angle center punch (Figure E-44). Layout of the drill and hole gage is now complete (Figure E-45).

Figure E-45 Completed layout for the drill and hole gage.

GEOMETRIC CONSTRUCTION LAYOUTS FOR BOLT CIRCLES

Certain types of layouts may be done using geometric constructions with only a divider, a scriber, a prick punch, and a rule. For example, to lay out four equally spaced locations:

1. Set the divider and scribe the bolt circle diameter (Figure E-46).
2. Prick punch the circle center and use the rule to scribe a horizontal centerline across the bolt circle diameter. Prick punch the points where the diameter scribe line intersects the circumference of the bolt circle (points 1 and 2).
3. Set one leg of the divider into one of these punch marks and adjust the divider spread so that it is spaced longer than the bolt circle radius. Using points 1 and 2 as center points, scribe arc 1 at the top and arc 3 at the bottom of the bolt circle. It is probably best to scribe

these construction arcs outside the circle, but they may be scribed inside the circle if necessary. The illustration shows the position of the divider so that the construction arcs may be scribed. Move the divider to the other side and at the same setting, scribe arcs 2 and 4 so that they intersect arcs 1 and 3 at points 3 and 4.

4. Use the rule and carefully line up the two points (3, 4) where the scribed arcs intersect. Scribe this line across the bolt circle diameter. If you do this carefully, the diameter scribe line should pass through the center of the bolt circle (Figure E-47).
5. Prick punch the two points where this perpendicular line intersects the bolt circle circumference at locations 2 and 4. Four equally spaced bolt circle hole center points have now been established (Figure E-48).

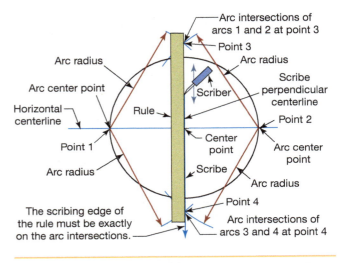

Figure E-47 Align the rule on the intersecting arcs and scribe the perpendicular centerline.

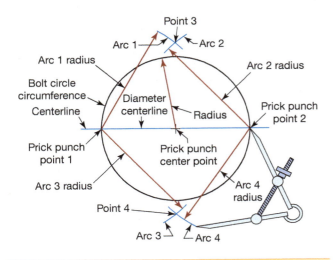

Figure E-46 Preliminary layout for four equally spaced locations.

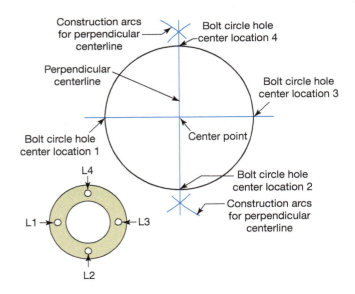

Figure E-48 Four equally spaced locations.

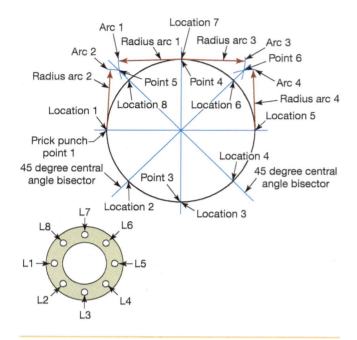

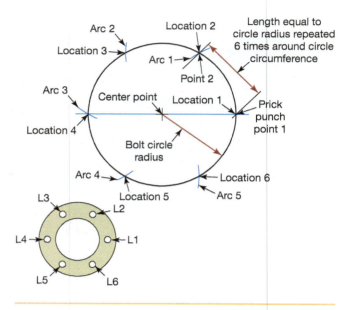

Figure E-50 Layout for six equally spaced locations.

Figure E-49 Bisecting the 90-degree central angles to establish four additional locations for a total of eight equally spaced locations.

To lay out an eight-hole pattern, follow the procedure just described for four holes:

1. When you are finished, bisect each 90-degree central angle to form two 45-degree central angles (Figure E-49).

2. Prick punch points 1 and 2 and also points 3 and 4. Set the divider leg in point 1 and adjust it so that the construction arcs fall outside the circle. Scribe arc 2. Set the divider tip in point 4 and scribe intersecting arc 1 at point 6. At the same setting using points 2 and 4 as centers, scribe intersecting arcs 3 and 4 intersecting at point 5.

3. Use the rule to line up the point of intersecting arcs (points 5 and 6) and the center of the bolt circle. Scribe completely across the circle, thus bisecting the opposite central angle. Where these lines intersect, the bolt circle circumference establishes eight equally spaced bolt circle hole centers (locations L1, L2, L3, L4, L5, L6, L7, and L8).

To lay out a six-hole pattern (Figure E-50), prick punch the center of the bolt circle and use the divider to scribe the bolt circle diameter.

Scribe the bolt circle diameter, then set the divider to the exact bolt circle radius of the circle. Prick punch a point on the bolt circle circumference where the diameter line intersects the circle circumference. Set one leg of the divider in the mark. Be sure that the divider is adjusted to the exact dimension of the bolt circle radius. Use this setting to scribe an arc intersecting the bolt circle circumference (arc 1). Prick punch this location, and holding the same divider setting, move the divider leg to this new location. Scribe the next arc on the bolt circle circumference. Prick punch the new location. Move around the circle until all six locations are marked. It is important to have an exact radius setting on the divider and to accurately position the divider tip, or small errors will crop up in the layout. This process will establish six equally spaced locations (locations L1, L2, L3, L4, L5, and L6).

SELF-TEST

1. How should a workpiece be prepared for layout?
2. What is the reason for placing a workpiece on a paper towel before applying layout die?
3. Describe the technique for using a layout punch.
4. Describe the use of the combination square and rule in layout.
5. Describe the technique of setting a divider to size using a rule.
6. List three type of scribers.
7. Why does a square tool bit make a good scriber?
8. Describe the use of an optical center punch.
9. Why are engraved rules best for setting dividers?
10. What is the difference between a center punch and a prick punch?

Basic Precision Layout Practice

Precision layout is more reliable and accurate than layout by semiprecision practice.

VERNIER HEIGHT GAGE

The main precision layout tool is the height gage. A height gage can typically discriminate to .001 inch. Layouts can also be made more quickly with a height gage. Major parts of the height gage include the base, beam, vernier slide, and scriber (Figure E-51). The size of height gages is measured by the maximum height gaging ability of the instrument. The typical range for a height gage is 18 inches.

Height gage scribers are made from tool steel or tungsten carbide. Carbide scribers are subject to chipping and must be treated gently. They do, however, retain their sharpness and scribe clean, narrow lines. Height gage scribers may be sharpened if they become dull. It is important that any sharpening be done on the slanted surface so that the scriber dimensions will not be changed.

The height gage scriber is attached to the vernier slide and can be moved up and down the beam. Scribers are either straight (Figure E-52) or offset (Figure E-53). The offset scriber permits direct readings with the height gage. The gage reads zero when the scriber rests on the

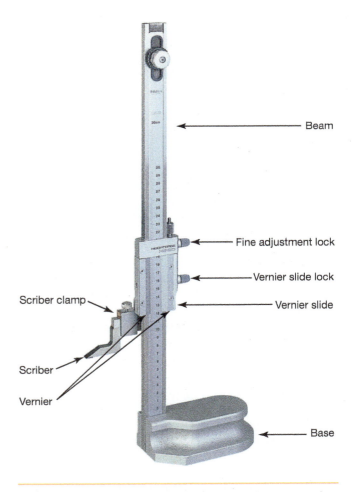

Figure E-51 Parts of the vernier height gage (*Mitutoyo America Corp.*).

Labels: Beam, Fine adjustment lock, Vernier slide lock, Vernier slide, Scriber clamp, Scriber, Vernier, Base

reference surface. With the straight scriber, the workpiece will have to be raised accordingly if direct readings are to be obtained. This type of height gage scriber is less convenient.

271

Figure E-52 Straight vernier height gage scriber.

Figure E-53 Offset height gage scriber.

Figure E-54 Mechanical dial type height gage.

Another popular type of height gage is the mechanical dial type (Figure E-54). These height gages are easy to use; they are direct reading and have no vernier scale. The dial height gage is also available as an electronic digital model.

CHECKING THE ZERO REFERENCE ON A HEIGHT GAGE

The height gage scriber must be checked against the reference surface before attempting to make any height measurements of layouts. Clean the surface of the layout table and the base of the gage. Slide the scriber down until it just rests on the reference surface (Figure E-55). Check the alignment of the zero mark on the scale. Adjust the dial to read zero.

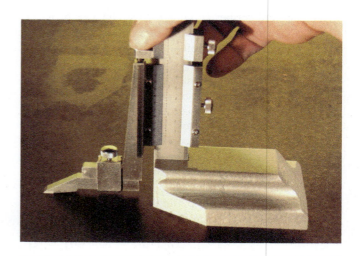

Figure E-55 Checking the zero reference.

APPLICATIONS OF THE HEIGHT GAGE IN LAYOUT

The primary function of the height gage in layout is to measure and scribe lines of known height on the workpiece (Figure E-56). Perpendicular lines may be scribed on the workpiece by the following procedure. First, clamp the work to a right-angle plate if necessary, and scribe the required lines in one direction. Set the height gage at an angle to the work, and pull the corner of the scriber across while keeping the height gage base firmly on the reference surface (Figure E-57). Apply only enough pressure with the scriber to remove the layout dye from the workpiece.

After scribing the required lines in one direction, turn the workpiece 90 degrees. Setup is critical if the scribe marks are to be perpendicular. A square (Figure E-58) or a dial test indicator may be used (Figure E-59) to establish the work at right angles. In both cases the edges of the workpiece must be machined smooth and square. After the clamp has been tightened, the perpendicular lines may be scribed at the required height (Figure E-60).

The height gage may be used to lay out centerlines on round stock (Figure E-61). The stock is clamped in a V block, and the correct dimension to center is determined.

Parallels (Figure E-62) are a valuable and useful layout accessory. Parallels are made from hardened steel or granite, and they are very accurate. Parallels are available in many sizes and lengths. In layout with the height gage, they can be used to support the workpiece (Figure E-63). Angles may be laid out by placing the workpiece on the sine bar (Figure E-64).

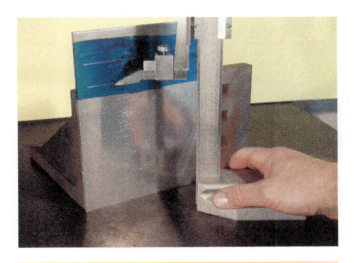

Figure E-56 Scribing height lines with the vernier height gage.

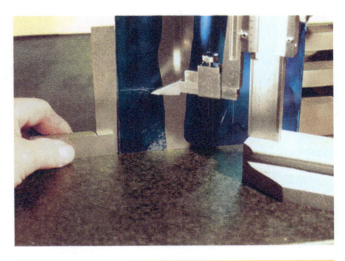

Figure E-58 After the workpiece is turned 90 degrees, it can be checked with a square.

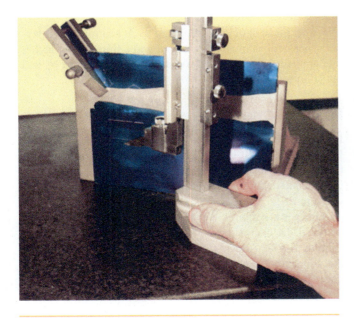

Figure E-57 Scribing layout lines with the workpiece clamped to a right-angle plate.

Figure E-59 Checking the work using a dial test indicator.

Figure E-60 Scribing perpendicular lines.

Figure E-61 Scribing centerlines on round stock clamped in a V block.

Figure E-62 Hardened-steel parallels.

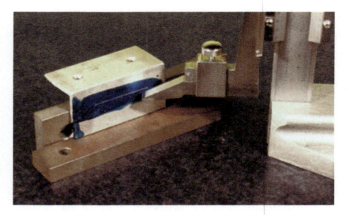

Figure E-63 Using parallels in layout.

Figure E-64 Laying out angle lines using the sine bar.

BASIC PRECISION LAYOUT PRACTICE

The workpiece should be prepared as in semiprecision layout. Sharp edges must be removed and a thin coat of layout dye applied. You will need a drawing of the part to be laid out (Figure E-65).

Position-One Layouts

In position one (Figure E-66), the clamp frame is on edge. In any position, all layouts can be defined as heights above the reference surface. Refer to the drawing on position-one layouts and determine all the layouts that can be done in this position.

Start by scribing the ¾ inch height that defines the width of the clamp frame. Set the height gage to .750 inch (Figure E-67). Attach the scriber (Figure E-68). Be sure that the scriber is sharp and properly installed for the height gage that you are using. Hold the workpiece and height gage firmly, and pull the scriber across the work in a smooth motion (Figure E-69). The height of the clamp screw hole can be laid out at this time. Refer to the part drawing and determine the height of the hole. Set the height gage at

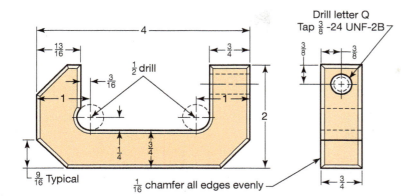

Drill letter Q
Tap $\frac{3}{8}$-24 UNF-2B

$\frac{1}{2}$ drill

$\frac{1}{16}$ chamfer all edges evenly

$\frac{9}{16}$ Typical

Figure E-65 Clamp frame layout.

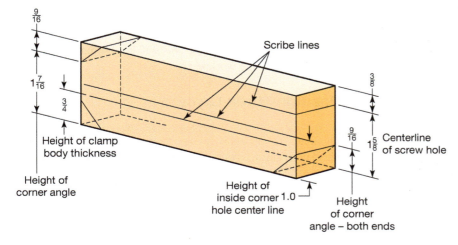

Scribe lines

Height of clamp
body thickness

Height of
corner angle

Height of
inside corner 1.0
hole center line

Height
of corner
angle – both ends

Centerline
of screw hole

Figure E-66 Clamp frame—position-one layouts.

Figure E-67 Setting the height gage to a dimension of .750 inch.

Figure E-68 Attaching the scriber.

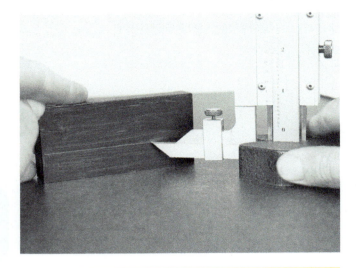

Figure E-69 Scribing the height equivalent of the frame thickness.

1.625 inches and scribe the line on the end of the workpiece (Figure E-70). The line may be projected around on the side of the part. This will facilitate setup in the drill press. Other layouts that can be done at position one include the height of the inside corner hole centerlines.

The starting points of the corner angles on both ends may also be laid out. Refer to the drawing on position-one layouts.

Position-Two Layouts

In position two, the workpiece is on its side (Figure E-71). Check the work with a micrometer to determine its exact thickness. Set the height gage to half this amount and scribe the centerline of the clamp screw hole. This layout will also establish the center point of the clamp screw hole. A height gage setting of .375 inch will probably be adequate provided the stock is .750 inch thick. However, if the thickness varies

Figure E-70 Scribing the height equivalent of the clamp screw hole.

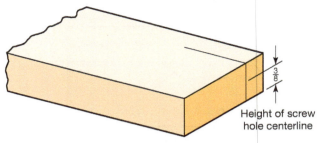

This layout establishes the hole centerpoint.

Figure E-71 Clamp frame—position-two layouts.

above or below .750 inch, the height gage can be set to half of whatever the thickness is. This will ensure that the hole is in the center of the workpiece.

Position-Three Layouts

In position three, the workpiece is on end and would be clamped to an angle plate. Perpendicularity should be checked using a square or dial test indicator. Set the height gage to .750 inch and scribe the height equivalent of the frame end thickness (Figure E-72). Other layouts possible at position three include the height equivalent of the inside corner hole centerlines. This layout will also locate the center points of the inside corner holes (Figure E-73). The height equivalent of the end thickness as well as the ending points of the corner angles can be scribed at position three. The completed layout of the clamp is shown in Figure E-74.

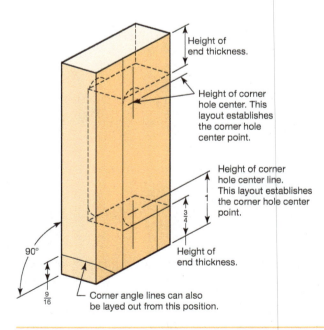

Figure E-72 Clamp frame—position-three layouts.

Figure E-73 Scribing the height equivalent of the end thickness.

Figure E-74 Completed layout of the clamp frame.

HEIGHT GAGE LAYOUT BY COORDINATE MEASURE

Many layouts can be done by calculating the coordinate position of the part features. Coordinate position simply means that each feature is located a certain distance from adjacent perpendicular reference lines. These are frequently known as the *X*- and *Y*-coordinates. You should begin to think of coordinates in terms of *X* and *Y*, as this terminology will be important, especially in computer numerically controlled machining (CNC). The *X*-coordinate on a two-dimensional drawing is horizontal. The *Y*-coordinate is perpendicular to *X* and in the same plane. On a drawing, *Y* is the vertical coordinate. The *X*- and *Y*-coordinate lines can be and often are the edges of the workpiece, provided the edges have been machined true and square to each other.

Coordinate lengths can be calculated by the application of appropriate trigonometric formulas. They may also be determined from tables of coordinate measure. Tables can be found in most machining handbooks.

Trigonometric Calculations in Layouts

Many layouts involve the use of trigonometry calculations. These include coordinate position calculations and the layout of angles. Trigonometry has many other applications in machining.

Trigonometry is based on the six trigonometric functions or ratios: the sine (sin), cosine (cos), tangent (tan), cotangent (cot), secant (sec), and cosecant (csc). These six trigonometric functions are ratios of the lengths of the right triangle's sides.

The length of each side of the right triangle may be compared with the length of the other two sides in several ways. For example, the side opposite an acute angle compared with the hypotenuse (the longest side) is defined as the sine (sin) ratio. The side adjacent to (next to or alongside of) an acute angle compared with the hypotenuse defines the cosine (cos) ratio. The side opposite an acute angle compared with the side adjacent defines the tangent (tan) ratio.

The three other functions, cotangent (cot), secant (sec), and cosecant (csc), are the inverse ratios of the sine, cosine, and tangent. The cotangent is the inverse of the tangent ratio, the secant is the inverse of the cosine ratio, and the cosecant is the inverse of the sine ratio. These inverse ratios or functions are less commonly used in routine machine shop trig calculations, but they may well be used to simplify a problem.

In the right triangle shown in Figure E-75, the following examples define the trigonometric functions of the acute angles *A* and *B*.

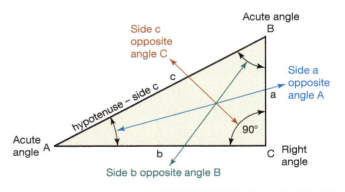

Figure E-75 Right-triangle nomenclature.

Functions for Acute Angle *A*:

$$\sin A = \frac{\text{opposite side}}{\text{hypotenuse}} = \frac{a}{c} \qquad \csc A = \frac{\text{hypotenuse}}{\text{opposite side}} = \frac{c}{a}$$

$$\cos A = \frac{\text{adjacent side}}{\text{hypotenuse}} = \frac{b}{c} \qquad \sec A = \frac{\text{hypotenuse}}{\text{adjacent side}} = \frac{c}{b}$$

$$\tan A = \frac{\text{opposite side}}{\text{adjacent side}} = \frac{a}{b} \qquad \cot A = \frac{\text{adjacent side}}{\text{opposite side}} = \frac{b}{a}$$

Functions for Acute Angle *B*:

$$\sin B = \frac{\text{opposite side}}{\text{hypotenuse}} = \frac{b}{c} \qquad \csc B = \frac{\text{hypotenuse}}{\text{opposite side}} = \frac{c}{b}$$

$$\cos B = \frac{\text{adjacent side}}{\text{hypotenuse}} = \frac{a}{c} \qquad \sec B = \frac{\text{hypotenuse}}{\text{adjacent side}} = \frac{c}{a}$$

$$\tan B = \frac{\text{opposite side}}{\text{adjacent side}} = \frac{b}{a} \qquad \cot B = \frac{\text{adjacent side}}{\text{opposite side}} = \frac{a}{b}$$

The key to performing a trig calculation is to select the correct function (ratio) based on what you know and what you need to determine. Appendix 2, Table 8, summarizes the trig formulas in their various transposed forms. Refer to this table and consult with your instructor as to how to select and apply the formula that will generate the information you need. The following examples of coordinate and angle layouts show the common applications of trigonometry to these layout requirements.

Calculating Coordinate Measurements

The drawing in Figure E-76 shows a five-hole equally spaced pattern centered on the workpiece. Because hole one is on the centerline, its coordinate position measured from

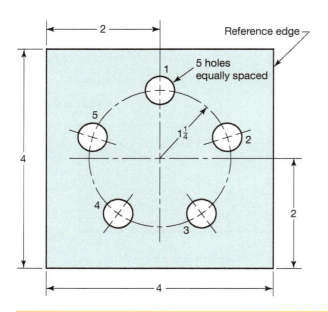

Figure E-76 Circle with five equally spaced holes.

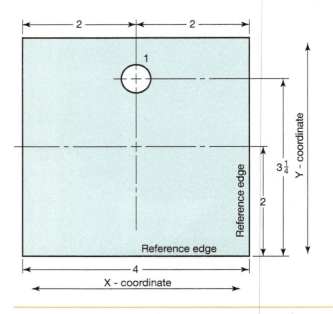

Figure E-77 Coordinate position of hole one.

the reference edges can easily be determined (Figure E-77). The *X*-coordinate (horizontal) is 2 inches. The *Y*-coordinate (vertical) is 2 inches plus the radius of the hole circle, or 3 ¼ inches.

The coordinate position of hole two can be calculated by the following: because there are five equally spaced holes, the center angle is 360/5, or 72 degrees. Right triangle *ABC* (Figure E-80) is formed by constructing a perpendicular line from point *B* to point *C*. Angle *A* equals 18 degrees (90 − 72 = 18). To find the *X*-coordinate, apply the following formula:

$$X_C = \text{circle radius} \times \cos 18°$$

$$= 1.250 \times .951$$

$$= 1.188 \text{ inches}$$

The *X*-coordinate length from the reference edge is found from

$$2.0 - 1.188 = .812 \qquad \text{(Figure E-80)}$$

The *Y*-coordinate is found by the following formula:

$$Y_C = \text{circle radius} \times \sin 18°$$

$$= 1.250 \times .309$$

$$= .386$$

The *Y*-coordinate length from the reference edge is found from

$$2.0 + .386 = 2.386 \text{ inches}$$

The coordinate position of hole three is calculated in a similar manner. Right triangle *AEF* is formed by constructing a perpendicular line from point *F* to point *E* (Figure E-78).

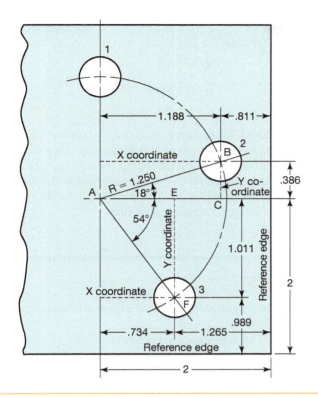

Figure E-78 Coordinate positions of holes two and three.

Angle *FAE* equals 54 degrees (72 − 18 = 54). To find the *X*-coordinate, apply the following formula:

$$X_C = \text{circle radius} \times \cos 54°$$

$$= 1.250 \times .587$$

$$= .734$$

The *X*-coordinate length from the reference edge is found from

$$2.0 - .734 = 1.266 \qquad \text{(Figure E-78)}$$

To find the *Y*-coordinate, apply the following formula:

$$Y_C = \text{radius} \times \sin 54°$$

$$= 1.250 \times .809$$

$$= 1.011 \text{ inches}$$

The *Y*-coordinate length from the reference edge is found from

$$2.0 - 1.011 = .989 \text{ inch} \qquad \text{(Figure E-78)}$$

The coordinate positions of holes four and five are the same distance from the centerlines as holes two and three. Their positions from the reference edges can be calculated easily.

Because this layout involves scribing perpendicular lines, the workpiece must be turned 90 degrees. If the edges of the work are used as references, they must be machined square. Either coordinate may be laid out first. The workpiece is then turned 90 degrees to the adjacent reference edge. This permits the layout of the perpendicular lines (Figure E-79).

LAYING OUT ANGLES

Angles may be laid out using the height gage by calculating the appropriate dimensions using trigonometry. In the example (Figure E-80), the layout of height *A* will establish angle *B* at 36 degrees. Height *A* is calculated by the following formula:

$$\text{Height } A = 1.25 \times \tan B$$

$$= 1.25 \times .726$$

$$= .908 \text{ inch}$$

After a height of .908 inch is scribed, the workpiece is turned 90 degrees, and the starting point of the angle is established at point *B*. Scribing from point *A* to point *B* will establish the desired angle.

Figure E-79 Height equivalents of coordinate positions for all holes.

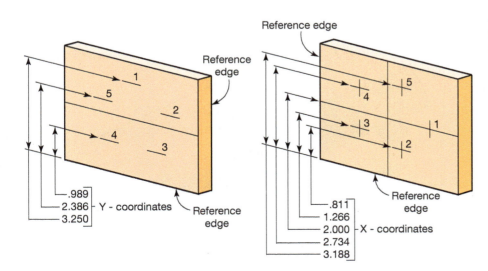

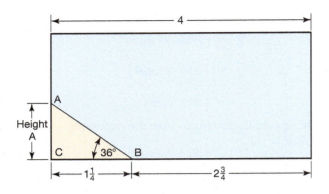

Figure E-80 Laying out a 36-degree angle.

The sine bar can also be used in angular layout. In the example (Figure E-81), the sine bar is elevated for the 25-degree angle. Sine bar elevation is calculated by the following formula:

Bar elevation = bar length × sine of required angle

If we assume a 5-inch sine bar,

$$Elevation = 5 \times \sin 25°$$

$$= 5 \times .422$$

$$= 2.113 \text{ inches}$$

A gage block stack is assembled and placed under the sine bar. Now that the bar has been elevated, the vertical distance *CD* from the corner to the scribe line *AB* must be determined. To find distance *DC*, a perpendicular line must be constructed from point *C* to point *D*. Angle *A* is

65 degrees (90 − 25 = 65). Length *CD* is found by the following formula:

$$CD = .500 \times \sin B$$

$$= .500 \times .906$$

$$= .453 \text{ inch}$$

The height of the corner must be determined and the length of *CD* subtracted from this dimension. This will result in the correct height gage setting for scribing line *AB*. The corner height should be determined using the height gage and dial test indicator. The corner height must not be determined using the height gage scriber.

SELF-TEST

1. Describe the procedure for checking the zero reference on a height gage.
2. How can the zero reference be adjusted?
3. How are perpendicular lines scribed with a height gage?
4. What is the range of a typical height gage?
5. When laying out angles, what tool is used in conjunction with the height gage?

INTERNET REFERENCES

Information on surface plates and height gages:

http://starrett.com

http://fvfowler.com

http://mitutoyo.com

Figure E-81 Laying out an angle using the sine bar.

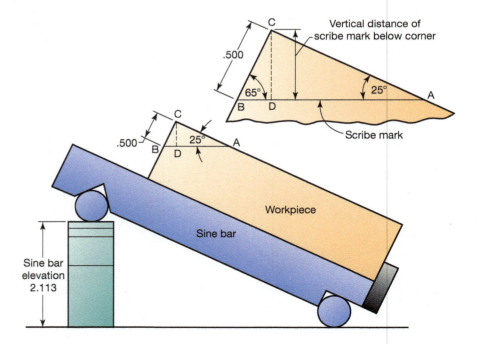

Preparation for Machining Operations

Cutting tools get hot when they are cutting. The heat rise on the cutting edge of a tool can be enough to burn the cutting edge and dull it (Figure F-1). The tool must then be replaced, resharpened, or a new sharp edge must be indexed into place. A certain amount of wear is to be expected, but lost machining time while replacing or sharpening the tool must be considered. There is a trade-off between maximum production and the cost of tooling. In many cases, it is cost effective to wear out tools more quickly. In these cases, the cost of tooling is more than offset by the increase in productivity.

Some tool materials can withstand higher temperature than others. The hardness and toughness of workpiece materials are also factors in tool breakdown caused by heating. Tool breakage can be caused by heavy cuts (excessive feed) and interrupted cuts.

Cutting speed is a major factor in controlling tool wear and breakdown. This is true of all machining operations—sawing, turning, drilling, milling, and grinding.

The *feed* on a machine is the movement that forces the tool into the workpiece. Excessive feed can result in broken tools (Figure F-2).

Cutting fluids are used in machining operations for two basic purposes: to cool the cutting tool and workpiece and to provide lubrication for easy chip flow across the face of the cutting tool.

SAFETY FIRST

Safety in Chip Handling   Certain metals, when divided finely as a powder or even as coarse as machining chips, can ignite with a spark or just by the heat of machining. Magnesium and zirconium are two such metals. These fires are difficult to extinguish, and if water or a water-based fire extinguisher is used, the fire will increase in intensity. The greatest danger of fire occurs when a machine operator fails to clean up zirconium or magnesium chips on a machine when the job is finished. The next operator may then cut alloy steel, which can produce high temperature in the chip or sparks that can ignite the magnesium chips. Such fires may destroy the entire machine or even the shop. Chloride-based powder fire extinguishers are commercially available. These are effective for such fires, because they prevent water absorption and form an air-excluding crust over the burning metal. Sand is also used to smother fires in magnesium.

Metal chips from machining operations are sharp, hot, and are a serious hazard. They should not be handled with bare hands. Gloves should only be worn when the machine is not running.

Some cutting fluids are basically coolants. These are usually water-based soluble oils or synthetics, and their main function is to cool the work–tool interface. Cutting oils provide some cooling action but also act as lubricants and are mainly used where cutting forces are high, as in die threading or tapping operations.

Tool materials range from high-speed steel to diamond, each type having its proper use, whether in traditional metal cutting in machine tools or for grinding operations. This section will enable you to choose and use cutting tools to their greatest advantage.

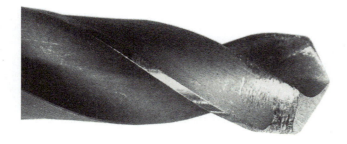

Figure F-1 The end of this twist drill has been burned due to excessive speed.

Figure F-2 The sharp corners of this end mill have been broken off because excessive feed was used.

Machinability and Chip Formation

Machinability is a measure of how easily a metal can be cut with regard to tool life, surface finish, and power consumption. Softer materials are easier to machine than harder ones. Chip removal and control are also major factors in machining. These and other considerations are covered in this unit.

OBJECTIVES

After completing this unit, you should be able to:

- Determine how metal cutting affects the surface structures of metals.
- Analyze chip formation, structures, and chip breakers.
- Explain machinability ratings and machining behavior of metals.

PRINCIPLES OF METAL CUTTING

In machining operations, the tool rotates or moves or the workpiece rotates or moves (Table F-1). The moving or rotating tool moves into the work material to cut a chip. This movement is called *feed*. The amount of feed controls the thickness of the chip. The depth of cut is called *infeed*. In addition to single-point tools with one cutting edge, as in lathe operations, there are multiple-point tools such as milling machine cutters, drills, and reamers. Grinding wheels could be considered to be multiple-point cutters, because small chips are removed by many tiny cutting points on grinding wheels.

One of the most important aspects of cutting tool geometry is rake (Figure F-3). Tool rake ranging from negative to positive has an effect on the formation of the chip and on the surface finish. Zero- and negative-rake tools are stronger and have a longer working life than positive-rake tools. Negative-rake tools produce poor finishes at low cutting speeds but give a good finish at high speeds. Positive-rake

tools are freer cutting at low speeds and can produce good finishes when they are properly sharpened.

A common misconception is that the material splits ahead of the tool, as wood does when it is being split with an axe. This is not true with metals; the metal is sheared off and does not split ahead of the chip (Figures F-4 to F-7). The metal is forced along in the direction of the cut, and the grains are elongated and distorted ahead of the tool and forced along a shear plane, as can be seen in the figures. The surface is disrupted more with a tool having negative rake than with a tool with positive rake. Negative-rake tools require more horsepower than positive-rake tools.

Higher speeds give better surface finishes and produce less disturbance of the grain structure, as can be seen in Figures F-8 and F-9. At a lower speed of 100 surface feet per minute (SFPM), the metal is disturbed to a depth of .005 to .006 inch, and the grain flow is shown to be moving in the direction of the cut. The grains are distorted, and in some places the surface is torn. This condition can later produce fatigue failures and a shorter working life of the part. Tool marks, rough surfaces, and an insufficient internal radius at shaft shoulders can also cause early fatigue failure where there are high stresses. At 400 SFPM, the surface is less disrupted and the grain structure is altered to a depth of only about .001 inch. When the cutting speeds are increased to 600 SFPM and above, little additional improvement is noted.

There is a great difference between the surface finish of metals cut with coolant or lubricant and metals that are cut dry. This is because of the cooling effect of the cutting fluid and the lubricating action that reduces friction between tool and chip. The chip tends to curl away from the tool more quickly and is more uniform when cutting fluid is used. Also, the chip becomes thinner and pressure welding is reduced. When metals are cut dry, pressure welding is a definite problem, especially in the softer metals such as 1100 aluminum and low-carbon steels. Pressure welding produces a built-up edge on the tool that causes a rough finish and

Table F-1 Machining Principles and Operations

Operation	Diagram	Characteristics	Type of Machines
Turning		Work rotates; tool moves for feed	Lathe and vertical turret lathe (VTL)
Milling (horizontal)		Cutter rotates and cuts on periphery; work feeds into cutter and can be moved in three axes	Horizontal milling machine
Face milling		Cutter rotates to cut on its end and periphery of vertical workpiece	Horizontal mill, profile mill, and machining center
Vertical (end) milling		Cutter rotates to cut on its end and periphery; work moves on three axes for feed or position; spindle also moves up or down	Vertical milling machine, die sinker, machining center
Shaping		Work is held stationary and tool reciprocates; work can move in two axes; toolhead can be moved up or down	Horizontal and vertical shapers
Planing		Work reciprocates while tool is stationary; tool can be moved up, down, or crosswise; worktable cannot be moved up or down	Planer
Horizontal sawing (cutoff)		Work is held stationary while the saw cuts in one direction, as in band sawing, or reciprocates while being fed downward into the work	Horizontal band saw, reciprocating cutoff saw

Table F-1 Machining Principles and Operations (*Continued*)

Operation	Diagram	Characteristics	Type of Machines
Vertical band sawing (contour sawing)		Endless band moves downward, cutting a kerf in the workpiece, which can be fed into the saw on one plane at any direction	Vertical band saw
Broaching		Workpiece is held stationary while a multitooth cutter is moved across the surface; each tooth in the cutter cuts progressively deeper	Vertical broaching machine, horizontal broaching machine
Horizontal spindle surface grinding		The rotating grinding wheel can be moved up or down to feed into the workpiece; the table, which is made to reciprocate, holds the work and can also be moved crosswise	Surface grinders, specialized industrial grinding machines
Vertical spindle surface grinding		The rotating grinding wheel can be moved up or down to feed into the workpiece; the circular table rotates	Blanchard-type surface grinders
Cylindrical grinding		The rotating grinding wheel contacts a turning workpiece that can reciprocate from end to end; the wheelhead can be moved into the work or away from it	Cylindrical grinders, specialized industrial grinding machines
Centerless grinding		Work is supported by a workrest between a large grinding wheel and a smaller feed wheel	Centerless grinder

Table F-1 Machining Principles and Operations (*Continued*)

Operation	Diagram	Characteristics	Type of Machines
Drilling and reaming		Drill or reamer rotates while work is stationary	Drill presses, vertical milling machines
Drilling and reaming		Work turns while drill or reamer is stationary	Engine lathes, turret lathes, automatic screw machines
Boring		Work rotates; tool moves for feed on internal surfaces	Lathes, and vertical turret lathe (VTL)

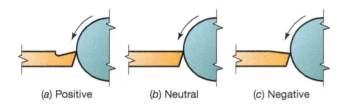

(a) Positive   (b) Neutral   (c) Negative

Figure F-3 Side view of rake angles.

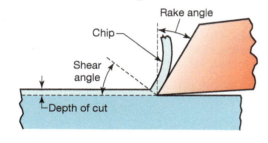

Figure F-4 Metal-cutting diagram.

Figure F-6 Point of a negative-rake tool magnified *100 at the point of the tool.

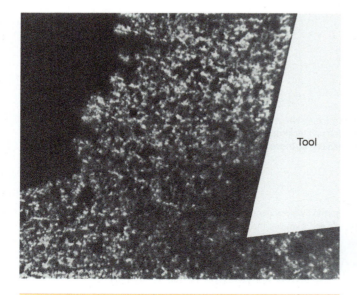

Figure F-5 A chip from a positive-rake tool magnified *100 at the point of the tool. The grain distortion is not as evident as in Figures F-6 and F-7. This is a continuous chip.

Figure F-7 A zero rake tool at magnified *100 shows grain flow and distortion similar to that of the negative-rake tool.

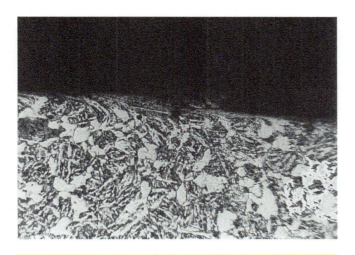

Figure F-8 This micrograph shows the surface of the specimen turned at 100 SFPM. The surface is irregular and torn. The grains are distorted to a depth of approximately .005 to .006 in. (magnification is 250*).

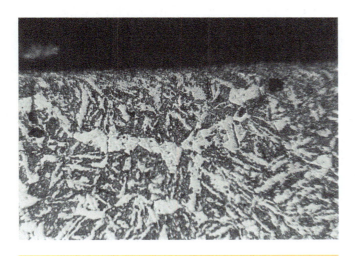

Figure F-9 At 400 SFPM, this micrograph reveals that the surface is fairly smooth and the grains are only slightly distorted to a depth of approximately .001 in. (magnification is 250*).

Figure F-10 A continuous-form chip is beginning to curl away from this positive-rake tool.

Figure F-11 A thick, discontinuous chatter chip being formed at slow speed with a negative-rake tool.

Figure F-12 A discontinuous, thick chip is being formed with a zero-rake tool.

tearing of the surface. A built-up edge is also caused by speeds that are too low. This often results in broken tools from excessive pressure on the cutting edge. Figures F-10 to F-12 show chips formed with tools having positive, negative, and zero rake. These chips were all formed at low surface speeds and consequently are thicker than they would have been at higher speeds.

At high speeds, cratering begins to form on the top surface of the tool (Figure F-13) because of chip wear against the tool; this causes the chip to begin to curl. The crater creates airspace between the chip and the tool, which is an ideal condition because it insulates the chip from the tool and allows the tool to remain cooler; the heat

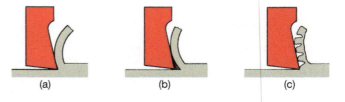

Figure F-14 Three types of chip formation: (*a*) continuous; (*b*) continuous with built-up edge; (*c*) discontinuous (segmented) (*Neely, John E.*, Basic Machine Tool Practices, *1st, ©2000. Electronically reproduced by permission of Pearson Education, Inc., Upper Saddle River, New Jersey*).

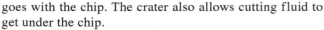

Figure F-13 The crater on the cutting edge of this tool was caused by chip wear at high speeds.

goes with the chip. The crater also allows cutting fluid to get under the chip.

Different metals cut in different ways. Softer, more ductile metals produce a thicker chip, and harder metals produce a thinner chip. A thin chip indicates a clean cutting action with a better finish. Speeds, feeds, tool shapes, and depth of cut have an effect on surface finish and the surface structure of the metal.

Tool materials include high-speed steel, carbide, ceramic, and diamond. Most manufacturing today is done with carbide tools. Greater amounts of material may be removed and tool life is extended considerably when carbides are used. Much higher speeds and feeds can be used with carbide tools than with high-speed tools.

ANALYSIS OF CHIPS

Machining produces chips of three basic types: the continuous chip, the continuous chip with a built-up edge on the tool, and the discontinuous chip. The formation of the three basic types of chips can be seen in Figure F-14. Several kinds of chips are shown in Figure F-15. High cutting speeds produce thin chips, and tools with a large positive rake angle favor the formation of a continuous chip. Any circumstances that lead to a reduction of friction between the chip–tool interface, such as the use of cutting fluid, tend to produce a continuous chip. The continuous chip usually produces the best surface finish and has the greatest efficiency in terms of power consumption. Continuous chips create lower temperatures at the cutting edge, but high speeds lead to higher cutting forces and extremely high tool pressures. Because there is less strength at the point of positive rake angle tools than with negative-rake tools, tool failure is more likely with large positive rake angles at high cutting speeds or with intermittent cuts.

Negative-rake tools tend to produce a built-up edge with a rough continuous chip and a rough surface finish, at lower cutting speeds and with soft materials. Positive rake angles with the use of cutting fluid and higher speeds

decrease the tendency for a built-up edge on the tool. Most carbide tools for turning machines have a negative rake for several reasons. Carbide is always used at high cutting speeds, lessening the tendency for a built-up edge. Better finishes are obtained at high speeds even with negative rake, and the tool can withstand greater shock loads than with positive-rake tools. Another advantage of negative-rake tools is that an indexable insert can have 90-degree angles between its top surface and flank. This means that the top edges of the insert can be used to cut and the bottom edges can also be used. This doubles the cutting edges on the insert. A negative rake holder gives the tool flank relief even though the tool has square edges. Negative-rake tools require more horsepower than positive-rake tools do when all other factors are the same.

The discontinuous or segmented chip is produced when a brittle metal, such as cast iron or hard bronze, is cut. Some ductile metals can form a discontinuous chip when the machine tool is old or loose and a chattering condition is present, or when the tool form is not correct. The discontinuous chip is formed as the cutting tool contacts the metal and compresses it. The chip then begins to flow along the tool, and when more stress is applied to the brittle metal, it tears loose and a rupture occurs. This causes the chip to separate from the work material. Then, a new cycle of compression, tearing away of the chip, and breaking off begins.

Low cutting speeds and zero or negative rake angle can produce discontinuous chips. The discontinuous chip is more easily handled on the machine, because it falls into the chip pan.

The continuous chip sometimes produces snarls or long strings that are not only inconvenient but dangerous to handle. The optimum chip for operator safety and a good surface finish is the figure 9-shaped chip. A figure 9-shaped chip is usually produced with a chip breaker (Figure F-16).

Most carbide tool holders have either an inserted chip breaker or one formed in the tool insert itself. A chip breaker is designed to curl the chip against the work and then to break it off to produce the proper type of chip.

When using high-speed tools, machinists must usually form the tool shape on a grinder. If they do not grind a chip

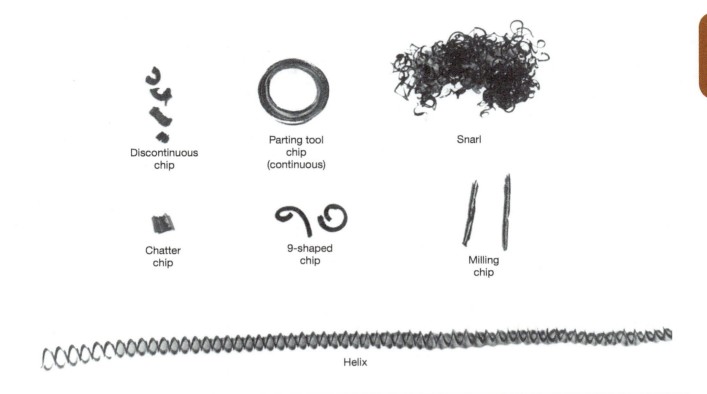

Discontinuous
chip

Parting tool
chip
(continuous)

Snarl

Chatter
chip

9-shaped
chip

Milling
chip

Helix

Figure F-15 Chips formed in machining operations.

breaker into the tool, a potentially long, dangerous, continuous chip will be formed. If they do grind a chip breaker in the tool, the tool will produce a more acceptable chip. A high rate of feed will produce a greater curl in the chip, and even without a chip breaker a curl can be produced by adjusting the feed of the machine properly. The depth of cut also has an effect on chip curl and breaking. A larger chip is produced when the depth of cut is greater. This heavier chip is less springy than light chips and tends to break up into small chips more readily. Figure F-17 shows how chip breakers may be formed in high-speed tools.

MACHINABILITY OF METALS

Machinability ratings by some manufacturers are now based on AISI B-1112 steel, which is assigned a machinability rating of 100 percent and at a cutting speed of 168 SFPM. If a metal is easier to machine than this, its machinability rating is higher than 100. If a metal is harder to machine than this, its rating is less than 100. Table F-2 gives the machinability of various alloys of steel based on AISI B-1112 as 100 percent. Machinability is greatly affected by the cutting tool material, tool geometry, and the use of cutting fluids.

(a)

(b)

Figure F-16 The action of a chip breaker to curl a chip and cause it to break off: (a) plain tool; (b) tool with chip breaker.

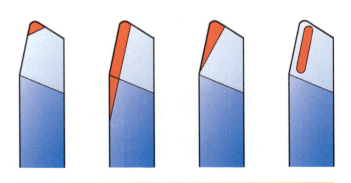

Figure F-17 Types of chip breakers in high-speed tools.

Table F-2 Machinability Ratings for Some Commonly Used Steels

AISI Number	Machinability Ratings %
B1112	100
C1120	81
C1140	72
C1008	66
C1020	72
C1030	70
C1040	64
C1060 (annealed)	51
C1090 (annealed)	42
3140 (annealed)	66
4140 (annealed)	66
5140 (annealed)	70
6120	57
8620	66
301 (stainless)	36
302 (stainless)	36
304 (stainless)	36
420 (stainless)	36
440A (stainless)	24
440B (stainless)	24
440C (stainless)	24

A machinist must select the type of tool, speeds, feeds, and cutting fluid for the material being cut. The most important material property is hardness. Hardness is also a factor in producing good finishes. Soft metals such as copper, 1100 series aluminum, and low-carbon, hot-rolled steel tend to have poor finishes. Cutting tools tend to tear the metal rather than cut these soft metals. Sharp tools and larger rake angles plus the use of cutting fluids will produce better finishes. Harder, tougher alloys almost always produce better finishes even with negative-rake carbide tools. For example, an AISI 1020 hot-rolled steel bar and an AISI 4140 steel bar of the same diameter cut at the same speed and fed with the same tooling will have very different finishes. The low-carbon, hot-rolled (HR) steel will have a poor finish compared with the alloy steel.

Cutting speeds can be different for materials in the same classification. Low-carbon steel, for example, can have a cutting speed of 70, 90, 100, or 120 depending on the hardness of the material.

The operator must also understand the effects of heat on metals such as alloy steel, tool steel, or gray cast iron. Welding on any of these metals may harden the metal near the weld and make it difficult to cut, even with carbide tools. The entire workpiece must be annealed to soften these hard areas.

Most alloy steels can be cut with high-speed tools but at relatively low cutting speeds. These low speeds produce a poor finish unless some back rake is used. These steels are best machined with carbide tools at higher cutting speeds.

Some nonferrous metals, such as aluminum and magnesium, can be machined at much higher speeds when the correct cutting fluid is used. Some alloy steels, however, are more difficult to machine because they tend to work harden. Examples of these are austenitic manganese steels, Inconel, stainless steels (SAE 301 and others), and some tool steels.

SELF-TEST

1. In what way do tool rakes, positive and negative, affect surface finish?
2. Soft materials tend to pressure weld on the top of the cutting edge of the tool. What is this condition called and what is its result?
3. Which type of chip indicates a greater disruption of the surface material: thin uniform chips or thick segmented chips?
4. Does the material split ahead of the tool when machining metal? If not, what does it do?
5. Which cutting edge is stronger, a negative- or positive-rake tool?
6. What effect does cutting speed have on surface finish? What effect on the disruption of the surface grain structure?
7. How can surface irregularities caused by machining affect the usefulness of the part?
8. Which property of metals is directly related to machinability? How can a machinist do a quick check to determine this property?
9. Describe the type of chip that is the safest and easiest to handle.
10. Define *machinability*.

INTERNET REFERENCE

Information on machinability of metals:

http://carbidedepot.com

Speeds and Feeds for Machine Tools

Modern machine tools are powerful, and with proper tooling they are designed for high production rates. However, a machinist who uses incorrect feeds and speeds will not be productive and will produce inferior parts. If a machinist does not have a good understanding of cutting tools, their cutting speeds and feeds may be too low to produce good surface finishes and achieve optimal metal removal. Labor and machine costs make time crucially important. Larger metal removal rates can shorten the time needed to produce a part and often improve its physical properties.

The importance of cutting speeds and feeds in machining operations cannot be overemphasized. The right speed can mean the difference between tool wear and breakage, or having many hours of cutting time between sharpenings or replacement. Excessive feeds often result in tool breakage. This unit will prepare you to choose correct feeds and speeds.

OBJECTIVES

After completing this unit, you should be able to:

- Calculate correct cutting speeds for various machine tools and grinding machines.
- Determine correct feeds for various machining operations.
- Define terms such as cutting speed, feed, chip load, SFPM, and so on.
- Locate cutting condition information using reference materials.
- Explain how the workpiece setup and machine rigidity affect cutting conditions.
- Calculate the proper RPM and feed rates for machining given the tool and material to be machined.

CUTTING SPEEDS

Different materials need to be cut at different speeds. The term used to describe a material's ideal machining speed is cutting speed. Cutting speed is the speed at the outside edge of the cutter as it is rotating. This is also known as surface speed. Surface speed, surface footage, and surface area are all directly related.

Two wheels can illustrate this (Figure F-18). Consider two wheels, one wheel is 3 feet in diameter and the other wheel is 1 foot in diameter. If each wheel is rolled one complete revolution, the 3-foot wheel moves about 9 feet (3 * PI) and the 1-foot diameter wheel moves about 3 feet (1 * PI). The larger wheel travelled farther because it has a larger circumference.

The same is true for a machine. Study Figure F-19. This figure represents two different diameter parts being cut on a lathe. As the spindle turns the part, the outside of the part (circumference) passes by the cutting tool. Loosely speaking, the cutter would make a 9-inch long chip on the 3-inch diameter part with each revolution of the spindle. In the bottom example, the lathe would only make a 3-inch long chip in one revolution.

Next imagine different types of materials: plastic, aluminum, brass, mild steel, alloy steel, hardened steel, and so on. Each of these materials would have an ideal speed at which they should be machined. For example, if alloy steel is being cut at an excessive speed, the tool's teeth will be subjected to excessive heat and will be damaged. But that speed might have been fine for brass. Every material has an ideal speed at which to be cut. These ideal speeds have been determined for all common materials. These ideal speeds are called cutting speeds. They can be found in cutting speed charts.

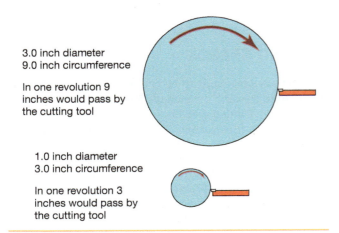

Length each wheel travels in one revolution

The 3 foot diameter wheel traveled about 9 feet

The 1 foot diameter wheel traveled about 3 feet

Figure F-18 An illustration comparing the circumference of two different diameter wheels and cutting speed.

3.0 inch diameter
9.0 inch circumference

In one revolution 9 inches would pass by the cutting tool

1.0 inch diameter
3.0 inch circumference

In one revolution 3 inches would pass by the cutting tool

Figure F-19 This figure shows the difference that diameter makes in cutting speed.

Cutting speeds are given in SFPM. Imagine that the parts in Figure F-20 are made of mild steel. The cutting speed for mild steel with a HSS tool is 100 SFPM. This would mean the ideal cutting condition would be 100 feet of material passing the tool point every minute. The 3-inch diameter part has a 9-inch circumference. That means that 9 inches go by the tool with every revolution. One hundred SFPM would be 1200 inches per minute (100 * 12).

To get 1200 inches of material to pass the tool with a 9-inch circumference, we would need to divide 1200 inches by 9 inches (1200"/9"). This would give us 133 for an answer. The spindle would have to turn at 133 revolutions per minute (RPM) to equal 100 SFPM. This would be the ideal speed to cut mild steel with a 3-inch diameter.

How about the 1-inch diameter mild steel part? The ideal cutting speed is still 100 SFPM, but the circumference of the part is only 3 inches. If we convert the 100 SFPM to inches it would be 1200 inches per minute. The part has a 3-inch circumference so 1200"/3" = 400. The ideal spindle speed for this smaller diameter part would be 400 RPM (Figure F-20).

Note that in this example that the smaller diameter part required a higher RPM to achieve the ideal cutting speed for the material. The larger diameter part required a lower RPM to achieve the ideal cutting speed.

In fact, the diameter of the larger part was three times as large as the 1-inch diameter part and the recommended RPM of the larger part was $\frac{1}{3}$ the RPM of the smaller part.

How is the ideal cutting speed of a material determined? Cutting speeds are determined by the machinability rating. Machinability is the ability of a material to be machined.

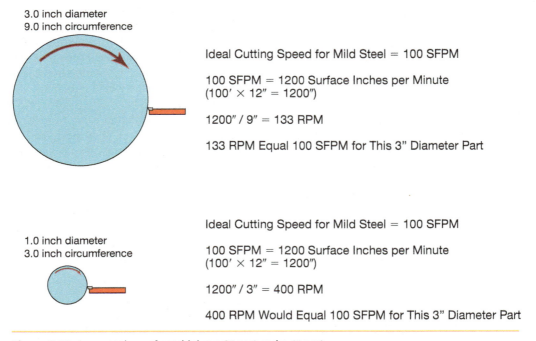

3.0 inch diameter
9.0 inch circumference

Ideal Cutting Speed for Mild Steel = 100 SFPM

100 SFPM = 1200 Surface Inches per Minute
(100′ × 12″ = 1200″)

1200″ / 9″ = 133 RPM

133 RPM Equal 100 SFPM for This 3" Diameter Part

1.0 inch diameter
3.0 inch circumference

Ideal Cutting Speed for Mild Steel = 100 SFPM

100 SFPM = 1200 Surface Inches per Minute
(100′ × 12″ = 1200″)

1200″ / 3″ = 400 RPM

400 RPM Would Equal 100 SFPM for This 3" Diameter Part

Figure F-20 A comparison of machining a 3" part and a 1" part.

Steel   Iron   Aluminum   Plastic

Increasing cutting speed

Figure F-21 The softer the material, the higher the cutting speed.

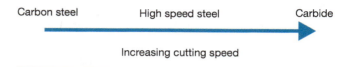

Carbon steel   High speed steel   Carbide

Increasing cutting speed

Figure F-22 The harder the cutting tool material, the higher the cutting speed.

Machinability ratings determine recommended cutting speeds. Recommended cutting speeds are given in charts.

The machinability of the material determines the recommended cutting speed. The harder the material, the slower the cutting speed. The softer the work material, the faster the recommended cutting speed (Figure F-21).

TYPE OF CUTTING TOOL AND ITS EFFECT ON CUTTING SPEED

The material the cutting tool is made of also determines the cutting speed. The two most common would be high-speed steel (HSS) and carbide. Cutting speed tables will give a cutting speed for HSS tooling and also for carbide tooling. The cutting speed for carbide is usually about four times faster. If the cutting speed for a steel is 80 SFPM for a HSS end mill, the SFPM would be approximately 320 for the same material with carbide.

The harder the cutting tool material, the higher the cutting speed (Figure F-22). The softer the cutting tool material, the lower the recommended cutting speed.

All cutting tools work on the surface footage principal. Cutting speeds depend primarily on the kind of material you are cutting and the kind of cutting tool you are using. Using the proper cutting speed for the material that is being machined will assure that cutting is optimal and is not too slow or fast for the tool.

OTHER FACTORS THAT AFFECT CUTTING SPEED

The depth of cut and the feed rate also affect the cutting speed. These three factors cutting speed, feed rate, and depth of cut are known as cutting conditions. Think of cutting conditions as being how much work is being done with the tool. A heavy roughing cut will create more heat and may require a lower cutting speed. A light finish cut would allow a higher cutting speed.

CALCULATING RPM

The spindle RPM must be set so that the part or cutter will be turning at the correct cutting speed for the material being machined. In the previous examples we used a lathe part. The same considerations apply to a cutting tool for a mill or other machine. For the rest of the RPM calculation examples

we will use rotating cutting tools, not a lathe tool. But the same RPM calculations apply to both.

To determine the proper machining speed, we need to calculate the RPM. The cutting speed or surface speed changes based on the size of the cutter. To keep the surface speed correct for different sizes of cutters, the formula must include the size of the cutter.

The correct RPM will change with the size of the cutter. As the milling cutter gets smaller, the RPM must increase to maintain the recommended cutting speed (SFPM). Remember the example of the wheel. Think of the cutter as a wheel and the cutting speed as a distance. A larger wheel (cutter) will need to turn fewer RPM to cover the same distance in the same amount of time than a smaller wheel (cutter). Therefore, to maintain the recommended cutting speed, larger diameter cutters must be run at slower speeds than smaller diameter cutters.

The machine must be set so that the cutter will be operating at the proper surface speed. Spindle speed settings on machines are in RPMs. To calculate the proper RPM for the tool, use the formula shown in Figure F-23.

A milling cut is to be taken with a 0.500-inch HSS end mill on a piece of 1018 plain carbon steel. Calculate the RPM for this cut (Figure F-24). Use the cutting speed charts in Figure F-25.

If the machine does not have variable speeds, choose the speed which is nearest to the calculated RPM.

There are additional considerations when choosing RPM. Is it a roughing or a finish cut? If you are roughing, go slower. If you are making a finish cut, go faster. What is your depth of cut? If it is a deep cut, go to the slower RPM setting. Go slower for setups that lack a great deal of rigidity. Are you using coolant? You may be able to go faster if you are using coolant.

$$\frac{\text{Cutting Speed (CS)} * 4}{\text{Diameter of Cutter (D)}} = \text{RPM}$$

Figure F-23 RPM formula.

$$\text{Cutting Speed} = 90\,\text{SFPM}$$

$$\text{Diameter of Cutter} = 0.500$$

$$\frac{\text{CS} * 4}{D} = \frac{90 * 4}{0.500} = \frac{360}{0.500} = 720\,\text{RPM}$$

Figure F-24 RPM calculation example.

Figure F-25 Recommended Cutting Speed for Milling in Surface Feet per Minute (SFPM)

Workpiece Material	Cutting Speed in SFPM	
	High-Speed Steel	**Carbide**
1018 Plain Carbon Steel	90–120	270–450
6061-T6 Aluminum	70–100	225–375
4140	30–70	145–300
303	100–150	325–500
11L17 Leaded Free Machining	170	400
304/316 Stainless Steel	60	230
A2 Tool Steel	50	250

The greatest indicator of using the correct cutting speed is the color of the chip. When using a HSS cutter, the chips should never be turning brown or blue. Straw colored chips indicate that you are on the maximum edge of the cutting speed for your cutting conditions. When using carbide, chip color can range from amber to blue, but should not be black. A dark purple color will indicate that you are on the maximum edge of your cutting conditions.

Let's try some examples:

A milling cut is to be taken with a 6.00-inch (HSS) face mill on a piece of 1045 steel (Cutting speed = 55 SFPM). Calculate the correct RPM for this cut (Figure F-26).

A 1-inch (HSS) drill is used on a piece of 1018 steel with a cutting speed of 100 SFPM. Calculate the RPM for this drilling operation (Figure F-27).

A cut is to be taken with a 3.00-inch carbide face mill on a piece of 4140 alloy steel with a cutting speed of 300 SFPM. Calculate the RPM for this cut (Figure F-28).

$$\text{Cutting Speed} = 55\,\text{SFPM}$$

$$\text{Diameter of Cutter} = 6.00$$

$$\frac{CS*4}{D} = \frac{55*4}{6.00} = \frac{220}{6.00} = 36\,\text{RPM}$$

Figure F-26 RPM Calculation example.

$$\text{Cutting Speed} = 100\,\text{SFPM}$$

$$\text{Diameter of Cutter} = 1.00$$

$$\frac{CS*4}{D} = \frac{100*4}{1.00} = \frac{400}{1.00} = 400\,\text{RPM}$$

Figure F-27 RPM Calculation example.

$$\text{Cutting Speed} = 300\,\text{SFPM}$$

$$\text{Diameter of Cutter} = 3.00$$

$$\frac{CS*4}{D} = \frac{300*4}{3.00} = \frac{1200}{3.00} = 400\,\text{RPM}$$

Figure F-28 RPM Calculation example.

FEED RATE CALCULATION

There are three factors that affect cutting conditions: cutting speed, depth of cut, and feed rate. The feed rate on milling machines is given in terms of inches per minute (IPM). IPM is the rate at which the tool will advance into the work.

The feed rate is determined by the RPM of the cutter, the cutter material, the number of teeth, and by the chip thickness that the cutter teeth can handle. The chip size is called the feed rate in inches per tooth or chip load.

Study Figure F-29. This figure shows chip load for conventional milling and for climb milling. If we set a feed with a chip load of .005″, it would mean that each tooth would take a .005″ thick cut every revolution.

Take a close look at the climb milling example. Note the direction of table feed and the direction of tool rotation. In climb milling the cutter tends to "pull" the table in the same direction as the feed. Note also that in climb milling each tooth takes the full chip load as it enters the material. Climb milling can produce better surface finishes if the machine is rigid and there is very little backlash in the table. If there is backlash or the machine is not rigid, chatter will result.

Next study the conventional milling example in the figure. The direction of the cutter rotation opposes the direction of the table feed. The chip load is very light as each tooth starts its cut. The thickness of the chip does not reach the full chip load until the end of the cut. Conventional milling will work better on machines that are not rigid or those with more backlash.

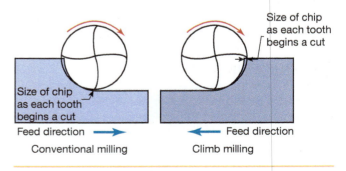

Figure F-29 Conventional milling and climb milling.

Figure F-30 Recommended feed in inches per tooth for high-speed steel milling cutters

Material	Chip Load for End Mills		Chip Load for HSS Drill $\frac{1}{6}$ to $\frac{3}{4}$ Diameter
	HSS	**Carbide**	
1018 Mild Steel	.001–.005	.006–.015	.001–.015
6061-T6 Aluminum	.002–.006	.002–.010	.001–.016
4140 Steel	.001–.004	.0015–.006	.001–.014
303 Stainless Steel	.001–.005	.001–.005	.001–.014
11L17 Free Machining Steel	.001–.005	.001–.007	.001–.018
304/316 Stainless Steel	.001–.005	.0005–.003	.001–.010
A2 Tool Steel	.0005–.003	.001–.004	.001–.007

The recommended values for chip load are based on the cutting tool material, the cutting tool size, and the hardness or machinability rating of the workpiece material. The recommended values for feed per tooth (chip load) can be found in charts, reference books, or on the internet. A typical feed in inches per tooth chart is shown in Figure F-30.

While the recommended feed rates found in these charts represent good machining practice, they are only recommended values. Deviations from these recommended values may be necessary due to machining conditions. The feed rate used on small or thin work may need to be reduced. The work holding technique has a great deal to do with the feed rate. Setups, which lack rigidity, may require a slower feed rate. Feed rates on rigid milling machines can be much heavier than those feed rates used on lighter duty milling machines. When using large carbide tools, a machine's horsepower and the rigidity of the machine influence the feed rate.

The feed rate in inches per tooth (chip load) must be converted into feed rate in IPM to set the feed rate on a machine. The formula for converting feed rate in inches per tooth into inches per minute is shown in Figure F-31.

These feed rates are a starting point. You will typically be given a range of chip load factors to use. A good rule of thumb is to start out at the middle of the range of feed per tooth and increase the feed rate to the capacity of the machine tool, the setup, and the desired surface finish. It must also be mentioned that using a chip load that is too small will cause excessive tool wear.

Let's try some feed rate calculations. Us the recommended feed rate chart in Figure F-30. A four-flute 0.5-inch HSS end mill is to be used on a piece of 1018 mild steel with a cutting speed of 80. The RPM setting to perform this cut is 320 rpm. Look up the feed per tooth in the charts and calculate the feed rate in IPM (Figure F-32).

$$\frac{CS*4}{D} = \frac{80*4}{0.50} = \frac{320}{0.50} = 640\,RPM$$
$$RPM = 640$$

$$Feed\,in\,Inches\,per\,Tooth\,(Chip\,Load) = 0.005"$$

$$Number\,of\,Flutes = 2$$

$$Feed\,(Inches\,per\,Minute) = RPM*Chip\,Load\,(CL)* \\ Number\,of\,Teeth$$
$$Feed\,(Inches\,per\,Minute) = 640*0.005*2$$

$$Feed = 6.4\,Inches\,per\,Minute\,(IPM)$$

Figure F-32 Feed rate calculation.

Some judgment must be used when selecting feed rates. The calculated feed rate is a good place to start. One must also consider the machining conditions. What are the surface finish requirements? A larger feed rate will leave a rougher finish. What is the depth of cut? If it is a deep cut, use a slower feed rate setting. Is the setup rigid? Go slower for setups that lack rigidity.

You may be able to have a higher feed if you are using coolant. Remember that the calculated feed rate is a good place to start. You may have to increase or decrease the feed rate to find the optimal feed rate.

Let's try another example. A two-flute .250-inch HSS end mill will be used to machine 8620 alloy steel with a cutting speed of 80 SFPM. Calculate the proper RPM. Also, calculate the feed rate in IPM using a chip load of .004 (Figure F-33).

DEPTH OF CUT

The depth of cut also affects speeds and feeds. The deeper the cut, the more work is being done. More work creates more heat in the tool and workpiece. Excessive heat can destroy tools. Therefore, to determine the depth of cut, we must first select the proper cutting tool, the proper machine, and a suitable setup. There is a term called the metal removal rate. Metal removal rate is the number of cubic-inches-per minute of material that can be removed from the part.

$$Feed\,Rate\,(inches/min.) = RPM*Chip\,Load* \\ Number\,of\,Teeth$$

Figure F-31 Feed rate formula.

$$\frac{CS*4}{D} = \frac{80*4}{0.25} = \frac{320}{0.25} = 1280 \, RPM$$
$$RPM = 1280$$

$$\text{Feed in Inches per Tooth (Chip Load)} = 0.008"$$

$$\text{Number of Flutes} = 2$$

$$\text{Feed (Inches per Minute)} = RPM * Chip Load (CL) * \\ \text{Number of Teeth}$$
$$\text{Feed (Inches per Minute)} = 1280 * 0.004 * 2$$

$$\text{Feed} = 10.24 \, \text{Inches per Minute (IPM)}$$

Figure F-33 Feed rate calculation.

TOOL TYPE

Roughing tools should be used when large amounts of material need to be removed. The serrated flute design enables a roughing end mill (Figure F-34) to remove three times as much material as a plain end mill.

You should always use the largest tool that will perform the job. Larger tools are capable of larger metal removal rates. In some cases, more than one tool will need to be used. One tool used for roughing cuts and a different tool for finish cuts. The profile of the part and efficiency may require that you need to rough with a larger cutting tool and finish with a smaller tool.

SELECTING THE PROPER MACHINE

The machine type and size will influence the depth of cut. How rigid is the machine? The easiest way to determine the rigidity of the machine is to look at the size of the spindle taper. The larger the spindle taper, the greater the rigidity of the machine. Make sure your metal removal rate decisions account for the style and size of the machine.

Available Horsepower

Does the machine have enough horsepower for the desired metal removal rate with the chosen cutting tool and conditions? Roughing cuts with carbide on tough materials require higher horsepower.

The horsepower that is available at the spindle is not the same as the horsepower of the main motor. The available horsepower at the spindle may be as low as 50 to 80 percent of the main motor horsepower. For maximum metal removal

Figure F-34 Roughing end mills can dramatically increase metal removal rates.

rates, you may need to calculate horsepower requirements. There are charts that show the required horsepower for various metal removal rates.

Rigidity of the Setup

How securely the part is clamped to the table and how far the tool extends from the spindle are the major considerations. If the tool is extended and the workpiece can only be clamped in a limited number of places, the depth of cut and the feed rate may have to be reduced.

The more rigid the setup is, the higher the metal removal rate can be. The shorter the tool, the higher the metal removal rate can be. The rigidity of the setup should always be maximized for productivity and safety.

Depth of Cut

When using a HSS end mill, the rule of thumb is the depth of cut should not exceed $\frac{1}{2}$ the diameter of the cutter. An operator should maximize the feed rate and the depth of cut while keeping the RPM in the calculated range. When maximizing the depth of cut, use a slower feed rate with an acceptable chip load factor.

If the cut you are taking is too large for the machine or conditions, it is better to decrease the depth of cut, rather than decreasing the feed rate below the recommended chip load. A feed rate that is too low will prematurely dull the end mill. Milling cutters with fewer teeth will allow a greater depth of cut. Fewer teeth make each tooth larger and stronger. Fewer teeth also allow more chip clearance. Use cutting fluids whenever possible. Cutting fluids dissipate heat. When a smooth, accurate finish is needed, take a roughing cut first to remove the majority of the material and then a finish cut.

SELF-TEST

1. If the cutting speed (CS) for low-carbon steel is 90, and the formula for rpm is $(CS \times 4)/D$, what should the rpm of the spindle of a drill press be for a $\frac{1}{8}$-inch-diameter high-speed twist drill? For a .750-in. diameter drill?

2. If the two drills in question 1 are used in a lathe instead of a drill press, what should the rpm of the lathe spindle be for both drills?

3. If, in question 2, the small drill could not be set at the calculated rpm because of machine limitations, what could you do?

4. An alloy-steel 2-inch-diameter cylindrical bar having a cutting speed of 50 is to be turned on a lathe using a high-speed tool. At what rpm should the lathe be set?

5. A formula for setting the safe rpm for an 8-inch grinding wheel, when the cutting speed is known, is rpm = $(CS \times 4)/D$. If the safe surface speed of a vitrified wheel is 6000 SFPM, what should the rpm be?

6. Are feeds on a drill press based on inches per minute or inches per revolution of the spindle?

7. Which roughing feed on a small lathe would work best, .100 or .010 inch per revolution?

8. What kind of machine tool uses inches per minute instead of inches per revolution?

INTERNET REFERENCES

Information on feeds and speeds for machining:

http://mrainey.freeservers.com

http://keyseaters.com

Cutting Fluids

Cutting fluids have been used in machining operations since the beginning of the Industrial Revolution. Lubricants in the form of animal fats were first used to reduce friction and cool the workpiece. These fatty oils tended to become rancid, had a disagreeable odor, and often caused skin rashes. Although lard oil alone is no longer used, it is still used as an additive in cutting oils. Plain water was sometimes used to cool workpieces, but water alone tends to rust machine parts and workpieces. Water can be combined with oil in an emulsion that cools but does not cause rust or corrosion. Many new chemical and petroleum-based cutting fluids in use today have increased efficiency in machining. Cutting fluids reduce machining time by allowing higher cutting speeds. Cutting fluids also reduce tool wear. This unit will prepare you to use cutting fluids properly.

OBJECTIVES

After completing this unit, you should be able to:

- List the various types of cutting fluids.
- Explain the correct uses and care in using several cutting fluids.
- Describe several methods of cutting fluid application.

EFFECTS OF CUTTING FLUIDS

Cutting fluid is a generic term that covers various products used in cutting operations. Cutting and grinding fluids serve two functions, cooling and lubrication. Water-based fluids are most effective for cooling, and oil-based fluids are better lubricants. Fortunately, there is a considerable overlap between the two.

Cutting force and temperature rise are generated not so much by the friction of the chip sliding over the surface of the tool as by the shear flow of the metal just ahead of the tool (Figure F-35). Some tool materials, such as carbide and ceramic, can withstand high temperatures. When these materials are used in high-speed cutting operations, the use of cutting fluids may actually be counterproductive. The increased temperatures tend to promote an easier shear flow and thus reduce cutting force. High-speed, high-temperature cutting also produces better finishes and disrupts the surface less than does lower-speed machining. For this reason, machining with extremely high cutting speeds using carbide tools is often done dry. The use of coolant may cause a carbide tool to crack and break up due to thermal shock, because the flooding coolant can never reach the hottest point at the tip of the tool and so cannot maintain a consistent temperature on the tool. Cutting fluids are a necessity in most precision grinding operations. Most milling machine, drilling, and many lathe operations require the use of cutting fluids.

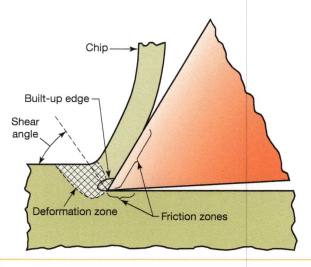

Figure F-35 Metal-cutting diagram showing the shear flow ahead of the tool.

TYPES OF CUTTING FLUIDS

Cutting fluids reduce cutting forces at lower cutting speeds. This enables somewhat higher speeds to be used. Cutting fluids also reduce the heat that tends to break down high-speed tools. Cutting fluids also extend tool life and improve workpiece finish. Another advantage in using cutting fluids is that the workpiece dimensions can be accurately measured on a cool workpiece. A hot workpiece must be cooled to room temperature to obtain a correct measurement.

Synthetic Fluids   These can be solutions of inorganic substances, which are dissolved in water. These include nitrites, nitrates, borates, or phosphates. These chemical substances are water soluble and vary in color from milky to transparent when mixed with water. This type is superior to others for machining and grinding titanium and for grinding operations in general. The surface-active type is a water solution that contains additives that lower the surface tension of the water. Some types have good lubricity and corrosion inhibitors. Lubricity is the measure of the reduction in friction of a lubricant.

Advantages of these chemical fluids include resistance to becoming rancid, good detergent (cleansing) properties, rapid heat dissipation, and the fact that they are easy to mix, requiring little agitation. There are also disadvantages. A lack of lubrication (oiliness) in most types may cause sticking in machine parts that depend on the cutting fluid for lubrication. Also, the high detergency has a tendency to remove skin oils and irritate workers' hands with long continual exposure. Synthetic fluids generally provide less corrosion resistance than oilier types, and some tend to foam. When improved lubricating qualities are needed for synthetic fluids, sulfur, chlorine, or phosphorus are added.

Semisynthetic Fluids   These cutting fluids are essentially a combination of synthetic fluids and emulsions. They have lower oil content than the straight emulsion fluids, and the oil droplets are smaller because of a higher content of emulsifying molecules, thus combining the best qualities of both types. Because a small amount of mineral oil is added to these semisynthetic fluids, they possess enhanced lubrication qualities. In addition, they may contain other additives such as chlorine, sulfur, or phosphorus.

Emulsions   Emulsions are also called *water-miscible fluids* or *water-soluble oils*. Ordinary soap is an emulsifying agent that causes oil to combine with water in a suspension of tiny droplets. Special soaps and other additives are blended with a naphtha-based or paraffin mineral oil to emulsify it. Other additions, such as bactericides, help reduce bacteria, fungi, mold, and extend emulsion life. Without these additives, an emulsion tends to develop a strong, offensive odor because of bacterial action and must be replaced with a new batch.

These emulsified oils contain oil particles large enough to reflect light, and therefore they appear opaque or milky when combined with water. In contrast, many of the synthetic and semisynthetic emulsion particles are so small that the water remains clear or translucent when mixed with the chemical. The mixing ratio of oil and water vary with the job requirement and can range from 5 to 70 parts water to 1 part oil. However, for most machining and grinding operations, a mixture of 1 part oil to 20 parts water is generally used. A mixture that is too rich for the job can be needlessly expensive and a mixture that is too lean may cause rust to form on the workpiece and machine parts. Be sure to check the instructions for the specific product you are using to find the correct mixing ratio.

Some emulsions are designed for greater lubrication, with animal or vegetable fats or oils providing a "superfatted" condition. Sulfur, chlorine, and phosphorus provide even greater lubricating value for metal cutting operations where extreme pressures are encountered in chip forming. These fluids are mixed in somewhat rich ratios: 1 part oil with 5 to 15 parts water. All these water-soluble cutting fluids are considered coolants even though they do provide lubrication.

Cutting Oils   Cutting oils may be animal, vegetable, petroleum, or marine oils (fatty tissue of fish and other marine animals), or a combination of them. The plain cutting oils (naphthenic and paraffinic) are considered lubricants and are useful for light-duty cutting on metals with high machinability, such as free-machining steels, brass, aluminum, and magnesium. Water-based cutting fluids should never be used for machining magnesium because of the fire hazard. The cutting fluid recommended for magnesium is anhydrous (without water) oil that has a low acid content. Fine magnesium chips can burn if ignited, and water tends to make it burn even more fiercely.

Where high cutting forces are encountered and extreme pressure lubrication is needed, as in threading and tapping operations, certain oils, fats, waxes, and synthetic materials are added. The addition of animal, marine, or vegetable oil to petroleum oil improves the lubricating quality and the wetting action. These chlorinated or sulfurized oils are dark in color and are commonly used for thread-cutting operations.

Cutting oils also tend to become rancid and develop disagreeable odors unless germicides are added to them. Cutting oils tend to stain metals and, if they contain sulfur, may severely stain nonferrous metals, such as brass and aluminum. In contrast, soluble oils generally do not stain workpieces or machines unless they are trapped for long periods of time between two surfaces, such as the base of a milling vise bolted to a machine table.

Coolant tanks containing soluble oil on lathes, milling machines, and grinding machines tend to collect oil and dirt or grinding particles. For this reason, the fluid should be removed periodically, the tank cleaned, and a clean solution put in the tank. Water-based cutting fluid can be contaminated quickly when a machinist uses an oilcan to apply cutting oil to the workpiece instead of using the coolant pump and nozzle on the machine. The oil goes into the tank and settles on the surface of the coolant, creating an oil seal, where bacteria can grow. This causes an odorous scum to form that quickly contaminates the entire tank.

Special vegetable-based lubricants have been developed that boost machining efficiency and tool life. Only small amounts are used, and messy coolant flooding is unnecessary. They are used in milling, sawing, reaming, and grinding operations.

Air Compressed air is sometimes used to prevent contamination on some workpieces, for example, reactive metals such as zirconium may be contaminated by water-based cutting fluids. Air can provide better cooling when a jet of compressed air is directed at the point of the cut.

METHODS OF APPLYING CUTTING FLUID

The simplest method of applying a cutting fluid is by using an oilcan. A brush can also be used to apply cutting oil to the workpiece, as in lathe threading operations.

Most milling and grinding machines, large drill presses, and lathes have a built-in tank or reservoir containing a cutting fluid and a pump system to deliver it to the cutting area. The cutting fluid used in these machines is typically a water-soluble synthetic or soluble oil mix. The cutting fluid in the tank is picked up by a low-pressure pump and delivered through a tube or hose to a nozzle where the machining or grinding operation is taking place (Figure F-36).

When a pump system is used, the most common method of application is by flooding. This is done by simply aiming the nozzle at the work–tool area and applying large amounts of cutting fluid. This system usually works well for most turning and milling operations, but not so well for grinding, since the rapid spinning of the grinding wheel tends to blow the fluid away from the work–tool interface. Special wraparound nozzles (Figure F-37) are often used on surface grinders to ensure complete flooding and cooling of

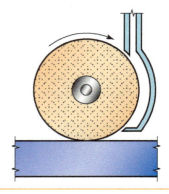

Figure F-37 Special wraparound nozzle used for surface grinding operations.

the workpiece. On cylindrical grinders, a fan-shaped nozzle the width of the wheel is normally used, but a specially designed nozzle (Figure F-38) is better.

In lathe operations, the nozzle is usually above the workpiece pointed downward at the tool. A better method of application would be above and below the tool, as shown in Figure F-39. Internal lathe work, such as boring operations and close chuck work, cause the coolant flow to be thrown outward by the spinning chuck. In this case, a chip guard is needed to contain the spray. Although a single nozzle is normally used to direct the coolant flow over the cutter and on the workpiece on milling machines, a better method is to use two nozzles (Figure F-40) to make sure the cutting zone is completely flooded. Some lathe toolholders are now provided with through-the-tool pressurized coolant systems. Also, some end mills for vertical milling machines have coolant channels that direct the fluid at the cutting edge under pressure. This is an extremely efficient way to apply coolant and it also forces the chips away from the cutting area.

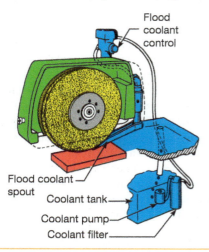

Figure F-36 Fluid recirculates through the tank, piping, nozzle, and drains in a floodgrinding system (*Courtesy of DoALL Company*).

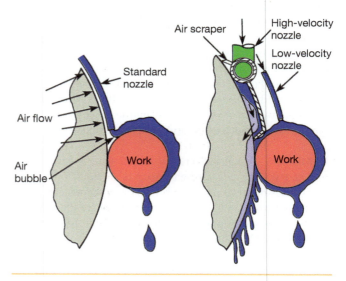

Figure F-38 This specially designed nozzle helps keep the fanlike effect of the rapidly rotating wheel from blowing cutting fluid away from the wheel–work interface (*Courtesy of MAG Industrial Automation Systems, LLC*).

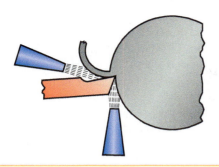

Figure F-39 Although not always possible, this is the ideal way of applying cutting fluid to a cutting tool.

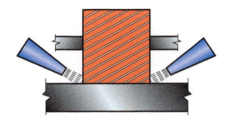

Figure F-40 The use of two nozzles ensures flooding of the cutting zone.

Figure F-41 Mist grinding fluid application on a surface grinder (*Courtesy of DoALL Company*).

The intermittent application of cutting fluids using a squirt bottle, pump, or brush should be avoided when using carbide cutting tools. The rapid and irregular cooling of the carbide tools may cause thermal cracking.

When a cutting fluid is properly applied, it can have a large effect on finishes and surfaces. A finish may be rough when machined dry but may be dramatically improved with lubrication. When carbide tools and coolant are used, the work–tool area must be flooded to avoid intermittent cooling and heating cycles that create thermal shock and cracking of the tool. The operator should not shut off the coolant while the cut is in progress to "see" the cutting operation.

There is no good way to cool and lubricate a twist drill, especially when drilling horizontally in a lathe. In deep holes, cutting forces and temperatures increase, often causing the drill to expand and bind. Even in vertical drilling on a drill press, the helical flutes, designed to lift out the chips, also pump out the cutting fluid. Some drills are made with oil holes running the length of the drill to the cutting tips to help offset the problem. The fluid is pumped through the flutes under relatively high pressure. The cutting oil lubricates and cools the drill and workpiece, and also flushes out the chips. Gun drills designed to make accurate and deep holes have a similar arrangement, but the fluid pressures are as high as 1,000 psi. Shields must be provided for operator safety when using these extremely high-pressure coolant systems.

Air-carried mist systems (Figure F-41) are used for some drilling and milling operations. Compressed air is used to create a spray of coolant drawn from a small tank on the machine. Because only a small amount of liquid ever reaches the cutting area, little lubrication is provided. These systems are chosen for their ability to cool rather than to lubricate. This factor may reduce tool life. However, mist cooling has many advantages. No splash guards, chip pans, and return hoses are needed, and only small amounts of liquid are used. The high-velocity air stream also cools the spray further by evaporation. There are safety hazards in using mist spray equipment from the standpoint of the operator's health. Conventional coolants are not highly toxic, but when sprayed in a fine mist they can be inhaled; this could affect some people over time. Many operators find that breathing any of the mist is uncomfortable and offensive. Good ventilation systems or ordinary fans can remove the mist from the operator's area.

COOLANT SEPARATORS

Cutting chips and abrasive materials become intermingled with the coolant in collection tanks. Oil and dirt are also contaminants. Discarding the old coolant and using new is expensive and is also an environmental disposal problem. Therefore, separators and filtration methods are often used to reclaim the old coolant.

Magnetic separation has been in use for many years for operations such as grinding, honing, broaching, and other machining of ferrous metals. These devices can remove as much as 98 percent by volume of ferrous solids larger than 20 or 30 μm. However, nonferrous solids and other contaminants must be removed by other processes, such as filtration. Oil skimmers are used to remove oils that collect in the coolant tank. This process reduces the need for frequent coolant maintenance. Portable coolant reconditioning systems are often used in place of large permanent stationary systems.

Wastewater treatment is another aspect of manufacturing today. Parts washers and other cleaning operations contaminate the water with oily waste. Shops normally have to pay specialized disposal companies to get rid of used coolants and solvents.

SELF-TEST

1. What are the two basic functions of cutting fluids?

2. Name at least three types of cutting fluids (*liquids*).

3. When soluble oil coolants become odorous or cutting oil becomes rancid in the reservoir, what can be done to correct the problem?

4. Why shouldn't a machinist use a pump can or brush containing cutting oil on a machine that has a coolant pump and tank?

5. Some of the synthetic cutting fluids tend to irritate workers' hands. What causes this?

6. How is cutting fluid typically applied on a small drill press without a coolant pump?

7. Why is a single-round nozzle rather inefficient when used on a grinding wheel?

8. Spray-mist cooling systems work well for cooling purposes but do not lubricate the work area and tool very well. Why is this?

Carbide Tooling Specification and Selection

The constant demand for higher productivity has led to the need for faster stock removal. A machinist must learn to achieve maximum productivity at minimum cost. A machinist must have a good understanding of carbide cutting tools and the ability to select them for specific machining tasks in order to be productive.

OBJECTIVES

After completing this unit, you should be able to:

- List six different cutting tool materials and compare some of their machining properties.
- Select a carbide tool for a job by reference to operating conditions, carbide grades, nose radii, tool style, rake angles, shank size, and insert size, shape, and thickness.
- Identify carbide inserts and toolholders by number systems developed by the American Standards Association.

The various tool materials used in today's machining operations are HSS, cemented carbides, ceramics, diamond, and cubic boron nitride (CBN).

FUNDAMENTALS OF CARBIDE TOOLING

Tooling is crucial to the success of a machine shop. A great machine with poor tooling will perform very poorly. This unit will concentrate on the fundamentals of carbide cutting tools and their selection, because good tooling techniques are vital to productivity.

Cemented Carbide

Cemented carbide, or tungsten carbide, is a form of powdered metallurgy. Fine powders consisting of tungsten carbide and other hard metals bonded with cobalt are pressed into required shapes and then sintered.

Sintering is the heating of the carbide materials to approximately 2,500°F. At this temperature, the cobalt melts and flows around the carbide materials. Cobalt acts as the binder that holds the carbide particles together. After the carbide insert cools, the insert is almost as hard as a diamond.

The hardness and physical properties of cemented carbides enable them to operate at high cutting speeds and feeds. The great hardness of carbide is also its Achilles' heel. Extremely hard materials are also very brittle, and this can cause problems under certain machining conditions. Carbide manufacturers have come up with different grades of carbide materials by the use of different mixtures of materials. Selecting the proper grade for the machining application is important for economy and productivity.

Selection of Carbide Tool Grade

Carbide tools come in a variety of grades. The grade is based on the carbide's wear resistance and toughness. As an insert becomes harder or more wear resistant, it becomes more brittle (less tough).

If you use a very hard, wear-resistant insert on a material that has an uneven or interrupted surface (interrupted cut), the insert will most likely break. The ANSI/ISO standards organizations have devised systems of grading carbide based on the carbide insert's application and physical makeup (Figure F-42).

Classification systems differ and can be quite confusing. There are over 5,000 different grades of carbide under 1,500 trade names. Carbide manufacturers help clarify matters with cross-reference charts in their catalogues (Figure F-43).

Coatings for Carbide Inserts

Carbide is a very hard, durable cutting tool, but it still wears. The wear resistance of cemented carbide can be greatly increased by using coated carbide inserts.

Wear-resistant coatings can be applied to the carbide substrate (base material) through the use of plasma coating or vapor deposition. The coating is very thin but very hard.

Figure F-42 ISO Grade Designations

ISO Grade	Material to Be Machined
P01	Steel, Steel Castings
P10	Steel, Steel Castings
P20	Steel, Steel Castings, Ductile Cast Iron with Long Chips
P30	Steel, Steel Castings, Ductile Cast Iron with Long Chips
P40	Steel, Steel Castings with Sand Inclusions and Cavities
P50	Steel, Steel Castings of Medium or Low Tensile Strength, with Sand Inclusions and Cavities
M10	Steel, Steel Castings, Manganese Steel, Gray Cast Iron, Alloy Cast Iron
M20	Steel, Steel Castings, Austenitic Steel or Manganese Steel, Gray Cast Iron
M30	Steel, Steel Castings, Austenitic Steel or Manganese Steel, Gray Cast Iron, High Temperature-Resistant Alloys
M40	Mild, Free Cutting Steel, Low-Tensile Steel, Non-Ferrous Metals and Light Alloys
K01	Very Hard Cast Iron, Chilled Castings over 85 Shore Hardness, High Silicon Aluminum Alloys, Hardened Steel, Highly Abrasive Plastics, Hard Cardboard, Ceramics
K10	Gray Cast Iron over 220 Brinell Hardness, Malleable Cast Iron with Short Chips, Hardened Steel, Silicon, Aluminum and Copper Alloys, Plastic, Glass, Hard Rubber, Hard Cardboard, Porcelain, Stone
K20	Gray Cast Iron over 220 Brinell Hardness, Non-Ferrous Metals, Copper, Brass, Aluminum
K30	Low-Tensile Gray Cast Iron, Low-Tensile Steel, Compressed Wood
K40	Soft Wood or Hard Wood, Non-Ferrous Metals

The most common types of coatings include titanium carbide (TiC), titanium nitride (TiN), and aluminum oxide (AlO). Aluminum oxide is a very wear-resistant coating used in high-speed finish cuts and light roughing cuts on most steels and all cast irons. TiN coatings are very hard and have the strength characteristics to perform well under heavy rough-cutting conditions. All three coatings will perform well on most steels, as well as on cast iron.

Diamond-Coated Inserts

Coated cutting tools have been around for years. Titanium and boron nitride materials have driven the coated cutting tool industry. One of the materials is the polycrystalline diamond, or PCD. PCD tools are becoming widely accepted as tooling solutions for difficult-to-machine materials. PCD material has the hardness of a diamond and the friction coefficient of Teflon. This combination has resulted in a remarkable increase in tool life.

INSERT CLASSIFICATION

Insert Shape

Carbide inserts come in many shapes. Inserts are clamped in toolholders and provide cutting tools with multiple cutting edges. After the cutting edges are worn to a point where they can no longer be used, they are discarded or saved and recycled. To correctly select the insert, the machinist must be able to answer a few specific questions about the job.

- What geometric features are required on the workpiece?
- What are the characteristics of the material that will be cut?
- How much material is being removed?
- How rigid is the work holding setup?

Tool selection and identification of carbide tools are done using a standard set of identifying letters and numbers known as ANSI standard tool nomenclature. An example of a typical insert style, shape, and size is a CNMG432. The first identifying feature found on all carbide inserts is shape. Insert shape is dictated by the part shape required. Position 1 on the chart identifies the shape of the insert (see Figure F-44). The C in this example would mean that we have an 80-degree diamond-shaped insert.

Figure F-45 shows different-shaped inserts in order of their strength. Round inserts have the greatest strength, as well as the greatest number of cutting edges, but the round configuration limits the operations that can be performed.

Square or 90-degree inserts have less strength and fewer cutting edges than round inserts but are a little more versatile.

Triangular inserts (Figure F-45) are more versatile than square inserts, but as the included angle is reduced from 90 degrees to 60 degrees it becomes weaker and more likely to break under heavy machining conditions.

Diamond-shaped inserts are probably the most commonly used shape. Diamond-shaped inserts range from a 35-degree to an 80-degree included angle. Diamond-shaped inserts are much more versatile than square and round inserts. It is good machining practice to select the largest included angle insert that will cut the shape of the part because the insert will be stronger.

Figure F-43 Typical grade system cross-reference chart gives grade selection choices for different tool manufacturers.

ISO/ANSI Grade	Valenite	Iscar	Sandvik	Mitsubishi	kennametal	Walter	Seco	Toshiba
P50–P40 C5	VP5535	IC635	GC4235		KC8050			
					KC9040			
					KC9240			
					KC9140			
P40–P30 C5–C6	VP5535	IC9025	GC4235	UE6035	KC9040	WPP30	TP400	T9035
	VPUP30	IC3028			KC8050	WAP30	TP3000	T9025
		IC635			TN7035			TD930
					KC5025			
P30–P20 C6–C7	VP5525	IC9025	GC4225	UE6020	KC9125	WPP20	TP2500	T9025
		IC3028		UE6035	TN7025	WAP30	TP2000	TD930
		IC9015		UP20M	KC8050	WAP20	CP500	T7020
		IC50M		VP15TF	KC5025			T725X
P20–P10 C5	VP5515	IC8048	GC4215	UE6010	KC9110	WPP10	TP2000	T9015
		IC570	GC1525	UE6110	KC9010	WAP20	TP1000	AT530
		IC9015		AP20N	TN7010	WAP10	TP200	T7010
		IC907		UP35N	KC5010			T715X
P10–P01 C8	VP1510	IC9015	GC4005	UE6005	KC9110	WPP01	TP1000	T9005
	VP1505	IC428	CT5015	UE6010	KC9010	WAP10	TP100	TD905
	VPUP10	IC520N		AP25N	TN7005	WAP01		AT520
		IC8048			KC9315	WPP05		
M40–M30	VP5535	IC9025	GC2035	US735	KC9240	WSM30	TP400	T6030
	VP9625	IC3028			KC9245	WAM30	CP500	T725X
	VPUS10	IC635			CL4			J740
					KC8050			
					TN7035			
M30–M20	VP8525	IC3028	GC2025	US7020	KC9225	WAM20	TP3000	T6030
	VP9625	IC9025		UP20M	KC8050		TP2500	T6020
	VPUS10	IC08		VP15F	KC5020		CP500	AH120
					TN7025		CP200	
					TN8025			
M20–M10 C8	VP8515	IC907	GC2025	US7020	KC5010	WAM10	TP100	T6020
	VP9610	IC570	GC1025	UP20M	KT315	WXN10	TP200	J530
	VPUS10	IC507		AP25N	KC5510		CP200	
		IC520						

Lead Angle

Lead angle, or side-cutting edge angle, is the angle at which the cutting tool enters the work (Figure F-46). The lead angle can be positive, neutral, or negative. The tool holder always dictates the amount of lead angle a tool will have.

Toolholders should be selected to provide the largest lead angle that the job will permit. There are two advantages to using a large lead angle. First, when the tool initially enters the work, it is at the middle of the insert where it is strongest, instead of at the tool tip, which is the weakest point of the tool (Figure F-47). Second, the cutting forces are spread over a wider area, reducing the chip thickness.

Clearance Relief Angle

The second identifying feature found on carbide insert identification charts is the relief angle (see Figure F-48). For our CNMG432 insert the second letter is an N. An N in the chart in Figure F-48 would be a 0-degree relief angle.

Figure F-44. Insert shape.

ANSI Insert Identification System

Shape			
Symbol Shape			Nose Angle
S		Square	90
T		Triangular	60
C			80
D			55
E		Diamond	75
F			50
M			86
V			80
W		Trigon	80
H		Hexagonal	120
O		Octagonal	135
P		Pentagonal	108
L		Rectangular	90
A			85
B		Parallelogram	82
N/K			55
R		Round	

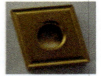

CNMG432

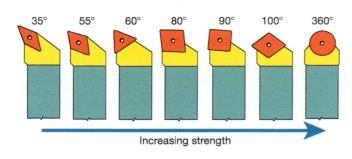

35° 55° 60° 80° 90° 100° 360°

Increasing strength

Figure F-45 The shape of the insert will have a great effect on the strength of the tool. Select the largest included angle that will cut the part.

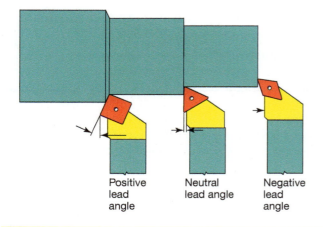

Positive lead angle

Neutral lead angle

Negative lead angle

Figure F-46 Lead or side-cutting edge angle is determined by the toolholder type. The lead angle can be positive, neutral, or negative.

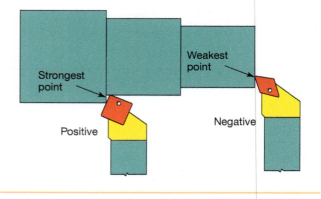

Strongest point

Positive

Weakest point

Negative

Figure F-47 The effect of the lead angle on the strength of the insert. Increasing the lead angle will greatly reduce tool breakage when roughing or cutting interrupted surfaces.

Figure F-48 Relief angle.

ANSI Insert Identification System

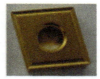

CNMG432

Clearance
Relief Angle
N - 0°
A - 3°
B - 5°
C - 7°
P - 11°
D - 15°
E - 20°
F - 25°
G - 30°

The side relief angle, also known as the side rake angle, is formed by the top face of the cutting tool and side cutting edge (see Figure F-49). The angle is measured in the amount of relief under the cutting edge.

Top view

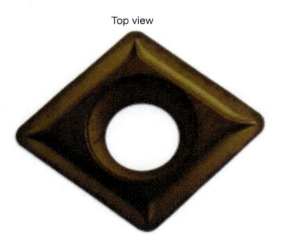

Side view

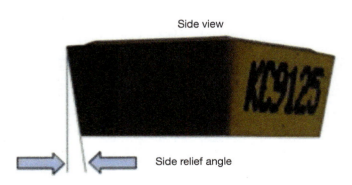

Side relief angle

Figure F-49 Relief angle.

Neutral rake inserts have an included angle of 90 degrees between the top rake and the end clearance angle (see Figure F-50).

This creates a 0-relief condition. Zero relief inserts have a letter designation of N or neutral. Relief under the cutting edge is essential. Negative rake holders must be used when using neutral rake inserts. The combination of the relief angle on the insert and the rake angle created by the holder is known as the effective rake angle. There are three principal rake angles: neutral, positive, and negative (see Figure F-51).

It is essential to look at the machining conditions when selecting the proper rake. Negative rake holders are a good, economical choice because they hold neutral rake inserts. Neutral rake inserts have twice as many cutting edges as positive rake inserts because the insert can be turned over and used.

Another advantage is that negative rake tool holders provide more support for the cutting edges of the insert. Under normal operating conditions, negative rake inserts are also a little stronger because of the compressive strength of carbide. Negative rake holders should be used when the tool and the

Top view

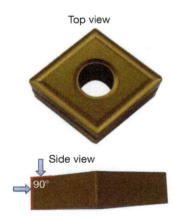

Side view

90°

Figure F-50 Neutral rake insert.

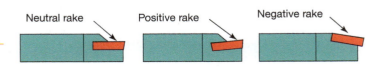

Neutral rake   Positive rake   Negative rake

Figure F-51 Side view of back rake angles.

work are held very rigidly and when high machining speeds and feeds can be maintained. More horsepower is required to cut with negative rake tool holders, which is why there is an increasing trend toward the use of positive rake cutting.

Positive rake cutting is more of a shearing effect than the pushing effect generated by negative rake. Positive rake holders generate less cutting force and have less of a tendency to chatter. Horsepower requirements are greatly reduced with positive-rake cutting tools.

The only drawback to positive-rake cutting tools is their inability to stand up to harder materials. Recent advances in carbide technology have produced tougher substrate materials that provide greater edge strength. Some carbide companies recommend positive rake holders whenever possible.

Positive rake should be used when machining softer materials because the chips are able to flow away from the cutting edge freely and the cutting action is more of a peeling effect. Positive rake cutting can be very successful on long slender parts or other operations that lack rigidity.

Insert Size Tolerance

The third identifying feature found on carbide insert identification charts is the insert size tolerance (see Figure F-52). For our example the third letter is an M. From the chart you will see that there are three tolerances specified for an M.

This letter designation states how much size variation is allowed from one insert to the next. The tolerance that is described by this letter designation includes the nose radius, the insert thickness, and the inscribed circle.

Hole and Chip Breaker Configuration

The fourth identifying feature on the ANSI insert identification chart designates the shape of the chip breaker that is molded into the insert and whether the insert has a hole (see Figure F-53). Our insert example is a G.

If the insert has a hole, the insert holder uses a lock pin to locate the insert. If there is no letter designation in the fourth field of the chart, then the insert doesn't have a hole. Inserts without holes are held only by a clamp on the holder.

Chip Breaker

The depth of cut and the feed rates must be taken into consideration to select the proper chip breaking geometry. If a roughing insert chip breaker is selected, the proper feed rate and depth of cut must be used. Using too light a feed rate or too small a depth of cut will result in long, stringy chip. An insert chip breaker that is designed for finish cuts will fail if it is used with a large depth of cut or heavy feed. Chip breaker configuration is covered later in this chapter.

Insert Size

The fifth identifying feature on the ANSI insert identification chart designates the size of the insert (see Figure F-54).

Figure F-52 Insert size tolerance.

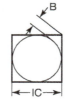

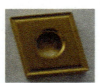
CNMG432

Tolerance Class	Tolerance on "B"		Tolerance on "IC"		Tolerance on "T"	
	INCH	MM	INCH	MM	INCH	MM
A	±.0002	±.0005	±.001	±.025	±.001	±.025
C	±.0005	±.013	±.001	±.025	±.001	±.025
E	±.001	±.025	±.001	±.025	±.001	±.025
F	±.0002	±.005	±.0005	±.025	±.001	±.025
G	±.001	±.025	±.001	±.13	±.005	±.13
H	±.0005	±.013	±.0005	±.025	±.001	±.025
J	±.002	±.005	±.002−.005	±.025	±.001	±.025
K	±.0005	±.013	±.002−.005	±.025	±.001	±.025
L	±.001	±.025	±.002−.005	±.025	±.001	±.025
M	±.002−.005	±.05−.13	±.002−.005	±.13	±.005	±.025
U	±.005−012	±.06−.25	±.005−.010	±.13	±.005	±.13
Tolerance						

ANSI Insert Identification System C N M G 4 3 2

Figure F-53 Hole and chip breaker configuration.

ANSI Insert Identification System C N M G 4 3 2

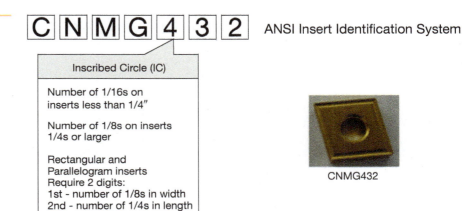

CNMG432

Geometry

A=		*K=	
B=		*L=	
C=		M=	
*D=		N=	
*E=		Q=	
F=		R=	
G=		T=	
H=		U=	
J=		W=	

* = Special Design

Figure F-54. Insert size.

C N M G 4 3 2 ANSI Insert Identification System

Inscribed Circle (IC)

Number of 1/16s on inserts less than 1/4″

Number of 1/8s on inserts 1/4s or larger

Rectangular and Parallelogram inserts Require 2 digits:
1st - number of 1/8s in width
2nd - number of 1/4s in length

CNMG432

An inscribed circle is the largest circle that will fit inside an insert. Our example has a four in the fifth position. This would mean that the inscribed circle would be .5″ $(4*1/8)$. The size of the insert is based on the inscribed circle, the insert thickness, and the tool-nose radius (see Figure F-55).

The depth of cut possible with an insert depends greatly on the insert size. The depth of cut should always be as great as the conditions will allow. A good rule of thumb is to select an insert with an inscribed circle at least twice that of the depth of cut.

Insert Thickness

The number in the sixth position of the ANSI insert identification system represents the number of sixteenths of thickness. For our example we have a three in the sixth position. That

would mean that the insert is $\frac{3}{16}$ of an inch thick (see Figure F-56). As the thickness of the insert increases, so does its strength. Remember, larger inserts can take deeper cuts. Insert thickness and inscribed circle size increase and decrease proportionally.

Insert Corner Geometry (Tool-Nose Radius)

The seventh position on the ANSI insert identification system is the size of the corner or nose radius on the insert (see Figure F-57). In our example we have a two in the seventh position. This would mean that this insert would have a $\frac{1}{32}$ radius.

Figure F-55 Diagram of a typical triangular insert.

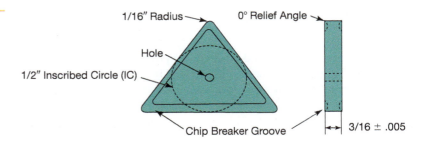

1/16″ Radius

0° Relief Angle

Hole

1/2″ Inscribed Circle (IC)

Chip Breaker Groove

3/16 ± .005

Figure F-56 Insert thickness.

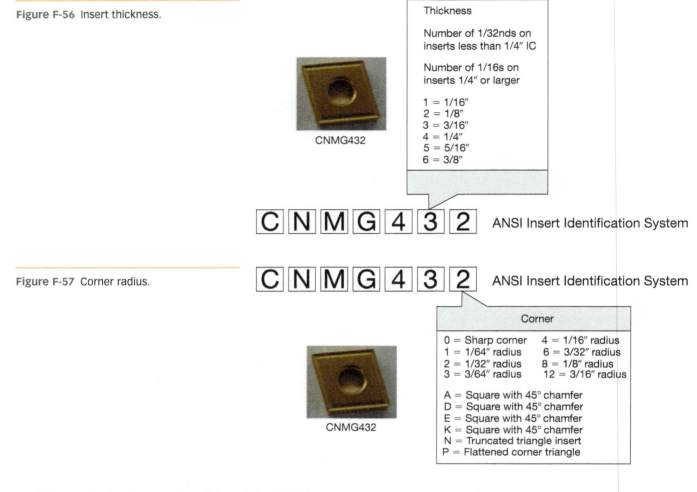

Thickness

Number of 1/32nds on inserts less than 1/4" IC

Number of 1/16s on inserts 1/4" or larger

1 = 1/16"
2 = 1/8"
3 = 3/16"
4 = 1/4"
5 = 5/16"
6 = 3/8"

CNMG432

C N M G 4 3 2   ANSI Insert Identification System

Figure F-57 Corner radius.

C N M G 4 3 2   ANSI Insert Identification System

Corner	
0 = Sharp corner	4 = 1/16" radius
1 = 1/64" radius	6 = 3/32" radius
2 = 1/32" radius	8 = 1/8" radius
3 = 3/64" radius	12 = 3/16" radius
A = Square with 45° chamfer	
D = Square with 45° chamfer	
E = Square with 45° chamfer	
K = Square with 45° chamfer	
N = Truncated triangle insert	
P = Flattened corner triangle	

CNMG432

The number in the seventh position of the ANSI insert identification system represents the number of sixty-fourths of nose radius. Although selecting the proper grade of insert is probably the most important, other factors such as nose radius are very important when selecting the proper tool for the application.

The nose radius of the tool directly affects tool strength and surface finish, as well as cutting speeds and feeds. The larger the nose radius, the stronger the tool. If the tool radius is too small, the sharp point will make the surface finish unacceptable, and the life of the tool will be shortened. Larger nose radii will give a better finish and longer tool life and will allow for higher feed rates. If the tool nose is too large, it can cause chatter. It is usually best to select an insert with a tool-nose radius as large as the machining operation will allow.

Insert Identification

Let's take one more look at our CNMG432 insert. What does this insert look like and what are the possible uses for this insert? Figure F-58 shows a CNMG432 insert.

The insert shape is designated as a C. This means that the shape of the insert is an 80-degree diamond. The clearance or relief angle designation is an N. This specifies that the insert has a zero or neutral relief angle and must be used

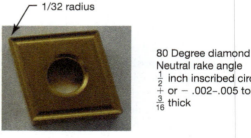

1/32 radius

80 Degree diamond
Neutral rake angle
$\frac{1}{2}$ inch inscribed circle (IC)
+ or − .002–.005 tolerance
$\frac{3}{16}$ thick

Figure F-58 A CMNG432 insert.

with a negative relief holder. The third letter in the designation indicates the size tolerance of the insert. The M guarantees that the inserts repeatability on the inscribed circle is ±.002–.005 of an inch and the thickness repeatability is ±.005 of an inch. The G in the fourth position of the chart signifies that the insert has a molded chip breaker on both sides of the insert and that this insert would be held in a holder which has a lock pin. The four in the fifth position of the chart states that the insert has an inscribed circle size of $\frac{4}{8}$ or one-half of an inch. The thickness of the insert is $\frac{3}{16}$ of an inch as indicated by the three in the sixth position of the chart. The last identifying number in the

CNMG432 insert is a two. This indicates the edge or nose of the insert has a radius of $\frac{2}{64}$ ($\frac{1}{32}$) of an inch.

Possible Usage

The 80-degree diamond shape and the neutral relief of this insert would indicate that it will be used primarily for roughing. In most cases, finish stock of 0.02 to 0.03 is left on the part after roughing, so the size tolerance and repeatability wouldn't be an issue. A molded chip breaker would help with chip control when roughing. An inscribed circle of one-half inch and an insert thickness of $\frac{3}{8}$ of an inch would allow us to take large roughing cuts. Keep in mind that the rigidity of the setup and the horsepower of the machine would also dictate the depth cut capabilities of this insert. The nose radius of this insert is $\frac{1}{32}$ ($\frac{2}{64}$s) of an inch. This size nose radius is certainly capable of handling roughing feed rates and heavy depths of cuts.

Insert Selection

Now that we have covered some aspects of carbide tool selection, let's look at the questions that we need to answer when selecting the proper insert grade and style. One of the first considerations is the material to be machined.

Machinability of Metals

Machinability describes the ease or difficulty with which a metal can be cut. Machining involves removing metal at the highest possible rate and at the lowest cost per piece. Different materials' structures pose different problems for the machinist. Materials that are easy to machine have high machinability ratings and therefore cost less to machine. Materials that are difficult to machine have lower machinability ratings and cost more to machine.

The machinability of a material is directly related to the material's hardness. A number of tests measure the hardness of a material, but the most common test for machinability is the Brinell test. Brinell hardness, or BHN, is stated as a number: the higher the BHN number, the harder the material. Hardness, although a major factor affecting machinability, is not the only factor that determines machinability.

Steels

Steels are classified based on their carbon content and their alloying elements. Plain carbon steels have only one alloy, carbon, mixed with iron. Carbon has a direct effect on steel's hardness. Plain carbon steel's machinability is directly related to its carbon content. Alloy steels, on the other hand, have carbon and other alloying elements mixed with iron. These alloying elements can give steel the characteristic of not only being hard but also tough. The major concern with machining alloy steels is their tendency to work harden, a phenomenon that occurs when too much heat from the cutting process is developed in the steel. The heat changes the properties of the steel, making it harder and difficult to machine. Great care must be taken when machining some alloy steels.

Plain carbon steel is divided into three categories: low carbon, medium carbon, and high carbon. Low-carbon steels have a carbon content of 0.10 to 0.30 percent and are relatively easy to machine. Medium-carbon steels have a carbon content of 0.30 to 0.50 percent. Medium-carbon steels are relatively easy to machine, but because of the higher carbon content, they have a lower cutting speed than that of low-carbon steel. High-carbon steels have a carbon content of 0.50 to 1.8 percent. When the carbon content exceeds 1.0 percent, high-carbon steel becomes quite difficult to machine.

Stainless Steel

Stainless steels have carbon, chromium, and nickel as alloys. Stainless steels are a very tough, shock-resistant material and are difficult to machine. Work hardening can be a problem when machining stainless steels. To avoid work hardening, use lower speeds and increased feed rates. Chip control is sometimes a problem when machining stainless because of its toughness and the chips' unwillingness to break.

Cast Iron

Cast iron is a broad classification for gray, malleable, nodular, and chilled-white cast iron. This grouping is in order of its machinability. Gray cast iron is relatively easy to machine, while chilled-white cast iron is sometimes unmachinable. Cast iron does not produce a continuous chip because of its brittleness.

The machinability of any material can be affected by factors such as heat treatment. Heat-treating can be used to harden or soften a material. The condition of the material at the time of machining should be taken into consideration when deciding a material's machinability.

Toolholder Style and Identification

Carbide manufacturers and the American Standards Association have created a toolholder identification system for indexable carbide toolholders. Figure F-59 shows ANSI designation for external toolholders with an example of the designation for typical toolholder. Figure F-60 shows the ANSI designation for boring bar toolholders with an example of the designation for typical boring bar.

Qualified Tooling

Tools that are used in CNC machines are machined to a high level of accuracy. Qualified tools are typically guaranteed to be within .003 of an inch.

The accuracy of the cutting tip is referenced to specific points or datums located on the holders. The higher level of accuracy enables the operator to change inserts without having to remeasure the tools. Figure F-61 shows an example of where the measurements are qualified from.

Figure F-59 ISO Tool holder identification system.

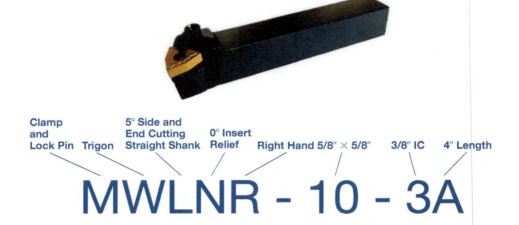

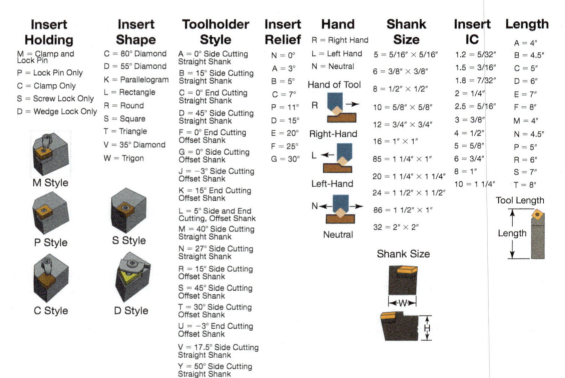

| Clamp and Lock Pin | Trigon | 5° Side and End Cutting Straight Shank | 0° Insert Relief | Right Hand 5/8″ × 5/8″ | 3/8″ IC | 4″ Length |

MWLNR - 10 - 3A

Insert Holding	Insert Shape	Toolholder Style	Insert Relief	Hand	Shank Size	Insert IC	Length
M = Clamp and Lock Pin	C = 80° Diamond	A = 0° Side Cutting Straight Shank	N = 0°	R = Right Hand	5 = 5/16″ × 5/16″	1.2 = 5/32″	A = 4″
P = Lock Pin Only	D = 55° Diamond	B = 15° Side Cutting Straight Shank	A = 3°	L = Left Hand	6 = 3/8″ × 3/8″	1.5 = 3/16″	B = 4.5″
C = Clamp Only	K = Parallelogram	C = 0° End Cutting Straight Shank	B = 5°	N = Neutral	8 = 1/2″ × 1/2″	1.8 = 7/32″	C = 5″
S = Screw Lock Only	L = Rectangle	D = 45° Side Cutting Straight Shank	C = 7°		10 = 5/8″ × 5/8″	2 = 1/4″	D = 6″
D = Wedge Lock Only	R = Round	F = 0° End Cutting Offset Shank	P = 11°		12 = 3/4″ × 3/4″	2.5 = 5/16″	E = 7″
	S = Square	G = 0° Side Cutting Offset Shank	D = 15°		16 = 1″ × 1″	3 = 3/8″	F = 8″
	T = Triangle	J = −3° Side Cutting Offset Shank	E = 20°		85 = 1 1/4″ × 1″	4 = 1/2″	M = 4″
	V = 35° Diamond	K = 15° End Cutting Offset Shank	F = 25°		20 = 1 1/4″ × 1 1/4″	5 = 5/8″	N = 4.5″
	W = Trigon	L = 5° Side and End Cutting, Offset Shank	G = 30°		24 = 1 1/2″ × 1 1/2″	6 = 3/4″	P = 5″
		M = 40° Side Cutting Straight Shank			86 = 1 1/2″ × 1″	8 = 1″	R = 6″
		N = 27° Side Cutting Straight Shank			32 = 2″ × 2″	10 = 1 1/4″	S = 7″
		R = 15° Side Cutting Offset Shank					T = 8″
		S = 45° Side Cutting Offset Shank					
		T = 30° Side Cutting Offset Shank					
		U = −3° End Cutting Offset Shank					
		V = 17.5° Side Cutting Straight Shank					
		Y = 50° Side Cutting Straight Shank					

M Style, P Style, C Style, S Style, D Style

Hand of Tool: R — Right-Hand; L — Left-Hand; N — Neutral

Shank Size, Tool Length, Length

Tool Insert and Toolholder Selection Practice

Figure F-62 shows a typical lathe part. An appropriate insert grade could be chosen to machine the part using the ISO grade designation chart shown in Figure F-42, the insert identification system shown in Figures F-63 and F-64, and an appropriate toolholder from Figure F-59.

- What type of material is being cut? Would you use a cast iron or a steel grade? Answer: The material used for the part is 1018 cold rolled steel, so a steel cutting grade would be appropriate (see Figure F-42).
- How hard is the material? How does this affect the grade? Answer: The material is a low-carbon alloy steel of only 200 Brinell hardness. A moderate hardness grade would be a good choice.

- What is the condition of the material? Does the surface show evidence of scale or hard spots? How does this affect the selection of the grade, insert shape, rake angle, and nose radius? Answer: The material is cold rolled steel, which has little or no scale. Again, a general-purpose insert of moderate hardness and strength would be applicable.
- What shape insert do we need to perform this job? Answer: For roughing this part, we would like to use a larger angled insert such as an 80-degree diamond. The finish tool needs to have a little smaller angle to cut the radius. A 55-degree diamond or triangular insert would be a good choice for finish cuts.
- How rigid is the machining setup? How does this affect the rake angles and nose radius? Answer: As you can see from the part print, the part has a small turned diameter

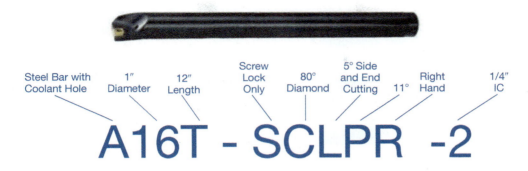

Steel Bar with Coolant Hole	1" Diameter	12" Length	Screw Lock Only	80° Diamond	5° Side and End Cutting	11°	Right Hand	1/4" IC

A16T - SCLPR -2

Boring Bar Type	Diameter	Length	Insert Holding	Insert Shape	Style	Insert Relief	Hand	Insert IC
S = Solid Steel Bar	03 = 3/16"	F = 3"	M = Clamp and Lock Pin	C = 80° Diamond	F = 0° End Cutting	N = 0°	R = RH	1.2 = 5/32"
	04 = 1/4"	G = 3 1/2"		D = 55° Diamond	J = −3° Side Cutting	A = 3°	L = LH	1.5 = 3/16"
A = Steel Bar with Coolant Hole	05 = 5/16"	H = 4"	P = Lock Pin Only	K = Parallelogram	K = 15° End Cutting	B = 5°		1.8 = 7/32"
B = Solid Steel Anti-Vibration Bar	06 = 3/8"	J = 4 1/2"	C = Clamp Only	L = Rectangle	L = −5° Side and End Cutting	C = 7°		2 = 1/4"
	08 = 1/2"	K = 5"	S = Screw Lock Only	R = Round		Right-Hand	2.5 = 5/16"
C = Carbide Bar (Steel Head)	10 = 5/8"	L = 5 1/2"		S = Square	Q = −17.5° End Cutting	P = 11°		3 = 3/8"
	12 = 3/4"	M = 6"		T = Triangle	S = −45° End Cutting	D = 15°		4 = 1/2"
D = Solid Steel Anti-Vibration Bar with coolant Hole	16 = 1"	N = 6 1/2"		V = 35° Diamond	U = −3° End Cutting	E = 20°	Left-Hand	5 = 5/8"
	20 = 1 1/4"	P = 6 3/4"		W = Trigon	W = −30° End Cutting	F = 25°		6 = 3/4"
E = Carbide Bar (Steel Head) with Coolant	24 = 1 1/2"	Q = 7"			Y = 5° End Cutting	G = 30°		8 = 1"
	28 = 1 3/4"	R = 8"					Neutral	10 = 1 1/4"
	32 = 2"	S = 10"						
F = Carbide Anti-Vib Bar (Steel Head)	36 = 2 1/4"	T = 12"						
	40 = 2 1/2"	U = 14"						
G = Carbide Anti-Vib Bar (Steel Head) with Coolant Hole		V = 16"						
		W = 18"						
H = Heavy Meatal Bar		Y = 20"						
J = Heavy Meatal Bar with coolant Hole								

Figure F-60 Boring bar designation.

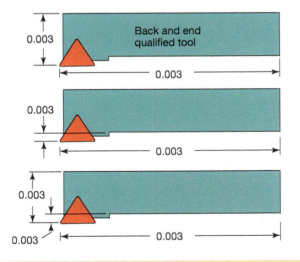

Figure F-61 Qualification of tool holders.

on the end. This small diameter may tend to deflect and chatter. A positive rake insert with a $\frac{1}{32}$ or $\frac{1}{16}$ tool-nose radius would probably be the best choice.

- What are the surface finish requirements of the part? How does this affect the nose radius? Answer: The surface finish requirement of the part is 125. A 125 finish is a standard machine finish that can be held using a $\frac{1}{32}$ or $\frac{1}{16}$ tool-nose radius. Slowing down the feed rate will also help acquire the 125-finish requirement.

CHIP CONTROL

Chip control refers to the ability to control the chip formation. Chip control is important to operator safety, tool life, and chip handling. CNC machines typically have chip conveyers to automatically deposit chips into recycling containers. If chips are long and stringy, they will clog up the chip conveyers. Long, stringy chips will also wrap around the tool, the workpiece, and the work-holding device, which can cause tool breakage and an especially dangerous situation for machine operators. Soft, gummy, and tough materials can wrap around spinning tools and workpieces. The chips begin whipping around, sending sharp, hot chips in every direction. For this reason, there has been considerable research in the area of chip control.

Molded Chip Breakers

Molded chip breakers use a molded groove to change the direction of the chip. Molded chip breakers are available in many different configurations (see Figure F-65). Some

Figure F-62 Lathe part.

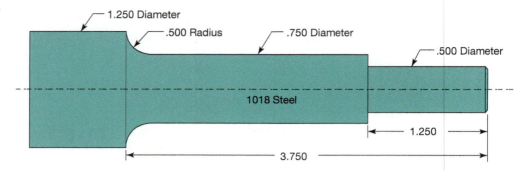

molded chip breakers are designed for certain materials, and others are designed for different feed rates and depths of cut.

The chip breaker is designed to redirect the flow of the chip, causing it to curl into a figure 6 or figure 9 (see Figure F-66). A chip of this configuration is said to be the perfect chip for steel cutting. When cutting cast iron, chip control is not a problem because iron is brittle and does not flow away from the cutting edge the way steel does.

Factors Affecting Chip Formation

Chip breakers have greatly increased our ability to control chips. There are factors that need to be considered no matter which type of chip breaker is used. The three factors that affect chip control the most are feed rate, cutting speed, and tool shape. One of the quickest ways to eliminate long, stringy chips is to increase the feed rate: a thicker chip will curl and break easier than thin chips.

Decreasing the side-cutting edge angle or lead angle will also create a thicker chip. Sometimes increasing the speed will help the chip flow and curl easier. Long, stringy chips are not the only chips that can cause problems in machining. Conduit chips are chips that are almost ready to break (see Figure F–66). A conduit chip is a long curly chip, which is common when machining soft ductile materials. To remedy the problem, try increasing the feed rate.

Corrugated chips (see Figure F-66) have a very tight curl and represent the opposite problem from stringy chips. These types of chips are being bent too much and are usually caused by excessive feed rates. Corrugated chips do not pose a chip control problem, but they are a sign of improper cutting action and should be dealt with immediately. If slowing the feed rate doesn't change the chip formation, use a narrower chip breaker to allow the chip to make a wider curl.

Chip Color

As a machine operator, you should always be aware of the chips you are producing. Analyze the chip's shape and color. A deep blue steel chip indicates that the heat of the cutting action is being drawn away from the workpiece, as it should. A dark purple or black chip indicates excessive heat. In this case, reduce the cutting speed and any other machining conditions until the color of the chip is acceptable. Chips should always be clean and smooth on the underside, not torn and ragged. Proper chip formation is a

balancing act of speeds, feeds, and chip breaker formation. Look to the chip for the information you need to balance these factors.

TROUBLESHOOTING

Carbide cutting tools are consistent and durable cutting tools. Problems will, however, sometimes result when using carbide tools. The CNC operator will then need to change cutting conditions to address the problems.

The first step is to diagnose the problem. Possible problems include premature failure of the insert, edge wear, crater wear, edge buildup, depth of cut notching, chipping, thermal cracking, or thermal deformation.

Catastrophic Breakage

Premature failure or insert breakage is a problem that will be apparent even to the least experienced machine operator. If the insert breaks and continues to break after being changed, there is a problem.

One possible cause of tool breakage is that the operating conditions are excessive. Slow down the speed and especially the feed. If the grade that you have selected is too hard (brittle) for the material or the condition of the material, select a tougher grade of insert. The lead angle may be too small. Select a toolholder that lends more support to the tool tip.

Edge Wear

Edge wear is more difficult to diagnose. Excessive edge wear is the unnatural wearing away of the insert along the side or flank of the cutting edge (see Figure F-67). The ability to recognize excessive edge wear comes with experience. If you believe that you are experiencing excessive edge wear, the probable cause is friction. Excessive friction causes heat to build along the cutting edge, which causes the binders to fail. One possible cause is that the lead angle is too great. Choose a holder that reduces the lead angle. Check the tool height. A crash or bump of the tool turret may be causing the tool to be too high. Another possible cause may be that the feed rate is too low. Increasing the feed rate will cause the chips to concentrate away from the cutting edge. Finally, it could be a grade selection problem.

Figure F-63 ANSI insert classification.

 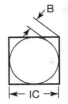

Tolerance Class	Tolerance on "B"		Tolerance on "IC"		Tolerance on "T"	
	INCH	MM	INCH	MM	INCH	MM
A	±.0002	±.0005	±.001	±.025	±.001	±.025
C	±.0005	±.013	±.001	±.025	±.001	±.025
E	±.001	±.025	±.001	±.025	±.001	±.025
F	±.0002	±.005	±.0005	±.025	±.001	±.025
G	±.001	±.025	±.001	±.13	±.005	±.13
H	±.0005	±.013	±.0005	±.025	±.001	±.025
J	±.002	±.005	±.002-.005	±.025	±.001	±.025
K	±.0005	±.013	±.002-.005	±.025	±.001	±.025
L	±.001	±.025	±.002-.005	±.025	±.001	±.025
M	±.002-.005	±.05-.13	±.002-.005	±.13	±.005	±.025
U	±.005-012	±.06-.25	±.005-.010	±.13	±.005	±.13

Tolerance

ANSI Insert Identification System **C N M G 4 3 2**

Shape			
Symbol Shape			Nose Angle
S	Square		90
T	Triangular		60
C	Diamond		80
D			55
E			75
F			50
M			86
V			80
W	Trigon		80
H	Hexagonal		120
O	Octagonal		135
P	Pentagonal		108
L	Rectangular		90
A	Parallelogram		85
B			82
N/K			55
R	Round		

Clearance
Relief Angle
N - 0°
A - 3°
B - 5°
C - 7°
P - 11°
D - 15°
E - 20°
F - 25°
G - 30°

Geometry

A=   *K=
B=   *L=
C=   M=
*D=   N=
*E=   Q=
F=   R=
G=   T=
H=   U=
J=   W=

* = Special Design

Figure F-64 ANSI insert classification.

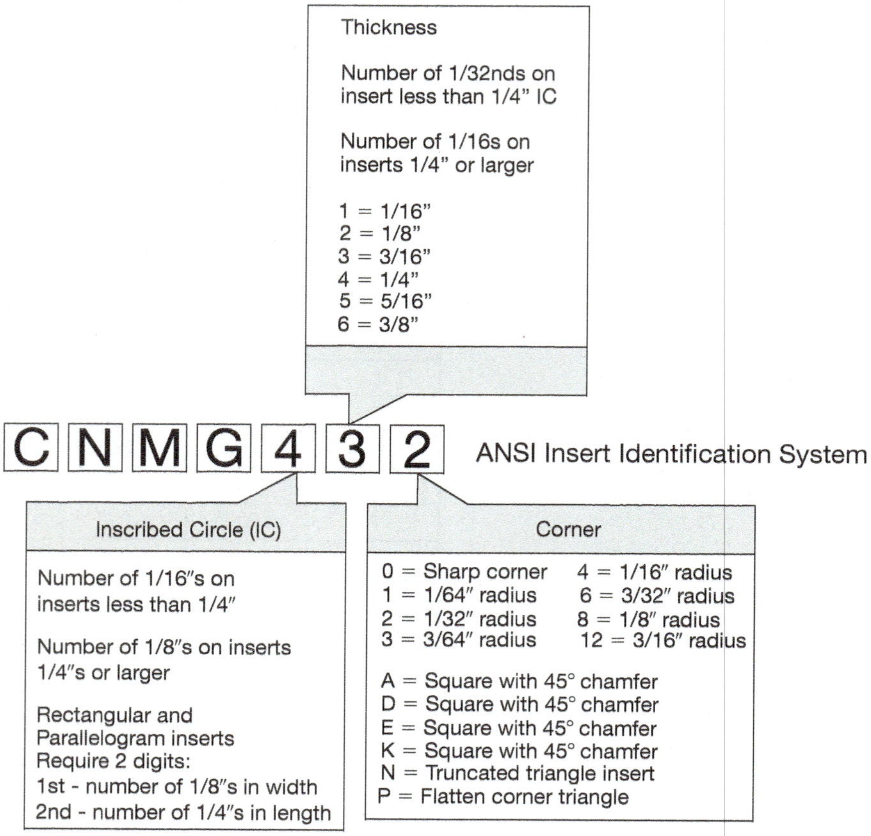

Thickness

Number of 1/32nds on insert less than 1/4" IC

Number of 1/16s on inserts 1/4" or larger

1 = 1/16"
2 = 1/8"
3 = 3/16"
4 = 1/4"
5 = 5/16"
6 = 3/8"

| C | N | M | G | 4 | 3 | 2 | ANSI Insert Identification System |

Inscribed Circle (IC)

Number of 1/16"s on inserts less than 1/4"

Number of 1/8"s on inserts 1/4"s or larger

Rectangular and Parallelogram inserts Require 2 digits:
1st - number of 1/8"s in width
2nd - number of 1/4"s in length

Corner

0 = Sharp corner   4 = 1/16" radius
1 = 1/64" radius   6 = 3/32" radius
2 = 1/32" radius   8 = 1/8" radius
3 = 3/64" radius   12 = 3/16" radius

A = Square with 45° chamfer
D = Square with 45° chamfer
E = Square with 45° chamfer
K = Square with 45° chamfer
N = Truncated triangle insert
P = Flatten corner triangle

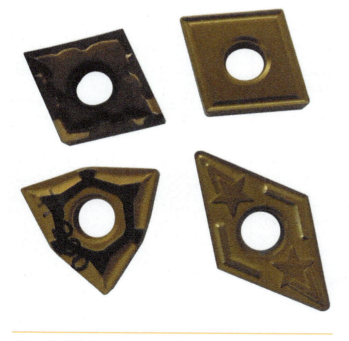

Figure F-65 Indexable inserts with molded chip breakers.

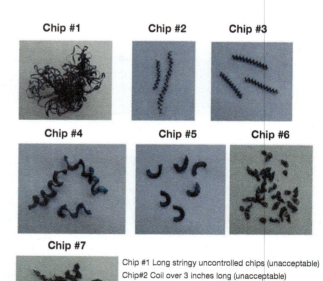

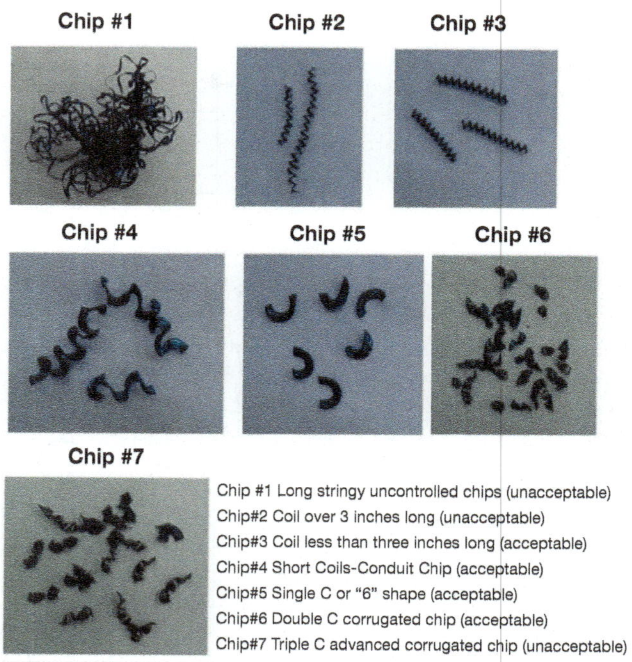

Chip #1   Chip #2   Chip #3

Chip #4   Chip #5   Chip #6

Chip #7

Chip #1 Long stringy uncontrolled chips (unacceptable)
Chip#2 Coil over 3 inches long (unacceptable)
Chip#3 Coil less than three inches long (acceptable)
Chip#4 Short Coils-Conduit Chip (acceptable)
Chip#5 Single C or "6" shape (acceptable)
Chip#6 Double C corrugated chip (acceptable)
Chip#7 Triple C advanced corrugated chip (unacceptable)

Figure F-66 Different chip configurations.

Figure F-67 Edge wear.

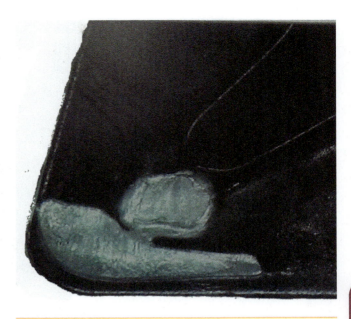

Figure F-68 Crater wear.

Crater Wear

Crater wear occurs when the binder is being replaced by the material you are cutting. When you are cutting steel, the constant passing of the chip over the insert causes the cobalt binder to be carried away by the chip, leaving the steel to act as the binder. The steel, not being a very good binder material, quickly wears away, leaving a crater (see Figure F-68). Cratering is usually a grade selection problem or an extreme heat problem, caused by cutting conditions that are too high. To minimize cratering, reduce the speed and feed, use a harder grade of carbide, or use a coated carbide insert.

Edge Buildup

Edge buildup or adhesion occurs when metal deposits build up on the cutting edge (see Figure F-69). Iron actually combines with the binder in the carbide substrate. Edge buildup occurs when the cutting conditions are too slow. Carbide cuts best at high temperatures and will rapidly wear if these temperatures are not reached. Machine operators can reduce edge buildup by increasing the speed and feed.

Depth-of-Cut Notching

Depth-of-cut notching is an unnatural chipping away of the insert right at the depth of cut line (see Figure F-70). Depth-of-cut notching is usually a grade selection problem. If you are using an uncoated insert, consider changing to a coated insert. If a coated insert is not available, try honing the edge of the insert. Honing should only be done on uncoated inserts. Honing is done at a 45-degree angle to the cutting edge. Proper honing just breaks the sharp edge of the insert.

Depth-of-cut notching may also be solved by lowering the feed rate and/or by reducing the lead angle.

Chipping

Chipping is a common insert problem. Chipping occurs along the cutting edge and is sometimes mistaken for edge wear (see Figure F-71). The major causes of insert chipping are lack of rigidity, an insert grade that is too hard, or low operating conditions. Carbide is very brittle and works best when it is well supported.

Many carbide cutting tool problems can be eliminated by decreasing the overhang of the tool and supporting the

Figure F-69 Edge buildup.

Figure F-70 Depth of cut notching.

Figure F-72 Thermal cracking.

Figure F-71 Chipping.

Figure F-73 Thermal deformation.

work better. If rigidity is not the problem, use a softer or tougher grade of insert. When making roughing cuts through hard spots or sand inclusions, use a tougher, not harder, grade of carbide. If the operating conditions are too low, abnormal pressures may build up, causing chipping. Increasing the cutting speed will sometimes eliminate chipping.

Thermal Cracking and Thermal Deformation

Two heat problems are commonly associated with carbide cutting tools: thermal cracking and thermal deformation (see Figures F-72 and F-73). Thermal cracking will show up as

small surface cracks along the cutting edge and tip of the insert. Cracking is caused by sudden changes in temperature. Thermal cracking can occur if coolant is being applied to the insert instead of in front of the insert. If coolant is applied in the middle of the cut, thermal cracking may occur.

The other heat-related problem associated with carbide cutting tools is thermal deformation. Thermal deformation is a melting away of the tool tip and is caused by operating conditions being too high (see Figure F-73). The excessive heat breaks down the binder materials in the carbide insert. There are two possible solutions to thermal deformation: reduce the cutting conditions or switch to a more heat-resistant grade of carbide.

Figure F-74 Troubleshooting chart. Find your machining problem on the left, and the right column lists potential cures for the problem.

Problem	Remedy
Tool life is too short due to excessive wear	1. Change to a harder, more wear-resistant grade. 2. Reduce the cutting speed. 3. Reduce the feed. 4. Increase the lead angle. 5. Increase the relief angle.
Excessive cratering	1. Use a harder, more wear-resistant grade. 2. Reduce the cutting speed. 3. Reduce the feed.
Cutting edge chipping	1. Increase the cutting speed. 2. Hone the cutting edge. 3. Change to a tougher grade. 4. Use a negative rake insert. 5. Increase the lead angle. 6. Reduce the feed.
Deformation of the cutting edge	1. Reduce the cutting speed. 2. Change to a grade with a higher red-hardness. 3. Reduce the feed.
Poor surface finish	1. Increase the cutting speed. 2. Increase the nose radius. 3. Reduce the feed. 4. Use positive rake inserts.

When diagnosing problems with carbide cutting tools, remember that troubleshooting is not a shot in the dark and should be done systematically. Troubleshooting must be a methodical procedure. The first step is to determine the problem. The second step is to arrive at all of the possible solutions. The third step is to examine each of the possible causes, changing only one condition at a time.

Use Figure F-74 to help diagnose your carbide cutting tool problems.

CERAMIC TOOLS

Ceramic or "cemented oxide" tools are made primarily from aluminum oxide. Some manufacturers add titanium, magnesium, or chromium oxides in quantities of 10 percent or less. The tool materials are molded at pressures over 4000 psi and sintered at temperatures of approximately 3000°F (1649°C). This process partly accounts for the high density and hardness of cemented oxide tools.

Ceramic tools are formed to shape by either cold pressing or hot pressing. Some hot-pressed, high-strength ceramics are termed *cermets* because they are a combination of ceramics and metals. They possess the high shear resistance of ceramics and the toughness and thermal shock resistance of metals. Cermets were originally designed for use at high temperatures such as those found in jet engines. These materials were subsequently adapted for cutting tools. A tool insert composed of titanium carbide and titanium nitride is an example. These tools have a low coefficient of friction and are less likely to form a built-up edge when machining steel. They are used at high cutting speeds. A common combination of materials in a cermet is aluminum oxide and titanium carbide. Another method of obtaining the high-speed cutting characteristics of ceramics and the toughness of metal tools is to coat a carbide tool with a ceramic composite. Ceramic tools should be used as a replacement for carbide tools that are wearing rapidly but not to replace carbide tools that are breaking.

CUBIC BORON NITRIDE TOOLS

CBN is next to diamond in hardness and therefore can be used to machine plain carbon steels, alloy steels, and gray cast irons with hardnesses of 45 Rc and above. Formerly, steels over 60 Rc would have to be abrasive machined, but with the use of CBN they can often be cut with single-point tools.

CBN inserts consist of a cemented carbide substrate with an outside layer of CBN formed as an integral part of the tool. Tool life, finishes, and resistance to cracking and abrasion make CBN a superior tool material to both carbides and ceramics.

DIAMOND TOOLS

Industrial diamonds are sometimes used to machine extremely hard workpieces. Only relatively small removal rates are possible with diamond tools, but high speeds are used and good finishes are obtained. Nonferrous metals are turned at 2000 to 2500 fpm, for example. Sintered polycrystalline diamond tools, available in shapes similar to those of ceramic tools, are used for materials that are abrasive and difficult to machine. Polycrystalline tools consist of a layer of randomly oriented synthetic diamond crystals brazed to a tungsten carbide insert.

Diamond tools are particularly effective for cutting abrasive materials that quickly wear out other tool materials. Nonferrous metals, plastics, and some nonmetallic materials are often cut with diamond tools. Diamond is not particularly effective on carbon steels or superalloys that contain cobalt or nickel. Ferrous alloys chemically attack single- or polycrystalline diamonds, causing rapid tool wear. Because of the high cost of diamond tool material, it is usually restricted to those applications where other tool materials cut poorly or break down quickly.

Diamond or ceramic tools should never be used for interrupted cuts such as on splines or keyways because they could chip or break. They must never be used at low speeds or on machines not capable of attaining the higher speeds at which these tools should operate.

Each of the cutting tool materials varies in hardness. The differences in hardness tend to become more pronounced at high temperatures (Figure F-75). Hardness is related to the wear resistance of a tool and its ability to machine materials that are softer than it is. Temperature change is important, because the increase in temperature during machining results in the softening of the tool material.

Figure F-75 Hardness of Cutting Materials at High and Low Temperatures

Tool Material	Hardness at Room Temperature	Hardness at 1400°F (760°C)
High-speed steel	RA 85	RA 60
Carbide	RA 92	RA 82
Cemented oxide (ceramic)	RA 93–94	RA 84
Cubic boron nitride		Near diamond hardness
Diamond		Hardest known substance

SELF-TEST

1. State the two main characteristics of carbide.
2. What is meant by *insert grade?*
3. What is a coated carbide?
4. Name two types of coating that are applied to carbide inserts.
5. Name three different insert shapes in order of increasing strength.
6. What is one of the most common binding materials that holds the carbide particles together?
7. State the purpose of tool-nose radius.
8. Name two factors to consider when selecting the proper shape of carbide insert.
9. What is meant by *inscribed circle?*
10. Describe the three types of rake angles.
11. What is lead angle?
12. What are qualified toolholders?
13. Describe the characteristics of a TPG 432 insert.
14. Describe the characteristics of a VNMG 332 insert.
15. Describe the characteristics of a CNMG 432 insert.
16. Describe the characteristics of a SPG 432 insert.
17. While face milling a previously drilled surface, you notice that the inserts start chipping as soon as the face-milling cutter enters the interrupted cut. What are some possible remedies for this type of tool failure?
18. After rough turning with a new insert for five minutes you notice that the insert is starting to spark and the part is getting very warm. What type of tool failure is this and what are some possible remedies for this type of tool failure?
19. While attempting to turn a piece of cast steel, the tool tip keeps breaking off. What are some possible remedies for this type of tool failure?
20. While turning tool steel, you notice that there is a groove appearing on the insert. What are some possible remedies for this type of tool failure?

INTERNET REFERENCE

Information on cutting tool materials:

http://kennametal.com

Sawing Machines

Sawing machines are some of the most important machine tools in the machine shop. These machines can be divided into two classifications. The first type is **cutoff machines**. The second type is the **vertical band saw**.

Cutoff machines are generally found near the stock supply area. The primary function of the cutoff machine is to reduce long lengths of bar stock material into lengths suitable for use in other machines.

Types of Cutoff Machines and Safety

Horizontal Band Saws

The band machine uses a steel band blade with the teeth on one edge. The band machine is very efficient because the band is cutting at all times with no wasted motion. Band saws are the workhorse of stock cutoff in machine shops (Figure G-1).

A modern band saw may be equipped with a variable-speed drive. This permits the most efficient cutting speed to be selected for the material being cut. The feed rate for cutting the material may also be varied. The size of the horizontal band saw is determined by the largest piece of square material that can be cut. Common horizontal band saws used for general stock cutoff in the machine shop may have a material capacity of 12 by 12 inches and range up to 24 or more inches.

Medium- and large-capacity horizontal band saws may be of the dual-column design (Figure G-2). On the dual-column machine, the saw frame moves vertically at both ends, enabling a larger workpiece to fit under the cutting band.

Figure G-2 Large dual-column horizontal band saw (*HE&M Saw*).

Universal Tilt Frame Band Saw

The universal tilt frame band saw is much like its horizontal counterpart. This machine has a vertical band blade and the frame can be tilted from side to side (Figure G-3). The tilt frame machine is particularly useful for making angle cuts up to 45 degrees both left and right on large structural shapes such as I-beams or pipe.

Cold Saw Cutoff Machines

A cold saw uses a circular metal saw. This machine tool can produce accurate cuts and is useful in applications where the length tolerance of the cut must be held as close as possible. A cold saw blade that is .040 to .080 inch thick can saw materials to a tolerance of plus or minus .002 inch. Large cold saws are used to cut structural shapes such as angle and flat bars. Cold saws are fast-cutting machines.

Figure G-1 Horizontal band cutoff saw.

Figure G-3 Universal tilt frame band saw (*Courtesy of Fox Valley Technical College*).

CUTOFF MACHINE SAFETY

Horizontal Band Saws

Safety regulations require that the blade of the horizontal band machine be fully guarded except at the point of cut (Figure G-4). When using a band saw make sure that the blade tension is correct. Check band tension, especially after installing a new band. New bands may stretch and loosen during their run-in period. Band teeth are sharp. When installing a new band, handle it with gloves. This is one of the few times when gloves may be worn around the machine shop. They must not be worn when operating any machine tool.

Figure G-4 The horizontal band blade is guarded except in the immediate area of the cut.

Endless band blades are often stored in double or triple coils. Be careful when unwinding them, as they are under tension. The coils may spring apart and can cause an injury. Make sure that the band is tracking properly on the wheels and in the blade guides. If a band should break, it could be ejected from the machine and cause an injury.

Make sure that the material being cut is properly secured in the vise. If you are cutting short pieces of material, the vise jaw must be supported at both ends (Figure G-5). It is poor practice to attempt to cut pieces of material that are very short. The stock cannot be secured properly and may be pulled from the vise by the pressure of the cut (Figure G-6). This can damage the saw band as well as possibly injure the operator. The stock should extend at least halfway through the vise at all times.

Many cutoff machines have a rollcase (stand with rollers) that supports long bars of material while they are being cut. Heavy stock should be brought to the saw on a rollcase (Figure G-7) or a simple rollstand. The rollcase makes it easy to move the material into the vise when positioning it for a cut. The pieces being cut off can sometimes be several feet long and should be similarly supported. Sharp burrs left from the cutting should be removed immediately with a file. You can acquire a nasty cut by sliding your hand over one of these burrs.

Figure G-5 Support both ends of the vise when cutting short material.

Figure G-6 Result of cutting stock that is too short.

Figure G-7 The material is brought into the saw on the rollcase (left side of the saw), and when pieces are cut off they are supported by the stand (this side of the saw). The stand prevents the part from falling to the floor.

Be careful around a rollcase, because bars of stock can roll, pinching fingers and hands. Also, be careful that heavy pieces of stock do not fall off the stock table or saw and injure feet or toes. Get help when lifting heavy bars of material. This will save your back and possibly your career.

VERTICAL BAND MACHINES

The vertical band saw (Figure G-8) is similar in construction to its horizontal counterpart. Basically, it consists of an endless band (saw blade) that runs on a drive and an idler wheel. The band blade runs vertically through a worktable on which the workpiece rests. The workpiece is pushed into the blade, and the direction of the cut is guided by hand or mechanical means.

ADVANTAGES OF BAND MACHINES

Shaping of material with the use of a saw blade is called **band machining**. A band machine can perform other machining tasks in addition to simple sawing. These include band friction sawing, band filing, and band polishing.

In any machining operation, a piece of material is cut by various processes to form the final shape and size of the part. In most machining operations, all the

Figure G-8 Vertical band saw (*Courtesy of Fox Valley Technical College*).

unwanted material must be reduced to chips to reveal the final shape and size of the workpiece. With a band saw, only a small portion of the unwanted material must be reduced to chips to reveal the final shape and size of the workpiece (Figure G-9). A piece of material can often be shaped to final size by one or two saw cuts. A further advantage is that the band saw cuts a narrow kerf, so a minimum amount of material is wasted.

A second important advantage of band sawing machines is **contouring ability**, that is, the capability to cut intricate curved shapes.

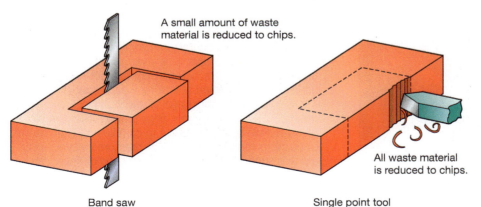

A small amount of waste material is reduced to chips.

All waste material is reduced to chips.

Band saw

Single point tool

Figure G-9 Sawing can reveal the workpiece shape in a minimum number of cuts.

Band sawing and band machining have several other advantages. There is no limit to the length, angle, or direction of the cut. However, the throat capacity of the sawing machine will be a factor depending on the dimension of the parts being sawed. Because the band tool is fed continuously into the work, the cutting efficiency is high. A band blade has a large number of cutting points passing the work. In most other machining operations, only one or a fairly low number of cutting points pass the work. With the band tool, wear is distributed over these many cutting points. This helps prolong the life of the saw band blade.

Water-jet machines and/or laser machines have largely taken over production-type cutting that may have been performed on band machines in the past. They are faster and can hold tight tolerances as well as cut complex shapes to finished size in many types of materials.

TYPES OF BAND MACHINES

General-Purpose Band Machine with Fixed Worktable

The general-purpose band machine is found in most machine shops. This machine tool has a non-power-fed worktable that can be tilted to make angle cuts. The table on this saw may be tilted 10 degrees left (Figure G-10). Tilt on this side is limited by the saw frame. The table on this saw may also be tilted 15 degrees right.

The workpiece is pushed into the blade by hand (Figure G-11). Mechanical or mechanical/hydraulic feeding mechanisms are also used on some saws.

High-Tool-Velocity Band Machines

High-velocity band machines are also available. Band speeds can range as high as 15,000 feet per minute (FPM). These machine tools are used in many band machining applications. They are frequently found cutting nonmetal products such as in trimming plastic laminates and cutting fiber materials.

Figure G-10 Vertical band machine worktable can be tilted 10 degrees left.

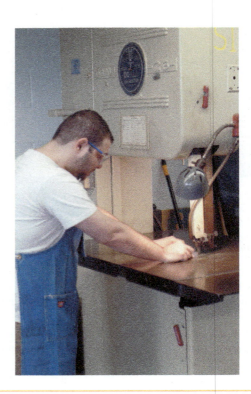

Figure G-11 Feeding the work into the blade by hand.

Conventional and Contour Sawing

Vertical band machines are used in many conventional sawing applications. They are found in the foundry trimming sprues and risers from castings. The band machine can accommodate a large casting and make widely spaced cuts. Castings can easily be trimmed with the high-velocity band machine. In the machine shop, the vertical band machine is used for general-purpose, straight-line, and contour cutting, mainly in sheet and plate stock.

SAFETY FIRST

The primary danger in operating a vertical band machine is accidental contact with the cutting blade. Workpieces are often hand guided. One advantage of sawing machines is that the pressure of the cut tends to hold the workpiece against the saw table. However, hands are often in close proximity to the blade. If you should contact the blade accidentally, an injury is almost sure to occur. You will not have time even to think about withdrawing your fingers before they are cut. Keep this in mind at all times when operating a band saw.

Always use a pusher against the workpiece whenever possible. This will keep your fingers away from the blade. Be careful when you are about to complete the cut: As the blade clears through the work, the pressure that you are applying is suddenly released, and your hand or finger can be carried into the blade. As you approach the end of the cut, reduce the feeding pressure as the blade cuts through.

VERTICAL BAND MACHINE SAFETY

The vertical band saw is not usually used to cut round stock. This can be dangerous and should be done on the horizontal band machine, where round stock can be secured in a vise. Handheld round stock will turn if it is cut on the vertical band machine. This can cause an injury and may damage the blade as well. If round stock must be cut on the vertical band saw, it must be clamped securely in a vise, V-block, or other suitable work-holding fixture.

Be sure to select the proper blade for the sawing requirements. Install it properly and apply the correct blade tension. Recheck band tension after a few cuts. New blades will tend to stretch during their break-in period. Band tension may have to be readjusted.

The entire blade must be guarded except at the point of the cut. This is accomplished by enclosing the wheels and blade behind guards that are easily opened for adjustments to the machine. Wheel and blade guards must be closed at all times during machine operation. The guidepost guard moves up and down with the guidepost (Figure G-12). For maximum safety, set the guidepost $\frac{1}{8}$ to $\frac{1}{4}$ inch above the workpiece.

Roller blade guides are used in friction and high-speed sawing. Depending on the material being cut, the entire cutting area may be enclosed, such as when cutting hard, brittle materials like granite and glass. Diamond blades are frequently used in cutting these materials. The clear shield protects the operator while permitting him or her to view the operation. Cutting fluids are also prevented from spilling on the floor. In any sawing operation making use of cutting fluids, see that they do not spill on the floor around the machine. Spilled coolant

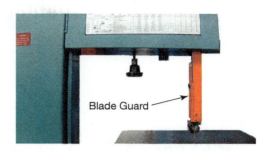

Figure G-12 Guidepost guard.

creates a dangerous situation, not only for you but for others in the shop as well.

Gloves should not be worn around any machine tool. An exception to this is when handling band blades. Gloves will protect hands from the sharp saw teeth.

SELF-TEST

1. What is the most common type of cutoff saw in a machine shop?
2. What extra capability does a tilt-frame saw have?
3. What is a cold saw?
4. What precautions should be taken when cutting round stock on a vertical band saw?
5. Which machines have taken over most of the contour cutting that used to be done on vertical band saws?

Using Horizontal Cutoff Saws

The horizontal band saw is the most common cutoff machine in a machine shop. The primary function is to cut long lengths of material into lengths suitable for other machining operations. The cutoff application is often the first step in machining a part to its final shape and size. In this unit you are introduced to saw blades and the operation of horizontal cutoff saws.

OBJECTIVES

After completing this unit, you should be able to:

- Use saw blade terminology.
- Describe the conditions that define blade selection.
- Identify the major parts of the reciprocating and horizontal band cutoff machine.
- Properly install blades on reciprocating and horizontal band machines.
- Properly use reciprocating and horizontal band machines in cutoff applications.

CUTTING SPEEDS

An understanding of cutting speeds is one of the most important aspects of machining. Cutting tools cut most effectively if fed through the workpiece at optimum speeds. If a tool feeds through the work too quickly, the heat generated by friction can rapidly dull the tool or cause it to fail completely. Too slow of a feed rate can result in premature dulling and low productivity.

Cutting speed refers to the amount of workpiece material that passes by a cutting tool in a given amount of time. Cutting speeds are measured in surface feet per minute (SFPM). In some machining operations, the tool passes the work, as in sawing; in other operations the work passes the tool, as with the lathe. In both cases, the SFPM is the same.

In sawing, SFPM is simply the speed of each saw tooth as it passes through a given length of material in one minute. If one tooth of a band saw travels one hundred feet in one minute, the cutting speed is one hundred SFPM. Cutting speeds are a critical factor in tool life. Productivity will be poor if the sawing machine is stopped often because a dull or damaged blade must be replaced. The additional cost of replacement of cutting tools must also be considered. Keep cutting speeds in mind for all machining operations.

Blade SFPM is a function of rpm (revolutions per minute) of the saw drive. That is, the setting of a specific rpm on the saw will produce a specific SFPM of the blade. SFPM also relates to the material being cut. Generally, hard, tough materials have low cutting speeds. Soft material has higher cutting speeds. In sawing, cutting speeds are affected by the material, size, and cross section of the workpiece.

SAW BLADES

The blade is the cutting tool of the sawing machine. At least three teeth on the saw blade must be in contact with the work at all times. This means that thin material requires a blade with more teeth per inch, whereas thick material can be cut with a blade having fewer teeth per inch. A machinist should be familiar with the terminologies of saw blades and saw cuts.

Blade Specifications

Blade Materials   Saw blades for band saws are made from carbon steels and high-speed alloy steels. Blades may also have tungsten carbide–tipped teeth. Some blades are bimetallic.

Blade Kerf   The kerf of a saw cut is the width of the cut as produced by the blade (Figure G-13).

Blade Width   The width of a saw blade is the distance from the tip of the tooth to the back of the blade (Figure G-14).

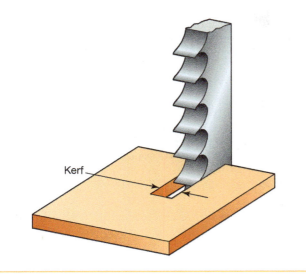

Figure G-13 Kerf.

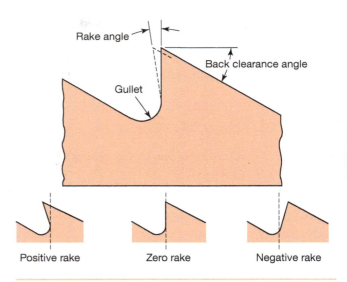

Figure G-15 Saw tooth terminology.

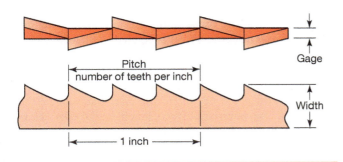

Figure G-14 Gauge, pitch, and width.

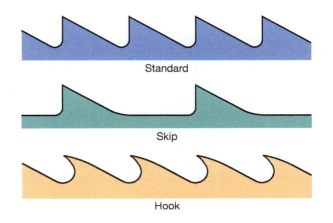

Figure G-16 Tooth forms.

Blade Gauge   Blade gauge is the thickness behind the set of the blade (Figure G-14). Reciprocating saw blades on large machines can be as thick as .250 inch. Common band saw blades are .025 to .035 inch thick.

Blade Pitch   The pitch of a saw blade is the number of teeth per inch (Figure G-14). An eight-pitch blade has eight teeth per inch (a tooth spacing of $\frac{1}{8}$ inch). Blades of variable pitch are also used.

SAW TEETH

Saw tooth rake is shown in Figure G-15.

Saw Tooth Specifications

Tooth Forms   Tooth form is the shape of the saw tooth. Saw tooth forms are standard, skip, or hook (Figure G-16). Standard form gives accurate cuts with a smooth finish. Skip tooth gives additional chip clearance. Hook form provides faster cutting because of the positive rake angle, especially in soft materials.

Set   The teeth of a saw blade must be offset on each side to provide clearance for the back of the blade. This offset is called **set** (Figure G-17). Set is equal on both sides of the blade. The set dimension is the total distance from the tip of a tooth on one side to the tip of a tooth on the other side.

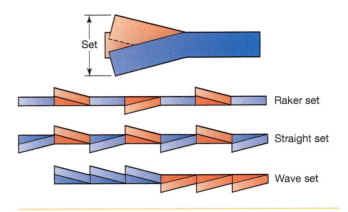

Figure G-17 Set and set patterns.

Set Patterns Set forms include raker, straight, and wave (Figure G-17). Raker and wave are the most common. Raker set is used in general sawing. Wave set is useful when the cross-sectional shape of the workpiece varies.

USING CUTTING FLUIDS

Cutting fluids are an important aid to sawing. The heat produced by the cutting action can affect the metallurgical properties of the blade teeth. Cutting fluids will dissipate much of this heat and greatly prolong the life of the blade. Besides functioning as a coolant, they also lubricate the blade. Sawing with cutting fluids will produce a smoother finish on the workpiece. One of the most important functions of a cutting fluid is to transport chips out of the cut. This allows the blade to work more efficiently. Common cutting fluids are oils, oils dissolved in water, and synthetic chemical cutting fluids.

Making the Cut

If you are cutting material with a sharp corner (Figure G-18), begin the cut on a flat side if possible. Note that angular material presents a sharp corner to the blade. Start the saw gently until a small flat is established. Bring the saw gently down until the blade has a chance to start cutting. Apply the proper feed. If you replace the blade after starting a cut, turn the workpiece over and begin a new cut (Figure G-19). Do not attempt to saw through the old cut, because this will damage the new blade. The kerf from the old blade will be narrower than the new blade and the teeth on the new blade will be immediately damaged.

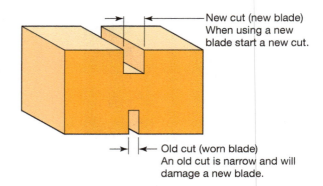

New cut (new blade) When using a new blade start a new cut.

Old cut (worn blade) An old cut is narrow and will damage a new blade.

Figure G-19 If the blade is changed, begin a new cut on the other side of the workpiece.

OPERATING THE HORIZONTAL BAND SAW

The horizontal band saw (Figure G-20) is the most common stock cutoff machine found in the machine shop. This machine tool uses an endless steel band blade with teeth on one edge. Because the blade passes through the work continuously, there is no wasted motion.

The size of the horizontal band saw is determined by the largest piece of square material that can be cut. Speeds on the horizontal band machine may be set by manual belt change, or a variable-speed drive on some saws. The variable-speed drive permits an infinite selection of band speeds within the capacity of the machine. Cutting speeds can be set precisely. Many horizontal band machines are of the hinge design. The saw head, containing the drive and idler wheels, hinges on a joint at the back of the machine.

The saw head may be raised and locked in the up position while stock is being placed into or removed from the machine. The down feed on many machines is

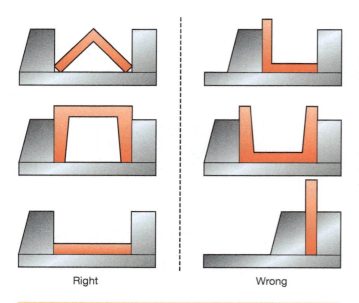

Right   Wrong

Figure G-18 Cutting workpieces with sharp corners.

Figure G-20 Horizontal endless band cutoff machine (*Courtesy of Fox Valley Technical College*).

Modern, Large Saw Control with
Variable Speed and Feed Controls

Small Saw with Valve to control Feed

Figure G-21 Feed controls.

hydraulically controlled. Some sawing machines use a hydraulic cylinder to regulate the feed rate. The head is held in the up position by the cylinder. A control valve permits oil to flow into the reservoir as the saw head descends. This permits the feed rate to be regulated. More advanced machines such as the ones shown on the left of Figure G-21 allow the operator to control the blade speed and down-feed from the control panel.

Cutting fluid is pumped from a reservoir and flows on the blade at the forward and rear guide on the saw shown in Figure G-21. Additional fluid is permitted to flow on the blade at the point of the cut. Cutting fluid is controlled by a control valve. Chips are cleared from the blade by a rotary brush that operates as the blade is running (Figure G-22).

Accessories available on horizontal band saws include programmable length feeds and roller stock tables. Figure G-23 shows a chip auger on a large saw that is used to automatically remove chips.

Figure G-22 Rotary chip brushes on the band saw blade.

Figure G-23 Chip auger.

Work Holding on the Horizontal Band Saw

The vise is the most common work-holding device on saws. Rapid-adjusting vises are common. These vises have large capacity and are quickly adjusted to the workpiece. After the vise jaws have contacted the workpiece, the vise is locked by the lock handle. The vise may be swiveled on some saws for miter or angle cuts. On some horizontal band cutoff machines, the entire top saw frame swivels for making angle cuts (Figure G-24).

Installing Blades on the Horizontal Band Machine

Blades for the horizontal band saw may be ordered prewelded in the proper length for the machine. Band blade may also be obtained in long rolls. The required length is then cut and welded.

To install the band, shut off power to the machine and open the wheel guards. Release the tension by turning the tension wheel. Place the blade around the drive and idler wheels. Be sure that the teeth are pointed in the direction of the cut. The blade will have to be twisted slightly to fit the guides. Guides should be adjusted so that they have .001– to .002-inch clearance with the blade. Adjust the blade tension. Reinstall the guards.

Figure G-24 Band saw swiveled for angle cutting.

Making the Cut

Speed of the saw should be set according to the blade type and material to be cut. The blade guides must be set so that they are as close to the work as possible (Figure G-25). This ensures maximum blade support and maximum accuracy of the cut. Sufficient feed should be used to produce a good chip. Excessive feed can cause blade failure. Too little feed can dull the blade prematurely. Go over the safety checklist for horizontal band saws. Release the head and lower it by hand until the blade starts to cut. Most saws are equipped with an automatic shutoff switch. When the cut is completed, the machine will shut off automatically.

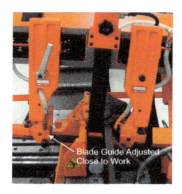

Figure G-25 The blade guide should be set as close to the workpiece as possible.

Figure G-26 Stock length stop in position.

SAWING PROBLEMS ON THE HORIZONTAL BAND MACHINE

The stock stop is used to gauge the length of material when multiple pieces are cut (Figure G-26). It is important to swing the stop clear of the work after tightening the vise and before beginning the cut. A cutoff workpiece can bind between the stop and the blade. This may destroy the blade set (Figure G-27). A blade with a tooth set worn on one side will drift in the direction of the side that has a set still remaining (Figure G-28). This is the principal cause of band breakage. As the saw progresses through the cut, the side draft of the blade will place the machine under great stress.

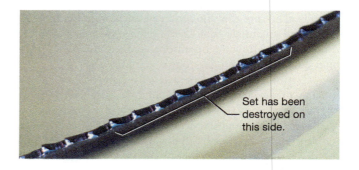

Figure G-27 Band saw blade with the set worn on one side.

Figure G-28 Using a band with the set worn on one side will cause the cut to drift toward the side of the blade with the setting remaining.

SHOP TIP

As the band set wears, the cut will drift toward the side of the band with the most remaining set. Although it is desirable to install a new band in this situation, you may be able to complete the started cut successfully by releasing the vise and allowing the workpiece to move a bit, thus relieving the side pressure on the band. Raise the blade a bit so that the cutting is stopped but the band is still in the kerf. Release the vise and allow the workpiece to "float" slightly so that the side pressure against the band is relieved. If the cut has drifted, the band tension will often push the workpiece slightly one way or the other. Reclamp the vise and complete the cut. This procedure will often allow an extended use of a somewhat dull band and also relieve the stress on the band saw caused by a drifting cut.

Improper feeds, band speeds, and types of materials being sawed will occasionally strip teeth from the band. These teeth are often deposited in the kerf and are difficult or impossible to remove. A band with a few teeth missing may still have some useful cutting life, assuming that the remaining teeth are good and equal set remains. However, running the old band or a new one into a cut where stripped-off saw teeth are lodged will be devastating to the saw band. If you suspect that broken saw teeth are lodged in the kerf, start a new cut on the workpiece. The use of cutting oils or other fluids during band sawing will flush chips and sometimes even broken teeth from the kerf. Sawing fluids will provide smoother, cooler, and faster cutting; greatly extend band life; and improve productivity.

SELF-TEST

1. Name the most common saw blade set patterns.
2. What considerations determine blade selection.
3. Describe what cutting speed is for a band cutoff saw.
4. What is set, and why is it necessary?
5. What are common tooth forms?
6. What can happen if the stock stop is left in place during the cut?
7. What type of cutoff saw will most likely be found in the machine shop?
8. What benefits do cutting fluids provide?
9. What can happen if chips are not properly removed from the cut?
10. If a blade is replaced after a cut has been started, what must be done with the workpiece?

INTERNET REFERENCES

Information on horizontal band saws:

http://www.marvelsaws.com/

http://www.dakecorp.com

Preparing a Vertical Band Saw for Use

A machine tool can only perform at maximum efficiency if it has been properly maintained, adjusted, and set up. Before the vertical band machine can be used, several important preparations must be made. These include welding saw blades into bands and making several adjustments to the machine tool.

OBJECTIVES

After completing this unit, you should be able to:

- Weld band saw blades.
- Prepare the vertical band machine for operation.

WELDING BAND SAW BLADES

Band saw blade stock can be purchased in rolls. The required length is measured and cut, and the ends are welded together to form an endless band. Most band machines are equipped with band welding attachments. These are frequently attached to the saw (Figure G-29). They may also be separate pieces of equipment.

The **band welder** is a resistance-type butt welder. It is often called **flash welder** because of the bright flash and shower of sparks created during the welding operation. The metal in the blade materials has a certain resistance to the flow of an electric current. This resistance causes the blade metal to heat as the electric current flows during the welding operation. The blade metal is heated to a temperature that permits the ends to be forged together under pressure. When the forging temperature is reached, the ends of the blade are pushed together by mechanical pressure. They fuse, forming a resistance weld. The band weld is then annealed or softened and ground to the correct thickness.

Welding band saw blades is a fairly simple operation. Sawing operations in which totally enclosed workpiece

Figure G-29 Band blade welder.

features must be cut require that the blade be inserted through a starting hole in the workpiece and then welded into a band. After the enclosed cut is made, the blade is broken apart and removed.

PREPARING THE BLADE FOR WELDING

The first step is to cut the required length of blade stock for the saw. Blade stock can be cut with snips or with a blade shear (Figure G-30). Start the cut on the side of the band opposite the teeth. Many band machines have a blade shear near the welder. The required length of the blade will usually be marked on the saw frame. Blade

Figure G-30 Blade shear.

length, B_L, for two-wheel sawing machines can be calculated by the formula

$$B_L = \pi D + 2L$$

where

D = the diameter of the band wheel

L = the distance between band wheel centers

Set the tension adjustment on the idler wheel about mid-range so that the blade will fit after welding. Many machine shops will have a permanent reference mark, probably on the floor, that can be used for measuring blade length.

After cutting the required length of stock, grind the ends of the blade so that they are square when positioned in the welder. Place the ends of the blade together so that the teeth are opposed (Figure G-31). Grind the blade ends in this position. The grinding wheel on the blade welder may be used for this operation. Blade ends may also be ground on a pedestal grinder (Figure G-32). Grinding the blade ends with the teeth opposed will ensure that the ends of the blade are square when the blade is positioned in the welder. Any

Figure G-31 Placing the blade ends together with the teeth opposed.

Figure G-32 End grinding the blade on the pedestal grinder.

small error in grinding will be canceled out when the teeth are placed in their normal position.

Proper grinding of the blade ends permits correct tooth spacing to be maintained. After the blade has been welded, the tooth spacing across the weld should be the same as at any other place on the band. The tooth set should be aligned as well. A certain amount of blade material is consumed in the welding process. Therefore, the blade must be ground correctly if correct tooth spacing is to be maintained. The amount of the blade material consumed by the welding process may vary with different blade welders. You will have to determine this by experimentation. For example, if ¼ inch of blade length is consumed in welding, this will amount to about one tooth on a four-pitch blade. Therefore, one tooth should be ground from the blade. This represents the amount lost in welding (Figure G-33). Be sure to grind only the tooth and not the end of the blade. The number of teeth to grind from a blade will vary according to the pitch and amount of material consumed by a specific welder. The weld should be made at the bottom of the tooth **gullet**. Exact tooth spacing can be somewhat difficult to obtain. This is much less of a consideration on finer pitch saw blades. You may have to practice end grinding and welding several pieces of scrap blades until you are familiar with the proper welding and tooth grinding procedure.

Make sure the jaws of the blade welder are clean before attempting any welding. Position the blade ends in the welder jaws (Figure G-34). The saw teeth should point toward the back. This prevents scoring of the jaws when welding blades of different widths. A uniform amount of the blade should extend from each jaw. The blade ends must contact squarely in the center of the gap between the welder jaws. Be sure that the blade ends are not offset or overlapped. Tighten the blade clamps.

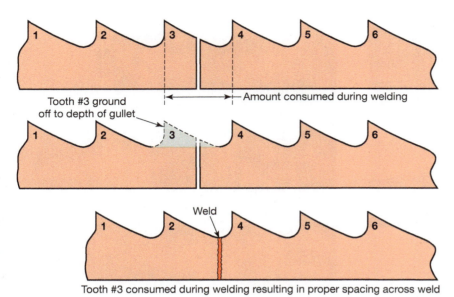

Tooth #3 ground off to depth of gullet

←— Amount consumed during welding —→

Weld

Tooth #3 consumed during welding resulting in proper spacing across weld

Figure G-33 The amount of blade lost in welding.

WELDING THE BLADE INTO AN ENDLESS BAND

SAFETY FIRST

Wear eye protection and stand to one side of the welder during the welding operation. Adjust the welder for the proper width of the blade to be welded. Depress the weld lever. A flash with a shower of sparks will be created (Figure G-35). Then the movable jaw of the welder moves toward the stationary jaw. The blade ends are heated to forging temperature by a flow of electric current, and the molten ends of the blades are pushed together and welded, forming a solid joint.

Loosen the blade clamps before releasing the weld lever. This prevents scoring of the welder jaws by the now-welded band. On correctly welded band, the weld flash will be evenly distributed across the weld zone (Figure G-36). Tooth spacing across the weld should be the same as on the rest of the band.

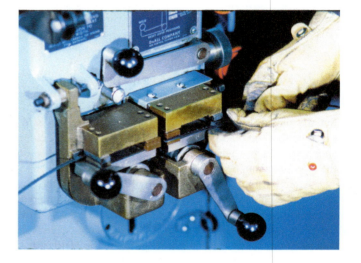

Figure G-34 Placing the blade in the welder.

ANNEALING THE WELD

The metal in the weld zone is hard and brittle after welding. For the band to function, the weld must be **annealed**, or **softened**. This process improves the strength of the weld. Place the band in the annealing jaws with the teeth pointed out (Figure G-37). This will concentrate the annealing heat away from the saw teeth. Compress the movable welder jaw slightly prior to clamping the band. This permits the jaw to move as the annealing heat expands the band.

It is most important not to overheat the weld during the annealing process. Overheating can destroy an otherwise good weld, causing it to become brittle. The correct

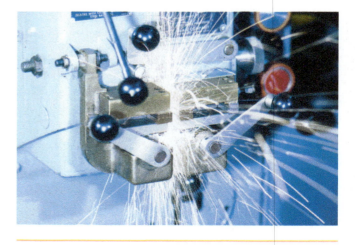

Figure G-35 Welding the blade into a band (*Courtesy of DoALL Company*).

Figure G-36 Weld flash evenly distributed across the weld.

Figure G-37 Positioning the band for annealing.

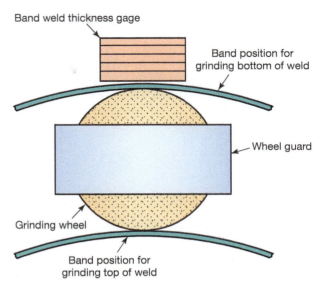

Figure G-38 Grinding the band weld.

Figure G-39 The saw teeth must not be ground while grinding the band weld.

annealing temperature is determined by the color of the weld zone during annealing. This should be a dull red color. Depress the anneal switch and watch the band heat. When the dull red color appears, release the anneal switch immediately and let the band begin to cool. As the weld cools, depress the anneal switch briefly several times to slow the cooling rate. Cooling the blade too rapidly can result in a band weld that is not properly annealed.

GRINDING THE WELD

Some machinists prefer to grind the band weld prior to annealing. This permits the annealing color to be seen more easily. More commonly the weld is ground after the annealing process. However, it is good practice to anneal the blade weld again after grinding. This will eliminate any hardness induced during the grinding operation. The grinding wheel on the band welder is designed for this operation. The top and bottom of the grinding wheel are exposed so that both sides of the weld can be ground (Figure G-38). Be careful not to grind the teeth when

grinding a band weld (Figure G-39). This will destroy the tooth set. Grind the band weld evenly on both sides. The weld should be ground to the same thickness as the rest of the band. If the weld area is ground thinner, the band will be weakened at that point. As you grind, check the band thickness in the gauge (Figure G-40) to determine when you have the proper thickness.

PROBLEMS IN BAND WELDING

Several problems may be encountered in band welding (Figure G-41). These include misaligned pitch, blade misalignment, insufficient welding heat, or too much welding heat. You should learn to recognize and avoid these problems. The best way to do this is to obtain some scrap blades and practice the welding and grinding operations.

Figure G-40 Band weld thickness gauge.

Figure G-41 Problems in band welding.

BAND GUIDES ON THE VERTICAL BAND MACHINE

Band guides must be properly installed if the saw is to cut accurately and if damage to the band is to be prevented. The band must be fully supported except for the teeth. Using band guides that are too thick and contact, the teeth will destroy the tooth set as soon as the machine is started.

Install the right-hand band guide and tighten the lock screw just enough to hold the guide insert in place

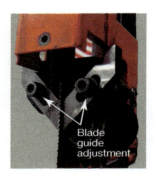

Figure G-42 Band guides must fully support the band but not extend over the saw teeth.

Figure G-43 Installing and adjusting the lower guides.

(Figure G-42). Check the backup bearing at this time. Clear any chips that might prevent it from turning freely. If the backup bearing cannot turn freely, it will be scored by the band and damaged permanently.

Install the left guide. Adjust the lower band guides in the same way (Figure G-43).

Roller band guides are used on some high-speed saws that are used for applications in which band velocities exceed 2,000 SFPM. The roller guides on these machines should be adjusted so that they have .001 to .002 inch clearance with the band.

ADJUSTING THE COOLANT NOZZLE

A band machine may be equipped with flood or mist coolant. Mist coolant is liquid coolant mixed with air. Certain sawing operations may require only small amounts of coolant. With the mist system, liquid coolant is conserved and is less likely to spill on the floor. When cutting with flood coolant, be sure that the runoff returns to the reservoir and does not spill on the floor. Flood coolant should be directed ahead of the band (Figure G-44).

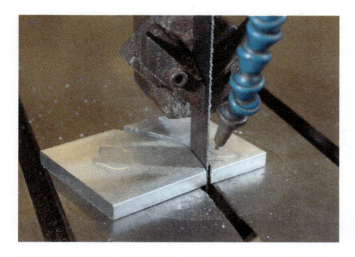

Figure G-44 Coolant should be directed ahead of the band.

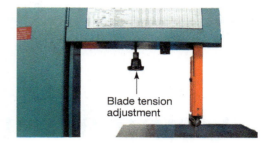

Figure G-45 Band tension crank.

INSTALLING THE BAND ON THE VERTICAL BAND MACHINE

Open the upper and lower wheel covers and remove the blade filler plate for the worktable. Wear gloves to protect your hands from the saw teeth. The hand tension crank is attached to the upper idler wheel (Figure G-45). Turn the crank to lower the wheel to a point where the band can be placed around the drive and idler wheels. Be sure to install the band so the teeth point in the direction of the cut. This is always in a down direction toward the worktable. If the saw teeth seem to be pointed in the wrong direction, you may have to turn the band inside out. Place the band around the drive and idler wheels and turn the tension crank so that tension is placed on the band. Be sure that the band slips into the upper and lower guides properly. Replace the filler plate in the worktable.

Adjusting Band Tension

Proper **band tension** is important to accurate cutting. Adjust the tension for the width of the band that you are using. After a new band has been run for a short time, recheck the tension. New bands tend to stretch during their initial use.

Adjusting Band Tracking

Band tracking refers to the position of the band as it runs on the idler wheels. On the vertical band machine, the idler wheel can be tilted to adjust the tracking position. The band tracking position should be set so that the back of the band just touches the backup bearing in the guide assembly. Generally, you will not often have to adjust band tracking. After you have installed a blade, check the tracking position. If it is incorrect, consult your instructor for help in adjusting the tracking position.

The tracking adjustment is made with the motor off and the speed range transmission in neutral. This permits

Figure G-46 Adjusting the band tracking position by tilting the idler wheel.

the band to be rolled by hand. Two knobs are located on the idler wheel hub. The outer knob (Figure G-46) tilts the wheel. The inner knob is the tilt lock. Loosen the lock knob and adjust the tilt of the idler wheel while rolling the band by hand. When the correct tracking position is reached, lock the inner knob. If the band machine has three idler wheels, adjust band tracking on the top wheel first and then adjust the tracking position on the back wheel.

SELF-TEST

1. Describe how blade ends should be ground to prepare for welding.
2. Describe the band welding procedure.
3. Describe the weld grinding procedure.
4. What is the purpose of the band blade guides?
5. Why is it important to use a band guide of the correct width?
6. What is the function of annealing the band weld?
7. Describe the annealing process.
8. What is band tracking?
9. How is band tracking adjusted?

Using a Vertical Band Saw

After a machine tool has been properly adjusted and set up, it is ready for machining. In the preceding unit you prepared the vertical band machine for use. In this unit you will learn to operate this versatile machine tool.

OBJECTIVES

After completing this unit, you should be able to:

- Use the vertical band machine job selector.
- Operate the band machine controls.
- Perform typical sawing operations on the vertical band machine.

SELECTING A BLADE FOR THE VERTICAL BAND MACHINE

Blade materials include carbon steel, high-speed steel, and bimetal blades. Band saw blades are available in a wide variety of widths, sets, pitches, and gauges.

The high-speed steel blade has hardened teeth and a hardened back. The harder back permits sufficient flexibility of the blade, but because of increased tensile strength, a higher band tension may be used, improving cutting accuracy.

High-speed steel and bimetallic high-speed steel blade materials are used in high-production and sawing applications where blades must have long-wearing characteristics. The high-speed steel blade can withstand much more heat than the carbon or carbon alloy materials. The cutting edge on a bimetallic blade is made from one type of high-speed steel, and the back is made from another type of high-speed steel that has been selected for high flexibility and high tensile strength. High-speed steel and bimetallic high-speed blades can cut longer, faster, and more accurately.

Band blade selection will depend on the sawing task. You should review saw blade terminologies discussed in Unit 1. The first consideration is blade pitch. The pitch of the blade should be such that at least two to three teeth are in contact with the workpiece. This generally means that fine-pitch blades with more teeth per inch will be used in thin materials. Thick material requires coarse-pitch blades so that chips will be more effectively cleared from the kerf.

Remember that machinists choose from three tooth sets (Figure G-47). Raker and wave set are the most common in the metalworking industries. Straight set may be used for cutting thin materials. Wave set is best for accurate cuts through materials with variable cross sections. Raker set may be used for general-purpose sawing.

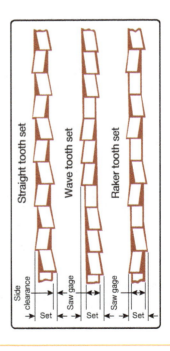

Figure G-47 Blade set patterns.

You also have a choice of **tooth forms** (Figure G-48). **Precision** or **regular** tooth form is best for accurate cuts where a good finish may be required. **Hook** form is fast cutting but leaves a rougher finish. **Skip** tooth is useful on deep cuts where additional chip clearance is required.

Several special bands are also used. **Straight, scalloped,** and **wavy edges** are used for cutting nonmetallic substances that regular saw teeth would tear (Figure G-49). **Continuous** (Figure G-50) and **segmented** (Figure G-51) **diamond-edged bands** are used for cutting hard nonmetallic materials.

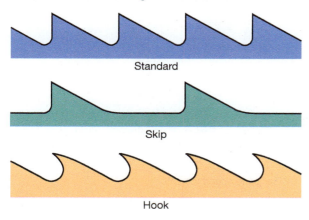

Figure G-48 Saw tooth forms.

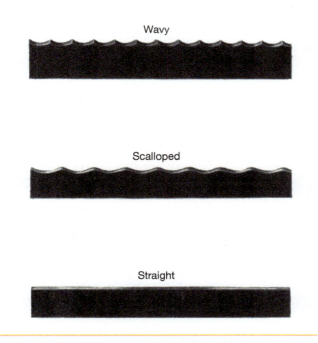

Figure G-49 Straight, scalloped, and wavy-edge bands (*Courtesy of DoALL Company*).

Figure G-50 Continuous edge diamond band (*Courtesy of DoALL Company*).

Figure G-51 Segmented diamond edge band (*Courtesy of DoALL Company*).

USING THE JOB SELECTOR ON THE VERTICAL BAND MACHINE

Many vertical band machines are equipped with a job selector. Job selectors are usually attached to the machine tool and are frequently arranged by material. On this selector the material to be cut is located on the rim. The selector disk is then turned until the sawing data for the material can be read (Figure G-52).

The job selector provides valuable information. Sawing velocity in FPM is the most important. The band must be operated at the correct cutting speed for the material. If it is not, the band may be damaged or productivity will be low. Saw velocity is read at the top of the column and is dependent on the material thickness. The job selector also indicates recommended pitch, set, feed, and temper. The job selector will also provide information on sawing of nonmetallic materials (Figure G-53).

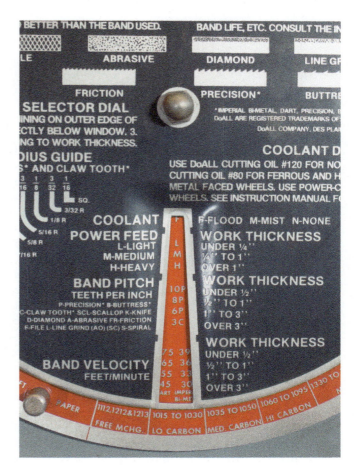

Figure G-52 Job selector on the vertical band machine.

Figure G-53 The job selector set for a nonferrous material.

SETTING SAW VELOCITY ON THE VERTICAL BAND MACHINE

Most vertical band machines are equipped with a variable-speed drive that permits a wide selection of band velocities. This is one of the factors that makes the band machine such a versatile machine tool. Saw velocities can be selected that enable efficient cutting of most materials.

The typical variable-speed drive uses a split-flange pulley to vary the speed of the drive wheel. As the flanges of the pulley are spread apart by adjusting the speed control, the belt runs deeper in the pulley groove. This is the same as running the drive belt on a smaller-diameter pulley. As the flanges of the pulley are adjusted for less spread, the belt runs toward the outside. This is equivalent to running the belt on a larger-diameter pulley: faster speeds are obtained.

Setting Band Velocity

Band velocity is indicated on the **band velocity indicator** (Figure G-54). Remember that band velocity is measured in SFPM. The inner scale indicates band velocity in the low-speed range. The outer scale indicates velocity in the high-speed range. Band velocity is regulated by adjusting the speed control. Adjust this control only while the motor is running, as this adjustment moves the flanges of the variable-speed pulley.

Figure G-54 Band speed indicator.

Setting Speed Ranges

Some vertical saws have both a high- and low-speed range, selected by operating the speed range shift lever (Figure G-55). This setting must be made while the band is stopped or is running at the lowest speed in the range. If the machine is set in high range and you want to go to low range, turn the band velocity control wheel until the band has slowed to the lowest speed possible. The speed range shift may now be changed to low

Figure G-55 Speed range and band speed controls.

speed. If the machine is in low speed and you want to shift to high range, slow the band to the lowest speed before shifting speed ranges. A speed range shift made while the band is running at a fast speed may damage the speed range transmission gears. Some band machines are equipped with an interlock to prevent speed range shifts except at low band velocity.

STRAIGHT CUTTING ON THE VERTICAL BAND SAW

Adjust the upper guidepost so that it is as close to the workpiece as possible (Figure G-56). This will maximize safety by properly supporting and guarding the band. It will also help make the cut more accurate. Adjust the guidepost by loosening the clamping knob and moving the post up or down according to the workpiece thickness.

Be sure to use a band of the proper pitch for the thickness of the material to be cut. If the band pitch is too fine, the teeth will clog (Figure G-57). This can strip and break the saw teeth by overloading them (Figure G-58). Cutting productivity will also be reduced. Using a fine-pitch band on thick material will result in slow cutting. The correct pitch for thick material will assure more efficient cutting.

As you begin a cut, feed the workpiece gently into the band. A sudden shock will cause the saw teeth to chip or fracture (Figure G-59), quickly reducing band life. See that chips are cleared from the band guides. These can score the band (Figure G-60), making it brittle and subject to breakage.

Figure G-57 Using too fine a pitch blade results in clogged teeth.

Figure G-56 Adjusting the upper guidepost (*Courtesy of Fox Valley Technical College*).

Figure G-58 Stripped and broken teeth resulting from overloading the saw.

Figure G-59 Chipped and fractured teeth resulting from shock and vibration.

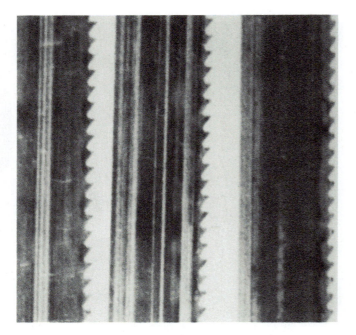

Figure G-60 Scored bands can become brittle and lose flexibility (*Courtesy of DoALL Company*).

Cutting Fluids

Cutting fluids are an important aid to sawing many materials. They cool and **lubricate** the band and **remove chips** from the kerf. Some band machines are equipped with a mist coolant system. Liquid cutting fluids are mixed with air to form a mist. If your band machine uses mist coolant, set the liquid flow first and then add air to create a mist. Do not use more coolant than is necessary. Overuse of air may cause a mist fog around the machine.

CONTOUR CUTTING ON THE VERTICAL BAND MACHINE

Contour cutting is cutting around corners to produce intricate shapes. The ability of the saw to cut a specific radius depends on the width of the blade. Charts are available that provide information on the minimum radius that can be cut

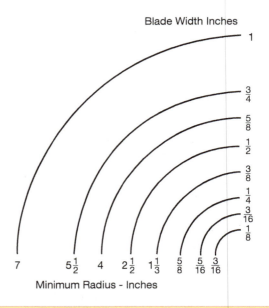

Blade Width Inches

Minimum Radius - Inches

Figure G-61 Minimum radius per saw width chart on the job selector (*Courtesy of Olson Saw Company*).

with a blade of a given width (Figure G-61). As you can see, a narrow band can cut a smaller radius than a wide band.

Whenever possible, use the widest band available for straight cuts. A narrow band will tend to wander, making straight cuts more difficult. For small-radius cutting, a narrow band will be required to follow the required radius. The band must have a full tooth set on each side for maximum efficiency in both cutting and following the desired radius curvature.

The set of the saw needs to be adequate for the corresponding band width. It is a good idea to make a test contour cut in a piece of scrap material. This will permit you to determine whether the saw set is adequate for cutting the desired radius. If the saw set is not adequate, you may not be able to keep the saw on the layout line as you complete the radius cut (Figure G-62).

Figure G-62 Band set must be adequate to cut the desired radius.

SELF-TEST

1. Name three saw blade sets and describe the applications of each.
2. When might scalloped or wavy-edged bands be used?
3. In what units of measure are band velocities measured?
4. Describe the upper guidepost adjustment.
5. What does band pitch have to do with sawing efficiency?

INTERNET REFERENCE

Information on band saw welders and band welding:

http://www.simonds.cc

SECTION H
Drilling Machines

INTRODUCTION

A machinist must be able to use various types of drilling machines. Safety should always be the most important consideration. Chips are produced in great quantities and must be safely handled. The operator must be protected from these chips as they fly from the machine. The rotating spindle and drill are also safety hazards. You should become familiar with the major parts of the drilling machine before you operate one. Work-holding devices must be used to secure the work to the drill press table to keep the operator safe.

In the following units, you will learn about tooling and how to operate a drill press.

DRILLING MACHINE SAFETY

People tend to dismiss the dangers involved in drilling, because most drilling is performed on relatively small drill presses with small-diameter drills. Dismissing the dangers of drill presses increases the potential danger to the operator. One example of a safety hazard is the drill "grabbing" when it breaks through the hole. If you are holding the workpiece with your hands, and the piece suddenly begins to spin, your hand will probably be injured, especially if the workpiece is thin. The sharp chips that turn with the drill could also cut the hand that holds the workpiece. Clamps should be used to hold down workpieces and vises, because hazards are always present even with small-diameter drills. Always clamp your work or the vise to the table of the drill press.

Figure H-1 Drilling operation on upright drill press with tools properly located on adjacent worktable (*Confederation College*).

SAFETY FIRST

Poor work habits produce many injuries. Chips that fly into unprotected eyes, dropped tools or parts falling from the drill press onto toes, slips on oily floors, and hair or clothing caught in a rotating drill are all hazards avoidable by safe work habits.

Safety Rules for All Types of Drill Presses

1. Never leave tools on the drill press table while drilling. Put them on an adjacent worktable (Figure H-1).
2. Get help when lifting heavy vises or workpieces.
3. Always secure workpieces with strap clamps, C-clamps, or fixtures. Use a vise when drilling small parts (Figure H-2). Place the clamp so that if the part or vise comes loose it will hit the clamp first. If a clamp should come loose and the part or vise begins to spin, do not try to stop them with your hands. Turn off the machine quickly; if the drill breaks or comes out of the chuck, the workpiece may fly off the table.

Figure H-2 A properly clamped drill press vise. Note that the vise would hit the clamp first if it became loose (*Courtesy of Fox Valley Technical College*).

4. Never clean the taper in the spindle when the spindle is rotating, because this can result in broken fingers or worse injuries.

5. Always remove the chuck key immediately after using it. A key left in the chuck will be thrown out at high velocity when the machine is turned on. It is a good practice to never let the chuck key leave your hand when you are using it. Do not leave it in the chuck even for a moment. Some keys are spring-loaded so that they will automatically be ejected from the chuck when released.

6. Never stop the drill press spindle with your hand after you have turned the machine off. Sharp chips often collect around the chuck or spindle. Do not reach around, near, or behind a revolving drill.

7. When removing taper shank drills with a drill drift, use a piece of wood under the drills so they will not drop on the table or your toes.

8. Interrupt the feed occasionally when drilling to break the chip so that it will not be a hazard.

9. Use a brush instead of your hands to clean chips off the machine. Never use an air jet for removing chips, as this will cause the chips to fly at a high velocity, and cuts or eye injuries may result. Do not clean up chips or wipe up oil while the machine is running.

10. Keep the floor clean. Immediately wipe up any oil that spills and sweep up chips, or the floor will be slippery and unsafe.

11. Remove burrs from a drilled workpiece as soon as possible, because any sharp edges or burrs can cause severe cuts.

12. When you are finished with a drill or other cutting tool, wipe it clean with a shop towel and store it properly.

13. Place oily shop towels in a closed metal container so they will not be a fire hazard.

Drill Press Fundamentals

Before operating any machine, a machinist must know the names and functions of its parts so that the machine can be operated safely. Familiarize yourself with the operating mechanisms of the drilling machines in this unit.

OBJECTIVES

After completing this unit, you should be able to:

- Identify two basic drill press types and explain their differences and primary uses.
- Identify the major parts of the radial arm drill press.

Drilling holes is one of the most basic machining operations. Metal cutting generates considerable pressure on the cutting edge. A drill press provides the necessary feed pressure either by hand or power feed. The primary use of the drill press is to drill holes, but it can also be used for other operations such as countersinking, counterboring, spotfacing, reaming, and tapping.

There are two major types of drilling machines: the upright drill press and the radial arm drill press. The **upright drill press** is used for small to medium size parts (Figure H-3). Drill press capacity is measured by the diameter of the work that can be drilled (Figure H-4).

A drill press has four major parts, not including the motor: the **head**, **column**, **table**, and **base**. Figure H-5 labels the man parts of the drill press. The spindle rotates within the quill, which does not rotate but carries the spindle up and down. The spindle speeds on drill presses can be adjusted by a V-pulleys and belt (Figure H-6), by a belt-driven variable-speed drive, or more commonly now, an electronic variable speed drive (Figure H-7).

The spindle drive on many upright drill presses are gear driven, making them very powerful. The motor must be stopped when changing speeds on a gear-driven drill press.

Figure H-3 Upright drill press (*Courtesy of Fox Valley Technical College*).

If drill press won't shift into the selected gear, turn the spindle by hand until the gears mesh. Some upright drill presses have power feeds that can be adjusted by the operator. The operator may either feed manually with a lever or hand wheel or engage the power feed.

Figure H-8 shows a radial arm drill press. The radial arm drill press is the most versatile type of drilling machine. Its size is determined by the diameter of the column and the

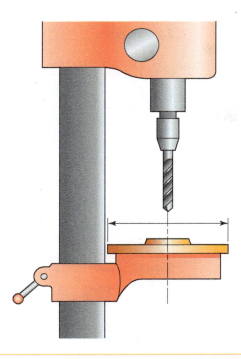

Figure H-4 Drill presses are measured by the largest diameter of a circular piece that can be drilled.

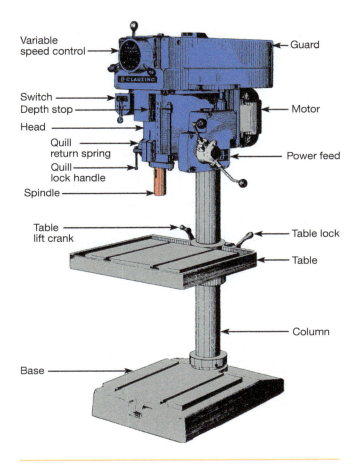

Variable speed control

Switch

Depth stop

Head

Quill return spring

Quill lock handle

Spindle

Table lift crank

Base

Guard

Motor

Power feed

Table lock

Table

Column

Figure H-5 Drill press showing the names of major parts (*Used with permission of Clausing Industrial, Inc.*).

Figure H-6 View of a V-belt drive (*Used with permission of Clausing Industrial, Inc.*).

Figure H-7 Variable-speed control (*Courtesy of Fox Valley Technical College*).

length of the arm measured from the center of the spindle to the outer edge of the column. It is useful for operations on large castings that are too heavy to be repositioned by the operator for drilling each hole. The work is clamped to the table or base, and the drill can then be positioned where it is needed by swinging the arm and moving the head along the arm. The arm and head can be raised or lowered on the column and then locked in place. The radial arm drill press is used for drilling small to very large holes and for reaming, counterboring, and countersinking. Radial arm drill presses have power and manual feeds.

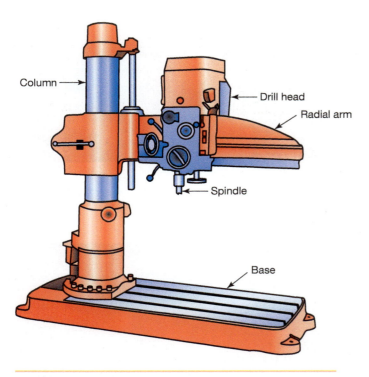

Figure H-8 Radial drill press.

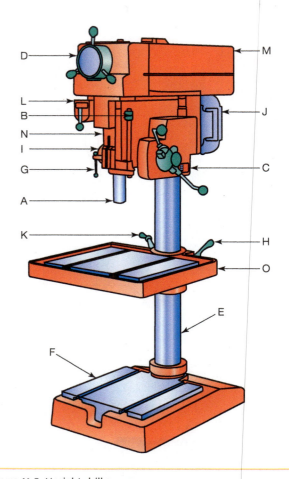

Figure H-9 Upright drill press.

SELF-TEST

1. List the basic types of drill presses and briefly explain their differences. Describe how the primary uses differ in each of these three drill press types.

2. *Sensitive drill press.* Match the correct letter from Figure H-9 with the name of that part.

Spindle	Base
Quill lock handle	Power feed
Column	Motor switch
Depth stop	Variable-speed control
Head	Table lift crank
Table	Quill return spring
Table lock	Guard

3. *Radial drill press.* Match the correct letter from Figure H-10 with the name of that part.

Column	Base
Radial arm	Drill head
Spindle	

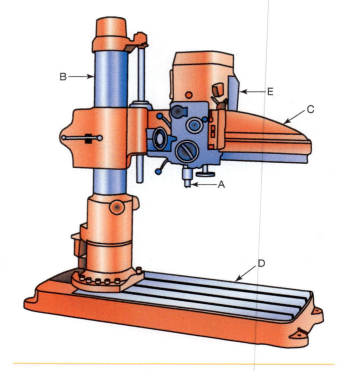

Figure H-10 Radial arm drill press.

Drilling Tools

There are great variety of drills and tooling available to the machinist. This unit will acquaint you with these tools how to select one for a given operation.

OBJECTIVES

After completing this unit, you should be able to:

- Identify the various features of a twist drill.
- Given 10 decimal drill sizes, identify the series and size of each.

DRILL BASICS

The drill is an end-cutting tool having one or more cutting lips and one or more flutes for the removal of chips and the passage of coolant. Drilling is the most efficient method of making a hole in metals softer than Rockwell 30. Harder metals can be successfully drilled, however, by using special drills and techniques. In the past, all drills were made of carbon steel and would lose their hardness if they became too hot from drilling. Today, however, most drills are made of high-speed steel or carbide. High-speed steel drills can operate at up to 1100°F (593°C) without breaking down. Cobalt steel alloys that contain more cobalt are variations of high-speed steel. Their main advantage is that they hold their hardness at much higher temperatures. A coating such as titanium nitride (TiN) is a very hard ceramic material that when used to coat a high-speed steel bit (usually a twist bit) can extend the cutting life by three or more times. Titanium aluminum nitride (TiAlN) is another coating that is used frequently. Titanium carbon nitride (TiCN) coating is superior to TiN. Solid carbide or carbide-tipped drills are used for special applications such as drilling abrasive materials and hard steels.

TWIST DRILLS

The twist drill is by far the most common type of drill. These drills are made with two or more flutes and cutting lips and in many varieties. Figure H-11 illustrates several of the most commonly used types of twist drills. The names of parts and features of a twist drill are shown in Figure H-12.

Figure H-11 Various types of twist drills used in drilling machines: (*a*) high-helix drill; (*b*) low-helix drill; (*c*) left-hand drill; (*d*) three-flute drill; (*e*) taper shank twist drill; (*f*) standard helix jobber drill; (*g*) center or spotting drill (*Courtesy of DoALL Company*).

Figure H-12 Features of a twist drill (*Used with Permission of Besly Cutting Tools, Inc.*).

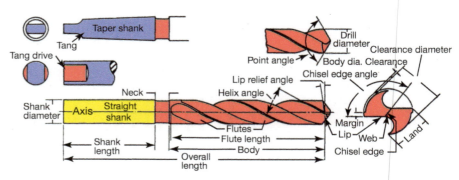

The twist drill has either a straight or tapered shank. The tapered shank drill has a **Morse taper**, a standard taper of about $\frac{5}{8}$ inch per foot, which has more driving power and greater rigidity than the straight shank drills. Straight shank drills are typically held in drill chucks (Figure H-13). A drill chuck is a friction drive, and the drill shank slipping in the chuck is a common problem. Tapered shank drills are held in special drill chucks with a Morse taper (Figure H-14). **Jobber's drills** have two flutes, a straight shank design, and a relatively short length-to-diameter ratio that helps maintain rigidity. These drills are used for drilling steel, cast iron, and nonferrous metals. **Center drills** and **spotting drills** are used for accurately starting holes in workpieces. **Oil-hole drills** are made so that coolant can be pumped through the drill to the cutting lips. This not only cools the cutting edges but also forces out the chips along the flutes. **Core drills** have from three to six flutes, making heavy stock removal possible. They are generally used for roughing holes to a larger diameter or for drilling out cores in castings. **Step drills** have a flat or angular cutting edge and can produce a hole with several diameters in one pass with either flat or countersunk shoulders.

Morse taper shanks on drills and Morse tapers in drill press spindles come in multiple sizes and are numbered from 1 to 6. For example, a smaller light-duty drill press typically has a No. 2 taper. Steel sleeves (Figure H-14) have a Morse taper inside and outside with a slot provided at the end of the inside taper to facilitate removal of the drill shank. A sleeve is used for enlarging the taper end on a drill to fit a larger spindle taper. In contrast, steel sockets (Figure H-15) adapt a smaller spindle taper to a larger drill. The tool used to remove a tapered shank drill is called a **drift** (Figure H-16), which is made in several sizes and is used to remove drills or sleeves. The drift is placed round side up and flat side against the drill and is struck a light blow with a hammer. A block of wood should be placed under the drill to keep it from being damaged and from being a safety hazard.

High- and Low-Helix Drills

High-helix drills, sometimes called *fast spiral drills*, are designed to remove chips from deep holes. The large rake angle (Figure H-17) makes these drills suitable for soft metals such

Figure H-14 Morse taper drill sleeve.

Figure H-15 Morse taper drill socket (*Used with permission of DoALL Company*).

Figure H-13 Drill chucks are used to hold straight shank drills.

Figure H-16 A drift being used to remove a sleeve from a drill (*Courtesy of Fox Valley Technical College*).

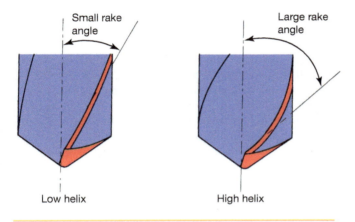

Figure H-17 Low and high rake angles on drills.

Figure H-18 Spotting drill (*Used with permission of DoALL Company*).

Size	Diameter A	Length L	Diameter D	Length C
00	1/8	1 1/8	0.025	0.030
0	1/8	1 1/8	1/32	0.038
1	1/8	1 1/4	3/64	3/64
2	3/16	1 7/8	5/64	5/64
3	1/4	2	7/64	7/64
4	5/16	2 1/8	1/8	1/8
5	7/16	2 3/4	3/16	3/16
6	½	3	7/32	7/32
7	5/8	3 1/4	1/4	1/4
8	3/4	3 1/2	5/16	5/16

Figure H-19 Center drill and size chart.

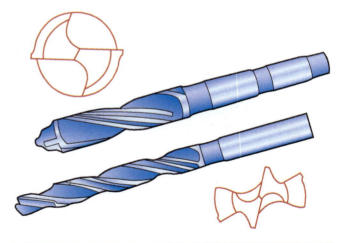

Figure H-20 Step and subland drills.

as aluminum and mild steel. Straight-fluted drills are used for drilling brass and other soft materials because the zero-rake angle eliminates the tendency for the drill to "grab" when breaking through the material. Low-helix drills, sometimes called *slow spiral drills*, are more rigid than standard helix drills and can stand more torque. Like straight-fluted drills, they are less likely to grab when emerging from a hole because of the small rake angle. Low-helix and straight-fluted drills are used primarily for drilling brass, bronze, and other nonferrous metals. Because of the low-helix angle, the flutes do not remove chips well from deep holes, but the large chip space allows maximum drilling efficiency in shallow holes.

Spotting Drills and Center Drills

Spotting drills (Figure H-18) are used to position holes accurately for further drilling with regular drills. Spotting drills are short and have little or no dead center. In spot drilling, only a spot is necessary to prevent the drill from wandering from the initial point of contact with the workpiece. Center drills (Figure H-19) are often used as spotting drills. Center drills come in several standard sizes. They are also available in high-speed steel or carbide. Center drills are short and stubby and so do not have a tendency to wander off location as they drill.

Step Drills

Step and subland drills (Figure H-20) are used when making a hole with two or more different diameters in one drilling operation. Often, one diameter makes a tap-drill hole, and the larger diameter makes a countersink or counterbore for a bolt or screw head.

Spade and Gun Drills

Special drills such as spade and gun drills are used in many manufacturing processes. A **spade drill** is simply a flat blade with sharpened cutting lips. The spade drill blade, which is clamped in the spade drill holder (Figure H-21), is replaceable. Some types provide for the coolant flow to the cutting edge through a hole in the holder or shank for the purpose of deep drilling. These drills are made with large diameters of 12 inches or more but can also be found as microdrills, for drilling holes smaller than a hair. By comparison, twist drills are rarely found with diameters over $3\frac{1}{2}$ inches.

Spade drills are usually ground with a flat top rake and with chip-breaker grooves on the end. A chisel edge and thinned web are ground in the dead center (Figure H-22). Some twist drills have carbide inserts for the cutting edges. These carbide-insert drills require a rigid drilling setup. **Gun drills** (Figure H-23) are also carbide tipped and have a single V-shaped flute in a steel tube through which the coolant is pumped under pressure. These drills are used in horizontal machines that feed the drill with a positive guide. Extremely deep precision holes are produced with gun drills, with a .0005-inch or better straightness if accurate setup and proper operating procedures are followed.

Carbide Drills

Figure H-24 shows three types of carbide drills. The top two are solid carbide. The bottom drill is a carbide insert drill. Carbide drills can be used to drill extremely hard material. They can also be run at much higher cutting speed and dramatically increase productivity.

DRILL SELECTION

The correct type of drill for a particular task depends on several factors. The type of machine being used, the rigidity of the workpiece, the setup, and the size of the hole to be drilled are all important factors. The composition and hardness of the workpiece are especially critical. The job may require a starting drill or one for secondary operations such as counterboring or spotfacing, and it may need to be a drill for a deep hole or a shallow one. If the drilling operation is too large for the size or rigidity of the machine, there will be chatter and the work surface will be rough or distorted. A machinist also must select the size of drill to be used. Twist drills are measured across the margins near the drill's point (Figure H-25). Worn drills measure slightly smaller here. Drills are normally tapered back along the margin so that they will measure a few thousandths of an inch smaller at the shank; this reduces friction and heat buildup.

There are four drill size series: fractional, number, letter, and metric. The fractional divisions are in $\frac{1}{64}$-inch increments, and the number, letter, and metric series have drill diameters that fall between the fractional inch measures. Together, these four series make up a series in decimal equivalents, as shown in Table H-1. Identification of a small drill is simple enough as long as the number or letter remains on the shank. Most shops, however, have several series of drills, and individual drills often become hard to identify, because the markings become worn off by the drill chuck. The machinist must then use a decimal equivalent table such as Table H-1. The drill in question is first measured with a micrometer; the decimal reading is then located in the table; and the equivalent fraction, number, letter, or metric size is found and noted.

Figure H-21 Spade drill clamped in holder (*Courtesy of DoALL Company*).

TiN (Titanium Nitride) coated
C2 carbide for high alloy steels,
stainless steel, etc.

TiAlN (Titanium Nitride) coated
C5 carbide for carbon steels, alloy steels,
and tool steel

Figure H-22 Spade drill blades.

Figure H-23 Single-flute gun drill with insert of carbide cutting tip (*Used with permission of DoALL Company*).

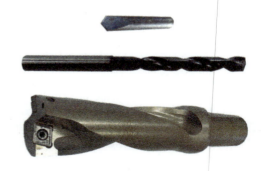

Figure H-24 Carbide drills.

Figure H-25 Drill being measured across the margins.

Table H-1 Decimal Equivalents for Drills

Decimal Size	Inch	Number	Millimeter	Decimal Size	Inch	Number	Millimeter	Decimal Size	Inch	Number	Millimeter
.0135		80		.0630			1.6	.1299			3.3
.0145		79		.0635		52		.1339			3.4
.0156	$\frac{1}{64}$			.0650			1.65	.1360		29	
.0157			.4	.0669			1.7	.1378			3.5
.0160		78		.0670		51		.1405		28	
.0180		77		.0689			1.75	.1406			
.0197			.5	.0700		50		.1417			3.6
.0200		76		.0709			1.8	.1440		27	
.0210		75		.0728			1.85	.1457			3.7
.0217			.55	.0730		49		.1470		26	
.0225		74		.0748			1.9	.1476			3.75
.0236			.6	.0760		48		.1495		25	
.0240		73		.0768			1.95	.1496			3.8
.0250		72		.0781	$\frac{5}{64}$			.1520		24	
.0256			.65	.0785		47		.1535			3.9
.0260		71		.0787			2	.1540		23	
.0276			.7	.0807			2.05	.1563	$\frac{5}{32}$		
.0280		70		.0810		46		.1570		22	
.0293		69		.0820		45		.1575			4
.0295			.75	.0827			2.1	.1590		21	
.0310		68		.0846			2.15	.1610		20	
.0313	$\frac{1}{32}$			.0860		45		.1614			4.1
.0315			.8	.0866			2.2	.1654			4.2
.0320		67		.0886			2.25	.1660		19	
.0330		66		.0890		43		.1673			4.25
.0335			.85	.0906			2.3	.1693			4.3
.0350		65		.0925			2.35	.1695		18	
.0354			.9	.0935		42		.1719	$\frac{11}{64}$		
.0360		64		.0938	$\frac{3}{32}$			.1730		17	
.0370		63		.0945			2.4	.1732			4.4
.0374			.95	.0960		41		.1770		16	
.0380		62		.0966			2.45	.1772			4.5
.0390		61		.0980		40		.1800		15	
.0394			1	.0984			2.5	.1811			4.6
.0400		60		.0995		39		.1820		14	
.0410		59		.1015		38		.1850		13	
.0413			1.05	.1024			2.6	.1850			4.7
.0420		58		.1040		37		.1870			4.75
.0430		57		.1063			2.7	.1875	$\frac{3}{16}$		
.0433			1.1	.1065		36		.1890			4.8
.0453			1.15	.1083			2.75	.1890		12	
.0465		56		.1094	$\frac{7}{64}$			.1910		11	
.0469	$\frac{3}{64}$			.1100		35		.1929			4.9
.0472			1.2	.1102			2.8	.1935		10	
.0492			1.25	.1110		34		.1960		9	
.0512			1.3	.1130		33		.1969			5
.0520		55		.1142			2.9	.1990		8	
.0531			1.35	.1160		32		.2008			5.1
.0550		54		.1181			3	.2010		7	
.0551			1.4	.1200		31		.2031	$\frac{13}{64}$		
.0571			1.45	.1220			3.1	.2040		6	
.0591			1.5	.1250	$\frac{1}{8}$			.2047			5.2
.0595		53		.1260			3.2	.2055		5	
.0610			1.55	.1280			3.25	.2067			5.25
.0625	$\frac{1}{16}$			.1285		30		.2087			5.3

Table H-1 Decimal Equivalents for Drills (*Continued*)

Decimal Size	Inch	Number Letter	Millimeter	Decimal Size	Inch	Letter	Millimeter	Decimal Size	Inch	Millimeter
.2090		4		.3150			8	.5156	$\frac{33}{64}$	
.2126			5.4	.3160		O		.5313	$\frac{17}{32}$	
.2130		3		.3189			8.1	.5315		13.5
.2165			5.5	.3228			8.2	.5469	$\frac{35}{64}$	
.2188	$\frac{7}{32}$			.3230		P		.5512		14
.2205			5.6	.3248			8.25	.5625	$\frac{9}{16}$	
.2210		2		.3268			8.3	.5709		14.5
.2244			5.7	.3281	$\frac{21}{64}$			.5781	$\frac{37}{64}$	
.2264			5.75	.3307			8.4	.5906		15
.2280		1		.3320		Q		.5938	$\frac{19}{32}$	
.2283			5.8	.3346			8.5	.6094	$\frac{39}{64}$	
.2323			5.9	.3386			8.6	.6102		15.5
.2340		A		.3390		R		.6250	$\frac{5}{8}$	
.2344	$\frac{15}{64}$			.3425			8.7	.6299		16
.2362			6	.3438	$\frac{11}{32}$			.6406	$\frac{41}{64}$	
.2380		B		.3445			8.75	.6496		16.5
.2402			6.1	.3465			8.8	.6563	$\frac{21}{32}$	
.2420		C		.3480		S		.6693		17
.2441			6.2	.3504			8.9	.6719	$\frac{43}{64}$	
.2460		D		.3543			9	.6875	$\frac{11}{16}$	
.2461			6.25	.3580		T		.6890		17.5
.2480			6.3	.3583			9.1	.7031	$\frac{45}{64}$	
.2500	$\frac{1}{4}$	E		.3594	$\frac{23}{64}$			.7087		18
.2520			6.4	.3622			9.2	.7188	$\frac{23}{32}$	
.2559			6.5	.3642			9.25	.7283		18.5
.2570		F		.3661			9.3	.7344	$\frac{47}{64}$	
.2598			6.6	.3680		U		.7480		19
.2610		G		.3701			9.4	.7500	$\frac{3}{4}$	
.2638			6.7	.3740			9.5	.7656	$\frac{49}{64}$	
.2656	$\frac{17}{64}$			.3750	$\frac{3}{8}$			.7677		19.5
.2657			6.75	.3770		V		.7812	$\frac{25}{32}$	
.2660		H		.3780			9.6	.7874		20
.2677			6.8	.3819			9.7	.7969	$\frac{51}{64}$	
.2717			6.9	.3839			9.75	.8071		20.5
.2720		I		.3858			9.8	.8125	$\frac{13}{16}$	
.2756			7	.3860		W		.8268		21
.2770		J		.3898			9.9	.8281	$\frac{53}{64}$	
.2795			7.1	.3906	$\frac{25}{64}$			.8438	$\frac{27}{32}$	
.2810		K		.3937			10	.8465		21.5
.2812	$\frac{9}{32}$			.3970		X		.8594	$\frac{55}{64}$	
.2835			7.2	.4040		Y		.8661		22
.2854			7.25	.4063	$\frac{13}{32}$			.8750	$\frac{7}{8}$	
.2874			7.3	.4130		Z		.8858		22.5
.2900		L		.4134			10.5	.8906	$\frac{57}{64}$	
.2913			7.4	.4219	$\frac{27}{64}$			.9055		23
.2950		M		.4331			11	.9063	$\frac{29}{32}$	
.2953			7.5	.4375	$\frac{7}{16}$			.9219	$\frac{59}{64}$	
.2969	$\frac{19}{64}$			.4528			11.53	.9252		23.5
.2992			7.6	.4531	$\frac{29}{64}$			.9375	$\frac{15}{16}$	
.3020		N		.4688	$\frac{15}{32}$			.9449		24
.3031			7.7	.4724			12	.9531	$\frac{61}{64}$	
.3051			7.75	.4844	$\frac{31}{64}$			.9646		24.5
.3071			7.8	.4921			12.5	.9688	$\frac{31}{32}$	
.3110			7.9	.5000	$\frac{1}{2}$			.9843		25
.3125	$\frac{5}{16}$			.5118			13	.9844	$\frac{63}{64}$	

Figure H-26 Split-point design of a drill point (*Used with permission of Besly Cutting Tools, Inc.*).

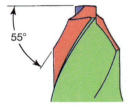

55°

Figure H-27 Drill parts (*Used with permission of Besly Cutting Tools, Inc.*).

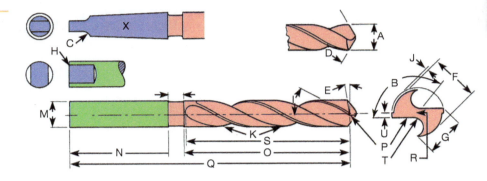

Accuracy of Drilled Holes

When twist drills have been sharpened by hand on a pedestal grinder, holes drilled by them are often rough and oversized. This is because it is very difficult to make precision angles and dimensions on a drill point by offhand grinding. However, rough drilled holes are often sufficient for such purposes as bolt holes and preparation for reaming and for tap drills. Drills almost never make a hole smaller than their measured diameter, but it is possible if the drill and workpiece become hot during the drilling operation. When the work cools, the hole becomes slightly smaller. A drilled hole may be 5 to 10 percent oversize if the drill has been hand-sharpened.

Drill Pointing Machines

A more precise method of sharpening drills is by using a drill grinding machine. Drill point grinding machines are available in several varieties and levels of complexity, ranging from those used for microdrilling to the automatic machines used by drill manufacturers. Some machines use a pivot so that the operator can shape the drill point to the proper geometry, and others use cams to produce the drill point geometry. One kind uses the side of a grinding wheel, whereas another uses the circumference of the wheel. All of these point grinding machines have one thing in common: precision. Precision is the major advantage of the drill sharpening machine. Additional advantages include: drill failures are reduced, drills stay sharp longer because both flutes cut evenly, and the hole size is more accurate. Split-point design (Figure H-26) is often used for drilling tough alloy steels. The shape of the point is critical and too difficult to achieve by hand grinding. Large shops and factories that have toolrooms are more likely to use drill grinding machines than are small job shops and maintenance machine shops. Drills used for CNC machines should be machine sharpened. Step and subland drills used in production

work should be machine-sharpened. Split-point or four-facet drill points used for deep-hole drilling as well as three- and four-flute core drills should be ground on a machine.

SELF-TEST

1. Match the correct letter from Figure H-27 to the list of drill parts.

Web	Body
Margin	Lip relief angle
Drill point angle	Land
Cutting lip	Chisel edge angle
Flute	Body clearance
Helix angle	Tang
Axis of drill	Tapered shank
Shank length	Straight shank

2. Determine the letter, number, fractional, or metric equivalent of each of the 10 following decimal measurements of drills:

Decimal Diameter	Fractional Size	Number Size	Letter Size	Metric Size
a. .0781				
b. .1495				
c. .272				
d. .159				
e. .1969				
f. .323				
g. .3125				
h. .4375				
i. .201				
j. .1875				

Hand Grinding of Drills on the Pedestal Grinder

Drill sharpening machines are not always available, so it is advantageous for a machinist to learn the art of offhand drill grinding.

OBJECTIVE

After completing this unit, you should be able to:

- Properly hand-sharpen a twist drill on a pedestal grinder so that it will drill a hole not more than .005 to .010 inch oversize.

BASICS OF HAND GRINDING DRILLS

One of the advantages of hand grinding drills is that special alterations of the drill point such as web thinning and rake modification can be done quickly. The greatest disadvantage of this method of drill sharpening is the possibility of producing inaccurate, oversize holes (Figure H-28). If the drill has been sharpened with unequal angles, the lip with the large angle will do most of the cutting (Figure H-28a) and will force the opposite margin to cut into the wall of the hole. If the drill has been sharpened with unequal lip lengths, the drill will wobble, and one margin will cut into the hole wall (Figure H-28b). When both conditions exist (Figure H-28c), holes drilled may be out of round and oversize. Drilling with inaccurate points places a great strain on the drill and the drill press spindle bearings. The frequent use of a drill point gage (Figure H-29) during the sharpening process will help keep the point accurate and prevent such drilling problems. The web of a twist drill (Figure H-30) is thicker near the shank. As the drill is ground shorter, a thicker web results near the point. Also, the dead center or chisel point of the drill is wider and requires greater pressure to force it into the workpiece, thus generating heat. Web thinning (Figure H-31) is one method of narrowing the dead center to restore the drill to its original efficiency.

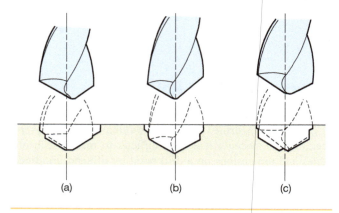

Figure H-28 Causes of oversize drilling: (*a*) drill lips ground to unequal lengths; (*b*) drill lips ground to unequal angles; (*c*) unequal angles and lengths.

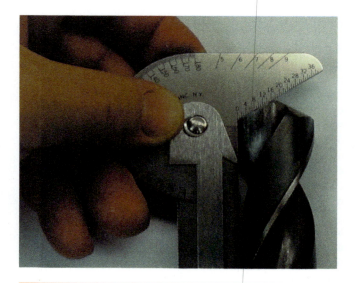

Figure H-29 Using a drill point gage.

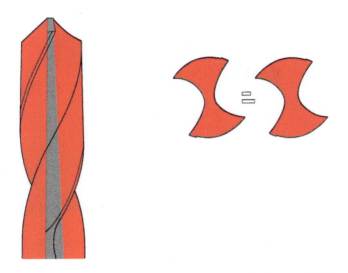

Figure H-30 Tapered web of twist drills (*Used with permission of Besly Cutting Tools, Inc.*).

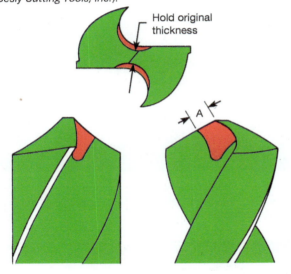

Figure H-31 The usual method of thinning the point on a drill. The web should not be made thinner than it was originally when the drill was new and full length (*Used with permission of Besly Cutting Tools, Inc.*).

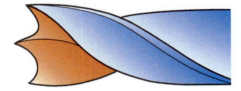

Figure H-32 Sheet metal drill point.

A sheet metal drill point (Figure H-32) may be ground by an experienced machinist. The rake angle on a drill can be modified for drilling brass as shown in Figure H-33. The standard drill point angle is a 118-degree included angle. Drill point angles should be from 135 to 150 degrees for drilling hard materials. A drill point angle from 60 to 90 degrees should be used when drilling soft materials, cast iron, abrasive

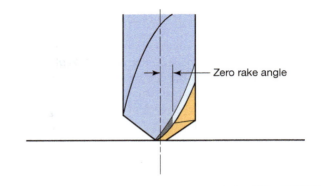

Figure H-33 Modification of the rake angle for drilling brass.

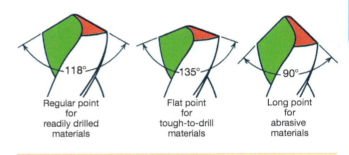

Figure H-34 Drill point angles (*Used with permission of Besly Cutting Tools, Inc.*).

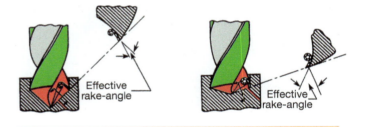

Figure H-35 Effective rake angles (*Used with permission of Besly Cutting Tools, Inc.*).

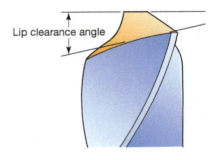

Figure H-36 Clearance angles on a drill point.

materials, plastics, and some nonferrous metals (Figure H-34). Too great a decrease in the included point angle is not advisable, however, because it will result in an abnormal decrease in the effective rake angle (Figure H-35). This will increase the required feed pressure and change the chip formation and chip flow in most steels. Clearance angles (Figure H-36) should be 8- to 12-degrees for most drilling applications.

DRILL GRINDING PROCEDURE

Check to make sure the roughing and finishing wheels are true (Figure H-37). If not, true the wheel(s) with a wheel dresser (Figure H-38). If the end of the drill is badly damaged, use the coarse wheel to remove the damaged part. If you overheat the drill, let it cool in air; do not cool high-speed steel drills in water.

The following method of grinding a drill is suggested:

Step 1 Hold the drill shank with one hand and the drill near the point with the other hand. Rest two fingers on the grinder tool rest. Hold the drill lightly at this point so that you can manipulate it from the shank end with the other hand (Figure H-39).

Step 2 Hold the drill approximately horizontal to the cutting lip (Figure H-40) level. The axis of the drill should be at 59 degrees from the face of the wheel.

Figure H-37 Grinding wheel dressed and ready to use.

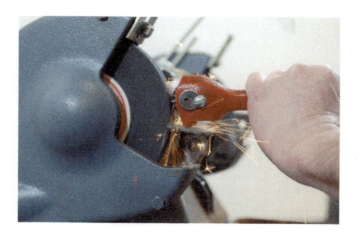

Figure H-38 Truing up a grinding wheel with a wheel dresser. The method shown here is often used in shops because it takes less time. The preferred method is to move the tool rest outward and to hook the lugs of the wheel dresser behind the tool rest, using it as a guide.

Figure H-39 Starting position showing 59-degree angle with the wheel and the cutting lip horizontal.

Figure H-40 Drill being held in the same starting position, approximately horizontal.

Figure H-41 The drill shank is moved downward.

Step 3 Using the tool rest and fingers as a pivot, slowly move the shank downward (Figure H-41). The drill must be free to slip forward slightly to keep it against the wheel (Figure H-42). Rotate the drill slightly. The beginner's most common mistake is to rotate the drill until the

Figure H-42 This figure shows that the drill shifts up and slightly forward to keep the drill against the wheel.

Figure H-43 Drill is almost to the final position of grinding. The shank has been rotated downward slightly from the starting position.

opposite cutting edge has been ground off. Do not rotate small drills at all—only larger ones. As you continue the downward movement of the shank, crowd the drill into the wheel so that it will grind lightly all the way from the lip to the heel (Figure H-43). This should all be one smooth movement. It is important at this point to allow proper clearance (8–12 degrees) at the heel of the drill.

Step 4 Without changing your body position, pull the drill back slightly and rotate 180 degrees so that the opposite lip is now in a level position. Repeat step 3.

Step 5 Check both cutting lips with the drill point gage (Figure H-44):

a. Check for correct angle.

b. Check for equal length.

c. Check lip clearances visually.

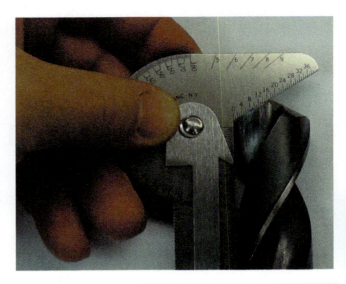

Figure H-44 The are cutting lips are checked with a drill point gage for length and angle.

The lip clearance angles should be between 8 and 12 degrees. If errors are found, adjust and regrind until correct.

Step 6 When you are completely satisfied that the drill point angles and lip lengths are correct, drill a hole in a scrap of metal that has been set aside for this purpose. Consult a drill speed table to select the correct rpm. Use cutting fluid.

Step 7 Check the condition of the hole. Did the drill chatter and cause the start of the hole to be misshapen? If so, this could be caused by too much lip clearance. Is the hole oversized more than .005–010 inches? Are the lips uneven or the lip angles off, or both? Running the drill too slowly for its size will cause a rough hole; too fast will overheat the drill. If the hole size is more than .010 inch over the drill size, resharpen the drill and try it again.

Step 8 When you have a correctly sharpened drill, show the drill and the workpiece to your instructor for his or her evaluation.

SELF-TEST

Sharpen drills on the pedestal grinder and submit them to your instructor for approval.

Operating Drilling Machines

You have already learned many things about drilling machines and tooling. You should now be ready to learn some important facts about the use of these machines. How fast should the drill run? How much feed should be applied? Which kind of cutting fluid should be used? Workpieces of a great many sizes and shapes are drilled by machinists. Several types of work-holding devices are used to hold parts safely and securely. In this unit you will learn how to correctly operate drilling machines.

OBJECTIVES

After completing this unit, you should be able to:

- Determine the correct drilling speeds for five given drill diameters.
- Determine the correct feed in steel by chip observation.
- Set up the correct feed on a machine by using a feed table.
- Identify and explain the correct uses for several work-holding and locating devices.
- Set up and drill holes in two parts of a continuing project; align and start a tap using the drill press.

CUTTING SPEEDS

Spindle speeds (rpm) for drilling are calculated using the following simplified formula:

$$rpm = \frac{CS \times 4}{D}$$

where

CS = an assigned surface speed for a given material
D = the diameter of the drill.

For metric units, use the following inputs:

$$m/min = \text{meters per minute}$$
$$rpm = \text{revolutions per minute}$$
$$D = \text{drill diameter}$$
$$\pi = 3.1416$$

To find:	Given:	Formula:
m/min	D (mm) and rpm	$m/min = \dfrac{\pi \times D \times rpm}{1000}$
rpm	D (mm) and m/min	$rpm = \dfrac{m/min \times 1000}{D \times \pi}$

For a $\frac{1}{2}$-inch drill in low-carbon steel, the cutting speed would be 90 rpm. See Table H-2 for cutting speeds for some metals. Cutting speeds/rpm tables for various materials are available in handbooks and as wall charts. The rpm for a $\frac{1}{2}$-inch drill in low-carbon steel would be

$$\frac{90 \times 4}{\frac{1}{2}} = 720\,rpm$$

To calculate rpm using a metric cutting speed, add the multiplier 3.28 to the formula:

$$3.28 \times CS \times 4/D$$

Table H-2 Drilling Speed Table

Material	Cutting Speed (ft/min)	Cutting Speed (m/min)
Low-carbon steel	60–100	20–30
Aluminum alloys	100–300	30–120
Cast iron	60–200	20–60
Alloy steel	40–90	15–25
Brass and bronze	50–200	15–60

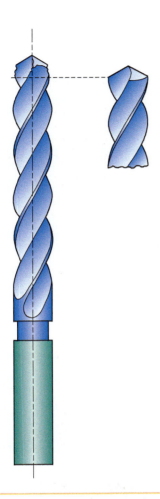

Figure H-45 Broken-down drill corrected by grinding back to full-diameter margins and regrinding cutting lips.

Excessive speeds can cause the outer corners and margins of the drill to bind in the hole, even if the speed is corrected and more cutting oil is applied. The only cure is to grind the drill back to its full diameter (Figure H-45).

If blue chips are produced when an HSS drill is drilling steel, the speed is too high. The tendency with small drills, however, is to set the rpm of the spindle too low. This gives the drill a low cutting speed, and little chips are formed unless the operator forces it with an excessive feed. The result is often a broken drill.

CONTROLLING FEEDS

The feed may be controlled by the "feel" of the cutting action and by observing the chip. A long, stringy chip indicates too much feed. The proper chip in soft steel should be a tightly rolled helix in both flutes (Figure H-46). Materials such as cast iron will produce a granular chip. Drilling machines that have power feeds are arranged to advance the drill a given amount for each revolution of the spindle. Therefore, .006-inch feed means that the drill advances .006 inch every time the drill makes one full revolution. The amount of feed varies according to the drill size and the work material. See Table H-3.

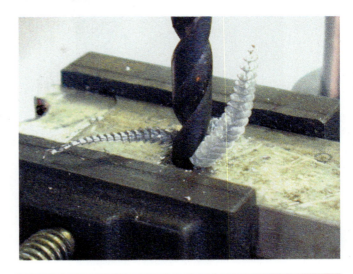

Figure H-46 Properly formed chips.

It is a good practice to start with smaller feeds than those given in tables. Materials and setups vary, so it is safer to start low and work up to an optimum feed. You should stop the feed occasionally to break the chip and allow the coolant to flow to the cutting edge of the drill. There is generally no breakthrough problem when using power feed, but when hand feeding, the drill may catch and grab while breaking through the last $\frac{1}{8}$ inch or so of the hole. Therefore, the operator should reduce the feed handle pressure near this point and ease the drill through the hole. This grabbing tendency is especially true of brass and some plastics, but it is also a problem in steels and other materials. Large upright drill presses and radial arm drills have power feed mechanisms with feed clutch handles (Figure H-47) that also can be used for hand feeding

Table H-3 Drilling Feed Table

Drill Size Diameter (inch)	Feed per Revolution (inch)
Under $\frac{1}{8}$	.001 to .002
$\frac{1}{8}$ to $\frac{1}{4}$	.002 to .004
$\frac{1}{4}$ to $\frac{1}{2}$	.004 to .007
$\frac{1}{2}$ to 1	.007 to .015
Over 1	.015 and over

Figure H-47 Feed clutch handle. The power feed is engaged by pulling the handles outward. When the power feed is disengaged, the handles may be used to hand feed the drill.

Figure H-48 Speed and feed control dials.

Figure H-49 Large speed and feed plates on the front of the head of the upright drill press can be read at a glance.

when the power feed is disengaged. Both feed and speed controls are set by levers or dials (Figure H-48). Speed and feed tables on plates are often found on large drilling machines (Figure H-49).

Some drill presses have tapping capability. Figure H-50 shows the control for a drill press. Note the drilling/tapping selector switch. To use this drill press to tap, the machinist would set the drilling/tapping switch to tapping. The machinist would also set the depth stop for the required thread depth. When the correct depth is reached the electronic switch in the depth stop will stop and reverse the spindle and the tap will reverse back out of the hole.

Tapping with small taps is often done on a drill press with a tapping attachment (Figure H-51) that has an adjustable friction clutch and reverse mechanism that screws the tap out when you raise the spindle. Large-size taps are power driven on upright or radial drill presses. These machines provide for spindle reversal (sometimes automatic) to reverse the spindle to back the tap out.

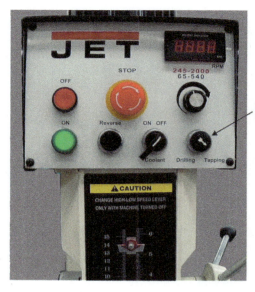

Drilling or Tapping Selector Switch

Figure H-50 Drill press with tapping selector switch.

Figure H-51 Tapping attachment.

CUTTING FLUIDS

A large variety of coolants and cutting oils are used for drilling operations. Emulsifying or soluble oils (either mineral or synthetic) mixed with water are used for drilling holes. Operations that tend to create more friction and need more lubrication to prevent **galling** (abrasion due to friction), require cutting oil. Organic or mineral oils with sulfur or chlorine added are often used. Reaming, counterboring, countersinking, and tapping all create friction and require the use of cutting oils, of which the sulfurized type is most frequently used. Cast iron and brass are usually drilled dry, but water-soluble oil can be used for both. Aluminum can be drilled with water-soluble oil or kerosene for a better finish. Both soluble and cutting oils are used for steel.

DRILLING PROCEDURES

Pilot Holes

When holes larger than $\frac{1}{2}$ inch are drilled a pilot-hole is usually drilled first (Figure H-52). The size of the pilot hole drill should be selected so that its diameter is slightly larger than the thickness of the dead center (thickness of web) of the larger drill. The pilot hole provides relief for the dead center of the larger hole and also reduces the pressure required to feed the drill through the hole. The pilot hole can also reduce the tendency of the larger drill to wander as it drills the hole.

SHOP TIPS—CRAFTSMANSHIP

When soft metals such as brass, aluminum, and plastics are being drilled, a drill may grab and be pulled into the work, possibly causing the tool to break or its shank to spin in the chuck. Such drilling problems may be solved by hand grinding a small flat area in the flute immediately behind and perpendicular to the drill's lip. This eliminates lip hook-over and lessens the tendency of the drill to grab.

DRILL DEPTH

Drill depth is measured from the top of the workpiece to the full diameter depth of the drilled hole (Figure H-53). Depth specified on blueprints is normally specified to the depth of the full diameter of the drill. Blind holes (holes that do not go through the piece) are measured from the edge of the drill margin to the required depth. But the tip of the drill drills deeper than the full diameter. The formulas to calculate how much deeper the center of the drill drills than full diameter depth are shown in Figure H-54. Note that the depth is

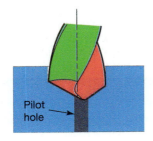

Figure H-52 A pilot hole for a large drill.

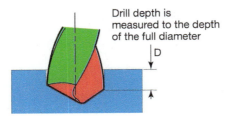

Figure H-53 Measuring the depth of a drilled hole.

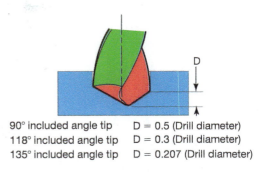

90° included angle tip	D = 0.5 (Drill diameter)
118° included angle tip	D = 0.3 (Drill diameter)
135° included angle tip	D = 0.207 (Drill diameter)

Figure H-54 Drill tip depth.

Figure H-55 Adjusting the depth stop.

dependent on the diameter of the drill and the included angle of the drill. CNC machines are able to automatically compensate for this difference in depth. A CNC control knows the length of the drill and the included angle.

A depth stop is provided on drilling machines to limit the travel of the quill so that the drill can be made to stop at a predetermined depth (Figure H-55). The use of a depth stop makes drilling several holes to the same depth quite easy. Spotfacing and counterboring should also be set up with the depth stop. Once measured, the depth can be set with the stop and drilling can proceed.

Deep-hole drilling requires sufficient drill length and quill stroke to complete the needed depth. A high-helix drill helps remove the chips, but sometimes the chips bind in the flutes of the drill and, if drilling continues, cause the drill to jam in the hole (Figure H-56).

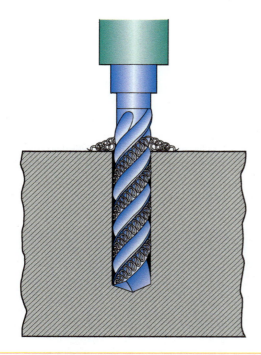

Figure H-56 Drill jammed in hole because of packed chips.

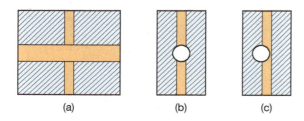

Figure H-57 Hole drilled at 90-degree angle into existing hole. Cross-drilling is done off center as well as on center: (*a*) side view; (*b*) end view drilled on center; (*c*) end view drilled off center.

A method of preventing this problem is called **pecking**; that is, the hole is drilled a short distance, then the drill is taken out from the hole, allowing the accumulated chips to fly off. The drill is again inserted into the hole, a similar amount is drilled, and the drill is again removed. Pecking is repeated until the required depth is reached.

Holes that must be drilled partly into or across existing holes (Figure H-57) may jam or bind the drill unless special precautions are taken. The hole can successfully be drilled without it wandering if a tight plug made of the same material as the work is first tapped into the cross hole (Figure H-58). The hole may then be drilled in a normal manner, and the plug can be removed.

Holes that overlap may be made on the drill press if care is taken and a set of counterbores with interchangeable pilots is available. First, pilot drill the holes with a size drill that does not overlap. Then, counterbore to the proper size with the appropriate pilot (Figure H-59). However, such an operation is best done with an end mill on a milling machine. Heavy-duty drilling should be done on an upright or radial drill press. The workpiece should be made secure, because high drilling forces are used with the larger drill sizes.

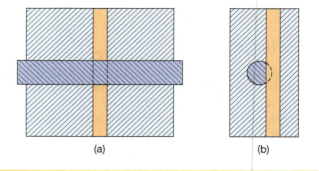

Figure H-58 Existing hole is plugged to allow the hole to be more easily drilled: (*a*) Hole is plugged with the same material as the workpiece; (*b*) end view showing hole drilled through plug.

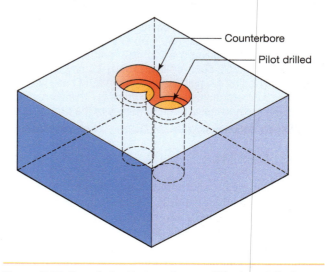

Counterbore

Pilot drilled

Figure H-59 Deep holes that overlap are difficult to drill. The holes are drilled alternately with a smaller pilot drill and a counterbore for final size.

The work should be well clamped or bolted to the worktable (Figure H-60). The head and column clamp should always be locked when drilling on a radial drill press. Cutting fluid is necessary for all heavy-duty drilling.

SHOP TIPS—CRAFTSMANSHIP

When you have several holes to machine involving multiple tools such as center drills, taps, and countersinks, it is more productive to do each similar operation on each hole. For example, you can center drill at each position, then tap drill at each position, and then tap each position. This procedure will involve fewer tool changes for each feature, but you must also precisely position the workpiece multiple times. In some cases, it may be more convenient to finish all machining operations on each hole before moving to the next feature. Although many manual tool changes can be tedious, completing a sequence of multiple operations may be a better procedure than risking a poorly positioned workpiece when trying to do similar operations on several features individually.

Figure H-60 Heavy-duty drilling on an upright drill press. Note how the vise is securely clamped to the table with the clamp oriented to stop the rotation of the vise if it should come loose (*Courtesy of Fox Valley Technical College*).

WORK HOLDING

Because of the great force applied by machines in drilling, some means must be provided to keep the workpiece from turning with the drill or from climbing up the flutes after the drill breaks through. This is necessary not only for safety but also for workpiece rigidity and good workmanship. One method of work holding is to use strap clamps (Figure H-61) and T-bolts (Figure H-62). The clamp should be parallel to the table or slightly higher than the part at the step block (Figure H-63). The T-bolt (clamping stud) should be kept as close to the workpiece as possible. Parallels (Figure H-64) are placed under the work at the point where the clamp is holding. This provides a space for the drill to break through without making

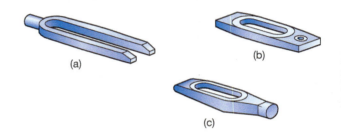

(a)

(b)

(c)

Figure H-61 Strap clamps: (*a*) U-clamp; (*b*) straight clamp; (*c*) finger clamp.

Figure H-62 T-bolt.

Figure H-63 Adjustable step blocks. This figure also shows the use of step blocks and the correct and incorrect setup for strap clamps. The clamp bolt should be as close to the workpiece as possible. The strap should also be level to prevent bending of the bolt.

Clamping force

Incorrect

Back end of clamp is lower (Exaggerated)

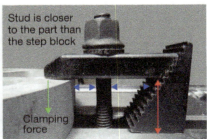

Stud is closer to the part than the step block

Clamping force

Correct

Clamping stud is closer to the part and the back end of the clamp is slightly higher

Figure H-64 Parallels of various sizes.

Figure H-66 C-clamp being used on an angle plate to hold work that would be difficult to support safely in other ways.

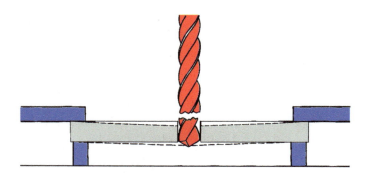

Figure H-65 Thin, springy material is supported too far from the drill. Drilling pressure forces the workpiece downward until the drill breaks through, relieving the pressure. The work then springs back, and the remaining "fin" of material is more than the drill can cut in one revolution. The result is drill breakage.

Figure H-67 Drill press vise. Small parts are held for drilling and other operations with the drill press vise.

a hole in the table. A thin or narrow workpiece should not be supported too far from the drill, since it will spring down under the pressure of the drilling. This can cause the drill on breakthrough to suddenly grab more material than it can handle. The result is often a broken drill (Figure H-65). Thin workpieces or sheet metal should be clamped over a wooden block to prevent this problem.

C-clamps of various sizes are used to hold workpieces on drill press tables and on angle plates (Figure H-66). **Angle plates** facilitate the holding of odd-shaped parts for drilling. The angle plate is either bolted or clamped to the table, and the work is fastened to the angle plate. For example, a gear or wheel that requires a hole to be drilled into a projecting hub can be clamped to an angle plate.

Drill press vises (Figure H-67) are frequently used for holding small workpieces with parallel sides. Vises provide the quickest and most efficient setup method for parallel work but should not be used if the work does not have parallel sides. The groove in the vise jaws is used to hold small cylindrical parts, whereas larger cylindrical parts can rest on

the bottom surface of the vise. The workpiece must be supported so that the drill will not go into the vise. If precision parallels are used for support, they and the drill can easily be damaged, because they are both hardened (Figure H-68). For rough drilling, however, keystock is sufficient for supporting the workpiece.

Angular vises can pivot a workpiece to a given angle so that angular holes can be drilled (Figure H-69). Another method of drilling angular holes is by tilting the drill press table. If there is no angular scale on the vise or table, a protractor head with a level may be used to set up the correct angle for drilling. Of course, this method will not be accurate if the drill press is not level. A better way is to measure the angle between the side of the quill and the table. Angle plates are also sometimes used for drilling angular holes (Figure H-70). The drill press table must be level (Figure H-71).

V-blocks come in sets of two, often with clamps for holding small-sized rounds (Figure H-72). Larger-sized round stock is set up with a strap clamp over the v-blocks (Figure H-73).

A wiggler is a tool that can be put into a drill chuck to locate a punch mark to the exact center of the spindle. Clamp the wiggler into a drill chuck and turn on the machine (Figure H-74). Push on the end of the pointer with a 6-inch rule or other piece of metal until it runs with no wobble (Figure H-75). With the machine still running, bring the pointer down into the punch mark. If the pointer begins

Figure H-68 Part set up in vise with parallels under it.

Figure H-69 Angle vise. Parts that must be held at an angle to the drill press table while being drilled are held with this vise.

to wobble again, the mark is not centered under the spindle, and the workpiece will have to be shifted. When the wiggler enters the punch mark without wobbling, the workpiece is centered. Once the drill is centered in this fashion, the workpiece is clamped and the hole is drilled.

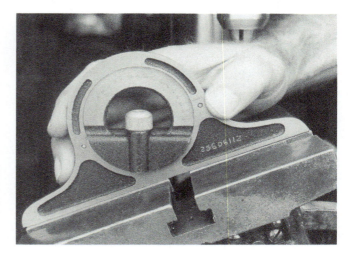

Figure H-70 View of an adjustable angle plate on a drill press table using a protractor for set up.

Figure H-71 Checking the level of the table.

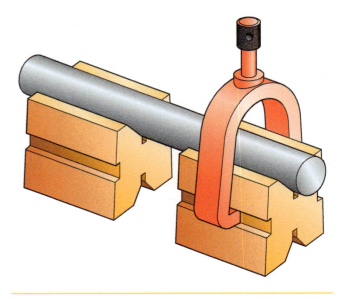

Figure H-72 Set of v-blocks with a v-block clamp.

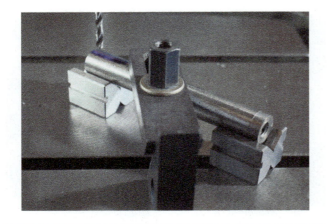

Figure H-73 Setup of two v-blocks and round stock with strap clamp.

Figure H-75 Wiggler centered.

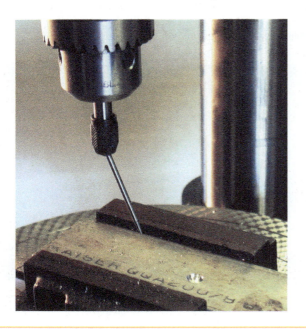

Figure H-74 Wiggler set in offset position.

After the work is centered, use a spotting or center drill to start the hole. Then, for larger holes, use a pilot drill, which is always a little larger than the web of the next drill size used. Pilot-drilled holes are made to reduce the feeding pressure caused by the web. When drilling alloy steels or work-hardened materials, drilling a pilot hole may be detrimental.

Use the correct cutting speed to set the rpm on the machine. Chamfer both sides of the finished hole with a countersink or chamfering tool. You may start a tap straight in the drill press by hand. After drilling the workpiece, and without removing any clamps, remove the tap drill from the chuck and replace it with a straight shank center. (An alternative method is to clamp the shank of the tap directly in the

Figure H-76 Typical box jig. Hardened bushings guide the drill to precise locations on the workpiece.

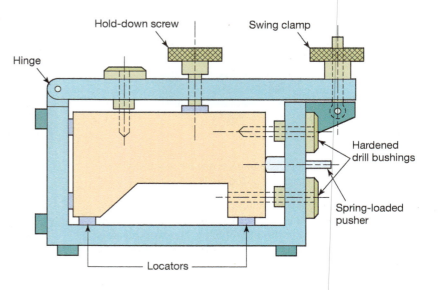

drill chuck and turn the spindle by hand two or three turns.) Insert a tap in the work and attach a tap handle. Then, put the center into the tap, but do not turn on the machine. Apply cutting oil and start the tap by turning the tap handle a few turns with one hand while feeding down with the other hand. Release the chuck while the tap is still in the work and finish the job of tapping the hole.

Jigs and **fixtures** are specially made tooling for production work. In general, fixtures reference a part to the cutting tool. Drill bushings guide the cutting tool (Figure H-76). The use of a jig ensures exact positioning of the hole pattern and eliminates laying out every part.

DRILLING PROCEDURE, C-CLAMP BODY

Given a combination square, wiggler, Q drill, countersink, $\frac{3}{8}$-24 tap, cutting oil, set of v-blocks with clamps, drill press vise, the C-clamp body as it is finished up to this point, and a piece of $\frac{1}{2}$-inch-diameter CR round stock $4\frac{1}{8}$ inches long:

Step 1 Set up a workpiece square with the drill press table.

Step 2 Locate the punch mark and center drill in the mark.

Step 3 Pilot drill and then tap drill.

Step 4 Hand start the tap in the drill press.

Step 5 Set up round stock in v-blocks, locate the center, and then clamp in place and drill a $\frac{3}{16}$-inch hole.

Figure H-77 Setup of C-clamp project in vise by squaring it with combination square.

SHOP TIPS—CRAFTSMANSHIP

When using a drill press, complete all operations before removing the workpiece from the vise or clamps. If you remove the workpiece before all operations are complete, you must then take extra time to set it up again and make sure that the feature to be machined is exactly centered under the cutting tool.

Drilling the Clamp Body

Step 1 Clamp the C-clamp body in the vise as shown in Figure H-77 so that the back side extends from the vise jaws about $\frac{1}{16}$-inch. Square it with the table by using the combination square. Tighten the vise.

Step 2 Put a wiggler into the chuck and align the center as explained previously. Clamp the vise to the table, taking care not to move it.

Step 3 Using the center drill or spotting drill, start the hole (Figure H-78). Change to a $\frac{1}{8}$- to $\frac{3}{16}$-inch pilot drill and drill the hole clear through. Now, change to the Q drill and enlarge the hole to size (Figure H-79). Chamfer the drilled hole with a countersink tool (Figure H-80). The chamfer should measure about $\frac{3}{8}$ inch across. Use cutting oil or coolant for drilling.

Figure H-78 Using a center drill to start the hole.

Figure H-79 Drilling the hole to the correct diameter for tapping. Note the correct chip formation.

Figure H-81 Hand tapping in the drill press to ensure good alignment.

Step 4 Place a straight shank center in the chuck and tighten it. Insert a $\frac{3}{8}$-24 tap and tap handle in the tap-drilled hole and support the other end on the center. Apply cutting oil. Do not turn on the machine. Feed lightly downward with one hand while hand turning the tap handle with the other hand (Figure H-81). The tap will be started straight when partway into the work. Release the chuck and finish tapping with the tap handle.

Figure H-80 Chamfering the drilled hole.

SELF-TEST

1. If rpm $= \dfrac{\text{CS} \times 4}{D}$, and the cutting speed for low-carbon steel is 90, what would the rpm be for the following drills: $\frac{1}{4}$, $\frac{1}{2}$, $\frac{3}{4}$, $\frac{3}{8}$, $1\frac{1}{2}$ inches in diameter?

2. What are some results of excessive drilling speed? What corrective measures can be taken?

3. What do chips look like if the feed is correct?

4. How are power feeds designated for drilling?

5. Counterboring, reaming, and tapping create friction that can cause heat. This can ruin a cutting edge. How can this situation be avoided?

6. How can jamming of a drill be prevented when drilling deep holes?

7. Name three uses for the depth stop on a drill press.

8. What is the main purpose for using work-holding devices on drilling machines?

9. List as many work-holding devices as you can.

10. Explain the uses of parallels for drilling setups.

11. Why should the support on a narrow or thin workpiece be as close to the drill as possible?

12. Angle drilling can be accomplished in several ways. Describe two methods.

13. What is the purpose of a wiggler?

14. What are angle plates used for?

15. What is the purpose of starting a tap in a drill press?

INTERNET REFERENCE

Twist drill troubleshooting guide:

http://www.carbidedepot.com/formulas-drills-troubleshoot.htm

Countersinking and Counterboring

In drill press work, it is often necessary to make a recess that will permit a bolt head to be below the surface of the workpiece. These recesses are made with countersinks or counterbores. When holes are drilled into rough castings or angular surfaces, a flat surface square to these holes is needed, and spotfacing is the operation used. This unit will familiarize you with these drill press operations.

OBJECTIVES

After completing this unit, you should be able to:

- Identify tools for countersinking and counterboring.
- Select speeds and feeds for countersinking and counterboring.

COUNTERSINKS

A countersink is a tool that is used to chamfer a hole (Figure H-82). Figure H-83 shows a common countersink designed to produce a smooth chamfer, free from chatter marks. A countersink is used as a chamfering or deburring tool to prepare a hole for reaming or tapping. Unless a hole requires a sharp edge, it should be chamfered to protect the end of the hole from nicks and burrs. A chamfer from $\frac{1}{32}$ to $\frac{1}{16}$ inch wide is sufficient for most holes. A hole made to receive a flathead screw or rivet should be countersunk deep enough for the head to be flush with the surface or up to .015 inch below the surface. A flathead fastener should never project above the surface. The included angles on commonly available countersinks are 60, 82, 90, and 100 degrees. Most flathead fasteners used in metal working have an 82-degree head angle, except in the aircraft industry, where the 100-degree angle is prevalent.

The cutting speed used when countersinking should always be slow enough to avoid chattering. A combination drill and countersink with a 60-degree angle (Figure H-84) is used to make center holes in workpieces for machining on lathes and grinders. The illustration shown is a bell-type

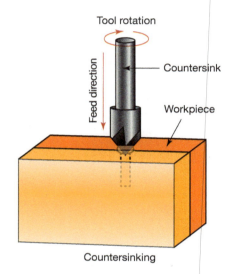

Figure H-82 Countersinking.

Figure H-83 Single-flute countersink (*Used with permission of DoALL Company*).

Figure H-84 Center drill or combination drill and countersink (*Used with permission of DoALL Company*).

center drill that provides an additional angle for a chamfer of the center, protecting it from damage. The combination drill and countersink, known as a *center drill*, is also used for spotting holes when using a drill press or milling machine, since it is extremely rigid and will not bend under pressure. By centerdrilling and providing a chamfer for the hole to be drilled, productivity is increased.

COUNTERBORES

Counterbored holes have flat bottoms, unlike the angled edges of countersunk holes, and are often used to recess a bolt head below the surface of a workpiece. Figure H-85 shows a counterboring operation. The pilot guides the counterbore tool into the work. The cutting edges cut a hole that the head of the screw will fit into. The diameter of the counterbore is usually $\frac{1}{32}$ inch larger than the head of the bolt.

Counterbores are designed to enlarge previously drilled holes and are guided into the hole by a pilot to ensure the concentricity of the two holes. A multi-flute counterbore is shown in Figure H-86, and a two-flute counterbore is shown in Figure H-87. The two-flute counterbore has more chip clearance and a larger rake angle than the multi-flute counterbore. When a variety of counterbore and pilot sizes are necessary, a set of interchangeable pilot counterbores is available. Figure H-88 shows a counterbore in which a number of standard or specially made pilots can be used. A pilot is illustrated in Figure H-89.

Counterbores are made with straight or tapered shanks and are used in drill presses and milling machines. For most counterboring operations the pilot should have from .002 to .005 inch of clearance in the hole. If the pilot is too tight in the hole, it may seize and break. If there is too much clearance between the pilot and the hole, the counterbore will be out of round and will have an unsatisfactory surface finish. It is important that the pilot be lubricated while counterboring. Usually, this lubrication is provided if a sulfurized cutting oil or soluble oil is used. When cutting dry, which is often

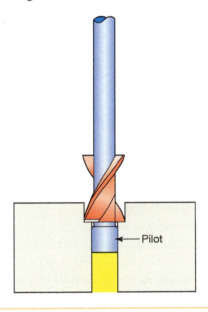

Figure H-85 A counterbore enlarges a portion of an already drilled hole.

Figure H-86 Multi-flute counterbore.

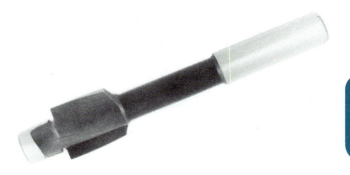

Figure H-87 Two-flute counterbore.

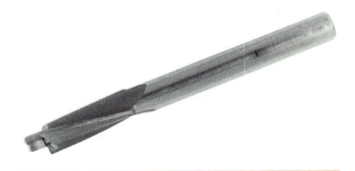

Figure H-88 Interchangeable pilot counterbore.

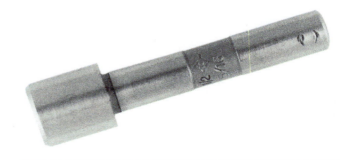

Figure H-89 Pilot for interchangeable pilot counterbore.

the case with brass and cast iron, lubricate the hole and pilot with a few drops of lubricating oil.

Counterbores or spotfacers are often used to provide a flat bearing surface for nut or bolt heads on rough castings or a raised boss (Figure H-90). This operation is called spotfacing. Because these rough surfaces may not be at right angles to the pilot hole, great strain is put on the pilot and counterbore and can cause breakage of either one. To avoid breaking the tool, be careful when starting the cut, especially when hand feeding. Prevent hogging into the work by tightening the spindle clamps slightly.

A back spotfacing tool (Figure H-91) is used when the back side of a hole is inaccessible to a standard spotfacer. The tool is fastened in a drill press chuck and put through the hole with the tool retracted. The tool swings out when

Table H-4 Feeds for Counterboring

$\frac{3}{8}$ inch diameter up to .004 inch per revolution
$\frac{5}{8}$ inch diameter up to .005 inch per revolution
$\frac{7}{8}$ inch diameter up to .006 inch per revolution
$1\frac{1}{4}$ inch diameter up to .007 inch per revolution
$1\frac{1}{2}$ inch diameter up to .008 inch per revolution

the spindle is rotated, and the spotface is then cut. The spindle is reversed to close and retract the tool from the hole. Some deburring tools operate on the same principle.

Recommended feed rates for counterboring are shown in Table H-4. The feed rate should be large enough to get under any surface scale quickly, preventing rapid dulling of the counterbore. The speeds used for counterboring are one-third less than the speeds used for twist drills of corresponding diameters. The choice of speeds and feeds is affected by the condition of the equipment, the power available, and the material being counterbored.

Before counterboring, the workpiece should be securely fastened to the machine table or tightly held in a vise because of the high cutting pressures. Workpieces also should be supported on parallels to allow for the protrusion of the pilot. To obtain several equally deep countersunk or counterbored holes on the drill press or milling machine, the spindle depth stop can be set.

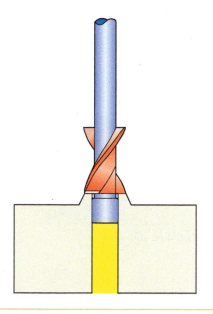

Figure H-90 Spotfacing on a raised boss.

SELF-TEST

1. What is a countersink used for?
2. Why are countersinks available in various angles?
3. What is a center drill used for?
4. What is a counterbore used for?
5. What is the relationship between pilot size and hole size?
6. Why is lubrication of the pilot important?
7. As a rule, how do the cutting speeds of an equal-sized counterbore and a twist drill compare?
8. What affects the selection of feed and speed when counterboring?
9. What is spotfacing?
10. What important points should be considered when making a counterboring setup?

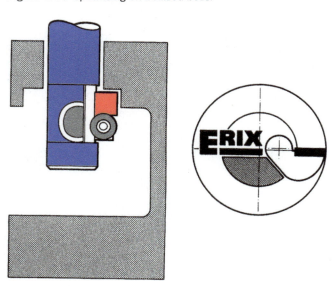

Figure H-91 Back spotfacing tool (*Based on ERIX TOOL AB*).

Reaming in the Drill Press

Many parts require the production of holes with smooth surfaces and accurate size. In many cases, holes produced by drilling alone do not entirely satisfy these requirements. The reamer was developed for enlarging or finishing previously formed holes. This unit will help you properly identify, select, and use machine reamers.

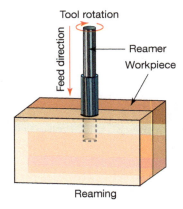

Figure H-92 Reaming operation.

OBJECTIVES

After completing this unit, you should be able to:

- Identify commonly used machine reamers.
- Select the correct feeds and speeds for commonly used materials.
- Determine appropriate amounts of stock allowance.
- Identify probable solutions to reaming problems.

REAMING

Figure H-92 shows an illustration of reaming. Note that a hole must be drilled before a reamer is used. A drilled hole may not be accurate enough in size and may not have a smooth surface finish. If a more accurate hole is required in size, roundness, and surface finish, the hole can be drilled slightly undersize and then reamed. A reamer can produce a very accurate hole size and much improved surface finish. There is a rule of thumb for reaming. If the finished size of the hole is $\frac{1}{2}$ inch or less, the hole should be drilled $\frac{1}{64}$ inch smaller than the finished size and then reamed. If the hole is larger than $\frac{1}{2}$ inch, the hole should be drilled $\frac{1}{32}$ smaller than the finished size and then reamed.

COMMON MACHINE REAMERS

Reamers are used to precisely finish holes, but they are also used in the heavy construction industry to enlarge or align existing holes. Machine reamers have straight or tapered shanks; the taper usually is a standard Morse taper. The parts of a machine reamer are shown in Figure H-93. The cutting end of a machine reamer is shown in Figure H-94.

Chucking and taper-shank reamers (Figures H-95 to H-97) are efficient in reaming a wide variety of materials and are commonly used in drilling machines. **Helical-flute reamers** have an extremely smooth cutting action that finishes holes accurately and precisely. Chucking reamers cut on the chamfer at the end of the flutes. This chamfer is usually at a 45-degree angle. **Jobber's reamers** (Figure H-98) are used where a longer flute length than that of chucking reamers is needed. The additional flute length gives added guide to the reamer, especially when reaming deep holes.

The **rose reamer** (Figure H-99) is primarily a roughing reamer used to enlarge holes to within .003 to .005 inch of finish size. The rose reamer is typically followed by a fluted reamer to bring the hole to finished size. The teeth are slightly backed off, which means that the reamer diameter is smaller toward the shank end by approximately .001 inch of flute length. The lands on these reamers are ground cylindrically without radial relief, and all cutting is done on the end of the reamer. This reamer will remove a considerable amount of material in one cut.

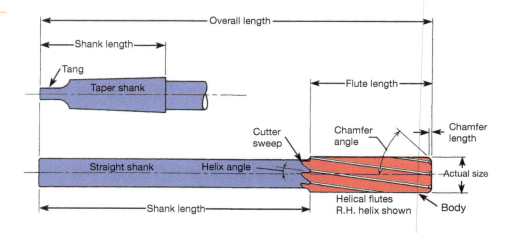

Figure H-93 The parts of a machine reamer (*Used with permission of Besly Cutting Tools, Inc.*).

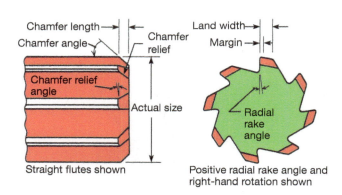

Figure H-94 Cutting end of a machine reamer (*Used with permission of Besly Cutting Tools, Inc.*).

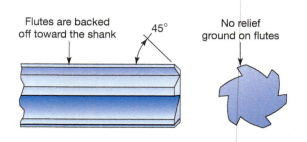

Figure H-99 Rose reamer.

Shell reamers (Figure H-100) are finishing reamers. Two slots in the shank end of the reamer fit over matching driving lugs on the shell reamer or box (Figure H-101). The hole in the shell reamer has a slight taper of $\frac{1}{8}$ inch per foot in it to ensure exact alignment with the shell reamer arbor. Shell reamers are made with straight or helical flutes and are commonly produced in sizes from $\frac{3}{4}$ inch to $2\frac{1}{2}$ inches in diameter. Shell reamer arbors come with matching straight or tapered shanks and are made in designated sizes from numbers 4 to 9. Morse taper reamers (Figure H-102), with straight or helical flutes, are used to finish ream-tapered holes in drill sockets, sleeves, and machine tool spindles.

Figure H-95 Straight shank straight-flute chucking reamer.

Figure H-96 Straight shank helical-flute chucking reamer.

Figure H-97 Taper shank straight-flute jobber's reamer.

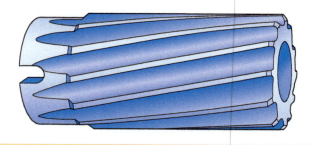

Figure H-100 Shell reamer helical flute.

Figure H-98 Taper shank straight-flute jobber's reamer.

Figure H-101 Tapered shank shell reamer arbor.

Helical taper pin reamers (Figure H-103) are especially suitable for machine reaming of taper pin holes. Chips do not get packed in the flutes, as the chips are pushed forward through the hole during the reaming operation. These reamers have a free-cutting action that produces a good finish at high cutting speeds. Taper pin reamers have a taper of $\frac{1}{4}$ inch per foot of length and are manufactured in 18 different sizes ranging from smallest number $\frac{8}{0}$ to the largest, number 10.

Taper bridge reamers (Figure H-104) are used in structural iron or steel work, bridge work, and ship construction where extreme accuracy is not required. They have long tapered pilot points for easy entry in the out-of-line holes often encountered in structural work. Taper bridge reamers are made with straight and helical flutes to ream holes with diameters from $\frac{1}{4}$ to $1\frac{5}{16}$ inches.

Carbide-tipped chucking reamers are often used in production, particularly where abrasive materials or sand and scale (as in castings) are encountered. The right-hand helix chucking reamer (Figure H-105) is recommended for ductile or highly abrasive materials, or when machining blind holes. The carbide-tipped left-hand helix chucking reamer (Figure H-106) will produce good finishes on heat-treated steels up to RC 40. These reamers should only be used for through holes, because the chips push out through the hole ahead of the reamer (Figure H-107). After becoming worn, expansion

Figure H-107 Carbide-tipped expansion reamer.

reamers can be expanded and resized by grinding. This feature offsets normal wear from abrasive materials and provides for a long tool life. These tools should not be adjusted for reaming size by loosening or tightening the expansion plug.

Reaming is intended to produce accurate, straight holes of uniform diameter. The required accuracy depends on a fine surface finish, close tolerance on diameter, roundness, straightness, and absence of bellmouth at the ends of holes. Machine reamers are often made of either high-speed steel or cemented carbide. Reamer cutting action is controlled by the cutting speed and feed used.

SPEEDS

The most efficient cutting speed for machine reaming depends on the type of material being reamed, the amount of stock to be removed, the cutting tool material, the finish required, and the rigidity of the setup. A good starting point is to use one-third to one-half of the cutting speed used for drilling the same material. Table H-5 may be used as a guide.

FEEDS

Feeds for reaming are usually two-to-three times greater than those used for drilling. The amount of feed may vary with different materials, but a good starting point is between .0015 and .004 inch per revolution. A feed that is too low may "glaze" the hole. This can work harden the material, causing chatter and excessive wear on the reamer.

Figure H-102 Morse taper reamer.

Figure H-103 Helical taper pin reamer.

Figure H-104 Helical-flute taper bridge reamer.

Figure H-105 Carbide-tipped helical-flute chucking reamer, right-hand helix.

Figure H-106 Carbide-tipped helical-flute chucking reamer, left-hand helix.

Table H-5 Reaming Speeds

Aluminum and its alloys	130–200
Brass	130–200
Bronze, high tensile	50–70
Cast iron	
Soft	70–100
Hard	50–70
Steel	
Low-carbon	50–70
Medium-carbon	40–50
High-carbon	35–40
Alloy	35–40
Stainless steel	
AISI 302	15–30
AISI 403	20–50
AISI 416	30–60
AISI 430	30–50
AISI 443	15–30

A feed that is too high tends to reduce the accuracy of the hole and the quality of the surface finish. Generally, it is best to use as high a feed as possible to produce the required finish and accuracy. The feed rate for reaming should be about twice the feed rate that would be used for drilling.

Tale H-5 shows cutting speeds in surface feet per minute (SFPM) for reaming with a HSS (high-speed steel) reamer. If carbide reamers are used, the speeds can be increased over those recommended for HSS reamers. The limiting factor as to whether carbide can be used is usually an absence of rigidity in the setup. Any chatter, which is often caused by excessive speed, is likely to chip the cutting edges of a carbide reamer. Always select a speed that is slow enough to eliminate chatter. Close tolerances and fine finishes often require the use of considerably lower speeds than those recommended in Table H-5.

STOCK ALLOWANCE

The stock removal allowance should be sufficient to assure good cutting action. Too small of a stock allowance results in burnishing (a slipping or polishing action) or wedges the reamer in the hole, causing excessive wear or breakage of the reamer. The condition of the hole before reaming also has an influence on the reaming allowance, because a rough hole will need a greater amount of stock removed than an equal-sized hole with a fairly smooth finish. See Table H-6 for commonly used stock allowances for reaming. When materials that work-harden easily are reamed, it is important to have adequate material for reaming. If too little material is left for reaming, the reamer tends to rub the material and create heat that can work harden the surface of the hole.

CUTTING FLUIDS

Cutting fluid is required to ream a hole with a good surface finish. Cutting fluid will cool the workpiece and the tool and will also act as a lubricant between the chip and the tool to reduce friction and heat buildup. Cutting fluids should be applied in sufficient volume to flush the chips away. Table H-7 lists some cutting fluids used for reaming different materials.

Table H-6 Stock Allowance for Machine Reaming

Reamer Size (inch)	Allowance (inch)
$\frac{1}{32}$ to $\frac{1}{8}$	.003 to .006
$\frac{1}{8}$ to $\frac{1}{4}$	.005 to .009
$\frac{1}{4}$ to $\frac{3}{8}$	.007 to .012
$\frac{3}{8}$ to $\frac{1}{2}$	.010 to .015
$\frac{1}{2}$ to $\frac{3}{4}$	$\frac{1}{64}$ to $\frac{1}{32}$
$\frac{3}{4}$ to 1	$\frac{1}{32}$

Table H-7 Cutting Fluids Used for Reaming

Material	Dry	Soluble Oil	Kerosene	Sulfurized Oil	Mineral Oil
Aluminum		X	X		
Brass	X	X			
Bronze	X	X			X
Cast iron	X				
Steels					
Low-carbon		X		X	
Alloy		X		X	
Stainless		X		X	

REAMING PROBLEMS

Chatter is often caused by the lack of machine rigidity, the workpiece, or the reamer itself. This condition may be corrected by reducing the speed, increasing the feed, putting a chamfer on the hole before reaming, using a reamer with a pilot (Figure H-108), or reducing the clearance angle on the cutting edge of the reamer. Carbide-tipped reamers cannot tolerate even momentary chatter at the start of a hole. Chatter is likely to chip the cutting edges of the reamer. Oversized holes can be caused by inadequate workpiece support, worn guide bushings, worn or loose spindle bearings, or bent reamer shanks. When reamers start cutting increasingly larger holes, it may be because the work material is galling or forming a built-up edge on reamer cutting surfaces (Figure H-109). Mild steel and some aluminum alloys are particularly troublesome. Changing to a different coolant may help. Reamers with highly polished flutes, margins, and relief angles or reamers that have special surface treatment may also improve the cutting action.

Bellmouthed holes are caused by the misalignment of the reamer with the hole. The use of accurate bushings or pilots may correct bellmouth, but in many cases the only

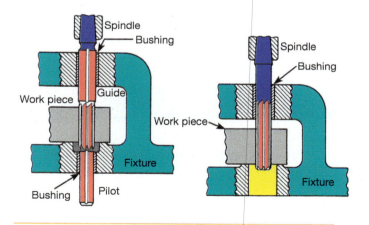

Figure H-108 Use of pilots and guided bushings on reamers. Pilots are provided so that the reamer can be held in alignment and can be supported as close as possible while allowing for chip clearance (*Used with permission of Besly Cutting Tools, Inc.*).

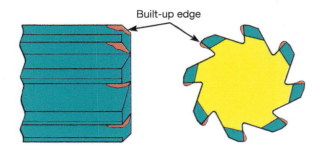

Figure H-109 Reamer teeth having built-up edges.

solution is the use of floating holders. A floating holder will allow movement in the horizontal direction while keeping the tool oriented vertically. A poor finish can be improved by decreasing the feed, but this will also increase the wear and shorten the life of the reamer. A worn reamer will never leave a good surface finish, as it will score or groove the finish and often produce a tapered hole. Too fast a feed will cause a reamer to break. Leaving an excessive amount of stock for finish reaming will produce a large volume of chips with heat buildup and will result in a poor hole finish. Too little stock allowance will cause the reamer teeth to rub as they cut, and not cut freely, which will produce a poor finish and cause rapid reamer wear. Insufficient cutting fluid may also cause rough surface finishes when reaming.

SELF-TEST

1. How is a machine reamer identified (versus a hand reamer)?
2. What is a jobber's reamer?
3. How does the surface finish of a hole affect its accuracy?
4. How do cutting speeds compare between drilling and reaming?
5. How do the feed rates compare between drilling and reaming?
6. How much reaming allowance would you leave on a $\frac{1}{2}$-inch hole?
7. What is the purpose of using cutting fluid while reaming?
8. What can be done to overcome chatter?
9. What causes a bellmouthed hole?
10. How can poor surface finish be improved?
11. When are carbide-tipped reamers used?
12. Why is vibration harmful to carbide-tipped reamers?

INTERNET REFERENCE

Additional machine reamer information:

http://www.gammons.com/reamers.html

The engine lathe is so named because it was originally powered by Watt's steam engine. The engine lathe may be used for turning, threading, boring, drilling, reaming, facing, spinning, and grinding. Lathe sizes range from the smallest jeweler's or precision lathes to the massive lathes used for machining huge forgings.

Engine lathes (Figure I-1) are used by machinists to produce one-of-a-kind parts or a few pieces for a short-run production. They are also used for toolmaking, machine repair, and maintenance.

Some lathes have a vertical spindle instead of a horizontal one, with a large rotating table on which the work is clamped. **Vertical turret lathes** (Figure I-2) can be relatively small up to among the largest machine tools. Huge cylindrical parts, weighing many tons, can be placed on the table and clamped in position to be machined. The machining of such castings would be impractical on a horizontal spindle lathe.

Lathes that are computer numerically controlled (CNC) produce workpieces such as shafts with tapers, threads, and precision diameters. CNC chuckers (Figure I-3) are high-production automatic lathes designed for chucking operations. Similar bar-feeding CNC types take a full-length bar through the spindle and automatically feed it in as needed. Some automatic lathes operate as either chucking machines or bar-feed machines.

Digital readout systems for manual machine tools are common. A digital readout uses linear encoders to accurately measure the axis movement. The machine axes movement is tracked and shown on a display (Figure I-4).

In ordinary turning, metal is removed from a rotating workpiece with a single-point tool. The tool must be harder than the workpiece. Chips formed from the workpiece slide across the face of the tool. This essentially is the way chips are produced in all metal-cutting operations. The pressures used in metal cutting can be as much as 20 tons per square inch. Tool geometry, therefore, is quite important for maintaining the strength at the cutting edge of the tool bit. In this section you will learn how to use common lathes, how to grind high-speed tools, and how to select tools for lathe work.

Figure I-2 Vertical turret lathe.

Figure I-1 A 15-inch manual lathe (*Courtesy of Fox Valley Technical College*).

Figure I-3 CNC chucking lathe with turret.

Figure I-4 Digital readout system (*Courtesy of Fox Valley Technical College*).

ON-SITE MACHINING

Most machining operations are performed at a permanent location such as in a machine shop. Some heavy machinery such as that used in earthmoving, logging, or mining operations cannot readily be moved from locations that are often remote. Also, disassembly and shipping costs along with the downtime can often be bypassed by using portable machine tools.

Portable boring bars can perform precision boring operations on-site. They are quite versatile tools that can be adapted to most job situations. Portable boring bars can produce hole diameters from 6 to 36 inches. These are hydraulically driven machines that can provide a range of rpm and torque.

Although milling operations are not necessarily lathe functions, they are used in on-site machining. Keyways can quickly be cut into shaft of any size with the portable key mill. This machine uses an end mill or other milling tools. It has many applications such as cutting motor mount slots. An adapter makes cross-milling possible. Some machinists are routinely sent, along with their on-site machining equipment, to remote, out-of-the-way locations to perform their jobs.

TURNING MACHINE SAFETY

SAFETY FIRST

The lathe can be a safe machine only if the machinist is aware of the hazards involved in its operation. You must always keep your mind on your work, to avoid accidents. Develop safe work habits in the use of setups, chip breakers, guards, and other protective devices. Standards for safety have been established as guidelines to help you eliminate unsafe practices and procedures on lathes.

Hazards in Lathe Operations

1. *Pinch points due to lathe movement.* A finger may get caught in gears or between the compound rest and a chuck jaw. The rule is to keep your hands away from such dangerous positions when the lathe is operating.
2. *Hazards associated with broken or falling components.* Heavy chucks or workpieces are dangerous when dropped. Care must be used when handling them. If a threaded spindle is suddenly reversed, the chuck can come off and fly out of the lathe. A chuck wrench left in the chuck can become a missile when the machine is turned on. Always remove a chuck wrench immediately after using it (Figure I-5).
3. *Hazards resulting from contact with high-temperature components.* Burns usually result from handling hot chips (up to $800^{O}F$ or even more) or a hot workpiece. Gloves may be worn when handling hot chips or workpieces. Gloves should never be worn when you are operating the machine.

Figure I-5 Always remove the chuck wrench when you finish using it (*Courtesy of Fox Valley Technical College*).

4. *Hazards resulting from contact with sharp edges, corners, and projections.* These are perhaps the most common cause of hand injuries in lathe work. Dangerous sharp edges are found in many places: on a long stringy chip, on a tool bit, or on a burred edge of a turned or threaded part. Shields should be used for protection from flying chips and coolant. These shields usually are made of clear plastic and are hinged over the chuck or clamped to the carriage of engine lathes. Even when shields are in place, safety glasses must be worn. Do not remove stringy chips with bare hands; wear heavy gloves and use hook tools or pliers. Never wear gloves while the machine is running. Always turn off the machine before attempting to remove chips. Chips should be broken and nine-shaped rather than in a stringy mass or a long wire (Figure I-6). Chip breakers on tools and correct feeds will help produce safe, easily handled chips. Burred edges must be removed before the workpiece is removed from the lathe. Always remove the tool bit when setting up workpieces on or removing workpieces from the lathe.

5. *Hazards of work-holding or driving devices.* When workpieces are clamped, their components often extend beyond the outside diameter of the holding device. Guards, barriers, and warnings such as signs or verbal instructions are all used to make you aware of the hazards. On power chucking devices you should be aware of potential pinch points between the workpiece and the work-holding device. Make certain that sufficient gripping force is exerted by the jaws to hold

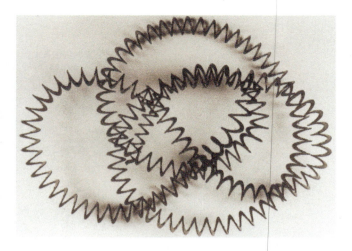

Figure I-6 Unbroken lathe chips are sharp and hazardous to the operator.

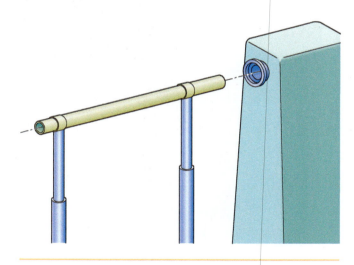

Figure I-7 Stock tube is used to support long workpieces that extend out of the headstock of a lathe.

the work safely. Keep tools, files, and micrometers off the machine. They may vibrate off into the revolving chuck or workpiece.

6. *Spindle braking.* The spindle or workpiece should never be slowed or stopped by hand. Always use machine controls to stop or slow it.

7. *Hazards associated with workpieces that extend out of the lathe.* Workpieces should be supported by a stock tube. If a slender workpiece is allowed to extend beyond the headstock spindle a foot or so without support, it can fly outward from the centrifugal force. The piece will not only be bent, but it will present a great danger to anyone standing nearby (Figure I-7).

Other Safety Considerations

Hold one end of the abrasive cloth strip in each hand when polishing rotating work. Don't let either hand get closer than a few inches to the work (Figure I-8).

Figure I-8 Polishing in the lathe with abrasive cloth (*Courtesy of Fox Valley Technical College*).

Figure I-9 A skyhook in use bringing a large chuck into place for mounting (*Courtesy of ATTCO, Inc.*).

Figure I-10 Left-hand filing in the lathe (*Courtesy of Fox Valley Technical College*).

Keep rags, brushes, and fingers away from rotating work, especially when knurling. Roughing cuts tend to quickly drag in and wrap up rags, clothing, abrasive cloth, and hair. Move the carriage back out of the way and cover the tool with a cloth when checking work. When removing or installing chucks or heavy workpieces, use a board on the ways so that the chuck can be slid into place. To lift a heavy chuck or workpiece (larger than an 8-inch diameter chuck), get help or use a crane (Figure I-9). Remove the tool or turn it out of the way while changing a chuck. Do not shift gears or try to take measurements while the machine is running and the workpiece is in motion.

Never use a file without a handle, as the file tang can quickly cut your hand or wrist if the file is struck by a spinning chuck jaw or lathe dog. Left-hand filing is considered safest; that is, the left hand grips the handle while the right hand holds the tip of the file (Figure I-10).

Engine Lathe Fundamentals

Modern lathes are highly accurate machines capable of performing a great variety of operations. Before attempting to operate a lathe, you should familiarize yourself with its principal parts and their operation. Good maintenance is important to the life and accuracy of machine tools. A machinist depends on the lathe to make precision parts. A poorly maintained machine loses its usefulness to a machinist. This unit will show you how to adjust, lubricate, and properly maintain lathes.

OBJECTIVES

After completing this unit, you should be able to:

- Identify the most important parts of a lathe and their functions.
- List all the lubrication points for one lathe in your shop.
- Determine the type of lubrication needed.
- Adjust the cross slide, compound slide, and tailstock, and clamp the compound after rotating it.

Figure I-11 An engine lathe (*Courtesy of Fox Valley Technical College*).

LATHE BASICS

A lathe is one of the most important machine tools (Figure I-11). A lathe is a device in which the work is rotated against a cutting tool. The shape of the workpiece is generated as the cutting tool is moved lengthwise and crosswise to the axis of the workpiece.

Figure I-12 shows a lathe and its most important parts. A lathe consists of the following major component groups: **headstock**, **bed**, **carriage**, **tailstock**, **quick-change gearbox**, and a **base** or **pedestal**. The headstock is fastened on the left side of the bed. It contains the spindle that drives the various work-holding devices.

The spindle is supported by spindle bearings on each end. If they are sleeve-type bearings, a thrust bearing is also used to take up end play. Tapered roller spindle bearings are often used on modern lathes. Spindle speed changes are also made in the headstock, either with belts or with gears.

Figure I-13 shows a gear-type headstock. Speed changes are made in these lathes by shifting gears.

Most belt-driven lathes have a low speed range when back gears are engaged. Figure I-14 shows a back-geared headstock. Smaller lathes have V-belt drives and back gears. Belt-driven lathes have mostly disappeared from machine shops.

A **feed reverse lever**, also called a *lead screw direction control,* is located on the headstock. Its function is simply to control the lead screw direction of rotation. This rotation determines the direction of feed and whether a thread cut on the lathe is left-hand or right-hand. The threading and feed mechanisms of the lathe are powered through the headstock.

The spindle is hollow to allow long, slender workpieces to pass through. The spindle end facing the tailstock is called the **spindle nose**. Spindle noses usually are one of three designs: a long taper key drive (Figure I-15), a camlock type (Figure I-16), or a threaded spindle nose (Figure I-17). Lathe chucks and other work-holding devices are fastened to and

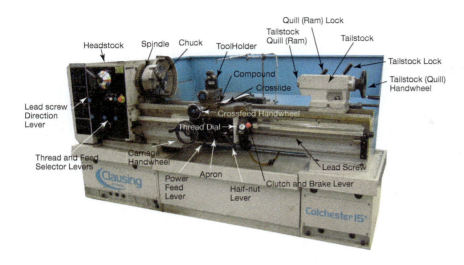

Headstock   Spindle   Chuck   ToolHolder

Quill (Ram) Lock
Tailstock
Quill (Ram)   Tailstock

Tailstock Lock

Compound

Tailstock (Quill)
Handwheel

Crosslide

Lead screw
Direction
Lever

Crossfeed Handwheel

Thread Dial

Thread and Feed
Selector Levers

Carriage
Handwheel

Apron   Lead Screw

Power
Feed
Lever

Half-nut
Lever

Clutch and Brake Lever

Colchester 15"

Figure I-12 Engine lathe with the parts identified (*Courtesy of Clausing Industrial, Inc.*).

Figure I-13 Geared headstock for heavy-duty lathe (*Courtesy of Fox Valley Technical College*).

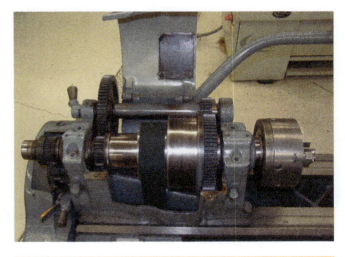

Figure I-14 Speeds are changed on this lathe by moving the belt to various steps on the pulley (*Courtesy of Fox Valley Technical College*).

driven by the spindle nose. The hole in the spindle nose typically has a standard Morse taper. The size of this taper varies with the size of the lathe.

The bed (Figure I-12) is the foundation and backbone of a lathe. Its rigidity and alignment affect the accuracy of the parts machined on it. Therefore, lathe beds are designed to withstand the stresses created by heavy cuts. On top of the bed are the **ways**, which usually consist of two inverted Vs and two flat bearing surfaces. The ways of the lathes are accurately machined by grinding or by milling and hand scraping. Wear or damage to the ways will affect the accuracy of the lathe. A gear rack is fastened below the front way of the lathe. Gears that link the carriage handwheel to this rack make possible the lengthwise movement of the carriage by hand.

The carriage is made up of the **saddle** and **apron** (Figure I-18). The apron is the part of the carriage facing the operator; it contains the gears and feed clutches that transmit motion from the feed rod or lead screw to the carriage and cross slide. The saddle slides on the ways and supports the cross slide and compound rest. The cross slide is moved crosswise at 90 degrees to the axis of the lathe by manually turning the crossfeed screw handle or by engaging the crossfeed lever (also called the *power feed lever* or, on some lathes, the *clutch knob*), which is located on the apron for automatic feed. On some lathes, a feed change lever (or plunger) on the apron directs power from the feed mechanism to either the longitudinal (lengthwise) travel of the carriage or the cross slide. On other lathes, two separate levers or knobs transmit motion to the carriage and cross slide.

Figure I-15 Long-taper key drive spindle nose.

Figure I-17 Threaded spindle nose (*Courtesy of Fox Valley Technical College*).

Figure I-16 Camlock spindle nose (*Courtesy of Fox Valley Technical College*).

Figure I-18 Lathe carriage (*Courtesy of Fox Valley Technical College*).

A thread dial on the apron (usually on the right side) indicates the exact place to engage the half nuts while cutting threads. The half-nut lever is used only for thread cutting and never for feeds for general turning. The entire carriage can be moved along the ways manually by turning the carriage handwheel or under power by engaging the power feed controls on the apron. The carriage can be clamped to the bed by tightening the carriage lock screw.

The compound rest is mounted on the cross slide and can be swiveled to any angle to produce bevels and tapers. The compound rest can be moved manually by turning the compound rest feed screw handle. Cutting tools are fastened on a tool post that is located on the compound rest.

The tailstock (Figure I-19) is used to support one end of a workpiece for machining or to hold various cutting tools such as drills, reamers, and taps. The tailstock slides on the ways and can be clamped in any position along the bed. The tailstock has

Figure I-19 Tailstock (*Courtesy of Fox Valley Technical College*).

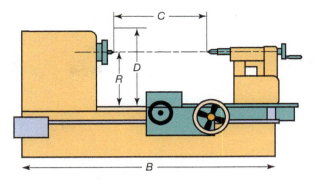

Figure I-21 Measuring the size of a lathe. *C* is the maximum distance between centers. *D* is the maximum diameter of workpiece over the ways (swing of lathe); *R* is the radius (one-half swing). *B* is the length of the bed.

Figure I-20 Quick-change gearbox showing index plate (*Courtesy of Fox Valley Technical College*).

a sliding spindle that is operated by a handwheel and locked in position with a spindle clamp lever. The spindle is bored to receive a standard Morse taper shank. The tailstock consists of an upper and a lower unit and can be adjusted to make tapered workpieces by adjustment screws in the base.

The quick-change gearbox (Figure I-20) is the link that transmits power between the spindle and the carriage. Different feeds can be selected by using the gear shift levers on the quick-change gearbox. Power is transmitted to the carriage through a feed rod or, as on smaller lathes, through the lead screw. The index plate on the quick-change gearbox indicates the feed in thousandths of an inch or as threads per inch for each lever position.

The base of the machine is used to level the lathe and to secure it to the floor. The motor of the lathe is usually mounted in the base. Figure I-21 shows how the size of a lathe is measured.

ENGINE LATHE MAINTENANCE AND ADJUSTMENTS

The engine lathe is a precision machine tool. With proper care a lathe will be accurate for many years, but its life will be shortened if it is misused. Even small nicks or burrs on the ways can prevent the carriage or tailstock from seating properly. Fine chips, filings, or grindings combine with the oil to form an abrasive mixture that can score and wear sliding surfaces and bearings. Frequent cleaning of way surfaces is helpful, but do

Figure I-22 A wood lathe board is being used for handling a chuck (*Courtesy of Fox Valley Technical College*).

not use an air jet, because this will blow the abrasive sludge into the bearing surfaces. Use a brush to remove chips, wipe off with a cloth, and then apply a thin film of oil.

Nicks or burrs are often caused by dropping chucks and workpieces on the ways or by laying tools such as files across them. Tools should not be placed on the ways. Wood lathe boards are used for handling chucks and heavy work (Figure I-22). Larger boards are often used for tool trays placed on an unused portion of the lathe ways (Figure I-23).

Figure I-23 Precision tools are protected from damaging the ways by keeping them in a safe place such as a tool tray.

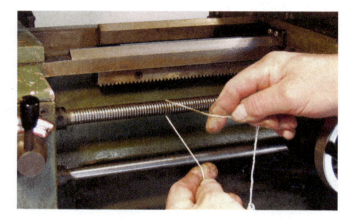

Figure I-24 Cleaning the lead screw using a piece of string.

The lead screw should be cleaned occasionally. To do this, loop a piece of string behind the lead screw and hold each end of the string (Figure I-24). With the machine on and the lead screw turning, draw the string along the threads. Never hold the string by wrapping it around your fingers; if the string grabs and begins winding on the lead screw, let it go. Lathes should be lubricated daily or before using. Oil cups or oil holes should be given a few drops of oil (too much just runs out). Apron and headstock reservoirs should be checked to be sure the oil level is correct. If they are low, use the oil recommended by the manufacturer, or its equivalent.

Before operating a lathe, wipe the way surfaces, even if they look clean, as gritty dust can settle on them when the machine has been idle for a few hours. Then, give the ways a thin film of way oil, which is specially compounded to remain on the way surfaces for a longer time than ordinary lubricating oil. After using the lathe, clean it so it is free of chips and grit, and lightly oil the ways to prevent rusting. Sweep up the chips and dirt in the surrounding area.

Lathe maintenance that requires extensive disassembly should be done by qualified personnel. A machinist should, however, be able to perform minor maintenance and make minor adjustments. The gibs on the cross slide and compound slide may need to be adjusted if there is excess clearance between the gib and the dovetail slide. Adjust the gibs

Figure I-25 Gibs are adjusted by tightening screws while the compound is centered over its slide (*Courtesy of Fox Valley Technical College*).

only when the slide is completely over its mating dovetail. Gibs adjusted without this backing may bend. Straight gibs are adjusted by tightening setscrews (Figure I-25) until the slide operates with just a slight drag. Overtightening will cause binding. The gibs on the compound slide should be kept fairly snug when the compound is not being used. Locknuts should be tightened after making the adjustment. Tapered gibs are adjusted by first loosening the lock screw and then adjusting the thrust screw. When the proper fit has been achieved, tighten the lock screw (Figure I-26).

NEW TECHNOLOGY

Traditional flat or tapered gibs on dovetail slides lack the high degree of stiffness required for some high-performance machine tools. The ideal condition is the least amount of clearance possible without limiting free movement.

Figure I-26 Method of adjustment of tapered gibs. The thrust screw is being tightened. The lock screw is on the other end of the gib.

Lathes have slack or backlash in the crossfeed screw and compound screw. This slack must be compensated for before starting a cut by backing away from the work two turns and then bringing the tool to the work. Some lathes have backlash-compensating crossfeed nuts (Figure I-27) that can be adjusted to remove most of the end play.

A machinist must occasionally interchange gears in the gear train on the headstock end of the lathe. Gears must sometimes be changed to cut various threads. When the proper gears have been set in place, the mounting or clamping bolts should be tightened lightly and a strip of paper should be placed between the gears. The gears should then be pushed

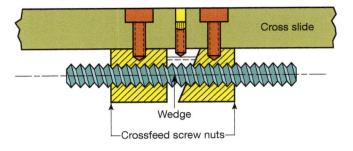

Figure I-27 Backlash compensating crossfeed nuts. This is one kind of backlash compensator in which a wedge is forced between two nuts on the same screw.

Figure I-28 Adjustment of feed change gears. Gear backlash could be approximately .005 inch (*Courtesy of Fox Valley Technical College*).

Figure I-29 Tailstock is clamped to the ways by means of a bolt and crossbar (*Courtesy of Fox Valley Technical College*).

Figure I-30 Camlock-type tailstock with lever being tightened (*Courtesy of Fox Valley Technical College*).

together against the paper shim. This spaces the gears approximately .005 inch apart. The clamping bolts should then be tightened (Figure I-28). If the gears are noisy, the adjustment must be made with more clearance between the gears.

The method by which the tailstock is clamped to the ways is simple and self-adjusting in many lathes, especially older ones. A large bolt and nut draws a crossbar up into the bed when the nut is tightened (Figure I-29).

A camlock type with a lever (Figure I-30) is sometimes used alone or on larger lathes in conjunction with a standard bolt and crossbar clamp. The clamping lever should never be forced. If it gets out of adjustment, it can be readjusted by means of a nut underneath the tailstock (Figure I-31).

The compound rotates so that angles can be machined. Degrees marked on its base are read against an index mark on the carriage. Various methods are used to clamp the compound. There are usually one or two bolts on each side (Figure I-32). Some types have setscrews.

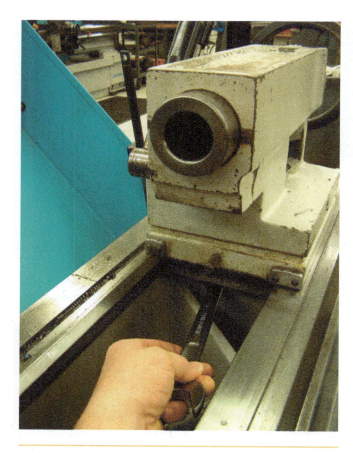

Figure I-31 Adjustment of camlock clamp. When the bolt on the crossbar is tightened, the cam will lock with the lever in a lower position (*Courtesy of Fox Valley Technical College*).

SELF-TEST

Part A At a lathe in the shop, identify the following parts and describe their functions. **Do not** turn on the lathe until you get permission from your instructor.

The Headstock

1. Spindle
2. Spindle speed-changing mechanism
3. Spindle nose
4. The kind of spindle nose on your lathe
5. Lead screw direction control

The Bed

1. The ways
2. The gear rack

The Carriage

1. Cross slide
2. Compound rest
3. Saddle
4. Apron
5. Power feed lever
6. Feed change lever
7. Half-nut lever

Figure I-32 Compound locknut.

8. Thread dial
9. Carriage handwheel
10. Carriage lock

The Tailstock

1. Spindle and spindle clamp lever
2. Tapered spindle hole and the size of its taper
3. Tailstock adjusting screws

The Quick-Change Gearbox

1. Lead screw
2. Shift the levers to obtain feeds of .005 and .013 inch per revolution. Rotating the lead screw with your hand may help when shifting feed levers.
3. Set the levers to obtain 4 threads per inch and then 12 threads per inch.
4. Measure the lathe and record its size.

Part B

1. Why should fine chips, filings, and grindings be cleaned from the ways and slides frequently?
2. How should the ways and slides be cleaned?
3. Nicks and scratches are damaging to lathe ways. What means can be employed to prevent them?
4. How often should a lathe be lubricated?
5. Should you begin work immediately on a lathe that looks perfectly clean? Explain.
6. What should you do when you are finished using the lathe?
7. Name two types of gibs used on lathes.
8. How tightly should the gibs be adjusted on the cross slide and on the compound?

INTERNET REFERENCES

Information on engine lathes:

http://www.clausing-industrial.com

http://www.americanmachinetools.com

Toolholders and Toolholding

Lathe cutting tools must be supported and fastened securely in the toolholder. There are many different types of toolholders available. Anyone working with a lathe should be able to select the best toolholding device for the operation.

OBJECTIVES

After completing this unit, you should be able to:

- Identify standard, quick-change, and turret-type toolholders mounted on a lathe carriage.
- Identify toolholding for the lathe tailstock.

Figure I-33 Quick-change tool post, dovetail type.

TOOLHOLDING

One of the most important considerations in lathe toolholding is rigidity. If tool bits or inserts are not well supported or the tool overhang is too great, vibration and tool chatter will often be the result. Tool posts are mounted on the compound rest and held securely by a T-bolt. The most common reason that tools turn and gouge into the workpiece is that the T-bolt is not tightened enough by the operator.

Because of the need for greater rigidity when using carbide cutting tools, several types of tool posts have been developed. The most commonly used are the quick-change types.

A **quick-change toolpost** (Figure I-33) is so named because of the speed with which tools can be interchanged. A quick-change toolpost is also more versatile than a standard toolpost. The toolholders used on a quick-change toolpost are accurately held in a dovetail on the post. This accuracy makes for more exact repetition of toolchanges. Tool height adjustments are made with a knurled adjustment collar, and the height alignment remains constant through repeated tool changes.

A three-sided quick-change tool post (Figure I-34) has the added ability to mount a tool on the tailstock side of the

tool post. These tool posts are securely clamped to the compound rest. The tool post in Figure I-34 uses a dovetail to locate the toolholders, which are clamped and released from the post by turning the top lever.

Toolholders for the quick-change tool posts include those for turning (Figure I-35), threading (Figures I-36 and I-37), and holding drills (Figure I-38). The drill holder makes it possible to use the carriage power feed when drilling holes instead of the tailstock hand feed.

Figure I-39 shows a boring bar toolholder in use. The boring bar is rigidly supported. The boring bar toolpost can be used to hold various boring bar sizes by using bushings. Quick-change toolposts are very rigid for boring.

An advantage of quick-change toolpost holders is that cutting tools of various shank thicknesses can be mounted in the toolholders (Figure I-40). The height adjustment knurled nut and locknut keep the tool at the correct height. Tools can be removed and replaced and maintain the correct height.

Figure I-41 shows a cutoff tool mounted in a toolholder.

A four-tool turret toolholder can be set up with several different tools such as turning tools, facing tools, or threading or boring tools. Often, one tool can perform two or

Figure I-34 Quick-change tool post (*Courtesy of Fox Valley Technical College*).

Figure I-36 Threading is accomplished with the bottom edge of the blade with the lathe spindle in reverse. This ensures cutting right-hand threads without hitting the shoulders (*Courtesy of Aloris Tool Technology Co. Inc.*).

Rocker-Arm Toolposts

An older-type tool post is shown in Figure I-45. These so-called standard tool posts are now obsolete but are still used on some older lathes and in some school shops. The two greatest disadvantages are their lack of rigidity and the tendency of the toolholder to turn in the tool post and gouge the workpiece.

Tool height adjustments are made by swiveling the rocker in the tool post ring. Making adjustments in this manner changes the effective back rake angle and also the front relief angle of the tool.

more operations, especially if the turret can be indexed in 30-degree intervals. A facing operation (Figure I-42), a turning operation (Figure I-43), and the chamfering of a bored hole (Figure I-44) are all performed from this turret. Tool height adjustments are made by adjusting a knurled nut.

Figure I-35 Quick-change toolholder being used for turning (*Courtesy of Fox Valley Technical College*).

Figure I-37 Quick-change toolholder being used for threading a left-hand thread. Note the bottom of the cutting tool is being used and the spindle would be running in reverse (*Courtesy of Fox Valley Technical College*).

Figure I-38 Drill toolholder in the tool post. Mounting the drill in the tool post makes drilling with power feed possible (*Courtesy of Aloris Tool Technology Co. Inc.*).

Many types of toolholders are used with the rocker-arm tool post. A straight shank turning toolholder (Figure I-46) is used with high-speed tool bits. The tool bit is held in the toolholder at a $16\frac{1}{2}$-degree angle, which provides a positive back rake angle for cutting. Straight shank toolholders are used for general machining on lathes. The type shown in Figure I-47 is used with carbide tools.

Figure I-48 shows an offset toolholder with no rake. It is typically used for carbide tools. Figure I-49 shows a right-hand offset toolholder.

Offset toolholders (Figures I-50) allow machining close to the chuck or tailstock of a lathe without tool post interference. The left-hand toolholder is intended for use with tools cutting from right to left or toward the headstock of the lathe.

A toolholder should be selected to match the application. The setup should be rigid, and the toolholder overhang should be kept to a minimum to prevent chatter. A variety of cutoff toolholders (Figure I-50) are used to cutoff or make grooves in workpieces. Cutoff tools are available in different thicknesses and sizes.

Figure I-51 shows a rocker-arm toolholder for boring. The cutting tool for this boring bar is inserted in the tool pocket and the front end of the boring bar is tightened to lock the cutting tool in the bar.

TOOLHOLDING FOR THE LATHE TAILSTOCK

The toolholders studied so far are all used on the carriage of a lathe. Toolholding is also done in the tailstock. Figure I-52 shows a Morse tapered drill that would fit into the tail stock spindle. One of the most common toolholding devices used in a tailstock is the **drill chuck** (Figure I-53). A drill chuck is used for holding straight shank drilling tools. When a series of operations must be performed and repeated on several workpieces, a **tailstock turret** (Figure I-54) can be used. The tailstock turret has six tool positions, one of which is used as a work stop. The other positions are for center drilling, drilling, reaming, counterboring, and tapping.

Figure I-39 Boring toolholder. This setup provides good boring bar rigidity (*Courtesy of Fox Valley Technical College*).

Height adjustment setting

Figure I-40 Tool height adjustment on a quick-change holder.

Figure I-41 Quick-change cutoff toolholder.

Figure I-42 Facing cut with a four-tool turret-type toolholder.

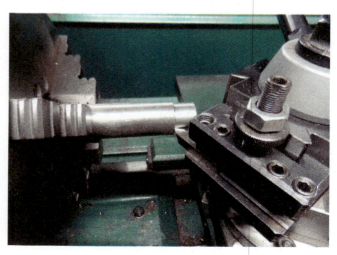

Figure I-43 Turning cut with a turret-type toolholder.

Figure I-44 Chamfering cut with a turret-type toolholder. Shown from the backside.

Figure I-45 Standard-type tool post with ring and rocker.

Figure I-46 Straight shank toolholder with a built-in back rake holding a high-speed right-hand tool.

Figure I-47 Right-hand toolholder for carbide tool bits without back rake.

Figure I-48 Left-hand toolholder.

Figure I-49 Right-hand toolholder.

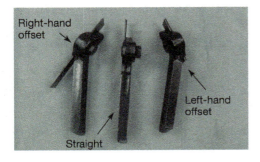

Figure I-50 Three kinds of cutoff toolholders with cutoff blades (*Courtesy of Fox Valley Technical College*).

Figure I-51 Boring bar tool post and bars with a special wrench.

Figure I-52 Taper shank drill with sleeve ready to insert in tailstock spindle.

Figure I-53 Drill chuck with Morse tapered shank.

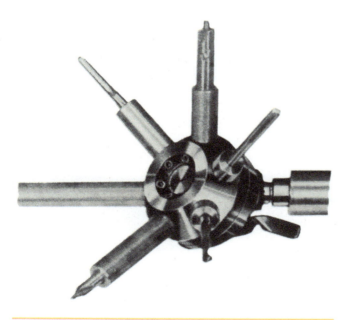

Figure I-54 Tailstock turret (*Enco Manufacturing Company*).

SELF-TEST

At a lathe in the shop, identify various toolholders and their functions.

1. If a tool turns into the workpiece and gouges it, what is usually the cause?
2. How are tool height adjustments made on a quick-change toolholder?
3. How are tool height adjustments made on a rocker-arm toolholder?
4. How does the toolholder overhang affect the turning operation?
5. What kinds of tools are used in the lathe tailstock?
6. How are tools fastened in the tailstock?

INTERNET REFERENCE

Information on quick-change toolholders for lathe work:

www.aloris.com

Cutting Tools for the Lathe

A machinist must fully understand cutting tool geometry. Whether a lathe operation can be done safely, economically, and with a good surface finish depends on the shape of the point, the rake and relief angles, and the nose radius of the tool. In this unit, you will learn about tool geometry and how to grind a lathe tool.

CUTTING TOOL GEOMETRY

On a lathe, metal is removed from a workpiece by turning it against a single-point cutting tool. The tool should not lose its hardness from the heat generated by machining. High-speed steel is used for many tools, as it fulfills these requirements and is easily shaped by grinding. High-speed steel is used in this unit to demonstrate tool geometry (see Table I-1). High-speed tools have been largely replaced in production machining by carbide tools (Figure I-55) because of their

Figure I-55 Carbide insert and toolholder.

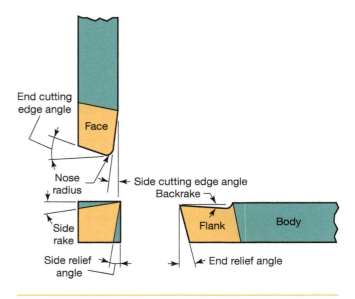

Figure I-56 The parts and angles of a tool.

higher metal-removal rates. However, in general machining operations high-speed tools are still used for special tooling. High-speed steel tools are required for older lathes equipped with only low-speed ranges. A good understanding of the tool geometry of high-speed tools will help you understand carbide tooling for CNC machining.

The most important characteristic of a lathe tool is its geometric form: the side and back rake, front and side clearance or relief angles, and chip breakers. Figure I-56 shows

Table I-1 Tool Geometry

Tool	Abbreviation	Angle Recommended
Back rake	BR	12°
Side rake	SR	12°
End relief	ER	10°
Side relief	SRF	10°
End cutting edge angle	ECEA	30°
Side cutting edge angle	SCEA	15°
Nose radius	NR	$\frac{1}{32}$ inch

the parts and angles of the typical lathe turning tool. The terms and definitions follow (the angles given are only examples and will vary according to the application).

1. The **tool shank** is that part held by the toolholder.
2. **Back rake** is important to smooth chip flow, which is needed for a uniform chip and a good finish, especially in soft materials.
3. The **side rake** directs the chip flow away from the point of cut and provides for a keen cutting edge.
4. The **end relief angle** prevents the front edge of the tool from rubbing on the work.
5. The **side relief angle** provides for cutting action by allowing the tool to feed into the work material.
6. The **side cutting edge angle** may vary considerably. For roughing, it should be almost square to the work, usually about 5 degrees. Tools used for squaring shoulders or for other light machining can have angles from 5 to 32 degrees, depending on the application. This angle may be established by turning the toolholder or by grinding it on the toolbit or both. In finishing operations with a large nose radius and light cut, side cutting edge angle is not an important factor. The side cutting edge angle directs the cutting forces back into a stronger section of the tool point and helps direct the chip flow away from the workpiece. It also affects the thickness of the cut (Figure I-57).
7. The nose radius will vary according to the finish required. The smallest nose radius that will give the desired finish should be used.

Grinding a tool provides both a sharp cutting edge and the shape needed for the cutting operation. Rake and relief angles must be clearly understood to successfully grind a tool. Left-hand tools are shaped just the opposite of right-hand tools (Figure I-58). The right-hand tool has the cutting edge on the left side and cuts to the left or toward the headstock. The hand of the lathe tool can easily be determined by looking at the end from the opposite side of the lathe; the cutting edge is to the right on a right-hand tool.

Tools are given a slight nose radius to strengthen the tip. A larger nose radius will produce a better finish but will also promote chattering (vibration) in a nonrigid setup. Figure I-59

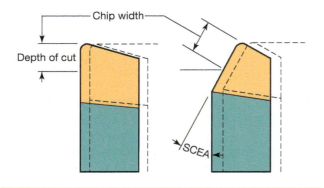

Figure I-57 Illustration of the change in chip width with an increase of the side cutting edge angle. A large edge angle can sometimes cause chatter (vibration of work or tool).

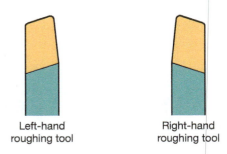

Figure I-58 Left- and right-hand roughing tools.

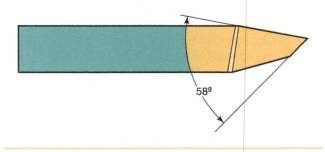

Figure I-59 One method of grinding the nose radius on the point of a tool.

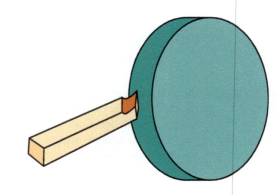

Figure I-60 Right-hand facing tool showing point angles. This tool is not suitable for roughing operations because of its acute point angle.

shows how a nose radius can be ground. All lathe tools require some nose radius, however small. A sharp-pointed tool is weak at the point and will usually break off in use, causing a rough finish on the work. A facing tool (Figure I-60) for shaft end-facing and mandrel work has little nose radius and an included angle of 58 degrees.

A right-hand or left-hand roughing or finishing tool is often used for facing chuck-mounted workpieces. Useful tool shapes are shown in Figure I-61. These are used for general lathe work.

Form Tools

Tools that have specially shaped cutting edges are called **form tools** (Figure I-62). These tools are plunged directly into the cut in one operation. Form tools include those used

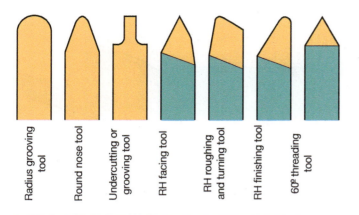

Figure I-61 Tool shapes most often used. The most commonly used tool shapes are the three on the right, which are the roughing or general turning tool, finishing tool, and threading tool.

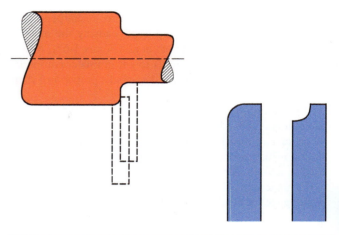

Figure I-63 Form tools are used to produce the desired shape in the workpiece. External radius tools, for example, are used to make outside corners round, whereas fillet radius tools are used on shafts to round the inside corners on shoulders.

Figure I-62 Form tool being checked with radius gage.

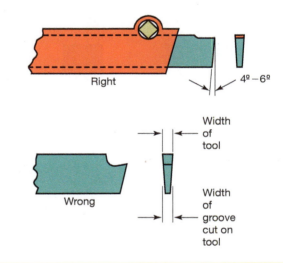

Figure I-64 Correct and incorrect methods of grinding a cutoff tool for deep parting.

for grooving, threading, and cutting internal and external radii and any other special shape needed for a particular operation. In contrast with single-point tools, these broad face tools have a tendency to chatter, so a lower cutting speed and cutting fluid are usually required when they are used.

The shape of any tool is dictated by the application. When cutting tool geometry is clearly understood, specially shaped tools can be made for specific purposes. For example, a tool with a nose radius can be used for making a fillet radius on a shaft (Figure I-63). Round corners can be formed with an external radius tool. A relief angle is ground on any kind of form tool to allow it to cut; side and back rakes are generally zero.

Cutoff Tools

Parting or cutoff tools are used for necking or undercutting, but their main function is cutting off material to the correct length. The correct and incorrect ways to grind a cutoff tool are shown in Figure I-64. Note that the width of the cutting edge becomes narrower than the blade as it is ground deeper, which causes the blade to bind in a groove deeper than the sharpened end. Too much end relief may cause the tool to "dig in" and jam in the cut.

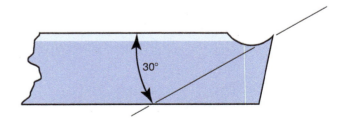

Figure I-65 Cutoff tools are sometimes ground with large back rake angles for aluminum and other soft metals.

Tools are sometimes specially ground for parting soft metals or specially shaped grooves (Figure I-65). The end is sometimes ground on a slight angle when a series of small hollow pieces are being cut off (Figure I-66). This helps eliminate the burr on small parts. This procedure is not recommended for deep parting.

Tools that have been ground back for resharpening too many times often form a "chip trap," causing the metal to be torn off or the tool not to cut at all (Figure I-67). A good machinist will never allow tools to get in this condition but will grind off the useless end and regrind a proper tool shape (Figure I-68).

Most toolholders hold the tool horizontally. Some lathe toolholders have a built-in back rake, so it is not necessary to grind one into the tool, as in Figure I-69. The tool in Figure I-70, however, is ground with a back rake and can be used in a toolholder that does not have a built-in back rake.

Threading tools (Figure I-71) should be checked with a center gage for 60-degree angle while they are being ground

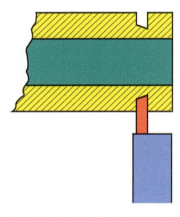

Figure I-66 Parting tool ground on an angle to avoid burrs on the cutoff pieces.

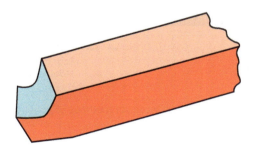

Figure I-67 Deformed tool caused by many resharpenings. The chip trap should be ground off and a new point ground on the tool.

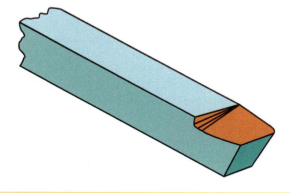

Figure I-68 Properly ground right-hand roughing tool.

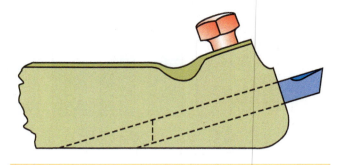

Figure I-69 Toolholder with back rake.

Figure I-70 Right-hand roughing tool with back rake.

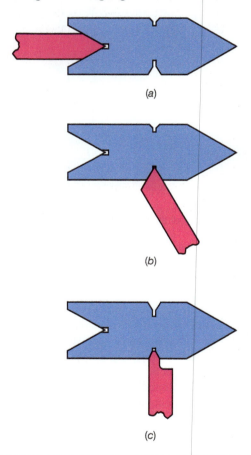

(a)

(b)

(c)

Figure I-71 *(a)* Checking a threading tool with a center gage; *(b)* checking a threading tool used for threading to a shoulder; *(c)* variation of the tool in *(a)* used for making fine threads to a shoulder.

or sharpened. Various kinds of threading tools can be made to suit the application. When you are threading close to a shoulder, a tool such as that in Figure I-71*b* or *c* should be used. Relief should be ground on each side of the tool. A slight flat should be honed on the end with an oilstone.

Tools for brass or plastics should have zero to negative rake to keep the tools from "digging in." Side rake, back rake, and relief angles are given for tools used for machining various metals in Table I-2.

Chip Control

It is important to make tools that will not produce hazardous chips. Long, unbroken chips are extremely dangerous. Tool geometry, especially side and back rake, has a considerable effect on chip formation. Side rakes with smaller angles tend to curl the chip more than those with large angles, and the curled chips are more likely to break up. Coarse feeds for roughing and maximum depth of cut also promote chip breaking.

Chip breakers are extensively used on both carbide and high-speed tools to curl the chip as it flows across the face of the tool. Because the chip is curled back into the work, it can go no farther and break (Figure I-72). A C-shaped chip is often the result, but a figure-9 chip is considered ideal (Figure I-73). This chip should drop safely into the chip pan without flying out.

Grinding the chip breaker too deep will form a chip trap that may cause binding of the chip and tool breakage (Figure I-74). The correct depth to grind a chip breaker is approximately $\frac{1}{32}$ inch. Chip breakers are typically of the parallel or angular types (Figure I-75). More skill is needed to offhand grind a chip breaker on a high-speed tool than to

Figure I-73 A figure-9 chip is the safest kind of chip.

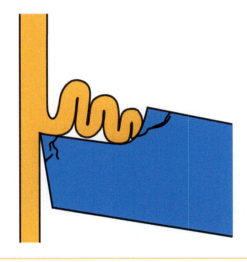

Figure I-74 Crowding of the chip is caused by a chip breaker that is ground too deeply.

Table I-2 Angle Degrees for High-Speed Steel Tools

Material	End Relief	Side Relief	Side Rake	Back Rake
Aluminum	8–10	12–14	14–16	30–35
Brass, free cutting	8–10	8–10	1–3	0
Bronze, free cutting	8–10	8–10	2–4	0
Cast iron, gray	6–8	8–10	10–12	3–5
Copper	12–14	13–14	18–20	14–16
Nickel and Monel	12–14	14–16	12–14	8–10
Steels, low carbon	8–10	8–10	10–12	10–12
Steels, Alloy	7–9	7–9	8–10	6–8

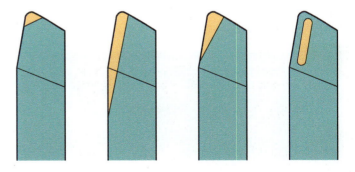

Figure I-75 Four common types of chip breakers.

grind the basic tool angles. Therefore, you should grind the basic tool and make an effort to produce safe chips by using correct feeds and depth of cut before grinding a chip breaker.

Care must be exercised while grinding on high-speed steel. A glazed wheel can generate heat up to 2000°F (1093°C) at the grinder–tool interface. Do not overheat the tool edge,

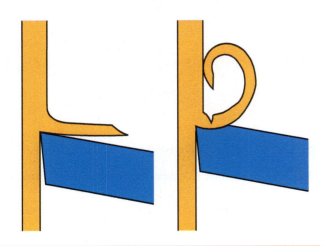

Figure I-72 Chip flow with a plain tool and with a chip breaker.

as this will cause small surface cracks that can result in the failure of the tool. Frequent cooling in water will keep the tool cool enough to handle. Do not quench the tool in water, however, if you have overheated it. Let it cool in air.

Because the right-hand roughing tool is most commonly used and the first you will need, you should begin with it. Use a piece of keystock the same size as the tool bit for practice until you can grind an acceptable tool. The tool gage used in the following tool grinding operation has an angle of 10 degrees on one side and $26\frac{1}{2}$ degrees on the other side. The Vs are 60 and 70 degrees. The larger angle of $26\frac{1}{2}$ degrees should be used only if the tool is to be used in a holder with a built-in back rake. If the tool is to be used in a horizontal position, both end and side relief angles should be checked with the 10-degree gage angle.

Grinding Procedure for a Lathe Tool Bit

Given a practice tool blank (a piece of keystock about 3 inches long), a tool gage, and a toolholder:

Step 1 Grind one acceptable practice right-hand roughing tool. Have your instructor evaluate your progress.

Step 2 Grind one acceptable right-hand roughing tool from a high-speed tool blank.

Wear goggles and make certain the tool rest on the grinder is adjusted properly (about $\frac{1}{16}$ inch from the wheel). True up the wheels with a wheel dresser if they are grooved, glazed, or out of round.

Step 3 Using the roughing wheel, grind the side relief angle and the side cutting angle about 10 degrees by holding the blank and supporting your hand on the tool rest (Figure I-76).

Step 4 Check the angle with a tool gage (Figure I-77). Correct if necessary.

Step 5 Rough out the end relief angle about 10 or 16 degrees, depending on the type of toolholder and the end cutting edge angle (Figure I-78).

Step 6 Check the angle with the tool gage (Figure I-79). Correct if necessary.

Figure I-76 Roughing the side relief angle and the side cutting edge angle.

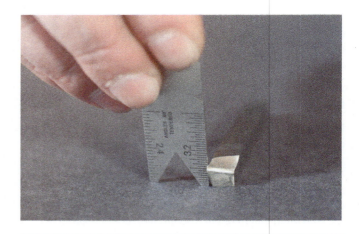

Figure I-77 Checking the side relief angle with a tool gage.

Figure I-78 Roughing the end relief angle and the end cutting edge angle.

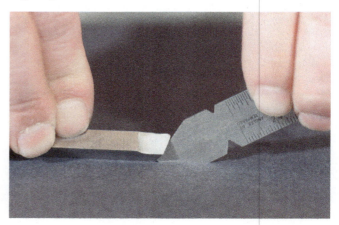

Figure I-79 Checking the end relief angle with a tool gage.

Step 7 Rough out the side rake. Stay clear of the side cutting edge by $\frac{1}{16}$ inch (Figure I-80).

Step 8 Check for wedge angle (Figure I-81). Correct if necessary.

Step 9 Change to a finer grit wheel and gently finish grinding the side and end relief angles. Try to avoid making

Figure I-80 Roughing the side rake.

Figure I-81 Checking for wedge angle with a tool gage.

several facets or grinds on one surface. A side-to-side oscillation will help produce a good finish.

Step 10 Grind the finish on the side rake as in Figure I-80 and bring the ground surface just to the side cutting edge—avoid going deeper.

Step 11 Grind a slight radius on the nose of the tool on the circumference of the wheel (Figure I-82) and all the way from the nose to the heel of the tool.

Step 12 Use a medium to fine oilstone to remove the burrs from the cutting edge (Figure I-83). The finished tool is shown in Figure I-84.

SELF-TEST

1. List two advantages of using high-speed steel for tools.
2. Other than hardness and toughness, what is the most important aspect of a lathe tool?
3. How do form tools work?
4. A tool that has been reground too many times can form a chip trap. Describe the problems that result from this.
5. Why is it not always necessary to grind a back rake into the tool?
6. When should a zero or negative rake be used?

Figure I-82 One method of grinding the nose radius is on the circumference of the wheel.

Figure I-83 Using an oil stone to remove the burrs from the cutting edge.

Figure I-84 The finished tool.

7. Explain the purpose of the side and end relief angles.
8. What is the function of the side and back rakes?
9. How can the side and back rake angles be checked?
10. Why should chips be broken?
11. In what ways can chips be broken?
12. A high-speed tool bit can easily be overheated by using a glazed wheel that needs dressing or by exerting too much pressure. What does this cause in the tool?

Lathe Spindle Tooling

There are several types of work-holding and driving devices that can be fastened to the spindle nose for machining on lathes. Various types of these work-holding devices, their uses, and proper care are detailed in this unit.

OBJECTIVES

After completing this unit, you should be able to:

- Explain the uses and care of independent and universal chucks.
- Explain the limitations and advantages of collets and describe a collet setup.
- Explain the use of a face driver or drive center.
- Explain the uses and differences of drive plates and faceplates.

LATHE SPINDLE NOSE

The lathe spindle nose can hold a variety of work-holding devices. The spindle is hollow and has an internal Morse taper at the nose end, which makes possible the use of taper shank drills or drill chucks (Figure I-85). This internal taper is also used to hold live centers, drive centers, or collet assemblies.

The outside of the spindle nose can have a threaded nose (Figure I-86), a long taper with key drive (Figure I-87), or a camlock (Figure I-88).

Threaded spindle noses are mostly found on older lathes. The chuck or faceplate is screwed on a coarse right-hand thread until it is forced against a shoulder on the spindle that aligns it. Two disadvantages of the threaded spindle nose are that the spindle cannot be rotated in reverse against a load and that it is sometimes difficult to remove a chuck or faceplate (Figure I-89).

The **long taper key drive spindle nose** relies on the principle that a tapered fit will always repeat its original position. The key gives additional driving power. A large nut

Figure I-85 Section view of the spindle.

Figure I-86 Threaded spindle nose (*Courtesy of Fox Valley Technical College*).

having a right-hand thread is turned with a spanner wrench. It draws the chuck into position and holds it there.

Camlock spindle noses use a short taper for alignment. Studs arranged in a circle fit into holes in the spindle nose. Each stud has a notch into which a cam is turned to lock it in place.

All spindle noses and their mating parts must be carefully cleaned before assembly. Small chips or grit will cause a work-holding device to run out of true and be damaged. A spring cleaner (Figure I-90) is used on mating threads for threaded spindles. Brushes and cloths are used for cleaning. A thin film of light oil should be applied to threads and mating surfaces.

Figure I-87 Long taper with key drive spindle nose (*Courtesy of Fox Valley Technical College*).

Figure I-88 Camlock spindle nose (*Courtesy of Fox Valley Technical College*).

Figure I-89 The chuck can be removed from a threaded spindle by using a large monkey wrench on one of the chuck jaws while the spindle is locked in a low gear. A long steel bar may also be used between the jaws (*Courtesy of Fox Valley Technical College*).

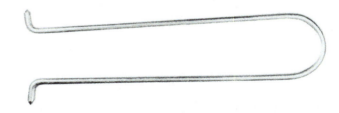

Figure I-90 A spring cleaner is used to clean internal threads on chucks.

INDEPENDENT AND UNIVERSAL CHUCKS

Independent four-jaw and universal three-jaw chucks and, occasionally, drive or faceplates are mounted on the spindle nose of engine lathes. Each of the four jaws of the **independent chuck** moves independently of the others, making it possible to set up oddly shaped pieces (Figure I-91).

The concentric rings on the chuck face help roughly center the work. Precise setups can be made with the four-jaw chuck by using a dial indicator, especially on round material. Each jaw of the chuck can be removed and reversed to accommodate irregular shapes. Some types are fitted with top jaws that can be reversed when bolts on the jaw are removed. Jaws in the reverse position can grip larger-diameter workpieces (Figure I-92). The independent chuck will hold work more securely for heavy cutting than will the three-jaw universal chuck.

Universal chucks usually have three jaws, but some have two jaws (Figure I-93) or six jaws (Figure I-94). The jaws are moved in or out equally in their slides by means of a scroll plate located at the back of the jaws. The scroll plate has a bevel gear on its reverse side that is driven by a pinion gear. This gear extends to the outside of the chuck body and is turned with the chuck wrench (Figure I-95). Universal chucks provide quick chucking and centering of round stock.

The jaws of universal chucks will not reverse, as independent chucks do, so a separate set of reverse jaws is used (Figure I-96) to hold pieces with larger diameters. The chuck and each of its jaws are stamped with identification numbers. Do not interchange any of these parts with another chuck; otherwise both will be inaccurate. Each jaw is also stamped

Figure I-91 Four-jaw independent chuck holding an offset rectangular part.

Figure I-92 Four-jaw chuck in reverse position holding a large-diameter workpiece (*Courtesy of Fox Valley Technical College*).

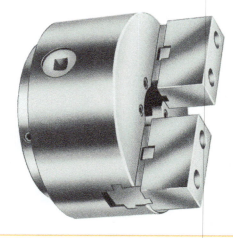

Figure I-93 Two-jaw universal chuck (*Used with permission of Hardinge Inc.*).

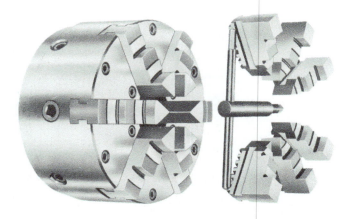

Figure I-94 Six-jaw universal chuck (*Courtesy of Buck Chuck Company*).

Figure I-95 Exploded view of universal three-jaw chuck (Adjust-tru) showing the scroll plate and gear drive mechanism (*Courtesy of Buck Chuck Company*).

1, 2, or 3 to correspond to the same number stamped by the slot on the chuck. Remove the jaws from the chuck in the order 3, 2, 1 and return them in the reverse order 1, 2, 3.

A universal chuck with top jaws (Figure I-97) is reversed by removing the bolts in the jaws and by reversing the jaws. The jaws must be carefully cleaned when this is done. Soft top jaws are frequently used when special gripping problems arise. Because the jaws are machined to fit the shape of the part (Figure I-98), they can grip it securely for heavy cuts.

One disadvantage of most universal chucks is that they lose their accuracy when the scroll and jaws wear, and normally there is no compensation for wear other than regrinding the jaws. The three-jaw adjustable chuck in Figure I-99 has a compensating adjustment for wear or misalignment.

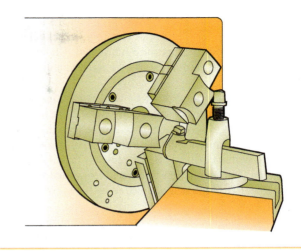

Figure I-98 Machining soft jaws to fit an oddly shaped workpiece on a jaw-turning fixture.

Figure I-96 Universal three-jaw chuck (Adjust-tru) with a set of outside jaws (*Used with permission of Buck Chuck Company*).

**1 PINION (C)
MOVES JAWS**

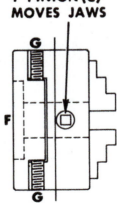

Figure I-97 Universal chuck with top jaws (*Courtesy of Haas Automation, Inc.*).

OTHER CHUCKS AND CHUCK COMPONENTS

Combination universal and independent chucks have independent adjustment for each jaw. These chucks are like the universal type, because three or four jaws move in or out equally, but each jaw can be adjusted independently as well.

Figure I-99 Universal chuck (Adjust-tru) with special adjustment feature (G) makes it possible to compensate for wear (*Courtesy of Buck Chuck Company*).

All chucks need frequent cleaning of scrolls and jaws. These should be lightly oiled after cleaning, and chucks with grease fittings should be pressure lubricated. Chucks come in several diameters and are made for light-, medium-, and heavy-duty uses.

Drive plates are used together with lathe dogs to drive work mounted between centers (Figure I-100). The live center fits directly into the spindle taper and turns with the spindle. A sleeve is sometimes used if the spindle taper is too large in diameter to fit the center. The live center is usually made of soft steel so the point can be machined as needed to keep it running true. Live centers are removed by means of a knock-out bar (Figure I-101).

When a machinist needs to machine the entire length of work mounted between centers without the interference of a lathe dog, special drive centers or face drivers (Figure I-102) can be used to machine a part without interference. Quite heavy cuts are possible with these drivers.

Faceplates are used for mounting workpieces or fix-tures. Unlike drive plates that have only slots, faceplates

Figure I-100 Drive plate for turning between centers (*Courtesy of Fox Valley Technical College*).

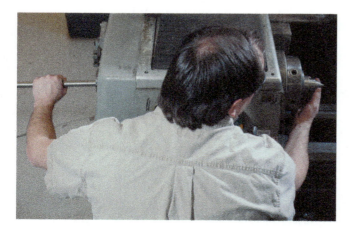

Figure I-101 Knockout bar is used to remove centers.

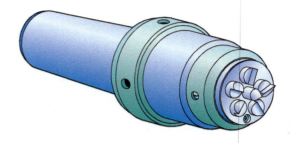

Figure I-102 Face driver is mounted in headstock spindle, and work is driven by the drive pins that surround the dead center.

Figure I-103 T-slot faceplate. Workpieces are clamped on the plate with T-bolts and strap clamps.

Figure I-104 Side and end views of a spring collet for round work.

have threaded holes and slots and are sturdier (Figure I-103). Faceplates are made of cast iron and so must be operated at relatively slow speeds. If the speed is excessive, the faceplate could fly apart.

Collet chucks are accurate work-holding devices and are used in producing small, high-precision parts. **Steel spring collets** (Figure I-104) are available for holding and turning hexagonal, square, and round workpieces. They are made in many different sizes, with a range of only a few thousandths of an inch. Workpieces to be gripped in a collet should not vary more than +.002 to −.003 inch from the collet size if the collets are to remain accurate. New collets can be expected to hold cylindrical work to .0005-inch runout. Rough and inaccurate workpieces should not be held in collets. The contact area in a collet would then be at a few points instead of along the entire length of the collet. The piece would not be held firmly and the collet could be damaged. To use collets in a lathe, an adapter called a **collet sleeve** is fitted into the spindle taper, and a draw bar is inserted into the spindle at the opposite end (Figure I-105). The collet is placed in the adapter and the draw bar is rotated, which threads the collet into the taper and closes it. Never tighten a collet without a workpiece in its jaws, as this can damage the collet. Before collets and adapters are installed, they should be cleaned to ensure accuracy.

Rubber flex collets (Figure I-106) have a set of tapered steel bars mounted in rubber. Rubber flex collets have a much wider range than the spring collet, each collet having a range of about $\frac{1}{8}$ inch. A large handwheel is used to open and close the collets instead of a draw bar. The concentricity that you can expect from each type of work-holding device is shown in Table I-3.

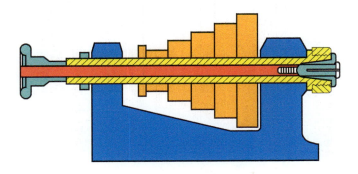

Figure I-105 Cross section of spindle showing construction of draw-in collet chuck attachment.

Table I-3 Accuracy of Holding Devices

Device	Centering Accuracy in Inches (Indicator Reading Difference)
Centers	Within .001
Four-jaw chuck	Within .001 (depending on the ability of the machinist)
Collets	.0005–.001
Three-jaw chuck	.001–.003 (good condition)
	.005 or more (poor condition)

SELF-TEST

1. Briefly describe the lathe spindle. How does the spindle support chucks and collets?
2. Name three spindle nose types.
3. What is an independent chuck, and what is it used for?
4. What is a universal chuck, and what is it used for?
5. Workpieces mounted between centers are driven with lathe dogs. Which type of plate is used on the spindle nose to turn the lathe dog?
6. Why is a spindle live center typically made from soft steel and how does it fit in the spindle nose?
7. On which type of plate are workpieces and fixtures mounted? What type of slots are used on it?
8. Name one advantage of using steel spring collets. Name one disadvantage.

INTERNET REFERENCES

Information on lathe spindle tooling:

http://www.workholding.com

http://www.royalprod.com

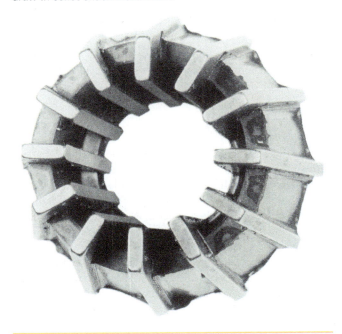

Figure I-106 Rubber flex collet.

Operating Lathe Controls

Before using any machine, know what the controls are for, and understand how they work. You must also be aware of the potential hazards for you and for the machine if it is misused. This unit prepares you to operate lathes.

OBJECTIVES

After completing this unit, you should be able to:

- Explain drives and shifting procedures for changing speeds on lathes.
- Describe the use of various feed control levers.
- Explain the relationship between longitudinal feeds and crossfeeds.
- State the differences in types of crossfeed screw micrometer collars.

Most lathes have similar control mechanisms and operating handles for feeds and threading.

Figure I-107 Speed-change levers and feed selection levers on a geared head lathe (*Courtesy of Fox Valley Technical College*).

DRIVES

Most lathes have gear drives; some are direct drive and some are variable speed. Geared head lathes are shifted with levers on the headstock (Figure I-107). Some of the levers are used to set up the various speeds within the range of the machine. The gears will not mesh unless they are perfectly aligned so that it is sometimes necessary to rotate the spindle by hand. **Never try to shift gears with the motor running and the clutch lever engaged.**

Older lathes used flat belts and stepped pulleys to provide different spindle speeds. The low-speed range on a flat-belt lathe relied on a back-gear drive system. Flat-belt-driven lathes have almost disappeared from school shops.

FEED CONTROL LEVERS

The carriage is moved along the ways by means of the lead screw when threading, or by a separate feed rod when using feeds. On very small lathes, however, a lead screw–feed rod combination may be used. To make left-hand threads by reversing the feed, the feed reverse lever is used. This lever reverses the lead screw. It should never be changed when the machine is running.

The quick-change gearbox in Figure I-108 typically has two or more sliding gear shift levers. These are used to select feeds or threads per inch. On those lathes also equipped with metric selections, the threads are expressed in pitch

Figure I-108 Quick-change gearbox with index plate.

Figure I-109 View of carriage apron with names of parts (*Courtesy of Fox Valley Technical College*).

(measured in millimeters). The carriage apron (Figure I-109) contains the handwheel for hand feeding and a power feed lever that engages a clutch to a gear drive train in the apron.

Hand feeding should not be used for long cuts, as the feed rate will not be consistent, and a poor finish will result. When using power feed and approaching a shoulder or the chuck jaws, disengage the power feed and hand feed the carriage for the last $\frac{1}{8}$ inch or so. Use the handwheel to finish the cut before returning to the starting point and beginning a new cut.

A feed change lever engages the feed to the carriage for longitudinal movement or to the cross slide. There is always some slack or backlash in the crossfeed and compound screws. As long as the tool is being fed to one direction against the workload, there is no problem, but if the screw is slightly backed off, the readings will be in error. Because of the backlash, the cross slide will not actually move even though the micrometer dial reading has been changed. When it is necessary to back the tool away from the cut a few thousandths of an inch, back off two full turns and then come back to the desired position on the micrometer dial.

Crossfeeds are geared differently than longitudinal feeds. On some lathes the crossfeed is approximately one-third to half that of the longitudinal feed, so a facing job (Figure I-110) with the quick-change gearbox set at about .012-inch feed will actually be only .004 inch for facing. The crossfeed ratio for each lathe is usually found on the quick-change gearbox index plate.

The half-nut or split-nut lever on the carriage engages the thread on the lead screw directly and is only used for threading. It cannot be engaged unless the feed change lever is in the neutral position.

Both the crossfeed screw handle and the compound rest feed screw handle are fitted with micrometer collars (Figure I-111). These collars traditionally have been graduated in English units, but some lathes have metric-conversion collars (Figure I-112). The dial in Figure I-112 can be rotated to change between English and metric.

Some micrometer collars are graduated to read single depth; that is, the tool moves as much as the reading shows. When turning a diameter such as a shaft, dials that read single depth will remove twice as much from the diameter

Figure I-110 Facing on a lathe. The guard is in place at left.

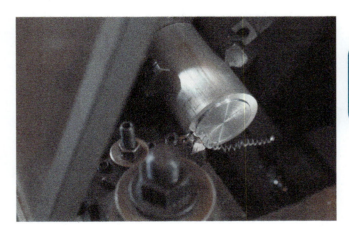

Figure I-111 Micrometer collar on the crossfeed screw that is graduated in English units. Each division represents .002 inch on this particular machine. Every division on this dial would reduce the diameter of the part by .002 inch.

Figure I-112 Crossfeed and compound screw handles with metric–English conversion collars (*Courtesy of Fox Valley Technical College*).

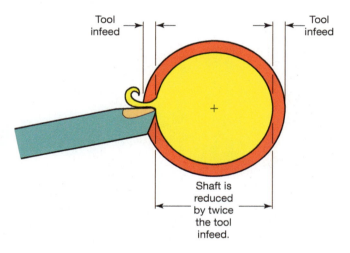

Figure I-113 The diameter of the workpiece is reduced by twice the amount the tool is moved.

Figure I-114 Clutch rod is actuated by moving the spindle clutch lever. This disengages the motor from the spindle and activates a brake on some lathes.

(Figure I-113). For example, if the crossfeed screw is turned in .020 inch and a cut is taken, the diameter will have been reduced by .040 inch. Many lathes, however, are graduated on the micrometer dial to compensate for double depth on cylindrical turning. On this type of lathe, if the crossfeed screw is turned in until .020 inch shows on the dial and a cut is taken, the diameter will have been reduced .020 inch. The tool will actually have moved into the work only .010 inch. This is sometimes called radius or diameter reduction.

To determine which type of graduation you are using, measure the diameter with a micrometer. Move the tool in a set amount such as .050 inch on the dial and take a cut. Measure the part diameter again after the cut. Check to see if the part is .050 smaller or .100 smaller.

Some lathes have a brake and clutch rod the same length as the lead screw. A clutch lever connected to the carriage apron rides along the clutch rod (Figure I-114). The spindle can be started and stopped without turning off the motor by using the spindle clutch lever. Many lathes also have a spindle brake that quickly stops the spindle when the spindle clutch lever is moved to the stop position.

An adjustable automatic clutch kick out is also a feature of the clutch rod.

When starting a lathe for the first time, use the following checkout list:

Step 1 Move the carriage and tailstock to the right to clear the work-holding device.

Step 2 Locate the feed clutches and half-nut lever, and disengage before starting spindle.

Step 3 Set up the lathe to operate at low speed.

Step 4 Read any machine information panels that may be located on the machine, and observe precautions.

Step 5 Note the feed direction; there are no built-in travel limits or warning devices to prevent feeding into the chuck or against the end of the slides.

Step 6 When you are finished with a lathe, disengage all clutches, clean up chips, and remove any attachments or special setups.

SELF-TEST

1. How are speeds set on geared head lathes?
2. What lever is shifted to reverse the lead screw?
3. The sliding gear shifter levers on the quick-change gearbox are used for just two purposes. What are they?
4. Why will the surface finish not be the same (tool marks per inch) on the face of a workpiece as on the outside diameter if using the same feed rate?
5. What is the half-nut lever used for?
6. How are micrometer collars on the crossfeed handle and compound handle graduated?
7. How can you determine if the lathe you are using is calibrated for single or double-depth?

Facing and Center Drilling

Facing and center drilling the workpiece are often the first steps taken in a lathe project. Much lathe work is done in a chuck, requiring facing and in some cases center drilling.

SETTING UP FOR FACING

Facing is done to obtain a flat surface on the face of parts clamped in a chuck, faceplate, or between centers (Figures I-115 and I-116).

Figure I-116 Facing the end of a shaft.

Figure I-117 Removing a camlock chuck mounted on a lathe spindle.

Figure I-115 Facing a workpiece in a chuck.

The work is typically held in a three- or four-jaw chuck. If the chuck is to be removed from the lathe spindle, a lathe board must be used to protect the ways. Figure I-117 shows a camlock-mounted chuck being removed. The correct procedure for installing a chuck on a camlock spindle nose is shown in Figures I-118 to I-122. All of the cams should be tightly snugged (Figure I-123).

Setting up work in an independent chuck is simple, but being efficient at it takes practice. Round stock can be set up using a dial indicator. Square or rectangular stock can be set up either with a toolholder turned backward or a dial indicator.

Figure I-118 Chips are cleaned from spindle nose with a brush.

Figure I-119 Cleaning chips from the chuck with a brush.

Figure I-120 Spindle nose is thoroughly cleaned with a soft cloth.

Figure I-121 Chuck is thoroughly cleaned with a soft cloth.

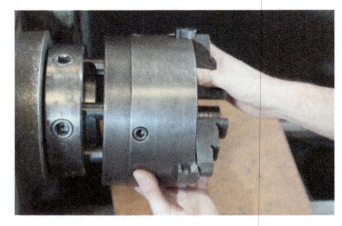

Figure I-122 Chuck is mounted on spindle nose.

Figure I-123 All cams are turned clockwise until locked securely.

Begin the setup by aligning two opposite jaws with the same concentric ring marked in the face of the chuck. This will roughly center the work. Set up the other two jaws with a concentric ring also. Next work on two opposite jaws first. Move the crossline dial until the toolholder just

Figure I-124 Rectangular stock being set up by using a toolholder turned backward.

Figure I-125 Adjusting the rectangular stock at 90 degrees from Figure I-124.

touches the part (Figure I-124). Read the dial. Then rotate the chuck 180 degrees. Move the cross slide in and read the dial again. The difference between the two readings will tell you how much these two jaws need to be adjusted. Loosen the low reading jaw and tighten the jaw with the high reading to move the part in the jaws. Repeat the process until the readings on the two jaws are the same. Next, repeat the process for the other two jaws (Figure I-125). Check and repeat until the piece is closely centered. If enough material is to be removed, this may be close enough. If it needs to be more perfectly centered, an indicator can be used. When you are using the dial indicator, zero the bezel at the lowest reading (Figure I-126). Then, rotate the chuck to the opposite jaw with the high reading and tighten it half the amount of the runout. It may be necessary to loosen the jaw on the low side slightly. Always tighten the jaws at the position where the dial indicator contacts the work, because any other location will give erroneous readings. Chalk is sometimes used for setting up rough castings and other work too irregular to be measured with a dial indicator. Workpieces can be chucked either internally or externally (Figures I-127 to I-129).

FACING

The material to be machined usually has been cut off in a power saw, so the piece is not square on the end or cut to the exact length. Facing from the center out (Figure I-130) produces a better finish, but it is difficult to cut on a solid face in the center. Facing from the outside (Figure I-131) is more convenient, because heavier cuts may be taken, and it is easier to work to the scribed lines on the circumference of the work. When you are facing from the center out, a right-hand turning tool in a left-hand toolholder is the best arrangement, but when facing from the outside to the center, a left-hand tool in a right-hand or straight toolholder can be used if you are using old-style toolholders.

With quick-change toolholders and light cuts, you can place the tool in any convenient position. When you are making heavy cuts, the tool pressure should be against the inside of the quick-change tool block to prevent the tool from moving. Facing or other machining should not be done on workpieces extending more than five diameters from the chuck jaws.

The tool point should be set to the dead center (Figure I-132). This is done by setting the tool to the tailstock center

Figure I-126 Setting up round stock in an independent chuck with a dial indicator.

Figure I-127 Normal chucking position.

Figure I-128 Internal chucking position.

point or by making a trial cut to the center of the work. If the tool is below center, a small uncut stub will be left. The tool can then be reset to the center of the stub.

The carriage can be locked (Figure I-133) when taking facing cuts, as the cutting pressure can cause the tool and carriage to move away, which will make the faced surface curved rather than flat. Finer feeds should be used for finishing than for roughing. Remember, on some lathes, the cross-feed is half to one-third that of the longitudinal (carriage) feed. On other lathes they are equal.

The ratio is usually listed on the index plate of the quick-change gearbox. A roughing feed can range from .005 to .015 inch, and a finishing feed from .003 to .005 inch. Use of cutting oils will help produce better finishes on finish facing cuts.

Facing to length may be accomplished by trying a cut and measuring with a hook rule (Figure I-134) or by facing to a previously made layout line. A more precise method is to use the graduations on the micrometer collar of the compound. The compound is set so that its slide is parallel to the ways (Figures I-135 and I-136). The carriage is locked in place, and a trial cut is taken with the micrometer collar set on the zero index. The workpiece is measured with a micrometer, and the desired length is subtracted from the measurement. The remainder is the amount you should remove by facing. If more than .015–.030 inch (depth left for finish cut) has to be removed, it should be taken off in two or more cuts by moving the compound micrometer dial the desired amount. A short trial cut (about $\frac{1}{8}$ inch) should again

be taken on the finish cut and adjustment made if necessary. Roughing cuts should be approximately .060 inch in depth, but they can vary considerably, depending on the machine size, horsepower, tooling, and setup.

The compound is often kept at 29–30 degrees for threading purposes (Figures I-137 and I-138). Note that the compound has been turned to 29.5 degrees in Figure I-138 but the reading shows 60.5 degrees. On some lathes the reading would be 29.5 and on other lathes it would be 60.5.

Figure I-129 External chucking position.

Figure I-130 Facing from the center to the outside of the workpiece.

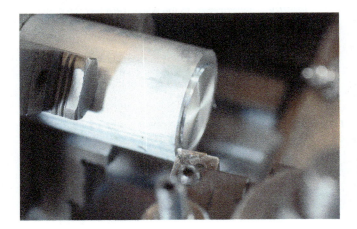

Figure I-131 Facing from the outside toward the center of the workpiece.

Figure I-132 Setting the tool to the center of the workpiece using the tailstock center.

Figure I-133 Carriage can be locked before taking a facing cut.

Figure I-134 Facing to length using a hook rule for measuring.

At this angle the tool feeds into the face of the work .001 inch for every .002 inch that the slide is moved. For example, if you wanted to remove .015 inch from the workpiece, you would turn in .030 inch on the compound micrometer dial (assuming it reads single depth).

Turning to size and facing on a shoulder requires a tool that can cut on both the end and the side (Figure I-139). Roughing should be done on both the diameter and the face before finishing to size. The diameter is usually the critical dimension and should therefore be finished to size after the face is finished.

Figure I-135 The compound is set at 90 degrees for facing operations.

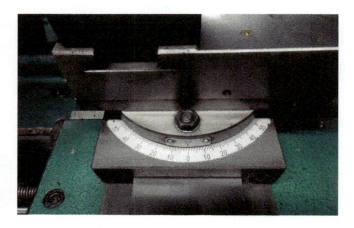

Figure I-136 Close-up of the compound aligned with the spindle. Note on this lathe the compound reads 0 degrees. On some lathes it might read 90 degrees.

Figure I-137 The compound set at 30 degrees.

Figure I-138 Close-up of the compound set at 29.5 degrees. Note however that the compound reading is about 60.5. On this lathe it would be 90 degrees minus 29.5 degrees to get the compound set correctly. This is because if you start with the compound perpendicular to the ways it reads 90 degrees. The compound must then be rotated 29.5 degrees which would then show 60.5 degrees.

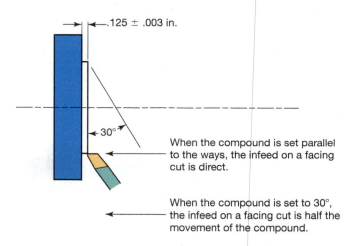

Figure I-139 Turning and facing an offset with a tool that cuts on both the end and the side.

Figure I-140 Half-centers make facing shaft ends easier.

A specially ground tool is used to face the end of a workpiece that is mounted between centers. The right-hand facing tool is shaped to fit in the angle between the center and the face of the workpiece. Half-centers (Figure I-140) are made to make tool access easier, but they should be used only for facing and not for general turning.

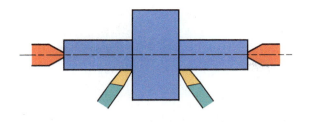

Figure I-141 Work held between centers on a mandrel can be faced on both sides with right- and left-hand facing tools.

Both right-hand and left-hand facing tools are used for facing work held on mandrels (Figure I-141). Care should be taken when machining pressure is toward the small end of a tapered mandrel (usually toward the tailstock). Excessive pressure may loosen the workpiece on the mandrel.

SPEEDS

Speeds (rpm) for a lathe are determined in the same way as speeds for drills. The only difference is that the diameter of the work is used instead of the diameter of the drill. In facing operations, the outside diameter of the workpiece has greater surface speed than its center. Ideally the rpm should vary as the tool is moved in or out. This can be done on CNC machines, but is difficult to accomplish on manual lathes. On a manual lathe a happy medium is usually chosen for the RPM. For facing work, the outside diameter is always used to determine maximum rpm. Thus

$$\text{rpm} = \frac{\text{CS} \times 4}{D}$$

where

D = diameter of workpiece (where machining is done)
rpm = revolutions per minute
CS = cutting speed, in surface feet per minute

EXAMPLE 1

The cutting speed for low-carbon steel is 90 SFPM (surface feet per minute), and the workpiece diameter to be faced is 6 inches. Find the correct rpm.

$$\text{rpm} = \frac{90 \times 4}{6} = 60$$

EXAMPLE 2

A center drill has a $\frac{1}{8}$-inch drill point. Find the correct rpm to use on low-carbon steel (CS 90).

$$\text{rpm} = \frac{90 \times 4}{\frac{1}{8}} = \frac{360}{1} \times \frac{8}{1} = 2880$$

These are approximate speeds and will vary according to the conditions. If chatter marks (vibration marks) appear on the workpiece, the rpm should be reduced. If this does not help, ask your instructor for assistance.

CENTER DRILLS AND DRILLING

When work is turned between centers, a center hole is required on each end of the workpiece. The center hole must have a 60-degree angle to conform to the center and have a smaller drilled hole to clear the center's point. This center hole is made with a center drill, often referred to as a *combination drill and countersink*. These drills are available in a range of sizes from $\frac{1}{8}$- to $\frac{3}{4}$-inch body diameter and are classified by numbers from 00 to 8, which are stamped on the drill body. Facing the workpiece is almost always necessary before center drilling, because an uneven surface can push sideways on the center-drill point and break it.

Center drills are usually held in a drill chuck in the tailstock. The workpieces are turned in a lathe chuck for center drilling (Figure I-142). Center holes are drilled by rotating the work in the chuck and feeding the center drill into the work by using the feed handle on the tailstock spindle. Long cylindrical pieces often need to be turned between the chuck and a tailstock center with a steady rest (Figure I-143). The end of the piece must first be center drilled. Long workpieces are generally faced by chucking one end and supporting the other in a steady rest. Because the end of stock is not always sawed square, it should be center drilled only after spotting a small hole with the lathe tool. A slow feed is needed to protect the small, delicate drill end. Cutting fluid should be used, and the drill should be backed out frequently to remove chips. The greater the work diameter and the heavier the cut, the larger the center hole should be. Table I-4 shows the suggested center-drill size for various workpiece diameters.

The size of the center hole is selected by the center-drill size and then controlled by the depth of drilling. You must be careful not to drill too deeply (Figure I-144), as this causes the center to contact only the sharp outer edge of the hole, which is a poor bearing surface. It soon becomes loose and out of round and causes such machining problems as chatter and roughness. Center drills often are broken from feeding the drill too fast with the lathe speed too slow or with the tailstock off center.

Figure I-142 Center drilling a workpiece held in a chuck.

Figure I-143 Long material supported in a steady rest (*Courtesy of Fox Valley Technical College*).

Table I-4 Center-Drill Sizes

Center-Drill Number	Drill Point Diameter (in.)	Body Diameter (in.)	Work Diameter (in)
1	$\frac{3}{64}$	$\frac{1}{8}$	$\frac{3}{16}-\frac{5}{16}$
2	$\frac{5}{64}$	$\frac{3}{16}$	$\frac{3}{8}-\frac{1}{2}$
3	$\frac{7}{64}$	$\frac{1}{4}$	$\frac{5}{8}-\frac{3}{4}$
4	$\frac{1}{8}$	$\frac{5}{16}$	$1-1\frac{1}{2}$
5	$\frac{3}{16}$	$\frac{7}{16}$	$2-3$
6	$\frac{7}{32}$	$\frac{1}{2}$	$3-4$
7	$\frac{1}{4}$	$\frac{5}{8}$	$4-5$
8	$\frac{5}{16}$	$\frac{3}{4}$	6 and over

Center drills are often used as starting or spotting drills when drilling is to be performed (Figures I-145 and I-146). This keeps the drill from "wandering" off center and making the hole eccentric. Spot drilling is done with the work chucked or supported in a steady rest. Care must be taken to center the workpiece properly in the steady rest, or the center drill will be broken.

SHOP TIPS—CRAFTMANSHIP

Round bar stock is neither perfectly round nor perfectly straight. If a workpiece is center drilled close to the chuck and then moved out and rechucked, the center-drilled hole might not run true. Placing the tailstock center in the center hole may bend the workpiece slightly as the center pressure forces it back into center. This will cause eccentric turning geometry on the part when it is removed from the lathe. In such cases, try facing the work close to the chuck and then moving it out and rechucking before center drilling. The center hole will be centered, and the tailstock center will not force the workpiece out of position. Exercise caution so that a slender workpiece does not start to whip and possibly bend during this procedure. It may be necessary to center drill at a lower spindle speed for safety during the center-drilling operation.

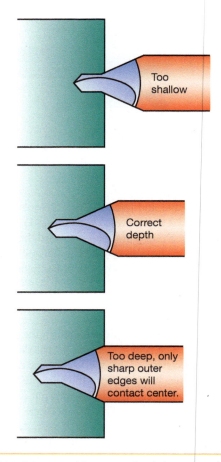

Figure I-144 Correct and incorrect depth for center drilling.

Figure I-145 Center drill is brought up to work and lightly fed into material.

Figure I-146 Center drill is fed into work with a slow, even feed (*Courtesy of Fox Valley Technical College*).

SELF-TEST

1. You have a rectangular workpiece that needs facing and center drilling. A three-jaw chuck is mounted on the lathe spindle. Describe what you need to do to prepare for machining?

2. Should the point of the tool be set above, below, or at the center of the spindle axis when taking a facing cut?

3. What is the drill point diameter for a number 4 center drill?

4. If the cutting speed of aluminum is 300 SFPM and the workpiece diameter is 4 inches, what is the rpm? The formula is

$$\text{rpm} = \frac{\text{CS} \times 4}{D}$$

5. Give two reasons for center drilling a workpiece in a lathe.

6. Name two causes of center-drill breakage.

7. What happens when you drill too deeply with a center drill?

8. What is the drill point diameter of a number 5 center drill?

INTERNET REFERENCE

Information on lathe operations:

http://en.wikipedia.org/wiki/Metal_lathe

Turning between Centers

Turning between centers is a good way for you to learn the basic principles of lathe operation.

OBJECTIVES

After completing this unit, you should be able to:

- Describe the correct setup procedure for turning between centers.
- Select correct feeds and speeds for a turning operation.
- Detail the steps necessary for turning to size predictably.
- Turn a $1\frac{1}{4}$-inch diameter shaft with six shoulders to a tolerance of +.000 or –.002 inch.

SETUP FOR TURNING BETWEEN CENTERS

To turn a workpiece between centers, support it between the dead center (tailstock center) and the live center in the spindle nose. A lathe dog (Figure I-147) clamped to the workpiece is driven by a drive plate (Figure I-148) mounted on the spindle nose. Machining with a single-point tool can be done anywhere on the workpiece except near the lathe dog.

Turning between centers has disadvantages. A workpiece cannot be cut off with a parting tool while being supported between centers, as this will bind and break the parting tool and ruin the workpiece. For drilling, boring, or machining the end of a long shaft, a steady rest is normally used to support the work, but these operations cannot be done when the shaft is supported only by centers.

The advantages of turning between centers are many. A shaft between centers can be turned end for end to continue machining without eccentricity if the live center runs true (Figure I-149). If a partially threaded part is removed from between centers for checking and everything is left the same on the lathe, the part can be returned to the lathe and the threading can be resumed where it was left off.

Figure I-147 Lathe dog (*Courtesy of Fox Valley Technical College*).

Figure I-148 Dog plate (drive plate) on the spindle nose of the lathe (*Courtesy of Fox Valley Technical College*).

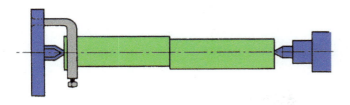

Figure I-149 Eccentricity in the center of the part because the live center is off center.

Lathe Centers

The center for the headstock spindle is sometimes referred to as a *live* center because it rotates, in contrast with a nonrotating dead center in the tailstock spindle. The live center rotates, and the dead center does not. To avoid difficulties in terminology in this text, the center in the headstock is referred to as a live center, and a nonrotating tailstock center as a dead center.

The center for the headstock spindle might not be hardened. This makes it possible to re-machine the tip when it is installed in the headstock spindle making it almost perfectly true. Thoroughly clean the inside of the spindle with a soft cloth and wipe off the live center. If the center is too small for the lathe spindle taper, use a tapered bushing that fits the lathe. Seat the bushing firmly in the taper (Figure I-150) and install the center (Figure I-151). Set up a dial indicator on the end of the center (Figure I-152) to check for runout. If there is runout, remove the center by using a knockout bar through the spindle. Be sure to catch the center with one hand. Check the outside of the center for nicks or burrs. These can be removed with a file. Check the inside of the spindle taper with your finger for nicks or grit. If nicks are found, **do not** use a file. Check with your instructor to see what to do next. After removing nicks, if the center still runs out more than the acceptable tolerances (usually .0001–.0005 inch), a light cut by tool or grinding can be taken with the compound set at 30 degrees.

A chuck center is often machined from a short piece of soft steel mounted in a chuck (Figure I-153). It is then left in place, and the workpiece is mounted between it and

Figure I-150 Make sure that the bushing is firmly seated in the taper.

Figure I-151 Installing the center.

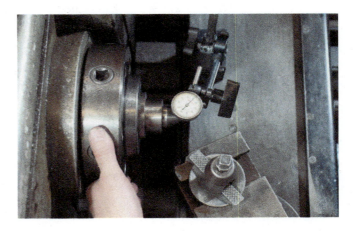

Figure I-152 Checking the live center for runout with a dial indicator.

Figure I-153 A live center being machined in a four-jaw chuck. Note that it is called a live center because it will turn with the work. The lathe dog on the workpiece will be driven by one of the chuck jaws.

the tailstock center. A lathe dog with the bent tail against a chuck jaw is used to drive the workpiece. This procedure sometimes saves time on large lathes where changing from the chuck to a drive plate is cumbersome, and the amount of work to be done between centers is small.

The nonrotating tailstock (dead) center (Figure I-154) is hardened to withstand machining pressures and friction.

Clean inside the taper and the center before installing. Ball bearing antifriction centers are often used in the tailstocks, as they will withstand high-speed turning without the overheating problems of dead centers. Ball bearing centers are used extensively in machine shops. Dead centers are virtually obsolete but have the advantage of greater rigidity. Rolling pipe centers are used for turning tubular material (Figure I-155). Ball bearing centers are shown in Figures I-156 and I-157.

To set up a workpiece that has been previously center drilled, slip a lathe dog on one end with the bent tail toward the drive plate. Do not tighten the dog yet. Put antifriction compound into the center hole toward the tailstock and then place the workpiece between centers (Figure I-158). The tailstock spindle should not extend out too far, as the machine will lose rigidity, and chatter or vibration may result. Set the dog in place and avoid any binding of the bent tail (Figure I-159). Tighten the dog and then adjust the tailstock so there is no end play and the bent tail of the dog moves freely in its slot. Tighten the tailstock binding lever. The heat of machining will expand the workpiece and cause the dead center to heat from friction. If overheated, the center may

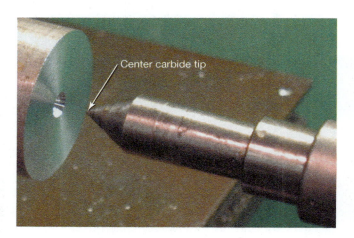

Center carbide tip

Figure I-154 This dead center is hardened to resist wear. It is made of high-speed steel with a carbide insert.

Figure I-155 The pipe center is useful for supporting a tubular workpiece. (*Courtesy of Fox Valley Technical College*).

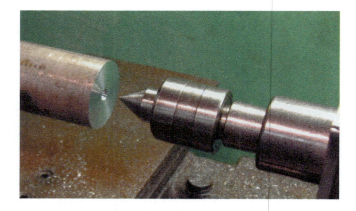

Figure I-156 Antifriction ball bearing center.

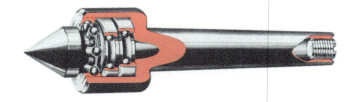

Figure I-157 Cutaway view of a ball bearing tailstock center (*Used with permission of DoALL Company*).

Figure I-158 Antifriction compound put in the center hole before setting the workpiece between centers. This step is not necessary when using an antifriction ball bearing center.

be ruined and may even be welded into your workpiece. Periodically, or at the end of each heavy cut, check the adjustment of the centers and reset if necessary.

When a tool post and toolholder are used, the toolholder must be positioned so that it will not contact the work. When heavy cuts are taken with this type of toolholder, it is more likely to move and dig into the workpiece than more modern rigid toolholders (Figure I-161). If the older style toolholder is used, Figure I-162 shows how it should be positioned so that the toolholder will swing away from the roughing cut rather than digging into the work. The tool and toolholder should not overhang too far for rough turning. The tool and holder should be close to

Figure I-159 Lathe dog in position.

Ram Lock Handle

The tailstock handle is positioned on the right side so that its weight will help keep the center tight.

Figure I-160 The tailstock handwheel handle (A) is positioned over center to the right side so that its weight will keep the center tight during turning operations.

OPERATING TIP

Avoiding Loose Tailstock Centers When tailstock centers become loose, the workpiece can fly out of the machine. This can happen if the tailstock ram lock is not sufficiently tight. One method to lessen the chance of a loose center is to position the crank handle (A) so that its weight will help keep the center tight, as shown in Figure I-160. If the handle is the other way, vibration may have a tendency to loosen the lock handle.

the tool post (Figure I-163). Tools should be set on or slightly above the center of the workpiece for roughing and on center for finishing. The tool may be set to the dead center or to a steel rule on the workpiece (Figure I-164). This method should not be used with carbide tools because of their brittleness.

Figure I-161 This quick-change toolholder can make heavy cuts without the tool loosening and moving if the clamp bolt is sufficiently tight.

Figure I-162 Correct position of toolholder for roughing. The toolholder will swing away from the cut with excessive feeds.

Figure I-163 Tool and toolholder in the correct position.

Figure I-164 Centering a tool by means of a steel rule. When the rule is vertical, the tool is centered.

TURNING BETWEEN CHUCK AND CENTER

SAFETY FIRST

Straight turning on shafts is often done with the workpiece held between the chuck and the tailstock center (Figure I-165). The advantages of this method are quick, solid setup, and a positive drive.

One disadvantage is that runout of the jaws can cause runout in the shaft. Another problem is the tendency for the workpiece to slip into the chuck jaws with heavy cuts, allowing the workpiece to come out of the tailstock center. This can present a hazardous situation; the workpiece can fly out of the machine, or if it is long and slender, it can act like a whip. Endwise movement of the shaft can be detected by watching the tailstock center closely or by making a mark next to the jaws. If the workpiece begins to loosen at the center, quickly shut off the machine and readjust the center. One way to prevent this movement is to first machine a shoulder on the shaft end that will be in the chuck (Figure I-166). Because the shoulder contacts the chuck jaws, the shaft cannot slide into the chuck. Machining pressure can also force the tailstock center to move away from the end of the part.

SPEEDS AND FEEDS FOR TURNING

Because machining time is an important factor in lathe operations, you must understand the principles of speeds and feeds to make the most economical use of your machine. Speeds are determined for turning between centers by using the same formula given for facing operations in the last unit:

$$rpm = \frac{CS \times 4}{D}$$

where

rpm = revolutions per minute

D = diameter of workpiece

CS = cutting speed, in surface feet per minute (sfpm)

Cutting speeds for various materials are given in Table I-5 using high-speed tools. When carbide tools are used, multiply these values by 3 or 4.

EXAMPLE

If the cutting speed is 40 for a certain alloy steel and the workpiece is 2 inches in diameter, find the rpm.

$$rpm = \frac{40 \times 4}{2} = 80$$

After calculating the rpm, use the nearest or next-lower speed on the lathe and set the speed.

Figure I-165 Work being turned between chuck and tailstock center.

Figure I-166 Shoulder turned on a shaft to prevent it from sliding into the chuck.

Feeds are expressed in inches per revolution (ipr) of the spindle. A .010-inch feed will move the carriage and tool .010 inch for one full turn of the headstock spindle. If the spindle speed is changed, the feed ratio still remains the same. Feeds are selected by means of an index chart (Figure I-167) found either on the quick-change gearbox or on the side of the headstock (Figure I-168). The sliding gear levers are shifted to different positions to obtain the feeds indicated on the index plate. The lower decimal numbers on the plate are feeds, and the upper numbers are threads per inch.

Feeds and the depth of cut should be as much as the tool, workpiece, and machine can stand without undue stress. The feed rate for roughing should be from one-fifth to one-tenth the depth of cut. A small 10- or 12-inch swing lathe should handle a $\frac{1}{8}$-inch depth of cut in soft steel, but in some cases, this may have to be reduced to $\frac{1}{16}$ inch. If .100 inch is selected as a trial depth of cut, the feed can be anywhere from .010 to .020 inch. If the machine seems to be overloaded, reduce the feed. Finishing feeds can be from .003 to .005 inch for steel. Use a tool with a larger nose radius for finishing.

Roughing and Finishing

Machining time is an exceedingly important aspect of lathe operations. The time required in finishing operations is generally governed by the surface finish requirements and tolerance limits. Shortcuts should never be taken when finishing

Figure I-168 Index chart for feed mechanism on a modern geared head lathe with both metric and inch thread and feed selections.

to size. Any time saved in machining a workpiece should be during the roughing operations. Coarser feeds, greater depth of cuts, higher rpm (when the tool material can withstand it), and cutting fluids all contribute to higher metal-removal rates. The formula to determine time for turning, boring, and facing is

$$T = \frac{L}{fN}$$

where

T = time, in minutes
L = length of cut, in inches
f = feed, in ipr
N = lathe spindle speed, in rpm

SHOP TIP

When making intermittent cuts on a lathe, as in turning a shaft with keyways or splines, use a greater depth of cut and normal speeds with smaller feeds. The edge angle of the tool should be larger than usual and the gibs should be tight. A tough grade of carbide insert such as C-5 or C-6 should be used.

Figure I-167 Index chart on the quick-change gearbox.

Table I-5 Cutting Speeds and Feeds for High-Speed Steel Tools

	Low-Carbon Steel	High-Carbon Steel Annealed	Alloy Steel Normalized	Aluminum Alloys	Cast Iron	Bronze
Speed (SFPM)						
Roughing	90	50	45	200	70	100
Finishing	120	65	60	300	80	130
Feed (ipr)						
Roughing	.010–.020	.010–.020	.010–.020	.015–.030	.010–.020	.010–
Finishing	.003–.005	.003–.005	.003–.005	.005–.010	.003–.010	.003–

EXAMPLE 1

A 3-inch diameter shaft is being turned at 130 rpm with an HSS (high-speed steel) tool. The feed is .020 ipr, and the depth of cut is .200 inch. The length of cut is 4 inches. The time required to make the cut is

$$\frac{4}{.02 \times 130} = 1.54 \text{ min}$$

EXAMPLE 2

The same 3-inch diameter shaft is being turned at 200 rpm with .005 ipr feed. The lathe horsepower limits the depth of cut to .050 inch because of the increased speed. The length of cut is 4 inches. The time required to make the cut is

$$\frac{4}{.005 \times 200} = 4 \text{ min}$$

These examples show that light cuts and small feeds often waste time in roughing operations, even when higher cutting speeds are used. The 4-inch long cut in Example 1 not only takes less time than the one in Example 2, but it is four times as deep.

PROFESSIONAL PRACTICE

A good machinist takes professional pride in his or her workmanship. When a person is new on the job, mistakes will be made, causing a part to be scrapped. Do not attempt to cheat by altering the part or by hiding it. It is better to keep an honest relationship with your employer and admit mistakes. Then you can make another part, profiting from your mistake, or the boss may find a way to save it for you.

Bosses know there will be failures at first. They would rather have an open, honest relationship with an employee than have to deal with an irate customer who has received substandard work.

When you have finished with a piece of work, be sure that it is clean, with no oil or fingerprints on it. Even if the work is within tolerances and the finishes are excellent, a dirty finished part looks unfinished.

Chip Safety

SAFETY FIRST

Chip formation and chip handling are important for safety. Coarser feeds, deeper cuts, and smaller rake angles all tend to increase chip curl, which breaks up the chip into small, safe pieces. Fine feeds and shallow cuts, on the other hand, produce a tangle of wiry, sharp, hazardous chips (Figures I-169 and I-170) even with a chip breaker on the tool. Long strings may come off the tool, suddenly wrap in the work, and be drawn back rapidly to the machine. The edges are like saws and can cause severe cuts.

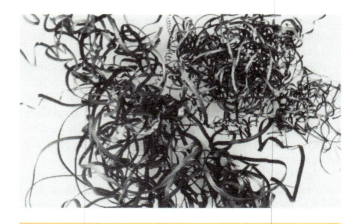

Figure I-169 Tangle of wiry chips. These chips can be hazardous to the operator.

Figure I-170 Better formation of chips. This type of chip will fall into the chip pan and is more easily handled.

Figure I-171 A trial cut is made to establish a setting of a micrometer dial in relation to the diameter of the workpiece.

TURNING TO SIZE

Get used to trusting the dials on the lathe. Use good methods so you do not introduce backlash. Take a cut (Figure I-171) and measure the diameter (Figure I-172). Take large cuts to get close to the finish size. Leave about .010–.015 for the finish cut. Remeasure and take the finish cut. If the machine has digital readouts for position, it is even more reliable. Digital readouts precisely measure the tool travel. Believe your dial or digital readout and take large roughing cuts to be efficient.

Finishing of machined parts with a file and abrasive cloth should not be necessary if the tools are sharp and honed, and if the feeds, speeds, and depth of cut are correct. A machine-finished part looks better and is more accurate than a part finished with a file and abrasive cloth.

If you do have to file on a lathe, use a low speed and long strokes, and file left-handed (Figure I-173). For polishing with abrasive cloth, set the lathe for a high speed and move the cloth back and forth across the work. Hold an end of the cloth strip in each hand (Figures I-174 and I-175).

Abrasive grains from abrasive cloth can damage the sliding surfaces of machinery. Some shops do not allow abrasive cloth to be used on machines. If you use it for polishing, always cover the ways and saddle with paper. Even if paper is used, abrasive cloth leaves grit on the ways of the lathe, so thoroughly clean the ways after polishing.

Figure I-174 Using abrasive cloth for polishing.

Figure I-175 Using a file for backing abrasive cloth for more uniform polishing.

Figure I-172 Measuring the workpiece with a micrometer.

SHOP TIPS—CRAFTSMANSHIP

At times, you must work with old and worn machine tools. In long turning operations on old and worn lathes, the workpiece diameter may taper considerably. This is often caused by wear in the ways, especially near the chuck, where most turning is done. Tapering can also be caused by tailstock centers out of alignment or a loose center that provides insufficient support for the workpiece. The problem may sometimes be solved by tightening up the gibs on the lathe carriage, realigning the tailstock center, and making sure that the tailstock center is tight in the workpiece center hole.

Turning to Length

Shoulders can be machined to specific lengths in several ways. Using a machinist's rule (Figure I-176) to measure workpiece length to a shoulder is a common but semiprecision method. Carriage micrometer stops (Figure I-177) can be used to limit carriage movement and establish a shoulder. This

Figure I-173 Filing in the lathe, left-handed.

Figure I-176 Measuring the workpiece length to a shoulder with a machinist's rule.

Figure I-177 Carriage micrometer stop set to limit tool travel to establish a shoulder.

method can be very accurate if it is set up correctly. The compound rest can also be used to machine a given distance to establish a shoulder. The compound rest is set parallel to the ways, and the micrometer collar indicates the distance moved. If the compound slide is not set accurately, a taper will be cut. Another accurate means of machining shoulders is using a dial indicator that has long travel. The indicators can show the amount of movement of the carriage. A digital readout system is the easiest and most precise method for turning to a specific length. Whichever method is used, the power feed should be turned off $\frac{1}{8}$ inch short of the workpiece shoulder and the tool should be hand-fed to the desired length. If the tool should be accidentally fed into an existing shoulder, the feed mechanism may jam and be difficult to release. A broken tool, toolholder, or lathe part may be the result.

MANDRELS

Mandrels, sometimes called **lathe arbors**, are used to hold work turned between centers (Figures I-178 and I-179). Tapered mandrels are made in standard sizes and have a taper of .006 inch per foot. A flat is milled on one end of the mandrel for the lathe dog setscrew. High-pressure lubricant is applied to the bore of the workpiece, and the mandrel is pressed into the workpiece with an arbor press. The assembly is mounted between centers, and the workpiece is turned or faced on either side. Care must be taken when the feed is toward the small end of the taper, because the workpiece may loosen on the mandrel.

Expanding mandrels (Figure I-180) have the advantage of providing a uniform gripping surface for the length of the bore (Figure I-181). A **tapered mandrel** grips tighter on one end than the other. **Gang mandrels** (Figure I-182) grip several pieces of similar size, such as disks, to turn their circumference. These are made with collars, thread, and a nut for clamping.

Figure I-178 Tapered mandrel or arbor.

Figure I-179 Tapered mandrel and workpiece set up between centers (*Courtesy of Fox Valley Technical College*).

Figure I-180 Expanding mandrel (*Courtesy of Fox Valley Technical College*).

Figure I-183 Stub mandrel being machined to size with a slight taper, about .006 inch per foot

Figure I-181 Expanding mandrel and workpiece set up between centers.

Figure I-184 Part to be machined being affixed to the mandrel. The mandrel is oiled so that the part can be removed easily.

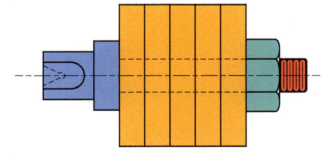

Figure I-182 Many similar parts can be machined at the same time with a gang mandrel.

Figure I-185 Part being machined after assembly on mandrel.

Stub mandrels (Figures I-183 to I-185) are used in chucking operations. These are often quickly machined for a single job and then discarded. Stub mandrels are used when production of many similar parts is carried out. Threaded stub mandrels are used for machining the outside surfaces of parts that are threaded in the bore (Figure I-186). An adjustable stub mandrel is shown in Figure I-187.

Coolants are used for heavy-duty and production turning. Oil–water emulsions and synthetic coolants are the most commonly used, whereas sulfurized oils usually are not used for turning operations except for threading. Some shop lathes do not have a coolant pump and tank, so if any cutting fluid is used, it is applied with a brush or pump oil can. Coolants and cutting oils for various materials are given in Table I-6.

Figure I-186 Threaded stub mandrel (*Courtesy of Fox Valley Technical College*).

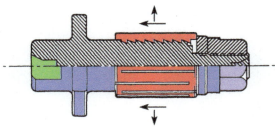

MC standard between centers nut actuated

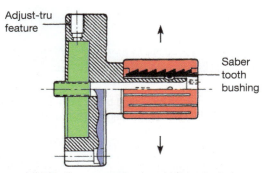

Adjust-tru feature

Saber tooth bushing

MLDR—flange mount—draw bolt—actuated stationary sleeve

Figure I-187 Precision adjustment can be maintained on the expanding stub mandrels with the Adjust-tru feature on the flange mount (*Used with permission of Buck Chuck Company*).

SELF-TEST

1. What other method besides turning between centers is used for turning shafts and long workpieces supported by the tailstock center?
2. What factors tend to promote or increase chip curl so that safer chips are formed?
3. Name three kinds of centers used in the tailstock and explain their uses.
4. How is the tailstock adjusted to the correct pressure when a dead center is used?
5. Why should the dead center be frequently adjusted when turning between centers?
6. Why should you avoid excess overhang with the tool and toolholder when roughing?
7. Calculate the rpm for turning a $1\frac{1}{2}$-inch diameter shaft of machine steel. Assume the cutting speed to be 90 SFPM.
8. How much stock should be left for finishing?

INTERNET REFERENCES

Information on lathe operations and specialty lathe centers:

www.riten.com

Table I-6 Coolants and Cutting Oils Used for Turning

Material	Dry	Water-Soluble Oil	Synthetic Coolants	Kerosene	Sulfurized Oil	Mineral Oil
Aluminum		X	X	X		
Brass	X	X	X			
Bronze	X	X	X			X
Cast iron	X					
Steel						
Low carbon		X	X		X	
Alloy		X	X		X	
Stainless		X	X		X	

Alignment of Lathe Centers

As a machinist, you must be able to check a workpiece for taper and properly adjust the tailstock of a lathe. This unit will show you how to align the centers of a lathe.

OBJECTIVES

After completing this unit, you should be able to:

- Check for taper with a test bar, and restore alignment by adjusting the tailstock.
- Check for taper by taking a cut with a tool and measuring the workpiece, and restore alignment by adjusting the tailstock.

TAILSTOCK ALIGNMENT

The tailstock will normally stay in alignment. If a lathe was used for taper turning with the tailstock offset, however, the tailstock may not have been realigned properly (Figure I-188). The tailstock also could be slightly out of alignment if an improper method of adjustment was used. It is therefore a good practice to check the center alignment of the lathe.

It is too late to save the workpiece if a taper is discovered while making a finish cut. A check for taper should be made on the workpiece while it is still in the roughing stage. You can do this by taking a light cut along the workpiece, then checking the diameter on each end with a micrometer. The difference between the two readings is the amount of taper in that distance.

Four methods are used for aligning centers on a lathe. In one method, the center points are brought together and visually checked for alignment (Figure I-189). This, of course, is not a precision method for checking alignment.

Another method of aligning centers is to use the tailstock witness marks. Adjusting the tailstock to the witness marks (Figure I-190), however, is only an approximate means of alignment.

The tailstock alignment can be adjusted by adjustment screws in the tailstock. A typical arrangement is shown in

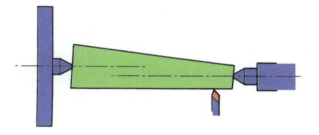

Figure I-188 Tailstock out of line, causing a tapered workpiece.

Figure I-189 Checking alignment by matching center points.

Figure I-191, where one setscrew is released and the opposite one is tightened to move the tailstock on its slide (Figure I-192). The tailstock clamp bolt must be released before the tailstock is adjusted.

USE OF A TEST BAR AND MACHINING

A more accurate means of aligning centers is to use a test bar. A test bar is simply a shaft that has true centers (is not off center) and has no taper. Some test bars are made with

Figure I-190 Adjusting the tailstock to the witness marks for alignment.

Figure I-191 Hexagonal socket setscrew that, when turned, moves the tailstock provided that the opposite one is loosened.

Figure I-192 The opposite setscrew being adjusted.

two diameters for convenience. No dog is necessary for checking alignment with a test bar, as the bar is not rotated. A dial indicator is mounted, preferably in the tool post, so that it will travel with the carriage (Figure I-193). Its contact point should be on the center of the test bar.

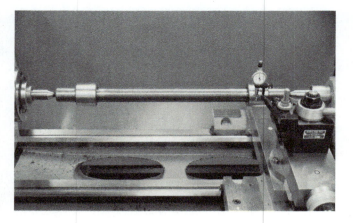

Figure I-193 Test bar setup between centers with a dial indicator mounted in the tool post.

Figure I-194 Indicator is moved to measuring surface at headstock end, and the bezel is set on zero.

Begin with the indicator at the headstock end, and set the indicator bezel to zero (Figure I-194).

Then, move the setup to the tailstock end of the test bar (Figure I-195) and check the dial indicator reading. If no movement of the needle has occurred, the centers are in line. If the needle has moved clockwise, the tailstock is misaligned toward the operator. This will cause the workpiece to taper with the smaller end near the tailstock. If the needle has moved counterclockwise, the tailstock is too far away from the operator, and the workpiece will taper with the smaller end at the headstock. In either case, move the tailstock until both diameters have the same reading.

Because only a minor adjustment is usually needed while a job is in progress, the most common method of aligning lathe centers is by cutting and measuring. It is also the most accurate. This method, unlike the bar test method, checks the workpiece while it is in the roughing stage (Figure I-196). A light cut is taken along the length of the test piece, and both ends are measured with a micrometer. If the diameter at the tailstock

Figure I-195 The carriage and the dial indicator are moved to the measuring surface near the tailstock. In this example, the dial indicator did not move, so the tailstock is on center.

Figure I-196 Checking for taper by taking a cut on a workpiece. After the cut is made for the length of the workpiece, a micrometer reading is taken at each end to determine any difference in diameter.

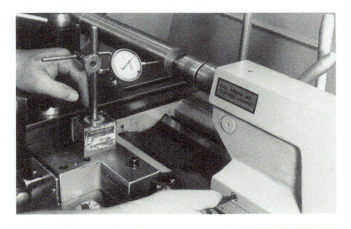

Figure I-197 Using a dial indicator to check the amount of movement of the tailstock when it is being realigned.

end is smaller, the tailstock is toward the operator, and if the diameter at the headstock end is larger, the tailstock is away from the operator. Set up a dial indicator (Figure I-197) and move the tailstock half the difference of the two micrometer readings. Make another light cut and check for taper.

SELF-TEST

1. What happens to the workpiece when the centers are out of line?
2. What happens to the workpiece when the tailstock is offset toward the operator?
3. Name three methods of aligning the centers.
4. Which measuring instrument is used when using a test bar?
5. Name a measuring tool that can be used to check a taper by taking a cut?

Other Lathe Operations

The lathe owes much of its versatility to the variety of tools and work-holding devices that can be used. This tooling makes many special operations possible.

OBJECTIVES

After completing this unit, you should be able to:

- Explain the procedures for drilling, boring, reaming, knurling, recessing, parting, and tapping in the lathe.
- Set up to drill, ream, bore, and tap on the lathe, and complete each of these operations.
- Set up for knurling, recessing, die threading, and parting on the lathe, and complete each of these operations.

DRILLING

Boring, tapping, and reaming usually begin with spotting and drilling a hole. A workpiece that requires a bore is mounted in a chuck, collet, or faceplate, and the drill is typically mounted in the tailstock spindle. If there is a slot in the tailstock spindle, the drill tang must be aligned with it when inserting the drill.

Drill chucks with Morse taper shanks are used to hold straight shank drills and center drills (Figure I-198). Center drills are used to accurately start a hole for drilling (Figure I-199). When large-size drills are used, a pilot drill should be used before the larger drill is used. The pilot drill's diameter should be a little larger than the web thickness of the larger drill.

Taper shank drills (Figure I-200) are inserted directly into the tailstock spindle. The friction of the taper is usually all that is needed to keep the drill from turning while a hole is being drilled (Figures I-201 and I-202), but when large drills are used, friction may not be enough. A lathe dog is sometimes clamped to the drill just above the flutes (Figure I-203) with the bent tail resting on the compound. Hole depth can be measured with a rule or by using the graduations on the tailstock spindle. The alignment of the tailstock with the lathe centerline should be checked before drilling or reaming.

Figure I-198 Mounting a straight shank drill in a drill chuck in the tailstock spindle.

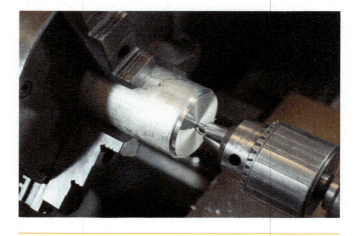

Figure I-199 Center drilling precedes drilling, reaming, or boring.

Drilled holes are not accurate enough for some applications, such as for gear or pulley bores, which should not be made more than .001–.002 inch over the nominal size. Drilling typically produces holes that are oversize and run eccentric to the center axis of the lathe.

Figure I-200 A drill sleeve is placed on the drill so that it will fit the taper in the tailstock spindle.

Figure I-201 The drill is then firmly seated in the tailstock spindle.

Figure I-202 Feed pressure is usually sufficient to keep the drill seated in the tailstock spindle, thus keeping the drill from turning.

Drilling produces holes with rough finishes, which along with size errors can be corrected by boring or reaming. The hole must first be drilled slightly smaller than the finish diameter to leave material for finishing by either of these methods.

Figure I-203 A lathe dog is used if the drill has a tendency to turn in the tailstock spindle.

Figure I-204 Small forged boring bar made of high-speed steel.

BORING

Boring is the process of enlarging and trueing an existing or drilled hole. A drilled hole for boring can be from $\frac{1}{32}$ to $\frac{1}{16}$ inch undersize, depending on the situation. Speeds and feeds for boring are determined in the same way as for external turning. Boring to size predictably is done in the same way as in external turning except that the crossfeed screw is turned counterclockwise to move the tool into the work.

Calipers are used by machinists for internal measuring, though the telescoping gage and outside micrometer are most commonly used to precisely measure small bores because they can take a more accurate measurement. Inside micrometers can be used for bores over $1\frac{1}{2}$ inches. Precision bore gages are used when tolerances are very tight.

A boring bar is clamped in a holder mounted on the carriage compound. Several types of boring bars and holders are used. Forged boring bars are sometimes used for small holes (Figure I-204). The forged end is sharpened by grinding. When the bar gets ground too far back, it must be reshaped or discarded. Some boring bars for holes with diameters over $\frac{1}{2}$ inch (Figure I-205) use high-speed tool inserts, which are typically hand ground in the form of a left-hand turning tool. These tools can be removed from the bar for resharpening when needed. The cutting tool can be held at various angles to obtain different results, which makes the boring bar useful for many applications. Standard bars generally come with a tool angle of 30, 45, or 90 degrees. Some boring bars are designed to use carbide inserts (Figure I-206).

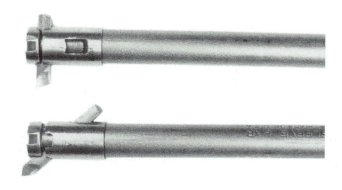

Figure I-205 Two boring bars with inserted tools set at different angles.

Figure I-206 Boring bar with carbide insert. When one edge is dull, a new one is selected.

SHOP TIP

Chatter can be caused by many things. Sometimes there is insufficient lubricant on the compound and cross-slide ways. This can cause chatter. Back off the slides and apply heavy way oil to them. The oil acts as a damper on the vibration.

There is always a certain amount of clearance on machine slides, and when they are dry, a chattering movement is possible. Machine slides that are too loose can cause chatter, even with sufficient lubricant.

Chatter is the vibration between a workpiece and a tool because of the lack of rigid support for the tool. Chatter is a problem in boring operations, because the bar must extend away from the support of the compound (Figure I-207). For this reason, boring bars should be kept back into their holders as far as practicable. Tuned boring bars can be adjusted so that their vibration is dampened (Figure I-208). If chatter occurs when boring, one or more of the following may help eliminate the vibration of the boring tool.

1. Shorten the boring bar overhang, if possible.
2. Increase feed.
3. Make sure that the tool is on center.
4. Reduce the spindle speed.
5. Use a boring bar as large in diameter as possible without its binding in the hole.
6. Reduce the nose radius on the tool.
7. Apply cutting oil to the bore.

Small diameter boring bars may spring away from the cut and cause bellmouth: a slight taper at the front edge of a

Figure I-207 Boring bar setup with a large overhang for making a deep bore. It is difficult to avoid chatter with this arrangement.

Figure I-208 Tuned boring bars contain dampening slugs of heavy material that can be adjusted by applying pressure with a screw (*Kennametal, Inc.*).

Figure I-209 Boring tools must have sufficient side relief and side rake to be efficient cutting tools. Back rake is not normally used.

bore. One or two extra cuts (called *free cuts or spring cuts*) taken without increasing the depth of cut will usually eliminate the problem.

Boring tools are made with side and end relief but usually with zero back rake (Figure I-209). Insufficient end relief will allow the heel of the tool to rub on the workpiece. Figures I-210 to I-214 are views of bores looking outward from inside the chuck. The end relief should be between 10 and 20 degrees. The machinist must use judgment when grinding the end relief because the larger the bore, the less end relief is required. If the end of the tool is relieved too much, the cutting edge will be weak and break down.

The point of the cutting tool should be positioned on the centerline of the workpiece (Figures I-211 to I-214). There must be a space to allow the chips to pass between the bar and the surface being machined, or the chips will wedge and bind on the backside of the bar, forcing the cutting tool deeper into the work (Figure I-214).

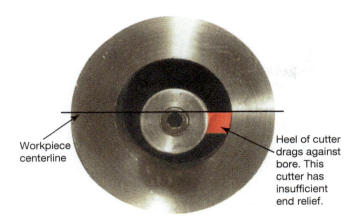

Figure I-210 A tool with insufficient end relief will rub on the heel of the tool (arrow) and will not cut.

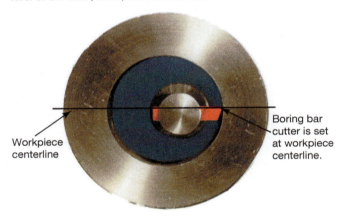

Figure I-211 The point of the boring tool must be positioned on the centerline of the workpiece.

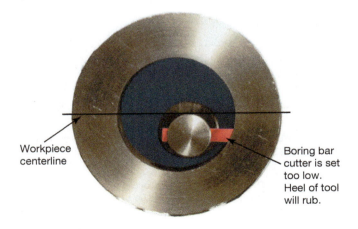

Figure I-212 If the boring tool is too low, the heel of the tool will rub and the tool will not cut, even if the tool has the correct relief angle.

Figure I-213 If the tool is too high, the back rake becomes excessively negative, and the tool point is likely to be broken off. A poor-quality finish is the result of this position.

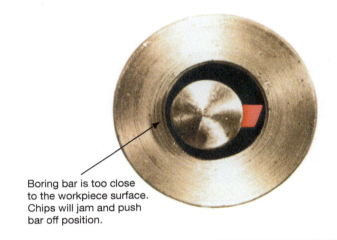

Figure I-214 Allowance must be made so that the chips can clear the space between the bar and the surface being machined. This setup has insufficient chip clearance (where the arrow is pointing).

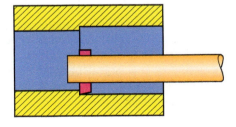

Figure I-215 Through boring, showing bar and tool arrangement.

Through boring is the boring of a workpiece from one end to the other or all the way through it. When through boring, the tool is held in a bar perpendicular to the axis of the workpiece. A slight side cutting edge angle is often used for through boring (Figure I-215). The surface on the backside of a through bore is sometimes back faced (Figure I-216) to

true it up. This facing is done with a specially ground right-hand tool, also held in a bar perpendicular to the workpiece. The amount of facing that can be done in this way is limited to the movement of the bar in the bore.

A **blind hole** is one that does not go all the way through the part to be machined (Figure I-217). Machining the bottom

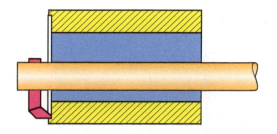

Figure I-216 Back facing, using a boring bar.

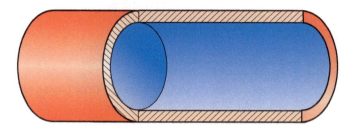

Figure I-217 A blind hole machined flat in the bottom.

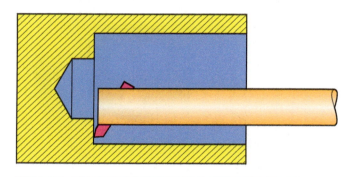

Figure I-218 A bar with an angled tool used to square the bottom of a hole with a drilled center.

or end of a blind hole to a flat is easier when the drilled center does not need to be cleaned up. A bar with the tool set at an angle, usually 30 or 45 degrees, is used to square the bottom of a hole with a drilled center (Figure I-218).

Most boring is performed on workpieces mounted in a chuck, but it is also done on the end of workpieces supported by a steady rest. Boring and other operations can be done on workpieces set up on a faceplate (Figure I-219).

A **thread relief** is an enlargement of a bore at the bottom of a blind hole. The purpose of a thread relief is to allow a threading tool to disengage the work at the end of a pass (Figure I-220).

When the work will allow it, a hole can be drilled deeper than necessary. This will give the end of the boring bar enough space so that the tool can reach into the area to be relieved and still be held at a 90-degree angle (Figure I-221). When the work will not allow for the deeper drilling, a special tool must be ground (Figure I-222).

Grooves in bores are made by feeding a form tool (Figure I-223) straight into the work. Snap ring, O-ring, and oil grooves are made in this way. Cutting oil should always be used in these operations.

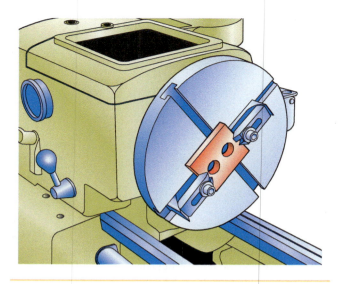

Figure I-219 Workpiece clamped on faceplate has been located, drilled, and bored.

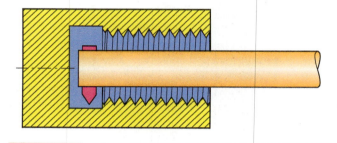

Figure I-220 Ample thread relief is necessary when making internal threads.

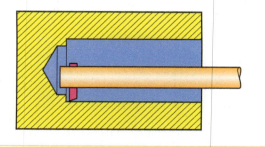

Figure I-221 The hole is drilled deeper than necessary to allow room for the boring bar.

Counterboring in a lathe is the process of enlarging a bore for definite length (Figure I-224). The shoulder produced in the end of the counterbore is usually made square (90 degrees) to the lathe axis. Boring and counterboring are also done on long workpieces supported in a steady rest. The edges and corners of all boring work should be broken or chamfered.

REAMING

Reaming is done in the lathe to quickly and accurately finish drilled or rough-bored holes to size. Machine reamers, like drills, are held in the tailstock spindle of the lathe. Floating

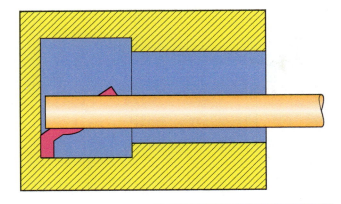

Figure I-222 A special tool is needed when the thread relief must be next to a flat bottom.

Figure I-223 A tool ground to the exact width of the desired groove can be moved directly into the work to the correct depth.

Figure I-224 A square shoulder is made with a cutting tool.

reamer holders are sometimes used to ensure alignment of the reamer, because the reamer follows the eccentricity of drilled holes. This helps eliminate bellmouth bores that result from reamer wobble but does not eliminate hole eccentricity. Only boring will remove runout in a hole.

Figure I-225 Hand reaming in the lathe.

Roughing reamers (rose reamers) are often used in drilled or cored holes, followed by machine or finish reamers. When drilled or cored holes have excessive eccentricity, they are bored .010–.015 inch undersize and machine reamed. If a greater degree of accuracy or a better finish is required, the hole can be bored to within .003–.005 inch of finish size and hand reamed (Figure I-225). For hand reaming, the machine is shut off and the hand reamer is turned with a tap wrench. Holes with diameters greater than $\frac{1}{2}$ inch that require precision are usually bored. An experienced machinist can make bores on a lathe within a tolerance of + or −.0002. Machine reamers may or may not produce an accurate hole, depending on their sharpness and whether they have a built-up edge. However, small hand reamers can produce accurate holes.

Cutting fluids used in reaming are similar to those used for drilling holes (see Table I-6). Cutting speeds are dependent on machine and workpiece material finish requirements. A rule of thumb for reaming speeds is to use half the speed used for drilling (see Table I-5).

Feeds for reaming are about twice that used for drilling. The cutting edge should not rub without cutting, as it causes glazing, work hardening, and dulling of the reamer.

A simple machine reaming sequence follows:

Step 1 Assuming that the hole has been drilled $\frac{1}{64}$ inch undersize, seat a taper shank machine reamer in the taper by hand pressure (Figure I-226). To insure seating use a mallet to seat the taper shank in the tailstock spindle.

Step 2 Apply cutting fluid to the hole, and start the reamer into the hole by turning the tailstock handwheel (Figure I-227). Complete the hole and remove the reamer from the hole. Never reverse the machine when reaming.

Step 3 Remove the reamer from the tailstock spindle and clean it with a cloth (Figure I-228). Then, return the reamer to the storage rack.

Step 4 Clean the lathe with a brush (Figure I-229).

Figure I-226 The reamer must be seated in the taper.

Figure I-227 Starting the reamer in the hole.

Figure I-228 The reamer should be cleaned and put away after use.

TAPPING

Tapping is a quick and accurate means of producing internal threads in a chucked part. Tapping in the lathe is similar to tapping in the drill press but is generally reserved for small-size holes, as tapping is the only way they can be internally threaded. Large internal threads can be made in the lathe with

Figure I-229 Chips are brushed into the chip pan.

Figure I-230 Hand tapping in the lathe by turning the tap wrench (*Courtesy of Fox Valley Technical College*).

a single-point threading tool held in a boring bar. A large tap requires considerable force or torque to turn, more than can be provided by hand turning. A tap aligned by the dead center will produce a tapped hole in line with the lathe axis.

A plug tap or spiral point tap may be used for tapping through holes. When tapping blind holes, a plug tap can be followed by a bottoming tap if threads are needed to the bottom of the hole. A good practice is to drill a blind hole deeper than the required depth of threads.

Two approaches may be used for hand tapping. Power is not used in either case. One method is to turn the tap by means of a tap wrench or adjustable wrench with the spindle engaged in a low gear so it will not turn (Figure I-230). The other method is to disengage the spindle and turn the chuck by hand while the tap wrench handle rests on the compound (Figure I-231). In both cases the tailstock is clamped to the ways, and the center is kept in the center of the tap by slowly turning the tailstock handwheel. The tailstock on small lathes need not be clamped to the ways for small taps but can be held firmly with one hand. Cutting oil should be used and the tap backed off every one or two turns to break chips unless it is a spiral point tap, sometimes called a gun tap.

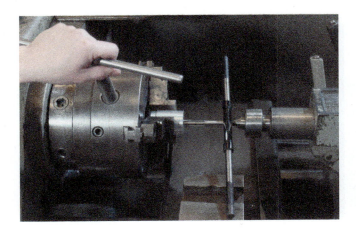

Figure I-231 Tapping by turning the chuck.

Figure I-232 Starting the die on a rod to be threaded in a lathe. The tailstock spindle (without a center) is used to start the die squarely onto the work.

The correct tap drill size should be found in a tap drill chart. Drills tend to drill slightly oversize, and tapping the oversize hole can produce poor internal threads with only a small percentage of thread cut. See Section B, Unit 7, for a discussion of percentage of thread. You can help make sure the drill produces the correct size hole by drilling first with a slightly smaller drill and then using the tap drill to finish the hole to the correct size.

Tapping can be done on the lathe with power, but it is recommended that it be done only if the spindle rotation can be reversed, if a spiral point tap is used, and if the hole is clear through the work. The tailstock is left free to move on the ways. Insert the tap in a drill chuck or tap driver in the tailstock and set the lathe on a low speed. Use cutting oil and slide the tailstock so that the tap engages the work. Reverse the rotation and remove the tap from the work every $\frac{3}{8}$ to $\frac{1}{2}$ inch. When reversing, apply light hand pressure on the tailstock to move it to the right until the tap is all the way out.

External threads cut with a die should be used only for nonprecision purposes, because the die may wobble, and the pitch (the distance from a point on one thread to the same point on the next) may not be uniform. Extend the rod to be threaded a short distance from the chuck, and start a die and diestock on the end (Figure I-232). Use cutting oil. Rest the handle against the compound. You may turn the chuck by hand, but if you use power, set the machine for low speed and reverse it every $\frac{3}{8}$ to $\frac{1}{2}$ inch to clear the chips. Finish the last $\frac{1}{4}$ inch by hand if approaching a shoulder. Reverse the lathe to remove the die.

Die holders are sometimes used to hold and guide the die from the tailstock spindle. The thread is much better aligned with less pitch error than it is with a diestock (though not as accurate as in single-point threading). The button die holder shown in Figure I-233 is hollow and is guided by a bar seated in the tailstock spindle. The lathe is running at low speed and the thread is cut when the knurled body is gripped, preventing its turning. As in other die cutting, the cutting action is stopped every two to five turns to break the chips. Fairly long threads on small diameter stock (not over $\frac{1}{2}$ inch) can be produced by this method.

When many threaded parts are needed, a die head such as that shown in Figure I-234 can be used. Die heads must be aligned accurately when used on an engine lathe to avoid

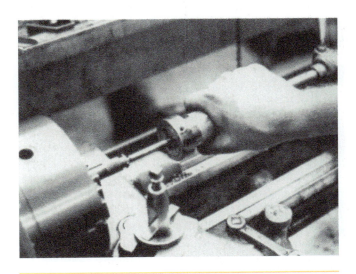

Figure I-233 Button die holder being used to cut a small-diameter thread on a parallel clamp screw.

Figure I-234 A self-opening die head can be quickly set up on any engine lathe for higher production of threads.

damage to the die. Most die heads are self-opening, and therefore the thread can be cut in one pass at high speed. The die automatically opens when the preset thread length is cut so that it can be retracted without stopping or reversing the machine.

RECESSING, GROOVING, AND PARTING

External diameters are recessed and grooved (Figures I-235 to I-237) to provide grooves for thread relief, snap rings, and O-rings. Special tools are ground for both external and internal grooves and recesses. Parting tools are sometimes used for external grooving and thread relief. Grooves are sometimes measured with dial, vernier, or electronic sliding-type calipers (Figure I-238).

Parting or cutoff tools are designed to withstand high cutting forces, but if chips are not sufficiently cleared or cutting fluid is not used, these tools can quickly jam and break (Figure I-239). Parting tools must be set on center and square with the work (Figure I-240). Lathe tools are often specially ground as parting tools for small or delicate parting jobs (Figure I-241). Diagonally ground parting tools leave no burr.

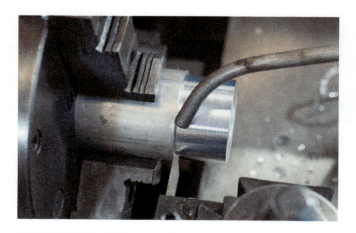

Figure I-235 The undercut tool is brought to the workpiece and the micrometer dial is zeroed. Cutting oil is applied to the work.

Figure I-236 The tool is fed to the single depth of the thread or the required depth of the groove. If a wider groove is necessary, the tool is moved over and a second cut is taken as shown.

Figure I-237 The finished groove.

Figure I-238 Groove being measured with a sliding caliper with digital readout.

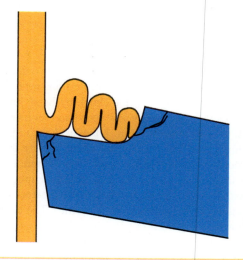

Figure I-239 Chips jamming and breaking the parting tool.

Parting alloy steels and other metals is sometimes difficult, and step parting (Figure I-242) may help in these cases. When deep-parting difficult material, extend the cutting tool from the holder a short distance and part to that depth. Then, back off the crossfeed and extend the tool a bit farther; part to that depth.

Figure I-240 Parting tools must be set to the center of the work.

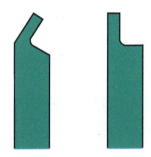

Figure I-241 Special parting tools ground from lathe tools for small or delicate parting jobs. These tools should be ground slightly wider at the cutting edge to clear the tool in the cut.

Figure I-242 Step parting.

Repeat the process until the center is reached. Heavy cutting oil works best for parting unless the lathe is equipped with a coolant pump and a steady flow of soluble oil is available. Parting tools are made in either straight or offset types. A right-hand offset cutoff tool is necessary when parting near the chuck.

All parting and grooving tools have a tendency to chatter; therefore, any setup must be as rigid as possible. A low speed should be used for parting; if the tool chatters, reduce the speed more. A feed that is too light can cause chatter, but a

feed that is too heavy can jam the tool. The tool should always be making a chip. Work should not extend far from the chuck when parting or grooving, and no parting should be done in the middle of a workpiece or at the end near the dead center. This is because the tool will bind in the cut when the material is almost cut through. When the workpiece is cut off near the dead center, the same binding problem exists, with the additional problem that the work will climb on the tool when it is cut off. This can break the tool and possibly damage the machine.

KNURLING

A **knurl** is a raised impression on the surface of the workpiece produced by two hardened rolls. Diamond or straight pattern knurling tools are available (Figure I-243). The diamond pattern is formed by a right-hand and a left-hand helix mounted in a self-centering head. The straight pattern is formed by two straight rolls. These common knurl patterns can be fine, medium, or coarse.

Diamond knurling is used to improve the appearance of a part and to provide a good gripping surface for levers and tool handles. **Straight knurling** is used to increase the size of a part for press fits in light-duty applications. A disadvantage to this use of knurls is that the fit has less contact area than a standard fit.

There are several types of tools used for knurling. Figure I-244 shows a few common types. The **straddle knurling tool** enables small diameters to be knurled with less distortion. The revolving head knurling tool can make three different size knurls.

Knurling works best on workpieces mounted between centers. When held in a chuck and supported by a center, the workpiece tends to crawl back into the chuck and out of the supporting center with the high pressure of the knurl. This is especially true when the knurl is started at the tailstock end and the feed is toward the chuck. Long, small diameter pieces push away from the knurling tool and will stay bent if the knurl is left on the work after the lathe is stopped.

Knurls do not cut. They displace the metal with high pressure. Lubrication is more important than cooling, so a cutting oil or lubricating oil is satisfactory. Low speeds (about the same as for threading) and a feed of about .010–.015 inch are used for knurling.

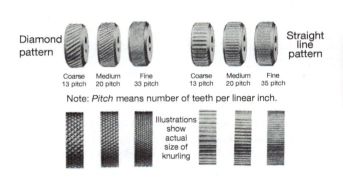

Figure I-243 Set of straight knurls and diagonal knurls.

Figure I-244 Knurling tools.

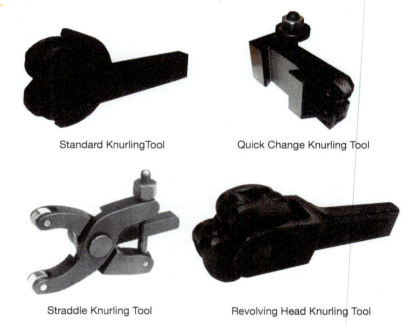

Standard KnurlingTool   Quick Change Knurling Tool

Straddle Knurling Tool   Revolving Head Knurling Tool

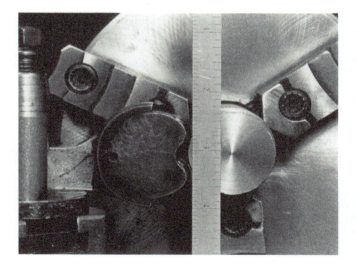

Figure I-245 Knurls are centered on the workpiece.

The knurls should be centered on the workpiece vertically (Figure I-245), and the knurl toolholder should be square with the work, unless the knurl pattern is difficult to establish, as it often is in tough materials. In that case, the toolholder should be angled about 5 degrees to the work so that the knurl can penetrate deeper (Figure I-246).

A knurl should be started in soft metal about half depth and the pattern checked. An even diamond pattern should develop. If one roll is dull or placed too high or too low, a double impression will develop (Figure I-247) because the rolls are not tracking evenly. If this happens, move the knurls to a new position along the workpiece, readjust up or down, and try again. If possible, the knurl should be made in one pass. However, this is not always possible with ordinary knurling toolholders because of the extreme pressure

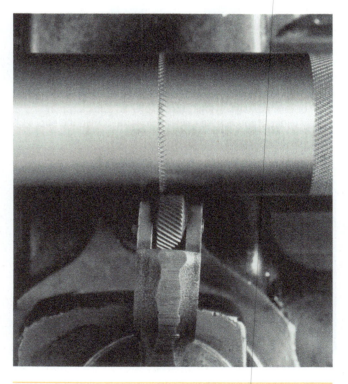

Figure I-246 Angling the toolholder 5 degrees often helps establish the diamond pattern.

bearing on one side of the workpiece. Several passes may be required on a slender workpiece to complete a knurl because the tool tends to push it away from the knurl. The knurls should be cleaned with a brush between passes.

Material that hardens as it is worked, such as high-carbon or spring steel, should be knurled in one pass if possible, and not more than two passes. Even in ordinary steel, the surface will harden after a diamond pattern has developed. It is best to stop knurling just before the points are sharp (Figure I-248). Metal flaking off the knurled surface is evidence that work hardening has occurred. Avoid knurling too deeply, as it produces an inferior knurled finish.

Figure I-247 Double impression on the left is the result of the rolls not tracking evenly.

Figure I-248 More than one pass is usually required to bring the knurl to full depth in hard or springy material.

SELF-TEST

1. Why isn't drilling used to produce bores in pulleys, gears, and bearing fits?

2. What is the chief advantage of boring over reaming in the lathe?

3. List five ways to eliminate chatter when using a boring bar.

4. How can grooves or thread reliefs be made in a bore?

5. State whether cutting speeds for reaming are twice or half that used for drilling; whether feeds used for reaming are twice or half that used for drilling.

6. Standard plug or bottoming taps can be used when hand tapping in the lathe. If power is used, what kind of tap works best?

7. How can you avoid chatter when cutting off stock with a parting tool?

8. State three reasons why knurling might be used on a part.

9. Ordinary knurls do not cut. How do they produce the diamond or straight pattern on the workpiece?

10. If a knurl is producing a double impression, what can you do to make it develop a diamond pattern?

Sixty-Degree Thread Information and Calculations

A machinist must know more than how to set up the lathe to cut threads. He or she must know the thread form, the class of fit, and thread calculation. This unit prepares you to cut threads, which you will do in the next unit.

OBJECTIVES

After completing this unit, you should be able to:

- Describe the several 60-degree thread forms, noting their similarities and differences.
- Calculate thread depth, infeeds, and minor diameters of threads.

SIXTY-DEGREE V THREAD FORM

Various screw thread forms are used for fastening or for moving or transmitting parts against loads. The most widely used of these forms are the 60-degree thread types. These are mostly used for fasteners. An early form of the 60-degree thread is the sharp V (Figure I-249). The sides of the thread form a 60-degree angle with each other. Theoretically, the sides and the base between two thread roots form an equilateral triangle, but in practice this is not the case; it is necessary to make a slight flat on top of the thread to deburr it. Also, the tool will always round off and leave a slight flat at the thread root.

The relationship between pitch and threads per inch is shown in Figure I-250. **Pitch** is the distance between a point on one screw thread and the corresponding point on the next thread, measured parallel to the thread axis.

Thread per inch means the number of threads in a 1 inch length. The pitch (P) may be calculated by dividing the number of threads per inch (tpi) into 1:

$$P = \frac{1}{\text{tpi}}$$

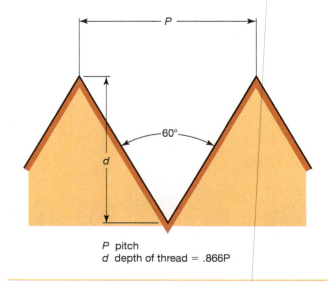

P pitch
d depth of thread = .866P

Figure I-249 The 60-degree sharp V-thread.

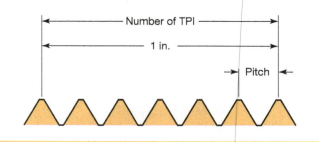

Figure I-250 Difference between threads per inch and pitch.

EXAMPLE

Find the pitch of a $\frac{1}{2}$-inch diameter 20-tpi machine screw thread.

$$P = \frac{1}{20} = .050 \text{ inch}$$

Figure I-251 Screw pitch gage for inch threads.

Figure I-252 Screw pitch gage for metric threads.

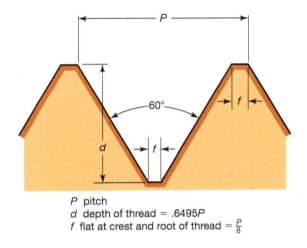

P pitch
d depth of thread = .6495*P*
f flat at crest and root of thread = $\frac{P}{8}$

Figure I-253 General dimensions of screw threads.

Pitch is checked on a screw thread with a screw pitch gage (Figures I-251 and I-252). General dimensions and symbols for screw threads are shown in Figure I-253.

UNIFIED AND AMERICAN NATIONAL FORMS

The American National form is now obsolete and has evolved into the Unified form. The basic geometry of the two systems is similar (Figure I-254).

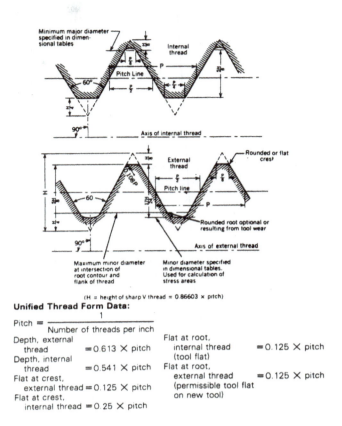

(H = height of sharp V thread = 0.86603 × pitch)

Unified Thread Form Data:

Pitch = $\dfrac{1}{\text{Number of threads per inch}}$

Depth, external thread	= 0.613 × pitch	Flat at root, internal thread (tool flat)	= 0.125 × pitch
Depth, internal thread	= 0.541 × pitch	Flat at root, external thread (permissible tool flat on new tool)	= 0.125 × pitch
Flat at crest, external thread	= 0.125 × pitch		
Flat at crest, internal thread	= 0.25 × pitch		

Figure I-254 Unified screw threads *(Reprinted from ASME B1.1–2003 by permission of The American Society of Mechanical Engineers. All rights reserved.).*

Taps and dies are marked with letter symbols to designate the series of the threads they form. For example, the symbol for American Standard taper pipe thread is NPT, for Unified coarse thread it is UNC, and for Unified fine thread the symbol is UNF.

Thread Depth

The American Standard for Unified threads (Figure I-254) is similar to the American National Standard with certain modifications. The thread forms are practically the same, and the basic 60-degree angle is the same. The depth of an external American National thread is .6495 × pitch, and the depth of the Unified thread is .6134 × pitch. The constant for American National thread depth may be rounded off from .6495 to .65, and the constant for Unified thread depth may be rounded to .613. The thread depth and the root truncation (tool flat) of the American National thread is fixed or definite, but these factors are variable within limits for Unified threads. A rounded root for Unified threads is desirable whether from tool wear or by design. A rounded crest is also desirable but not required. The constants .613 for thread depth and .125 for the flat on the end of the tool were selected for calculations on Unified threads in this unit.

Thread Fit Classes and Thread Designations

Unified and American National Standard form threads are interchangeable. An NC bolt will fit a UNC nut. The principal difference between the two systems is that of tolerances. The Unified system, a modified version of the old system, allows for more tolerances of fit. Thread fit classes 1, 2, 3, 4, and 5 were used with the American National Standard, 1 being a loose fit, 2 a free fit, 3 a close fit, 4 a snug fit, and 5 a jam or interference fit. The Unified system expanded this number system to include a letter, so the threads could be identified as class 1A, 1B, 2A, 2B, and so on. *A* indicates an external thread and *B* an internal thread. Because of this expansion in the Unified system, tolerances are now possible on external threads and are 30 percent greater on internal threads. These changes make easier the manufacturer's job of controlling tolerances to ensure the interchangeability of threaded parts. See *Machinery's Handbook* for tables of Unified thread limits. Limits are the maximum and minimum allowable dimensions of a part—in this case, internal and external threads.

Threads are designated by the nominal bolt size or major diameter, the threads per inch, the letter series, the thread tolerance, and the thread direction. Thus $1\frac{1}{4}$-inch–12 UNF-2BLH would indicate a $1\frac{1}{4}$-inch Unified nut with 12 threads per inch, a class 2 thread fit, and a left-hand helix.

Tool Flats and Infeeds for Thread Cutting

The flat on the crest of the thread on both the Unified and American National systems is P/8 or P * .125. The root flat (flat on the end of the external threading tool) is calculated as P/8 or P * .125 for the Unified thread system.

To cut 60-degree form threads, the tool is fed into the work with the compound set at 30 degrees (Figure I-255). However, in practice, the compound is actually set to 29 or 29.5 degrees to provide a light finish cut on the trailing side of the threading tool. The small difference will make only a slight difference in the calculation. The infeed depth along the flank of the thread at 30 degrees is greater than the depth at 90 degrees from the work axis. This depth may be calculated for the Unified form by the following formula:

For 29-degree compound angle cutting Unified form threads,

$$\text{Infeed} = .701/n \text{ or } .701\ P$$

where

n = number of threads per inch (tpi)

P = pitch, or $1/n$

Thus, for a Unified thread with 10 tpi:

$$\text{Infeed} = .701/10 = .0701$$

or

$$\text{Infeed} = .701 \times .100 = .0701$$

For external Unified threads the infeed at 29 degrees may be calculated by the formula

$$\text{Infeed} = \frac{.708}{n} \text{ or } .708P$$

Thus, for a thread with 10 threads per inch (.100-inch pitch),

$$\text{Infeed} = \frac{.708}{10} = .0708 \text{ in.}$$

The formula for compound infeed is derived mathematically by taking the constant for the thread (Unified form = .613) and dividing it by the cosine of the compound rest angle.

PITCH DIAMETER, LEAD OR HELIX ANGLE, AND PERCENT OF THREADS

The clearances and tolerances for thread fits are derived from the pitch diameter. The pitch diameter on a straight thread is the diameter of an imaginary cylinder that passes through the thread profiles at a point where the width of the groove and thread are equal. The mating surfaces are the flanks of the thread.

The percentage of thread has little to do with fit but refers to the actual minor diameter of the internal thread. The typical nut for machine screws has 75 percent threads, which are easier to tap than 100 percent threads and retain sufficient strength for most thread applications.

The **lead** or **helix angle** of a screw thread (Figure I-256) is larger for greater lead threads than for smaller leads; and the larger diameter of the workpiece, the smaller the helix angle for the same lead. Helix angles should be taken into account when grinding tools for threading.

The relief and helix angles must be ground on the leading or cutting edge of the tool. A protractor may be used to check this angle (Figure I-257).

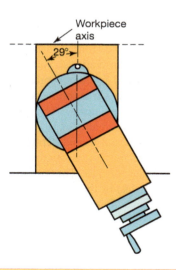

Figure I-255 Compound at 29 degrees for cutting 60-degree threads.

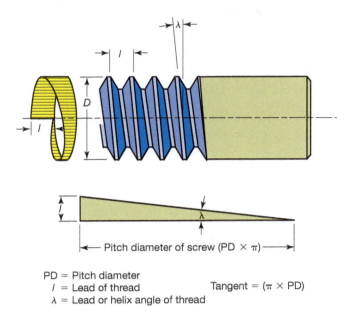

PD = Pitch diameter
 l = Lead of thread
 λ = Lead or helix angle of thread

Tangent = (π × PD)

Figure I-256 Screw thread helix angle.

Figure I-257 Checking the relief and helix angle on the threading tool with a protractor.

Lead or helix angles may be determined by the following formula:

$$\frac{\text{Tangent of the lead or}}{\text{helix angle}} = \frac{\text{lead of the thread}}{\text{pitch diameter of the thread}}$$

$$\text{Pitch diameter (PD) of the thread} = \frac{\text{lead of the thread}}{\pi \times \text{PD}}$$

where

$$\pi = 3.1416$$

Also note that pitch and lead are the same for single lead screws. Helix or lead angles are given for Unified and other thread series in handbooks such as *Machinery's Handbook*.

A taper thread is made on the internal or external surface of a cone. An example of a 60-degree taper thread is the American National Standard pipe thread (Figure I-258). A line bisecting the 60-degree thread is perpendicular to the

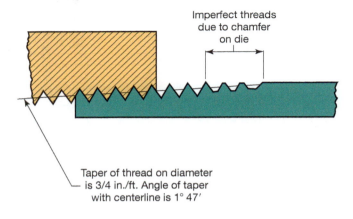

Imperfect threads due to chamfer on die

Taper of thread on diameter is 3/4 in./ft. Angle of taper with centerline is 1° 47′

Figure I-258 American National standard taper pipe thread.

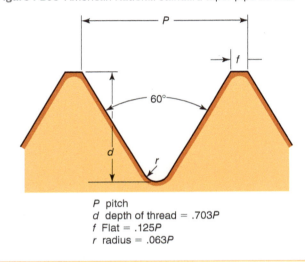

P pitch
 d depth of thread = .703*P*
 f Flat = .125*P*
 r radius = .063*P*

Figure I-259 SI metric thread form.

axis of the workpiece. On a taper thread the pitch diameter at a given position on the thread axis is the diameter of the pitch cone at that position.

METRIC THREAD FORMS

The Système International (SI) thread form (Figure I-259), adopted in 1898, is similar to the American National Standard. The British Standard for ISO (International Organization for Standardization) metric screw thread was set up to standardize metric thread forms. The basic form of the ISO metric thread (Figure I-260) is similar to the Unified thread form. These and other metric thread systems are listed in *Machinery's Handbook*. See Table 4 in Appendix 2 for ISO metric tap drill sizes.

SELF-TEST

1. Give one disadvantage of the sharp V-thread.
2. Explain the difference between the threads per inch and the pitch of the thread.

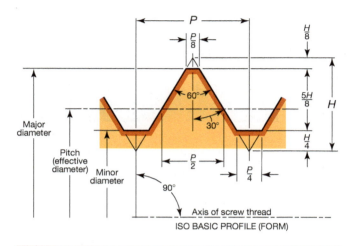

Figure I-260 ISO metric thread form.

3. Name two similarities and two differences between American National and Unified threads.

4. What is a major reason for thread allowances and classes of fits?

5. What does $\frac{1}{2}$–20 UNC-2A describe?

6. The root truncation (flat) for Unified threads and for American National threads is calculated by .125P. What should the flat on the end of the threading tools be for a $\frac{1}{2}$-20 thread?

7. Explain the difference between the fit of threads and the percent of thread.

8. How are Unified and metric thread forms similar?

Cutting Unified External Threads

A machinist is frequently called on to cut threads of various forms on the engine lathe. The threads most commonly made are the V-form, Unified Series. This unit will show you how to make these threads on a lathe. You will need to practice to gain confidence in your ability to make external Unified threads on any workpiece.

OBJECTIVES

After completing this unit, you should be able to:

- Detail the steps and procedures necessary to cut a Unified thread to the correct depth.
- Set up a lathe for threading and cutting several different thread pitches and diameters.
- Identify tools and procedures for thread measurement.

Figure I-261 An equal chip is formed on each side of the threading tool when the infeed is made with the cross slide.

HOW THREADS ARE CUT ON A LATHE

Threads are cut on a lathe with a single-point tool by taking a series of cuts. A direct ratio exists between the headstock spindle rotation, the lead screw rotation, and the number of threads on the lead screw. This ratio can be altered by the quick-change gearbox to make a variety of threads. When the half nut is clamped on the thread of the lead screw, the carriage will move a given distance for each revolution of the spindle. This distance is the lead of the thread.

If the infeed of a thread is made with the cross slide (Figure I-261), equal-sized chips will be formed on both cutting edges of the tool. This causes higher tool pressures, which can result in tool breakdown, and sometimes causes tearing of the threads because of insufficient chip clearance. A more accepted practice is to feed in with the compound, which is set at 29 degrees (Figure I-262) toward the right of the operator, for cutting right-hand threads. This ensures a cleaner cutting action than with 30 degrees, with most of the chip taken from the leading edge and a scraping cut from the following edge of the tool.

Figure I-262 A chip is formed on the leading edge of the tool when the infeed is made with the compound set at 29 degrees to the right of the operator to make right-hand threads.

SETTING UP FOR THREADING

Begin the setup by obtaining or grinding a tool for cutting Unified threads of the required thread pitch. The only difference in tools for various pitches is the flat on the end of the tool. For Unified threads this is $.125P$, as discussed in Unit 10. If the toolholder you are using has no back rake, no grinding on the top of the tool is necessary. If the toolholder does have back rake, the tool must be ground to provide zero rake (Figure I-263).

A center gage (Figure I-264) or an optical comparator may be used to check the tool angle. An adequate allowance for the helix angle on the leading edge will ensure sufficient side relief.

Set up the part to be threaded between centers, in a chuck or in a collet. Make an undercut of .005 inch less than the minor diameter at the end of the thread. Its width should be sufficient to clear the tool.

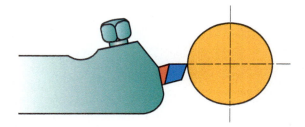

Figure I-263 The tool must have zero rake and be set on the center of the work to produce the correct form.

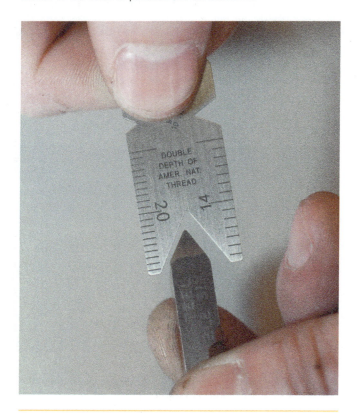

Figure I-264 Checking the tool angle with a center gage.

Figure I-265 The threading tool is placed in the holder and lightly clamped.

Figure I-266 The tool is adjusted to the dead center for height. A tool set too high or too low will not produce a true 60-degree angle in the cut thread.

Clamp the tool in the holder and set it on the centerline of the workpiece (Figures I-265 and I-266). Use a center gage to align the tool to the workpiece (Figure I-267). Clamp the toolholder tightly after the tool is properly aligned.

Setting Dials on the Compound and Crossfeed

Bring the point of the tool into contact with the work by moving the crossfeed handle, and set the micrometer collar on the zero mark (Figure I-268). Also set the compound micrometer collar on zero (Figure I-269), but first be sure that all slack or backlash is removed by turning the compound feed handle clockwise.

Setting Apron Controls

Some lathes have a feed change lever that selects between cross or longitudinal feeds and must be moved to a neutral position for threading. This action locks out the feed

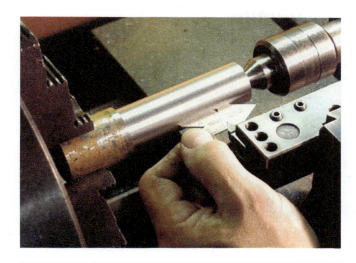

Figure I-267 The tool is properly aligned by using a center gage. The toolholder is adjusted until the tool is aligned. The toolholder is then tightened.

Figure I-268 After the tool is brought into contact with the work, the crossfeed micrometer collar is set to the zero index mark.

Figure I-269 The operator then sets the compound micrometer collar to the zero index.

Figure I-270 The threads per inch are selected on the quick-change gearbox.

mechanism so that no mechanical interference is possible. All lathes have some interlock mechanism to prevent interference when the half-nut lever is used. The half-nut lever causes two halves of a nut to clamp over the lead screw. The carriage will move the distance of the lead of the thread on the lead screw for each revolution of the lead screw.

Thread dials operate off the lead screw and continue to turn when the lead screw is rotating and the carriage is not moving. When the half-nut lever is engaged, the threading dial stops turning and the carriage moves. The marks on the dial indicate when it is safe to engage the half-nut lever. If the half nuts are engaged at the wrong place, the threading tool will not track in the same groove as before and may cut into the center of the thread and ruin it. With any even number of threads such as 4, 6, 12, or 20, the half nut may be engaged at any line. Odd-numbered threads such as 5, 7, or 13 may be engaged at any numbered line. With fractional threads it is safest to engage the half nut at the same line every time.

Quick-Change Gearbox

The settings for the gear shift levers on the quick-change gearbox are selected according to the desired threads per inch (Figure I-270). If the lathe has an interchangeable stud gear, be sure the correct one is in place.

Spindle Speeds

Spindle speeds for thread cutting are approximately one quarter of turning speeds. The speed should be slow enough so that you will have complete control of the thread cutting operation (Figure I-271).

Figure I-271 Setting the correct speed for threading.

CUTTING THE THREAD

The following is the procedure for cutting right-hand threads:

Step 1 Move the tool off the work and turn the crossfeed micrometer dial back to zero.

Step 2 Feed it in .002 inch on the compound dial.

Step 3 Turn on the lathe and engage the half-nut lever (Figure I-272).

Step 4 Take a scratch cut without using cutting fluid (Figure I-273). Disengage the half nut at the end of the cut. Stop the lathe and back out the tool using the crossfeed. Return the carriage to the starting position.

Step 5 Check the thread pitch with a screw pitch gage or a rule (Figure I-274). If the pitch is wrong, it can still be corrected.

Step 6 Apply appropriate cutting fluid to the work (Figure I-275).

Step 7 Feed the compound in .005–.020 inch for the first pass, depending on the pitch of the thread. For a coarse

Figure I-272 The half-nut lever is engaged at the correct line or numbered line on the thread dial depending on whether the thread is odd-, even-, or fractional-numbered.

Figure I-273 A light scratch cut is taken for the purpose of checking the pitch.

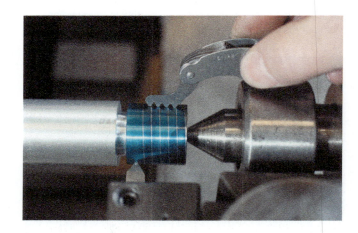

Figure I-274 The pitch of the thread is being checked with a screw pitch gage.

Figure I-275 Applying cutting fluid to the work.

thread, heavy cuts can be taken on the first few cuts. Reduce the depth of cut for each pass until it is about .002 inch at the final passes. Bring the crossfeed dial to zero.

Step 8 Make the second cut (Figure I-276).

Figure I-276 The second cut is taken after feeding in the compound .010 inch.

Figure I-278 The thread is checked with a ring gage.

Figure I-277 The finish cut is taken with infeed of .001 to .002 inch.

Figure I-279 A standard nut is often used to check a thread.

Figure I-280 The screw thread micro
check the pitch diameter of the thre

Step 9 Continue this process until the tool is within .010 inch of the finished depth (Figure I-277).

Step 10 Brush the threads to remove the chips. Check the thread fit with a ring gage (Figure I-278), standard nut

or mating part (Figure I-279)
(Figure I-280). You may re
centers and return withou
provided that the tail of th

Step 11 Continue to take cuts of .001 or .002 inch, and check the fit between each cut. Thread the nut or gage with your fingers; it should go on easily but without end play. A class 2 fit is desirable for most purposes.

Step 12 Chamfer the end of the thread to protect it from damage.

Left-Hand Threads

The procedure for cutting left-hand threads (Figure I-281) is the same as that used for cutting right-hand threads with two exceptions. The compound is set at 29 degrees to the left of the operator (Figure I-282) and the lead screw rotation is reversed so that the cut is made from the left to the right. The feed reverse lever is moved to reverse the lead screw. Sufficient undercut must be made for a starting place for the tool. Also, sufficient relief must be provided on the *right* side of the tool.

Methods of Terminating Threads

Undercuts are often used for terminating threads. They should be made the single depth of the threads. The undercut should have a radius to less .005 inch. of fatigue failure resulting from stress conc sibility sharp corners.

Machinists sometimes remove the tool qu the end of the thread while disengaging the half nut. chinist misjudges and waits too long, the point of the ing tool will be broken off. A dial indicator is some used to locate the exact position of the carriage for removi the tool. When this tool withdrawal method is used, an un dercut is not necessary.

Terminating threads close to a shoulder requires specially ground tools (Figure I-283). Sometimes, when cutting right-hand threads to a shoulder, it is convenient to turn the tool upside down and reverse the lathe spindle, which then moves the carriage from left to right. Some commercial threading tools are made for this purpose (Figure I-284).

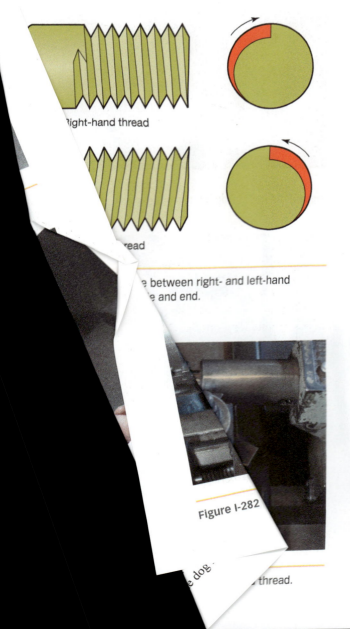

Right-hand thread

...read

...e between right- and left-hand ...e and end.

Figure I-282

...e dog... ...thread.

Figure I-283 A threading tool used for threading to a shoulder.

Figure I-284 By placing the blade of the threading tool in the upper position, this tool can be made to thread on the bottom side of the tool. The lathe is reversed and the thread is cut from left to right. Cutting threads from left to right when there is a shoulder makes the job easier.

Other Tool Types

Because the crest of the Unified thread form should be rounded, some commercial threading tools provide this and other advantages (Figures I-285 and I-286).

Picking up a Thread

It sometimes becomes necessary to reposition the tool when its position against the work has been changed during a threading operation. This is sometimes called *chasing a thread*. This position change can occur if the threading tool is removed for grinding, the work slips in the chuck or lathe dog, or the tool moves from the pressure of the cut.

To reposition the tool the following steps may be taken:

Step 1 Check the tool position with reference to the work by using a center gage. If necessary, realign the tool.

Figure I-285 Carbide threading insert for external threading. Inserts are made for many different thread forms and for boring bars. (*Used with permission of Kennametal, Inc.*)

Figure I-286 Threads being turned at high speed with carbide insert threading tool (*Kennametal, Inc.*).

Figure I-287 Repositioning the tool.

Step 2 With the tool backed away from the threads, engage the half nut with the machine running. Turn off the machine with the half nuts still engaged and the tool located over the partially cut threads.

Step 3 Position the tool in its original location in the threads by moving both the crossfeed and compound handles (Figure I-287).

Step 4 Set the micrometer dial to zero on the crossfeed collar and set the dial on the compound to the last setting used.

Step 5 Back off the crossfeed and disengage the half nuts. Resume threading where you left off.

BASIC EXTERNAL THREAD MEASUREMENT

The simplest method for checking a thread is to try the mating part for fit. The fit is determined solely by feel with no measurement involved. Although a loose, medium, or close fit can be determined by this method, the threads cannot be depended on for interchangeability with others of the same size and pitch.

Thread roll snap gages can be used to check the accuracy of external screw threads (Figure I-288). These common measuring tools are easier and faster to use than thread micrometers or ring gages. The part size is compared with a preset dimension on the roll gage. The first set of rolls are the go and the second the no-go rolls.

Thread roll snap gages, ring gages, and **plug gages** are used in production manufacturing where quick gaging methods are needed. These gaging methods depend on the operator's "feel," and the level of precision is only as good as the accuracy of the gage. The thread sizes are not measurable in any definite way.

The **thread comparator micrometer** (Figure I-289) has two conical points. This micrometer does not measure the pitch diameter of a thread. The micrometer is used to

Figure I-288   Thread roll snap gage.

Figure I-289   Thread comparator micrometer.

Figure I-290   Set of "best" three wires for various thread sizes.

make a comparison with a known standard. The micrometer is first set to the threaded part and then compared with the reading obtained from a thread plug gage.

ADVANCED METHODS OF THREAD MEASUREMENT

The most accurate place to measure a screw thread is on the flank or angular surface of the thread at the pitch diameter. The outside diameter measured at the crest, or the minor diameter measured at the root can vary considerably. Threads may be measured with standard micrometers and specially designated wires (Figure I-290) or with a screw thread micrometer. The pitch diameter is measured directly by these methods. The kit in Figure I-290 has wire sizes for most thread pitches. Most kits provide a number to be added to the nominal OD of the threads, which is the reading that should be seen on the micrometer if the pitch diameter of the thread is correct.

Three-Wire Method

The **three-wire method** of measuring threads is considered one of the best and most accurate. Figure I-291 shows three wires placed in the threads with the micrometer measuring over them. Different sizes and pitches of threads require different-sized wires. For greatest accuracy a wire size that will contact the thread at the pitch diameter should be used. This is called the "best" wire size.

The pitch diameter of a thread can be calculated by subtracting the wire constant (which is the single depth of a sharp V-thread, or .866 * P) from the measurement over the three wires when the best wire size is used. The wires used for three-wire measurement of threads are hardened and lapped steel, and are available in sets that cover a large range of threaded pitches.

A formula by which the best size wire may be found is

$$\text{Wire size} = \frac{.57735}{n}$$

where

n = the number of threads per inch

Figure I-291 Measuring threads with the three wire and micrometer

EXAMPLE

The preceding formula is used to find the best wire size for measuring a $1\frac{1}{4}$ inch-12 UNC screw thread:

$$\text{Wire size} = \frac{.57735}{12} = .048$$

Therefore, the best wire size to use for measuring a 12-pitch thread would be .048 inch.

If the best wire sizes are not available, smaller or larger wires may be used within limits. They should not be so small that they are below the major diameter of the thread or so large that they do not contact the flank of the thread. Subtract the constant $(.866025 \times P)$ for best wire size from the pitch diameter and add three times the diameter of the available wire when the best wire size is not available.

EXAMPLE

The best wire size for $1\frac{1}{4}$-12 2A is .048 inch, but only a $\frac{3}{64}$-inch diameter drill rod is available, which has a diameter of .0469 inch. The maximum PD of this thread is 1.1959, and the minimum PD is 1.1912. The fit classification must be considered because the PD varies from fit to fit.

$$\text{Best wire size} = .57735 \times .08333 = .048 \text{ inch}$$
$$\text{Smallest wire size} = .56 \times .08333 = .047 \text{ inch}$$
$$\text{Largest wire size} = .90 \times .08333 = .075 \text{ inch}$$

The following formula is used to determine the micrometer reading over the wires for Unified threads:

$$M = E - .0866025 \times P + 3W$$

where

M = measurement over wires

E = pitch diameter

W = wire size

P = pitch

For the $1\frac{1}{4}$-12 UNC 3A thread, the maximum and minimum PD dimensions are

$$\text{Max. PD} = 1.1959$$
$$\text{Min. PD} = 1.1913$$

The calculation for the maximum and minimum measurement over the wires is

$$\text{Max. } M \text{ dimension} = 1.1959 - .866025 \times .08333$$
$$+ 3 \times .055 = 1.2887$$

$$\text{Min. } M \text{ dimension} = 1.1913 - .866025 \times .08333$$
$$+ 3 \times .055 = 1.2841$$

This calculation determines the acceptable tolerance of the M dimension and is an industry standard.

The measurement over the wires will be slightly different from that of the best wire size because of the difference in wire size. After the best size wire is found, the wires are positioned in the threaded grooves as shown in Figure I-317. The anvil and spindle of a standard outside micrometer are then placed against the three wires, and the measurement is taken.

EXAMPLE

To find M for a $1\frac{1}{4}$-12 UNC thread, proceed as follows. If

$$W = .048$$
$$D = 1.250$$
$$n = 12$$

then

$$M = 1.250 + 3 \times .048 - 1.5156/12$$
$$= 1.250 + .144 - .126$$
$$= 1.268 \text{ (micrometer measurement)}$$

The same method and formula are used for measuring a Unified fine thread, except that the constant is 1.732 instead of 1.5155.

The wire method of thread measurement is also used for other thread forms such as Acme and buttress. Information and tables may be found in *Machinery's Handbook*.

The screw thread micrometer (Figure I-292) may be used to measure threads. The spindle is pointed to a 60-degree included angle. The anvil, which swivels, has a double-V shape to contact the pitch diameter. The thread micrometer measures the pitch diameter directly from the screw thread. This reading may be compared with pitch diameters given in handbook tables. Thread micrometers have interchangeable anvils that will fit a wide range of thread pitches. Some are made in sets of four micrometers that have a capacity up to 1 inch, and each covers a range of threads. The range of these micrometers depends on the manufacturer. Typical ranges are as follows:

No. 1	8–14 threads per inch
No. 2	14–20 threads per inch
No. 3	22–30 threads per inch
No. 4	32–40 threads per inch

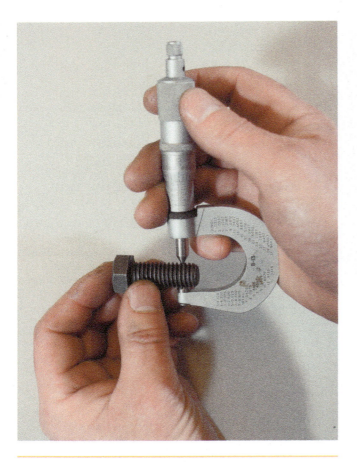

Figure I-292 Screw thread micrometer.

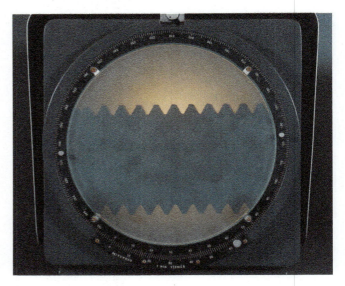

Figure I-293 Profile of thread as shown on the screen of the optical comparator.

The optical comparator is sometimes used to check thread form, helix angle, and depth of thread on external threads (Figure I-293). The part is mounted in a screw thread accessory that is adjusted to the helix angle of the thread so that the light beam will show a true profile of the thread.

SELF-TEST

1. How are threads cut or chased with a single-point tool in a lathe? How is the lathe set up to produce a specific thread lead?

2. Why is it better to feed the tool in with the compound set at 29 degrees rather than to feed with the cross slide when cutting threads?

3. How can the 60-degree angle be checked on a threading tool?

4. How can the number of threads per inch be checked?

5. How is a threading tool aligned with the work?

6. Is the carriage moved along the ways by gears when the half-nut lever is engaged? Explain.

7. Explain which positions on the threading dial are used for engaging the half nuts for even-, odd-, and fractional-numbered threads.

8. How fast should the spindle be turning for threading?

9. What is the procedure for cutting left-hand threads?

10. If for some reason it becomes necessary for you to temporarily remove the tool or the entire threading setup before a thread is completed, what procedure should you follow to finish the thread?

INTERNET REFERENCE

Thread measurement tooling:

www.threadcheck.com

Cutting Unified Internal Threads

Small internal threads are usually tapped. Internal threads larger than 1 inch are often cut in a lathe. The problems and calculations involved with cutting internal threads differ in some ways from those of cutting external threads. This unit will help you understand these differences.

OBJECTIVE

After completing this unit, you should be able to:

- Calculate the dimensions of and cut an internal Unified thread.

MAKING INTERNAL THREADS

Many of the same rules used for external threading apply to internal threading. The tool must be shaped to the exact form of the thread, and the tool must be set on the center of the workpiece. When you are cutting an internal thread with a single-point tool, the inside diameter (hole size) of the workpiece should be the minor diameter of the internal thread (Figure I-294). If the thread is made by tapping in the lathe, the hole size of the workpiece can be chosen to obtain the desired percent of thread.

Full-depth threads are difficult to tap in soft metals and impossible in tough materials. Tests have proved that above 60 percent of thread, little additional strength is gained. Lower percentages, however, provide less flank surface for wear. Most commercial internal threads in steel are about 75 percent. Therefore, the purpose of making an internal thread with less than 100 percent thread depth is only for easier tapping. No such problem exists when making internal threads with a single-point tool on a lathe.

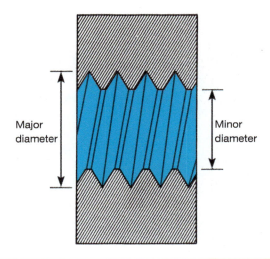

Figure I-294 View of internal threads showing major and minor diameters.

SINGLE-POINT TOOL THREADING

The advantages of making internal threads with a single-point tool are that large threads of various forms can be made and that the threads are concentric to the axis of the work. Threads may not be concentric if they are tapped. Difficulties can be encountered when making internal threads. The tool is often hidden from view, and tool spring must be taken into account.

The hole to be threaded is first drilled to a diameter $\frac{1}{16}$ inch less than the minor diameter. Then a boring bar is set up, and the hole is bored to the minor diameter of the thread. If the thread is to go completely through the work, no recess is necessary, but if threading is done in a blind hole, a recess must be made. The compound rest should be swiveled 29 degrees to the left of the operator for cutting right-hand threads (Figures I-295 and I-296).

Figure I-295 For right-hand internal threads, the compound is swiveled to the left.

Figure I-297 An alternative method of internal threading is shown here in which the tool is turned 180 degrees away from the operator. Note: this is a view from over the top of the setup.

Figure I-296 The compound rest is swiveled to the right for left-hand internal threads.

Figure I-298 Aligning the threading tool with a center gage.

An alternative is to turn the boring bar 180 degrees so that the tool bit is opposite and the cutting edge faces downward (Figure I-297).

With this method, the compound is swiveled to the right for right-hand internal threads. The advantages of this method are that the cutting operations can be more easily observed and the compound can remain in the same position as for external right-hand threads. For both methods, the threading tool is clamped in the bar in the same way and aligned by means of a center gage (Figure I-298).

The compound micrometer collar is moved to the zero index after the slack (backlash) has been removed by turning the screw outward or counterclockwise. The tool is brought to the work with the cross-slide handle and its collar is set on zero. Threading may now proceed in the same way as external threads. The compound is advanced outward a few thousandths of an inch, a scratch cut is made, and the thread pitch is checked with a screw pitch gage.

The cross slide is backed out of the cut and reset to zero before the next pass. Cutting oil is used. The compound is advanced a few thousandths of an inch (.001–.010 inch).

The infeed on the compound is calculated in the same way as with external threads; for Unified internal threads, use the formula for the depth of cut with the compound set at 29 degrees.

$$\text{Infeed} = P \times .619$$

For a pitch of 6 threads per inch,

$$\text{Infeed} = .1666 \times .619 = .102 \text{ inch}$$

Often, it is necessary to realign an internal threading tool with the thread when the tool has been moved for sharpening or when the setup has moved during the cut. The tool is realigned in the same way as is done for external threads: by engaging the half nut and positioning the tool in a convenient place over the threads, then moving both the compound and cross slides to adjust the tool position.

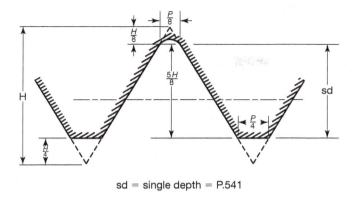

sd = single depth = P.541

Figure I-299 Single depth of the Unified internal thread.

The exact amount of infeed depends on how rigid the boring bar and holder are and how deep the cut has progressed. Too much infeed will cause the bar to spring away and produce a bellmouth internal thread.

If a slender boring bar is necessary or there is more than usual overhang, lighter cuts must be used to avoid chatter. The bar may spring away from the cut, causing the major diameter to be less than the calculated amount or that amount fed in on the compound. If several passes are taken through the thread with the same setting on the compound, this problem can often be corrected.

The single depth of the Unified internal thread (Figure I-299) equals $P * .541$. The minor diameter is found by subtracting the double depth of the thread from the major diameter. Thus, if

$$D = \text{major diameter}$$
$$d = \text{minor diameter}$$
$$P = \text{thread pitch}$$
$$P \times .541 = \text{single depth}$$

the formula is

$$d = D - (P \times .541 \times 2)$$

EXAMPLE

A $\frac{1}{2}$-6 UNC nut must be bored and threaded to fit a stud. Find the dimension of the bore.

$$P = \frac{1}{6} = .1666$$
$$d = 1.500 - (.1666 \times .541 \times 2)$$
$$= 1.500 - .180$$
$$= 1.320$$

The bore should be 1.320 inch.

Cutting the Internal Thread

The following is the procedure for cutting right-hand internal threads:

Step 1 After correctly setting up the boring bar and tool, touch the threading tool to the bore and set both crossfeed and compound micrometer dials to zero.

Step 2 Feed in counterclockwise .002 inch on the compound dial.

Step 3 Turn on the lathe and engage the half-nut lever.

Step 4 Take a scratch cut. Disengage the half nut when the tool is through the workpiece.

Step 5 Check the thread pitch with a screw pitch gage.

Step 6 Apply cutting fluid to the work.

Step 7 Feed the compound in an appropriate amount. Slightly less depth for cutting internal threads than for similar external threads may be necessary because of the spring of the boring bar.

Step 8 When nearing the calculated depth, test the thread with a plug gage between each pass. When the plug gage turns completely into the thread without being loose, the thread is finished.

Step 9 Take several free cuts (passes without infeed) when nearing the finish depth to compensate for boring bar spring. Use the test plug between free passes.

Step 10 The inside and outside edges of the internal thread should be chamfered.

BASIC INTERNAL THREAD MEASUREMENT

Internal threads cut with a single-point tool need to be checked. A precision thread plug gage (Figure I-300) is generally sufficient for most purposes. Thread plug gages are available for all standard thread sizes. An external screw thread is on each end of the handle. The longer threaded end is called the go gage, and the shorter end is called the no-go gage. The no-go end is made to a slightly larger dimension than the pitch diameter for the class of fit that the gage tests. To test an internal thread, both the go and no-go gages should be tried in the hole. If the part is within the range or tolerance of the gage, the go end should turn in flush to the bottom of the internal thread, but the no-go end should just start into the hole and become snug with no more than three turns. The gage should never be forced into the hole. If no gage is available, a shop gage can be made by cutting the required external thread to precise dimensions. If only one threaded part of a kind is to be made and no interchangeability is required, the mating part may be used as a gage.

Figure I-300 Thread plug gage.

SELF-TEST

1. What should the bore size be for making internal threads on a lathe?

2. What percent of thread are tap drill charts based on?

3. List two advantages of making internal threads with a single-point tool on the lathe.

4. When making internal right-hand threads, in which direction should the compound be swiveled?

5. After a scratch cut is made, what is the easiest method to measure the pitch of the thread?

6. What does deflection or spring of the boring bar cause when cutting internal threads?

7. Using P * .541 as a constant for Unified single-depth internal threads, calculate the minor diameter for a UNC 1-8 thread.

8. List two methods of checking an internal thread for size.

Cutting Tapers

Tapers often need to be machined on parts. The machinist should be able to calculate a specific taper and to set up a machine to produce it. The machinist should also be able to measure tapers accurately and determine proper fits. This unit will help you understand the various methods and principles involved in making a taper.

OBJECTIVE

After completing this unit, you should be able to:

■ Describe different types of tapers and the methods used to produce and measure them.

USE OF TAPERS

Tapers are used because of their ability to align and hold machine parts and to realign them when they are assembled and disassembled. This repeatability assures that tools such as centers in lathes, taper shank drills in drill presses, and arbors in milling machines will run in perfect alignment when placed in the machine. When a taper is slight, it is called a *self-holding taper,* because it is held in and driven by friction (Figure I-301). A Morse taper, which is about $\frac{5}{8}$-inch taper per foot, is an example of a slight taper. A steep taper, such as a quick-release taper of $3\frac{1}{2}$ inches per foot used on most milling machines, must be held in place with a draw bolt (Figure I-302).

A **taper** may be defined as a uniform increase in diameter on a workpiece for a given length. Internal or external tapers are expressed in taper per foot (tpf), taper per inch, or degrees. The tpf or tpi refers to the difference in diameters in the length of 1 foot or 1 inch, respectively (Figure I-303). This difference is measured in inches. Angles of taper, on the other hand, may refer to the included angles or the angles with the centerline (Figure I-304).

Some machine parts measured in tpf are mandrels (.006 inch per foot), taper pins and reamers ($\frac{1}{4}$ inch per ft), the Jarno

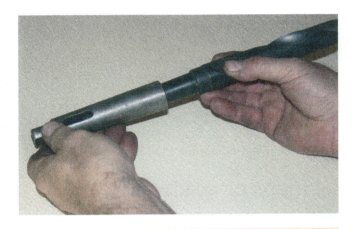

Figure I-301 The Morse taper shank on this drill and the tang keeps the drill from turning when the hole is being drilled.

Figure I-302 The milling machine taper is driven by lugs and held in by a draw bolt.

taper series (.006 inch foot), the Brown and Sharpe taper series (approximately $\frac{1}{2}$ inch per foot), and the Morse taper series (about $\frac{5}{8}$ inch per foot). Morse tapers include eight sizes that range from size 0 to size 7. Tapers and dimensions vary

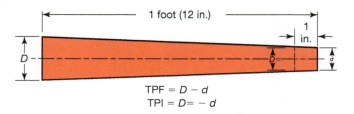

$$TPF = D - d$$
$$TPI = D = - d$$

Figure I-303 Difference between taper per foot (tpf) and taper per inch (tpi).

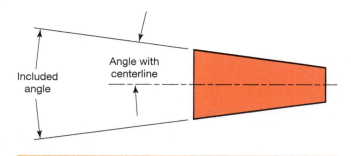

Figure I-304 Included angles and angles with centerline.

slightly from size to size in both the Brown and Sharpe and the Morse series. For instance, a No. 2 Morse taper has .5944 inch per foot taper and a No. 4 has .6233 inch per foot taper. See Table I-7 for more information on Morse tapers.

METHODS OF MAKING A TAPER

There are several methods of turning a taper on a lathe: the compound slide method, the offset tailstock method, the taper attachment method, or the use of a form tool. Each method has its advantages and disadvantages, so the kind of taper needed on a workpiece should determine which method will be used.

Compound Slide Method

Internal or external short, steep tapers can be turned on a lathe by hand feeding the compound slide (Figure I-305). The swivel base of the compound is divided into degrees.

When the compound slide is in line with the ways of the lathe, the 0-degree line will align with the index line on the cross slide (Figure I-306). When the compound is swiveled off the index, which is parallel to the centerline of the lathe, a direct reading may be taken for the half-angle or angle to the centerline of the machined part (Figure I-307). When a taper is machined off the lathe centerline, its included angle will be twice the angle that is set on the compound. Not all lathes are indexed in this manner.

When the compound slide is aligned with the axis of the cross slide and swiveled off the index in either direction, an angle is directly read off the cross-slide

Figure I-305 Making a taper using the compound slide.

Table I-7 Morse Taper Specification

Number of Taper	Taper per Foot	Taper per Inch	P, Standard Plug Depth	D, Diameter of Plug at Small End	A, Diameter at End of Socket	H, Depth of Hole
0	.6246	.0520	2	.252	.356	$2\frac{1}{32}$
1	.5986	.0499	$2\frac{1}{8}$	.396	.475	$2\frac{3}{16}$
2	.5994	.0500	$2\frac{9}{16}$	.572	700	$2\frac{5}{8}$
3	.6023	.0502	$3\frac{3}{16}$	.778	.938	$3\frac{1}{4}$
4	.6326	.0519	$4\frac{1}{16}$	1.020	1.231	$4\frac{1}{8}$
5	.6315	.0526	$5\frac{3}{16}$	1.475	1.748	$5\frac{1}{4}$
6	.6256	.0521	$7\frac{1}{4}$	2.116	2.494	$7\frac{3}{8}$
7	.6240	.0520	10	2.750	3.270	$10\frac{1}{8}$

Figure I-306 Alignment of the compound parallel with the ways.

Figure I-307 An angle may be set off the axis of the lathe from this index.

centerline (Figure I-307). In this example the compound was set to 14.5 degrees. That would cut a 29-degree included angle.

Tapers of any angle may be cut by this method, but the length is limited to the travel of the compound slide. Since tapers are often given in tpf, it is sometimes convenient to consult a tpf-to-angle conversion table, like Table I-8.

A more complete table may be found in the *Machinery's Handbook*.

If a more precise conversion is desired, the following formula may be used to find the included angle: Divide the taper in inches per foot by 24, find the angle that corresponds to the quotient in a table of tangents, and double this angle. If the angle with centerline is desired, do not double the angle.

Table I-8 Tapers and Corresponding Angles

Inches per Foot	Included Angle		Angle with Centerline		Taper per Inch
	Degrees	Minutes	Degrees	Minutes	
$\frac{1}{8}$	0	36	0	18	.0104
$\frac{3}{16}$	0	54	0	27	.0156
$\frac{1}{4}$	1	12	0	36	.0208
$\frac{5}{16}$	1	30	0	45	.0260
$\frac{3}{8}$	1	47	0	53	.0313
$\frac{7}{16}$	2	5	1	2	.0365
$\frac{1}{2}$	2	23	1	11	.0417
$\frac{9}{16}$	2	42	1	21	.0469
$\frac{5}{8}$	3	00	1	30	.0521
$\frac{11}{16}$	3	18	1	39	.0573
$\frac{3}{4}$	3	35	1	48	.0625
$\frac{13}{16}$	3	52	1	56	.0677
$\frac{7}{8}$	4	12	2	6	.0729
$\frac{15}{16}$	4	28	2	14	.0781
1	4	45	2	23	.0833
$1\frac{1}{4}$	5	58	2	59	.1042
$1\frac{1}{2}$	7	8	3	34	.1250
$1\frac{3}{4}$	8	20	4	10	.1458
2	9	32	4	46	.1667
$2\frac{1}{2}$	11	54	5	57	.2083
3	14	16	7	8	.2500
$3\frac{1}{2}$	16	36	8	18	.2917
4	18	56	9	28	.3333
$4\frac{1}{2}$	21	14	10	37	.3750
5	23	32	11	46	.4167
6	28	4	14	2	.5000

EXAMPLE

What angle is equivalent to a taper of $3\frac{1}{2}$ inch per foot?

$$\frac{3.5}{24} = .14583$$

The angle of this tangent is 8 degrees 18 minutes, and the included angle is twice this value, or 16 degrees 36 minutes.

When an internal or external taper is turned on a lathe, the cutting tool height must be on the centerline of the lathe spindle. If the tool is either too high or too low, the taper may be inaccurate.

OFFSET TAILSTOCK METHOD

Long, slight tapers may be produced on shafts and external parts between centers. Internal tapers cannot be made by this method. Power feed is used, so good finishes are obtainable. Because many lathes are equipped with taper attachments, the offset tailstock method is not commonly used. Its greatest advantage is that longer tapers can be made by this method than by any other. The tpf or taper per inch must be known so that the amount of offset for the tailstock can be calculated. Because tapers are of different lengths, they will not be the same tpi or tpf for the same offset (Figure I-308).

When the taper per inch is known, the offset calculation is as follows:

$$\text{Offset} = \frac{\text{tpi} \times L}{2}$$

where

tpi = taper per inch

L = length of workpiece

When the tpf is known, use the following formula:

$$\text{Offset} = \frac{\text{tpf} \times L}{24}$$

If the workpiece has a short taper in any part of its length (Figure I-309) and the tpi or tpf is not given, use the following formula:

$$\text{Offset} = \frac{L \times (D - d)}{2 \times L_1}$$

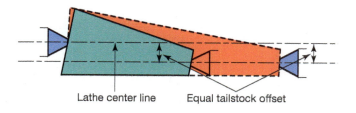

Figure I-308 When tapers are of different lengths, the tpf is not the same with the same offset.

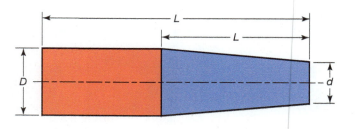

Figure I-309 Long workpiece with a short taper.

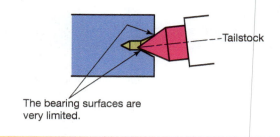

Figure I-310 The contact area between the center hole and the center is small.

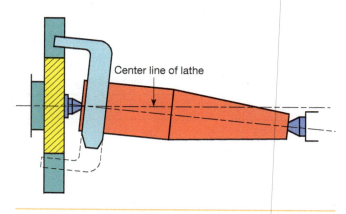

Figure I-311 The bent tail of the lathe dog should have adequate clearance.

where

D = diameter at large end of taper

d = diameter at small end of taper

L = total length of workpiece

L_1 = length of taper

When turning a taper between centers, the contact area between the center and the center hole is limited (Figure I-310). Frequent lubrication of the centers may be necessary.

You should also note the path of the lathe dog bent tail in the drive slot (Figure I-311). Check to see that there is adequate clearance.

To measure the offset on the tailstock, use either the centers and a rule (Figure I-312) or the witness mark and a rule (Figure I-313); both methods may be adequate for some applications. A more precise measurement is possible with a

Figure I-312 Measuring the offset on the tailstock by use of the centers and the rule.

Figure I-313 Measuring the offset with the witness mark and rule.

Figure I-314 Using a dial indicator to measure the offset.

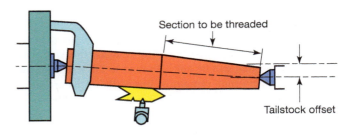

Figure I-315 Adjusting the threading tool for cutting tapered threads using the offset tailstock method. The tool is set square to the centerline of the work rather than the taper.

dial indicator, as shown in Figure I-314. The indicator is set on the tailstock spindle while the centers are still aligned. A slight loading of the indicator is advised, since the first .010 or .020 inch of indicator movement may be inaccurate, or the mechanism may be loose owing to wear, causing fluctuating readings. Set the bezel at zero, and move the tailstock toward the amount you calculated. Clamp the tailstock to the way. If the indicator reading changes, loosen the clamp and readjust.

When tapered threads such as pipe threads are being cut, the tool should be square with the centerline of the workpiece, not the taper (Figure I-315). When you have finished making tapers by the offset tailstock method, realign the centers to .001 inch or less in 12 inches. When more than one part must be turned by this method, all parts must have identical lengths and center hole depths if the tapers are to be the same.

Figure I-316 Plain taper attachment.

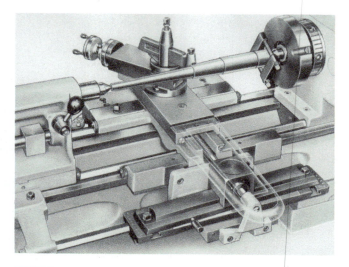

Figure I-317 Telescopic taper attachment (*Courtesy of Clausing Industrial, Inc.*).

Figure I-318 Adjusting the taper attachment to a given taper with a dial indicator.

Taper Attachment Method

The taper attachment has a slide independent that is independent to the ways that can be angled and will move the cross slide according to the angle that is set. Slight to fairly steep tapers ($3\frac{1}{2}$ inches per foot) may be made this way. The length of taper that can be cut is limited to the length of the slide bar. Centers may remain in line without distortion of the center holes. Work may be held in a chuck, and both external and internal tapers may be made, often with the same setting for mating parts. Power feed is used. Taper attachments are graduated tpf or in degrees.

There are two types of taper attachments: the **plain taper attachment** (Figure I-316) and the **telescoping taper attachment** (Figure I-317). The crossfeed binding screw must be removed to free the nut when the plain type is set up. The depth of cut must then be made by using the compound feed screw handle. The crossfeed may be used for depth of cut when using the telescoping taper attachment, because the crossfeed binding screw is not disengaged with this type.

When a workpiece is to be duplicated or an internal taper is to be made for an existing external taper, it is often convenient to set up the taper attachment by using a dial indicator (Figure I-318). The contact point of the dial indicator must be on the center of the workpiece. The workpiece is first set up in a chuck or between centers so that there is no

runout when it is rotated. With the lathe spindle stopped, the indicator is moved from one end of the taper to the other. The taper attachment is adjusted until the indicator does not change its reading when moved. A digital readout that reads in both X and Z axes can also be used.

The angle, the tpf, or the taper per inch must be known to set up the taper attachment to cut specific tapers. If none of these are known, proceed as follows: If the end diameters (D and d) and the length of taper (L) are given in inches, the following applies:

$$\text{Taper per foot} = \frac{D - d}{L} \times 12$$

If the tpf is given, but you want to know the amount of taper in inches for a given length, use the following formula:

$$\text{Amount of taper} = \frac{\text{tpf}}{12} \times \text{given length of tapered part}$$

When the tpf is known, to find tpi divide the tpf by 12. When the tpi is known, to find tpf multiply the tpi by 12.

Turning a Taper with a Taper Attachment

Step 1 Clean and oil the slide bar (*a*) (Figure I-319).

Step 2 Set up the workpiece and the cutting tool on center. Bring the tool near the workpiece and to the center of the taper.

Step 3 Remove the crossfeed binding screw (*b*) that binds the crossfeed screw nut to the cross slide. *Do not remove* this screw if you are using a telescoping taper attachment. The screw is removed *only* on the plain type. Put a temporary plug in the hole to keep chips out.

Step 4 Loosen the lock screws (*c*) on both ends of the slide bar and adjust to the required degree of taper.

Step 5 Tighten the lock screws.

Step 6 Tighten the binding lever (*d*) on the slotted cross-slide extension at the sliding block, *plain type only.*

Step 7 Lock the clamp bracket (*e*) to the lathe bed.

Step 8 Move the carriage to the right so that the tool is from $\frac{1}{2}$ to $\frac{3}{4}$ inch past the start position. Do this every pass to remove any backlash in the taper attachment.

Step 9 Feed the tool in for the depth of the first cut with the cross slide unless you are using a plain-type attachment. Use the compound slide for the plain type.

Step 10 Take a trial cut and check the diameter. Continue the roughing cut.

Step 11 Check the taper for fit and readjust the taper attachment if necessary.

Step 12 Take a light cut, about .010 inch, and check the taper again. If it is correct, complete the roughing and final finish cuts.

Internal tapers are best made with the taper attachment. They are set up in the same manner as prescribed for external tapers.

Other Methods of Making Tapers

A tool may be set with a protractor to a given angle (Figure I-320), and a single plunge cut may be made to produce a short taper (Figure I-321). This method is often used for chamfering a workpiece to an angle such as the chamfer

Figure I-320 Tool is set up with protractor to make an accurate chamfer or taper.

Figure I-319 Parts of the taper attachment (*Courtesy of Fox Valley Technical College*).

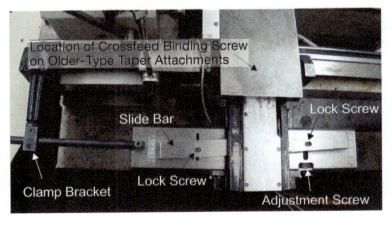

Location of Crossfeed Binding Screw on Older-Type Taper Attachments

Slide Bar

Lock Screw

Clamp Bracket

Lock Screw

Adjustment Screw

Figure I-321 Making the chamfer with a tool.

Figure I-322 Taper plug gage.

Figure I-323 Taper ring gage.

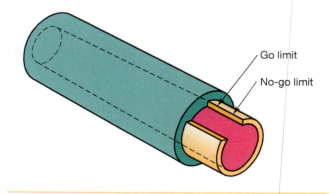

Figure I-324 Go/no-go taper ring gage.

Figure I-325 Chalk mark is made along a taper plug gage prior to checking an internal taper.

used for hexagonal bolt heads and nuts. Tapered form tools sometimes are used to make V-shaped grooves. Only short tapers can be made with form tools.

CNC machines can be programmed to make any taper, either external or internal. An advantage to this method is the precision repeatability when many tapers must be made with the same dimensions.

Tapered reamers are sometimes used to produce a specific internal taper such as a Morse taper. A roughing reamer is used first, followed by a finishing reamer. Finishing Morse taper reamers can also be used to clean up a nicked or scarred internal Morse taper.

METHODS OF MEASURING TAPERS

The most convenient and simple way to check an internal taper is to use a taper plug gage (Figure I-322). A taper ring gage (Figure I-323) is used for external tapers. Some taper gages have go and no-go limit marks on them (Figure I-324).

To check an internal taper, first make a chalk or Prussian blue mark along the length of the taper plug gage (Figures I-325 and I-326). Then, insert the gage into the internal taper and turn slightly. When you take the gage out, the chalk mark or Prussian blue will be partly rubbed off where contact was made. Adjust the taper until the chalk mark is rubbed off along its full length of contact, indicating a good fit. Mark an external taper with chalk and check it in the same way with a taper ring gage (Figures I-327 to I-329).

The taper per inch may be checked by scribing two marks 1 inch apart on the taper and measuring the diameters (Figures I-330 and I-331) at these marks. The difference is the taper per inch.

Perhaps an even more precise method of measuring a taper is with the sine bar and gage blocks on the surface plate (Figure I-332). When this is done, it is important to keep the centerline of the taper parallel to the sine bar and to read the indicator at the highest point. Tapers may be measured with a taper micrometer.

Figure I-326 The taper has been tested and the chalk mark has been rubbed off evenly, indicating a good fit.

Figure I-327 The external taper is marked with chalk or Prussian blue before being checked with a taper ring gage.

Figure I-328 The ring gage is placed on the taper snugly and is rotated slightly.

Figure I-329 The ring gage is removed and the chalk mark is rubbed off evenly for the entire length of the ring gage. This indicates a good fit.

Figure I-330 Measuring the taper per inch (tpi) with a micrometer. The larger diameter is measured on the line with the edge of the spindle and the anvil of the micrometer contacting the line.

SELF-TEST

1. State the difference in holding between steep tapers and slight tapers.
2. In what three ways are tapers expressed (measured)?
3. Briefly describe three methods of turning a taper in the lathe.
4. When a taper is produced by the compound slide method, is the reading in degrees on the compound swivel base the same as the angle of the finished workpiece? Explain.
5. If the swivel base is set to a 35-degree angle at the cross-slide centerline index, what will the reading be at the lathe centerline index?

Figure I-331 The second measurement is taken on the smaller diameter at the edge of the line in the same manner.

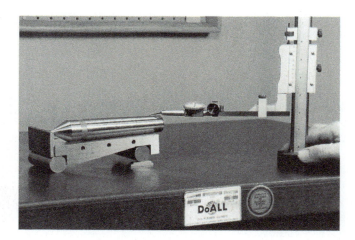

Figure I-332 Using a sine bar and gage blocks with a dial indicator to measure a taper.

6. Calculate the offset for the taper shown in Figure I-333. The formula is

$$\text{Offset} = \frac{L \times (D - d)}{2 \times L_1}$$

7. Name three methods of measuring tapers.

8. What is the most practical and convenient way to check internal and external tapers in the lathe?

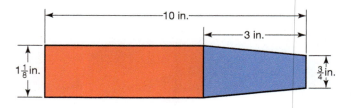

Figure I-333 Calculate the offset for this taper.

Using Steady and Follower Rests

Many lathe operations would not be possible without the use of steady and follower rests. These valuable attachments make internal and external machining operations on long workpieces possible.

OBJECTIVES

After completing this unit, you should be able to:

- Identify the parts and explain the uses of the steady rest.
- Explain the correct uses of the follower rest.
- Correctly set up a steady rest on a straight shaft.
- Correctly set up a follower rest on a prepared shaft.

STEADY REST

Long shafts tend to vibrate when cuts are made, leaving chatter marks. Even light finish cuts will often produce chatter when the shaft is long and slender. To help eliminate these problems, use a steady rest to support workpieces that extend out from a chuck more than four or five diameters of the workpiece.

The **steady rest** (Figure I-334) is made of a cast iron or steel frame that is hinged so that it will open to accommodate workpieces. It has three or more adjustable jaws that are tipped with bronze, plastic, or ball bearing rollers. The base of the frame is machined to fit the ways of the lathe, and it is clamped to the bed by means of a bolt and crossbar.

A steady rest is also used to support long workpieces for various other machining operations such as threading, grooving, and knurling (Figure I-335). Heavy cuts can be made by using one or more steady rests along a shaft.

Adjusting the Steady Rest

Workpieces should be mounted and centered in a chuck whether a tailstock center is used or not. If the shaft has centers and finished surfaces that turn concentric (have no

Figure I-334 A steady rest.

runout) with the lathe centerline, setup of the steady rest is simple. The steady rest is slid to a convenient location on the shaft, which is supported in the dead center and chuck, and the base is clamped to the bed. The two lower jaws are brought up to the shaft finger tight only (Figure I-336). A good high-pressure lubricant is applied to the shaft, and the top half of the steady rest is closed and clamped. The upper jaw is brought to the shaft finger tight, and then all

Figure I-335 A long, slender workpiece is supported by a steady rest near the center to limit vibration or chatter.

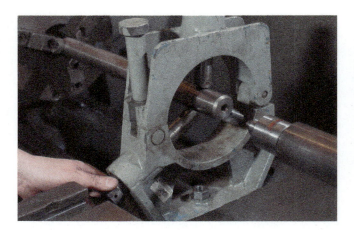

Figure I-336 The jaws are adjusted to the shaft.

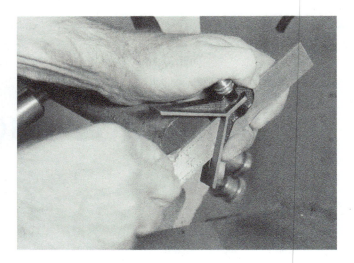

Figure I-337 Laying out the center of a shaft.

Figure I-338 Adjusting the steady rest jaws to a centered shaft. Note: the tailstock center is being moved close to the end of the piece so it can be used as a guide to help center the work in the steady rest.

three lockscrews are tightened. Some clearance is necessary on the upper jaw to avoid scoring of the shaft. As the shaft warms or heats up from friction during machining, readjustment of the upper jaw is necessary.

A finished workpiece can be scored if any hardness or grit is present on the jaws. To protect finishes, brass or copper strips are often placed between the jaws and the workpiece. Steady or follower rest jaws should never support rough surfaces because they will soon be worn away, even if lubricant is used. Also, any machining done will be inaccurate because of the surface irregularities of the workpiece.

When there is no center in a finished shaft of the same diameter, setup procedure is as follows:

Step 1 Position the steady rest near the end of the shaft with the other end lightly chucked in a three- or four-jaw chuck.

Step 2 (Note: If the piece has not been center drilled). You can scribe two cross-centerlines with a center head on the end of the shaft and prick punch the mark (Figure I-337). The dead center can then be used to align with the center lines.

Step 3 (If the end of the shaft was center drilled). Bring up the dead center near to the center drilled hole (Figure I-338).

Step 4 Adjust the lower jaws of the steady rest to center the center-drilled hole and the dead center.

Step 5 Tighten the chuck. If it is a four-jaw chuck, check for runout with a dial indicator.

Step 6 You may now remove the steady rest to any location along the shaft.

Step 7 If the shaft has not been center drilled, you may now drill a center hole in the end of the shaft.

Stepped shafts may be set up by using a similar procedure, but the steady rest must remain on the diameter on which it is set up.

Figure I-339 Turning a concentric bearing surface on rough stock for the steady rest jaws.

Figure I-340 The rough shaft can now rotate in the steady rest jaws, which are being adjusted to the bearing surface in this view.

Using the Steady Rest

Steady rest jaws should never be used on rough surfaces. When a forging, casting, or hot-rolled bar must be placed in a steady rest, a concentric bearing with a good finish must be turned (Figures I-339 and Figures I-340). Thick-walled tubing or other materials that tend to be out of round also should have bearing surfaces machined on them. The usual practice is to remove no more in diameter than necessary to clean up the bearing spot.

When the piece to be set up is irregular, such as a square (Figure I-341), hexagonal part, HR bar, or rough casting, a **cathead** is used.

Figure I-341 Using a cathead to support a square piece. Drilling and boring in the end of a heavy square bar requires the use of an external cathead.

Figure I-342 The cathead is adjusted with setscrews to center the shaft with the dead center.

Figure I-343 The cathead is adjusted over the irregular end of the shaft.

The procedure to set up a cathead is not difficult. The piece is placed in the cathead, and the cathead is mounted in the steady rest while the other end of the workpiece is centered in the chuck. The workpiece is made to run true

Figure I-344 Tubing being set up with a cathead using a dial indicator to true the inside diameter.

near the steady rest by adjusting screws on the cathead (Figures I-342).

In most cases the workpiece is given a center to provide more support for turning operations. A centered cathead (Figures I-343) is sometimes used when a permanent center is not required in the workpiece. An internal cathead (Figure I-344) is used for truing to the inside diameter of tubing.

FOLLOWER REST

Long, slender shafts tend to spring away from the tool, vary in diameter, chatter, and often climb the tool. To prevent these problems when machining a slender shaft, a follower rest (Figure I-345) is often used. Follower rests are bolted to the carriage and follow along with the tool. Most follower rests have two jaws placed to back up the work opposite the tool thrust. Some types are made with different-sized bushings to fit the work.

Using the Follower Rest

The workpiece should be 1 to 2 inches longer than needed to allow room for the follower rest jaws. The end is turned to smaller than the finish size. The tool is adjusted ahead of the

Figure I-346 Adjusting the follower rest.

Figure I-345 A follower rest is used to turn this long shaft.

Figure I-347 Long, slender Acme threaded screw being machined with the aid of a follower rest.

Figure I-348 Both steady and follower rests being used.

jaws about $1\frac{1}{2}$ inch, and a trial cut of 2 or 3 inches is made with the jaws backed off. Then, the lower jaw is adjusted fingertight (Figure I-346) followed by the upper jaw. Both locking screws are tightened. Oil should be used to lubricate the jaws.

A follower rest is often used when cutting threads on long, slender shafts, especially when cutting square or Acme threads (Figure I-347). Burrs should be removed between passes to prevent them from cutting into the jaws. Jaws with rolls are sometimes used for this purpose. On long shafts, a steady rest and follower rest are often used (Figure I-348).

SELF-TEST

1. When should a steady rest be used?
2. How is the steady rest set up on a straight finished shaft with centers in the ends?
3. What precaution can be taken to prevent the jaws of the steady rest from scoring the finished shaft?
4. How can a steady rest be set up when there is no center hole in the shaft?
5. Is it possible to set up a steady rest by using a dial indicator on the rotating shaft to watch for runout?
6. Should a steady rest be used on a rough surface? Explain.
7. How can a steady rest be used on an irregular surface such as square or hex stock?
8. What lathe attachment could be used if a long, slender shaft needs to be turned or threaded for its entire length?

Additional Thread Forms

Many thread forms other than the 60-degree types are found on machines. Each of the forms is unique and has a special use. A machinist must be able to recognize these thread forms.

As a machinist, you may occasionally be called on to make a multiple-lead thread. This unit will acquaint you with the various methods and procedures for cutting multiple-lead threads.

OBJECTIVES

After completing this unit, you should be able to:

- Identify five different thread forms and explain their uses.
- Calculate the dimensions needed to machine the five thread forms.
- Describe the methods and procedures for machining multiple-lead threads.

THREAD FORM BASICS

Transmitting (translating) screw threads are used primarily to transmit or impart power or motion to a mechanical part. Often, these transmitting screws are of multiple lead to effect rapid motion. Bench vises and house jacks are familiar applications of single-lead transmitting screws. The lead screw on a lathe and the table feed screws on milling machines are examples of use of these screw threads to impart power along the axis of the screw to move a part.

The earliest type of transmitting screw was the square thread form (Figure I-349). Thrust on the flanks of the thread is fully axial, thus reducing friction to a minimum. The square thread is more difficult to produce than other types and is not now widely used. The thread depth and thickness are half the pitch. Clearance must be provided on the flanks and major diameter of the thread.

The modified square thread form (Figure I-350) was designed to replace the square thread. It is easier to produce

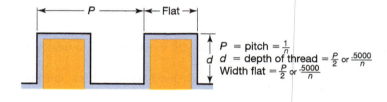

$$P = \text{pitch} = \frac{1}{n}$$
$$d = \text{depth of thread} = \frac{P}{2} \text{ or } \frac{.5000}{n}$$
$$\text{Width flat} = \frac{P}{2} \text{ or } \frac{.5000}{n}$$

Figure I-349 Square thread.

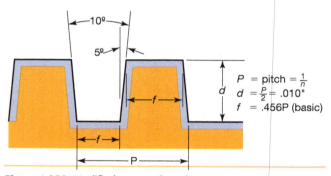

$$P = \text{pitch} = \frac{1}{n}$$
$$d = \frac{P}{2} = .010''$$
$$f = .456P \text{ (basic)}$$

Figure I-350 Modified square thread.

than the square thread, yet it has all the advantages—and some of the drawbacks—of the square thread. Like the square thread, it is not widely used.

The **Acme thread** (Figure I-351) is generally accepted throughout the mechanical industries as an improved thread form over the square and modified square. An Acme thread is easier to make and stronger than the square form thread, because the Acme root cross section is thicker than its root clearance. Acme thread screws are used on milling machines and lathes. Like the square thread, the Acme has a basic depth equal to half the pitch; however, clearance is added both at the crest and the root of the thread for the general-purpose fit. The Acme general-purpose threads bear on the flanks (Figure I-352). Centralizing fits bear at the major diameter (Figure I-353). For more detailed information on Acme thread fits, see the *Machinery's Handbook*.

Three classes of general-purpose Acme threads—2G, 3G, and 4G—are used. Class 2G is preferred for general-purpose

Figure I-351 Acme thread.

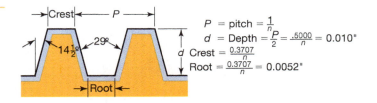

$$P = \text{pitch} = \frac{1}{n}$$
$$d = \text{Depth} = \frac{P}{2} = \frac{.5000}{n} = 0.010''$$
$$\text{Crest} = \frac{0.3707}{n}$$
$$\text{Root} = \frac{0.3707}{n} = 0.0052''$$

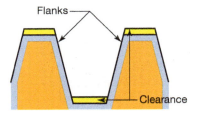

Figure I-352 General-purpose Acme threads bear on the flanks.

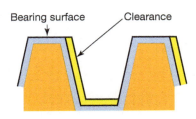

Figure I-353 Centralizing Acme threads bear at the major diameter.

assemblies. If less backlash or end play is desired, classes 3G and 4G are used. Internal threads of any class may be combined with any external class to provide other degrees of fit. The included angle of general-purpose Acme threads is 29 degrees. Depth of thread is half the pitch plus .010 inch. for 10 tpi and coarser. For finer pitches the depth of thread is half the pitch plus .005 inch.

Stub Acme threads (Figures I-354 and I-355) are used where a coarse pitch thread with a shallow depth is required. The depth for the stub Acme is only $.3P$, and $.433P$ for the American Standard stub, as compared with $.5P$ for the standard Acme threads.

The Buttress thread (Figure I-356) is not usually used for translating motion. It is often used where great pressures are applied in one direction only, such as on vise screws and the breech of large guns. Acme threads are gradually replacing square and Buttress thread forms.

Acme threads may be measured by using the one-wire method (Figure I-357). If a wire with a diameter equal to $.48725 * P$ is placed in the groove of an Acme thread, the wire will be flush with the top of the thread. For further information on one-wire and three-wire methods for checking Acme threads, see the *Machinery's Handbook*.

MULTIPLE-LEAD THREADS

Multiple threads, though not often used, are usually found on industrial machines, valves, fire hydrants, and aircraft landing gear. They are also used on jars and other containers.

Most screws and bolts have single-lead threads, which are formed by cutting one groove with a single-point tool. A double-lead thread has two grooves, a triple-lead thread has three grooves, and a quadruple-lead thread has four grooves. Double, triple, and quadruple threads are also known as multiple-lead threads. You may determine whether a bolt or screw is single or multiple threaded by looking at its end (Figure I-358) and counting the grooves that have been started. Multiple threads have less holding power, and less force is produced when these screws are tightened. A single-lead thread should be used for fasteners where locking power is required.

Multiple-lead threads offer several advantages:

1. They furnish more bearing surface area than single threads.
2. They have larger minor diameters; therefore, a bolt is stronger than one with a single thread.
3. They provide rapid movement.
4. They are more efficient, as they lose less power to friction than do single-lead threads.

The **lead** is the distance that a nut travels in one revolution. In one turn on a single-lead screw, a nut moves forward the distance (pitch) of one thread; in one turn on a double-lead thread, it moves twice as far; on a triple-lead thread, it moves three times as far; and on a quadruple-lend thread, it moves four times as far. On a single-lead screw the lead and

Figure I-354 American Standard Stub Acme thread.

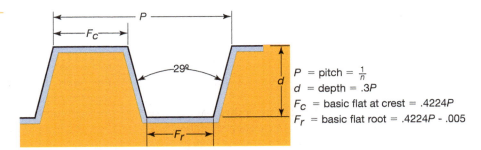

$$P = \text{pitch} = \frac{1}{n}$$
$$d = \text{depth} = .3P$$
$$F_C = \text{basic flat at crest} = .4224P$$
$$F_r = \text{basic flat root} = .4224P - .005$$

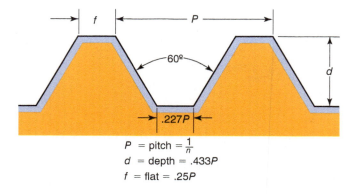

P = pitch = $\frac{1}{n}$
d = depth = .433P
f = flat = .25P

Figure I-355 Stub Acme thread.

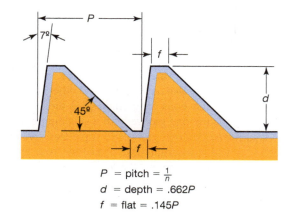

P = pitch = $\frac{1}{n}$
d = depth = .662P
f = flat = .145P

Figure I-356 Buttress thread.

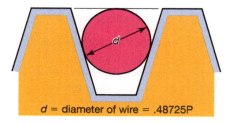

d = diameter of wire = .48725P

Figure I-357 One-wire method for measuring Acme threads.

pitch are the same, but on a two-lead screw, the lead is twice the pitch. Pitch is measured in the same way for single- and multiple-lead thread. It is the distance from a point on one thread to the corresponding point on the next thread.

The form and depth of thread for multiple-lead threads can be based on any recognized thread form: Buttress, square, sharp V, Acme, American National, metric, or Unified, both left- and right-hand. The thread depth is based on the pitch of the thread and not the lead.

CUTTING MULTIPLE-LEAD THREADS

Several methods are used for indexing or dividing multiple-lead threads. One method is to use an accurately slotted face-plate (Figure I-359). The lathe dog is moved 180 degrees for

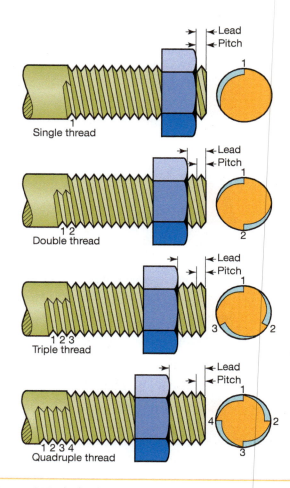

Figure I-358 Single, double, triple, and quadruple threads.

Figure I-359 Using the faceplate and lathe dog method of indexing multiple-lead threads.

two leads, 120 degrees for three leads, and 90 degrees for four leads. This method will only work for external thread-ing. Another method is to mark the stud gear at 180 degrees for two leads, disengage the gear, rotate it 180 degrees, and reset it (Figure I-360). This procedure will work only if the spindle and stud gear have a 1:1 (1 to 1) ratio.

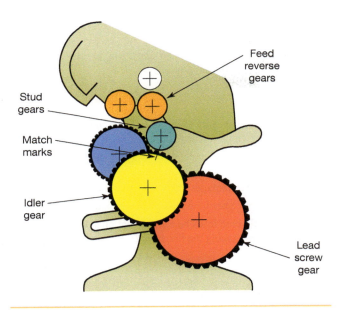

Figure I-360 Using the stud gear in the drive train for indexing.

Figure I-361 A thread chasing dial that may be used to cut double-lead threads.

A thread dial (Figure I-361) may be used to cut double threads if they are fractional or odd numbered. The first lead is cut on the numbered lines; the second lead is cut on the unnumbered lines. Some experimentation on various lathes will be useful to determine where the divisions will be found. If you cut the thread with the compound set at 29 degrees, be sure to back the compound out to its original position when you move to the next lead.

Many machinists prefer to index the thread with the compound set at 90 degrees, or parallel to the ways (Figure I-362). With this method, the tool must be fed straight into the work, so lighter cuts should be taken to keep from tearing the thread. Begin by taking up the slack in the compound feed screw and

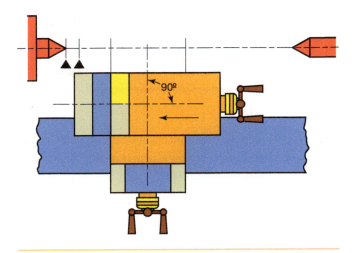

Figure I-362 Setting the compound parallel to the ways for cutting multiple-lead threads. To index, move the compound a distance equal to the pitch, always in the same direction.

setting the tool to cut the first lead. Set the compound and crossfeed micrometer dials to zero. Cut the first thread to the correct depth, and move the compound forward one pitch of the thread using the micrometer dial or a dial indicator. Be sure that the slack is out of the screw at all times. It is a good idea to tighten the compound gibs for this operation.

Make the cut for the second lead and check the fit. If it is a two-start thread, for example, use a gage or two-start nut to check it. If the thread is too tight, move the cross slide in a few thousandths of an inch, take a cut, then advance the compound *forward* to the other position and take the same depth cut. This method may be used for a thread with any number of starts. On coarse threads (under 12 threads per inch), rough out both threads before taking the finish cuts. Setups or tools may slip out of place slightly on heavy cuts, thus allowing for no adjustment if one thread is already finished.

An example of using this method to cut multiple-lead threads follows. If a two-start, double-lead unified screw thread with .100-inch pitch is required, the following steps can be used to make this thread.

Cutting a Double-Lead Thread

Step 1 The lead is .200 inch, or 5 threads per inch (tpi). Set the quick-change gearbox at 5 tpi.

Step 2 Set the compound at 90 degrees, or parallel to the axis of the lathe, and set up the tool.

Step 3 Zero the micrometer dials and take a scratch cut. Advance the compound .100 inch and take the second scratch cut. The pitch should now be .100 inch or 10 tpi when checked with a screw pitch gage.

Step 4 The depth of the external unified thread is $P * .613$.

Step 5 Feed in no more than .005 inch per pass but leave .010 inch for finishing. Repeat this process in the second lead.

Step 6 Cut both leads to finish diameter.

If you use a dial indicator instead of a micrometer dial to measure the movement of the compound, the two lead threads can be within .001 inch of true position.

SELF-TEST

1. What two purposes are translating-type screw threads used for?
2. Name at least 3 thread forms used as translating screws.
3. What is the depth of thread for a 4 tpi square thread?
4. What is the depth of thread for a general-purpose 4 tpi Acme thread?
5. What is the included angle for Acme threads?
6. Of the translating thread forms, which type is most used and is easiest to machine?
7. What are Buttress threads used for?
8. A $\frac{1}{8}$-inch-pitch single-lead thread will move .125 inch in one revolution of the nut. How far will a $\frac{1}{8}$-inch- pitch, three-start thread move in one revolution?
9. Define *pitch*.
10. Define *lead*.
11. Which thread has more force or holding power for fasteners, a single or a multiple lead?
12. Name at least three methods of indexing the lathe for multiple threads.
13. Name three advantages offered by multiple-lead threads.
14. How can you determine the number of leads?
15. Why should both leads be roughed out on a double-lead thread before taking the finish cut?

Cutting Acme Threads on the Lathe

Machinists are sometimes required to cut internal and external Acme threads. These threads are, in most cases, larger and coarser than 60-degree form threads, and greater skill is required to produce them well. In this unit you will learn the procedure for cutting these threads.

Cutting Acme threads is similar to cutting 60-degree threads in many ways. The threads per inch and infeeds are calculated in the same way. Some calculations, tool form and relief angles, and finishes, however, involve different problems and procedures.

OBJECTIVE

After completing this unit, you should be able to:

- Describe setup and procedure for making external and internal Acme threads.

GRINDING THE TOOL FOR ACME THREADS

The cutting tool form must be checked with the Acme tool gage (Figures I-363 and I-364) when the 29-degree included angle is ground. Side relief must be ground at the same time. The end of the tool is ground at the same time. The end of the tool is ground flat and perpendicular to the bisector of the angle. The flat is checked (Figure I-365) with the tool gage at the number corresponding to the threads per inch you will cut. It is important to have the flat the exact width needed for the particular thread being cut.

The relief angles on the tool (Figure I-366) are of greater importance when coarse threads are cut, because if the heel of the tool rubs on either side, a rough, inaccurate thread will result. When the helix angle has been determined (Figure I-367), it should be added to the relief angle (8 to 12 degrees) of the tool on the leading edge and similar relief provided on the trailing edge of the tool (Figure I-368). As

Figure I-363 Acme tool gage.

in other threading operations, the tool must be ground for 0-degree rake and set on the center of the work to maintain the correct thread form.

SETTING UP TO CUT EXTERNAL ACME THREADS ON THE LATHE

The threads per inch are set up normally, and the lead screw rotation is set for right- or left-hand threads. The gears, lead screw, and carriage should be lubricated before cutting coarse threads. The compound is most often set at $14\frac{1}{2}$ degrees

Figure I-364 Checking the tool angle with an Acme tool gage.

Figure I-365 Checking the flat on the end of the tool with an Acme tool gage.

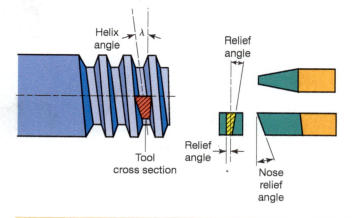

Figure I-366 Acme threads, showing the importance of the relief angles on both sides of the tool.

to the right for right-hand external threads. The workpiece must be set up and held securely in the work-holding device; a four-jaw chuck and dead center will be most secure. The tool is aligned with the work by using the Acme gage (Figure I-369). With this setup, the tool is fed into the work

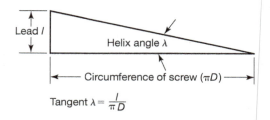

$$\text{Tangent } \lambda = \frac{l}{\pi D}$$

Figure I-367 The screw thread helix angle may be determined by dividing the lead by the circumference of the screw. The number thus obtained is the tangent of the helix angle, which can then be found in a table of tangents. D = diameter of screw, l = lead of thread, and λ = helix angle of thread.

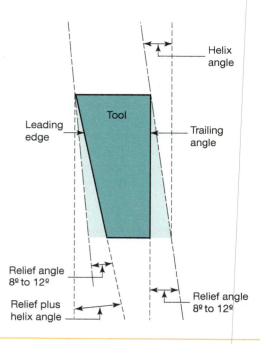

Figure I-368 The relationship of the helix angle of the thread to the relief angle of the tool.

Figure I-369 Aligning the tool with the work using an Acme gage.

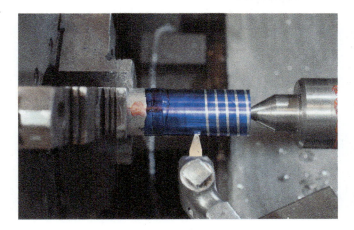

Figure I-370 Taking a scratch cut.

Figure I-371 Compound is set 14 1/2 degrees to the left of the operator for right-hand internal threads.

by advancing the compound in small steps, as with 60-degree threads. An undercut must be made at the end of the threads to clear the tool.

When Acme threads coarser than 5 tpi are cut, a square or round nose roughing tool smaller than the Acme tool should be used to remove up to 90 percent of the finished thread. The roughing tool does not cut to full depth or width, and the Acme form tool is used to finish the thread. This procedure is also used when making threads in tough alloy materials.

Some machinists prefer to set the compound parallel to the ways so they can "shave" both flanks of the thread for a good finish. When this is done, the tool must be made a few thousandths of an inch narrower to allow for the "shaving" operation.

Making the Cut

Take a scratch cut (Figure I-370) and measure. Move the cross slide out and return the carriage. Set the cross slide on zero and advance the compound .005–.010 inch, depending on what the lathe and setup will handle without chatter. Use heavy cutting oil. The total depth of the cut is $.5P + .010$ inch. Feed in on the cross slide for the last few thousandths of an inch so that the trailing flank will also receive a finish cut. For other Acme thread fits, see the *Machinery's Handbook*.

INTERNAL THREADS

The bore size for making a general-purpose Acme internal thread is the major diameter of the screw minus the pitch. The actual minor diameter of an Acme thread is the inside diameter of the bore. The minor diameter of an Acme 1″–5 thread is 1″–.200 = .800 inch. The internal major diameter should be the major diameter of the screw plus .010 inch for 10 or more tpi and .020 inch for pitches less than 10 tpi.

The compound is set $14\frac{1}{2}$ degrees to the left for cutting right-hand internal threads (Figure I-371). An internal Acme threading tool is ground, checked (Figure I-372) and set up, and then fed into the work with the compound .002–.005 inch for each pass. Figure I-373 shows the setup after

Figure I-372 Aligning the threading tool with an Acme gage to cut internal threads.

Figure I-373 Cutting internal Acme threads.

several cutting passes have been made. An Acme screw plug gage or the mating external thread should be used to check the fit as the internal thread nears completion.

When internal Acme threads are too small in diameter to be cut with a boring bar and tool, an Acme tap (Figure I-374)

Figure I-374 Acme taps.

Figure I-375 Tandem Acme tap (*Courtesy of Fox Valley Technical College*).

is used to make the thread. Acme taps are made in sets of two or three; each tap cuts more of the thread, and the last tap makes the finishing cut. Some Acme taps have roughing and finishing teeth on the same shank, as in Figure I-375. The part is drilled or bored to the minor diameter of the thread, and the Acme tap is turned in by hand. Cutting oil is used when tapping steel, but threads in bronze are cut dry.

External Acme threads must often have a good finish. A final honing of the tool before the last few shaving passes will help. The setup must be rigid and the gibs tight. Low speeds are essential. The grade of cutting oil is extremely important in this finishing operation.

The thread may be finished after it is cut by using a thin, safe edge file at low rpm and by using abrasive cloth at a higher rpm. A thin piece of wood is sometimes used to back up the abrasive cloth while each flank is being polished.

SELF-TEST

1. What are two major differences between V-form threads and Acme threads?
2. When grinding an Acme threading tool, what three things should be carefully measured?
3. What is the depth of thread for a $\frac{3}{4}$-6 external general-purpose Acme thread?
4. How is an ACME threading tool aligned with the workpiece?
5. Determine the bore size to make a $\frac{3}{4}$-6 general-purpose internal Acme thread.
6. Which is the best way to make small internal Acme threads?
7. What can you use to check internal threads for fit?
8. Explain how a good tool finish may be obtained on an Acme thread.

SECTION J

Vertical Milling Machines

The first vertical milling machines appeared in the 1860s. The next significant step in the development of the vertical milling machine came in the mid-1880s with the adaptation of the "knee and column" from the horizontal milling machine. This allowed the milling table to be raised and lowered in relation to the spindle. The spindle heads on some of these machines could be tilted at an angle to the table.

Just after the turn of the twentieth century, vertical milling machines began to appear with power feeds on the spindle. During that period, micrometers and vernier scales were used to make them suitable for precisely locating holes. Improvements in vertical milling machine design after 1910 related mostly to its drive and control mechanisms. By 1927, hydraulic tracing controls had been applied to vertical milling machines.

The vertical milling machine (Figure J-1) is one of the most versatile machine tools in a machine shop. This machine tool can accomplish a wide variety of machining tasks, including milling, drilling, and boring.

VERTICAL MILLING MACHINE SAFETY

Safe operation of the vertical milling machine requires that you learn about hazards so that you can protect yourself from them. You must always be alert when you are working on a machine tool. It is dangerous to operate equipment when sick, tired, or emotionally upset. Some prescription drugs should not be taken before driving a car because they affect reflexes; the same is true when operating machinery. A safe machine operator thinks before doing anything. You need to know what is going to happen before you turn a lever or operate a control.

SAFETY FIRST

Proper dress is important for safe machine operation. Short sleeves or tight-fitting sleeves are protection against being caught in a revolving spindle. Rings, bracelets, earrings, and wristwatches can become dangerous if worn around machinery and should be removed. Safety glasses should always be worn in a machine shop. Flying particles from the machine can blind you. Long hair should be safely tied back or covered under a cap. Heavy shoes should be worn to protect your feet from chips and from falling objects. Gloves should not be worn while operating machines because of the danger of being caught by the machine. All machine guards should be in place prior to starting up a machine. Unsafe equipment and practices should be immediately reported. A safe workplace depends on everyone. If you need to lift a heavy workpiece or machine attachment, use a hoist or have someone help you.

SHOP TIP

Chips are sharp, hot, and contaminated with cutting fluids. Never handle chips with bare hands. Remove chips with a brush or vacuum cleaner. Do not use compressed air to blow chips off a machine, or they will become small missiles that can injure someone. Compressed air will also force chips and dirt between the sliding surfaces of the milling machine, where they will cause scoring and rapid wear. Attempt to clean chips and cutting fluids from the machine or workpiece only after the cutter has stopped rotating. Do not use the machine table or ways as a worktable, because even small nicks and scratches will affect the accuracy of the machine. Keep the area around the machine clean of chips, oil spills, cutting fluids, and other obstructions to prevent slipping and stumbling. Use a shop towel to protect your hands when handling cutting tools or sharp-edged workpieces to avoid getting cut. Make sure that workpieces are rigidly supported and tightly clamped to withstand the high cutting forces encountered in machining. If a workpiece comes loose while it is being machined, it is often ruined, and so is the cutter. The operator is also in danger from flying particles from a broken cutter or the workpiece. The cutting tools need to be securely fastened in the machine spindle to prevent movement during cutting. Measurements are frequently made during machining operations. Do not take any measurements until the spindle has come to a complete standstill. **Never leave a running machine unattended.**

Figure J-1 Vertical milling machine.

Vertical Milling Machines

The first step in efficient and safe operation of any machine tool is to know the names of the machine's parts and its various controls. The next step is to know the function of each part and control so that you can operate the machine without damaging it or the workpiece, or injuring yourself. The purpose of this unit is to acquaint you with the vertical milling machine and to identify the controls and their functions. Before starting any milling job, take time to operate all the machine controls and observe their function.

OBJECTIVES

After completing this unit, you should be able to:

■ Identify the important components and controls on the vertical milling machine.
■ Describe the functions of machine parts and controls.
■ Perform routine maintenance on the machine.

IDENTIFYING MACHINE PARTS, CONTROLS, AND THEIR FUNCTIONS

The major assemblies of the vertical milling machine are the base, and column, knee, saddle, table, ram, and toolhead (Figure J-2).

Base and Column

The base and column are one piece and are the major structural components of a vertical mill. A dovetail slide is machined on the face of the column to provide a vertical guide for the knee. A similar slide is machined on the top of the column to provide a guide for the ram. The top column slide and ram can be swiveled right and left of center to permit wide positioning of the toolhead over the table.

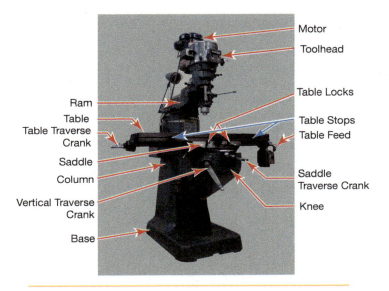

Figure J-2 The important parts of a vertical milling machine.

Knee

The knee engages the slide on the face of the column and is moved up and down by turning the vertical traverse crank. The knee supports the saddle and table. Knee locks secure the knee at any position.

Saddle

The saddle engages the slide on the top of the knee and can be moved in and out by turning the saddle-traverse crank. The saddle supports the table. Saddle locks lock the saddle at any position.

Table

The table engages the slide on the top of the saddle and is moved right and left by turning the table traverse handle. The workpiece or work-holding device is secured to the table. Table locks are provided so that the table may be locked at any position. Many milling machines have a power feed mechanism on the table. This enables the table to be fed at a variable feed rate in either direction.

Ram

The ram is mounted on the slide on the top of the column. The ram is moved in and out by turning the ram positioning pinion gear. The ram movement can extend the head to increase the working area. Ram locks secure the ram position.

Toolhead

The toolhead (Figure J-3) is attached to the end of the ram. The spindle motor is mounted on the toolhead. The spindle is turned on with a three-position switch (High Range/Off/Low Range). Be sure that the spindle is rotating in the proper direction when you turn on the spindle. Speed changes are made with V-belts, gears, or variable-speed drives. When you change speeds into the high- or low-speed range, the spindle has to be stopped. The same is true for V-belt or gear-driven speed changes. On variable belt drives, however, the spindle must be revolving while speed changes are being made.

The quill does not rotate. The quill contains the rotating spindle. The quill can be extended and retracted into the toolhead by a quill feed hand lever or handwheel. The quill feed hand lever is used to rapidly position the quill or to drill holes. Some vertical mills have a quill feed handwheel. A quill feed handwheel provides a controlled, slow manual feed.

Power feed to the quill is engaged by the feed control lever. Different quill feed rates, usually .0015, .003, and .006 inch per spindle revolution, are selected with the quill feed selector. The power feed is automatically disengaged when the quill dog contacts the adjustable micrometer depth stop (Figure J-4). When feeding upward, the power

feed disengages when the quill reaches its upper limit. The micrometer dial allows depth stop adjustment in .001-inch increments. The quill clamp is used to lock the quill in a fixed position. To obtain maximum rigidity while milling, you should tighten the quill clamp.

The spindle lock or spindle brake is used to keep the spindle from rotating when installing or removing tools. Tools are usually held in collets, which are secured in the spindle with a drawbolt. The drawbolt threads into the upper end of the collet. When the drawbolt is tightened, the collet is drawn into the taper in the spindle. This aligns and holds the tool. To release a tool from the spindle, use the following procedure:

Step 1 Use the spindle brake to lock the spindle.

Step 2 Raise the quill to its top position and lock.

Step 3 Unscrew the drawbolt one turn.

Step 4 Give the drawbolt a firm blow with a dead blow hammer. This should release the collet from the spindle taper. At the same time, the tool is released from the collet. You should be holding the tool with a shop towel during this step to prevent the tool from falling on the workpiece or the machine table.

Adjustable Quill Stop

Figure J-4 Quill stop.

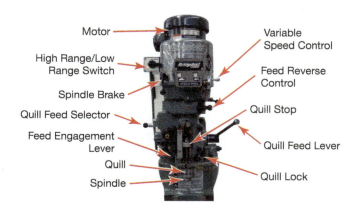

Motor
Variable Speed Control
High Range/Low Range Switch
Feed Reverse Control
Spindle Brake
Quill Feed Selector
Quill Stop
Feed Engagement Lever
Quill Feed Lever
Quill
Quill Lock
Spindle

Figure J-3 The toolhead.

LOCKS, ADJUSTMENTS, AND MAINTENANCE

The knee, saddle, table, and quill are equipped with locks that will prevent movement (Figure J-5). During machining, all axes except the moving one should be locked. This will increase the rigidity of the setup. Do not use these clamping devices to compensate for wear on the machine slides. If the machine slides become loose, make adjustments with the gib adjustment screws (Figure J-6). Turning this screw in will tighten a tapered gib. Note that in most cases there is also a screw at the opposite end of the gib. This one needs to be loosened to move the gib. When the gib is in the correct position, both screws should be tight to keep the gib in position. Partially turn the screw, and then try moving the unit with the handwheel. Repeat this operation until free but not loose movement is obtained. If it is adjusted too tightly lubricant is squeezed out from the slides, which results in rapid wear.

All machine tools require periodic adjustment and lubrication. Many mills are equipped with one-shot lubricators, often located on the side of the knee. Oil from the lubricator is pressure fed to the knee, saddle, and table slides. Any other oil cups on the machine should be kept filled with light oil as specified by the manufacturer.

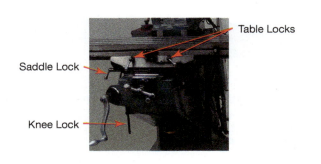

Figure J-5 Clamping devices (*Courtesy of Fox Valley Technical College*).

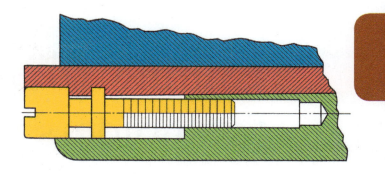

Figure J-6 Gib adjusting screw (*Courtesy of MAG Industrial Automation Systems LLC*).

SELF-TEST

1. Name the six major components of a vertical milling machine.
2. How can the table be moved longitudinally?
3. How can the saddle be moved?
4. How can the quill be moved manually?
5. What are the table clamp used for?
6. What is the purpose of the spindle brake?
7. What is important when changing the spindle speed range from high to low?
8. What additional capability does the ram provide?
9. How is a loose table movement adjusted?
10. What is the purpose of the quill clamp?

Cutting Tools and Cutting Toolholders for the Vertical Milling Machine

There are a wide variety of cutting tools available for vertical milling machines. This unit will cover common types of cutters and aid in the selection of tooling that best fits a given application.

OBJECTIVES

After completing this unit, you should be able to:

- Identify common cutters for the vertical mill.
- Select a cutter for a given machining task.

Milling and drilling tools make up the majority of the types of tools used on the vertical mill. This section will discuss standard types of tools and toolholders used on milling machines. High-speed steel (HSS) and carbide tools will be covered in this section.

High-Speed Steel Drills

The two basic types of high-speed drills are the twist drill and the spade drill. The HSS twist drill is the most commonly used tool for producing holes.

Twist drills are great for rapidly producing holes that do not have to be very accurate in size or position. If the holes must be very accurate in size they are drilled to a smaller size and reamed, milled, or bored to size. If the position of the hole must be very accurate, the drilled hole must be milled or bored on location.

Twist drills are made with two or more flutes and come in a variety of styles (Figure J-7).

Larger drills typically have a tapered shank with a tang on the end (Figure J-8). The tang keeps the tapered shank drill from slipping under the higher torque conditions associated with drilling large holes.

Figure J-7 Twist drills are the most common hole-producing tools in the machine shop. Twist drills have either a straight or tapered shank. Straight-shank drills are common up to 1/2 inch in diameter and are held in drill chucks.

Figure J-8 The tang on the end of the tapered shank drives the drill.

Center or Spotting Drills

When drilled holes need to be accurately located, you should center- or spot-drill the holes prior to drilling. This spotting or centering is done with center or spotting drills (Figure J-9). These drills are short, stubby, and rigid and do not flex or deflect, as longer drills have a tendency to. The spot drill

Figure J-9 Center and spotting drills come in a variety of sizes. The short, stubby design of the drills allows them to accurately locate holes.

produces a small start point that is accurately located. When the hole is drilled, the drill point will follow the starting hole that the spot drill made. This method can produce holes that are reasonably accurate in location.

Spade Drills

The spade drill has a flat blade with sharp cutting edges (Figure J-10). The spade cutting tool is clamped in a holder and can be resharpened many times. Spade drills typically are used for drilling very large diameter holes. They can lower tooling costs because the blade holders can hold a variety of sizes of blades. Spade drills are designed to drill holes in one pass. Spade drills require approximately 50 percent more horsepower than twist drills. Spade drilling also requires a rigid machine and setup.

Carbide Drills

Carbide-tipped twist drills have been around for many years. They are basically carbon steel drills with a piece of tungsten carbide brazed into them. They look similar to a spade drill but are usually made in smaller diameters. Solid carbide drills are just that, a solid carbide cutting tool. Solid carbide drills are typically found in small diameters because of the cost of the carbide materials.

One of the newer innovations in carbide drilling technology is carbide insert drills (Figure J-11). These drills incorporate indexable or replaceable inserts and can remove metal four to ten times faster than an HSS drill. Carbide insert drills require a rigid setup and a machine with substantial horsepower.

Figure J-10 Spade drills are a two-piece tool consisting of the blade and the holder. Spade blades and holders increase productivity because one holder can be used to drill many different diameter holes simply by changing blades.

Figure J-11 Carbide insert drills allow you to drill hard materials at feeds and speeds much higher than those of conventional drills. When the drill becomes dull, the carbide inserts can be indexed or replaced.

Auxiliary Hole-Producing Operations

Drilling may be the most common method of producing holes, but it is not the most accurate. In some cases, holes may need a very accurate size and/or finish. If an accurate-size hole is needed, reaming may be the quickest method.

A reamer is a cylindrical tool similar in appearance to the drill (see Figure J-12). Reamers produce holes to a tight tolerance with a smooth finish. A slightly smaller hole must be drilled in the part before the hole can be reamed. The reamer follows the drilled hole, so inaccuracies in location cannot be corrected by reaming.

Boring

If accurate location is needed as well as accurate size, boring may be necessary. Reamers are a quick way of producing accurately sized holes. Boring can produce holes of any size with a good finish and can locate them very accurately.

Boring is done with an offset boring head. Figure J-13 illustrates a boring head with a carbide insert. The offset boring head holds the tool and can be adjusted to cut any size hole within its range.

Most boring heads are made to hold various length boring bars. Figure J-14 shows a typical boring bar that could be used in an offset boring head.

Figure J-12 Reamers consistently produce accurately sized holes. The rule of thumbs for reamer feeds and speeds is that reamers should be run at half the speed and twice the feed of the same size drill.

Figure J-13 The offset boring head can be adjusted to cut any size hole within its size range.

Figure J-14 Boring bars are tools used to do the cutting in a boring operation.

Boring, similar to reaming, can be done only on an existing hole. As a general rule, the boring tool should be as short and as large in diameter as possible. When using an HSS tool, the length -to-diameter ratio should not be greater than 5 to 1. For example, if a 1-inch-diameter boring bar is used, no more than 5 inches should be sticking out. Carbide has a 3 to 1 recommended ratio. Reducing the ratio helps insure against chatter because as the tool overhang becomes shorter, the amount of force it takes to flex the boring bar increases.

Tapping

Tapping is the process of producing internal threads by using a tap. There are many different types of taps. The most common types of tap used on machines are the gun tap and the spiral-fluted tap. The gun tap is especially useful for tapping holes that go through the workpiece (Figure J-15 Top). Chip clearance is especially important when tapping. If chips clog the hole, a broken tap often results. The spiral-fluted tap brings the chips up and out of the hole (Figure J-15 Bottom).

END MILLS

End mills are the most commonly used tool on a vertical milling machine. End mills may have two, three, four, or more **flutes** and may be right- or left-hand cutting. To determine the cutting direction of an end mill, observe the cutter from its cutting end (Figure J-16). A right-hand cutter will cut while turning in a counterclockwise direction. A left-hand cutter will cut while turning in a clockwise direction.

Two-flute end mills can be used for plunge cutting. These are called center cutting because they can be plunged into the work without a starting hole (Figure J-17). Some four-flute end mills may also be center cutting. However, if these are center drilled or gashed on the end, they cannot start their own holes. This type of end mill will cut only on its periphery.

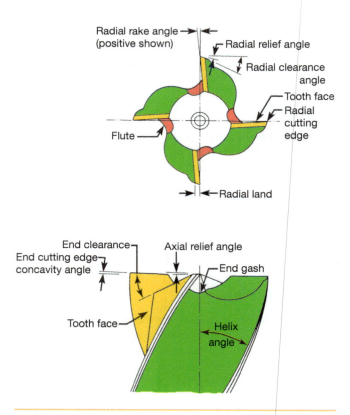

Figure J-16 End mill nomenclature (*Based on Regal Cutting Tools*).

Figure J-15 Gun taps (top) push the chips ahead of the tap, so you must consider the amount of chip clearance that is available. The spiral flutes on a spiral-flute tap (bottom) enable lubricant or coolant to reach the end of the tap.

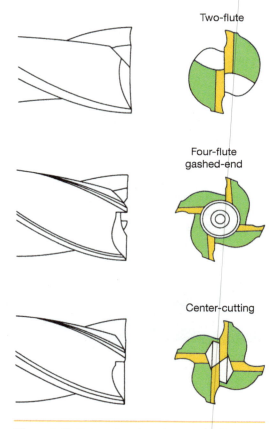

Figure J-17 Types of end teeth on end mills (*Based on Regal Cutting Tools*).

A three-flute end mill is a compromise between the bigger chip capacity of a two flute and the better finishing ability of the four-flute end mill. A three-flute end mill is an excellent choice for slotting. Often, one cut with a three-flute end mill will give an acceptable surface finish. Much of the time, two of the three teeth are in the cut, giving a smooth, chatterless cutting action. A four-flute end mill is stronger than two- or three-flute end mills; this enables higher feed rates. Multiple-flute end mills produce fine finishes following a roughing cut. A corner radius or chamfer on an end mill increases cutter life by preventing corner chipping. These end mills should be used whenever a fillet radius in a pocket is acceptable. These cutters are also called "bull-nose" end mills (Figure J-18).

HSS Helical and Straight-Fluted End Mills

The HSS end mills may be single ended (Figure J-19) or double ended (Figure J-20). They may also have straight flutes (Figure J-21). Slow, regular, and fast helix angles are

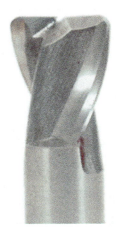

Figure J-18 End mill with corner radius.

Figure J-19 Single-end helical-tooth end mill (*Used with permission of Weldon Tool—A Dauphin Precision Tool Brand*).

Figure J-20 Two-flute double-end helical-tooth end mill (*Used with permission of Weldon Tool—A Dauphin Precision Tool Brand*).

Figure J-21 Straight-tooth double-end end mill (*Used with permission of Weldon Tool—A Dauphin Precision Tool Brand*).

Figure J-22 Forty-five-degree helix angle aluminum cutting end mill (*Used with permission of Weldon Tool—A Dauphin Precision Tool Brand*).

Figure J-23 Roughing mill (*Used with permission of Illinois Tool Works, Inc.*).

Figure J-24 Three-flute tapered end mill (*Used with permission of Weldon Tool—A Dauphin Precision Tool Brand*).

also available. An example of a slow helix is one in which the helix angle of the cutter is about 12 degrees. A regular helix angle may be 30 degrees, and a high helix angle is 40 degrees or more. Selection of helix angle will depend on the machining task. For example, aluminum can be machined efficiently with a high-helix-angle cutter (45 degrees) and with a highly polished cutting face to minimize chip adherence (Figure J-22). Chips welding to the cutting face can mar the surface finish of the part.

Roughing and Tapered End Mills

The roughing end mill (Figure J-23) is used when large amounts of material must be removed (roughed) quickly. These end mills are also called hogging end mills and have a wavy tooth form cut on their periphery. These wavy teeth form many individual cutting edges. The tip of each wave contacts the work and produces one short compact chip. Each succeeding wave tip is offset from the next one, which results in a relatively smooth surface finish. During the cutting operation, multiple teeth are in contact with the work. This reduces the possibility of vibration and chatter.

Tapered end mills (Figure J-24) are used in mold making, die work, and pattern making, where precise tapered surfaces need to be machined. Tapered end mills include tapers ranging from 1 degree to more than 10 degrees. Tapered end mills are also called diesinking mills.

Geometry-Forming, Dovetail, T-Slot, Woodruff Key, and Shell End Mills

Several types of end mills are used to form a particular geometry on a workpiece. Ball-end end mills (Figure J-25) have two or more flutes and form an inside radius (fillet) between surfaces. Ball-end end mills are used in mold making and in diesinking operations. Round-bottom grooves can also be machined with them. Precise convex radii can be machined on the edge of parts with corner-rounding end mills (Figure J-26). Dovetails are machined with single-angle milling cutters (Figure J-27). The two commonly available angles are 45 degrees and 60 degrees. **T-slots** in machine tables are machined with T-slot cutters (Figure J-28). T-slot cutters are made in sizes to fit standard T-nuts. Woodruff keyseats are cut into shafts to retain a Woodruff key. Woodruff keys are used to connect and drive shafts and pulleys. Woodruff keyseat cutters (Figure J-29) come in many different sizes.

Figure J-25 Two-flute single-end ball-end end mill (*Used with permission of Weldon Tool—A Dauphin Precision Tool Brand*).

Figure J-26 Corner-rounding milling cutter (*Used with permission of Illinois Tool Works, Inc.*).

Figure J-27 Single-angle, carbide-tipped dovetail cutter (*Used with permission of Illinois Tool Works, Inc.*).

Figure J-28 T-slot milling cutter (*Used with permission of Weldon Tool—A Dauphin Precision Tool Brand*).

Figure J-29 Woodruff keyslot milling cutter (*Used with permission of Illinois Tool Works, Inc.*).

Flycutters

A **flycutter** is a single-point tool with a high-speed or carbide tool secured in a special holder (Figure J-30). Flycutters are often used to take light face cuts from large surface areas. The tool bit in the flycutter must be properly ground to obtain the correct rake and clearance angles for the material being machined. Care must be exercised when using a flycutter. When the tool is revolving, the tip of the cutting tool becomes almost invisible and can injure the operator.

Carbide and Indexable End Mills

Carbide end mills are designed to outperform HSS cutters, especially when machining abrasive materials. Carbide end mills are used primarily because they can dramatically improve productivity. If these mills are made from fine-grain carbide, their toughness approaches that of HSS with the hardness of carbide. Indexable tooling has largely replaced the use of HSS tools in industrial machining. A major advantage of indexable tools is the ability to select tool grades that are the best for a particular material. Other advantages are the availability of many cutter types. Indexable cutters do not require resharpening: you just replace the inserts. Center-cutting indexable end mills are economical tools. One end mill with different insert grades can be used for cutting steel, cast iron, or aluminum, with each insert grade and style having the correct tool geometry for its application (Figure J-31). When deeper cuts have to be made, indexable end mills with additional inserts along the body are used. The end mill shown has a helical flute design, which gives a free-cutting

Figure J-30 Flycutter with an HSS tool installed.

Figure J-31 Indexable end mill.

action for all types of metals. The inserts enter the cut in overlapping succession; this minimizes shock and contributes to smooth cutting (Figure J-32). The first row of inserts is 80-degree parallelograms, which allows the machining of square shoulders. Indexable ball nose end mills can be used to plunge and side cut (Figure J-33). Indexable 45-degree chamfer-cutting end mills can be used to chamfer holes, bevel surfaces, and for flat surface milling (Figure J-34).

FACE MILLING CUTTERS

The cutter body on a carbide insert face milling cutter is made from steel. The cutting inserts are carbide and can be changed. When a face mill is mounted on the machine spindle, it is important that all mating surfaces between the cutter and the spindle be clean and free of any nicks or burrs. Face mills are used when flat surfaces have to be produced (Figure J-35).

Face mills are designed lighter-duty cutters for finishing cuts or heavy-duty cutters with fewer teeth and heavier bodies for roughing operations. Positive rake angles produce a good surface finish, increase cutter life, and use less power in cutting (Figure J-36). Positive rake angles are effective in machining tough and work-hardened materials. Cutting pressures, which may deflect thin-walled workpieces, are smaller with positive-rake cutters than with negative-rake cutters under the same conditions. Zero or negative-rake cutters are strong and will give good service under heavy impact or interrupted cutting conditions (Figure J-36). Negative-rake inserts create high cutting pressures, which

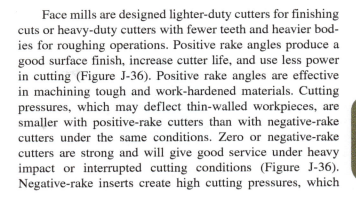

Figure J-34 Indexable 45-degree chamfer cutter for chamfering holes and cutting bevels.

Figure J-32 Indexable end mill with helical flutes and coolant holes.

Figure J-35 Inserted carbide tooth shell mill.

Figure J-33 Indexable ball nose end mill used to plunge and side cut.

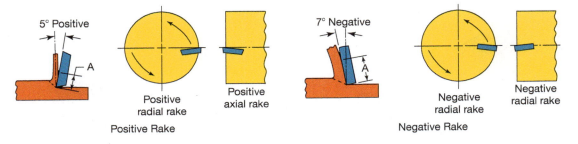

Figure J-36 Positive and negative rake angles (*Used with permission of Kennametal, Inc.*).

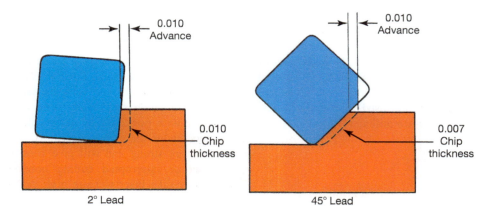

Figure J-37 Small lead angle (*Used with permission of Kennametal, Inc.*).

tend to force the workpiece away from the cutter. Negative-rake inserts should not be used on work-hardened materials or ductile materials such as aluminum or copper.

USE OF FACE MILLING CUTTERS

The **lead angle**, which is the angle of the cutting edge measured from the periphery of the cutter, varies from 0 to 45 degrees, depending on the application. Small lead angles of 1 to 3 degrees can be used to machine close to square shoulders. A small lead angle cutter, when used with a square insert, will have sufficient clearance on the face of the cutter to prevent its rubbing. Figure J-37 shows the effect of a small lead angle on chip thickness. With a .010 inch feed per insert, the chip is also .010 inch thick and as long as the depth of cut. Figure J-37 uses the same feed and depth of cut, but the chip is quite different. The chip now is longer than the depth of cut, and its thickness is only .007 inch. A thinner chip gives an increased cutting-edge life but limits the effective depth of cut. Feed rates of less than .004 inch per insert should be avoided. Figure J-38 shows another beneficial effect of a large lead angle: The cutter contacts the workpiece at a point away from the tip of the cutting edge. The cutter does not cut a full-size chip on initial impact but eases into the cut. The same thing happens on completion of the cut as the cutter eases out of the work.

The surface finish produced by a cutter depends largely on nose radius of the insert and the feed used. Figure J-39 is an exaggerated view of the ridges left by the cutter. The finish can be improved by the use of an insert with a wiper flat (Figure J-39).

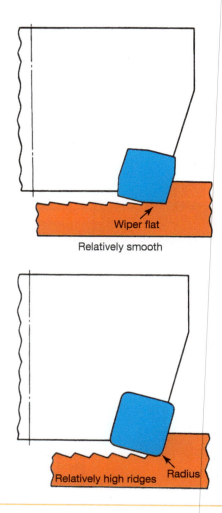

Figure J-39 Feed lines on surface (*Used with permission of Kennametal, Inc.*).

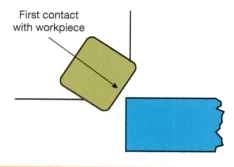

Figure J-38 Lead angle effect.

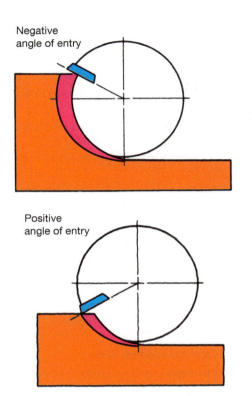

Figure J-40 Work contacts tool away from cutting tool tip (*Used with permission of Kennametal, Inc.*).

For finishing operations use one wiper insert; the rest are regular inserts in the cutter body. The maximum feed rate should not exceed two-thirds of the width of the wiper flat. As an example, let's use a wiper insert with a .300-inch-wide flat and the cutter revolving at 200 rpm. Two-thirds of .300 inch equals .200 inch per spindle revolution, and .200 inch times .200 rpm gives 40 ipm. The wiper insert also extends axially .004 inch beyond the other inserts—this makes the wiper insert the finishing insert.

The cutting action of a face mill is also affected by its diameter in relation to the width of the workpiece. When it is practical, a cutter should be chosen that is larger than the width of cut. A good ratio is obtained when the cut is two-thirds as wide as the cutter diameter. Cuts as wide as the cutter diameter should not be taken, because of the high friction and rubbing of the cutting edge before the material starts to shear. This rubbing action results in rapid cutting-edge wear. Tool life can be extended when the cutting tool enters the workpiece at a negative angle (left side of Figure J-40). The work makes its initial tool contact away from the cutting tip at a point where the insert is stronger. When possible, arrange the width of cut to obtain the negative entry angle cutting action in Figure J-40.

The graphic on the bottom of Figure J-40 illustrates a workpiece where the cutter enters at a positive angle. Contact takes place at the weakest point of the insert, and the cutting edge may chip.

Cutting fluid is important in milling. A good cutting fluid absorbs heat generated in cutting and provides a lubricant between the chips and the cutting tool. Cutting fluid should be applied where the cutting is taking place, and it should flood the cutting area. Cooling carbide tolling is difficult because of the intermittent nature of the cutting process and the fanning action of the revolving cutter. If the cooling of the carbide insert is intermittent, thermal cracking, which causes rapid tool failure, takes place. Unless a sufficient coolant flow at the cutting edge is maintained, it is better to machine dry. A mist-type coolant application is sometimes used to provide lubrication between the chip and cutting tool. This reduces friction and generates less heat.

To machine a flat surface with a face mill, it is crucial that the cutter spindle is perpendicular to the table. If the spindle is not exactly perpendicular, the cutter will exhibit back drag or produce a convex surface. Back drag appears on the workpiece as light cuts by the trailing cutting edges of the cutter. A concave surface is generated when the cutter is tilted in relation to the table, so that the trailing cutting edges are not in contact with the work surface. A slight tilt, where the trailing cutting edges are .001 to .002 inch higher than the leading cutting edges, gives satisfactory surfaces in most applications. This is generally only a problem on milling machines that have tilting heads.

Effective face milling operations depend on several factors. Rigidity of the setup is the main concern. The workpiece must be supported where the cutting takes place. Rigid stops prevent the workpiece from being moved by the cutting pressures. The table should be as close to the spindle as possible. The gibs on the machine slides need to be adjusted to prevent looseness. Locking devices should be used on machine slides. Coarse-pitch cutters usually have an uneven spacing of the teeth, which helps eliminate or reduce vibrations. Efficient machining requires sharp cutting edges. Dull tools produce higher cutting pressures, which may deflect the workpiece. Dull tools also require more power. When dull tools are used on work-hardened materials, both the tool and workpiece may be ruined. The speed of the cutter affects the surface finish of the workpiece.

When a carbide face mill is used, the color of the chips produced is an indicator of the machining conditions. When steel is milled, chips should be straw colored or lightly blue. A dark blue to black color indicates that the surface speed is too high. When chips show no color, the surface speed is too low. If the feed is too low, the rubbing action wears down the cutting edge. If the feed is too high, the cutting edges will chip or break. Milling cutters often have high power requirements.

CUTTER HOLDING ON THE VERTICAL MILL

No matter which type of milling cutter you are using, it must be securely mounted in the machine spindle. Solid holders and collet holders are the most widely used on vertical milling machines. The solid holder (Figure J-41) is more rigid and holds tools more securely than a collet holder. The solid

Figure J-41 Solid holder.

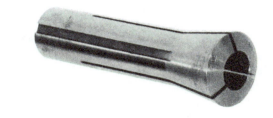

Figure J-42 Split collet.

holder has a precision-ground shank that fits the spindle on the milling machine. Many vertical milling machines have an R-8 spindle, meaning they will accept all standard R-8 tooling. The end mill is secured with setscrews that are tightened against a flat on the cutter shank. The split collet (Figure J-42) is also used to hold cutters. When the tapered part of the collet is pulled into the spindle taper by the drawbolt, the split in the collet permits it to squeeze tightly against the shank of the end mill. Although split collets are effective cutter-holding devices, it is possible for a cutter to be pulled from the collet because of heavy feed rates or a dull cutting tool. Helical flute end mills may tend to be pulled from the collet. Solid holders have an advantage over the split type, since the setscrews prevent slippage of the cutter.

SELF-TEST

1. How is a right-hand-cut end mill identified?
2. What characteristic of end mills allows them to be used for plunge cutting?
3. What is the main difference between a general-purpose end mill and one designed to cut aluminum?
4. When are carbide-tipped end mills chosen over HSS end mills?
5. What type of end mill is designed to be used when a large amount of material needs to be removed?
6. What are tapered end mills used for?
7. What are some advantages of carbide insert tools?
8. Name two types of toolholders for straight shank tools?
9. Among a two-, three-, and four-flute end mill, which is the strongest?

INTERNET REFERENCE

Indexable milling troubleshooting guide:

http://www.carbidedepot.com/formulas-mc-troubleshoot.htm

Setups on the Vertical Milling Machine

Before any machining can be done on the vertical mill, the toolhead must be squared to the table and saddle axes. The workpiece or work-holding device must also be secured to and aligned with the table and saddle axes. This unit will describe the procedure for squaring the toolhead and discuss common workpiece setups.

OBJECTIVES

After completing this unit, you should be able to:

- Square the toolhead.
- Set up and align a workpiece on the table.
- Set up and align a mill vise.
- Locate the edges of a workpiece relative to the spindle and position the spindle over a hole center.

SQUARING THE TOOLHEAD

On many vertical mills, the toolhead can be swiveled relative to the table and saddle axes (Figure J-43).

This enables drilling and milling at an angle. The large majority of milling and drilling operations, however, are done with the toolhead set perpendicular to the table and saddle axes. This alignment is critical, as it is directly responsible for square mill cuts and perpendicular drilling. The following procedure is used to square the toolhead.

Step 1 Fasten a dial indicator in the machine spindle (Figure J-44). The indicator should sweep a circle slightly smaller than the width of the table.

Step 2 Lower the quill until the indicator contact point is depressed .015 to .020 inch. Lock the quill in this position. The spindle with attached indicator will be turned by hand and used to determine the position of the toolhead relative to the table and saddle. There is a tendency for the indicator tip to catch in the table T-slot as it is turned. To prevent this, an accurately machined ring or a flat plate can be used as an indicating surface. The swivel base from a precision mill vise can be used for this purpose. Be sure that the table is clean and that there are no burrs on the ring or plate. Note that if a plate is used it must be flat and the top must be parallel to the bottom. Note that you do not have to use a ring or a plate. If you indicate on the table, you must be careful as the indicator needle drops into the table slots.

Step 3 Tighten the knee clamp locks. If this is neglected, the knee will sag in the front and introduce an error in the indicator reading. That would cause an error if you did lock the knee for machining. If there is very little play in the ways of the knee the error will be small.

Step 4 Loosen the front toolhead clamping bolts (Figure J-45) one at a time and retighten them to provide a slight drag on the toolhead. Fine adjustments will be easier if the toolhead is just loose enough to be moved by slight pressure.

Step 5 Rotate the spindle by hand until the indicator is to the left or right of the spindle and in line with the table axis.

Step 6 Set the indicator bezel to zero. The example shown will aid you in the alignment procedure (Figure J-46).

Step 7 Rotate the spindle 180 degrees so that the indicator is positioned on the opposite side and in line with the table axis. Note the reading at this position.

Step 8 Tilt the toolhead using the tilt screw (Figure J-47) until the indicator moves back toward zero, with half the amount showing.

Step 9 Turn the indicator 180 degrees and see whether the reading varies. If both readings are the same, tighten the toolhead clamping bolts and recheck the readings. Tightening the toolhead clamps will sometimes move the head slightly, requiring a small additional adjustment.

Step 10 Repeat this procedure for the toolhead alignment relative to the saddle axis. Begin by loosening the side head clamping bolts and retighten them to provide a slight drag on the toolhead (Figure J-47).

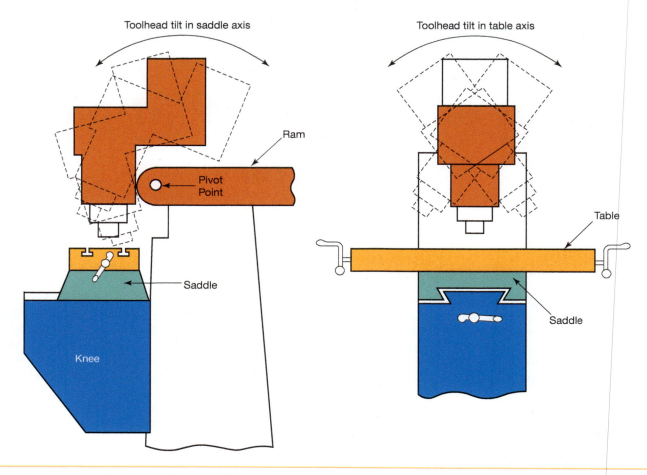

Figure J-43 Vertical mill toolhead tilting capability.

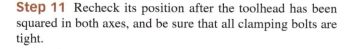

Figure J-44 Aligning the toolhead square to the table using the swivel base from a precision mill vise and a dial indicator.

Step 11 Recheck its position after the toolhead has been squared in both axes, and be sure that all clamping bolts are tight.

Figure J-45 Front toolhead clamping bolts (*Courtesy of Fox Valley Technical College*).

SHOP TIP

Good lighting is a major factor in quality. Proper lighting is absolutely necessary to achieve quality work. Workers are more alert in a bright shop environment and are less likely to make mistakes. Working under proper lighting makes for a safer work area and reduces accidents where high-speed machinery is in use.

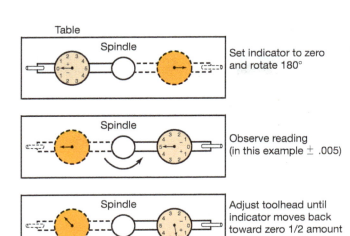

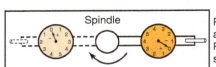

Table

Spindle — Set indicator to zero and rotate 180°

Spindle — Observe reading (in this example ± .005)

Spindle — Adjust toolhead until indicator moves back toward zero 1/2 amount shown (± .0025)

Spindle — Reset indicator to zero and move back 180°. Readings should be the same on both sides.

Figure J-46 Indicator used for alignment.

Tilt Adjustment Screw

Toolhead Clamping Bolts

Figure J-47 Tilt adjustment screw and toolhead side clamping bolts (*Courtesy of Fox Valley Technical College*).

WORK-HOLDING DEVICES

Work-holding techniques are very important in the setup and operation of a machining center. Before any machining can be done, the operator must make sure that the part or work-holding device is properly positioned and fastened to the table. Some setups may be as simple as placing the part in a vise, but some setups may take considerable ingenuity and time. Whatever the case, it is important to remember to make your setup as safe as possible.

Vises

The vise is the most common work-holding device (Figure J-48). The plain vise is used for holding work with parallel sides and is bolted directly to the table using the T-slots in the machine table. This vise has a swivel so that the vise can be set at any angle. Air or hydraulically operated vises can be used in high-production operations to increase productivity.

Angle Plates

Work that needs to be held at a 90-degree angle to the table can be held on an angle plate (Figure J-49). An angle plate is an L-shaped piece of cast iron or steel that has tapped holes or slots.

Direct Workpiece Mounting

Work that is too big or has an odd configuration can be bolted directly to the table (Figure J-50). This method of work holding takes the most ingenuity and expertise. There are a number of accessories that can be used to aid the setup person. A variety of clamp styles are commercially available for directly mounting workpieces to the machine table (Figure J-51). Figure J-52 shows some examples of good

Swivel

Figure J-48 The standard milling machine vise is used to hold relatively small parts that have a square or rectangular shape.

Figure J-49 Angle plates come in a variety of sizes and are typically bolted directly to the machine table.

Figure J-50 Parts that are clamped directly to the table are typically of an odd configuration, such as a weldment. Setting up these types of workpieces takes some ingenuity.

Figure J-51 Strap clamps are used to fasten work to the machine table, fixture, or angle plate. Strap clamps are usually supported by step blocks. T-bolts should be placed as close to the workpiece as possible.

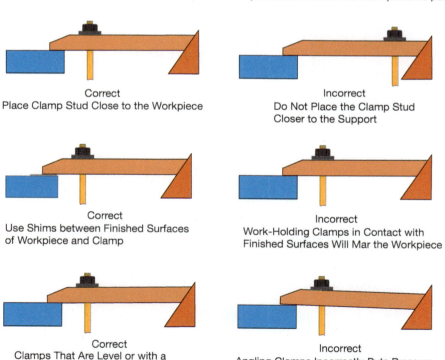

Correct
Place Clamp Stud Close to the Workpiece

Incorrect
Do Not Place the Clamp Stud
Closer to the Support

Correct
Use Shims between Finished Surfaces
of Workpiece and Clamp

Incorrect
Work-Holding Clamps in Contact with
Finished Surfaces Will Mar the Workpiece

Correct
Clamps That Are Level or with a
Slight Decline toward the Workpiece
Equalize Pressure on the Workpiece

Incorrect
Angling Clamps Incorrectly Puts Pressure
on the Support, Not on the Workpiece and It
Tends to Push the Workpiece Out of Position

Figure J-52 Study these clamping practices carefully.

and bad clamping technique. Following are some tips for clamping work directly to the table:

1. Tables should be protected from abrasive materials, such as cast iron, by placing plastic or aluminum shims between the work and the table.
2. Clamps should be located on both sides of the workpiece if possible.
3. Clamps should always be located over supports to prevent distortion or breakage of parts.
4. Clamps and supports should be placed at the same height.
5. Screw jacks should be placed under parts for support to prevent vibration and distortion.

OTHER WORK-HOLDING ACCESSORIES FOR THE MILLING MACHINE

ROTARY TABLE

The rotary table is a useful accessory for the milling machine (Figure J-53). The rotating motion is controlled by a precision worm and worm gear assembly. Common ratios of rotary tables are 40:1 and 80:1. A 40:1 ratio means that it takes 40 complete revolutions of the hand crank to revolve the rotary table once. The common rotary table positions the workpiece by degrees of arc. Discrimination on rotary tables varies from 1 minute to fractions of a second. Full degrees are graduated on the circumference of the table. Fractions of a degree are read on the worm crank scale, usually with the aid of a vernier. Rotary tables are designed so that they can be mounted on a machine tool in either a vertical or horizontal position. Three-jaw chucks can also be fastened to the table of a rotary table to hold and machine round parts.

When the rotary table is mounted in a horizontal position on the machine table, circular and linear cuts can be made without reclamping the workpiece. Figure J-54 shows an example of typical part that could be machined on a

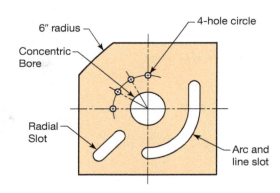

Figure J-54 Machining operations using a rotary table.

rotary table. Before CNC machines the rotary table was the only way to make a part like this. Always have the workpiece securely fastened and supported on parallels if holes or other features go completely through the workpiece. Every machine axis not moving during a cut, linear (machine table) or rotary (rotary table), should be securely locked. Remember that metal removal operations create tremendous cutting pressures, which may move your workpiece. When the rotary table is in a vertical position, operations can be performed on the circumference of a workpiece.

SHOP TIP

Rotary tables are used to machine arcs, angular cuts, or bolt circles on manual machines. To make accurately spaced holes on a circle, the rotary table must be precisely positioned on the machine table. Always check the surfaces of the machine table and any accessories for nicks or burrs before mounting them on the table. The center of the rotary table also must be aligned with the centerline of the machine spindle. Using a dial test indicator mounted in the machine spindle, align the hole in the center of the rotary table with the spindle. Once the alignment is made in both the table and saddle axes, the machine dials or the digital readout should be set to zero. Mount the workpiece on the rotary table and align it as needed. Next, move the table axis a distance equal to the radius value of the desired hole circle. Lock any movable axes before starting the machining operation.

Dividing Head

Figure J-55 shows a dividing (indexing) head. Work can be held between centers, in collets, or in a chuck. The indexing head, also called the dividing head, is a useful milling machine accessory. It is used to rotate the workpiece a full or partial turn to machine a specific number of divisions, an angle, or a circular feature. The indexing head gets its rotational movement from a worm and worm gear assembly. The index crank is geared to the spindle worm gear, most commonly in a 40:1 ratio. This means that 40 complete turns

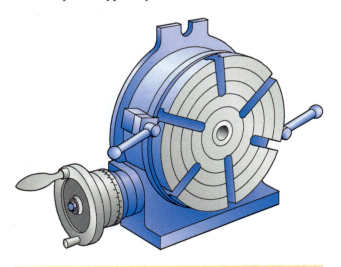

Figure J-53 A rotary table can be used in a horizontal or vertical position.

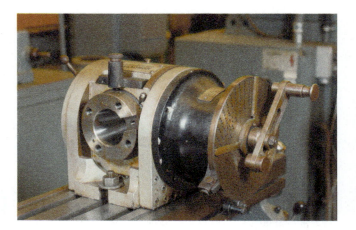

Figure J-55 Typical dividing head.

Figure J-56 Dividing head sector arms set for indexing 11 spaces.

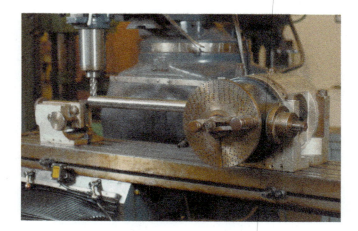

Figure J-57 Dividing head in use.

of the index crank will result in 1 complete revolution of the spindle. A lock is provided so that the spindle may be secured at a particular setting to increase the rigidity of a setup.

The index crank contains a pin on a spring-loaded plunger that engages the holes in the index plate. Hole circles in the index plate are used for the indexing function. The pattern on the plate consists of circles of holes, and the crank pin is adjusted in and out radially so that the desired hole circle may be used. Two sector arms rotate about the crank hub and can be adjusted to indicate the correct number of holes for a partial turn of the crank. For example, if you are indexing one turn plus 11 holes for each division, the sector arms can be set so that you will not have to count the 11 extra holes each time around (Figure J-56). The sector arms rotate about the index crank hub and are locked in place by a lock screw. The facing edges of the arms are beveled, and the number of holes required for fractional turns is established between the beveled edges. Remember that the number of holes required for a fractional turn will be new holes. The hole in which the crank pin is engaged is not counted.

The index head is secured to the mill table and aligned with the alignment keys that fit the table T-slots. The end of a workpiece can be supported by a footstock (Figure J-57).

Common machining operations that involve indexing include cutting splines, keyways, hexagons, octagons, squares, and other geometric shapes. The indexing calculation to machine a square, when using an indexing head with a 40:1 gear ratio, is

$$\text{Index crank turns} = 40/N = 40/4 = 10$$

where

$$N = \text{the required number of divisions}$$

The result 10 means that 10 full revolutions (turns) of the index crank are required from one division to the next. The calculation for a hexagon would be

$$\text{Index crank turns} = 40/N = 40/6 = 6\tfrac{4}{6}$$

This fraction shows that six complete index crank turns plus 4/6 turn are needed. To achieve this partial turn of 4/6,

reduce this fraction to its lowest common denominator, which is 2/3. The available index plate has 8 circles of 24, 25, 28, 30, 34, 37, 38, and 39 holes. Three of these hole circles (24, 30, and 39) can be divided by the lowest common denominator of 3. Always use the hole circle with the greatest number of holes that is divisible by the denominator to achieve the greatest accuracy. The result is six turns plus 2/3 * 13/13 = 26/39 = 26 holes in the 39-hole circle.

To operate the index head, unlock the spindle and pull back the crank pin plunger. Rotate the crank the required number of turns plus the fraction of a turn. Set the sector arms to keep track of the additional holes beyond the full turn of the crank. Bring the crank pin around until it just drops into the required hole. If you overshoot the position, back off well past the required hole and come up to it once again. This will eliminate any backlash in the worm and worm gear assembly. Always index in only one direction, and always lock the spindle before beginning the machining operation. After each indexing move, the sector arms must be rotated so that they will be properly positioned for the next move.

Fixtures

Fixtures are tools that are designed to accurately position and hold a specific part. They are typically found in a production shop and can be built to hold one part or many parts. The fixtured part is usually one that cannot be held in a vise (Figure J-58). Fixtures are used quite extensively in the machining industry. They should be kept simple to allow for quick loading and unloading of parts. Fixtures also need to be designed and built in a foolproof manner so that the part can be loaded in only one way. A well-designed fixture will lower the cost of producing parts.

WORK HOLDING ON THE VERTICAL MILL

The most common work-holding methods on a vertical mill are securing the part directly to the table or by holding the part in a mill vise.

Mounting Directly to the Table

Mounting a workpiece directly to the machine table is an excellent work-holding method. The same clamping techniques that you learned in drilling are applied in milling. Be careful not to distort the workpiece when applying clamping pressure. In many cases, the workpiece must be aligned with the table or saddle axis to ensure parallel or perpendicular cuts. A workpiece can be aligned quite closely by placing it against stops that just fit the table T-slots (Figure J-59). Another method for roughly aligning the workpiece is to measure from the edge of the table to the workpiece (Figure J-60). The most accurate method of alignment is to use a dial indicator fastened in the machine spindle or to the toolhead (Figure J-61). The workpiece is brought into contact with the indicator, and the table or saddle is run back and forth while the workpiece position is adjusted until the indicator reads zero for the entire length.

Using a Mill Vise

A precision mill vise may also be used to hold the part. The vise must be aligned with the table or saddle axis. Use a dial indicator for alignment and always indicate a vise on its solid jaw.

Figure J-58 Fixtures are used to hold and accurately position workpieces. Repeatability in locating the part is very important.

Figure J-59 Work aligned by locating against stops in T-slots

Mill vises are precision tools. Always treat them gently when setting them down on the machine table. Be sure that the table is clean and that there are no burrs on the bottom of the vise.

Soft-Jaw Vises   Sometimes a mill vise may have soft steel or aluminum jaws instead of the regular hardened steel types. After the vise has been bolted to the mill table and

Figure J-60 Measuring the distance from the edge of the table to the workpiece.

Figure J-61 Aligning a workpiece with the aid of a dial indicator.

roughly aligned, a light cut can be taken on the soft jaws. The result of this procedure is a vise jaw that has been machined true to the axis of the saddle or table. Soft jaws are often used in production machining operations or where it might be desirable to shape the vise jaw in a certain way to hold a particular part to be machined.

WORK EDGE AND HOLE CENTERLINE LOCATING

After a workpiece has been set up on the mill, it may be necessary to position the machine spindle relative to the edge of the part or to center the spindle over an existing hole. Edge finding and centerline finding are common operations that you will need to perform.

Using an Edge Finder

The edge finder is a very useful tool for location (Figure J-62). The edge finder consists of a shank with a floating tip that is retained by an internal spring. The edge finder tip is accurately machined to a known diameter, usually .200 or .500 inch. Follow this procedure for using an edge finder.

Step 1 Secure the edge finder in a collet or chuck in the machine spindle.

Step 2 Set the spindle speed to 600 to 800 rpm, and nudge the edge finder tip so that it is off center.

Step 3 Start the spindle and lower the quill or raise the knee so that the edge finder tip can contact the edge of the part to be located.

Step 4 Turn the table or saddle crank and move the workpiece until it contacts the rotating edge of the edge finder tip. Continue to slowly advance the workpiece against the edge finder tip until the tip runs perfectly true and then suddenly moves sideways. Stop movement at this moment (Figure J-63). The machine spindle is now positioned a distance equal to half of the edge finder tip diameter from the edge of the workpiece. If you are using a .200-inch-diameter tip, the centerline of the spindle is .100 inch from the edge of the workpiece. Repeat the edge finder-to-workpiece approach at least two times to confirm the readings of the machine dials or digital display.

Step 5 When you are sure that the positioning is correct, lower the workpiece or raise the spindle and set the digital display or the table and saddle dials to zero. Then, move the table or saddle the additional .100 inch in the same direction that it was moving as it approached the workpiece and reset the digital readout dial to zero. This will prevent introduction of a backlash error owing to slack in the table or saddle nut.

Locating Hole Centers   Many vertical milling machine operations require that the machine spindle be positioned over the center of an existing hole in the workpiece. Use the following procedure to locate a hole center.

Figure J-62 Offset edge finder (*Courtesy of the L.S. Starrett Co.*).

Figure J-63 Work approaching the tip of the offset edge finder.

SHOP TIP-CRAFTSMANSHIP

When work requires accurate table movements, back-lash must be considered. When a digital readout is not available on your milling machine, use a dial indicator with a measuring capacity of 1 inch or more. When this dial indicator is fitted with a magnetic base and clamped on the machine, it will measure exact table movement regardless of the play (backlash) existing between the nut and leadscrew of the machine. Watch the dial while tightening the machine locking levers to tell if unwanted table movements are happening, and make corrections before making a wrong cut.

Step 1 Mount a dial test indicator in the machine spindle.

Step 2 Move the indicator contact point so it touches the side of the hole (Figure J-64). Set the indicator bezel to zero.

Step 3 Rotate the spindle 180 degrees. Compare the two indicator readings and split the difference between them by moving the machine table. Locate the spindle center first in one axis, and then in the second axis.

Step 4 Even if the hole is not round or cylindrical, the spindle still has to be centered in both the table and saddle axes. If the hole is not round, the indicator readings will be identical, 180 degrees apart, but will vary from axis to axis.

Step 5 Always double-check the readings on the indicator and the machine dial settings before performing any additional machining.

Figure J-64 Dial indicator used to locate the center of a hole.

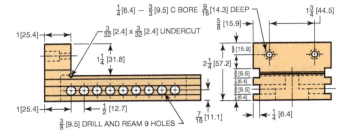

Figure J-65 Machining the holes in the vise body.

Figure J-66 Drilling 11/32-inch holes in the vise body.

Example of Setup and Machining Holes in Vise Body (Figure J-65)

Step 1 Align the workhead square to the machine table.

Step 2 Align and fasten a machine vise on the table so that its jaw is parallel to the long axis of the table.

Step 3 Mount the vise body in the machine vise with the bottom surface against the solid jaw of the machine vise (Figure J-66).

Step 4 Mount an edge finder in a spindle collet and align the spindle axis with the base surface of the vise body.

Step 5 Move the table the required .452-inch distance and lock the table cross slide.

Step 6 Pick up the outside of the solid jaw of the vise body.

Step 7 Move to the first hole location 1.015 inches from the outside edge.

Step 8 Center drill this hole.

Step 9 Use a 1/4-inch drill and drill this hole 1 ½ inch deep.

Step 10 Repeat steps 8 and 9 for the remaining eight holes. Accurate positioning is done with the micrometer dials or digital readout.

Step 11 Remove the workpiece from the machine vise. Turn it over so that the just-drilled holes are down, and the bottom surface of the vise body is again against the solid jaw.

Step 12 Use the edge finder to pick up the two sides, as for the first drilling operation.

Step 13 Position the spindle over the first-hole location, again with the first hole on the solid jaw side.

Step 14 Center drill this hole.

Step 15 Drill this hole with a ¼-inch drill deep enough to meet the hole from below.

Step 16 Switch to a 22/32-inch drill and drill completely through the vise body. The ¼-inch hole acts as a pilot hole to let the 11/32-inch drill come out in the correct place on the bottom side.

Step 17 Change from the 11/32-inch drill to a ⅜-inch machine reamer and ream this hole completely through also (Figure J-67).

Step 18 Repeat steps 14 to 17 for the remaining eight holes.

Step 19 Reposition the workpiece so that it is upright in the machine vise with the solid jaw of the vise body up.

Step 20 Locate the edges of the workpiece with an edge finder.

Step 21 Position for the two-hole locations and drill the 11/64 holes with their 13/32 counterbores (Figure J-68).

Step 22 Remove all burrs.

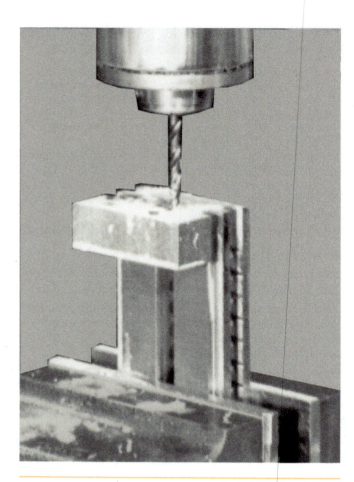

Figure J-68 Drill and counterbore holes in the solid jaw of the vise body.

SELF-TEST

1. How can workpieces be aligned when they are clamped to the table?

2. How is a vise aligned on a machine table?

3. When is the toolhead alignment checked?

4. Why is it important that the knee clamping bolts should be tight before aligning a toolhead?

5. Why does the toolhead alignment need to be checked again after all the clamping bolts are tightened?

6. How can the machine spindle be located exactly over the edge of a workpiece?

7. With an edge finder, when do you know that the spindle axis is .100 inch away from the edge of the workpiece?

8. What is the recommended rpm to use with an edge finder?

9. When locating multiple positions on a workpiece, how can you eliminate the backlash in the machine screws?

10. How is the center of an existing hole located?

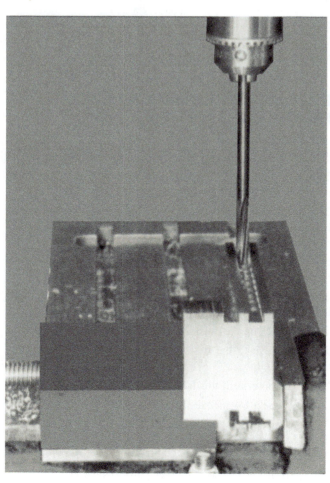

Figure J-67 Reaming ⅜-inch holes in the vise body.

Vertical Milling Machine Operations

The vertical milling machine is one of the most versatile machine tools in a machine shop. The purpose of this unit is to explore some of this versatility.

After completing this unit, you should be able to:

- Calculate cutting speeds and feeds for end milling operations.
- Identify and select vertical milling machine setups and operations for a variety of machining tasks.

CLIMB AND CONVENTIONAL MILLING

In milling, the direction that the workpiece is being fed can be either the same as the direction of cutter rotation or opposed to the direction of cutter rotation. When the direction of feed is opposed to the direction of rotation, this is said to be *conventional* or *up milling* (Figure J-69). When the direction of feed is the same as the direction of cutter rotation, this is said to be *climb* or *down milling*, because the cutter is attempting to climb onto the workpiece as it is fed into the cutter. If there is a large amount of backlash, the workpiece can be pulled into the cutter during climb milling. This can result in a broken cutter, damaged workpiece, and possible injury from flying metal. Climb milling should be avoided in most cases on older manual machines. In certain situations, it is desirable to climb mill. For example, if the milling machine has ball screws, and backlash is virtually eliminated, climb milling is a very good technique. Even on conventional machines, climb milling with a light cut can result in a better surface finish, because chips are not swept back through the cut. During any milling operation, all table movements should be locked except the one that is moving. This will ensure the most rigid setup possible. Spiral-fluted end mills may work their way out of a split collet when deep

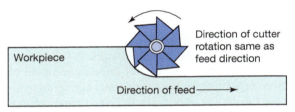

Climb or down milling

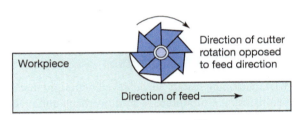

Conventional or up milling

Figure J-69 Climb and conventional milling.

heavy cuts are made or when the end mill gets dull. As a precaution, to warn you that this is happening, you can make a mark with a felt-tip pen on the revolving end mill shank where it meets the collet face. Observing this mark during the cut will give you an early indication if the end mill is changing its position in the collet.

CUTTING PERFORMANCE FACTORS

Detailed information on machinability, cutting speeds, feeds, and cutting fluids is found in Section F.

Cutting Speeds

Cutting speed in milling is the rate at which a point on the cutter passes by a point on the workpiece in a given period of time. Cutting speed is an extremely important factor in all machining operations. For rotating milling cutters, cutting speed is expressed as a function of cutter rpm by the formula

$$rpm = \frac{CS \times 4}{D}$$

where

rpm = revolutions per minute of the cutter

CS = cutting speed of the material being machined, in feet per minute

D = diameter of the cutter, in inches

Cutting speed constants (Table J-1) are influenced by the cutting tool material, workpiece material, rigidity of the machine setup, and use of cutting fluids. As a rule, lower cutting speeds are used to machine hard or tough materials or when heavy cuts are taken, and it is desirable to minimize tool wear and to extend tool life. Higher cutting speeds are used in machining softer materials to achieve better surface finishes. Higher speeds also apply when using small-diameter cutters for light cuts on fragile workpieces and in delicate setups. Table J-1 gives starting values for common materials. These values may have to be varied higher or lower depending on the specific machining task. Always observe the cutting action carefully and make appropriate speed corrections as needed. Until you gain some experience in milling, use the lower values in the table when selecting cutting speeds. Ball-end end mills, corner rounding mills, and angle milling cutters should be operated at half to two-thirds the speed of a comparable end mill.

Advanced Tool Materials

The best machining results are achieved with rigid machines and setups using a minimum of tool overhang and machine spindle extension. Vibration-causing chatter will rapidly destroy tools. Many carbide grades are made for general-purpose machining within a range of different work materials. Others are designed for one specific application and will fail if used for anything else. When using carbide, cermet, ceramic, CBN, or diamond inserts, it is absolutely essential to refer to the cutting tool manufacturer's catalog for specific application recommendations for a specific insert.

Feed Rates

Another equally important factor in safe and efficient machining is the feed rate. Feed rate is the rate at which the cutter is advanced into the work material. Because each tooth of a milling cutter is cutting, a chip of a given thickness will be removed depending on the rate of feed. Chip thickness affects the life of the milling cutter. Excessive feed rates can cause a chipped cutting edge or a broken cutter. On the other hand, the highest practical feed rate per tooth will give the longest tool life. Feed rate in milling is measured in inches per minute (ipm) and is calculated by the formula

$$ipm = F \times N \times rpm$$

where

ipm = feed rate, in inches per minute

F = feed per tooth, in inches

N = number of teeth on the cutter being used

rpm = revolutions per minute of the cutter

Table J-2 gives starting values for chip loads for common materials. Information on how the feed per tooth compares with the actual chip thickness per tooth is presented in Section M, Unit 3.

Table J-1 Cutting Speeds for Some Commonly Used Materials

Work Material	Tool Material						
	High-Speed Steel	Uncoated Carbide	Coated Carbide	Cermet	Ceramic	CBN	Diamond
Aluminum							
Low silicon	300–800	700–1400					1000–5000
High silicon							500–2500
Bronze	65–130	500–700					1000–3000
Gray cast iron	50–80	250–450	350–500	400–1000	700–2000	700–1500	
Chilled cast iron					250–600	250–500	
Low-carbon steel	60–100	250–350	500–900	500–1300	1000–2500		
Alloy steel	40–70		350–600	100–300	500–1500	250–600	
Tool steel	40–70		250–500		500–1200	150–300	
Stainless steel							
200 and 300 series	30–80	100–250	400–650		300–1100		
400 and 500 series			250–350		400–1200		
Nonmetallics		400–600					400–2000
Superalloys		70–100	90–150		500–1000	300–800	

Table J-2 Feeds for High-Speed Steel End Mills (feed per tooth in inches)

Cutter Diameter (in.)	Aluminum	Brass	Bronze	Cast Iron	Low-Carbon Steel	High-Carbon Steel	Medium Alloy Steel	Stainless Steel
$\frac{1}{8}$	.002	.001	.0005	.0005	.0005	.0005	.0005	.0005
$\frac{1}{4}$	.002	.002	.001	.001	.001	.001	.0005	.001
$\frac{3}{8}$	.003	.003	.002	.002	.002	.002	.001	.002
$\frac{1}{2}$	.005	.002	.003	.0025	.002	.002	.001	.002
$\frac{3}{4}$	.006	.004	.003	.003	.004	.003	.002	.003
1	.007	.005	.004	.0035	.005	.003	.003	.004
$1\frac{1}{2}$	.008	.005	.005	.004	.006	.004	.003	.004
2	.009	.006	.005	.005	.007	.004	.003	.005

Cutting Speed and Feed Rate Calculations

The first step is to calculate the correct rpm for the cutter. Refer to Table J-1 for the starting value of the cutting speed.

EXAMPLE

Calculate the rpm for a ½-inch-diameter HSS end mill machining aluminum.

$$\text{rpm} = \frac{\text{CS} \times 4}{D} = \frac{300 \times 4}{1/2} = \frac{1200}{.5} = 2400$$

The next step is to calculate the feed rate. Refer to Table J-2 for the starting values.

$$\text{ipm} = F \times N \times \text{rpm}$$

$$= .005 \text{ (feed per tooth, Table J-2)}$$

$$\times \ N(\text{number of teeth}) \ \times \ \text{rpm}$$

$$= .005 \times 2 \times 2400 = 24 \text{ ipm}$$

Therefore, the cutter should revolve at 2400 rpm, and the feed rate should be 24 ipm.

Depth of Cut

The third factor to consider in using end mills is the depth of cut. The depth of cut is limited by the amount of material that needs to be removed from the workpiece, by the power available at the machine spindle, and by the rigidity of the workpiece, tool, and setup. As a rule, the depth of cut for an end mill should not exceed half the diameter of the tool in steel. In softer metals, the depth of cut can be larger. With roughing end mills, the depth of cut can be as much as one and a half times the cutter diameter and the width of cut half the cutter diameter. If deeper cuts need to be made, the feed rate needs to be reduced to prevent tool breakage. The end mill must be sharp. The end mill should be mounted with no more tool overhang than necessary to do the job.

Another factor to consider is the horsepower rating of the machine. For horsepower calculations, see Section K, Unit 5.

Cutting Fluids

Milling is often done with **cutting fluids**. Cutting fluids dissipate heat generated by the friction of the cutter against the workpiece. They also lubricate the interface between the cutting edge and work and also flush the chips away from the cutting area. Machining operations are greatly improved by using cutting fluids. Cutting fluids increase productivity, extend tool life, and improve the surface finish of the workpiece. Cutting fluids are applied to the cut in a flood stream or by an air/cutting fluid mix mist. Cutting fluids generally reduce carbide tool life in milling cast iron and steel. The cutting fluid cannot reach the cutting edge while the insert is in the cut. After the extremely hot insert leaves the cut, the cutting fluid then has an extremely shocking cooling effect. These excessive temperature variations result in minute cracks along the cutting edge and premature tool failure. Cutting fluid recommendations of cutting tool manufacturers should be followed to prevent tool failure caused by incorrect coolant use.

Materials such as cast iron, brass, and plastic are often machined dry. A stream of compressed air can cool tools and keep the cutting area clear of chips. A safety shield and protective clothing should always be used when compressed air is applied, for protection from flying hot chips.

SHOP TIP

After selecting an end mill, check the condition of its cutting edges. If the cutting edges are dull and show indication of wear, the end mill needs to be sharpened or you need to use a different one. If it is an indexable tool, you must rotate the insert to expose a new cutting edge, or change the end mill. End mills are sharpened on a tool and cutter grinder. Dull tools create excessive cutting pressures that can break the end mill or leave an unsatisfactory surface finish. Worn end mills will also produce undersized grooves or slots.

COMMON MILLING OPERATIONS

Squaring a Workpiece

Often, a rough piece of stock (Figure J-70) must be machined square. In this figure the raw stock is gray and the finished part is shown in blue. The sides are also numbered and will be machined in the order shown. This can be accomplished by machining all four sides in the correct sequence. The workpiece should be set up in a mill vise. The following procedure may be used.

Step 1 Mill side 1 (Figure J-71). In this example side 4 was a flat surface. Side 4 is placed against the solid jaw of the vise on one parallel adjacent to the solid jaw. A round steel rod is placed between the movable jaw and the workpiece. This assures that side 4 will be flat against the solid jaw. The vise is tightened and a dead blow hammer is used to seat the back edge of the workpiece onto the parallel. The top (side 1) is then milled.

Step 2 Mill side 2 (Figure J-72). Deburr and place this surface (side 1) against the solid jaw of the vise with one parallel under the back side. Place a short length of soft round rod between the work and the movable jaw of the vise. Tighten the vise and seat the workpiece down on the parallel with a dead blow hammer. The round rod ensures that the reference surface is pressed flat against the solid vise jaw. Mill the top (side 2).

Step 3 Mill side 3 (Figure J-73) Next clamp the part with side 2 against the solid back jaw of the vise and on top of the parallel. Place a short length of soft round rod between the work and the movable jaw of the vise. Tighten the vise. Check to make sure the part is tight on the parallel. Mill Side 3. Measure the

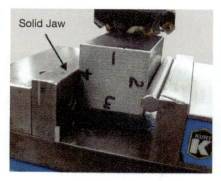

Figure J-70 Machining a square from rough stock.

After the top is milled, side 1 will be square with side 4

Figure J-71 Setup to machine side 1.

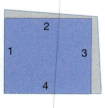

After side 2 is milled it will be square with side 1 and parallel to side 4

Figure J-72 Setup to machine side 2.

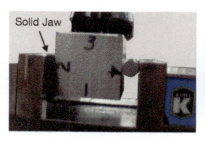

After side 3 is milled it will be square with sides 2 and 4 and parallel to side 1

Figure J-73 Milling side 3.

dimension between side 1 and side 3. Take additional cuts if the piece is not to the specified size.

Step 4 Mill side 4 (Figure J-74). Next clamp the part with side 3 against the solid back jaw of the vise and on top of the parallel. Place a short length of soft round rod between the work and the movable jaw of the vise. Tighten the vise. Check to make sure the part is tight on the parallel. Mill Side 4. Measure the dimension between side 4 and side 2. Take additional cuts if the piece is not to the specified size.

Step 5 To mill side 4, clamp the part with side 2 against the solid jaw and tight on the parallels (Figure J-74). Note that at this point, two parallels are used because sides 2 and 3 are parallel and square with side 1. Mill side 4 and then measure the distance between sides 1 and 4. If it is not to the specified size, take additional cuts.

Step 6 To square the ends of the part, line up the part sing a solid square in the vise (Figure J-75). Then repeat the process for the other end.

Step 7 Next clamp the part with side 4 on the solid back jaw of the vise and on top of the parallel. Check to make sure the part is tight on the parallel. Mill Side 3. Measure the dimension between side 1 and side 3. Take additional cuts if the piece is not to the specified size.

Step 8 To mill side 4, clamp the part with side 2 against the solid jaw and tight on the parallels. Note that at this point, 2 parallels are used because sides 2 and 3 are parallel and square with

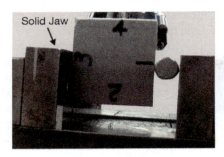

After a skim cut is taken on side 4 it will be square with sides 3 and 1 and parallel to side 2

Figure J-74 Setup to machine side 4.

Figure J-75 Setup of the workpiece to machine the end square.

side 1. Mill side 4 and then measure the distance between side 1 and side 4. If it is not to the specified size, take additional cuts.

Step 9 To square the ends of the part, line up the part using a solid square in the vise (Figure J-75). Then repeat the process for the other end.

Machining Steps and Squaring the End of a Workpiece

Common milling operations on the vertical mill include machining steps (Figure J-76) and squaring or machining a surface perpendicular to another surface (Figure J-77). The ends of the workpiece can be machined square and to a given length by using an end mill. If a large amount of material has to be removed, it is best to use a roughing end mill first (Figure J-78), and then finish to size with a regular end mill. On low-horse power vertical mills, plunge cutting is an efficient method of removing material quickly (Figure J-79). In this operation, the end mill is plunged a predetermined width and depth of cut, retracted, and then advanced and plunged again repeatedly. In plunging, the maximum cutting force is in the direction in which the machine is the strongest—in the vertical direction.

Figure J-76 Using an end mill to mill steps.

Figure J-77 Using an end mill to square stock.

Figure J-78 Roughing end mill used to remove a large volume of material.

Milling a Cavity

Center-cutting end mills make their own starting hole when used to mill a pocket or cavity (Figure J-80). Prior to making any mill cuts, the outline of the cavity should be laid out on

Figure J-79 Plunge cutting with an end mill.

SHOP TIP

Plunge cutting can be an efficient method of removing material quickly. In this operation, an end mill is plunged and then fed vertically into the workpiece with the quill feed hand lever. The depth of the plunge cut can be limited with the depth stop. After each plunge cut, the tool is retracted and the table is advanced for the next cut. This sequence is repeated until the roughing is completed. In plunge cutting, the highest cutting forces are in the direction of the greatest strength of the machine, which is the vertical spindle direction.

the workpiece. Only when finish cuts are made should these layout lines disappear. Good milling practice is to rough out the cavity to within .030 inch of finished size before making any finish cuts.

When you are milling a cavity on a manual machine, the direction of the feed should be against the rotation of the cutter (Figure J-81). This ensures positive control over the distance the cutter travels and prevents the workpiece from being pulled into the cutter because of backlash. When you

reverse the direction of table travel, you will have to compensate for the backlash in the table feed mechanism.

Newer machines may have ball screws which minimize backlash. If the machine you are using does not have much backlash, climb milling is always the preferred method for machining. This is especially true when using carbide inserts. Figure J-82 shows an end mill that uses inserts, including an insert on the bottom that allows plunge cutting. Different insert corner radii make machining a pocket with a bottom radius easy.

It is important that the chips be removed from the cavity during the milling operation. Chips can be removed with a brush. If the chips are left to accumulate, the cutter will jam and likely break.

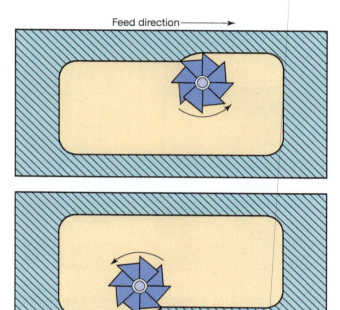

Figure J-81 Feed direction is against cutter rotation.

Figure J-80 Using an end mill to machine a pocket.

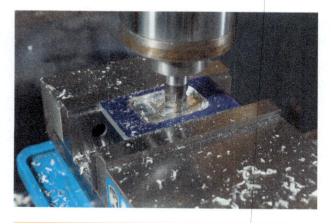

Figure J-82 Plunge cutting and milling a cavity with a positive-rake center-cutting end mill.

SHOP TIP

The size of an end mill is normally marked on the tool shank. Assuming that this marking is correct may lead you into trouble. Often, a resharpened end mill is undersize because both the diameter and the end have been reground. It is good practice to check the diameter of a cutter before using it.

End Milling a Shaft Keyway

A common operation on the vertical mill is milling keyways. A shaft keyway must be centered on the shaft and cut to the correct depth. The following procedure may be used.

Step 1 Secure the workpiece to the machine table or in an aligned mill vise.

Step 2 Select the correct size of cutter either in a two-flute or center-cutting multi-flute type. Install the cutter in the machine spindle.

Step 3 Move the workpiece aside and lower the cutter beside the part. With the spindle motor off, insert a slip of paper between cutter and workpiece. An edge finder may also be used for this procedure.

Step 4 Move the workpiece toward the cutter until the paper feeler is pulled between cutter and work as you rotate the spindle by hand. At this point, the cutter is about .002 inch from the shaft (Figure J-83).

Step 5 Set the saddle micrometer collar to zero, compensating for the .002 of paper thickness.

Step 6 Add the diameter of the shaft and cutter, and divide by 2. This is the total distance to move the workpiece for centering.

Step 7 Raise the cutter clear of the work and move the workpiece over the correct amount in the same direction that it was moving as it approached the cutter.

Step 8 Raise the quill to its top position and lock the quill and the saddle in place.

Figure J-83 Moving an end mill to the side of a shaft with the aid of a paper feeler.

Figure J-84 Cutter centered over the shaft and lowered to make a circular mark.

Step 9 Move the table crank to position the cutter at the point where the keyway is to begin.

Step 10 Start the spindle and raise the knee until the cutter contacts the shaft (Figure J-84). Set the knee micrometer collar to zero. Raise the knee and move the cutter off the end of the shaft. Raise the knee a distance equal to half the cutter diameter plus .005 inch and lock the knee in this position. Using correct speeds and feeds, mill the keyway to the required length.

Machining T-Slots and Dovetails, Angle Milling, and Drilling

Two operations are performed to machine a **dovetail** into a workpiece (Figure J-85). First, a slot is cut with a regular end mill. The slot reduces the amount of material that the dovetail cutter has to remove. There are three dovetails to be cut

Figure J-85 Milling a dovetail.

Figure J-86 Formulas for calculating distance across pins.

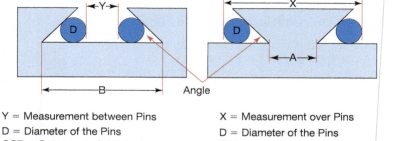

Y = Measurement between Pins
D = Diameter of the Pins
COT = Cotangent of the Angle
B = Desired Distance across Angle Points
Y = B − D(1 + COT ½ Angle)

X = Measurement over Pins
D = Diameter of the Pins
COT = Cotangent
A = Desired Distance across Angle Points
X = D(1 + COT ½ Angle) + A

into the part shown in Figure J-85. Three slots were cut first with an end mill. The dovetail cutter is cutting the middle slot. An initial pass would be taken with the dovetail cutter through the part and then it would be measured. Dovetails are measured by using pins (Figure J-86). The machinist would choose an appropriate size for pins and then calculate the distance to be measured. The machinist would then measure the distance and move the table the required amount and take another cut. If the dovetail needs to be centered on a specific dimension, the machinist would move half of the distance one direction, take a cut, re-measure and then move the opposite direction for the final cut. Remember to take backlash into account. Note that there are also free dovetail calculators on the internet that can do the calculation for you.

Two operations are performed to machine a T-slot into a workpiece. First, a slot is cut with a regular end mill, and then a T-slot cutter is used to finish the machining (Figure J-87).

Angular cuts on workpieces can be made by tilting the workpiece in a vise with the aid of a protractor (Figure J-88) and its built-in level. The angle can be machined with an end mill (Figure J-88). Another way to machine angles is to tilt the workhead (Figures J-89 and J-90). The head can be tilted so that angular holes can be drilled. These holes can be drilled by using the sensitive quill feed lever or the power feed mechanism, or in the case of vertical holes, the knee can be raised. For work involving compound angles, a universal vise (Figure

Figure J-88 Setting up a workpiece for an angular cut with a protractor.

J-91) is used. This vise can be swiveled 90 degrees in the vertical plane and 360 degrees in the horizontal plane.

Other Vertical Mill Operations and Accessories

Rotary tables are designed so that they can be mounted on a machine tool in either a vertical or horizontal position. Figure J-92 shows the use of a rotary table to mill a circular clot in a workpiece.

Figure J-87 First a slot is milled and then the T-slot cutter makes the T-slot.

Figure J-89 Cutting an angle by tilting the workhead and using the end teeth of an end mill.

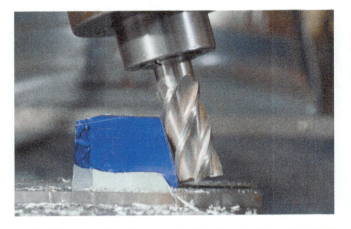

Figure J-90 Cutting an angle by tilting the workhead and using the peripheral teeth of an end mill.

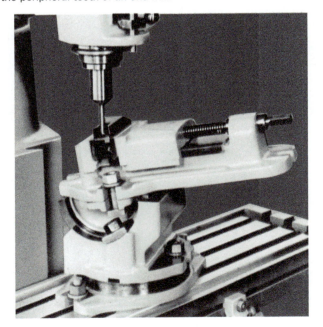

Figure J-91 Universal vise (*Courtesy of MAG Industrial Automation Systems LLC*).

Figure J-92 Using a rotary table to mill a circular slot.

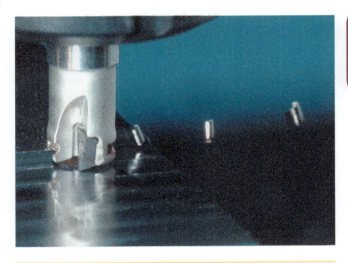

Figure J-93 Small face mill with double positive rake angles removes material efficiently (*Ingersoll Cutting Tools*).

Face Milling

On small machines it is best to use a small-diameter cutter and take multiple passes. A large-diameter cutter requires too much horsepower and leads to undesirable small feed rates and depth of cuts. If the cutter has double positive (axial and radial) rake angles, cutting forces are reduced significantly and productivity goes up (Figure J-93). Coarse-pitch cutters have bigger chip pockets and fewer teeth in the cut. A differential-pitch cutter has unequally spaced teeth, which cuts down on vibration and chatter. Climb milling is always recommended, because the chip is thick as the cutter enters the workpiece and thin on exit. The thin exit chip gives extended tool life.

Optimum width of cut is two-thirds the diameter of the cutter. Centrally positioned cutters result in an alternating cutting force, which can cause vibration. To obtain a better surface finish when using a carbide insert tool, exchange one of the regular inserts with a wiper insert. A wiper insert protrudes below the other inserts by approximately .002 inch. The width of this insert should be approximately one-third greater than the distance the cutter travels per revolution. This wider insert then removes the tool marks left by the other inserts.

Difficult-to-machine materials such as high-alloy steels, austenitic stainless steels, titanium alloys, and heat-resistant superalloys (nickel, cobalt, and iron based) often machine the easiest when round inserts are used. The round insert gives maximum edge strength and limits the chip thickness. Start with a low cutting speed and increase it if possible. The more difficult a material is to machine, the lower the cutting speed should be. Use flood coolant. Round inserts not only are effective cutting tools for exotic materials but can be used on all materials. Round inserts with a positive rake have an effective shearing action. The large radius also provides an outstanding surface finish. For machining hardened steel, using a CBN polycrystalline insert in a face mill gives excellent results.

SELF-TEST

1. When are lower cutting speeds recommended?
2. When are higher cutting speeds used?
3. Should you always use calculated rpm?
4. How is the tool life of an end mill affected by the chip thickness of a cut?
5. What is normally considered the maximum depth of cut for an HSS end mill?
6. Calculate the feed rate for a two-flute ¼-inch carbide end mill to machine low-carbon steel.
7. How is an end mill centered over a shaft prior to cutting a keyway?
8. Why should the feed direction be against the cutter rotation when milling a cavity?
9. What can cause an end mill to work itself out of a collet while cutting?
10. Describe two methods to cut angular surfaces in a vertical milling machine.
11. What device can be used to mill circular slots?
12. What milling machine attachment can be used to mill a precise square or hexagon on a shaft?
13. Which type of insert can be used to make a better surface finish with a face mill?
14. Why are two operations necessary to mill a T-slot or dovetail?

INTERNET REFERENCE

End mill troubleshooting guide:

http://www.carbidedepot.com/formulas-endmill-troubleshoot.htm

Using an Offset Boring Head

The offset boring head is used to machine precisely controlled diameters at accurate locations. Boring heads can also be used to cut grooves and recesses. This unit describes the setup and operation of offset boring heads.

OBJECTIVE

After completing this unit, you should be able to:

- Set up and use the offset boring head in common boring operations.

OFFSET BORING HEAD

Most holes in workpieces are machined by drilling. When a better surface finish and better diameter accuracy are required, reaming may follow drilling. However, drilling and reaming are limited to standard sizes in which these tools are available. In addition, drilled holes may drift off position during a machining operation.

To machine holes of any size and at exact locations, the offset boring head is used. The workpiece must be predrilled on the mill or drill press to use an offset boring head. Some operations that can be performed with an offset boring head are illustrated in Figure J-94.

Parts of the Offset Boring Head

The offset boring head consists of the body with shank and the tool slide (Figure J-95). The shank permits the body to be secured in the machine spindle. The tool slide contains several holes that will accommodate several sizes of boring bars. Boring bar materials include HSS, brazed carbide, and disposable carbide inserts. Boring bars are made with different shapes of inserts, from square to diamond, or triangular. The shape of the insert and how the insert is positioned in the bar determine whether the lead angle is positive, neutral, or negative. A 5-degree negative lead angle is shown on the

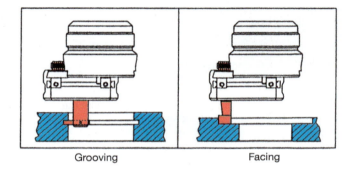
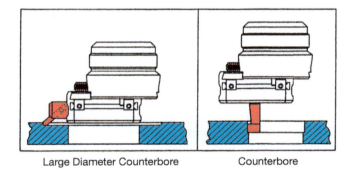

Grooving   Facing

Large Diameter Counterbore   Counterbore

Figure J-94 Offset boring head operations (*Used with permission of Criterion Machine Works*).

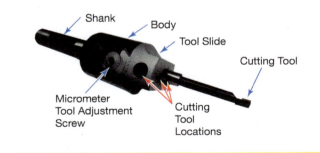

Figure J-95 Offset boring head.

Figure J-97 Workpiece supported on parallels. Note the clearance underneath for the boring tool.

boring bar in Figure J-96. A negative lead angle is needed to allow the tool to machine a square shoulder.

WORKPIECE PREPARATION AND SETUP

The workpiece should be predrilled or otherwise machined to within about 1/16 inch of finished size. If the hole in the part is rough and or off center, leave additional material so that cleanup machining will be ensured. A workpiece to be bored may be clamped directly to the machine table by standard methods or held in a vise or other work-holding fixture. If the part is to be bored through, it must be supported on parallels so that the boring tool will not cut into the machine table (Figure J-97). The parallels must also be set far enough apart so that the tool will clear as it turns.

Positioning the Boring Head

Before boring can begin, the boring head must be positioned over the hole to be bored. An edge finder can be used to position the axes correctly to get the boring head in the correct position to bore the hole.

USING THE OFFSET BORING HEAD

Bar Setup

A boring tool's cutting edge must be on the centerline of the boring head and in line with the axis of the tool slide movement (Figure J-98). The rake angles and clearance angles are then correct. This is also the position in which the tool's cutting edge

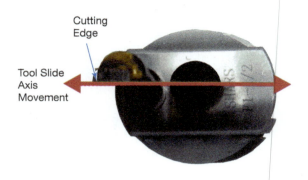

Figure J-98 Boring tool cutting edge is on the centerline of the boring head.

moves the same distance as the tool slide when adjustments are made. When selecting a boring bar, always pick the largest-diameter bar with the shortest shank that will fit the hole to be bored. This ensures maximum rigidity of the setup.

When using any boring head, it is important to determine the amount that tool slide advances when the micrometer adjusting screw is rotated one graduation. On some boring heads, the tool slide advances the same amount as on the micrometer dial. This will increase the diameter of the bore by twice the movement on the dial. Other boring heads are direct reading. The tool slide advance will increase the bore diameter by the same amount indicated on the dial. Movement of the tool slide is in thousandths of an inch, and the ratio of tool slide advance to micrometer graduations will usually be indicated on the boring head.

Feeds and Speeds in Boring

The tool material and workpiece material determine the cutting speed that should be used, but the rigidity of the machine spindle and the setup often require a lower-than-calculated rpm because the imbalance of the offset boring head creates heavy machine vibrations.

The quill feed on many vertical milling machines is limited to .0015, .003, and .006 inch of feed per spindle revolution. Roughing cuts should be made at the higher figure, and finishing cuts should be made at the two lower values. Roughing cuts are usually made with the tool feeding down into the hole. Finishing cuts are made with the tool feeding down, and often the tool is fed back up through the hole by changing the feed direction at the bottom of the hole. Because of the tool deflection, a light cut will be made without resetting the tool on that second cut. When cuts are made with the tool feeding only down but not up, the spindle rotation is stopped before the tool is withdrawn from the hole. If the spindle rotates while the quill is raised, a helical groove will be cut into the wall of the just-completed hole, possibly spoiling it.

Controlling the Bore Diameter

To obtain a predictable change in hole size for a given adjustment, certain conditions must be met. The depth of cut needs to be the same around the circumference, not like the varying

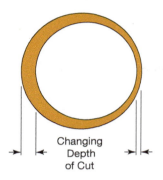

Figure J-99 When the hole is eccentric to the spindle centerline, it will cause a variable depth of cut for the boring tool.

Figure J-100 A radius is machined on a workpiece with the offset boring head.

depth of cut illustrated in Figure J-99. Roughing cuts should be taken until the hole is round and concentric with the spindle centerline. The depth of cut needs to be equal for successive cuts. As an example, assume that a hole has been rough bored to be concentric with the spindle axis. The tool is now resharpened and fastened in the tool slide. The tool slide is advanced until the tool just touches the wall of the hole. After the tool is raised above the work, the tool is moved 20 graduations, or a distance that should increase the hole diameter by .020 inch. The spindle is turned on, and with a feed of .003 inch per revolution, the cut is made through the hole. The spindle is stopped and the tool is withdrawn from the hole.

Measuring the hole shows the diameter to have increased by only .015 inch. What happened is that the tool was deflected by the cutting pressure to produce a hole .005 inch smaller than expected. The tool is now advanced to increase the hole diameter by .020. The hole is bored again.

The hole is measured and it is .020 larger. Additional cuts made with the same depth of cut and the same feed will give additional .020-inch-diameter increases. If the depth of cut is increased, more tool deflection will take place and may result in a smaller-than-expected diameter increase. If the depth of cut is decreased, the tool will be deflected less, which may result in a larger-than-expected diameter.

When the same depth of cut is maintained but the feed per revolution is increased, higher cutting pressures will produce more tool deflection and a smaller-than-expected hole diameter. With an equal depth of cut and a smaller feed per revolution, less cutting pressure may produce a larger-than-expected hole diameter. Another factor that affects the hole diameter with a given depth of cut is tool wear. As a tool cuts, it becomes dull. A dull tool will produce higher cutting pressures with a resultant larger tool deflection.

Machining Radii and Facing with the Boring Head

An offset boring head can also be used to machine a precise radius on a workpiece (Figure J-100). The workpiece is positioned the specified distance from the spindle axis. A scrap piece of the same kind of material is clamped to the table opposite the workpiece. As cuts are made with the offset boring head, the tool cuts on the workpiece and the scrap piece. The radius can be measured by the diameter dimension between the pieces being machined.

SELF-TEST

1. What is an offset boring head used for?
2. Why is a workpiece placed on parallels?
3. Why are the locking screws snugged after tool slide adjustments have been made?
4. Why does the tool slide have multiple holes to hold boring tools?
5. Why is it important to determine the amount of tool movement for each graduation on the adjustment screw?
6. What is the most important consideration when selecting a boring tool to use on a job?
7. How important is the alignment of the tool's cutting edge with the axis of the tool slide?
8. What factors affect the size of the hole obtained for a given amount of tool adjustment?
9. Name three causes for boring tool deflection.
10. What determines the cutting speed for boring?

Horizontal Spindle Milling Machines

In the preceding section you studied the vertical milling machine. This section deals with the horizontal milling machine, in which the spindle is in the horizontal axis. The horizontal milling machine, like its vertical counterpart, is an extremely versatile machine capable of accomplishing a variety of machining tasks.

HORIZONTAL MILLING MACHINE SAFETY

Before operating the rapid traverse control on a milling machine, loosen the locking devices on the machine axis to be moved. Check that the handwheels or hand cranks are disengaged, or they will spin and injure anyone near them when the rapid traverse is engaged. The rapid traverse control will move more than one axis at a time. Make sure you engage only one axis feed lever at a time. Do not try to position a workpiece too close to the cutter with the rapid feed. Stop a couple inches before the desired position and move the final distance by using the handwheels or hand cranks.

Measurements should be taken on a milling machine only after the cutter has stopped rotating and after the chips have been cleared away.

Many milling machine attachments and workpieces are heavy; use a hoist to lift them on or off the table. Do not walk under a hoisted load. The hoist may release and drop the load on you. If a hoist is not available, ask for assistance.

Injuries can be caused by improper setups or the use of the wrong tools. Use the correct-size wrench when loosening or tightening nuts or bolts, preferably a box wrench or a socket wrench. An oversized wrench will round off the corners on bolts and nuts and prevent sufficient tightening or loosening. A slipping wrench can cause smashed fingers, severe cuts, or other injuries to the hands or arms.

Milling machine cutters have sharp cutting edges. Handling cutters carefully and with a cloth will prevent cuts on the hands. Check all machine guards to see that they are in good condition and in place. Clean the working area after a job is completed. A clean machine is safer than one buried under chips.

SAFETY FIRST

To be safe an operator must be properly dressed for the job. Loose-fitting clothing is dangerous—it can catch in rotating machinery. Sleeves should be rolled up. Rings, bracelets, and watches should be removed before operating machinery. Long or loose hair can be caught in a cutter or even be wrapped around a rotating smooth shaft. Persons with long hair should tie it back or wear a cap in the machine shop. A milling machine should not be operated while wearing gloves because of the danger of their getting caught in the machine. When gloves are needed to handle sharp-edged materials, the machine has to be stopped. Eye protection should be worn at all times in the machine shop. Eye injuries can be caused by flying chips, tool breakage, or cutting-fluid sprays. Keep your fingers away from the moving parts of the milling machine such as the cutter, gears, spindle, or arbor. Never reach over a rotating spindle or arbor.

Safe operation of a machine tool requires that you think before you do something. Before starting up a machine, know the location and operation of its controls. Operate all controls on the machine yourself. Do not have another person start or stop the machine for you. Chances are good that they may turn a control at the wrong time. While operating a milling machine, observe the cutting action at all times so that you can stop the machine immediately if you see or hear something out of the ordinary. Always stay within reach of the controls while the machine is running. An unexpected emergency may require quick action on your part. Never leave a running machine unattended.

Horizontal Spindle Milling Machines

Before operating a horizontal mill, you should know the names of the machine parts, controls, and their functions. You must also know how to perform the routine maintenance on the machine required to preserve its accuracy and provide ease of operation. The purpose of this unit is to identify the parts, controls, and control functions of the horizontal mill and to describe routine maintenance procedures.

OBJECTIVES

After completing this unit, you should be able to:

- Identify the important components and controls on the horizontal milling machine.
- Describe the functions of machine parts and controls.
- Perform routine maintenance on the machine.

HORIZONTAL MILLING MACHINE

Identifying Major Parts, Controls, and Their Functions

The major assemblies of the horizontal mill are the base and column, knee, saddle, table, spindle, and overarm (Figure K-1).

Base and Column   The base along with the column forms the one-piece major structural component of the machine tool. A dovetail slide is machined on the vertical face of the column, providing an accurate guide for the vertical travel of the knee. A dovetail slide is also machined on the top of the column, providing a guide for the overarm. The column also contains the machine spindle, main drive motor, spindle speed selector mechanism, and spindle rotation direction selector.

Knee   The knee engages the slide on the face of the column and is moved vertically by turning the vertical hand feed crank. A slide on the top of the knee provides a guide for the saddle.

Saddle   The saddle engages the slide on the top of the knee and is moved horizontally toward or away from the face of the column by turning the cross feed handwheel. The saddle supports the table.

Table   The table engages the slide on the top of the saddle and can be moved horizontally right and left by turning the table handwheel. The table is equipped with T-slots for direct mounting of the workpiece, vise, or other fixture.

Spindle   The machine spindle is located in the upper part of the column and is used to hold, align, and drive cutters, chucks, and arbors. The front end or spindle nose has a tapered socket with a standard milling machine taper. This taper aligns the milling machine adapter or cutter arbor. Driving force is provided by two keys located on the spindle nose. These engage slots on the adapter or arbor. Arbors and adapters are held in place by means of a drawbolt extending through the hollow center of the spindle to the rear of the machine. As in the vertical mill, the drawbolt is threaded on one end and is designed to screw into the thread in the end of the arbor or adapter shank. Tightening the drawbolt lock nut draws the taper shank of the arbor into the spindle taper.

Overarm and Arbor Support   The overarm engages the slide on the top of the column and may be moved in and out by loosening the overarm clamps and sliding this part to

Figure K-1 Horizontal milling machine.

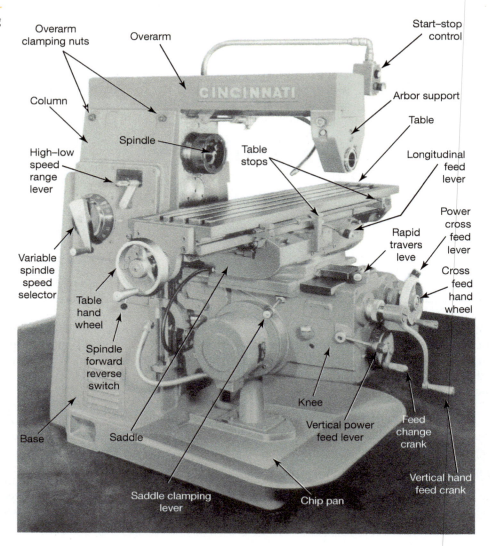

the desired position. The arbor support engages the dovetail on the overarm. The arbor support contains a bearing that is exactly in line with the spindle of the mill. The arbor support provides a rigid bearing support for the outer end of the mill arbor.

Machine Controls

Most horizontal milling machines are equipped with power feeds for the table, saddle, and knee. Horizontal milling machines are also equipped with a rapid traverse feature that permits rapid positioning of a workpiece without the need to turn table, saddle, and knee cranks by hand.

Controls for Manual Movements   Cranks and crank handwheels are provided to elevate the knee and move

table and saddle. All these controls are equipped with a micrometer collar graduated in .001-in. increments.

Feed Rate Selector and Feed Engage Controls

The feed rate selector is located on the knee (Figure K-2) and is used to select the power feed rate for table, saddle, and knee in inches per minute (ipm). Power feeds are engaged by individual controls on the table, saddle, and knee. However, engaging all these at the same time will cause them to move at the same time. On many mills, power feeds will not function unless the spindle is turning. Two safety stops at each axis travel limit prevent accidental damage to the feed mechanism by providing automatic kickout of the power feed. Adjustable power feed trip dogs are also

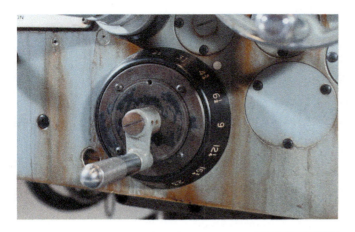

Figure K-2 Feed change crank.

Figure K-3 Rapid traverse lever.

provided so that you may preset the point at which the feed is to be disengaged.

Using the Rapid Traverse

To expedite the positioning of the knee, saddle, and table to rapidly move the workpiece up to the cutter or clear of the overarm, a rapid traverse feature is provided. When the rapid traverse control is engaged (Figure K-3), it overrides the feed rate selector and rapidly moves the table, saddle, or knee, depending on which feed control is engaged. The direction of rapid traverse is in the same direction as that of the feed. Furthermore, the rapid traverse will rapidly move the knee, saddle, and table all at the same time if they happen to have their feed controls engaged. Be careful when using the rapid traverse function not to run the work or table into the

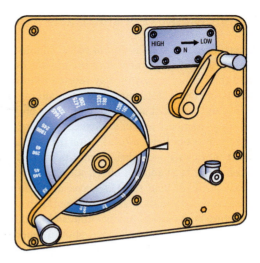

Figure K-4 Speed change levers.

cutter or overarm. This can damage the work, cutter, and machine and can cause an injury.

Spindle Controls

Spindle controls include the main motor switch, clutch, spindle speed, and speed range controls. The motor switch will usually reverse the motor and spindle direction electrically. On some mills, spindle rotation is changed mechanically. A clutch is sometimes used to connect spindle and motor.

Spindle speeds are selected from the control on the side of the column (Figure K-4). Several spindle speeds are available in both a low- and a high-speed range. Variable-speed controls are also used. The speed range selector is adjacent to the speed selector. This control has a neutral position between high- and low-speed settings. In the neutral position, the spindle may be turned by hand during machine setups. Spindle speed and direction must be selected and set while the motor and spindle are stopped. Shifting gears while the spindle is turning can damage the drive mechanism. However, on variable-speed drives, speeds must be set while the spindle is in motion.

Locks

Locks are provided on the table, saddle, and knee. These permit the components to be locked in place during a machining operation to increase the rigidity of the setup. All locks must be released before moving any part either by hand or under power. During machining, locks should be snugged, except the one on the moving axis. Locks should not be used to compensate for wear in the machine slides.

ROUTINE MAINTENANCE ON MILLS

Before any machine tool is operated, it should be lubricated. A good starting point is to wipe all sliding surfaces clean and then apply a coat of way lubricant to them. Way lubricant is specially formulated oil for sliding surfaces. Dirt, chips, and dust will act like an abrasive compound between sliding members and cause excessive machine wear. Most machine tools have a lubrication chart that outlines the correct lubricants and lubrication procedures. When no lubrication chart is available, check all oil sight gages for the correct oil level, and refill if necessary. Too much oil causes leakage. Lubrication should be performed progressively, starting at the top of the machine and working down. Machine points that are hand oiled should receive only a small amount of oil at any one time, but this should be repeated at regular intervals.

SELF-TEST

1. On a horizontal milling machine, locate the following parts:

Overarm	Switch for coolant pump
Column	Knee
Saddle clamping lever	Table
Speed change lever	Feed change lever
Power feed levers for longitudinal feed, cross feed, and vertical feed	Spindle nose
	Saddle
Rapid traverse lever	Knee clamping lever
Switch for spindle on–off	Spindle forward–reverse switch
Arbor support	Trip dogs for all three axes

2. Lubricate a horizontal milling machine.

Types of Spindles, Arbors, and Adapters

Several different devices hold and drive cutters on the horizontal mill. As a machinist, part of your job is to know what these are and to select the one that best fits your needs. The purpose of this unit is to identify and describe types of spindles, arbors, and adapters used on the horizontal mill.

OBJECTIVE

After completing this unit, you should be able to:

■ Identify machine spindles and set up different cutting tool mounting systems used to drive milling cutters.

MILL SPINDLE TAPERS

The milling machine spindle provides the driving force for the milling cutter. A milling cutter may be attached directly to the spindle nose (Figure K-5).

Most modern milling machines have self-releasing tapers in their spindle sockets. These permit quick and easy installation and removal of tooling. The standard national

milling machine taper is $3\frac{1}{2}$ inch per foot, which is an included angle of approximately $16\frac{1}{2}$ degrees. Because the standard milling machine taper is a locating or aligning taper only, tooling must be held in the spindle by a draw-in bolt or drawbolt extending through the center of the spindle. National milling machine tapers are available in four different sizes and are identified by the numbers 30, 40, 50, and 60. Positive drive of tapered shank tooling is provided through two keys on the spindle nose that engage keyways in the flange on the arbor or adapter.

TYPES OF MILLING MACHINE ARBORS

Two common arbor styles are shown in Figure K-6. A style A arbor has a cylindrical pilot on the end opposite the shank. The pilot is used to support the free end of the arbor. Style A arbors are used mostly on small milling machines, but they are also used on larger machines when a style B arbor support cannot be used because of a small-diameter cutter or interference between the arbor support and the workpiece. Style B arbors are supported by one or more bearing collars and arbor supports. Style B arbors are used to produce rigid setups in heavy-duty milling operations.

Figure K-5 Mounting a face mill on the spindle nose of a milling machine (*Courtesy of MAG Industrial Automation Systems, LLC*).

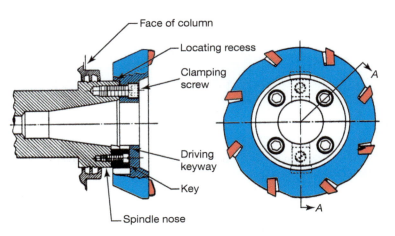

Face of column
Locating recess
Clamping screw
Driving keyway
Key
Spindle nose
A
A

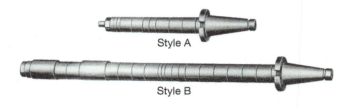

Style A

Style B

Figure K-6 Arbors, styles A and B (*Courtesy of MAG Industrial Automation Systems, LLC*).

Arbors have one or two full-length keyways machined on them. Square keys are inserted into these keyways to drive the cutter. These keys should always be long enough to extend through the cutter and both adjacent collars to provide the driving force. Without these keys, the cutter is driven only by frictional forces. Arbors are best stored in a vertical position to keep them from bending.

Spacing and Bearing Collars

Precision spacing collars take up the space between the cutter and the ends of the arbor. Shims may also be used to obtain exact cutter spacings in straddle milling operations. Bearing collars are larger in diameter than spacing collars. These collars ride in the arbor support bearing. On style A arbors, the end of the arbor rides in the arbor support bearing.

All collars are manufactured to close tolerances with their ends or faces parallel and also square to the hole. It is important that the collars and other parts fitting on the arbor be handled carefully to prevent damaging the collar faces. **Any nicks, chips, or dirt between the collar faces will misalign the cutter or deflect the arbor and cause cutter runout.**

Arbor Support Bearings

The arbor support bearing supports the outer end of the arbor (Figure K-7). The arbor bearing collar or arbor pilot fits this bearing. The arbor support bearing may be a sleeve bushing or a sealed ball bearing. On mills where a sleeve bushing is used for the arbor support bearing, the fit of the arbor bearing collar or pilot is important.

A provision for adjusting the fit of the arbor support bushing to the arbor collar or pilot is provided. A loose fit will cause inaccuracies in cuts or permit chatter. A fit that is too tight will cause frictional heating and can damage the arbor collar, pilot, or the arbor support bushing. An arbor turning at high revolution per minute (rpm) will require more clearance than at slow rpm.

Arbor support bushings must be lubricated properly. Some mills have an oil reservoir in the arbor support for supplying oil to the arbor bushing. Check the oil level and method of lubrication before operating the machine, and consult with your instructor regarding the adjustment of the bushing to fit the arbor you are using.

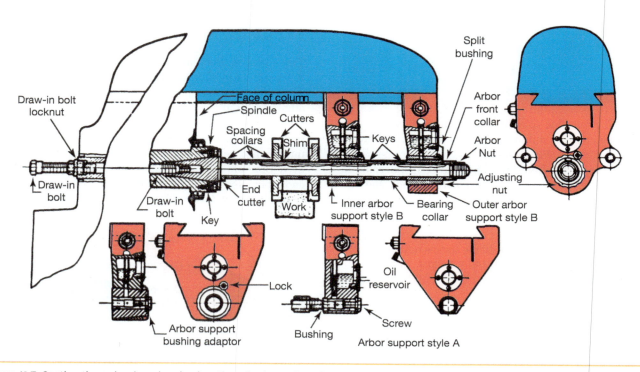

Figure K-7 Section through arbor showing location of arbor collars, keys, bearing collars, and various arbor supports (*Courtesy of MAG Industrial Automation Systems, LLC*).

SELF-TEST

1. What type of cutters are mounted directly on the spindle nose?
2. What makes a milling machine taper self-releasing?
3. What is the amount of taper on a national milling machine taper?
4. When is a style A arbor used?
5. When is the style B arbor used?
6. Is the friction between the taper shank tooling and the spindle nose taper sufficient to drive the cutter?
7. Describe the function and care of spacing and bearing collars.
8. What is the result of dirty or nicked spacing collars?
9. What is important in the use of arbor support bearings?
10. How are the arbor support bushings oiled?

Arbor-Driven Milling Cutters

Much of the versatility of the horizontal mill is due to the many types of milling cutters that are available. The purpose of this unit is to describe many of these common milling cutters and to aid you in selecting the one that best fits the needs of a specific machining task.

OBJECTIVE

After completing this unit, you should be able to:

- Identify common milling cutters, list their names, and select a suitable cutter for a given machining task.

CLASSIFYING MILLING CUTTERS

Milling cutters are designed to perform specific machining operations. A machinist should be able to identify common types by sight and know their capabilities and limitations. Many milling cutters for a horizontal mill are made from high-speed steel (HSS). Special purpose form cutters like gear cutting tools are made of HSS. Carbide tools are very prevalent, especially for shank driven tooling. Solid carbide and carbide insert cutters have much higher metal removal rates.

Arbor-Driven or Shank Types

Milling cutters for the horizontal mill are designed to be driven by the mill arbor, or they may have their own tapered shanks for direct mounting in the machine spindle socket. Although in many instances you will use a tapered shank mounted cutter in the horizontal mill, probably the most common cutters you will be using are arbor-mounted types.

PLAIN ARBOR-DRIVEN CUTTERS

Plain milling cutters are designed for milling plain surfaces where the width of the work is narrower than the cutter (Figure K-8). Plain milling cutters less than $\frac{3}{4}$-inch wide have

Figure K-8 Light-duty plain milling cutter (*Used with permission of Illinois Tool Works, Inc.*).

Figure K-9 Heavy-duty plain milling cutter (*Used with permission of Illinois Tool Works, Inc.*).

straight teeth. On straight-tooth cutters, the cutting edge will cut along its entire width at the same time. Cutting pressure increases until the chip is completed. At that time the sudden change in tooth load causes a shock that is transmitted through the drive and often leaves chatter marks or an unsatisfactory surface finish. Light-duty milling cutters have many teeth, which limit their use to light or finishing cuts because of insufficient chip space. Heavy-duty plain mills (Figure K-9) have fewer teeth, which make for strong teeth with ample chip clearance. The helix angle of heavy-duty mills is about 45 degrees. The helical form enables each tooth to take a cut gradually, which reduces shock and lowers the tendency to chatter. Plain milling cutters are also called slab mills. Plain milling cutters do not have side cutting teeth and should not be used to mill shoulders or steps on workpieces.

SIDE MILLING CUTTERS

Side milling cutters are used to machine steps or grooves. These cutters are available in various sizes. Figure K-10 shows a straight-tooth side milling cutter. To cut deep slots or grooves, a staggered-tooth side milling cutter (Figure K-11) is preferred because the alternate right-hand and left-hand helical teeth reduce chatter and give more chip space or higher feeds than possible with straight-tooth side milling cutters. To cut slots over 1 inch wide, two or more side milling cutters may be mounted on the arbor simultaneously. Shims between the hubs of the side mills can produce any precise width cut.

Inserted-tooth carbide cutters (Figure K-12) are used to cut slots and grooves. Because of the inserted carbide teeth, these cutters can be set to cut variable-width slots. The common range of width adjustments on one cutter is .060 inch. One big advantage of this type of cutter is that the inserts can quickly be changed when dull. The great number of carbide grades available allows the correct grade of carbide to be selected for the material and the machining conditions.

Figure K-11 Staggered-tooth milling cutter (*Used with permission of Illinois Tool Works, Inc.*).

Half-side milling cutters are designed for heavy-duty milling where only one side of the cutter is used (Figure K-13). For straddle milling, a right-hand and a left-hand cutter combination is used. Plain metal slitting saws are designed for slotting and cutoff operations (Figure K-14). Their sides are slightly relieved or dished to prevent binding in a slot. Their use is limited to a relatively shallow depth of cut. These saws are made in widths from $\frac{1}{32}$ to $\frac{5}{6}$ inch. To cut deep slots or when many teeth are in contact with the work, a side-tooth

Figure K-10 Side milling cutter.

Figure K-12 Inserted-tooth carbide cutter (*Used with permission of Kennametal, Inc.*).

Figure K-13 Half side milling cutter (*Used with permission of Illinois Tool Works, Inc.*).

Figure K-15 Side-tooth metal slitting saw (Used with permission of *Illinois Tool Works, Inc.*).

Figure K-14 Plain metal slitting saw (*Used with permission of Illinois Tool Works, Inc.*).

Figure K-16 Staggered-tooth metal slitting saw (*Used with permission of Illinois Tool Works, Inc.*).

metal slitting saw will perform better than a plain metal slitting saw (Figure K-15). These saws are made from $\frac{1}{16}$- to $\frac{3}{16}$-inch wide. Extra-deep cuts can be made with a staggered-tooth metal slitting saw (Figure K-16). Staggered-tooth saws have greater chip-carrying capacity than other saw types. All metal slitting saws have a slight clearance ground on the sides toward the hole to prevent binding in the slot and scoring of

the walls of the slot. Staggered-tooth saws are made from $\frac{3}{16}$- to $\frac{5}{16}$-inch wide.

Indexable tooth slotting cutters are an excellent choice of cutting tool because of the great number of tool grades available (Figure K-17). Angular milling cutters are used for angular milling such as cutting of dovetails, V-notches, and serrations. Single-angle cutters (Figure K-18) form an included angle of 45 to 60 degrees, with one side of the angle at 90 degrees to the axis of the cutter. Double-angle milling cutters (Figure K-19) usually have an included angle of 45, 60, or 90 degrees. Angles other than those mentioned

Figure K-17 Indexable tooth slotting cutters provide for efficient machining (*Used with permission of Lovejoy Tool Company, Inc.*).

are special milling cutters. Convex milling cutters (Figure K-20) produce concave bottom grooves, or they can be used to make a radius in an inside corner. Concave milling cutters (Figure K-21) make convex surfaces. Corner rounding milling cutters (Figure K-22) make rounded corners. Involute gear cutters (Figure K-23) are commonly available in a set of eight cutters for a given pitch, depending on the number of teeth for which the cutter is to be used. These eight cutters are designed so that their forms are correct for the lowest number of teeth in each range. An accurate tooth form near the upper end of a range requires a special cutter. The ranges for the individual cutters are shown in Table K-1.

Figure K-20 Convex milling cutter (*Used with permission of Illinois Tool Works, Inc.*).

Figure K-18 Single-angle milling cutter (*Used with permission of Illinois Tool Works, Inc.*).

Figure K-19 Double-angle milling cutter (*Used with permission of Illinois Tool Works, Inc.*).

Figure K-21 Concave milling cutter (*Used with permission of Illinois Tool Works, Inc.*).

Figure K-22 Corner rounding milling cutter (*Used with permission of Illinois Tool Works, Inc.*).

Figure K-23 Involute gear cutter.

Table K-1 Involute Gear Cutters

Cutter Number	Range of Teeth
1	135 to rack
2	55 to 134
3	35 to 54
4	26 to 34
5	21 to 25
6	17 to 20
7	14 to 16
8	12 and 13

SELF-TEST

1. What are the two basic kinds of milling cutters based on their tooth shape?
2. What is the difference between a light- and a heavy-duty plain milling cutter?
3. Why are plain milling cutters not used to mill steps or grooves?
4. What kind of cutter is used to mill grooves?
5. How does the cutting action of a straight-tooth side milling cutter differ from that of a staggered-tooth side milling cutter?
6. Give an example of an application of half side milling cutters.
7. What are metal slitting saws used for?
8. Give two examples of form-relieved milling cutters.
9. What are angular milling cutters used for?
10. When facing the spindle, in which direction should a right-hand cutter be rotated to cut?

Work-Holding Methods and Standard Setups

Before any machining can be done on the horizontal mill, the workpiece or work-holding fixture must be secured and aligned on the machine table. A specific setup will depend on the particular workpiece and the machining task to be done. Considerable ingenuity on your part is required to make a safe and secure setup that will not distort or damage the workpiece. Milling machine setups are almost infinite in number, and it is not possible to discuss them all. The purpose of this unit is to introduce you to the equipment and techniques for common setups on this machine.

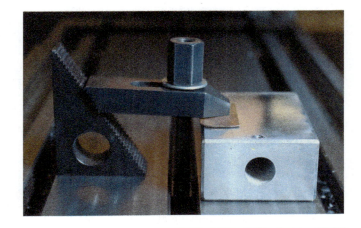

Figure K-24 Highly finished surfaces should be protected from clamping damage.

OBJECTIVES

After completing this unit, you should be able to:

- Select a work-holding method and device for common milling tasks.
- Safely set up a workpiece on the machine.

DIRECT MOUNTING TO THE TABLE

A large variety of milling clamps are available for direct table mounting of the workpiece. Always follow the rules of good clamping. Fully screw studs or bolts into their T-nuts. Clamp nut threads should be fully engaged on studs. Locate studs or clamp bolts as close to the workpiece as possible and arrange clamp support blocks so that clamping pressure is applied to the workpiece. Locate clamps on both sides of the workpiece if possible. This will ensure maximum safety. **Clamp supports must be the same height as the workpiece. Never use clamp supports that are lower than the workpiece.** Adjustable step blocks are extremely useful for this, as the height of the clamp bar may be adjusted to ensure maximum clamping pressure.

To protect a soft or finished surface from damage by clamping pressure, place a shim between the clamp and work (Figure K-24). If you are machining a rough casting or weldment (Figure K-25), protect the machine table from

Figure K-25 Protect the machine table surface from rough workpieces.

damage by using parallels or a shim under the workpiece. Brass and metal are shim materials.

Workpieces can easily be distorted, broken, or otherwise damaged by excessive or improper clamping. One solution to this problem is to use a screw jack to support the workpiece (Figure K-26). The jack should be placed directly under the clamp. In lieu of a screw jack, solid blocks of the correct size may be used as both workpiece supports and clamp supports.

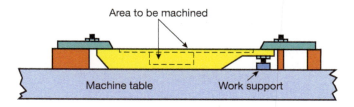

Figure K-26 Workpiece supported under the clamp.

In machining operations where heavy cuts are involved, top clamps alone may not be sufficient to restrain the workpiece. A stop block clamped to the machine table will offset the cutting pressures and prevent slippage of the part being machined.

Toe clamps (Figure K-27) may be used to secure workpieces directly to the machine table. Because the workpiece is clamped from the sides, the complete top surface is accessible for machining. One advantage of these clamps is that they exert pressure in both downward and horizontal directions. The clamping action is independent of the clamp mounting, permitting the workpiece to be clamped or released without actually moving the clamp itself.

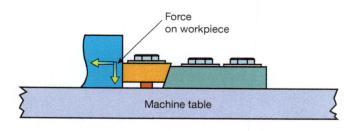

Figure K-27 Toe clamps hold workpiece securely to the machine table.

MILL VISES

The most common method of work holding on a milling machine is a vise. Vises are simple to use and can be adjusted quickly to the size of the workpiece. A vise should be used to hold work with parallel sides if it is within the size limits of the vise. This is the quickest and most economical work-holding method. The plain vise is bolted to the machine table. Alignment with the table is provided by two slots at right angles to each other on the underside of the vise. These slots are fitted with removable keys that align the vise with the table T-slots either lengthwise or crosswise. A swivel-vise allows the vise to be rotated so that angles can be cut on workpieces.

The strongest setup is one in which the workpiece is clamped close to the table surface.

Air- or hydraulically-operated vises are often used in high-volume production work, but in general, work vises are opened and closed by cranks or levers.

SELF-TEST

1. In relation to the workpiece, where should the clamping bolt be located?
2. What precautions should be taken when clamping on a finished surface?
3. What are screw jacks used for?
4. What is the reason to clamp a stop block to the table?
5. Why are toe clamps used?
6. What is a swivel vise used for?
7. Why clamp a workpiece directly to a table?

INTERNET REFERENCE

Information on work holding:

www.kurtworkholding.com/news/tips.http

Machine Setup and Plain Milling

Plain milling is the operation of milling a flat surface parallel to the cutter axis and machine table. Milling on a horizontal milling machine involves setting up the machine, selecting an appropriate arbor and cutter, and calculating correct speeds, feeds, and depth of cuts. The purpose of this unit is to describe the procedure for preparing the machine tool and plain milling processes.

OBJECTIVES

After completing this unit, you should be able to:

- Set up the mill for plain milling.
- Select and set up a work-holding system.
- Select and set up an appropriate cutter and arbor.

Climb and Conventional Milling

When milling, the direction that the workpiece is being fed can either be in the same direction as the cutter rotation or opposed to the direction of cutter rotation (Figure K-28). When the direction of feed is opposed to the rotation direction of the cutter, this is said to be *up* or *conventional milling*. When the direction of feed is the same as the rotation direction, this is said to be *down* or *climb milling* because the cutter is attempting to climb onto the workpiece. If excessive backlash exists in the table or saddle, the workpiece may be pulled into the cutter during climb milling. This can result in a bent arbor, broken cutter, damaged workpiece, and possible injury to the operator. Climb milling should be avoided unless the mill is equipped with adequate backlash control. Remember that any cutter can be operated in an up-milling or down-milling mode depending only on which side of the workpiece the cut is started.

Selecting and Setting Up Mill Arbors

When selecting a mill arbor, use one that has minimum overhang beyond the outer arbor support. Excessive overhang can cause vibration and chatter. After selecting the proper arbor, insert the tapered shank into the spindle socket. Be sure that the socket is clean and free from burrs or nicks (Figure K-29). Large mill arbors are heavy, and you may need help holding them in place until the drawbolt is engaged. Do not let the arbor fall out onto the machine table. Thread the drawbolt into the arbor shank all the way and then draw the arbor into

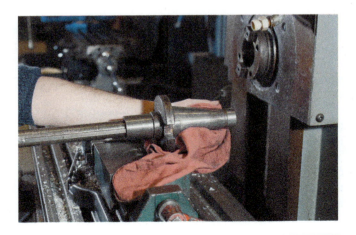

Figure K-29 Cleaning the external and internal taper prior to installation.

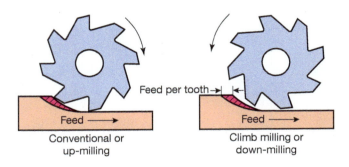

Feed per tooth

Conventional or up-milling

Climb milling or down-milling

Figure K-28 Conventional and climb milling.

Figure K-30 Tightening the drawbolt locknut with a wrench.

Figure K-31 Tightening the arbor nut.

the spindle taper by turning the drawbolt locknut. Tighten the locknut with a wrench (Figure K-30).

Remove the arbor nut and spacing collars. Place these on a clean surface so that their precision surfaces are not damaged. Position the cutter on the arbor as close to the spindle as the machining task will permit. Place a sufficient number of spacing collars on either side of the cutter to position it correctly. The cutter, spacing collars, and bearing collar should be a smooth sliding fit on the arbor.

A key is generally used to ensure a positive drive between cutter and arbor. Consult with your instructor and follow instructions regarding the use of keys.

Place the bearing collar on the arbor and locate it as close to the cutter as the machining task will permit. Place the arbor support on the overarm and slide it in until the arbor bearing collar slips through the arbor support bearing. Tighten overarm and support clamps. Tighten the arbor nut only after the support is in place (Figure K-31). Tightening the arbor nut before the support is in place may bend the arbor. Do not overtighten the arbor nut.

Removing and Storing Mill Arbors

Exercise care in removing the arbor from the spindle. Loosen the drawbolt locknut about one turn (Figure K-32). You may have to tap the drawbolt lightly to release the arbor

Figure K-32 Loosening the locknut on the drawbolt.

Figure K-33 Tapping the end of the drawbolt with a lead hammer.

Figure K-34 Have another person hold heavy arbors when you are loosening them.

taper shank from the spindle socket (Figure K-33). Hold the arbor in place or get help while you unscrew the drawbolt from the arbor shank (Figure K-34). Remove the arbor from the machine and store it in an upright position.

PLAIN MILLING

Good milling practice is to take roughing cuts and then a finish cut. Better surface finish and higher dimensional accuracy are achieved when roughing and finishing cuts are made. The machining time of a workpiece is dependent to a large extent on how efficiently material is removed during the roughing operation. The depth of the roughing cut is often limited by the horsepower of the machine or the rigidity of the setup. A good starting depth for roughing is .100 to .200 inch.

The finishing cut should be .015 to .030 inch deep. Cuts less than .015 inch deep should be avoided because a milling cutter, especially in conventional or up milling, has a strong rubbing action before the cutter actually starts cutting. This rubbing action causes a cutter to dull rapidly.

Making a Typical Roughing Cuts

To take a roughing cut .100 inch deep, use the following procedure.

Step 1 Loosen the knee locking clamp and the cross-slide lock.

Step 2 Turn on the spindle and check its rotation.

Step 3 Position the table so that the workpiece is under the cutter.

Step 4 You may use a paper strip to determine when the cutter is about .002 inch away from the workpiece (Figure K-35). **This should be done with the spindle off and turned by hand.**

Step 5 Set the micrometer dial on the knee crank to zero.

Step 6 Lower the knee approximately half a revolution of the hand feed crank. If the knee is not lowered, the cutter will leave tool marks on the workpiece in the following operation.

Step 7 Move the table right or left until the cutter is clear of the workpiece. Start the cut on whichever side of the part results in up or conventional milling.

Step 8 Raise the knee past the zero mark to .100-inch depth.

Step 9 Tighten the knee lock and the cross-slide lock. Prior to starting a machining operation, snug all locking clamps except the one that would restrict table movement while cutting. This helps prevent chatter.

Step 10 The machine is now ready for the cut. Turn on the coolant and engage the power feed.

Step 11 When the cut is completed, disengage the power feed, stop the spindle rotation, and turn off the coolant before returning the table to its starting position. If the revolving cutter is returned over the newly machined surface, it will leave cutter marks and mar the finish.

Step 12 After brushing the chips off and wiping the workpiece clean, measure the part while it still fastened in the machine. If the workpiece is parallel, additional cuts can be made until the desired size is reached.

To square the end of a workpiece, tighten the part in the vise lightly and tap the part with a soft hammer to bring it into a square position. You can check it with a precision square if the tolerance is not too close. A dial indicator can also be used to position the workpiece in a perpendicular position. Attach the indicator to the overarm or arbor with the tip in contact with the workpiece. Crank the knee up and down and adjust the workpiece until a zero-indicator reading is obtained. If a workpiece needs to be fastened in a vise off center (Figure K-36), it is important that the opposite end of the vise jaw be supported with another piece of material the same thickness as the workpiece. Without this supporting spacer the vise cannot securely clamp the workpiece.

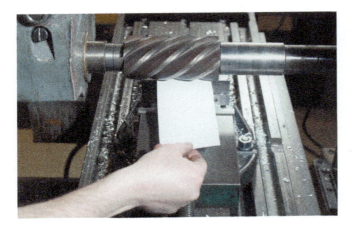

Figure K-35 Using a paper strip to position the cutter on the top of the workpiece.

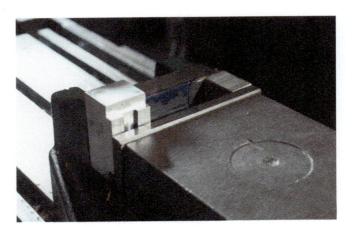

Figure K-36 Work clamped off center needs a spacer.

SELF-TEST

1. Describe climb versus conventional milling advantages and disadvantages.
2. What is one of the main considerations when choosing a milling arbor?
3. How are arbors secured in the spindle?
4. Why should the cutter and overarm be located as close as possible to the spindle?
5. Why are all table axis locks snugged except the one being used during machining?

INTERNET REFERENCE

Cutting fluids and coolant selection and application:

http://www.substech.com/dokuwiki/doku.php?id=cutting_fluids_coolants

Using Side Milling Cutters

Machining a flat surface is only one of many machining operations performed on the horizontal mill. Steps and grooves, straddle milling, and gang milling all involve the use of side milling cutters. The purpose of this unit is to introduce you to these useful milling cutters and their applications.

OBJECTIVES

After completing this unit, you should be able to:

- Set up side milling cutters and cut steps and grooves.
- Use side milling cutters for straddle milling.
- Use side milling cutters in gang milling.

SIDE MILLING CUTTERS AND SIDE MILLING

Milling cutters with side teeth are called **side milling cutters**. These are used to machine steps and grooves or, when only the sides of the workpiece are to be machined, in straddle setups. An example is cutting hexes on bolt heads. Grooves are best machined in the workpiece with full side mills that have cutting teeth on both sides (Figure K-37). Steps may be cut with half side mills having cutting teeth only on one side (Figure K-38).

The size, type, and diameter of side milling cutter to be used will depend on the machining task. As a rule, the smallest-diameter cutter that will do the job should be used as long as sufficient clearance is maintained between the arbor support and work or vise.

Preparing and Setting Up the Workpiece

A machinist may mark the workpiece with layout lines before securing it in the machine. The layout should be an exact outline of the final part shape and size. The reason for

Figure K-37 Full-side milling cutter machining a groove.

making the layout prior to beginning work is that reference surfaces are often removed during machining. After the layout has been made, diagonal lines should be chalked on the workpiece indicating the portions to be cut away. This helps identify on which side of the layout lines the cut is to be made (Figure K-39). The workpiece may be mounted by any

Figure K-38 Half-side milling cutter machining a step.

of the traditional methods discussed previously. If you are using a vise, be sure it is aligned along with the table. If you are clamping directly to the table, follow the rules of good clamping.

Before you use a milling cutter, inspect the cutting edge for sharpness. If the cutting edge of the tool shows any sign of wear or dullness, it should be sharpened. Milling cutters need to be sharpened on a tool and cutter grinder so that every cutting edge is exactly in the same cylindrical plane

as every other one. If one cutting edge extends beyond any other, it will produce a wavy or scalloped surface on the machined workpiece. Dull milling cutters create excessive cutting pressures that may distort the workpiece or move it from the work-holding device. Milling cutters leave a rough surface finish if they need to be sharpened.

Frequently, side milling cutters will cut a slot or groove slightly wider than the nominal width of the cutter. The causes of this can be cutter wobble due to small chips or dirt between the cutter and spacing collars or multiple passes through the workpiece. Other factors influencing the width of cut are the feed rate and/or the type of material being machined. A slow feed rate will allow the cutter to make a wider slot; this can also happen when softer materials are machined. If the drawing calls for a slot width of .375 inch, using a .375-inch-wide cutter may result in a slot .3755- to .376-inch wide. To achieve the correct slot width, a 250-inch-wide cutter may have to be used and multiple passes made.

Positioning a Side Milling Cutter

To machine a slot or groove at a particular location on the workpiece, the side mill must be positioned both horizontally (for location) and vertically (for depth of cut). To position the cutter for location, lower below the top surface of the workpiece. **With the spindle off and free to turn by hand,** insert a paper strip between cutter and work (Figure K-40) and move the workpiece toward the cutter until the paper is pulled between the work and cutter. At this point, the cutter is about .002 inch from the workpiece. Set the saddle micrometer collar to zero, compensating for the .002 inch of paper thickness.

Position the depth by lowering the knee and moving the workpiece under the cutter. Then, raise the knee and use the paper strip gage to determine when the cutter is about .002 inch above the workpiece. Set the knee micrometer collar to zero. Move the table until the cutter is clear of the workpiece, and then raise the knee to the amount required for the depth of the feature.

Figure K-39 Work laid out for milling.

Figure K-40 Setting cutter height by using a paper strip.

Figure K-41 Positioning a cutter using a steel rule.

Figure K-43 Measuring a slot with an adjustable parallel and a micrometer.

Figure K-42 Taking a trial cut.

A less accurate but quicker method of cutter alignment is direct measurement (Figure K-41). A rule may be used to measure the position of the cutter relative to the edge of the workpiece, after which the saddle micrometer collar should be set to zero.

Making the Cut

After the cutter has been positioned, raise the knee the amount required for the depth of the cut. Machining to full depth may be accomplished in one pass, depending on the width of the cut and the material being machined. A deep slot may have to be machined in more than one pass, each pass somewhat deeper than the one before. Until you gain experience in milling, hold the depth setting to about .100 inch.

Set proper feeds and speeds and turn on the spindle. Approach the workpiece in an up-milling mode. Hand feed the cutter into the work until a small nick is machined on the corner (Figure K-42, point X). Stop the spindle, back away, and check the dimension relative to the edge of the workpiece. If you are machining a slot, check the width with an adjustable parallel and a micrometer (Figure K-43), or use a dial/vernier caliper. If the dimensions are correct, complete

Figure K-44 Measuring depth of a step.

the required cuts. Precise depth can be measured with a depth micrometer (Figure K-44).

Milling Keyways

Side milling is an excellent way to machine a shaft **keyway**, especially if the keyway is long. The procedure for keyway milling is much the same as in vertical milling. The cutter must be centered over the shaft and set for depth. After

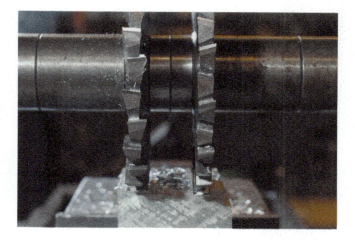

Figure K-45 Straddle milling.

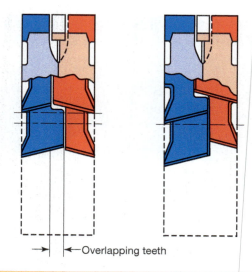

←Overlapping teeth

Figure K-46 Interlocking side milling cutters.

selecting and mounting a cutter of the proper width, raise the workpiece beside the cutter and use the paper feeler technique to position the cutter alongside the shaft. Lower the knee and move the workpiece over a distance equal to half the total of cutter width plus shaft diameter. Raise the knee until the cutter contacts the shaft and cuts a full width cut. Set the knee micrometer collar to zero. Lower the knee and move the cutter clear of the workpiece in the table axis only. Raise the knee to obtain the correct keyway depth. Using proper feeds and speeds, approach the workpiece in an up-milling mode, and mill the keyway to the proper length.

Straddle Milling

Side milling cutters are combined to perform straddle milling. In straddle milling, two side milling cutters are mounted on an arbor and set at an exact spacing (Figure K-45). Two sides of the workpiece are machined simultaneously, and final width dimensions are closely controlled. Straddle milling has many useful applications in production machining. Parallel slots of equal depth can be milled by using straddle mills of equal diameter.

Interlocking Tooth and Right- and Left-Hand Helical Side Mills

Interlocking side mills are used when grooves of a precise width are machined in one operation (Figure K-46). Shims inserted between individual cutters make precise adjustment possible. The overlapping teeth leave a smooth finish in the bottom of the groove. Cutter combinations that have become thinner from resharpenings can also be adjusted to their full width by adding shims.

SELF-TEST

1. What are full-side milling cutters used for?
2. What are half-side milling cutters used for?
3. How is the diameter of a cutter chosen?
4. Is a groove the same width as the cutter that produces it?
5. Why should a layout be made on workpieces?
6. How can a side milling cutter be positioned for a cut without marring the workpiece surface?
7. Why should measurements be made before removing a workpiece from the work-holding device?
8. How is the width of a workpiece controlled in a straddle milling operation?
9. How can you check the accuracy of a cut before the cut is completed?
10. What are interlocking side mills used for?

Grinding and Abrasive Machining Processes

Abrasive materials are the cutting tools in grinding processes. Grinding can produce very precise and accurate work. Grinding offers great challenges to machinists. Consequently, many people consider grinding more of an art than other machining practices. Grinding processes are typically finish processes. Furthermore, grinding processes can produce very smooth surface finishes on the workpiece. In this section text you will study several types of grinding machines and types of abrasives used. You will also learn to set up and operate grinding machines that are commonly found in machine shops.

Types of Grinders

TYPES OF GRINDING MACHINES

There are many types of grinding machines. Machinists should be familiar with the types of grinding machines and the operations they can be used for.

Surface Grinders

One of the most common grinding machines is the surface grinder. There are three basic types of surface grinders:

- Horizontal spindle with reciprocating table
- Horizontal spindle with rotary table
- Vertical spindle with either reciprocating or rotary table

The first type of surface grinder has the grinding wheel on a horizontal spindle and the table reciprocates (Figure L-1). The face (periphery) of the wheel performs the grinding. The workpiece, if it is made of magnetic material, is usually held on a magnetic chuck and moved back and forth under the rotating grinding wheel. The table (or the wheelhead) may also be moved in and out. The depth of cut is controlled by raising and lowering the wheelhead. Figure L-2 shows a typical horizontal spindle surface grinder. A surface grinder can be expected to produce a flat surface on a workpiece

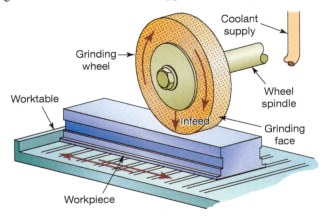

Figure L-1 Principle of the horizontal reciprocating surface grinder.

Figure L-2 Horizontal spindle reciprocating surface grinder (*Courtesy of Maximum Advantage—Carolinas*).

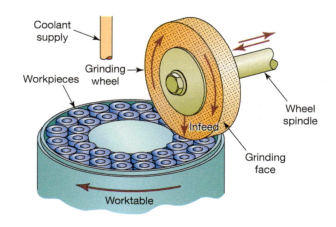

Figure L-3 Horizontal spindle, rotary table surface grinder for production work.

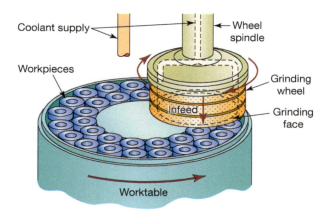

Figure L-4 Principle of the vertical spindle rotary grinder.

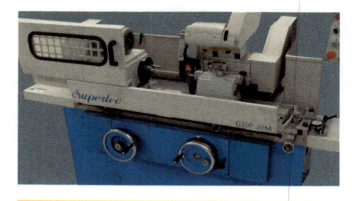

Figure L-5 Universal cylindrical grinder (*Courtesy of MAG Industrial Automation Systems, LLC*).

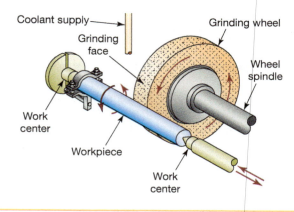

Figure L-6 Principle of the cylindrical grinding machine showing the workpiece and wheel motions.

within a tolerance of less than ±.0001 inch. The horizontal spindle reciprocating surface grinder is the most common type of surface grinder.

The second type of surface grinder has a horizontal spindle but the table is rotary (Figure L-3). These are often used for higher production of parts. Many workpieces can be ground at one time.

The third type of surface grinder has a vertical spindle, and the face of the wheel is brought into contact with the workpiece. The worktable is typically rotary (Figure L-4). The rotary table type is typically called a Blanchard grinder. Blanchard grinders are excellent for grinding large plates, flat and parallel and for grinding many parts at once.

Universal Cylindrical Grinders

Cylindrical grinders are used to grind the diameters of cylindrical workpieces. The universal cylindrical grinder is a very versatile machine (Figure L-5). A universal cylindrical grinder can be used to grind the outside or inside cylindrical surfaces of parts. The head and table can typically be rotated to an angle to grind tapers. Cylindrical tapers can be ground

internally or externally on parts. Work pieces can be held in a chuck on the headstock or between centers. Many universal cylindrical grinders also have a powered internal grinding attachment that can be rotated down to grind internal surfaces.

The workpiece is traversed past the grinding wheel while being rotated from 50 to 100 SFPM (Figure L-6). The table can be swiveled to enable tapers to be ground. This machine is also used for plunge grinding, in which the work and wheel are brought together without table traverse. Plunge grinding can be used to quickly grind a diameter on a part. The width of the area to be ground must be narrower than the wheel width in plunge grinding. If the area to be ground is wider that the wheel width, the table is traversed.

Roll Grinder   Roll grinders are used to grind and resurface large steel rolls for the paper or steel industry (Figure L-7). Because these rolls are very large, long, and heavy, they are supported in bearings (rather than on centers) while in the grinding machine. The wheelhead is traversed along the rotating workpiece to accomplish the grinding.

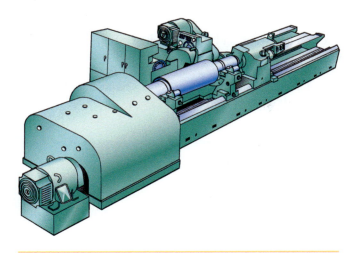

Figure L-7 Roll grinding machine for grinding steel mill rolls.

Internal Cylindrical Grinding

Universal cylindrical grinders are used to grind cylindrical parts. They are very versatile machines. The principle of a cylindrical grinder is shown in Figure L-8. Universal cylindrical grinders often have an internal grinding attachment (Figure L-9). This is a spindle that can be swiveled down to be in line with the headstock centerline. Mounted abrasive wheels are used for internal grinding. These are grinding wheels attached to a shank for mounting into the machine's grinding spindle. Internal grinding can be done on concentric workpieces

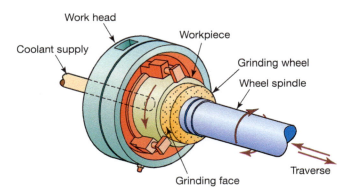

Figure L-8 Principle of internal cylindrical grinding.

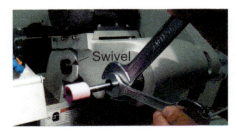

Figure L-9 Grinding attachment swiveled into position with an internal grinding wheel being tightened in the spindle.

Figure L-10 Internal cylindrical grinding.

parallel to the machine spindle axis (Figure L-10). Internal tapers may also be ground.

Miscellaneous Grinding Machines

Centerless Grinder   A centerless grinder is used to grind cylindrical workpieces without the use of centers. It can be used for short or long cylindrical workpieces, such as lengths of drill rod.

The workpiece is fed between the grinding wheel and regulating wheel (Figure L-11). The work rest located between the grinding and regulating wheel supports the workpiece. The regulating wheel is usually a rubber-bonded wheel that has the same grit characteristics as the grinding wheel. It rotates in the same direction as the grinding wheel but at a much slower speed, acting as a brake on the workpiece. The feeding action for moving the workpiece between the grinding wheel and the regulating wheel along the work rest blade is provided by a small tilt (typically 1 to 3 degrees) of the regulating wheel (Figure L-12). As shown in this figure, this feeding action may also be used with a stop for the workpiece. A common through-feed centerless grinding operation is the grinding of wrist pins for automotive pistons.

Tool and Cutter Grinder   The universal tool and cutter grinder (Figure L-13) is a type of cylindrical grinder. On these machines, the grinding head can be rotated around the vertical axis, and the workhead can be rotated and tilted

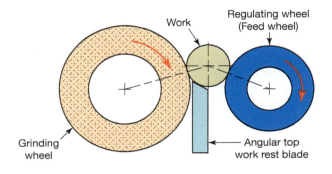

Figure L-11 Principle of the centerless grinder. The grinding wheel travels at normal speeds, and the regulating wheel travels at a slower speed to control the rate of spin of the workpiece.

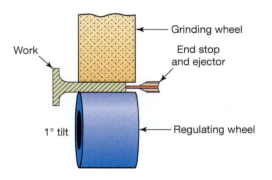

Figure L-12 End feeding in the centerless grinder. The regulating wheel is set at a small tilt angle to keep the workpiece against the end stop.

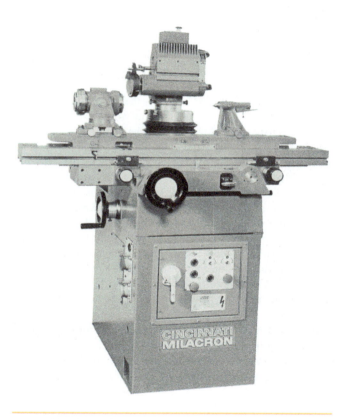

Figure L-13 Universal tool and cutter grinder (*Courtesy of MAG Industrial Automation Systems, LLC*).

in a variety of directions. The grinding spindle can also be tilted on many tool and cutter grinding machines. A variety of grinding wheel shapes and a variety of abrasives can be used with tool and cutter grinders. The primary application of tool and cutter grinders is in the re-sharpening of milling cutters and other cutting tools.

SELF-TEST

1. What are the three basic types of surface grinders?
2. A surface grinder can typically grind to a tolerance of ____?
3. What does the term universal mean when applied to a cylindrical grinder?
4. What is a centerless grinder used for?
5. Describe how internal grinding can be done on a universal cylindrical grinder.

Selection and Use of Grinding Wheels

Selecting a grinding wheel is much like selecting any other cutting tool. Many of the same factors apply, such as size and shape. However, in grinding wheels, additional factors influence the selection process: types of abrasives, grit, grade, structure, and bond, as well as the particular grinding task to be done. The purpose of this unit is to familiarize you with the identification of grinding wheels and to assist you in selecting the one that best fits your needs.

OBJECTIVES

After completing this unit, you should be able to:

- List at least three principal abrasives with their general areas of best use.
- List at least three principal bonds with the types of applications where they are most used.
- Identify four commonly used shapes of grinding wheels.
- Interpret wheel shape and size markings together with five basic symbols of a wheel specification into a description of the grinding wheel.
- Given several standard, common grinding jobs, recommend the appropriate abrasive, approximate grit size and grade, and bond.
- Describe truing, dressing, and balancing of grinding wheels.
- Distinguish the difference between truing and dressing a grinding wheel.
- Correctly position a single-point diamond dresser in relation to the grinding wheel.

TYPES OF ABRASIVES

Grinding wheels have two major components—the abrasive grains and the bond. The abrasive grains do the cutting. The bond holds the grains together to form the wheel.

The bond also has another important function. The bond releases abrasive grains as they become dull. A grinding wheel's structure is determined by the percentages of grain and bond and their spacing.

The abrasive type used in a wheel is selected based on the work material. The proper abrasive will remain sharp during the grinding operation. If the correct abrasive and bond is chosen, the wheel will self sharpen during cutting. Dull grains are released by the bond and new ones are exposed. This ability for the bond to crumble and expose new grains is called friability.

ABRASIVE GRAIN TYPES

Aluminum Oxide

Aluminum oxide is the most widely used abrasive in manufacturing. The higher the level of purity, the more friable (less tough) the grit becomes. Aluminum oxide abrasive is usually chosen for grinding carbon steel, high-speed steel, alloy steels, annealed malleable iron, bronzes, and similar metals.

There are many types of aluminum-oxide abrasives. Each type is designed for particular types of applications. The purer grades are often white in color and are used mainly on hardened steels. The grinding of hardened steel causes a fairly rapid breakdown of the individual grains to expose new sharp edges. A sharp wheel cuts easily, with less heat being produced. Less pure aluminum oxide is usually gray in appearance. It has a tougher grain structure that resists fracturing, and it is better for applications such as offhand grinding on a pedestal grinder. For extremely harsh applications, such as rough grinding large part edges (snagging) in foundries on large grinders, a crystalline combination of aluminum and zirconium oxide was developed to result in a tough abrasive. Although aluminum oxide works well on steels, it may work poorly on cast iron.

Zirconia Alumina

Zirconia alumina is a combination of aluminum oxide and zirconium oxide. This results in durable, tough abrasive that works well for rough grinding and cutoff-type applications. There are several types of zirconia alumina.

Silicon Carbide

Silicon carbide is a somewhat harder abrasive than aluminum oxide but has a sharper, more friable, and quite brittle crystalline structure. It works well on cast iron and nonferrous materials such as aluminum and copper-based alloys. It also works on carbide cutting tools, but not as well as the cooler-cutting diamond abrasives, which have largely replaced silicon carbide for this application. Silicon carbide is useful in grinding many titanium alloys as well as austenitic stainless steels.

Ceramic Aluminum Oxide

Ceramic aluminum oxide is produced by a process similar to growing crystals. This method permits engineering the crystal shape for specific purposes. This results in an abrasive wheel with that fractures at a controlled rate. This exposes thousands of new cutting points.

By combining ceramic aluminum oxide with fused aluminum oxide, a wheel can be produced that has a combination of sharpness, aggressive cutting action, form retention, and resultant long life between dressings. This results in a three- to five-fold improvement in grinding ratio over straight fused aluminum oxide grinding wheels. This abrasive is very hard and strong. It performs well for precision grinding on difficult to grind steels and alloys.

Superabrasives

Superabrasives are designed to grind the hardest materials. Carbides, hardened steels, and other materials can be almost as hard as the abrasives used in grinding wheels. Superabrasives are used for grinding hard materials such as these. Superabrasive wheels have their abrasive on the outside cutting edge of the wheel bonded to a core material, which forms the shape of the wheel. Superabrasive wheels are available in the same standard grit range as conventional wheels. They are available in a range of grades and concentrations (the amount of diamond in the bond).

Cubic Boron Nitride

Cubic boron nitride (CBN) is much harder than silicon carbide. This abrasive was created by the General Electric Company and is called BorazonTM. It works best on hardened ferrous alloys, especially the difficult-to-machine cobalt and nickel superalloys used widely in jet engine applications. Like most milling chips, the chips produced by this superabrasive are well formed. This sharp cutting action results in a much cooler operating temperature, which is important in the grinding of heat-sensitive materials. At high temperatures CBN tends to react with water; hence CBN was initially applied with straight oil cutting fluids. Recent developments in water-based fluids have improved the CBN grinding ratio in these fluids to approach that of straight oil cutting fluids.

CBN is not competitive with diamond in the grinding of tungsten carbide, however. The abrasive is much more expensive than either aluminum oxide or silicon carbide. Machine requirements are also rigorous. However, time saved from faster cutting action, less dressing, consistent size control, and less-frequent wheel changing often makes cubic boron nitride the least expensive abrasive per part produced.

Diamond

Diamond is another superabrasive obtained in either natural or manufactured form. It is the hardest substance known. The single-crystal diamond nib used for the truing and dressing of grinding wheels is a common application. For several economic and strategic reasons, there was great interest in the development of synthetic diamond. This feat was accomplished by the General Electric Company in the early 1950s. Today, most diamonds used in precision grinding applications are manufactured. Manufactured diamonds are engineered to have controlled crystal configurations not duplicated in natural diamonds.

Diamond abrasive is especially useful in the grinding of ceramics and tungsten carbide. Diamond abrasives are not effective on steels or superalloys containing cobalt or nickel. At the elevated operating temperature of grinding these materials, they readily pick up carbon from diamond. Diamond grinding wheels are especially useful in sharpening carbide cutting tools.

Because of the cost of diamond, its use is usually restricted to those applications where no other abrasive will work effectively. If using diamond costs less per part, it should be the preferred tool for that operation.

Grit Size

After a grain type has been chosen, grit size must be selected. Every grinding wheel has a number that designates its grit size. Grit size is the size of the individual abrasive grains in the wheel. When grit is manufactured, it is run through a series of ever finer screens to establish various sizes of grains. Grit size is established by the number of openings per linear inch in the final screen used to size the grain. The higher the number, the smaller the openings in the screen. Coarse grains would have lower numbers such as 10, 16 or 24. Coarse grains are used for rapid stock removal where a fine finish is not required. Fine grit wheels have higher numbers such as 70, 100, and 180. They are used for fine finishes and for hard, brittle materials.

Bonds

The bond material holds the abrasive grains in a wheel together. The correct bond must be chosen if the abrasive is to cut efficiently. The bond is designed to wear away. As the bond wears away, abrasive grains wear and fall away so that new sharp grains are exposed.

There are three principal types of bonds used in conventional grinding wheels: vitrified, resinoid (organic), and rubber. Each type offers different characteristics to the grinding action of a wheel. Bond type is selected based on wheel operating speed, the material to be ground, the type of grinding operation, and the precision required in the application. Table L-1 shows various bonds and their symbols.

Vitrified bonds are the most common. Vitrified bond wheels are very rigid, strong, and porous. Vitrified bond wheels have high metal removal rates and can be used for precision grinding. Vitrified bonds are very hard, but at the same time they are brittle like glass. They are broken down by the pressure of grinding. Vitrified bonds are made with a mixture of carefully selected clays. Grinding wheels are made at very high temperatures in kilns. The high temperatures fuse the clays and the abrasive grains into a molten glass-like condition. As it cools, the glass attaches grains together. Vitrified wheels are rated up to 6,500 SFPM.

The second type of bond is made from organic substances. These bonds soften under the heat of grinding. The resinoid bond is the most common type of organic bond. The resinoid bond material is made from synthetic resin.

Resinoid bonds are good choices for rapid stock removal. They are also a good choice when better finishes are required. Resinoid bonds are designed to operate at higher speeds. Resinoid bond wheels are also used in rough-grinding operations where some flexibility is needed, such as snagging of castings in a foundry, with high wheel speeds and heavy stock removal. Resinoid bonds are also used with superabrasives for carbide grinding on tool and cutter grinders. Resinoid wheels are rated to 16,000 SFPM or higher.

Rubber is also an organic bond. Rubber bonds are often found in wheels used where a high quality of finish is required. Rubber-bonded grinding wheels are used in the finish grinding of bearing surfaces. Rubber-bonded wheels result in a smooth grinding action. They are also used for cut-off wheels.

Table L-1 Bond Types and Their Symbols

Bond Name	Symbol	Description
Vitrified	V	Glass-based; made via vitrification of clays and feldspars
Resinoid	B	Resin-based; made from plants or petroleum distillates
Silicate	S	Silicate-based
Shellac	E	Shellac-based
Rubber	R	Made from natural or synthetic rubber
Metal	M	Made from various alloys

Silicate Bond

Silicate bonds are made by mixing abrasive grains with silicate and soda or water glass. It is then molded into the desired shape, dried, and then removed from the mold. The raw wheel is then baked in a furnace.

Shellac Bond

Shellac bonded wheels are made by mixing abrasive with shellac and then molding by rolling and pressing. The wheels are then heated to 300°F for several hours. A shellac bond has greater elasticity than other bonds with good strength. Shellac bonds are recommended for cool cutting on materials such as hardened steel, thin sections, cast iron, and steel rolls.

Metal-bonds are used for diamond wheels. Diamond wheels are also used for grinding hard nonmetallics, such as ceramics and stone.

Grade

The grade of the grinding wheel designates the strength of the bond in the grinding wheel. The bond is a hard grade if the spans between each abrasive grain are very strong and securely holds the grains during grinding. A soft wheel releases the grains under small grinding forces. The relative amount of bond in the wheel determines the wheel's grade or hardness. Hard grade wheels have longer wheel life. They are used for jobs on high-horsepower machines and for jobs that have narrow areas of contact. Soft grade wheels are used for rapid stock removal and for jobs with large areas of contact. Soft grades are also used for hard materials like tool steels and carbides.

CONCENTRATION

Concentration is a term applied to superabrasive grinding wheels that refers to the amount of abrasive contained in a unit volume of usable grinding wheel. A concentration of 100 corresponds to 72 carats per cubic inch of wheel in the effective superabrasive volume of the wheel. The concentration values are usually between 30 and 175. The desirable concentration depends on the projected area of wheel contact with the workpiece material. A large area, as in flat surface grinding, would call for a low concentration. A limited contact area, such as is found in most tool and cutter grinding applications, would call for a relatively high concentration. It is important to match the concentration to the application. Workpiece materials, such as glass, that are likely to chip while being ground need a low concentration.

SIZE AND SHAPE OF WHEELS

Grinding wheels come in a variety of shapes. The straight wheel is the most common. The grinding face is on the periphery of a straight wheel.

The cutting face is on the side of some grinding wheels. These wheels are usually named based on their shapes, such as cup wheels, cylinder wheels, and dish wheels. Wheels with cutting faces on their sides are often used to grind the teeth of cutting tools and other hard-to-reach surfaces. Segmented wheels are also used. Segments are assembled in a segmented wheel to form a side grinding wheel. Segmented wheels are often used for large grinding wheels such as for Blanchard grinding machines.

Mounted wheels are small grinding wheels with special shapes that are mounted on steel shafts. They are used for a variety of off-hand and internal grinding jobs.

Figure L-14 shows a few of the standard wheel shapes. The standard coding system uses numbers 1 to 28 to indicate wheel size and shape.

A straight (type 1) wheel has its grinding face on the periphery. The grinding face is at right angles to the sides. This is called an "a" face. A cylinder (type 2) wheel has the grinding face on the rim or wall end of the wheel, and has three dimensions: diameter, thickness, and wall thickness. Straight cup (type 6) wheels have a grinding face that has a flat rim or wall on the end of the cup. The grinding face on a flaring cup (type 11) wheel is the flat rim or wall of the cup. Note that the wall of the cup is tapered. A dish (type 12) wheel is similar to type 11, but it has a narrow, straight peripheral grinding face in addition to the wall grinding face. This is the only wheel of those shown considered safe for both peripheral and wall (or rim) grinding.

STANDARD WHEEL MARKING SYSTEMS

A standard wheel marking system is used to identify the five major characteristics in grinding wheel selection (Figure L-15):

1. Type of abrasive
2. Grit size
3. Grade or hardness
4. Structure
5. Bond

These factors are indicated on the grinding wheel blotter by a numeric and letter identification code. For example, a wheel marked **A 60-J8V** indicates the following:

First Symbol: Abrasive Material

The A indicates that this wheel's abrasive material is fused aluminum oxide.

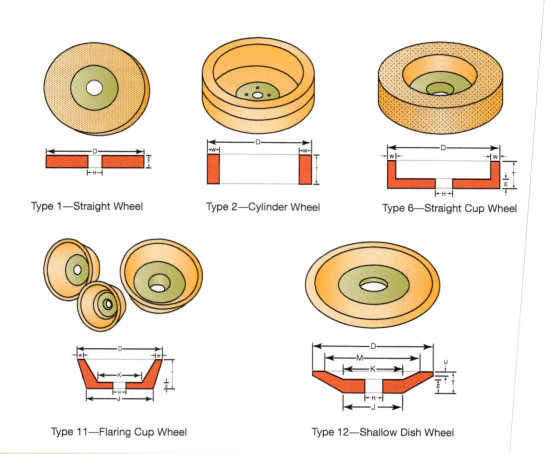

Type 1—Straight Wheel   Type 2—Cylinder Wheel   Type 6—Straight Cup Wheel

Type 11—Flaring Cup Wheel   Type 12—Shallow Dish Wheel

Figure L-14 A few wheel types.

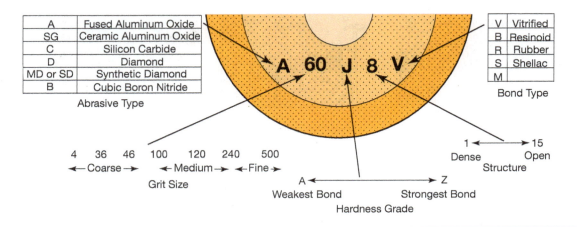

A	Fused Aluminum Oxide
SG	Ceramic Aluminum Oxide
C	Silicon Carbide
D	Diamond
MD or SD	Synthetic Diamond
B	Cubic Boron Nitride

Abrasive Type

V	Vitrified
B	Resinoid
R	Rubber
S	Shellac
M	

Bond Type

A 60 J 8 V

4   36   46   100   120   240   500
←— Coarse —→   ← Medium →   ← Fine →
Grit Size

A ←————————————→ Z
Weakest Bond   Strongest Bond
Hardness Grade

1 ←————→ 15
Dense   Open
Structure

Figure L-15 Wheel specification.

Second Symbol: Grain Size

The grain size of 60 indicates a coarse grain. Grain size for general purpose grinding is generally between 46 and 100.

Third Symbol: Grade of Hardness (A 60-J8V)

Grade of hardness (Figure L-16) is a measure of the bond strength of the grinding wheel. The bond material holds the abrasive grains together in the wheel. The stronger the bond, the harder the wheel. Precision grinding wheels tend to be softer grades, because it is necessary to have dull abrasive grains pulled from the wheel as soon as they become dulled, to expose new sharp grains to the workpiece. If this does not happen, the wheel will become glazed with dull abrasive. Cutting efficiency and surface finish will be poor. Later alphabet letters indicate harder grades. For example, F to G are soft, whereas R to Z are hard.

Fourth Symbol: Structure (A 60-J8V)

Structure, or the spacing of the abrasive grains in the wheel (Figure L-17), is indicated by the numbers 1 (dense) to 15 (open). Structure provides chip clearance so that chips may be thrown from the wheel by centrifugal force or washed out by the grinding fluid. If this does not happen, the wheel becomes loaded with workpiece particles and must be dressed.

Fifth Symbol: Bond (A 60-J8V)

The bond in this example is vitrified.
  Rules of thumb for grinding wheel selection:

  The harder the steel (material), the softer the wheel.

  The more stock to remove, the coarser the wheel.

  The finer finish required, the finer the grit.

  The sharper internal corner desired, the harder and finer the wheel to do just the corner.

GRINDING MACHINE SAFETY

Wheel Speed and Wheel Guards

To reduce the safety hazards from wheel explosions, wheel guards are used on nearly every type of grinding machine. All grinding wheels are rated at a specific maximum rpm. Exceeding rated speeds can cause a wheel to fly apart.

Weak Bond
Soft Wheel

Medium Bond
Medium Hard Wheel

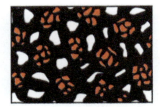

Strong Bond
Hard Wheel

Figure L-16 Three sketches illustrating a soft, a medium, and a hard wheel. This is the grade of the wheel. The white areas are voids with nothing but air, the black areas are the bond, and the others are the abrasive grain. The harder the wheel, the greater the proportion of bond and, usually, the smaller the voids.

Dense Grain Spacing
Dense Structure

Medium Grain Spacing
Medium Structure

Open Grain Spacing
Open Structure

Figure L-17 Three similar sketches showing structure. From left to right, dense, medium, and open structure or grain spacing. The proportions of bond, grain, and voids in all three sketches are about the same.

SAFETY FIRST

The same general rules of safety apply to grinding machines as apply to other machine tools, but there are additional hazards because of the typically high speed of grinding wheels, which can store a great deal of energy. If a grinding wheel becomes cracked, it can fly apart, ejecting chunks of wheel like missiles.

Always check rated speeds marked on the wheel blotter and never operate any grinding wheel beyond its stated maximum speed.

The American National Standards Institute has a document on grinding: ANSI B11.9-2010 "Safety Requirements for Grinding Machines." This document incorporates the earlier standards that specifically covered only grinding wheel safety.

Ring Testing a Grinding Wheel

A vitrified bond grinding wheel can be "ring tested" for possible cracks (Figure L-18). Hold the wheel on your finger or on a small pin. Tap the wheel lightly with a wooden mallet or screwdriver handle. A good wheel will give off a clear ringing sound, whereas a cracked wheel will sound dull. If you discover a cracked wheel, advise your instructor or supervisor immediately. Large grinding wheels may be ring tested while resting on the floor or being supported by a sling. Ring tests should be made a total of four times, indexing the wheel in 90-degree increments.

Figure L-18 Making a ring test on a small wheel.

Grinding Wheel Safety Rules

1. Inspect all wheels for cracks or chips before mounting. Ring test them.
2. Do not alter a wheel to fit the grinding machine, and do not force it onto the machine spindle.
3. Make sure that the operating speed of the grinder is not higher than the maximum allowable speed of the wheel.
4. Ensure that mounting flange bearing surfaces are clean and free of burrs (Figure L-19).
5. Use mounting blotters unless the wheel already has them.
6. Do not overtighten the mounting nut.
7. Do not grind on the side of a straight wheel. There are certain detailed exceptions to this rule. In applications such as shoulder and form grinding, some amount of side grinding takes place.
8. Use a safety guard that covers at least half of the grinding wheel.

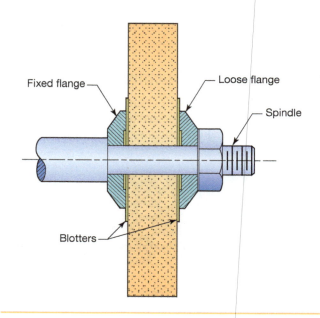

Fixed flange

Loose flange

Spindle

Blotters

Figure L-19 Typical set of flanges with flat rims and hollow centers, with blotters separating the wheel and the flanges. Overtightening the nut could spring the flanges and perhaps even crack the wheel. (Follow the wheel manufacturer's torque specifications.)

9. Allow the grinding wheel to run at least one minute before using it to grind and do not stand directly in line with the rotating grinding wheel.
10. Always wear safety glasses or other approved eye protection.
11. Use the correct wheel for the material you are grinding. Check the *Machining Data Handbooks* for guidance.
12. Be sure there is no possibility of a collision between the grinding wheel and the workpiece before starting the grinding machine. This is especially critical in cylindrical and surface grinding.

Grinding wheels should be stored so they are protected from damage from other wheels. The storage area should be free from extreme variations in temperature and humidity. Extreme variations in temperature or humidity can damage the bonds in some wheels.

Grinding wheels must be carefully handled. If a grinding wheel is too heavy to be carried safely, lifting equipment should be used. The wheel should be cushioned while moving it.

Before mounting a vitrified wheel, it must be ring tested. A cracked wheel cannot be used.

Make sure the spindle rpm of the machine doesn't exceed the maximum safe speed of the grinding wheel.

Always use a wheel with a hole size that fits snugly yet freely on the spindle without forcing it. Never alter the center hole. Use a matched pair of clean, recessed flanges at least one-third the diameter of the wheel. Flange bearing surfaces must be flat and free of any burrs or dirt.

Only tighten the spindle nut enough to firmly hold the wheel. Do not overtighten a wheel. If the grinding wheel is a directional wheel, look for an arrow on the wheel and make sure it points in the direction of spindle rotation.

Make sure that guards are in place before operating the machine. After the wheel is securely mounted and the guards are in place, turn on the spindle, step back out and off to the side, and let it run for at least one minute at operating speed before using the machine.

Grind only on the face of a straight wheel. Make grinding contact gently, without bumping the wheel. Do not overwork the grinder. If the motor slows, you are overworking the machine.

TRUING, DRESSING, AND BALANCING OF GRINDING WHEELS

A grinding wheel must run true with every point on its cutting surface concentric with the machine spindle. As the wheel becomes loaded with workpiece material, it must be dressed to restore sharpness. It must also run in balance because of its great speed. Truing, dressing, and balancing are important parts of grinding operations.

TRUING AND DRESSING

When a new wheel is installed on a grinder, it must be trued before it is used. The cutting surface of a new wheel will run out slightly due to the clearance between the wheel bore and

machine spindle. Truing a wheel will make every point on its outer cutting surface concentric with the machine spindle. This concentricity is important for achieving smooth and accurate grinding.

Dressing is the process of sharpening a grinding wheel. Truing and dressing are performed in the same manner.

Truing and dressing remove material from the grinding wheel. Wheels should be trued and dressed only enough to establish concentricity (truing) or to expose new sharp abrasive grains to the workpiece (dressing). If the correct wheel is chosen for the application, it should result in an essentially self-dressing condition, where the force of the grinding action is sufficient to release dull abrasive grains from the bond and keep the grinding wheel sharp.

Precision grinders are trued and dressed with single- or multiple-point diamond dressers (Figure L-20). The diamond dresser is mounted on the grinder chuck so that it can be traversed across the cutting surface of the wheel. The dresser must be positioned off center on the wheel on the outgoing rotation side (Figure L-21) to prevent the dresser from getting caught and being pulled under the wheel. Truing and dressing should always be done with coolant to cool the diamond dresser.

When truing, lower the wheelhead while traversing the diamond with the table travel until you contact the wheel. Continue to traverse the dresser across the wheel and feed down .001 inch after each pass. Each time, a little more abrasive will be removed from the wheel. When the dresser is cutting all around, the wheel has been fully trued. Do not remove any more abrasive than is necessary to achieve concentric running of the cutting surface. After truing and dressing, it is desirable to "break" the sharp corners of the wheel with a dressing stick, leaving a small radius. This will prevent the sharp corners from leaving undesirable feed-line scratches on the workpiece.

The speed of traverse can influence the surface finish that can be achieved on the workpiece. A slow dressing traverse will cause the diamond to machine the abrasive

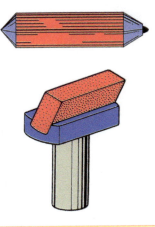

Figure L-20 Single-point diamond dresser is shown on the top of the figure. A multiple-point dresser is shown on the bottom of the figure (*Used with permission of Desmond-Stephan Manufacturing Company*).

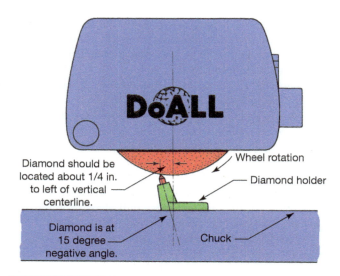

Figure L-21 The dresser with its diamond is placed and magnetically secured on a clean magnetic chuck. Note that the diamond is slanted at a 15-degree angle and slightly past the vertical centerline of the wheel (*Used with permission of DoALL Company*).

grit smoother, which in turn will result in a smoother workpiece finish but with less efficient grinding. A rapid dressing traverse will leave a sharper wheel, but the surface finish will be rougher. In most cases the same wheel can be used for both roughing and finishing: a coarse dress for roughing multiple parts, followed by a fine dress for finishing the parts.

The bore of a diamond wheel is machined so that it will closely fit the grinder spindle. When a resinoid-bonded diamond or CBN wheel is mounted, it may be adjusted to run true by using a dial indicator. The wheel is tapped lightly using a block of wood (Figure L-22). These wheels are not dressed normally.

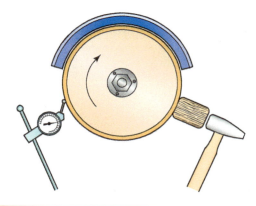

Figure L-22 Runout on a diamond wheel is checked with a dial indicator and must be within .0005 inch for resinoid wheels or .00025 inch (half as much) for metal-bonded wheels. (Some metal-bonded wheels are constructed with a concentric groove for indicating purposes.) Tapping a wooden block held against the wheel to shift the wheel on the spindle is often enough to bring it within limits. Otherwise, it will have to be trued.

RADIUS AND FORM DRESSING

Contours, radii, and other special shapes can be ground by forming the reverse geometry on the grinding wheel. Various methods are used for this. Using a radius dresser that swings the single-point diamond in a preset arc is one method.

BALANCING

Balancing is usually required on large wheels (over 14-inch diameter) but may not be required for smaller wheels. An out-of-balance wheel can cause chatter marks in the workpiece finish.

Wheels are balanced on a disk balancing tool (Figure L-23) or on parallel ways (Figure L-24). The parallel ways balancing tool must be leveled. The grinding wheel is mounted on a balancing arbor and placed on the ways. The heavy point rotates to the lowest position. Wheels are balanced by adjusting weights in the flanges (Figure L-25). Balancing should be done carefully, to the point where the weight of a postage stamp applied to the grinding surface in a horizontal position will cause the wheel to move. Balancing on ways or with the overlapping disk tool results is only static-balancing. Sometimes, even a carefully static-balanced wheel can be quite unbalanced in operation. Sometimes, it is necessary to measure the needed weight for static balance and use two smaller weights arrayed about 60 degrees from the light spot.

Some production grinders are equipped with devices used to balance a wheel while it is in motion. In one system, a cylindrical grinder has its wheel spindle fitted with computer-controlled motor-driven weights that continuously compensate for any detected imbalances. The process is termed *dynamic balancing*. Dynamic balancing adapts to changes that can occur during the working life of the grinding wheel. The result of automatic high-speed spindle and

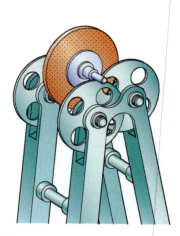

Figure L-23 This type of balancing device with overlapping wheels or disks is quite common. It has an advantage in that it need not be precisely leveled.

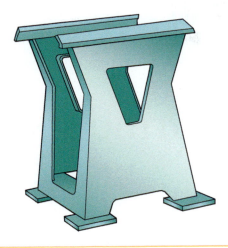

Figure L-24 Balancing a wheel on two knife edges, as on this unit, is accurate because there is minimal friction. Of course, the unit must be perfectly level and true. Otherwise, the wheel may roll from causes other than being out-of-balance.

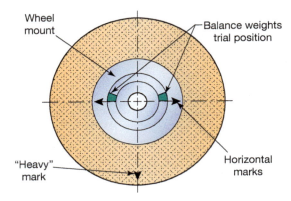

Figure L-25 With the weights between the vertical and the horizontal centerlines, the wheel should be in proper balance, stationary in any position. If not, a matched pair of differing balance weights should be used.

wheel balancing improves product quality by suppressing chatter. Chatter, which is caused by an out-of-balance condition, can result in metallurgical damage to heat-sensitive materials such as bearings.

SELF-TEST

1. A grinding wheel has two major components. What are they and what does each do?
2. How is the abrasive type chosen for a grinding operation?
3. What is friability?
4. What is meant by *wheel grade*?
5. What is the most widely used abrasive?
6. Describe grit size and its affect in grinding.
7. What are the names of five common shapes for grinding wheels?
8. A standard wheel marking system is used to identify the five major characteristics in grinding wheel. What are the five characteristics that are used?
9. What is the most important safety consideration in wheel section to avoid catastrophic failure?
10. Describe ring testing and its purpose.
11. Describe dressing and truing a grinding wheel and the purpose of each.
12. Explain the placement of a single-point dresser.
13. How is a diamond or CBN wheel trued?
14. Generally, what determines whether a wheel needs to be balanced?
15. Describe the balancing procedure for a grinding wheel.
16. What happens when grinding a wheel is chosen with a bond strength that is too hard?
17. You have been given a straight grinding wheel to identify with the markings C 80 J8B. Describe the structure, bond hardness, grit size, and abrasive material, and name the bonding material.

INTERNET REFERENCES

Further information on grinding wheels:

http://www.georgiagrindingwheel.com

http://en.wikipedia.org/wiki/Grinding_wheel

http://www.mmsonline.com

Setup of Surface Grinders

The horizontal spindle reciprocating table surface grinder is probably the most common precision grinder found in the machine shop. The primary use of this machine is surface grinding. The purpose of this unit is to familiarize you with the major parts of this machine, its controls, and available accessories.

Figure L-26 Surface grinder with direction and control of movements (*Courtesy of Okamoto Corporation*).

OBJECTIVES

After completing this unit, you should be able to:

- Name the components of the horizontal spindle surface grinder.
- Define the functions of the various component parts of the grinder.
- Name and describe the functions of at least two accessory devices used to increase the versatility of the surface grinder.
- List reasons for using grinding fluids.
- List three types of grinding fluid.
- Describe methods of grinding fluid application.
- Describe methods of cleaning grinding fluids.

IDENTIFYING MACHINE PARTS AND THEIR FUNCTIONS

The horizontal-spindle, reciprocating table, surface grinder (Figure L-26) is usually just called a *surface grinder*. The workpiece is mounted on a table that moves back and forth under the wheel. Major parts of this grinding machine include the following.

Wheelhead

The wheelhead contains the spindle, bearings, and drive motor. This assembly is mounted on the downfeed slide.

Table with Chuck

The table supports the chuck, which is the primary work-holding device on surface grinders. The table reciprocates right and left to move the workpiece under the grinding wheel. Table reciprocation may be done by hand or it may be a powered function. On powered tables, the length of stroke is preset by positioning the table reverse trip dogs. The table is then reversed at the end of each stroke.

Saddle

The table is supported by the saddle, which moves in or out to move the workpiece forward or backward after each left-to-right reciprocation. This action is termed *crossfeed*. This is also an automatic, powered function on many grinders.

Controls

The downfeed handwheel (Figure L-27) located on the wheelhead raises and lowers the wheel in reference to the workpiece. Depth of cut is controlled by the downfeed. The downfeed handwheel is graduated in .0001-inch increments and usually has an arrow indicating the direction of wheel feed. The downfeed handwheel is equipped with a micrometer collar (Figure L-27) that may be adjusted to zero at the point where the wheel just begins to cut the workpiece.

The table feed handwheel reciprocates the table back and forth. The crossfeed handwheel moves the saddle in and out. The crossfeed handwheel usually has a dial graduated in .001-inch increments, enabling the crossfeed to be regulated so that the correct amount of overlap is obtained on each grinding pass.

In general, large crossfeed movements combined with small downfeeding movements are preferred. This combination keeps more of the wheel working and tends to keep the wheel surface flat longer between dressings. Narrow crossfeed movements tend to round the edges of the wheel excessively.

WORK HOLDING ON THE SURFACE GRINDER

A fundamental consideration in any machining operation is work holding. The purpose of this unit is to examine the more common grinder work-holding devices.

TYPES OF GRINDER CHUCKS

Magnetic Chucks

The most common work-holding device for the surface grinder is the magnetic chuck. The main types are the permanent-magnet chuck (Figure L-28) and the electromagnetic chuck (Figure L-29). The permanent-magnet chuck

Figure L-28 Permanent-magnet chuck, the most widely used surface grinding chuck (*Mitutoyo America Corp.*).

Figure L-29 Electromagnetic chuck for reciprocating surface grinder. The stop strips at the back and left side are adjustable for height and can be used to prevent work from sliding off the chuck (*Courtesy of Walker Magnetics*).

Figure L-27 Closeup of down-feed handwheel. Moving from one graduation to the next lowers or raises the grinding wheel by .0001 inch.

Figure L-30 Magnetic sine chuck needed for grinding nonparallel surfaces (*Courtesy of Walker Magnetics*).

has a series of alternating plates of powerful magnets with pole pieces inside that can be moved to increase the magnetism for work holding or to diminish the magnetism to release the part.

The electromagnetic chuck is magnetized when electrically energized. In most cases, the grinding that is done on electromagnetic chucks is done with a substantial flow of grinding fluid, which carries away the generated heat. The chuck remains magnetized as long as the power is applied. When power is removed, the chuck is demagnetized so that the workpiece can be removed. Some residual magnetism often remains in the workpiece. An electrical demagnetizer can be used to remove any residual magnetism from the part.

Figure L-30 shows a magnetic sine plate being used to hold a part at the correct angle for grinding. Note that the magnetic sine plate is being held on the magnetic table of the grinder.

CARE OF GRINDER CHUCKS

Like much of the equipment found in the machine shop, grinder chucks are precision tools and deserve the same care and handling you would give your precision measuring instruments. With reasonable care they will maintain their accuracy over many years. Observe the following precautions when using grinder chucks.

Step 1 Carefully clean the chucking surface before mounting a workpiece.

Step 2 Be sure that the workpiece is free from burrs.

Step 3 On a magnetic chuck, be sure that small parts span as many magnetic poles as possible to ensure maximum holding power.

Step 4 Deburr the chucking surface from time to time (Figure L-31).

Figure L-31 Periodic deburring of the chuck with a granite deburring stone as shown, or with a fine-grit oilstone, is a good practice.

Step 5 Check the chucking surface with a dial indicator if the chuck has been removed and reinstalled on the machine or a new chuck is being used for the first time. It may be necessary to "grind in" the chuck before it is ready to use for precise work.

GRINDER CHUCKING SETUPS

Odd-Shaped Workpieces

A magnetic chuck will exert the most reliable holding power on an odd-shaped workpiece, if the area of contact with the chuck is large enough. Whenever possible, place the workpiece so that the largest surface area possible is on the chuck. Sometimes it is necessary to block a small or thin part with additional material to prevent the part from moving on the chuck during the grinding process. The blocking, of course, needs to be thinner than the workpiece. Sometimes a part may have a protruding part that makes it impossible to lay flat on the magnetic chuck (Figure L-32). In these cases, the workpiece can be supported with magnetic (laminated) parallels or V-blocks (Figure L-33). These are specially designed for grinder work. These laminated accessories are made with nonmagnetic and soft steel inserts so that the

Figure L-32 Chuck setup for a workpiece with a projection on chucking side. The work is supported on laminated magnetic parallels.

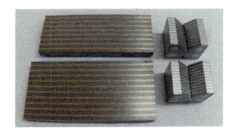

Figure L-33 A set of magnetic parallels and V-blocks can be useful.

lines of magnetic force will be conducted through to the workpiece. Laminated parallels must be treated with great care because the magnetically permeable materials are soft. They should be checked carefully for burrs each time they are used. Odd-shaped workpieces may also be held by traditional precision milling-type machine tool work-holding methods such as vises or clamps.

Figure L-34 Spring-tooth clamps.

Figure L-35 Spring-tooth clamps.

Figure L-36 Grinding thin warped parts.

Nonmagnetic Work

Nonmagnetic materials may be held on a magnetic chuck by blocking the parts with steel (magnetic) spring-tooth clamps (Figure L-34). The comb-like teeth will keep a nonmagnetic workpiece from sliding off the chuck. As the magnetism is applied, the comb-like teeth of the clamps are pulled down against the nonmagnetic workpiece to keep the part from sliding off the chuck (Figure L-35). If you are using spring-tooth clamps, be sure that they are lower than the workpiece so that they will not come in contact with the grinding wheel. Double-sided tape is also used in some applications. Care must be taken with all grinding operations as the pressures can throw parts out of the grinder if the holding or blocking strength is insufficient.

Thin Workpieces

Thin material that is already warped or twisted will probably be pulled flat on either a magnetic or vacuum chuck. These

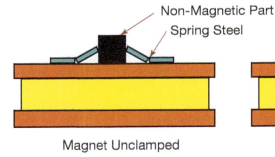

Non-Magnetic Part
Spring Steel

Magnet Unclamped   Magnet Clamped

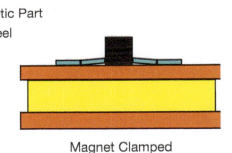

Grinding with no Shims

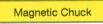

Warped Piece

Magnetic Chuck

The Part is Pulled Flat on the Magnetic Chuck

The Part Returns to Its Warped Shape when Removed from the Magnetic Chuck

Grinding with Shims

Magnetic Chuck

Part is Shimmed under the Warp

Part Top is Now Flat when Removed from the Magnetic Chuck

Magnetic Chuck

Flat Top of Part Placed down on the Magnetic Chuck. Red Line Shows that the Part Will Be Flat and Parallel after Grinding

workpieces will assume their original distorted shape after grinding and removal from the chuck. Figure L-36 shows an example of what happened to a thin warped piece when ground. The three illustrations on the top show what happens. The warp is pulled flat when the part is placed on the magnetic chuck. The top is then ground. When the part is removed from the chuck it returns to its warped shape. The bottom three illustrations show how a part can be shimmed for grinding. The shim prevents the part from being pulled down flat on the chuck. The top is then ground flat. The part is then removed, flipped over and put on the chuck to be ground. After the grind, the part will be flat and parallel. It is important to grind uniform amounts from both sides of materials and to be prepared to grind opposite sides multiple times to get a flat, dimensionally stable part.

Nonmagnetic material can be ground by using double-backed tape (with adhesive on both sides). One side of the tape is applied to the chuck while the other side holds the workpiece. This may be sufficient for light cuts. Avoid the use of grinding fluids when double-backed tape is used, as it can cause part slippage. The use of double-backed tape requires great care to avoid the risk of part slippage, which can cause significant damage to machine and/or operator.

GRINDING FLUIDS

A great deal of heat is generated from the friction between abrasive grains and the workpiece. Temperatures up to 2000°F (1093°C) are not uncommon at the point where the grain is cutting. Grinding fluids are essential in grinding.

PURPOSE OF GRINDING FLUIDS

The correct selection of grinding fluid can greatly affect the grinding process, but even the correct grinding fluid, poorly applied, can be worse than none at all. All these fluids serve a cooling function, but others emphasize lubrication, even at the expense of cooling capacity. Grinding fluids do the following:

1. Reduce the temperature produced by the grinding process in the workpiece, thus reducing warping in thin parts and aiding in size control of all shapes of parts.
2. Lubricate the contact area between wheel and workpiece, thus helping prevent chips from adhering to the wheel, which aids surface finish.
3. Flush chips and abrasive (swarf) from the workpiece.
4. Help control grinding dust, which can be hazardous to the health of the operator.

TYPES OF GRINDING FLUIDS

Water-Soluble Chemical Types

Water-soluble chemical types of fluids, called **synthetic fluids**, are typically mineral oil-free. They are usually transparent, which helps with visibility. Some have additives for

rust control. They are not prone to bacterial growth. They provide a high level of cooling capacity but have less lubricating ability than either the soluble or straight oils.

Semisynthetic grinding fluids have been developed to combine the benefits of synthetics and soluble oils. These contain less mineral oil in their content than soluble oils do. The semisynthetics offer good lubricity and better wetting and cooling properties than soluble oils.

Water-Soluble Oil Types

Water-soluble oil types use water and a water-soluble mineral oil. These water-soluble oil fluids are the common milky fluids often seen in the machine shop. For grinding, this fluid is good for medium stock removal operations. These should be mixed to the concentration specified by the manufacturer.

Bacterial growth, resulting in a foul-smelling fluid, often occurs when the fluid is idle or at slightly elevated temperatures, making it necessary to add bacterial growth inhibitor. The aeration that normally occurs during daily use helps inhibit bacterial growth.

METHODS OF APPLYING COOLANT

Most grinding fluids are applied in high-volume, low-velocity flood streams by the coolant pump. The grinding fluid is constantly recirculated and cleaned of swarf (Figure L-37). The supply must be replenished from time to time with makeup mixture and additives.

It is important to apply grinding fluid where the grinding is taking place. This is the area where the wheel contacts the workpiece. One of the greatest difficulties in both surface and cylindrical grinding is "grinding dry with fluid." On some heat-sensitive steels, inadequate grinding fluid

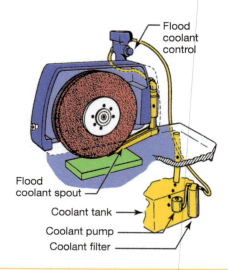

Flood coolant control

Flood coolant spout

Coolant tank

Coolant pump

Coolant filter

Figure L-37 Fluid recirculates through the tank, piping, nozzle, and drains in the flood grinding system (*Used with permission of DoALL Company*).

application can result in the development of grinding cracks from having the high-temperature workpiece contact area with the grinding wheel quenched an instant later with the grinding fluid. This type of grinding crack is parallel to the grinding wheel axis.

USING GRINDING FLUIDS

When using grinding fluids, be sure to observe the following:

- Maintain proper coolant level.
- Maintain proper balance of water and additives.
- Check for contamination by lubricating oils; skim off oil droplets.
- Make sure that coolant is supplied to the contact area in sufficient volume and at the right place.

SELF-TEST

1. Why are small down feed movements in combination with large cross feeding movements preferable in surface grinding?
2. What is the most widely used work-holding device on the surface grinder?
3. What are laminated accessories, and how can they be used in grinder work holding?
4. Name at least two methods of work holding on the surface grinder that can be used to hold nonmagnetic workpieces.
5. Explain how to grind a warped thin part to make it flat and parallel.
6. List at least three major functions of grinding fluids.
7. Where should the grinding fluid be applied?
8. What is mist coolant and how is it applied?
9. What is one of the advantages of using water-soluble chemical-type grinding fluids?
10. What is meant by the term *grinding dry with fluid*?

INTERNET REFERENCES

Further information on magnetic chucks and other magnetic work-holding tools:

http://www.mmsonline.com/articles

http://www.newmantools.com

Information on surface grinders:

http://www.clausing-industrial.com

http://en.wikipedia.org/wiki/Surface_grinder

Using a Surface Grinder

Using the surface grinder involves selecting and mounting the proper grinding wheel, surface grinding the chuck (grinding-in) if necessary, and using the proper combination of table speeds and crossfeeds. The surface grinder is a versatile and precision machine tool. Used correctly, it can produce very accurate parts. The purpose of this unit is to describe the surface grinder, familiarize you with its setup, and with the grinding processes required to finish common parts.

OBJECTIVES

After completing this unit, you should be able to:

- Prepare the surface grinder for a typical grinding job.
- Check and grind-in the chuck if necessary.
- Finish grind a set of V-blocks to required specifications.

SURFACE GRINDING BASICS

In surface grinding, it is important to coordinate the workpiece speed, crossfeed, and downfeed to keep the grinding action as consistent as possible. The example in this unit will be the grinding of a V-block, but before that, other basics about horizontal spindle surface grinding must be considered.

Starting with a broad surface of a soft steel like 1018, a suitable selection of wheel and grinding variables would be: an aluminum oxide abrasive of 46-grit size in a J-bond hardness. The bond itself is vitrified. The wheel speed for vitrified wheels is usually from 5,500 to 6,500 SFPM; the table speed would be about 50 SFPM. The crossfeed rate would be .050–.500 inches per pass. For roughing, the downfeed could be as much as .003 inch, and for finishing it should be less than .001 inch.

How would it change the wheel recommendation if the 1018 material was pack hardened (carburized) to about 55 Rockwell C? The wheel speed and the table speed would

remain the same. The abrasive chosen and the grit size would also remain the same, but the bond selected would be one grade softer, or I bond. The roughing downfeed would change to .002 inch and the finishing to .0005 inch. The greatest change in recommendation other than the grade of the wheel would be in the way that the crossfeeding is done. Here the recommendation would be .025–.250 inches with one-tenth of the wheel width as a maximum. All these recommendations are based on a sharp wheel.

A general observation about horizontal surface grinding is to keep the crossfeed travel as large as possible and the downfeed adjustments matched so that you cannot hear a significant change in grinding wheel speed during the work's traverse under the wheel.

Familiarize yourself with controls on the surface grinder you will be using. If someone else is using the machine, observe its operation for a while. Check the supply of grinding wheels and determine their possible uses.

TRUING A MAGNETIC CHUCK

The first step in surface grinding is to check, and if necessary true the chuck. Truing the chuck is usually not necessary unless it has been removed and reinstalled. To true a grinder chuck, the following procedure may be used:

Step 1 Mount the chuck and align it in both table and saddle axes using a dial indicator with .0001-inch resolution. Make sure the chuck and bed are clean before mounting the chuck.

Step 2 Use the indicator to check the chucking surface to determine if truing will be necessary.

Step 3 If truing is necessary, select and mount a proper wheel for the application.

Step 4 Mark the entire surface of the chuck with bluing or pencil lines. As you are grinding, the removal of the last marking will show you when the whole chuck surface has been trued.

Step 5 True and dress the wheel. Make the final dressing pass with a relatively fast cross movement of the diamond, for a free-cutting grinding action.

Step 6 Magnetize the chuck.

Step 7 Bring the wheel down toward the chuck, using a feeler gage to set the wheel a few thousandths of an inch above the chucking surface. Then, restart the grinding wheel. Set the table in motion with the table stop dogs set to overtravel about 1 inch at each end. Lower the wheel slowly while you are moving the table saddle back and forth cross-wise until you find and barely contact the highest point on the chuck's surface. Then, start the grinding fluid.

Step 8 Set the downfeed handwheel collar to zero.

Step 9 Using a downfeed of no more than .0002 inch, a fairly rapid table speed (50–100 SFPM), and a crossfeed of about one-fourth the wheel width on each pass, grind to a *cleanup* condition on the chucking surface. Remove only enough material to clean the surface completely. (The total amount is usually no more than .001 inch.)

GRINDING V-BLOCKS

The V-block (Figure L-38) is a common and useful machine shop tool. Precision types are finish machined by grinding after they have been rough machined oversize (about .015 inch) and heat-treated. Finish grinding V-blocks provides a variety of surface grinding setups and techniques.

Selecting the Wheel

For these V-blocks, the best abrasive will be a friable grade of aluminum oxide. A vitrified wheel of 46 grit with an I-bond hardness would be a reasonable selection.

Grinding Machine Setup

The following procedure may be used in setting up the grinding machine for V-block grinding.

Step 1 Select a suitable wheel.

Step 2 Clean the spindle. Use a cloth to remove any grit or dirt from the spindle.

Figure L-39 Diamond dresser in the correct position. Note it is placed so that if it were to be thrown out it would be free of the diameter of the rotating wheel.

Step 3 Ring test the grinding wheel.

Step 4 Mount the wheel. Blotters will probably be attached to the wheel. If not, place a correctly sized blotter between each of the flanges and the wheel.

Step 5 Install the spindle nut and tighten firmly. Do not overtighten.

Step 6 Replace or close the wheel guard.

Step 7 Place the diamond dresser on the magnetic chuck in the proper position (Figure L-39).

Step 8 Dress the wheel, using fluid and a relatively rapid crossfeed.

Step 9 Check the chucking surface for nicks and burrs. Use a deburring stone if necessary.

Side and End Grinding Procedure

The V-blocks should be ground together in pairs so that they are exactly the same dimensions when completed. The following steps and illustrations will describe the process for side and end grinding.

Step 1 Place the blocks with the large V-side up on the grinder chuck (Figure L-40).

Step 2 Magnetize the chuck. Remember to re-magnetize the chuck after each setup change.

Step 3 Lower the wheelhead until it is about an inch above the workpiece. Adjust table and saddle position so that blocks are centered.

Step 4 Adjust table feed reverse trips so that the workpiece has about 1 inch of overtravel at each end of the table stroke (Figure L-41).

Step 5 With the grinding wheel stopped, use a feeler gage or piece of paper and lower the wheelhead until it is a few thousandths of an inch above the surface to be ground.

Figure L-38 Finished, hardened, and ground precision V-block.

Figure L-40 The rough blocks in place with the large V side up ready to be ground.

Figure L-42 Grinding the first surface until the surface is "cleaned up."

Figure L-41 Setting the table trip dogs (stop dogs).

Figure L-43 The blocks are turned over and the opposite sides (small v) is ground.

Step 6 Start the wheel and grinding fluid. Start table crossfeed and table travel. Carefully lower the wheel until it just begins to contact the high point of the workpiece. Set the downfeed micrometer collar to zero at this point.

Step 7 The amount to grind from the workpiece will depend on the amount of extra material left from the original machining. Downfeed about .001 inch per pass and grind to a *cleanup* condition (Figure L-42). About .003 to .005 inches should be left for finish grinding on each surface.

Step 8 Turn the V-blocks over and surface grind the side with the small V (Figure L-43).

Step 9 The end of the block may be ground by clamping it to a precision angle plate (Figure L-44). Adjust the block into alignment using a test indicator. The end of the block should extend slightly beyond the angle plate.

Step 10 Set the angle plate on the chuck, and end grind the V-block (Figure L-45).

Step 11 For grinding the remaining side, clamp the ground end to the angle plate, leaving the side surface projecting above

the angle plate (Figure L-46). Grind one remaining side square to the end and square to the first surface ground in Step 7.

Step 12 End grind opposite ends (Figure L-47). The V-blocks may be set up on the magnetic chuck without further support.

Step 13 Grind the remaining sides leaving .003 to .005 inches for finishing to the specification (Figure L-48).

Step 14 Re-dress the wheel with a slow pass (or passes) of the diamond dresser for finish grinding. Use a light cut of about .0002 inch.

Step 15 Check all sides and ends for squareness. Small errors can be corrected by further grinding if necessary. Tissue paper shims (.001 inch) may be used to achieve squareness. Take as little material as possible if corrections are needed.

Step 16 Check all dimensions with a vernier micrometer (.0001 inch discrimination) or dial test indicator.

Step 17 Finish grind all sides and ends to final dimensions. Both blocks should be match ground.

Figure L-44 A ground side of the V-block is clamped to a precision angle plate, and the part is made parallel in preparation for grinding the end square.

Figure L-45 The precision angle plate and V-block setup is turned with the V-block end up on the magnetic chuck. The end of the V-block is ground square to a ground side.

Figure L-46 Setting up to grind the third square surface.

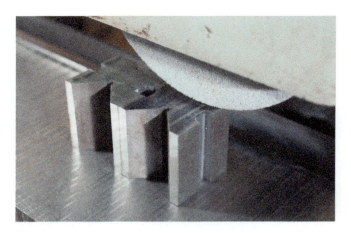

Figure L-47 Grinding the opposite ends of both blocks in one setup to make them parallel.

Figure L-48 The unground sides are ground in one setup to make them parallel to the other sides.

V-Grinding

Step 1 Grind one side of the large V (Figure L-49). Set the blocks in a magnetic V-block and carefully align with table travel. (Do not make contact with the side of the wheel.) Check that the workpiece is being firmly held by the magnetic V-blocks.

Step 2 After rough grinding one side of the V, note the number on the downfeed micrometer collar or set it to zero index. Raise the wheel about $\frac{1}{2}$ inch. Reverse the blocks in the magnetic V-block and grind the other side. Bring the wheel down to contact the work and make grinding passes until the micrometer collar is at the same position as it was when the first V-side was roughed. Then, dress the wheel for finishing and repeat the procedure for both Vs, removing only enough material necessary to obtain a finish. This procedure will ensure that the V is accurately centered on the blocks.

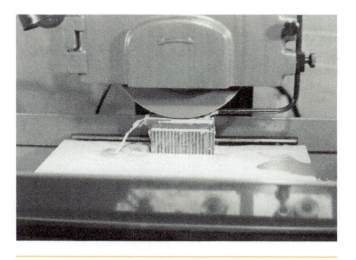

Figure L-49 Setting up the magnetic V-block to grind the angular surfaces on the large v's.

Step 3 Repeat the procedure for the remaining V grooves.

Step 4 Set up the parts again in the magnetic V-blocks to grind the external angular surfaces (chamfers). This operation must be done gently, as work holding from the accessory magnetic V-blocks is less secure than from the surface of the magnetic chuck itself.

PROBLEMS AND SOLUTIONS IN SURFACE GRINDING

In any machining operation, many factors influence the results. The experienced machinist can identify a problem and implement a solution. The purpose of this unit is to help you identify common problems in surface grinding and determine solutions.

Table L-2 Summary of Surface Grinding Defects and Possible Causes

Causes	Burning or Checking	Burnishing of Work	Chatter Marks	Scratches on Work	Wheel Glazing	Wheel Loading	Work Not Flat	Work Out of Parallel	Work Sliding on Chuck
Machine operation									
Dirty coolant				X		X			
Insufficient coolant	X						X	X	
Wrong coolant					X	X			
Dirty or burred chuck				X			X	X	
Inadequate blocking									X
Poor chuck loading							X	X	X
Sliding work off chuck				X					
Dull diamond					X				
Too fine dress	X				X	X	X		
Stroke is too long								X	
Loose dirt under guard				X					
Grinding wheel									
Too fine grain size	X				X	X			
Too dense structure					X	X			
Too hard grade	X	X	X		X	X	X		
Too soft grade			X	X					
Machine adjustment									
Chuck out of line								X	
Loose or cracked diamond				X			X		
No magnetism									X
Vibration			X						
Condition of work									
Heat-treat stresses							X		
Thin							X		

Source: *Used with permission of DoALL Company.*

General Causes of Grinding Problems

Surface grinding problems appear in a variety of forms. In many cases, they are related to acceptable surface finish, or the avoidance of metallurgical changes (burning) of the workpiece. In some cases, such as with safety-critical workpieces, grinding defects can be a serious matter. Surface grinding problems can usually be traced to the general factors of machine condition and machine operation.

Specific problems and solutions are shown in surface grinding Table L-2.

SELF-TEST

1. If a grinding-in or truing of the grinding chuck becomes necessary, how do you determine when the surface has been trued?

2. If chatter or vibration marks are observed on the workpiece, what steps can the operator take to correct the problem?

3. If the workpiece shows burn marks, what can the operator do?

4. What should the operator do if the work is not parallel?

UNIT FIVE

Cylindrical Grinding

The center-type cylindrical grinder, as its name implies, is used to grind the outside diameters of cylindrical (or conical) workpieces mounted on centers. Inside diameters can also be ground, if the proper accessories are available. The cylindrical grinder is a versatile machine tool capable of finish machining a cylindrical or a conical part to a high degree of dimensional accuracy. Various setups on the basic machine permit many different grinding tasks to be accomplished. The purpose of this unit is to familiarize you with the major parts of this machine and its general capabilities.

OBJECTIVES

After completing this unit, you should be able to:

- Identify the major parts of the cylindrical grinder.
- Describe the various movements of the major parts.
- Describe the general capabilities of this machine.

CYLINDRICAL GRINDING

Cylindrical grinders can be used to grind the outside or inside diameter of a cylindrical part. The rotating abrasive wheel, typically moving from 4,000 to 6,500 SFPM, is brought into contact with the rotating workpiece moving in the opposite direction at 50–200 SFPM. The workpiece is traversed lengthwise against the wheel to reduce the outside diameter, as in the case of external grinding. In the case of internal grinding, the workpiece inside diameter will be increasing. In cylindrical plunge grinding, the wheel is brought into contact with the work but without traverse. In all cylindrical grinding, the workpiece is rotated in the opposite direction of the rotation of the abrasive wheel.

IDENTIFYING MACHINE PARTS AND THEIR FUNCTIONS

On the center-type cylindrical grinder (Figure L-50), the workpiece is mounted between centers. A plain cylindrical grinder has a fixed wheelhead that cannot be swiveled, but can only be moved toward or away from the center axis of the workpiece. The table can be swiveled to permit the grinding of tapered workpieces. On the universal cylindrical grinder (Figure L-51), both the wheelhead and table may be swiveled to grind tapers. All possible motions are illustrated in Figure L-52.

Major Parts of the Universal Center-Type Cylindrical Grinder

Major parts of the machine include the bed, slide, swivel table, headstock, footstock, and wheelhead.

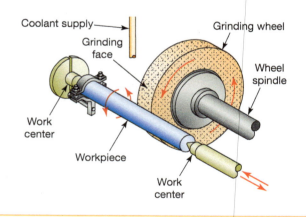

Figure L-50 Sketch of center-type cylindrical grinder set up for traverse grinding. Note particularly the direction of travel of the grinding wheel and the workpiece, and the method of rotating the workpiece.

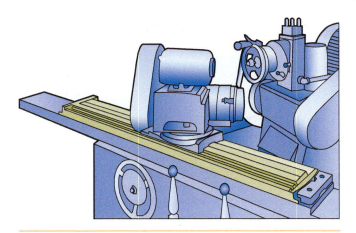

Figure L-51 Universal cylindrical grinder.

Bed   The bed is the main structural component and is responsible for the rigidity of the machine tool. The bed supports the slide, which in turn supports the swivel table.

Slide and Swivel Table   The slide carries the swivel table and provides the traverse motion to carry the workpiece past the wheel. The swivel table is mounted on the slide and supports the headstock and footstock. The swivel table has graduations for establishing taper angles.

Headstock   The headstock (Figure L-52) mounts on the swivel table and is used to support one end of the workpiece. The headstock also provides the rotating motion for the workpiece. The headstock spindle is typically designed to accept a chuck or face plate. The headstock center is used

when workpieces are mounted between centers. Variable headstock spindle speed selection is also available. For the most precise cylindrical grinding, the headstock center is held stationary while the driving plate that rotates the part rotates concentric to the dead center. This procedure eliminates the possibility of duplicating headstock bearing irregularities into the workpiece. For parts that can tolerate minor runout errors, it is preferable to have the center turn with the driving plate.

Footstock   The footstock is also mounted on the swivel table and supports the opposite end of a workpiece mounted between centers. The footstock center does not rotate. It is spring-loaded and retracted by a lever. This permits the easy installation and removal of the workpiece. Compression loading on the spring is adjustable, and the footstock spindle typically can be locked after it is adjusted. The center on this end should be lubricated with high pressure lubrication grease. The footstock assembly is positioned on the swivel table at whatever distance is needed to accommodate the length of the workpiece.

Wheelhead   The wheelhead, located at the back of the machine, contains the spindle, bearings, drive, and main motor.

WORK HOLDING

A workpiece that has a center on each end would be mounted between the head and footstock centers. This effectively provides a single point mounting on each end of the workpiece, permitting maximum accuracy to be achieved in the grinding operation.

CAPABILITIES OF THE CENTER-TYPE CYLINDRICAL GRINDER

Most common cylindrical grinding operation is simple traverse grinding (Figure L-53). This may be done on interrupted surfaces where the wheel face is wide enough to span two or more surfaces.

Figure L-53 Traverse grinding (*Courtesy of MAG Industrial Automation Systems, LLC*).

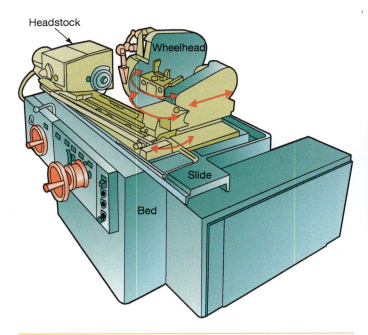

Figure L-52 View of universal cylindrical grinder with arrows indicating the swiveling capabilities of various major components.

Multiple-diameter-form grinding is a type of plunge grinding in which the wheel may be form dressed to a desired shape. In straight plunge grinding, the wheel is brought into contact with the workpiece, but the workpiece is not traversed (Figure L-54).

O.D. taper grinding (Figures L-55 and L-56) is done by swiveling the table. Steeper tapers may require that both table and wheelhead be swiveled in combination (Figure L-57).

Angular shoulder grinding (Figure L-58) and angular plunge grinding may require that the wheel be dressed at the appropriate angle.

Internal cylindrical grinding can be straight (Figure L-59) or tapered. The workpiece is held in a chuck or fixture so that the inside diameter can be accessed by the grinding wheel.

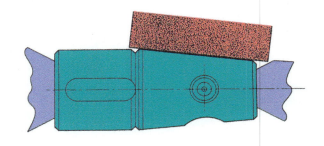

Figure L-56 Taper grinding with the workpiece swiveled to the desired angle. For a steeper taper, the wheel many also have to be dressed at an angle of less than 90 degrees (*Courtesy of MAG Industrial Automation Systems, LLC*).

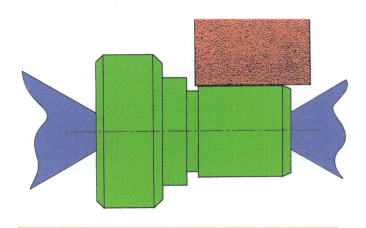

Figure L-54 Straight plunge grinding, in which the wheel is usually wider than the length of the workpiece feature (*Courtesy of MAG Industrial Automation Systems, LLC*).

Figure L-57 Steep taper grinding. Here the wheelhead has been swiveled to grind the workpiece taper (*Courtesy of MAG Industrial Automation Systems, LLC*).

Figure L-55 O.D. taper grinding, which may be done to produce a tapered (conical) finished workpiece or to correct a workpiece that was previously tapered.

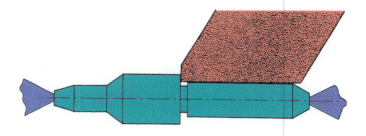

Figure L-58 The illustration shows angular plunge grinding with shoulder grinding. (*Courtesy of MAG Industrial Automation Systems, LLC*).

Figure L-59 Internal grinding, which requires a special high-speed attachment mounted on the wheelhead so that it can be swung up out of the way when not in use. It also requires either a chuck or a face plate on the headstock.

SELF-TEST

1. Explain the differences in construction between plain and universal cylindrical grinders.
2. Why is the most critical cylindrical grinding done on nonrotating centers?
3. Why should the footstock adjustment be locked after the workpiece has been positioned?
4. Can the "plain" cylindrical grinder produce a tapered (conical) surface?
5. Because many cylindrically ground workpieces are hardened, what sort of center-hole preparation is necessary to prevent heat-treatment scale from causing inaccurate cylindrical grinding?
6. What is traverse grinding?
7. What is plunge grinding?
8. Describe the basic methods by which a workpiece is held on a center-type cylindrical grinder.
9. If you are traverse grinding a workpiece with interrupted surfaces, what major limitation needs to be observed in the workpiece?
10. What two additional components are required to perform internal grinding on a universal cylindrical grinder?

INTERNET REFERENCES

Further information on cylindrical grinders:

http://en.wikipedia.org/wiki/Cylindrical_grinder

http://www.msgrinding.com

Using a Cylindrical Grinder

Using the cylindrical grinder involves many of the same steps that you learned in surface grinding. The cylindrical grinder is a versatile and precise machine tool. Used correctly, it will produce a very accurate part with a very fine surface finish. The purpose of this unit is to familiarize you with the preparation of the machine and the grinding processes required to finish a typical cylindrically ground workpiece.

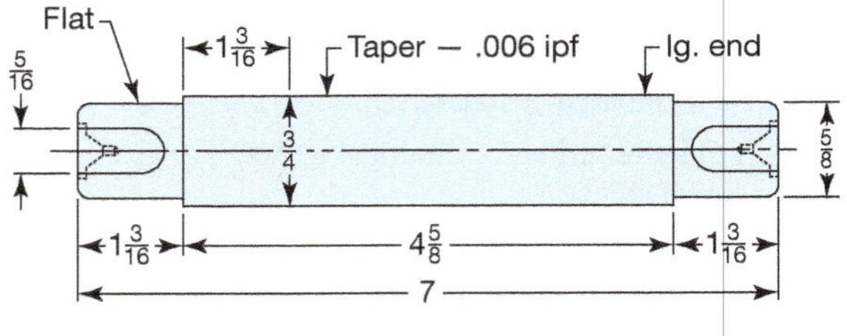

Figure L-60 Sketch of tapered lathe mandrel.

OBJECTIVES

After completing this unit, you should be able to:

- Prepare the cylindrical grinder for a typical grinding job.
- Finish grind a lathe mandrel to required specifications.

CYLINDRICAL GRINDING A LATHE MANDREL

A lathe mandrel is a common and useful machine shop tool. Precision mandrels are finished by cylindrical grinding after they have been machined and heat-treated. Finish grinding the mandrel will provide you with basic experience in cylindrical grinding. Refer to the working sketch shown in Figure L-60 to determine the required dimensions.

Figure L-61 Setting up the diamond for wheel dressing.

Grinding Machine Setup

Step 1 Set up the diamond dresser (Figure L-61). Diamond dressers have a height to match the wheel centerline. On many machines the dresser is built into the footstock assembly.

Step 2 Dress the wheel using a full flow of cutting fluid. If you cannot obtain continuous fluid coverage, then dress dry. Use a relatively rapid traverse of the diamond, with about

.001 inch of infeed per pass, until the wheel is true and sharp (Figure L-62).

Step 3 Place a parallel test bar between centers (Figure L-63). Mount a dial indicator on the wheelhead, and set it to zero at one end of the test bar.

Step 4 Move the table 12 inches and read the indicator at the other end of the bar. The indicator should read .003, or a total of .006 per foot of taper (Figure L-64).

Figure L-62 Dressing the wheel with the grinding fluid on.

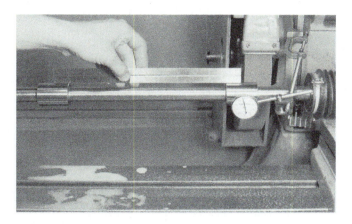

Figure L-63 Setting up the parallel test bar and taking the first reading.

Figure L-64 Second reading of the dial indicator.

Step 5 Adjust the swivel table to obtain the correct amount of taper (Figure L-65).

Step 6 Lubricate the center hole in each end of the mandrel, using a high-pressure lubricant specially prepared for use with centers.

Figure L-65 Adjusting the swivel table.

Step 7 Place the dog on the mandrel and insert it between the headstock and footstock centers. Move the footstock in so that some tension is on the footstock center (Figure L-66).

Step 8 The work must have no detectable axial movement but be free to rotate on the centers. The dog must be clamped firmly to the work and contact the drive pin (Figure L-67). If your machine has a slotted driver, be sure that the dog does not touch the bottom of the slot, or it could force the workpiece off its seating on the center.

Step 9 Set the table stops (Figure L-68) so that the wheel can be traversed some distance beyond the ground surface but have at least $\frac{1}{4}$-inch clearance from contacting the driving dog. If there is adequate clearance on the footstock center, set the stop to permit at least one-third of the wheel width to go beyond the surface being ground. This overtravel will ensure that the ground surface of the mandrel gets completely finished.

Mandrel Grinding Procedure

Step 1 Move the wheel close to what will become the smaller O.D. of the workpiece. (This is usually the footstock end.) With the wheel stopped, you may use a paper strip

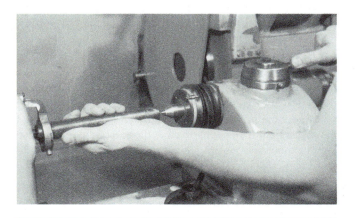

Figure L-66 Clamping the footstock.

Figure L-67 Checking the work dog clearance and drive pin contact.

Figure L-68 Adjusting the travel stops.

(about .003 inch thick) as a feeler gage to position the wheel close to the mandrel.

Step 2 Turn on the wheel and headstock spindle. Adjust the work speed to provide 70–100 SFPM.

Step 3 Start the table traverse. The table traverse rate should be about one-fourth of the wheel width for each revolution of the workpiece. (On unhardened work a more rapid traverse would be suitable, up to half the wheel width.) If your machine is equipped with a dwell (tarry) control, set the adjustment for a slight dwell at the end of the traverse, so that the table does not "bounce" on the table stops.

Step 4 Infeed the wheel until it contacts the rotating mandrel (Figure L-69). When the wheel contacts the work you will see sparks. Then, start the grinding fluid and infeed about .001 in. per traverse until the surface of the part is cleaned up. At this point, set the infeed dial to read zero. Then, retract the wheel.

Step 5 Stop the machine completely. Measure the mandrel in two places, 4 inches apart (Figures L-70 and L-71), to verify that the taper setting was correctly done. There should be a .002 inch difference in diameter between the larger and smaller diameters. Make an adjustment if necessary.

Figure L-69 Turning on the wheel and work rotation.

Figure L-70 Confirming the taper required.

Figure L-71 Measuring taper 4 inches down the part.

Be certain that the grinding head is retracted from the work if you need to make this adjustment.

Step 6 Rough and finish grind the mandrel to final size. The grinder may have an automatic infeed stop that will stop the infeed when final diameter is reached. On a manual machine, infeed the wheelhead to remove about .001 inch from the diameter per complete traverse cycle. The finishing passes should remove about .0002 inch from the diameter.

Step 7 Check the taper and diameter periodically while making the finishing passes. The table traverse rate should be reduced to about one-eighth the width of the wheel per workpiece revolution. When the final size is reached, allow the work to spark out by traversing through several cycles without additional infeed (Figure L-72).

Only after careful measurement should the workpiece be removed from between the centers (Figure L-73), as returning the work to the centers to remove small amounts of material is usually not successful, because of minor irregularities that occur. The smallest amount of grit can cause large differences.

Figure L-73 Finished mandrel.

SELF-TEST

1. Why is the wheel used on a cylindrical grinder typically of a denser structure than a surface grinding wheel for the same workpiece material condition?
2. What advantage is there in having the diamond wheel dresser integral with the footstock?
3. What is the purpose of a parallel test bar on a cylindrical grinder?
4. What amount of end play should there be when mounting workpieces on centers?
5. What special care must be taken when using a driver dog in cylindrical grinding?
6. When the traversing stops are being set, how much of the wheel width should be permitted to overtravel the ground part length?
7. What purpose is served by the tarry control?
8. For rough cylindrical grinding, about how much of the wheel width should be traversed for each rotation of the workpiece?
9. For finish grinding, about how much of the wheel width should be traversed for each rotation of the workpiece?
10. Finishing passes on the cylindrical grinder should remove about _____ inches from the diameter.

Figure L-72 Grinding the mandrel to dimension.

Universal Tool and Cutter Grinder

Like all cutting tools, milling cutters become dull or occasionally break during normal use. No production machine shop can function at peak efficiency unless it can keep its cutting tools sharp. The tool and cutter grinder is an essential tool for sharpening milling cutters and reamers.

The operation of the tool and cutter grinder in a large machine shop is often delegated to specialists in tool grinding. Even though tool grinding is a somewhat specialized area of machining, any well-rounded machinist should be familiar with the machines and processes used. A knowledge of tool and cutter grinding also makes it possible for the machinist to make special tools to meet the needs of unusual applications.

OBJECTIVES

After completing this unit, you should be able to:

- Identify a tool and cutter grinder and its parts.
- Briefly describe the function of this machine tool.
- Under guidance from your instructor, sharpen common cutting tools.

TOOL AND CUTTER GRINDER

The tool and cutter grinder is constructed much like a center-type cylindrical grinder (Figure L-74). It has the additional ability to swivel and tilt the workhead in two axes. Some also have the ability to tilt the wheelhead. Another unique capability is the raising and lowering of the wheelhead. The main use of the tool and cutter grinders is in sharpening cutters of all types.

Primary and Secondary Clearance Angles

A milling cutter must have clearance behind its cutting edge to cut. The required clearance actually consists of two angles that form the primary and secondary clearances.

The surface immediately behind the cutting edge is called the *land*. The angle formed by the land and a line tangent to the cutter at the tooth tip creates the primary clearance (Figure L-75). The angle between the back of the land and the heel of the tooth forms the secondary clearance.

Primary clearance is extremely important to the performance and life of the milling cutter. If this clearance is excessive, the cutting edge will be insufficiently supported and will chip from the pressure of the cut.

Primary and secondary clearance angles for machining various materials are detailed in Table L-3. When a milling cutter is sharpened, the primary clearance is ground first. The secondary clearance is ground later, so that the land may be ground to the recommended width. The correct land width will vary depending on the diameter of the cutter.

Measuring Clearance Angles

There are two common methods that can be used to determine the clearance angle of a milling cutter. The indicator drop method (Figure L-76) uses a dial test indicator to measure the difference in height across the width of the land (Figure L-77). Table L-4 shows the amount of indicated drop for various size cutters.

The cutter clearance gage (Figure L-78) is somewhat more convenient, as clearance angles can be measured directly in degrees.

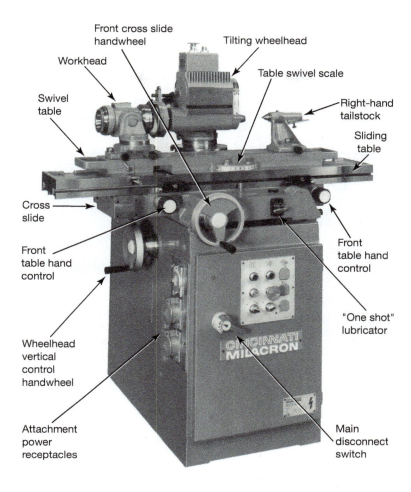

Figure L-74 Components of the cutter and tool grinder (*Courtesy of MAG Industrial Automation Systems, LLC*).

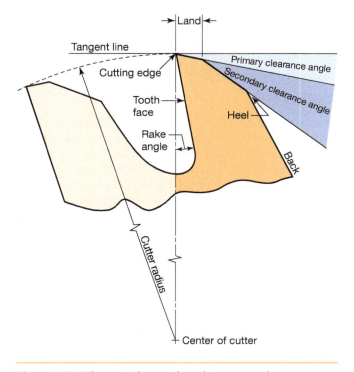

Figure L-75 Primary and secondary clearance angles.

Table L-3 Primary and Secondary Clearance Angles Suggested for High-Speed Steel Cutters

Material	Primary Clearance (degree)	Secondary Clearance (degree)
Carbon steels	3–5	8–10
Gray cast iron	4–7	9–12
Bronze	4–7	7–12
Brasses and other copper alloys	5–8	10–13
Stainless steels	5–7	11–15
Titanium	8–12	14–18
Aluminum and magnesium alloys	10–12	15–17

Wheel Styles for Cutter Sharpening

Milling cutters may be sharpened with a flaring cup wheel or straight wheel. The flaring cup wheel will produce a flat land (Figure L-79). A straight wheel will produce a slight

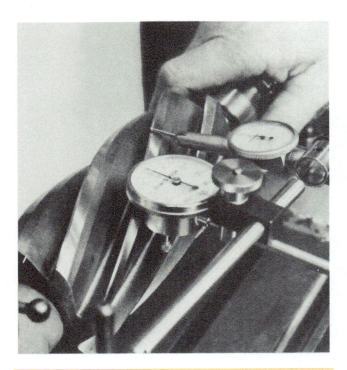

Figure L-76 Setting up indicator for the indicator drop method of checking clearances.

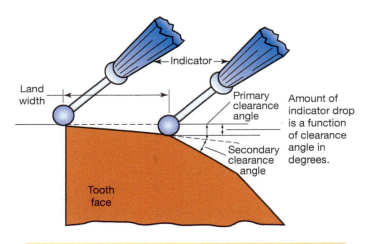

Figure L-77 Measuring primary clearance by indicator drop.

hollow grind behind the cutting edge. The resulting curvature is dependent on the diameter of the grinding wheel used. Form relieved cutters are sharpened on the tooth face using a shallow dish wheel.

Wheel Abrasive Selection

The grinding wheels selected for tool and cutter grinding should be biased toward a friable, medium-soft grade of about 60 grit. A wheel with grit that is too fine can result in slowed cutting action and the generation of excessive

Figure L-78 Checking the primary relief of a stagger-tooth milling cutter with a Starrett Cutter clearance gage (*Courtesy of Fox Valley Technical College.*).

heat, which can in turn result in metallurgical damage. Since so little material is ground in the reconditioning of most cutters, a softer grade of wheel can be utilized economically as compared with the one selected for surface grinding.

Work Holding

Arbor-driven cutters are held on grinding arbors for sharpening (Figure L-80). An adjustable grinding mandrel using an expanding slotted bushing may also be used (Figure L-81). (These must be used with great care to obtain the needed concentricity.) Cutters with shanks (such as end mills) are held in an accessory spindle or in the universal workhead.

Toothrests

Because the cutter is free to turn with the grinding arbor, a provision must be made so that each tooth to be sharpened can be solidly positioned. The toothrest is used for this purpose. Types of toothrests include the plain design (Figure L-82) and the micrometer adjustable type (Figure L-83). Various designs of toothrests are used (Figure L-84) depending on the type of cutter to be sharpened. An offset toothrest is used for helical milling cutters (Figure L-85).

Setting Up the Grinding Machine for Cutter Grinding

The centers (Figure L-86) are mounted on the swivel table so that they will accommodate the length of the cutter arbor. One center is spring loaded to facilitate installation and removal of the cutter arbor.

Table L-4 Indicator Drop Method of Determining Primary Clearance Angles

Diameter of Cutter (inches)	Average Range of Primary Clearance (degree)	Indicator Drop for Range of Primary Clearance Shown		Radial Movement for Checking
		Minimum	Maximum	
$\frac{1}{16}$	20–25	.0018	.0027	.010
$\frac{1}{8}$	15–19	.0021	.0032	.015
$\frac{3}{16}$	12–16	.0020	.0034	.020
$\frac{1}{4}$	10–14	.0019	.0033	.020
$\frac{5}{16}$	10–13	.0020	.0033	.020
$\frac{7}{16}$	9–12	.0025	.0038	.025
$\frac{1}{2}$	9–12	.0027	.0040	.025
$\frac{5}{8}$	8–11	.0028	.0045	$\frac{1}{32}$
$\frac{7}{8}$	8–11	.0033	.0049	$\frac{1}{32}$
1	7–10	.0028	.0045	$\frac{1}{32}$
$1\frac{1}{4}$	6–9	.0025	.0042	$\frac{1}{32}$
$1\frac{1}{2}$	6–9	.0026	.0043	$\frac{1}{32}$
$1\frac{3}{4}$	6–9	.0027	.0044	$\frac{1}{32}$
2	6–9	.0028	.0045	$\frac{1}{32}$

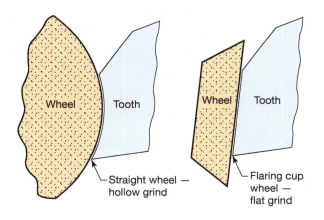

Figure L-79 The wheel type used will determine whether you get a flat or hollow-ground form.

Figure L-80 Slitting saw being mounted on grinding arbor.

Figure L-81 Adjustable grinding mandrel. The slotted bushing is moved along the mandrel to adjust for the I.D. of the tool to be mounted.

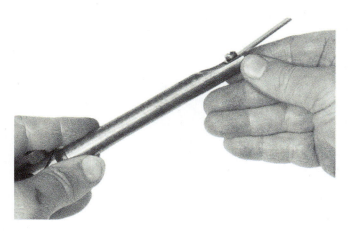

Figure L-82 Plain-type toothrest support.

Figure L-83 Micrometer toothrest support. This example also has provision for spring loading of the finger to permit ratcheting of the cutter tooth, called a flicker finger.

Figure L-85 Offset toothrest blade for use in grinding helical milling cutters.

Figure L-86 Tailstock centers are basic to a large portion of cutter grinding.

Figure L-84 Various designs of toothrests.

The arbor must be checked for runout, and the table must be aligned to travel without runout past the wheelhead. Attach a dial indicator to the wheelhead (Figure L-87) and use a parallel grinding arbor or test bar held between centers to determine table alignment. Make the necessary adjustments and lock the table in place (Figure L-88).

Establishing Clearance Angles

The type of cutter to be sharpened will determine where the toothrest is to be mounted. For a straight-tooth cutter such as a slitting saw, the toothrest will be mounted on the grinder table. When sharpening this type of cutter, it is necessary to rest only each tooth solidly against the toothrest blade.

A center height gage that is adjusted to the arbor centerline is used to set the toothrest blade and the grinder spindle to the arbor centerline height.

Figure L-87 After the grinding arbor is checked for runout on the centers at each end, the bezel should be zeroed and table alignment checked.

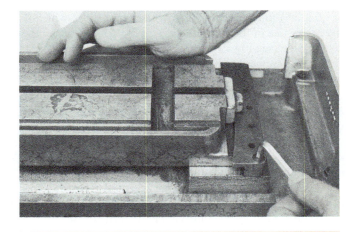

Figure L-88 Adjusting and locking the swivel table in alignment.

Because the toothrest is mounted to the table, clearance angles are established by raising the wheelhead above the center of the cutter. This will cause the grinding to occur behind the cutting edge, thus establishing the correct clearance angle. The amount to raise the wheelhead (W_r) is calculated by the formula:

$$W_r = \text{sine (clearance angle)} \times \text{radius of grinding wheel}$$

Helical cutters, because of their helix angle, must be rotated during grinding so that each point on the tooth is presented to the grinding wheel. It would not be possible to rotate the cutter if the toothrest were mounted on the table. However, mounting the toothrest on the wheelhead allows the cutter tooth constantly to be kept in contact with the toothrest blade as the cutter rotates through its helix.

The clearance angle for the cutting edge on the cutter is established by lowering the wheelhead. This causes the toothrest to lower also and causes the cutter tooth to be rotated below the cutter centerline. The grinding then occurs behind the cutting edge with the correct clearance angle. The amount to lower the wheelhead is calculated by the formula:

$$W_1 = \text{sine (clearance angle)} \times \text{radius of cutter}$$

Sharpening a Slitting Saw

When sharpening slitting saws, the toothrest is mounted on the swivel table, because it is necessary to hold only the cutter in a fixed position to grind each tooth. Rotation of the cutting edge on the toothrest during the grinding process is not necessary, as it is for helical cutters. A *flicker-type* toothrest is useful in this operation. After one tooth is ground, the table with the cutter is moved aside to clear the wheel, and the cutter is rotated up one tooth. The spring-loaded toothrest snaps aside to index for the next tooth (Figure L-89).

Staggered-Tooth Cutters

The staggered-tooth cutter is a helical cutter with the helix angle alternating direction each consecutive tooth. The grinding wheel must be dressed to a shallow V-form, and the toothrest must be mounted on the wheelhead and carefully centered (Figure L-90) to the apex of the V-form dressed into the wheel.

An additional procedure is required to grind the side teeth of a staggered-tooth cutter. Primary and secondary clearances are obtained by tilting the workhead (Figure L-91). On some machines, the wheelhead itself can be tilted. Note that the side cutting tooth must be in a horizontal orientation for grinding.

Form-Relieved Cutters

Form-relieved cutters are designed to machine a specific workpiece contour. Included among these are involute gear cutters and convex and concave milling cutters. Form cutters are sharpened only on the face of the tooth (Figure L-92) so that the particular geometry is maintained. It is important to have the face that is being ground in line with the cutter axis, unless marked otherwise on the side of the cutter. The shallow dish-shaped grinding wheel must be properly dressed and positioned relative to the face of the tooth.

Figure L-89 Grinding the primary relief (*Courtesy of K&M Industrial Machinery Co.*).

Figure L-90 Grinding two successive teeth to check for centering of the toothrest (*Courtesy of K&M Industrial Machinery Co.*).

Grinding End Mills with an Air Bearing Fixture

Figure L-93 shows an air bearing tool grinding fixture with the components and attachments identified.

Sharpening the Periphery of an End Mill Using the Cup Wheel Method

Bolt the fixture on the table of the grinding machine as shown in Figure L-94. Air pressure is used to act as the bearing between the spindle and the bore of the spindle housing. The escaping air also keeps grinding dust and contaminants out of the bearing area between the spindle and bore of the spindle housing. The air should be turned on before inserting the spindle. Air pressure should be no greater than 70 psi. Whenever the spindle is in the housing, the air must be on to keep from marring the precision bearing surfaces. Insert the spindle. Mount a cup wheel on the grinder. Absolute alignment with the table ways is unnecessary, since straightness comes from the stroke of the sliding spindle not the machine ways. The operator should stand behind the air-flow spindle while operating it.

The sliding spindle should be centrally balanced if the length of the end mill permits. On some grinders, a slight compounding of the table may be necessary to reach the wheel.

Select the collet for the end mill and secure it in the mouth of the spindle.

Position the clearance collar to the zero mark (see Figure L-95) and loosen the outboard finger arm so it can be moved along the dovetail.

Adjust the outboard arm and micrometer thimble adjustments to bring the finger blade rest (hereafter referred to as "finger") to the position shown in Figure L-96, which is the center line of the endmill. This can be done by using the center gage (see Figure L-96). When the finger is centered, secure the outboard arm and also the finger stem by tightening the knurled thumb screw.

Now center the end mill approximately to the grinder spindle so that the center of the end mill and the wheel spindle are about the same height (see Figure L-97).

Rotate the clearance collar clockwise the recommended number of degrees for the primary relief (as viewed for the operator's position) and re-secure the collar (see Figure L-98).

Dress the grinding area of the wheel to a narrow contact, but not too narrow as it will wear too quickly.

With the fixture rocking hand knob forward (see Figure L-99), in the grinding position, move the finger close to the wheel almost to the point of contact with the finger either exactly at the apex of the bevel or 1/16 nearer to the operator (see Figure L-100).

Next lock the longitudinal table movement; as you continue to grind, keep the wheel oriented the same way and the finger in the same relative position as the starting position. Positioning the finger nearer to the operator assures that as the end mill flute crosses the finger, as it leaves the wheel, it will not fall down suddenly and nick the end.

Figure L-91 The workhead is given additional tilt to provide for secondary clearance (*Courtesy of Fox Valley Technical College*).

Figure L-92 The wheel is positioned relative to the face of the tooth (*Courtesy of Fox Valley Technical College.*).

Figure L-93 Components of an air bearing fixture 1. Base 2. Swivel Housing Bracket 3. Spindle Bushing Housing 4. Tooth Rest Assembly 5. Spindle for 5C collet 6. Collet Drawbar 7. Cutter Arbor 8. Combination Tooth Rest Assembly 9. T Wrench 10. Spindle Stop Collar 11. 24-division Index Collar 12. Center Height Gage 13. Wide Tooth Rest 14. 5-C Collet 15. Indexing Lock Bracket (*Courtesy of Fox Valley Technical College*).

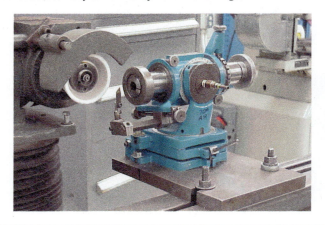

Figure L-94 Air bearing fixture mounted on the tool and cutter grinder (*Courtesy of Fox Valley Technical College*).

Commencing to Grind the Primary Clearance

Grasping the rear end of the spindle, slide and rotate the spindle forward until the finger rests on the flute at the end toward the shank (see Figure L-101). The stop collar is then secured on the spindle to prevent too much spindle movement.

Pull the spindle back causing the end mill flute to pass smoothly across the finger. Use gentle finger pressure clockwise as the spindle is pulled.

Now that the end mill is pulled back off the finger, rock the end mill away (see Figure L-102) from the grinding wheel by grasping the hand knob. Index the end mill to the next flute at the rear end of the flute, holding forward against the stop collar, resting securely on the finger.

Next let up on the rocker arm slowly and make contact with the grinding wheel on this flute. Do not slam the end mill

Figure L-95 Clearance collar set to zero (*Courtesy of Fox Valley Technical College*).

Figure L-96 Using the center gage to zero the finger (*Courtesy of Fox Valley Technical College*).

into the wheel. Now pull back on the spindle again for this flute and sharpen it. Continue grinding in .001 to .003 inch increments until the cutting edge is sharp and free of nicks.

Grinding the Secondary Clearance

Loosen and rotate the clearance collar further clockwise the recommended number of degrees for the secondary relief, as viewed for the operator's position, and re-secure the collar.

Notice how rotating the collar has lowered the finger height so that when one rests the end mill on the finger, the other has lowered the cutting edge and is now exposing the heel of the land to the grinding wheel. Now, move into the wheel and grind the heel until the recommended amount of primary margin (land) remains. Repeat for as many flutes as are on the end mill.

Figure L-97 Centering the end mill to the grinding wheel (*Courtesy of Fox Valley Technical College*).

Figure L-98 Setting the collar to grind primary clearance (*Courtesy of Fox Valley Technical College*).

Figure L-99 Fixture rocking hand knob forward (*Courtesy of Fox Valley Technical College*).

Sharpening the End Teeth Using the Indexing Stop Collar Method

Place the end mill in the collet. Adjust the front stop collars and the indexing stop collar on the spindle to keep the spindle from sliding in or out. Do not overtighten the set screws in the stop collars. With the plunger in one of the indexing

Figure L-100 Finger position for grinding (*Courtesy of Fox Valley Technical College*).

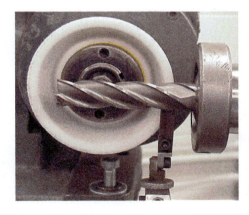

Figure L-101 Start grinding near the shank (*Courtesy of Fox Valley Technical College*).

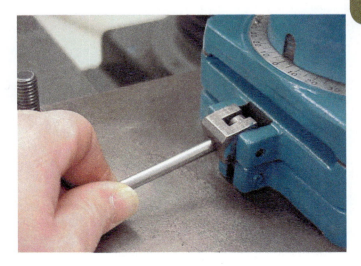

Figure L-102 Rock the end mill away to clear the wheel (*Courtesy of Fox Valley Technical College*).

Figure L-103 Tooth needs to be level (*Courtesy of Fox Valley Technical College*).

Figure L-104 Swivel the housing up for primary clearance grinding (*Courtesy of Fox Valley Technical College*).

Figure L-105 Swivel the base to grind dish on the cutter (*Courtesy of Fox Valley Technical College*).

slots, adjust the end mill so that the tooth to be ground is level (see Figure L-103).

Next, swivel the spindle housing up the recommended number of degrees for the primary clearance (see Figure L-104).

The swivel base is set at 1 or 2 degrees off of perpendicular to allow for center clearance grinding (Figure L-105).

The wheel is then raised above the center somewhat so the operator may sharpen the cutting edge right to the center

without colliding with the adjacent cutting edge. Set the table stop dog at the position where the grinding wheel edge sharpens to the center of the end mill without colliding with the adjacent cutting edge (see Figure L-106).

The longitudinal table feed is used to stroke across the cutting edge. After each stroke the end mill is indexed to the next tooth. When indexing, the operator pulls the plunger up a fraction of an inch to escape to the tooth on the indexing collar. Index the proper number of slots determined by the number of teeth on the end mill. Complete the grinding of the primary clearance in .001–.003 inch increments. Grind each tooth completely before setting a new depth of grind.

Grinding the Secondary Clearance

Grinding the secondary angle is identical to grinding the primary angle. The housing is swiveled downward the recommended number of degrees for the secondary clearance.

The end mill is then raised slightly above the center on the grinding wheel so the operator may sharpen the cutting edge right to the center without colliding with the adjacent cutting edge. Reset the table stop dog at the position where the grinding wheel edge sharpens to the center of the end mill without colliding with the adjacent cutting edge. Complete the grinding of the secondary clearance in .001–.003 inch increments until the primary land on the end is–the recommended specification.

Notching

When resharpening the end of the end mill, the end teeth may need to be notched for chip clearance (see Figure L-107).

Figure L-106 Sharpening to the center of the end mill (*Courtesy of Fox Valley Technical College*).

Figure L-107 Properly notched end mill (*Courtesy of Fox Valley Technical College*).

Figure L-108 General fixture position for notching (*Courtesy of Fox Valley Technical College*).

End mill notching is done in the general position as shown in Figure L-108 with a plunger pin cooperating with an index collar. Here, there is no finger on the end mill itself. The housing is tipped up, the wheel is dressed square, and the longitudinal table movement is used to plunge into the ends forming notches, the style of which can be seen on a new end mill.

A table stop is used to prevent over-stroking and control the depth of the plunge cut. End mill notching can also be done free hand, but this method should be done only under the supervision of an experienced tool grinder.

Maintenance and Care Instructions for an Air Bearing Fixture

1. Before using, remove the spindle carefully and wipe clean of any oil film, using acetone as cleaning agent. Insert spindle carefully in sleeve; use gentle alignment, not force.
2. Never use kerosene, oil or any lubricant. It is not necessary to use wax or other types of protective coatings.
3. Before operation, excessive moisture in air line should be blown out. It is recommended to use a dirt and moisture filter on air line, especially in extremely humid atmospheres.
4. After operation, the spindle may be removed from sleeve and wiped clean and dry for storing. Removal is recommended if work is in a dirty, oily, or moist atmosphere.
5. Do not operate except under approximately 70 psi of compressed air, CO_2 or other inactive gas.
6. Do not overtighten the split ring collar on the housing.

> **SAFETY FIRST**
>
> It should be noted that the gashing techniques shown present a significant hazard. The cutter is hand-held with fingers close to the high-speed wheel. Wrapping a cutter in a shop cloth also presents a significant hazard near the revolving equipment. **You must be extremely careful if you are gashing end mills by this technique.** Until you become experienced, ask your instructor for help in performing this cutter grinding operation.

Figure L-109 Center gashing an endmill (*Courtesy of Fox Valley Technical College*).

Center Gashing   Center gashing is necessary to permit a center-cutting end mill to enter the work axially. A straight wheel is dressed to a sharp beveled edge. The cutter is then gashed (Figure L-109) to the center. After each tooth is ground, burrs may be removed with a suitable dressing stick.

SELF-TEST

1. What machine capabilities differentiate the tool and cutter grinder from other designs of universal cylindrical grinders?
2. If the cutter clearances are excessive, what will result if the cutter is used?
3. Why is it desirable to sharpen different sets of milling cutters for aluminum alloys as compared with steels?
4. Why is the primary clearance ground first on a milling cutter?
5. Why is the toothrest attached to the grinding head when helical cutters are sharpened?
6. In grinding cutters, why should the second trial cut be made 180 degrees from the first grind?
7. In grinding staggered-tooth milling cutters, why must the toothrest be carefully centered to the grinding wheel?
8. On what surface are form-relieved cutters sharpened?
9. What grinding accessory is essential for the sharpening of the sides of end mills?
10. What tool and cutter grinder accessory is necessary for grinding the clearances on both the ends of end milling cutters and the side teeth of staggered tooth cutters?

INTERNET REFERENCES

Further information on universal tool and cutter grinders:

http://en.wikipedia.org/wiki/Tool_and_cutter_grinder

Computer Numerical Control and Other Advanced Machining Processes

Fundamentals of Computer Numerical Control (CNC)

INTRODUCTION

American manufacturers have worked diligently to increase productivity, reliability, and flexibility. In order to be competitive with foreign manufacturers, American manufacturers have had to improve quality and efficiency. Manufacturers have implemented many technologies that improve quality and productivity.

OBJECTIVES

After completing this unit, you should be able to:

- Describe various types of CNC machines.
- Identify various components of CNC machines.
- Describe the different axis coordinate systems.
- Identify positions on a Cartesian coordinate grid using absolute and incremental programming methods.
- Describe basic safe practices for CNC machines.

TYPES OF CNC MACHINES

The most common CNC machines are machining centers (milling machines) and turning centers (lathes). These two types represent more than half of the CNC machines on the market. The basic programming is very similar for all types of CNC machines.

Vertical Machining Centers

Vertical machining centers are vertical milling machines that use CNC controls for positioning and automatic tool changing to efficiently produce complex machine parts (Figure M-1).

Figure M-1 CNC vertical machining centers use automatic tool changing systems to complete multiple machining processes.

Machining centers can accomplish all milling tasks including straight-line, angular and circular milling, drilling, tapping, reaming, boring, chamfering, and face milling (Figure M-2). In addition to performing these common processes, a machining center has the capability to mill complex contours (Figure M-3). A machining center can also do thread milling (Figure M-4).

Figure M-2 Vertical machining center milling a part (*Courtesy of Haas Automation, Inc.*).

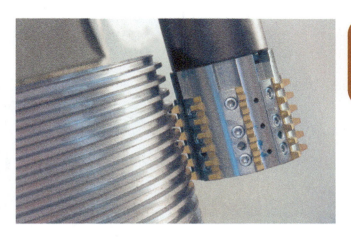

Figure M-4 Thread milling is another capability of the machining center (*Courtesy of Haas Automation, Inc.*).

Figure M-3 The CNC machining center can accomplish contour cutting in multiple axes (*Courtesy of Haas Automation, Inc.*).

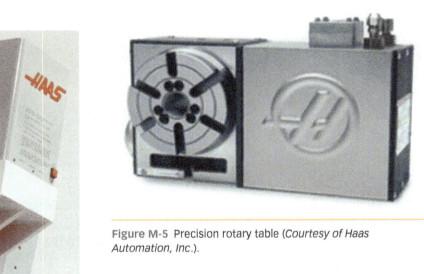

Figure M-5 Precision rotary table (*Courtesy of Haas Automation, Inc.*).

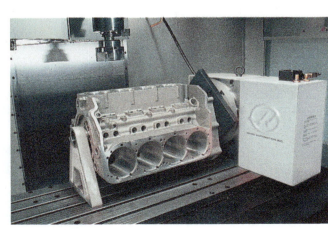

Figure M-6 The rotary table in the rotational axis of the vertical spindle machining center used for angular positioning of an engine block for a boring operation (*Courtesy of Haas Automation, Inc.*).

When the machining center is equipped with a programmable rotary table axis, its versatility is further enhanced (Figure M-5). The rotary axis may be used for angular workpiece positioning (Figure M-6), helical milling, or other multidimensional contour machining (Figure M-7).

Figure M-7 Helical milling is another application of the precision rotary table (*Courtesy of Haas Automation, Inc*.).

Horizontal Machining Centers

Horizontal machining centers are equipped with automatic tool changers along with a variety of other features to increase their versatility and production capabilities. Horizontal machining centers have a horizontal spindle. Horizontal machines are often larger and more rigid, enabling heavier, faster machining (Figure M-8).

CNC Turning Center

CNC turning centers are lathe-type machines and have between two and four axes (Figure M-9). The standard configuration consists of a two-axis lathe with one axis parallel to the spindle (*Z*) and one axis perpendicular to the spindle (*X*). A four-axis turning center has a rear or second tool turret that can be programmed independently of the master tool turret (Figure M-10).

Figure M-8 A CNC horizontal machining center (*Courtesy of Haas Automation Inc*.).

Figure M-10 Main turret tooling on the CNC turning center (*Courtesy of Haas Automation, Inc*.).

Figure M-9 CNC Turning center with a bar feeder (*Courtesy of Haas Automation Inc*.).

COMPONENTS OF CNC MACHINES

The CNC Machine

Modern CNC machine tools have been completely redesigned for CNC machining and bear little resemblance to their conventional counterparts. Requirements for new CNC machines include rigidity, rapid mechanical response, low inertia of moving parts, high accuracy, and low friction along the ways.

Tool Changing

Tool changers automatically change cutting tools without operator intervention. Machining centers can have anywhere from 16 to 100 tooling stations. Lathe or turning centers typically have 8–12 tools, which are indexed automatically. The speed of tool changing is an important factor for CNC machines used in production environments. Many machines have bidirectional turrets that can select the quickest route to the desired tool.

Rotary tool changers are very common (Figure M-11). Other types of vertical spindle machining centers include the gantry type on which the table is fixed and the spindle moves across on the cross rail, which is carried back and forth by twin gantries, one on each side of the machine. Large machining centers may have a movable table, with the tool head and spindle carried on the cross rail. These machine tools, popular in the aircraft manufacturing industry, may have tables 40 or more feet long and also have multiple tooling spindles.

AXES AND COORDINATE SYSTEMS

To fully understand CNC programming, you must understand axes and coordinates. Think about a simple part. You could describe the part to someone else by its geometry. For example, the part is a 4-inch by 6-inch rectangle. Any point on a machined part, such as a hole to be drilled, can be described in terms of its position. The system that allows us to do this, called the *Cartesian coordinate system* or *rectangular coordinate system,* was developed by a French mathematician, René Descartes.

The Cartesian Coordinate System

Consider one axis first. Imagine a line with zero marked in the center (Figure M-12). To the right of zero every inch has been marked with a positive number representing how far the mark is to the right of zero. Mark the inches to the left of zero beginning with −1, −2, −3, and so on. This line is called the X axis. We could describe a particular position on our line by giving its position in inches from zero. This would be called a position's *coordinate.*

Next a perpendicular line (axis) is added that crosses the first line (X axis) at zero. The horizontal line is the X axis and the vertical line is the Y axis. The point where the lines cross is the zero point, usually called the *origin*. Points are described by their distance along the axis and by their direction from the origin by a plus (+) or minus (−) sign (Figure M-13).

Quadrants

The axes divide the work envelope into four sections called *quadrants* (Figure M-14). The quadrants are numbered in a counterclockwise direction, starting from the upper right.

Points in the upper right, quadrant 1, have positive X (+X) and positive Y (+Y) values.

Points in quadrant 2 have negative X (−X) and positive Y (+Y) values.

Points in quadrant 3 have negative X (−X) and negative Y (−Y) values.

Points in quadrant 4 have positive X (+X) and negative Y (−Y) values.

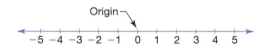

Figure M-12 Single-axis coordinate line.

Figure M-11 Rotary tool changer (*Courtesy Fox Valley Technical College*).

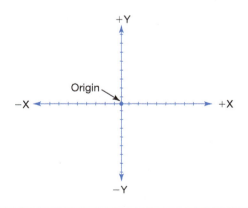

Figure M-13 Dual-axis coordinate grid.

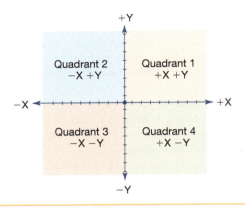

Figure M-14 The four quadrants of the Cartesian coordinate system.

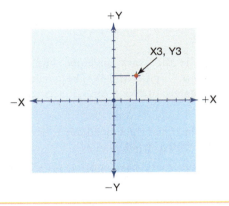

Figure M-15 Locating a position using the Cartesian coordinate system.

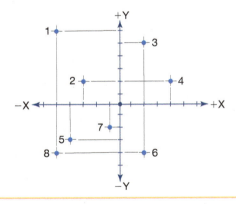

Figure M-16 Cartesian coordinate system and the XY coordinates for eight points.

Location	X	Y
1	-5	6
2	-3	2
3	2	5
4	4	2
5	-4	-3
6	2	-4
7	-1	-2
8	-5	-4

To locate a point such as (X3.0, Y3.0) in the two-axis system, start at the zero point and count to the right (+ move) three units on the *X* axis and up (+ move) three units on the *Y* axis.

Figure M-15 shows a point in the Cartesian coordinate system. The point's coordinates are identified as X3, Y3. Note that only one point would match these coordinates. Figure M-16 shows another Cartesian coordinate system with eight points identified with their coordinates. Add one more axis (Figure M-17) to represent depth. To drill a hole, the hole would be described by its location by its *X* and *Y* coordinates. A *Z* value would be used to represent the depth of the hole. The *Z* axis is added perpendicular to the *X* and *Y* axes.

Polar Coordinates

It is also possible to describe the position of points by stating angles and distances along the angles. The direction of the angular line is in relation to the *X* axis. A positive angular dimension runs counterclockwise from your present position. A negative angular dimension would run clockwise from your present position. Study Figure M-18.

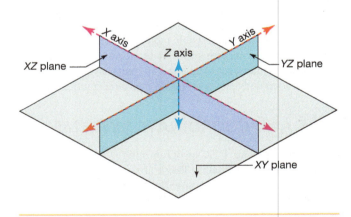

Figure M-17 Three-dimensional space defined by the *X–Y*, *X–Z*, and *Y–Z* planes.

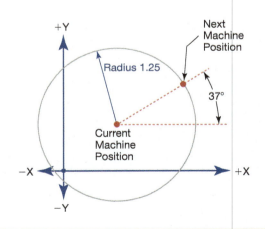

Figure M-18 Polar coordinates are described from the current position, not from the absolute axes origin (X0,Y0). In this example, the next position is 1.250″ on a line that is +37° through the current position from the *X* axis.

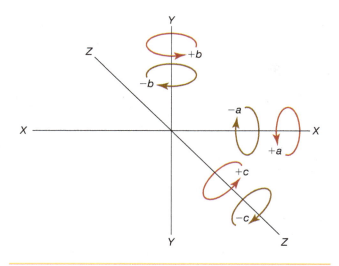

Figure M-19 The programmable rotational axes, *a, b,* and *c,* and directions of motions.

The desired position therefore is a point on the angular line, the desired radial distance from your present position (A+37, 1.250).

When using polar coordinates, the *X* value represents the radius distance and the *Y* value represents the angle from 0 degrees. In the example in Figure M-18, the *X* would be 1.25 and the *Y* would be 37 degrees in the polar coordinate mode.

These three axes, *X, Y,* and *Z,* are the fundamental programmable axes on all CNC machine tools. However, there may also be three programmable rotational axes: *a, b,* and *c* (Figure M-19). The a **axis** is rotational around the *X* axis, the b **axis** is rotational around the *Y* axis, and the c **axis** is rotational around the *Z* axis. Rotational axes permit the programming of accessories such as rotary

tables and indexing devices. With six axes available to be programmed and the addition of a rotary table, indexing, and swivel tool heads, the machining center has the capability to machine extremely complex multidimensional contours.

ABSOLUTE AND INCREMENTAL PROGRAMMING

Absolute coordinates specify the relative tool moving position in relation to the program zero. Incremental coordinates specify the tool moving distance and direction from its current position. CNC programs can be written in absolute, incremental, or may use both formats in the same program.

Absolute programming specifies a position or an end point from the workpiece coordinate zero (datum). It is an absolute position (Figure M-20).

Incremental programming specifies the movement or distances from the point where the spindle is currently located (Figure M-21). A move to the right or up from this position is always a positive move (+); a move to the left or down is always a negative move (−). With an incremental move, we are specifying how far and in what direction we want the machine to move.

Absolute positioning systems have a major advantage over incremental positioning. If the programmer makes a mistake when using absolute positioning, the mistake is isolated to the one location.

When the programmer makes a positioning error using incremental positioning, all future positions are affected. Most CNC machines allow the programmer to mix absolute and incremental programming. There are times when using both systems in one program will make the program easier to write.

Figure M-20 Absolute coordinate positions are always located from the program zero.

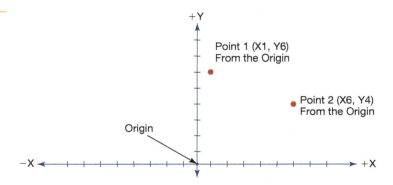

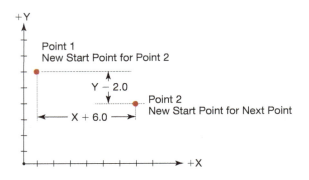

Figure M-21 In incremental coordinate positioning, your present position becomes the program start position. From the start point, point 1 is one position to the right on the *X* axis (*X*+1) and six positions up on the *Y* axis (*Y*+6). When we move to point 2, point 1 becomes our start point. Point 2 is five positions to the right of our present position on the *X* axis (*X*+5) and two positions down on the *Y* axis (*Y*−2). Remember, in incremental programming, moves down or moves to the left are negative moves.

CNC SAFETY

CNC operations represent unique safety issues beyond those of conventional machine tools. In conventional machines, your hands are typically on the machine cranks or other controls and you can hear, feel, and see what is happening. However, with CNC most of these functions are under the control of the CNC controller. You can see and hear what is happening, but you have no feel for it. Starting a machine function from the machine control happens so fast that you may not have time to react before the machine, tooling, or workpiece is damaged. To protect yourself and an expensive piece of equipment, double-check everything that you do, before you act.

CNC machine tools are equipped with several features that can increase safety. The feed hold control on the machine control will cause the machine to stop motion and go into a hold mode although on some CNCs, the feed hold mode will not turn off the spindle. The feed hold button is useful if you see or hear something that does not seem right. After you have checked things you can continue machining.

The single block feature is also useful for testing and program verification. With single block turned on, only one program block is executed at a time, permitting you to check each step. Most machining centers have a provision to inhibit the Z (spindle)-axis motion or to run all program blocks with no motion in any axis. These features can be useful during program verification. In case of a real problem, the emergency stop will halt all functions. When the emergency stop button is used however, you will have to restart the program.

Most problems in CNC arise from programming errors, so if you have a method to verify the program before going to the machine, it may avoid a crash, ruined part and/or damage to the machine. Most of the computer programming systems will enable complete program testing. This is a great method for eliminating problems before the program is run on the machine. You still must exercise caution the first time any program is run on a machine.

SELF-TEST

1. Describe the three axes that a mill typically has.
2. What direction is negative Z in terms of the spindle?
3. Make a sketch of the four Cartesian quadrants and identify the signs of each quadrant.
4. Describe the incremental positioning mode.
5. Name one disadvantage of the incremental positioning mode.
6. Identify the positions marked on the Cartesian coordinate grid below. Use absolute positioning.

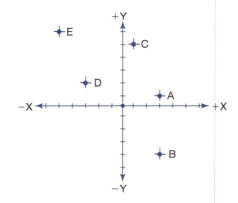

Point	X	Y
A		
B		
C		
D		
E		

Point	X	Y
Origin to Point A		
Point A to Point B		
Point B to Point C		
Point C to Point D		
Point D to Point E		

Using the Cartesian coordinate system, write down the *X* and *Y* values for the incremental moves.

Fundamentals of Machining Centers

INTRODUCTION

A CNC machining center is a CNC milling machine that is equipped with an automatic tool changer. The manual milling machine is a very versatile and productive machine tool, but when coupled with a computer control, it becomes the production center of the machine shop. Repetitive operations, such as milling, drilling, tapping, and boring, are perfect applications for the machining center.

OBJECTIVES

After completing this unit, you should be able to:

- Describe the purpose and function of the machining center.
- Identify the major components of the machining center.
- Identify the axes and directions of motion on a typical machining center.
- Describe the different methods of manually moving the machine axes.
- Define terms such as "MDI" and "conversational."

VERTICAL MACHINING CENTER

The vertical machining center is probably the most versatile and common CNC machine found in the machine shop (Figure M-22). The vertical configuration of the spindle lends itself to quick, easy workpiece setups.

PARTS OF THE MACHINING CENTER

Figure M-23 shows the main components of a vertical machining center.

Figure M-22 Vertical spindle machining centers are very versatile machines. The quick setup of these machines makes them very popular in the machine shop (*Courtesy of Haas Automation Inc.*).

Column

The column is the backbone of the machine. The column is typically mounted to the saddle and provides one of the axes or directions of travel. The rigid construction of the column will keep the machine from twisting during machining.

Bed

The bed is one of the more integral parts of the machining center. The bed is typically produced from high-quality cast iron, which absorbs the vibration of the machining operation. Hardened and ground slideways are mounted to the bed to provide alignment and support for the machine axes.

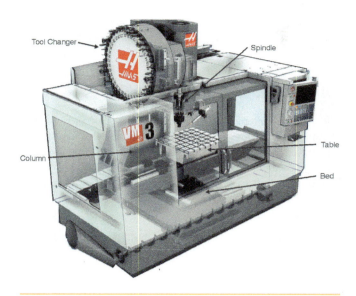

Figure M-23 The main components of the vertical machining center (*Courtesy of Haas Automation Inc*.).

Table

The table is mounted on the bed, and the work or a work-holding device is mounted to the table. The table has T-shaped slots milled in it for mounting the work or work-holding device.

Spindle

The spindle holds the cutting tool and is programmable in revolutions per minute.

Tool Changers

Tool changers are an automatic storage and retrieval system for the cutting tools. An automatic tool changer makes the CNC milling machine a machining center.

Carousel-Type Tool Changers

Carousel-type tool changers are spindle-direct tool changers, meaning they do not use auxiliary arms to change tools (Figure M-24). The carousel can be mounted on the back or side of the machine. Carousel tool changers are typically found on vertical machining centers. When a tool change is commanded, the machine moves to the tool change position and puts the current tool away. The carousel then rotates to the position of the new tool and loads it.

Arm-Type Tool Changers

The tool change arm rotates between the machine spindle and the tool magazine. After getting the tool change command, the tool that is in the spindle will come to a fixed position known as the "tool change position." The automatic tool change (ATC) arm will rotate and will pick up the tool

Figure M-24 Carousel-type tool changing systems are usually found on vertical machining centers (top figure). On a vertical machining center the spindle positions itself over the tool and then moves down and clamps the tool in the spindle. The photo on the bottom is (*Courtesy of Haas Automation Inc*.).

using a gripper. The arm has two grippers, one on each end of the arm. Each gripper can rotate through 90 degrees, to deliver tools to the spindle. One end of the arm will pick up the old tool from spindle and the other end will pick up the new tool from the tool magazine. The arm then rotates and places the old tool back in the tool magazine.

Tool carousels typically have bidirectional capabilities, which enable faster tool changes. Tool-changing cycle time is very important to a machining center's productivity. Tool-changing time is nonproductive time. No machining occurs while a tool is being changed.

AXES OF MOTION

The linear axes or directions of travel of the machining center are defined by the letters *X, Y,* and *Z* (Figure M-25). Axis designation letters appear with positive or negative signs for direction of travel. The *Z* axis always lies in the same direction as the spindle. A negative *Z* (−*Z*) movement always moves the cutting tool closer to the work. The *X* axis is usually the

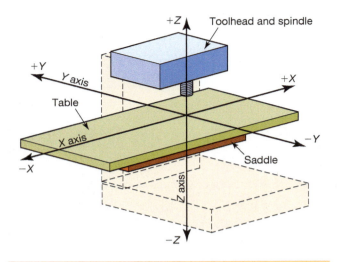

Figure M-25 The three linear axes of the vertical machining center are identified as X, Y, and Z. The Z axis always lies in the same plane as the spindle.

Figure M-26 The rotary table adds a fourth axis of travel to the machining center. The rotary table can be used to do helical-type machining or it can be used to reposition the workpiece (*Courtesy of Haas Automation Inc.*).

axis with the greatest amount of travel. On a vertical machining center, the X axis would be the left/right travel of the table. The Y axis on the vertical machining center would be the travel toward and away from the operator. On the horizontal machining center, the Y axis would be the up and down travel of the spindle head. It may be helpful to think of the motion in terms of tool position. If the table is moved to the left, the tool is positioned more in the +X direction. If the table is moved toward the front, the tool is positioned more in the +Y direction.

Rotational Axes

A horizontal machining center that is equipped with a rotary table is capable of four axes of motion. The fourth axis is known as the C-axis (Figure M-26). The C-axis is a rotational axis about the Z-axis. Machining centers that are equipped with a rotary table and tilting, contouring spindle

are said to have five axes: three linear and two rotary. Four- and five-axis machines are used to machine parts with complex surfaces such as mold cavities or rotary turbines.

MACHINE CONTROL FEATURES

Machining center controls come in all styles and levels of complexity, but they all have many of the same features. If you have a good understanding of one machine, the next control will be much easier to learn. Control functions are divided into two distinct areas: manual controls and program controls.

Manual Control

Manual control features are those buttons or switches that control machine movement (Figure M-27).

Emergency Stop Button

The emergency stop button is the most important component on the machine control. This button has saved more than one operator from disaster by shutting down all machine movement.

This big red button with the word "Reset" or "E-stop" on the front should be used when it is evident that a collision or tool breakage is going to occur.

MOVING THE AXES OF THE MACHINE

Manual movement of the machine axes is done in a number of ways. Most controls are equipped with a pulse-generating hand wheel (Figure M-28).

The hand wheel has an axis selection switch that allows the operator to choose which axis he or she wants to move.

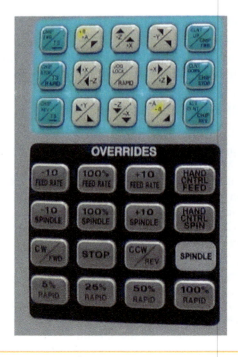

Figure M-27 The manual machine controls are located on the left-side of this control.

Figure M-28 Some controls have a hand wheel for each axis, while other controls have one hand wheel and a switch to select which axis the operator will move.

The handle sends a signal or electronic pulse to the motors that move the table or the spindle head. If the handle of the *Y* axis is turned in the negative direction, the tool moves toward the operator. If the *Z* axis hand wheel is moved in a negative direction, the tool moves toward the table.

Some machines are equipped with jog buttons (Figure M-29). When the jog button for a certain axis is pressed, the axis moves. The distance or speed at which the machine moves is selected by the operator prior to the move.

Spindle Speed and Feed Rate Override Switches

Spindle speed and feed rate overrides are used to speed up or slow down the feeds and speeds of the machine while machining (Figure M-30). The override controls are typically used by the operator to adjust to changes in cutting conditions, such as hard spots in the material. Spindle speeds can typically be adjusted from 0 to 200 percent of the programmed spindle speed.

Cycle Start/Feed Hold Buttons

The two most commonly used buttons on the control are the cycle start and feed hold buttons (Figure M-31). The cycle start button is used to start execution of the program. The feed hold

will stop execution of the program without stopping the spindle or any other miscellaneous functions. By pushing the cycle start, the operator can restart the execution of the program.

Home or Zero Return

The home or zero return button, when selected, will return all of the axes of the machine to the home position. Home position is usually defined as a position at the extreme travel limits of the three main axes. The zero or home position is

Figure M-30 The spindle speed and feed rate override switches give the operator a greater amount of control over the machine.

Figure M-31 The cycle start and feed hold buttons are located adjacent to each other. These buttons will typically light up when activated. The feed hold button will only stop the axis from traveling; it will not stop the spindle or any other miscellaneous function.

Figure M-29 Jog buttons are used for machine axis movement. The amount of movement per push of the jog button is controlled by the mode selector switch.

set by using switches and encoders. The home return button is used when the operator wants to load or unload parts or to start the program from the home position.

WORKPIECE COORDINATE SETTING

The workpiece coordinate or program zero is the point or position from which all of the programmed coordinates are established. For example, if the programmer looks at the part print and notices that all of the dimensions come from the center of the part, these datums are then used to establish the program zero or workpiece coordinate (Figure M-32).

The part origin is the X0, Y0, Z0 location of the part in the rectangular or Cartesian coordinate system. In absolute programming, all of the tool movements would be programmed with respect to this point. If all of the dimensions were located from the center of the bored hole, then that point would become the program zero.

During part setup, the X and Y zero position of the part has to be located. Using the hand wheels or other manual positioning devices and an edge finder or probe, the machinist locates the point at which the center of the spindle and the part origin are the same (Figure M-33).

The "home zero" is then entered as a G-code in the appropriate area of the program or in an offset table.

The machinist must then measure and enter the values of the tool lengths in the offset table for each tool being used in the program. Each actual tool used in the program probably has a different length.

The control must then be told to compensate for the difference in the lengths. The tool length offsets are typically the distance from Z at machine zero to the Z position of the part zero (Figure M-34). The tool length offsets are stored as an offset in the control (Figure M-35).

Single Block Operation

The single block option on the control is used by the operator to advance through the program one block or line at a time. When the single block switch is on, the operator presses the button each time he or she wants to execute a program block.

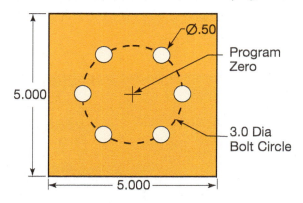

Figure M-32 The workpiece coordinate system can be set from any datum feature. Pick the feature that would allow the programmer to do the fewest calculations.

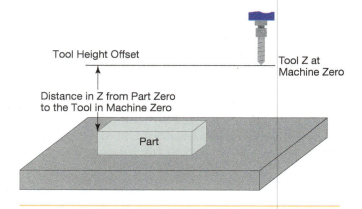

Figure M-33 It is extremely important to accurately locate the workpiece zero point; otherwise, all of the machined features will be shifted out of location. An edge finder (top) or probe (bottom) is used to position the center of the spindle at the part zero location.

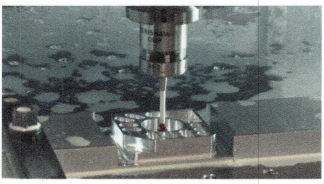

Figure M-34 The operator touches the tool off on the Z0 point of the part. He or she then takes the Z axis distance from the Z0 point of the part and the machine home position. The distance is recorded in the tool length offset table. This must be done for every tool that is used in the program.

When the operator wants the program to run automatically, he or she can turn the single block off and press cycle start, and the program will run through without stopping. The purpose of the single block switch is to allow the operator to watch each operation of the program carefully. It is typically used the first time a program is run.

Figure M-35 When the program calls for the tool length offset, the control accesses the register located in the tool offset area. The control uses this to compensate for the tool length.

<< PROBING		TOOL OFFSET			TOOL INFO >>	
TOOL 1	COOLANT	H(LENGTH)		D(DIA)		
OFFSET	POSITION	GEOMETRY	WEAR	GEOMETRY	WEAR	
1 SPINDLE	10	4.5680	0.	0.	0.	
2	0	0.	0.	0.	0.	
3	0	0.	0.	0.	0.	
4	0	0.	0.	0.	0.	
5	0	0.	0.	0.	0.	
6	0	0.	0.	0.	0.	
7	0	0.	0.	0.	0.	
8	0	0.	0.	0.	0.	
9	0	0.	0.	0.	0.	

<< WORK PROBE		WORK ZERO OFFSET				WORK PROBE >>
G CODE	X AXIS	Y AXIS	Z AXIS			
G52	0.	0.	0.			
G54	-12.5680	-8.4890	-23.1480			
G55	0.	0.	0.			
G56	0.	0.	0.			
G57	0.	0.	0.			
G58	0.	0.	0.			
G59	0.	0.	0.			
G154 P1	0.	0.	0.			
G154 P2	0.	0.	0.			
G154 P3	0.	0.	0.			
ENTER A VALUE						

Manual Data Input

Manual data input, or MDI, is a means of inputting commands and data. MDI can be used to enter a simple command, such as starting the spindle, or to enter an entire program and is done through the alphanumeric keyboard located on the control (Figure M-36).

Program Editing

After a part program is loaded into the control, it may need some modification. The machinist can make changes using the program edit mode. The machinist uses the display to find the program errors and the keyboard to correct the errors.

Display

The display shows information such as the written program or part graphics. Graphics can be displayed on the screen if the machine has graphics capabilities. Graphics are a representation of the part and the tool path, which would be generated by the program. Graphics are a safe way to test part programs.

Conversational Programming

Conversational programming is a built-in feature that allows the programmer to respond to a set of questions that are displayed on the graphics screen. The questions guide the programmer through each phase of machining. First, the operator might input the material to be machined. This information is then used by the control to calculate speeds and feeds. The operator can override these. Next, the operator would choose the operations to be performed and input the geometry. Operations such as pocket milling, grooving, drilling, and tapping can be performed.

Figure M-36 The alphanumeric keyboard lets the operator edit or enter a program at the machine control.

SELF-TEST

1. Describe the function of the tool changer.
2. What are the three major axes associated with the machining center?
3. Which axis always lies in the same plane as the spindle?
4. What piece of equipment gives a machining center a 4th axis of motion?
5. Describe two methods of manually moving the axes of the machining center.
6. Describe the workpiece coordinate or zero point.

Fundamentals of Programming Machining Centers

INTRODUCTION

CNC programming is the process of taking the information from a part print that would be used to do manual machining and converting it to a language that a CNC machine will understand. This chapter focuses on word address programming.

WORD ADDRESS PROGRAMMING

Word address programming precisely controls a machine's movements and functions by using short sentence-like commands. Consider a simple operation that could be performed on a manual milling machine and convert it to a word address command: turn the spindle on in a clockwise direction at a spindle speed of 600 RPM (revolutions per minute).

The command to do this on a CNC machine would be: M03 S600. The M03 commands the spindle to start in a clockwise direction. The S600 tells the spindle how fast to turn. This is one block of information.

Letter Address Commands

A CNC machine is controlled by the use of *letter address commands*. Following are abbreviated descriptions of the most common letter address commands.

N is used for the line number or sequence number for each block of program.

G is used for specific modes of operation. G-codes are also called preparatory functions. G-Codes set up the mode in which the machining operation(s) are to be executed.

F is the feed rate.

S is the spindle speed setting.

T is the letter address for a tool.

M is a miscellaneous function. Miscellaneous functions include coolant on/off, spindle forward, spindle reverse, and many others.

H and D are letter address codes used for tool height and tool diameter offsets.

Programming Terminology

A program word is composed of two parts: an address and a number. Study Figure M-37. In the first example, G is the address and 01 is the number. Together they are called a *word*. A G01 word commands the machine to make a linear move. The second example of a word (S800) is composed of the S address and 800. S800 would set a spindle speed of 800 RPM. The third example (M08) of a word is composed of an address of M and 08. An M08 word commands the control to turn flood coolant on.

A block is one line of code for a CNC machine. Think of a block as being one operation. It can consist of one or more words (see Figure M-38). Each block is ended with an end-of-block character (;). Note that the end-of-block

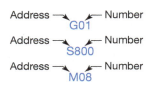

Figure M-37 Examples of word addresses.

Figure M-38 A block of code.

N20 M06 T01;   Block 1

N030 M03 S800;   Block 2

N040 G01 X1.0 F8.0;   Block 3

N050 G01 Y2.5;   Block 4

N060 G01 X5.25;   Block 5

Figure M-39 Order of execution.

character is usually generated automatically when the programmer hits the enter key at the end of a line. The end-of-block character does not normally show up on the machine display. When operating, a CNC control reads one block and executes it, it then reads the next block and executes it, and so on (Figure M-39).

The order of words in a block can vary. Figures M-40 and M-41 show the typical arrangement and order of words in blocks. Note that most blocks will contain far fewer words.

Program Numbers

Programs are stored in the controller by their program number. Controls can store many programs. Program numbers start with the letter O. For example, O5 would be program 5. The program number normally appears before the first line of code in a program (Figure M-42). In this figure, the program number is O100. Note also the use of sequence numbers to name blocks of code. Remarks can be put in parentheses on any line. The controller ignores remarks.

Part Programming

A part program is simply a series of blocks that execute motions and machine functions to make a part. Consider a simple program that cuts around the outside of a 3-inch by 4-inch block with a 1/2-inch diameter end mill (Figure M-43). The program must compensate for the radius of the cutter to make the part the correct size. The cutter radius is .250 so the

	N G X, Y, Z F S T M H, D	
N	Sequence (line) number	
G	Preparatory function. Specifies the mode of operation under which a command will be executed.	
X, Y, Z	Dimension words. Used to specify a position or distance to move.	
F	Feed rate	
S	Spindle speed	
T	Designates the tool to be used.	
M	Miscellaneous function code. Designates things such as spindle on/off, spindle direction, coolant on/off, etc.	
H,D	Used to specify tool offsets for height and/or diameter.	

Figure M-40 Typical arrangement of words in a block.

N050	G01	X3.250	Y1.750	F8.0	S800	M03	;
Sequence Number	Preparatory Function	Dimension Words		Feed Rate	Spindle RPM	Miscellaneous Function	End of Block

Figure M-41 Typical order for one block of code.

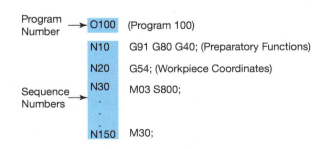

Figure M-42 Use of program number and sequence numbers.

Program Number → O100   (Program 100)
N10   G91 G80 G40; (Preparatory Functions)
N20   G54; (Workpiece Coordinates)
Sequence Numbers → N30   M03 S800;
.
.
.
N150   M30;

Figure M-43 Profile mill of a 3-inch by 4-inch block.

program must keep the center of the tool .250 to the left of the part as it moves in a clockwise direction around it.

N0010 M03 S800

N0020 G00 X-1.00 Y-1.00 (point 1)

N0030 G01 X-.25 Y-.25 F10.0 (point 2)

N0040 G01 Y3.25 (point 3)

N0050 G01 X4.25 (point 4)

N0060 G01 Y-.25 (point 5)

N0070 G01 X-.50 (point 6)

N0010 M03 S800

Line number 10 turns on the spindle in a clockwise direction at 800 RPM.

N0020 G00 X-1.00 Y-1.00

Line number 20 is a rapid feed move (G00) to position the tool just off the lower left-hand corner of the part (X-1, Y-1). This is position 1 in Figure M-43.

N0030 G01 X-.25 Y-.25 F10.0

Line number 30 feeds (G01) the tool at an angle to a position that is 1/2 of the tool diameter to the left and below the side of the part (point 2, X-.250). The tool will feed at 10 inches per minute (F10.0 IPM). The right edge of the cutter is now aligned with the left-hand side of the part. It is now ready to cut the left side of the part.

N0040 G01 Y3.25

Line number 40 cuts the left side of the part and positions the spindle center past the top of the part (Y3.25) by 1/2 the tool diameter (point 3). This positions the edge of the tool for the cut across the top of the part. The feed rate is still 10 inches per minute.

N0050 G01 X4.25

Line 50 cuts the top of the part. The feed rate is still 10 IPM, because we have not changed it since line number 20. The spindle center is now positioned .25 to the right of the part (point 4). This gets us ready to cut the right-hand side of the part.

N0060 G01 Y-.25

Line number 60 moves the tool to point 5. The right-hand side of the part is now complete. The tool center is also positioned 1/2 tool diameter below the bottom of the part, ready to cut.

N0070 G01 X-.50

Line number 70 cuts the bottom of the part to size and moves the tool completely off the part (point 6, X-.50).

Next, take a closer look at the individual parts of a word address part program.

Part Datum Location

To program a part, you need to determine where the workpiece zero (or part datum) should be located. The part datum is a feature of the part from which the majority of the dimensions of the part are located.

Because all of the dimensions of the part in Figure M-43 come from the lower left-hand corner, this was the logical choice for the workpiece zero point. It is good practice to choose a part feature that is easy to access with an edge finder and one that will involve the fewest number of calculations. This approach will help avoid errors.

Sequence Numbers (Nxxxx)

Sequence numbers are a way to identify blocks of information within a program. Line numbers are handy for the operator. The machine controller can be commanded to find blocks of information by their line numbers.

In addition, line or sequence numbers are needed in the use of some canned cycles, which will be covered later in this book.

G-Codes (Preparatory Functions)

G-codes are used to set modes such as linear interpolation (G01) or rapid traverse (G00). Linear interpolation means that the cutter moves on a controlled linear path (line). A "G" followed by a two-digit number determines the machining mode in that block or line.

G-codes or preparatory functions fall into two categories: modal or nonmodal. Nonmodal or "one-shot" G-codes are those command codes that are only active in the block in which they are specified.

Modal G-codes are those command codes that will remain active until another code of the same type overrides them.

For example, if you had five lines that were all linear feed moves, you would only have to put a G01 in the first line. The next four lines would be controlled by the previous G01 code. The feed rate was modal in the first example. The feed rate does not change unless a different feed rate is commanded.

The G-codes shown in Figure M-44 are commonly used machining center G-codes, but some of them may be slightly different from those used on your machine tool. Consult the manual for your machine to be sure. Note: Appendix A has a more extensive list of common G codes and examples of their use.

SPINDLE CONTROL FUNCTIONS

Spindle speeds are controlled with an "S" followed by up to four digits. When programming a machining center, the spindle speed is programmed in RPM. A spindle speed of 600 RPM would be programmed S600. Spindle speeds may also be programmed in surface feet per minute (SFPM) through the use of a G96 preparatory code. SFPM is the cutting speed of the material you are machining. Most turning center controls will typically program in SFPM. This allows the spindle speed to automatically change as the diameter of the workpiece changes, maintaining a constant surface speed. For example, a cutting speed for mild steel and a carbide tool might be 400 SFPM. A spindle speed of 400 SFPM would be programmed G96 S400. Then as the diameter of the turned part gets smaller, the RPM would increase to keep the cutting speed correct. The

G00	Rapid traverse (rapid move)	Modal
G01	Linear positioning at a feed rate	Modal
G02	Circular interpolation clockwise	Modal
G03	Circular interpolation counter-clockwise	Modal
G17	XY Plane Selection	Modal
G20	Inch Programming	Modal
G28	Zero or home return	Non-Modal
G40	Tool diameter compensation cancel	Modal
G41	Tool diameter compensation-left	Modal
G42	Tool diameter compensation-right	Modal
G43	Tool height offset	Modal
G49	Tool height offset cancel	Modal
G54	Workpiece coordinate preset	
G80	Canned cycle cancel	Modal
G81	Canned drill cycle	Modal
G83	Canned peck drill cycle	Modal
G84	Canned tapping cycle	Modal
G85	Canned boring cycle	Modal
G90	Absolute coordinate positioning	Modal
G91	Incremental positioning	Modal
G92	Workpiece coordinate preset	
G98	Canned cycle initial point return	Modal
G99	Canned cycle R point return	Modal

Figure M-44 Commonly used machining center G-codes.

M00	Program stop	Non-Modal
M01	Optional stop	Non-Modal
M02	End of program	Non-Modal
M03	Spindle start clockwise	Modal
M04	Spindle start counterclockwise	Modal
M05	Spindle stop	Modal
M06	Tool change	Non-Modal
M07	Mist coolant on	Modal
M08	Flood coolant on	Modal
M09	Coolant off	Modal
M30	End of program & reset to the top of program	Non-Modal
M40	Spindle low range	Modal
M41	Spindle high range	Modal
M98	Subprogram call	Modal
M99	End subprogram & return to main program	Modal

Figure M-45 Miscellaneous functions.

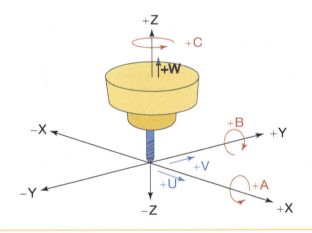

Figure M-46 Machine axes of motion.

spindle is turned on using a miscellaneous (M) code of either M03 or M04. An M03 will turn the spindle on in a clockwise direction, while an M04 will turn the spindle on in a counter-clockwise direction. A M05 turns the spindle off.

MISCELLANEOUS FUNCTIONS (M-CODES)

Miscellaneous functions or M-codes perform miscellaneous functions such as tool changes, coolant control, and spindle operations. An M-code is a two- or three-digit numerical value preceded by a letter address, "M." M-codes, similar to G-codes, can be modal or nonmodal. Figure M-45 lists commonly used machining center miscellaneous functions (M-codes).

Tool Calls

On a machining center, a tool change is commanded with a miscellaneous code of M06. Next you tell the control which tool to change to. A typical tool change block for a machining center would be M06 T02. This line of code would tell the CNC controller to change to tool 2.

AXES WORDS (*X, Y, Z*)

There are three axes on a typical milling machine: *X, Y,* and *Z.* Each axis is specified by a letter (*X, Y,* or *Z*), which may be preceded by a direction sign (+ or −). Unit 1 covered

the Cartesian coordinate system, which specifies how the axes of the machines are oriented as well as the direction of travel. A simple command block to rapid position the tool of the milling machine to 1 inch above the workpiece zero would be: N0010 G00 Z1.00. The G00 means a rapid move, and the Z1.0 means 1 inch above the workpiece zero.

There are additional axes on some machines. Figure M-46 shows additional axes of motion. If a machine has additional auxiliary axes that can move in the *X, Y,* and *Z* directions, they are called U, V, and W. Rotational extra axes around the *X, Y,* and *Z* are labeled A, B, and C.

MOTION BLOCKS

Motion can be controlled in three ways: rapid positioning, linear feed, or circular feed.

Rapid Traverse Positioning (G00)

A rapid positioning block consists of a preparatory G-code (G00) and the coordinate to which you want to move. A rapid move to a location of X10, Y5, and Z1 would be programmed:

G00 X10.0 Y5.0 Z1.0. This block would command the machine to move at a rapid traverse rate to this position, moving all of the axes simultaneously. An operator can choose to reduce the rapid rate from 0 to 100 percent on the CNC control. Rapid feeds should be reduced when testing a new program.

Linear Feed Mode (G01)

A G01 (or G1) linear interpolation code moves the tool to a commanded position in a straight line at a specific feed rate. The feed rate is the speed at which the machine axes move. Linear feed blocks are normally cutting blocks. The rate at which the metal is removed is controlled through a feed rate code (F). Machining centers can use feed rates in inches per minute (IPM), or inches per revolution of the spindle (IPR). The feed rate type selection of IPM or IPR is controlled by G codes. Machining centers are typically programmed in inches per minute, or IPM.

To make a straight-line cutting motion on a machining center, the block of information would look like this: G01 X10.00 F8.00. The tool would move to an *X*-axis position of 10.00 inches at a feed rate of 8 inches per minute. Remember, straight-line moves can also be angular. CNC machine controls are able to make simultaneous axes moves (Figure M-47).

The G00, G01, and F codes are all modal. Modal commands are active unless changed by another preparatory code. If you were programming a series of straight-line moves, you would only have to put the G01 and the feed rate in the first line. The lines that follow would be controlled by the previous G01 and feed rate. Note that it is OK to put the G01 or G1 in every line and it may make the program easier to understand.

The example shown in Figure M-48 uses the programming procedures that have been covered to this point. We

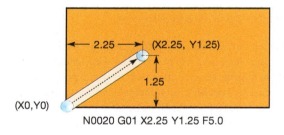

N0020 G01 X2.25 Y1.25 F5.0

Figure M-47 G01 linear interpolation example.

need to mill around the outside of the part, .250 inches deep, with a .50-inch end mill at a feed rate of 5 inches per minute. The first step is to set up the control by doing our preliminary procedures. The second step is the tool call. The third step sets the WPC or workpiece zero point. In our fourth step, we need to start the spindle and set the RPM. Next, we rapid position close to the part and start our linear cutting moves. After we have cut the profile of the part, we need to return to the home position and end the program.

O001 (Program name O0001)

N100 G20 G17 G40 G49 G80 G90 (Preparatory Information - G20 - input data in inches, G17 - X, Y plane, G40 - cancel cutter diameter compensation, G49 - cancel tool length compensation, G80 - cancel fixed cycles, G90 - absolute mode)

N101 G54 (Work piece fixture location 1)

N104 T1 M6 (T1- Tool 1- .5" end mill, M6 - tool change)

N105 S800 M3 (S800 - Spindle speed 800, M3 - spindle on clockwise)

N106 G00 X-1.0 Y-1.0 (G00 - Rapid move to X-1.0 Y-1.0)

N108 G00 Z.1 M8 (G00 - Rapid to Z. 1, M8 - flood coolant on)

N110 G01 Z-.25 F5.0 (G01 - linear move to Z-.25, F5.0 - Feed rate 5 inches per minute)

N111 G01 X-.25 Y- .25 (G01 - Linear move to X-.25 Y-.25 at same feed rate as the previous line)

N114 G01 Y2.75 (G01 - Linear move to Y2.75)

N116 G01 X4.75 (G01 - Linear move to X4.75)

N118 G01 Y.75 (G01 - linear move to Y.75)

N120 G01 X2.75 (G01 - Linear move to X2.75)

N122 G01 Y-.25 (G01 - Linear move to Y-.25)

N124 G01 X-1.00 (G01 - Linear move to X-1.00)

N132 G00 Z.1 (G00 - Linear move to Z.1)

N134 M05 (M05 - Spindle stop)

N136 G28 M09 (G28 - Return to reference point, M09 - flood coolant off)

N138 M30 (Program end, memory reset)

%

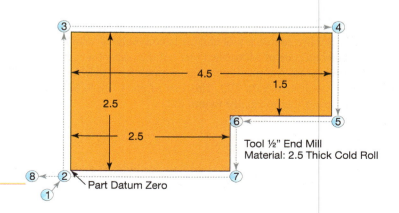

Figure M-48 Simple contour programming example.

PROGRAMMING

Programs can be thought of as having three sections. In fact, a CNC program is similar to a business letter. There is a heading, the body of the letter, and then an ending.

- The heading, or the introduction, consists of preparatory codes.
- The body consists of the machining operations.
- The ending consists of codes needed to end the program.

Preparatory Codes

Preparatory codes are used to set conditions or cancel conditions.

The program tells the machine how to operate in order to produce a quality part in a safe manner. Some preparatory G-codes are used to cancel modes. For example, if the machine just finished running a program that used offsets, the offsets should be canceled before running another program. If the same program is run again, the offsets are called at the beginning of the program.

A G40 would cancel diameter offsets and a G49 would cancel height offsets. Canned cycles that were active should be canceled at the end of a program. G40 and G49 are examples of canceling modes that may be active for safety and proper operation.

Other codes are used to set the mode we want to operate in. For example, if our program is written in inch mode (not metric), we would use a G20. If our program is developed in the XY plane, we would tell the control to use the XY plane by using a G17.

Study the lines of code below. There are three lines. Each line of code is considered one block of information to the control. The first line is O0001. This is the name of the program. Line N10 has a G20 code. This tells the control that our program is written in inch mode (not metric). Study line N020. The G17 tells the control that the program was written in the XY plane. The G40 tells the control to cancel diameter offsets. The G49 cancels height offset. The G80 cancels canned cycles. The G90 sets the control to absolute mode.

O0001

N10 G20

N20 G17 G40 G49 G80 G90

The order in which they are done and the line numbers are unimportant. In fact, many shops have somewhat standard header files they use to cancel all offsets and canned cycles and set the modes they normally use for programming.

CODES TO CONSIDER FOR PREPARATORY FUNCTIONS

Codes to cancel modes:

G40	Cancel diameter offset
G49	Cancel height offset

G80	Cancel canned cycles

Codes to set modes of operation:

G17	XY plane designation
G18	ZX plane selection
G19	YZ plane selection
G20	Inch mode
G21	Metric mode
G54	Workpiece coordinates 1 (Many machines also allow G55-G59 to be used for alternative workpiece coordinates.)
G90	Absolute programming
G91	Incremental programming
G94	Feet-per-minute
G95	Feed-per-revolution

Comments

Other information may also be included in the program. Many companies include setup information for the operator. Study the example that follows. Note the parentheses around the comments in each block. The control ignores text that is between parentheses. It is only for operator information.

O0002

N010 (X0 Y0 is the lower left-hand corner)

N020 (Tool 1: 1″ spot drill)

N030 (Tool 2: .25 drill)

N040 (Tool 3: .500 end mill)

N050 (Set the part up in a vise with 2-inch parallels)

Machining Operations

The body of the program contains the machining operations. The code in the body of the program is used to load the proper tools, control speeds, feeds, and motion of tools.

End the Program

This is the simplest part of the program. In this part of the program, you might want to make sure the tools have been put away, cancel canned cycles, and tell the control the program is complete.

WORKPIECE COORDINATE (WPC) SETTING

The machine must know where the part is on the table. Remember that the machine has a zero position. Figure M-49 shows that for this machine, the machine zero position is located at the left front and top of the machine table. The machine always knows where this is. The machine has no idea where the workpiece is until we tell it where it is. The part position can be called workpiece coordinates. This is called workpiece coordinate setting. The WPC tells the machine the position of the part datum (Figure M-49).

Figure M-49 Workpiece coordinates.

Figure M-50 Workpiece coordinates.

Figure M-51 Machine zero and part origin.

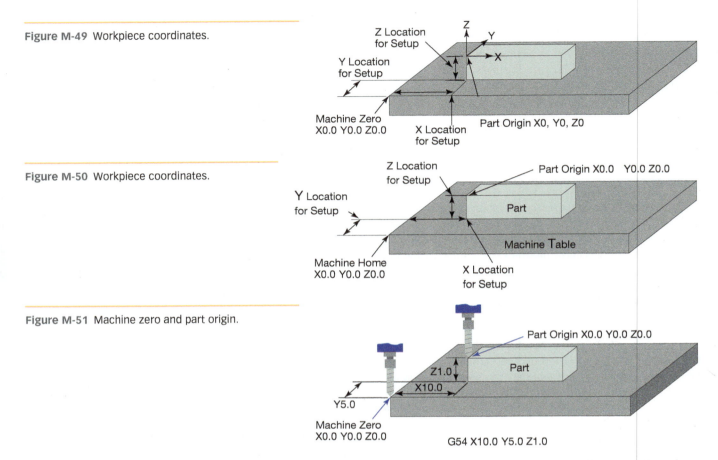

The workpiece part zero may be located on any part of the workpiece, but we have to tell the control where this point is on the machine table. The technique for locating the workpiece zero varies for each machine tool. Some controls use a button to set the zero point. The setup person or operator uses the jog buttons to position the spindle center over the part datum and then presses the zero set or zero shift button to set the coordinate system to zero. On other types of controls, the WPC is set with a G-code. There are two common WPC G-codes: G54 and G92. Machines use either a G54-G59 for stored values or a G92 followed by X, Y, and Z dimensions.

The G54-G59 workpiece coordinate is the absolute coordinate position of the part zero (Figure M-50). These are not available on all machines. Six are available, and all serve the same function. This allows the programmer to have six different workpiece coordinates established on a machine. This would be very beneficial for repetitive jobs that could be located at the same position on the machine table. For example, a job that is run once each week might use the G59. The G59 would be used to establish the location for that particular job.

To locate this position, the operator would position the center of the spindle directly over the part zero using an edge finder or probe and then take note of the machine position (Figure M-51). The coordinates of this position would be placed in the G54 line. A typical workpiece coordinate setting of this type would be written: N0010 G54. Note that the values for X, Y, and Z are stored in the CNC and don't appear in the program.

The G92 workpiece coordinate is the incremental distance from the workpiece datum (X, Y, and Z zero) to the center of

the spindle (Figure M-52). In effect, it tells the machine where the spindle is in relation to the workpiece. The spindle must then be in that position when you start to run the program.

When a G92 is called, the center of the spindle must be in the preprogramed position. If it is not, the control will start machining at the wrong position. If, for example, the center of the spindle at the home position is 10 inches to the right on the X axis, 5 inches back on the Y axis, and 8 inches above the part on the Z axis, the G92 would be written: G92 X10.00 Y5.00 Z8.00. If the center of the spindle were located any other distance away from the part datum, when the G92 was called, the tool would try to cut the part in the wrong location. This is why using a G92 can be very dangerous!

The G54 type of WPC setting is a lot safer than a G92. No matter where you are when the G54 is called, the control knows exactly where your part is located because it is an absolute position, not an incremental distance.

It is important to remember that a G54 or G92 will not move the machine tool to this point; it merely tells the control where the part (G54) or spindle (G92) is.

INCREMENTAL PROGRAMMING

Now that we have established the basics of absolute programming, we can look at another type of positioning, incremental positioning. Absolute programming is when all of the coordinates of the part program are related to an absolute zero point. Incremental programming defines the coordinates of the part in relationship to the present position.

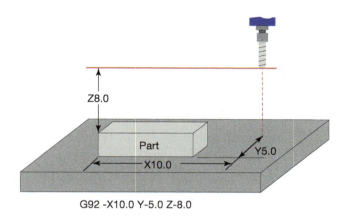

Figure M-52 G92 workpiece coordinates.

Incremental programming is also known as point-to-point positioning. The point where the spindle is presently, is the zero for the next coordinate position. Incremental positions are the direction (+ or −) and the distance to the next point. See Figure M-53 which contrasts absolute and incremental moves. For example, to move from point 3 to point 4 in incremental mode, X axis would not need to move and Y would need to move −3.5″. Incremental moves are simply the distance and direction needed to go from the current machine position to the next position.

Incremental programming can be used to program the whole part or just certain sections of the program. Incremental positioning is programmed with a G91 preparatory code and can be very useful when programming a series of holes that are incrementally located on the part print. Figure M-54 would be a typical application for incremental programming.

Most of the program to drill the holes could just be incremental .75″ moves in the X or Y direction. If you wished to switch back to absolute programming at any point in the program, you would use a G90 code.

CIRCULAR INTERPOLATION

Up to this point, we have discussed only straight-line moves. One of the most important capabilities of a CNC machine is the ability to do circular cutting motions. CNC machines are capable of cutting any arc of any specified radius value. Arc or radius cutting is known as *circular interpolation*. Circular interpolation is carried out with a G02 or G03 code.

Programming Circular Moves

There are two basic methods used to program circular moves: incremental arc center or radius.

Programming Circular Moves Using the Incremental Arc Center Method

When the machine starts cutting an arc, the tool is already positioned at the start point of the arc. To program a circular move, we first need to tell the direction of the arc. Is it a

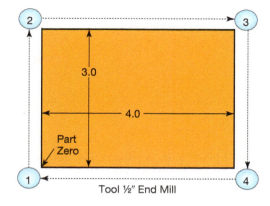

Tool Movement	Absolute Command	Incremental Command
Point 1 to Point 2	X-.25 Y3.25	X0.0 Y3.5
Point 2 to Point 3	X4.25 Y3.25	X4.5 Y0.0
Point 3 to Point 4	X4.25 Y-.25	X0.0 Y-3.5
Point 4 to Point 1	X-.25 Y-.25	X-4.5 Y0.0

Figure M-53 Comparison of absolute and incremental programming.

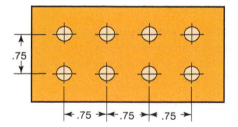

Figure M-54 Incremental programming example.

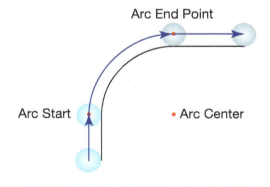

Figure M-55 The critical pieces of information needed to cut an arc are the arc start point, arc direction, arc end point, and arc centerpoint location.

clockwise (G02) or counterclockwise (G03) arc? The second piece of information the control needs is the end point of the arc. The last piece of information is the location of the arc center (Figure M-55).

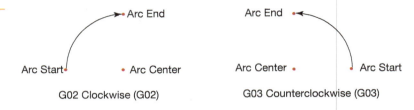

Figure M-56 G02 and G03 arc direction.

Arc Start Point

The arc start point is the coordinate location where the arc starts. The tool is moved to the arc start point in the line prior to the arc generation line. Simply stated, the start point of the arc is the point where the spindle is located when you want to start machining the arc (Figure M-55).

Arc Direction (G02, G03)

Circular interpolation can be carried out in two directions, clockwise or counterclockwise. There are two G-codes that specify arc direction (Figure M-56). The G02 code is used for circular interpolation in a clockwise direction. The G03 code is used for circular interpolation in a counterclockwise direction. Both G02 and G03 codes are modal and are controlled by a feed rate (F) code, just like a G01.

Arc End Point

The arc end point is the coordinate position for the end point of the arc. The arc start point and arc end point set up the tool path, which is generated according to the arc center position (Figures M-57 and M-58).

Arc Centerpoints

To generate a circular path, the controller must know where the center of the arc is. When using the IJ method, a particular problem arises. How do we describe the position of the arc center? If we use X, Y, Z coordinate position words to describe the end point of the arc, how will the controller discriminate between the end point coordinates and the arc center coordinates? The answer is that we use different letters to describe the same axes. Secondary axes addresses are used to designate arc centerpoints. The secondary axes addresses for the axes are:

I = X-axis coordinate of an arc centerpoint

J = Y-axis coordinate of an arc centerpoint

K = Z-axis coordinate of an arc centerpoint

Only two of three secondary addresses will be used to generate an arc, because we are only cutting the arc in the XY plane. When cutting arcs in the XY axes, the I/J letter addresses will be used. If we were cutting an arc on a turning center, the X/Z axes would be the primary axes, and the I/K letter addresses would be used to describe the arc centerpoint.

The type of controller you are using dictates how these secondary axes are located. With most controllers, the arc centerpoint position is described as the incremental distance from the arc start point to the arc center (Figure M-58).

While very uncommon, there are a few controls where the arc centerpoint position is described as the absolute location of the arc centerpoint from the workpiece zero point (Figure M-59). Note: this is very uncommon.

If the arc centerpoint is located down or to the left of the start point, a negative sign (–) must precede the coordinate dimension (Figure M-60).

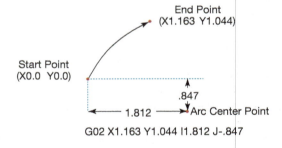

Figure M-58 The center point of the arc is the distance from the start point position to the arc center position.

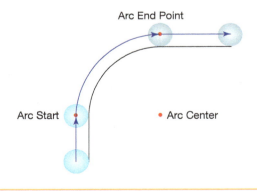

Figure M-57 Arc end point location.

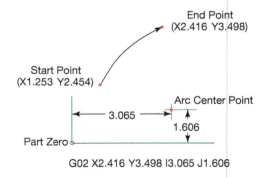

Figure M-59 When using absolute centerpoint positioning, the centerpoint is the coordinate position from the part zero. This method is very uncommon.

Figure M-60 Circular interpolation example. The incremental arc method was used.

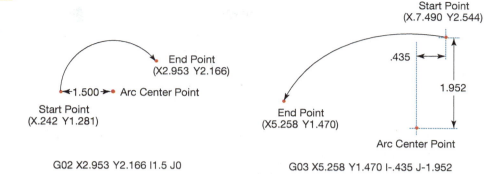

G02 X2.953 Y2.166 I1.5 J0

G03 X5.258 Y1.470 I-.435 J-1.952

Remember that I and J values are normally incremental. I and J values must be the incremental distance and direction. The direction is shown by the plus or minus sign. Also remember that the distance and direction is taken from the start of the arc to the center of the arc.

Circular Interpolation Using the Radius Method

Circular motion can also be programmed using the radius of the arc. The general format is G02 (or G03) X Y R. Figure M-61 shows an example of a circular cut using the radius method. The code for this cut would be G02 X4.000 Y3.000 R1.0. The G02 means that it will be a clockwise circular cut. The X value is 4.000 inches (absolute) in the X axis. The Y value is 3.000 inches (absolute) in the Y axis. The R value is the length of the radius value. In this example, it is +1.0 inches.

Study Figure M-62. Note that there are two possible arcs with the same start point, same end point, and radius of 1.0″. One is greater than 180 degrees and the second one is less than 180 degrees. We tell the controller which one to use by giving the radius value a plus or minus sign.

Figure M-63 shows an arc that is more than 180 degrees. The code to machine this arc would be G02 X4.0625 Y3.0625 R-2.5. G02 means that it is a clockwise arc. The X value is the

absolute X position of the end point. The Y value is the absolute position of the end point in the Y axis. The R value is the value of the radius of the arc. Note that it must be negative in this example because the arc is more than 180 degrees.

Figure M-64 shows an arc that is less than 180 degrees. The code to machine this arc would be G02 X4.0625 Y3.0625

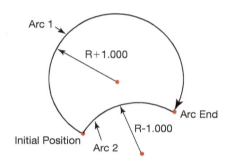

Figure M-62 Two possible arcs with the same start point, same end point, and same radius.

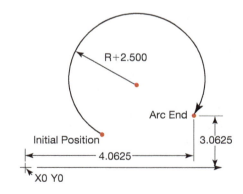

Figure M-63 Arc of more than 180 degrees.

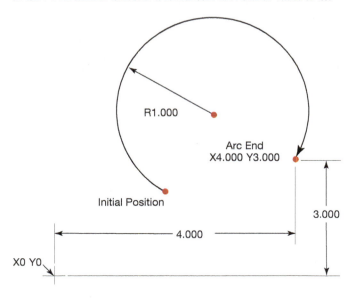

Figure M-61 Circular cut using the radius programming method.

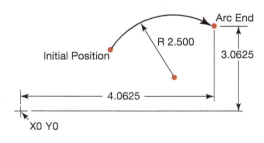

Figure M-64 Arc of less than 180 degrees.

R2.5. G02 means that it is a clockwise arc. The X value is the absolute position of the end point in the *X* axis. The Y value is the absolute Y position of the end point. The R value is the value of the radius of the arc. Note that it must be positive in this example because the arc is less than 180 degrees.

The radius method cannot be used on most CNC controls to program a full circle. The IJ method should be used for full circles.

COMPREHENSIVE PROGRAMMING EXERCISE

Figure M-65 shows a base plate that involves linear and circular cutting. To simplify the programming of the part, the holding device has been eliminated and programming is done at the center of the spindle (no tool offsets).

O0001 (Program name O0001)

N10 T2 M6 (Tool change to tool 2)

N20 G54 (Workpiece coordinates are in G54 register)

N30 M3 S800 (Turn spindle on clockwise direction 800 RPM)

N40 G00 X6.00 Y − .25 (Rapid position spindle to X6.0 Y-.25)

N50 G00 Z.100 (Rapid position to Z.100)

N60 G01 Z − .500 F6.0 (Linear feed to Z-.500 at 6.0 IPM)

N70 G01 X1.00 (Linear feed to X1.00 at 6.0 IPM)

N80 G02 X − .25 Y1.00 I0.0 J1.25 (Circular mill an arc in the clockwise direction at 6.0 IPM)

N90 G01 Y2.00 (Linear feed to Y2.00 at a feed rate of 6.0 IPM)

N100 G02 X1.00 Y3.25 I1.25 J0.0 (Circular mill an arc in the clockwise direction at 6.0 IPM)

N110 G01 X4.00 (Linear feed to X4.00 at 6.0 IPM)

N120 G02 X5.25 Y2.00 I0.0 J − 1.25 (Circular mill an arc in the clockwise direction at 6.0 IPM)

N130 G01 Y1.00 (Linear feed to Y1.00 at 6.0 IPM)

N140 G02 X4.00 Y − .25 I − 1.25 J0.0 (Circular mill an arc in the clockwise direction at 6.0 IPM)

N150 G01 Y − 1.00 (Linear feed to -Y1.00 at 6.0 IPM)

N160 M05 (Turn spindle off)

N170 M6 T1 (Tool change to tool 1)

N180 M03 S750 (Spindle on counterclockwise at 750 RPM)

N190 G00 X1.00 Y1.00 (Rapid to X1.00 Y1.00)

N200 G00 Z.100 (Rapid to Z.100)

N210 G01 Z − .525 F2.0 (Feed to Z-.525 at 6.0 IPM)

N220 G01 Y2.00 F5.0 (Feed to Y2.00 at 5.0 IPM)

N230 G01 Z.100 (Feed to Z.100 at 5.0 IPM)

N240 G00 X4.00 (Rapid to X4.00)

N250 G01 Z − .525 F2.0 (Feed to Z-.525 at 2.0 IPM)

N260 G01 Y1.00 F5.0 (Feed to Y1.00 at 5.0 IPM)

N270 G01 Z.100 (Feed to Z.100 at 5.0 IPM)

N280 T0 M6 (Put the tool away)

N290 M30 (Program end, memory reset)

%

TOOL LENGTH OFFSETS

Up to this point, we have not used tool length offsets. Length offsets make it possible for a CNC machine to adjust to different tool lengths. Every tool is going to be a different length, but CNC machines can deal with this quite easily. CNC controllers have a special area within the control to store tool length offsets.

The tool length offset is the distance from the tool tip at home position to the workpiece Z zero position (Figure M-66). This distance is stored in a table that the programmer can access using a G-code or tool code. On a machining center a G43 code is typically used. The letter address G43 code is accompanied by an "H" auxiliary letter and a two-digit number. The G43 tells the control to compensate the Z axis, while the H and the number tell the control which offset to call out of the tool offset table. The tool length offset typically needs to be accompanied by a Z axis move to activate it.

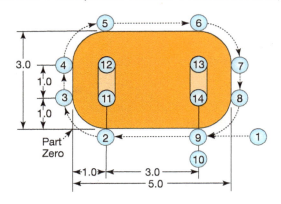

Figure M-65 Base plate.

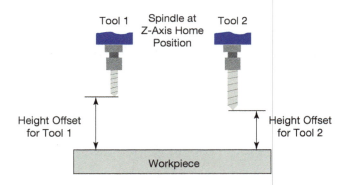

Figure M-66 Tool length offsets.

A typical tool-length offset block would look like this:

N0010 G43 H10;

The G43 calls for a tool length offset, and the H10 is the number of the offset, which is found in register 10 of the tool length offset file. It is a good idea to correspond the tool length offset register number to the tool number. For example, if you are using tool number 10 (T10), try to correspond the height offset by using height offset number 10 (H10).

On some other brands of CNC machines, the height offset is called up with the tool number. If the program calls for tool number 10, the control automatically accesses the tool file and offsets the tool according to the tool length registered in the tool file under the tool number 10.

Because machine controls vary, it is a good idea to find out how your control deals with variations in tool lengths.

TOOL DIAMETER OFFSETS

Tools also differ in diameter, and to compensate for this we use tool diameter offsets. Tool diameter offsets are also used to control the size of milled features. Tool diameter offsets allow you to program the part, not the toolpath.

In Figure M-67 we had to compensate for tool diameter by offsetting the tool path by the radius of the tool to the left or right. The control can offset the path of the tool automatically so we can program the part just as it appears on the part print. This eliminates having to mathematically calculate the cutter path. The diameter offset also allows the programmer to use the same program for any size cutter by just changing the offset. Without diameter offsets, the programmer would have to know the precise size of the tool to be used and program using the center of the spindle.

With cutter compensation, the cutter size can be ignored and the part profile can be programmed. The radius of the cutting tool is entered into the offset file, and when the offset is called, the tool path will automatically be offset by the tool radius. If the part is too big or too small, the offset can be changed so that the next part will be accurate.

Cutter compensation can be to the right or to the left of the part profile. To determine which offset to use, imagine yourself walking behind the cutting tool. Do you want the tool to be on the left of the programmed path or to the right (Figure M-67)?

When compensation to the left is desired, a G41 is used. When compensation to the right is desired, a G42 is used. When using the cutter compensation codes, you need to tell the controller which offset to use from the offset table. The offset identification is a number that is placed after the direction code. A typical cutter compensation line would look like this: G41 D12.

To initialize cutter compensation, the programmer has to make a move (ramp on). This additional move must occur before cutting begins. This move allows the control to evaluate its present position and make the necessary adjustment from centerline positioning to cutter periphery positioning. This initial move must be larger than the amount of the tool offset. The machine corrects for the offset of the tool during the ramp move. In Figure M-68 the machine compensates for the offset in the move between point 1 and point 2.

To cancel the cutter compensation and return to cutter centerline programming, the programmer must make a linear move (ramp off) to invoke a cutter compensation cancellation (G40). This is an additional move after the cut is complete (point 5 to point 6). Do not cancel the cutter compensation until the tool is away from the part by at least as much as the compensation amount. Figure M-68 shows a typical programming example that uses tool length and tool diameter compensation.

N10 G17 G20 G40 G80 G90 (G17 - X, Y plane, G20 - input data in inches, G40 - Cancel Cutter diameter compensation, G80 - cancel fixed cycles, G90 - Absolute mode)

N20 G54 (Work piece coordinates 1)

N30 T2 M06 (Tool change to tool 2- .375 end mill)

N40 S800 M03 (Turn spindle on clockwise at 800 RPM)

N50 G00 X-1.0 Y-1.0 (Rapid to X-1.0 Y-1.0 Position 1)

N60 G43 H2 Z.1 (Tool length compensation positive by the amount in H2, rapid to Z.1)

N70 G01 Z- .5 F5.0 (Linear move to Z-.5 at a feed rate of 5.0 inches per minute)

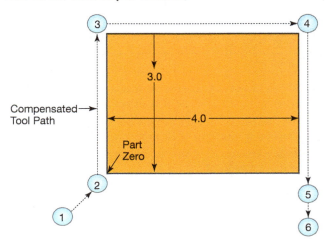

Figure M-68 Programming example using tool length and cutter diameter compensation.

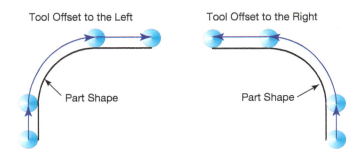

Figure M-67 Left (G41) and right (G42) tool compensation.

N80 G41 D2 X0.0 Y0.0 (Cutter diameter compensation left by the amount found in D2, Ramp compensation on during linear move to X0.0 Y0.0 Position 2)

N90 G01 Y3.0 (Linear move to Y3.0 Position 3)

N100 G01 X4.0 (Linear move to X4.0 Position 4)

N110 G01 Y-.50 (Linear move to Y-.25 Position 5)

N260 G40 Y-1.0 (Ramp off and cancel cutter compensation feed to position 6)

N270 G00 Z.1 (Rapid to Z.1)

N280 M5 (Spindle stop)

N290 T0 M06 (Return to tool change position)

N300 M30 (Program end, memory reset)

%

It is very important that the programmer's tooling intentions are communicated to the operator. This is usually done using the process plan (router) and setup sheets.

SELF-TEST

1. What primary role do preparatory functions serve?
2. Name three functions that miscellaneous codes control.
3. Name two considerations that must be taken into account when selecting a part datum location.
4. What does modal mean?
5. Complete the following table.

Code	Function
G00	
G01	
M03	
G54	
G92	
G02	
G03	
G70	
M08	

6. Write a line of code to move the spindle 10 inches to the right in the X axis at a feed rate of 10 inches per minute.
7. Write a line of code to move the Y axis 5 inches in a negative direction at a feed rate of 10 inches per minute.
8. Write a line of code to turn the spindle on in a clockwise direction at 800 RPM.
9. Write a line of code to make sure the control is in absolute and inch mode.
10. Write a line of code to change to tool 4.

11. Write a line of code to move the Z axis up 5 inches in rapid mode.
12. Describe the difference between G92 and G54 workpiece coordinate settings.
13. Write code to make the following move at a feed rate of 10 inches per minute. Write it in absolute mode. Move to X8.000 Y5.250
14. Write code to make the following move at a feed rate of 10 inches per minute. Write it in incremental mode. Move 5.000 inches to the right.
15. Write code to make the following move at a feed rate of 10 inches per minute. Write it in absolute mode. Move to X-7.500 Y-3.250.
16. Write a line of code using the incremental arc center method to machine the following arc at a feed rate of 5 inches per minute.

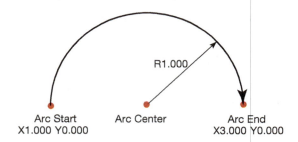

Arc Start
X1.000 Y0.000   Arc Center   Arc End
X3.000 Y0.000

17. Write a line of code using the incremental arc center method to machine the following arc at a feed rate of 8 inches per minute.

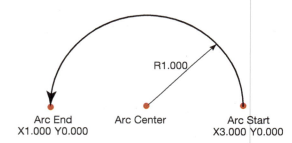

Arc End
X1.000 Y0.000   Arc Center   Arc Start
X3.000 Y0.000

18. Write a line of code using the incremental arc center method to machine the following arc at a feed rate of 5 inches per minute.

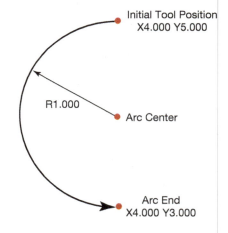

Initial Tool Position
X4.000 Y5.000

R1.000   Arc Center

Arc End
X4.000 Y3.000

19. Write a line of code using the incremental arc center method to machine the following arc at a feed rate of 7.5 inches per minute.

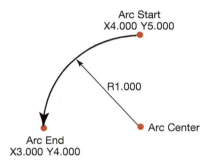

20. Write a line of code using the incremental arc center method to machine the following arc at a feed rate of 7 inches per minute.

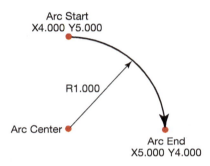

21. Write a line of code using the radius method to machine the following arc at a feed rate of 5 inches per minute.

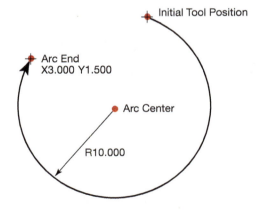

22. Write a line of code using the radius method to machine the following arc at a feed rate of 5 inches per minute.

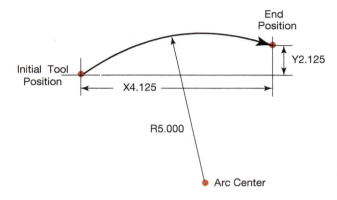

23. Write a line of code using the radius method to machine the following arc at a feed rate of 5 inches per minute.

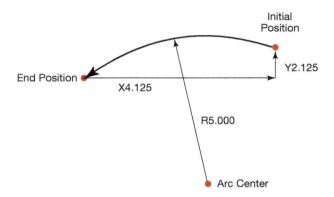

24. Write a line of code using the radius method to machine the following arc at a feed rate of 5 inches per minute.

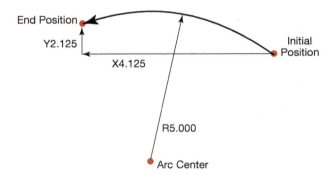

25. What purpose do tool length offsets serve?

26. What must be done to invoke tool diameter compensation?

27. Complete the blocks for the parts shown in the figure. Use a .50 end mill and cut to a depth of .25 around the part.

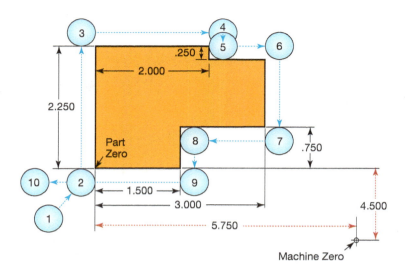

N10 G_ G_ G_ G_ G_; (Inch programming, absolute programming, cancel diameter compensation, cancel canned cycles, XY plane)

N20 M_ T02; (Tool change, tool #2)

N30 G54 X__ Y__; (Workpiece zero setting)

N40 M_ S800; (Spindle start clockwise, 800 RPM)

N50 G_ X-1.00 Y-1.00; (Rapid to position #1)

N60 G_ Z_; (Rapid down to .100 clearance above the part)

N70 G__ Z__ _5.0; (Feed down to depth at 5 inches per minute)

N80 G01 X__ Y__; (Feed to position #2, offsetting for the tool radius)

N90 G01 X__ Y__; (Feed to position #3)

N100 G01 X__ Y__; (Feed to position #4)

N110 G01 X__ Y__; (Feed to position #5)

N120 G01 X__ Y__; (Feed to position #6)

N130 G01 X__ Y__; (Feed to position #7)

N140 G01 X__ Y__; (Feed to position #8)

N150 G01 X__ Y__; (Feed to position #9)

N160 G01 X__ Y__; (Feed to position #10, 1 inch to the left of the part)

N170 G__; (Return all axes to home position)

N180 M__ T__; (Tool change, tool 0)

N190 M__; (End program, rewind program to beginning)

28. Program the part shown in the figure below. Use a .25 end mill to machine the outside shape of the part and machine to a depth of .35. Assume it is tool number 5. Make sure you use offsets. Program the part using the climb milling technique.

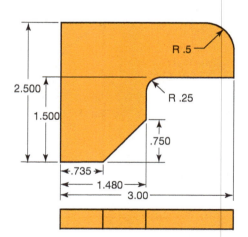

Programming Examples

This unit will examine sample programs with line-by-line explanations. The first program will be thoroughly explained. After the first program, the rest will have explanations for the program commands that are new or might be difficult to understand.

PROGRAMMING EXAMPLES

The first program we examine will drill several 1/4-inch holes in a part. This example will use basic G-codes to drill the parts. Canned drilling cycles will be covered later. There are two example programs for this part. The first program uses absolute mode programming. The second uses incremental programming. Both programs use several G- and M-codes. Tool length offsets will also be used. Figure M-69 shows the part.

Programming can be thought of as a three-step process. We will follow a three-step process for these examples.

Step 1 contains the preparatory functions to get the control into the proper control modes.

Step 2 will contain the actual machining operations. There may be several. For example, step 2a might be a milling operation, step 2b might be a drilling operation, step 2c might be a tapping operation, and so on.

Step 3 ends the program correctly after all machining operations have been completed.

PROGRAM 1: ABSOLUTE MODE PROGRAMMING

The first step in a program sets the control in the desired modes. Lines N10 through N20 are preparatory lines of code to get the control ready to run the part. Lines 10 and 20 set the desired modes for the control. The order of these codes at the beginning of a program does not matter.

O001 (Program name O001)
N10 G20

    Line N10 tells the control to operate in inch programming mode, not metric.

Figure M-69 The part print is shown on left and the order of drilling on the right.

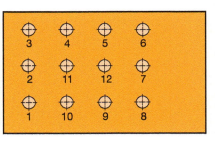

N20 G17 G40 G49 G80 G90

G17 tells the control to operate in the XY plane. G40 cancels diameter offsets that may have been in the control from a previous program. G49 cancels tool length compensation that may have been in the control from a previous program. G80 cancels canned cycles that may have been in the control from a previous program. G90 tells the control to operate in absolute mode.

The second part of a program performs operations. This part of the program drills 12 holes. Only one tool is used. Tool 1 is a 1/4-inch drill.

N30 T1 M6

First a drill change loads tool 1 (Line N30).

N40 G54

In line N40 the control is told that the workpiece coordinates (X, Y, Z) are located in the G54 register. Note that the operator must make sure that they measure and put the correct values in the G54 register.

N50 S3500 M3

Then the spindle is turned in a clockwise direction at 3,500 RPM. (Line N50).

N60 G00 X.5 Y.63

Next, a rapid move positions the tool over the location for the first hole X.5 Y.630 (Line N60).

N70 G43 H1 Z.1 (Tool length compensation positive amount found in H1, rapid to Z.1)

In line N70, positive tool length compensation is called (Figure M-70). The amount of the compensation is found in register H1. A rapid move to Z.1 inches then occurs because the G00 is modal and still active from line N60. The control implements the length offset during the move to Z.1.

N80 G01 Z-.75 F8.0 (Linear move to Z-.75 at a feed rate of 8 inches per minute)

In line N80, the hole is drilled as the tool is fed to a Z depth of -.75 inches at a feed rate of 8.0 inches per minute.

N90 G00 Z.1 (Rapid move up to Z.1)

In line 90, the drill is moved with a rapid feed to Z.1 inches.

The rest of the program repeats these actions. A rapid feed positions the tool to the next hole location, the drill is fed to a depth of -.75 inches, the drill is moved with a rapid feed, back to Z.1 inches, and the process is repeated for the next hole.

The whole program is shown next.

O001 (Program name O001)

N10 G20 (Tells the control to operate in inch programming mode, not metric)

N20 G17 G40 G49 G80 G90

N30 T1 M6 (Change to tool 1, 1/4 drill)

N40 G54 (Workpiece coordinates are found in G54 register)

N50 S3500 M3 (Turn on spindle CW at a speed of 3500 RPM)

N60 G00 X.5 Y.63 (Rapid move to X.5 Y.630, hole 1)

N70 G43 H1 Z.1 (Tool length compensation positive amount found in H1, rapid to Z.1)

N80 G01 Z-.75 F8.0 (Linear move to Z-.75 at a feed rate of 8 inches per minute)

N90 G00 Z.1 (Rapid move up to Z.1)

N100 Y1.38 (Rapid move to Y 1.38, hole 2)

N110 G01 Z-.75 F8.0 (Linear move to Z-.75 at a feed rate of 8 inches per minute)

N120 G00 Z.1 (Rapid move up to Z.1)

N130 Y2.13 (Rapid move to Y2.13, hole 3)

N140 G01 Z-.75 F8.0 (Linear move to Z-.75 at a feed rate of 8 inches per minute)

N150 G00 Z.1 (Rapid move up to Z.1)

N160 X1.25 (Rapid move to X1.25, hole 4)

N170 G01 Z-.75 F8.0 (Linear move to Z-.75 at a feed rate of 8 inches per minute)

N180 G00 Z.1 (Rapid move up to Z.1)

N190 X2. (Rapid move to X2, hole 5)

N200 G01 Z-.75 F8.0 (Linear move to Z-.75 at a feed rate of 8 inches per minute)

Figure M-70 Tool length compensation amount for T1 (drill).

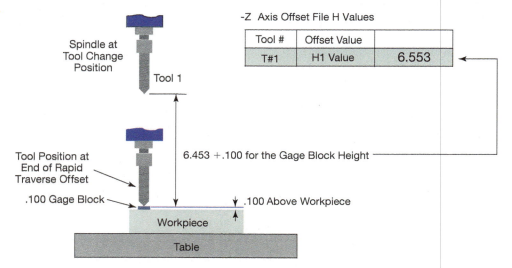

-Z Axis Offset File H Values

Tool #	Offset Value	
T#1	H1 Value	6.553

Spindle at Tool Change Position

Tool 1

Tool Position at End of Rapid Traverse Offset

.100 Gage Block

6.453 +.100 for the Gage Block Height

.100 Above Workpiece

Workpiece

Table

N210 G00 Z.1 (Rapid move up to Z.1)

N220 X2.75 (Rapid move to X2.75, hole 6)

N230 G01 Z-.75 F8.0 (Linear move to Z-.75 at a feed rate of 8 inches per minute)

N240 G00 Z.1 (Rapid move up to Z.1)

N250 Y1.38 (Rapid move to Y1.38, hole 7)

N260 G01 Z-.75 F8.0 (Linear move to Z-.75 at a feed rate of 8 inches per minute)

N270 G00 Z.1 (Rapid move up to Z.1)

N280 Y.63 (Rapid move to Y.63, hole 8)

N290 G01 Z-.75 F8.0 (Linear move to Z-.75 at a feed rate of 8 inches per minute)

N300 G00 Z.1 (Rapid move up to Z.1)

N310 X2. (Rapid move to X2, hole 9)

N320 G01 Z-.75 F8.0 (Linear move to Z-.75 at a feed rate of 8 inches per minute)

N330 G00 Z.1 (Rapid move up to Z.1)

N340 X1.25 (Rapid move to Y1.25, hole 10)

N350 G01 Z-.75 F8.0 (Linear move to Z-.75 at a feed rate of 8 inches per minute)

N360 G00 Z.1 (Rapid move up to Z.1)

N370 Y1.38 (Rapid move to Y1.38, hole 11)

N380 G01 Z-.75 F8.0 (Linear move to Z-.75 at a feed rate of 8 inches per minute)

N390 G00 Z.1 (Rapid move up to Z.1)

N400 X2. (Rapid move to X2, hole 12)

N410 G01 Z-.75 F8.0 (Linear move to Z-.75 at a feed rate of 8 inches per minute)

N420 G00 Z.1 (Rapid move up to Z.1)

All machining is done at this point, and the program is ended in lines N430-N460.

N430 M5 (Stop spindle rotation)

N450 G28 (Return to reference point)

N460 M30 (Program end, memory reset)

%

Remember to think of a program as a three-step process.

Step 1: Get the control into the desired modes through the use of preparatory functions.

Step 2a … 2n: Perform the machining operations.

Step 3: End the program.

PROGRAM 2: INCREMENTAL MODE PROGRAMMING

The second example uses the same part (Figure M-71). This program will use incremental mode, however.

Lines N10 through N20 are preparatory lines of code to get the control ready to run the part. Lines 10 and 20 set the desired modes for the control. Remember that the order of these preparatory codes does not matter.

O002 (Program name O002)

N10 G20 (Inch programming)

N20 G17 G40 G49 G80 G90 (Rapid, XY Plane, cancel diameter offsets, cancel tool length compensation, cancel canned cycles, absolute mode)

The second part of a program performs operations. This part of the program drills 12 holes. Only one tool is used. Tool 1 is a 1/4-inch drill. First a drill change loads tool 1 in line N30. Then the control is told that the workpiece coordinates (*X, Y, Z*) are located in the G54 register (line N40). Note that the operator must make sure that they measure and put the correct values in the G54 register. The spindle is turned on in Line N50.

In line N60 a rapid move positions the tool to the first hole location. In line N70, positive tool length compensation is called. The amount of the compensation is found in register H1. A rapid move to Z.1 then occurs. The control implements the length offset during the move to Z.1.

Line N80 puts the control into incremental mode. In line N90, the hole is drilled as the tool is fed to a Z depth of -.75 at a feed rate of 8.0 inches per minute. In line 100, the drill is moved with a rapid feed to Z.1. The rest of the program repeats these actions. A rapid move positions the tool to the next hole location, the drill is fed to a depth of -.85 (-.85 incremental depth, absolute depth of -.75), the drill is moved with a rapid back to Z.1, and the process is repeated for the next hole.

Figure M-71 The part print is shown on the left and the drill order on the right.

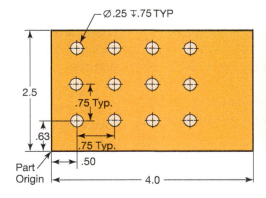
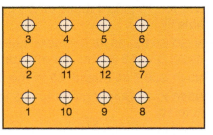

N30 T1 M6 (Change to tool 1, 1/4 drill)

N40 G54 (Workpiece coordinates are found in G54 register)

N50 S3500 M3 (Turn spindle on CW at 3500 RPM)

N60 G00 X.5 Y.63 (Rapid move to X.5 Y.63, hole 1)

N70 G43 H1 Z.1 (Positive tool height offset, value of offset in D1, move to Z.1)

N80 G91 (Incremental mode)

N90 G01 Z-.85 F8.0 (Linear move to Z-.85 at a feed rate of 8 inches per minute)

N100 G00 Z.85 (Rapid .85 in the Z+ direction)

N110 Y.75 (Rapid move .75 in the Y+ direction, hole 2)

N120 G01 Z-.85 F8.0 (Linear move to -.85 in the Z- direction at 8 inches per minute)

N130 G00 Z.85 (Rapid .85 in the Z+ direction)

N140 Y.75 (Rapid move .75 in the Y+ direction, hole 3)

N150 G01 Z-.85 F8.0 (Linear move to -.85 in the Z- direction at 8 inches per minute)

N160 G00 Z.85 (Rapid .85 in the Z+ direction)

N170 X.75 (Rapid move .75 in the X+ direction, hole 4)

N180 G01 Z-.85 F8.0 (Linear move to -.85 in the Z- direction at 8 inches per minute)

N190 G00 Z.85 (Rapid .85 in the Z+ direction)

N200 X.75 (Rapid move .75 in the X+ direction, hole 5)

N210 G01 Z-.85 F8.0 (Linear move to -.85 in the Z- direction at 8 inches per minute)

N220 G00 Z.85 (Rapid .85 in the Z+ direction)

N230 X.75 (Rapid move .75 in the X+ direction, hole 6)

N240 G01 Z-.85 F8.0 (Linear move to -.85 in the Z- direction at 8 inches per minute)

N250 G00 Z.85 (Rapid .85 in the Z+ direction)

N260 Y-.75 (Rapid -.75 in the Y- direction, hole 7)

N270 G01 Z-.85 F8.0 (Linear move to -.85 in the Z- direction at 8 inches per minute)

N280 G00 Z.85 (Rapid .85 in the Z+ direction)

N290 Y-.75 (Rapid move to -.75 in the Y- direction, hole 8)

N300 G01 Z-.85 F8.0 (Linear move to -.85 in the Z- direction at 8 inches per minute)

N310 G00 Z.85 (Rapid .85 in the Z+ direction)

N320 X-.75 (Rapid move to -.75 in the X- direction, hole 9)

N330 G01 Z-.85 F8.0 (Linear move to -.85 in the Z- direction at 8 inches per minute)

N340 G00 Z.85 (Rapid .85 in the Z+ direction)

N350 X-.75 (Rapid move to -.75 in the X- direction, hole 10)

N360 G01 Z-.85 F8.0 (Linear move to -.85 in the Z- direction at 8 inches per minute)

N370 G00 Z.85 (Rapid .85 in the Z+ direction)

N380 Y.75 (Linear move to .75 in the Y+ direction, hole 11)

N390 G01 Z-.85 F8.0 (Linear move to -.85 in the Z- direction at 8 inches per minute)

N400 G00 Z.85 (Rapid .85 in the Z+ direction)

N410 X.75 (Rapid move .75 in the X+ direction, hole 12)

N420 G01 Z-.85 F8.0 (Linear move to -.85 in the Z- direction at 8 inches per minute)

N430 G00 Z.85 (Rapid .85 in the Z+ direction)

All machining is done at this point, and the program is ended in lines N440–N460

N440 M5 (Spindle stop)

N450 G28 (Return to reference point)

N460 M30 (Program end, memory reset)

%

PROGRAM 3: MILLING A GROOVE

This program mills a triangular groove in a part. No tool diameter offsets will be used in this program. Positive-tool height compensation will be used. The part is shown in Figure M-72.

O003 (Program name O003)

Lines N10 through N20 are preparatory lines of code to get the control ready to run the part. Lines 10 and 20 set the desired modes for the control. Remember that the order of these at the beginning of a program does not matter.

N10 G20 (Inch mode)

N20 G17 G40 G49 G80 G90 (Rapid, XY Plane, cancel diameter offsets, cancel tool length compensation, cancel canned cycles, absolute mode)

Line N30 tells the control that the workpiece coordinates (X, Y, Z) are located in the G54 register. Note that the operator must make sure that they put the correct values in the G54 register.

N30 G54 (Workpiece coordinates are found in G54 register)

N40 T1 M6 (Change to tool 1)

N50 S5000 M3 (Turn on spindle clockwise at a speed of 5000 RPM)

N60 G00 X1.5 Y.625 (Linear move to X1.5 Y.625, point 1)

Line N70 calls for positive tool length compensation by the amount in H1 and rapids the tool to .1 above the workpiece. The control implements the length offset during the move to Z.1.

N70 G43 H1 Z1.

N80 G00 Z.1 (Rapid to Z.1)

Figure M-72 The part print is shown on the right and the machining path on the left. The part will be machined with a 1/4-inch end mill.

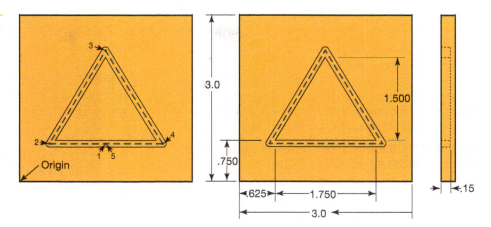

N90 G01 Z-.15 F6. (Linear move to Z-.15 at a feed rate of 6.0 inches per minute)

N100 G01 X.4074 F10. (Linear move to X1.5 Y.625 at a feed rate of 10.0 inches per minute, point 2)

N110 G01 X1.5 Y2.4981 (Linear move to X1.5 Y2.4981, point 3)

N120 G01 X2.5926 Y.625 (Linear move to X2.5926 Y.625, point 4)

N130 G01 X1.5 (Linear move to X1.5, point 5)

N140 G01 Z.1 F6. (Linear move to Z.1 at a feed rate of 6.0 inches per minute)

N150 G0 Z1. (Rapid to Z.1)

The rest of the lines stop the spindle, return to reference, and end the program.

N160 M5 (Spindle stop)

N170 G28 (Return to reference point)

N180 M30 (Program end, memory reset)

%

PROGRAM 3: MILLING WITH CIRCULAR INTERPOLATION

This part will involve milling a groove in a part that involves circular interpolation. The part is shown in Figure M-73. The cutter will be a 1/4-inch end mill. The depth of cut should be .13.

Lines N10 and N20 are the preparatory codes to set the correct modes for the control. Line N30 tells the control that the coordinates for the workpiece are in register G54. Note that the operator must make sure that they measure and put the correct values in the G54 register.

O004 (Program name O004)

N10 G20 (Inch mode)

N20 G17 G40 G49 G80 G90 (Rapid, XY Plane, cancel diameter offsets, cancel tool length compensation, cancel canned cycles, absolute mode)

N30 G54 (Workpiece coordinates are found in G54 register)

N40 T1 M6 (Load tool 1, 1/4 end mill)

N50 S5000 M3 (Turn on spindle clockwise at a speed of 5000 RPM)

N60 G00 X.625 Y1. (Rapid move to X.625 Y1, point 1)

Line N70 calls for positive tool length compensation by the amount in H1 and rapids the tool to 1.0 inches above the workpiece. The control implements the length offset during the move to Z.1.

N70 G43 H1 Z1. (Positive tool height offset, value of offset in D1, move to Z.1)

N80 G01 Z.1 (Linear move to Z.1)

N90 G01 Z-.13 F6.0 (Linear move to Z-.13 at a feed rate of 6.0 inches per minute)

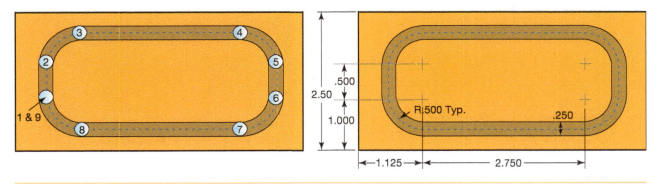

Figure M-73 The part print is shown on the right and the machining path order on the left.

Line N100 is a clockwise circular interpolation to point 3. The X1. and the Y2. are the end points of the circular move. The I value is the value in the X direction between the start point and the center of the arc. An I value of .5 means that the arc center is .5 inch to the right (the plus direction) the start point in the X direction. The J value is not listed, which means that the arc center location in the Y direction is the same as the start point Y location value.

N100 G01 Y1.5 F20. (Linear move to point 2 at a feed rate of 20 inches per minute)

Line N100 is a clockwise circular interpolation to point 3. The X1. and the Y2. are the end points of the circular move.

The I value is the value in the X direction between the start point and the center of the arc. An I value of .5 means that the arc center is .5 inch to the right (the plus direction) the start point in the X direction. The J value is not listed, which means that the arc center location in the Y direction is the same as the start point Y location value.

N110 G02 X1.125 Y2. I.5 (Clockwise circular interpolation to point 3)
N120 G01 X3.875 (point 4)

Line N130 is a clockwise circular interpolation to point 5. The X4.375 and the Y1.5 are the end points of the circular move. The I value is not listed, which means that the arc center in the X direction is the same as the start point X value. The J value is the value in the Y direction between the start point and the center of the arc. A J value of -.5 means that the arc center is -.5 below (the minus direction) the start point in the Y direction.

N130 G02 X4.375 Y1.5 J-.5 (Clockwise circular interpolation to point 5)
N140 G01 Y1 (point 6)

Line N150 is a clockwise circular interpolation to point 7. The X3.875 and the Y.5 are the end points of circular move. The I value is the value in the X direction between the start point and the center of the arc. A I value of -.5 means that the arc center is -.5 inch to the left (the minus direction) the start point in the X direction. The J

value is not listed, which means that the arc center location in the Y direction is the same as the start point Y location value.

N150 G02 X3.875 Y.5 I-.5 (Clockwise circular interpolation to point 7)
N160 G01 X1.1125 (Linear move to point 8)

Line N170 is a clockwise circular interpolation to point 9. The X.63 and the Y1. are the end points of the circular move. The I value is not listed, which means that the arc center in the X direction is the same as the start point X value. The J value is the value in the Y direction between the start point and the center of the arc. A J value of .5 means that the arc center is .5 above (the plus direction) the start point in the Y direction.

N170 G02 X.625 Y1. J.5 (Clockwise circular interpolation to point 9)
N180 G01 Z.1 F6.0 (Linear move to Z.1 at a feed rate of 6. inches per minute)
N190 G00 Z1. (Rapid to Z1.)

The next lines stop the spindle, return to the reference position, and end the program.

N200 M5 (Spindle stop)
N210 G28 (Return to reference point)
N220 M30 (Program end, memory reset)
%

PROGRAM 4: CIRCULAR INTERPOLATION USING THE RADIUS PROGRAMMING METHOD

The next part will involve tool compensation and circular interpolation. This program will use the radius method of circular interpolation. The part is shown in Figure M-74.

Lines N10 through N20 are the preparatory lines. Line N10 sets the desired modes for the controller. Line N20 sets the workpiece coordinates.

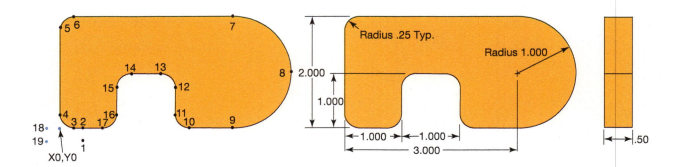

Figure M-74 The part print is shown on the right and the machining path order on the left.

O006 (Program name O006)

N10 G17 G20 G40 G80 G90 (Rapid, XY Plane, cancel diameter offsets, cancel tool length compensation, cancel canned cycles, absolute mode)

N20 G54 (Workpiece coordinates are found in G54 register)

N30 T1 M06 (Tool Change to tool 1, .375 END MILL)

N40 S4000 M3 (Turn on spindle clockwise at a speed of 4000 RPM)

N50 G00 X.375 Y-.375 (Rapid move to X.375 Y-.375, point 1)

Line N60 invokes positive tool length compensation by the amount in register H1. The control implements the compensation in this line as it moves the cutter to Z.1.

N60 G43 H1 Z.1

N70 G01 Z-.5 F6.0 (Linear move to Z-.5 at a feed rate of 6.0 inches per minute, point)

Line N80 tells the control to offset the tool to the left of the cutter path by the amount found in D1. In this example, the cutter is .375 diameter, so the offset would be 1/2 of that size (.1875). The operator should make sure that D1 contains .1875. Line N80 also performs a linear move to Y0. at a feed rate of 15.0 inches per minute. During this move, the control implements the offset. When the cutter arrives at Y0., the control has offset the cutter to the left by .1875.

N80 G41 D1 Y0. F15.0 (Linear move to Y0., point 2)

N90 G01 X.25 (Linear move to X.25, point 3)

Line N100 is a clockwise circular interpolation. The end point for the move will be at X0. Y.25. The r value of .25 is the radius of the arc.

N100 G02 X0. Y.25 R.25 (Clockwise circular move to point 4)

N110 G01 Y1.75 (Linear move to Y1.75, point 5)

Line N120 is a clockwise circular interpolation. The end point for the move will be at X.25 Y2. The r value of .25 is the radius of the arc.

N120 G02 X.25 Y2. R.25 (Clockwise circular move to point 6)

N130 G01 X3. (Linear move to X3., point 7)

N140 G02 X4.0 Y1.0 R1.0 (Clockwise circular move to point 8)

N150 G02 X3. Y0. R1. (Clockwise circular move to point 9)

N160 G01 X2.25 (Linear move to X2.25, point 10)

N170 G02 X2. Y.25 R.25 (Clockwise circular move to point 11)

N180 G01 Y.75 (Linear move to Y.75, point 12)

N190 G03 X1.75 Y1. R.25 (Counterclockwise circular move, point 13)

N200 G01 X1.25 (Linear move to Y1.25, point 14)

N210 G03 X1. Y.75 R.25 (Counterclockwise circular move, point 15)

N220 G01 Y.25 (Linear move to Y.25, point 16)

N230 G02 X.75 Y0. R.25 (Clockwise circular move to point 17)

N240 G01 X-.375 (Linear move to X-.375, point 18)

N250 G01 Y-.375 (Linear move to Y-.375, point 19)

Note that line N260 cancels the diameter offset that was called for this tool.

N260 G40 (Cancel tool diameter compensation)

N270 G01 Z.1 F6.0 (Linear move to Z.1 at a feed rate of 6.0 inches per minute)

Lines N280 through N310 stop the spindle, return to reference position, and perform a program end and memory reset.

N280 M5 (Stop spindle)

N290 G28 (Return to reference point)

N300 M30 (Program end, memory reset)

%

PROGRAM 5: HEIGHT AND DIAMETER COMPENSATION EXAMPLE

The next program will be another milling example. The part is shown in Figure M-75. It will use height and diameter compensation. There is nothing new introduced in this program. The only explanation for this program will be in each line of code.

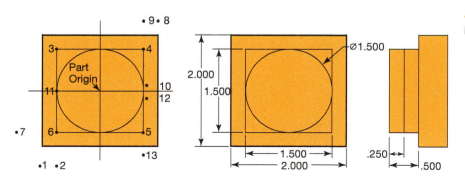

Figure M-75 Milling example.

O008 (Program name O008)

N10 G20 (Inch mode)

N20 G17 G40 G49 G80 G90 (Rapid, XY Plane, cancel diameter offsets, cancel tool length compensation, cancel canned cycles, absolute mode)

N30 G54 (Workpiece coordinates are found in G54 register)

N40 T1 M6 (Load tool 1, 3/4 end mill)

N50 S2000 M3 (Turn on spindle clockwise at a speed of 2000 RPM)

N60 G00 X-1.25 Y-1.75 (Rapid move to X-1.25 Y-1.75., point 1)

N70 G43 H1 Z1. (G43 H1 – Positive tool length compensation amount in H1, rapid to Z1.)

N80 G01 Z-.5 F6. (Linear feed to Z-.5 at a feed rate of 6.)

N90 G41 D1 X-.75 F24. (Cutter diameter compensation left amount in D1, linear move to X-.75, point 2)

N100 G01 Y.75 (Linear move to Y.75, point 3)

N110 G01 X.75 (Linear move to X.75, point 4)

N120 G01 Y-.75 (Linear move to Y-.75, point 5)

N130 G01 X-.75 (Linear move to X-.75, point 6)

N140 X-1.875 (Linear move to X-1.875, point 7)

N150 G01 G40 Z.1 F6.

N160 G00 Z1. (Rapid to Z1.)

N170 G00 X1.00 Y1.125 (Rapid move to X1.00 Y1.125, point 8)

N180 G00 Z.1 (Rapid move to Z.1)

N190 G01 Z-.25 (Linear feed to Z-.25)

N200 G41 D1 X.75 (Left diameter cutter compensation by amount in D1, linear move to X.75, point 9)

N210 G01 Y0. F24. (Linear move at 24. inches per minute, point 10)

N210 G02 X-.75 Y0. I-.75 (Clockwise circular interpolation, point 11)

N220 G02 X.75 Y0. I.75 (Clockwise circular interpolation, point 12)

N230 G01 Y-1.125 (Linear move to Y.75, point 13)

N240 G01 G40 Z.1 F6. (G40 – Cancel tool diameter compensation, G01 Z.1 F6. – Linear move to Z.1 at a feed rate of 6. inches per minute)

N250 G00 Z1. (Rapid to Z1.)

N260 M5 (Spindle stop)

N270 G28 (Return to reference point)

N280 M30 (Program end, memory reset)

%

SELF-TEST

1. Program the part. Use a .25 end mill to machine the outside shape of the part and machine to a depth of .35. Assume it is tool number 5. Make sure you use height and diameter offset compensation. Program the part using the conventional milling technique (counter-clockwise).

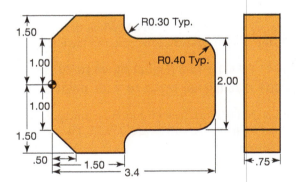

2. Program the part. Use a .5-end mill to machine the outside shape of the part and machine to a depth of .50. Assume it is tool number 5. Make sure you use height and diameter offset compensation.

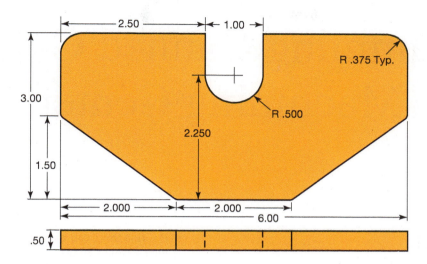

3. Program the part. Use a .5-inch end mill. Mill the contour to a .50-depth.

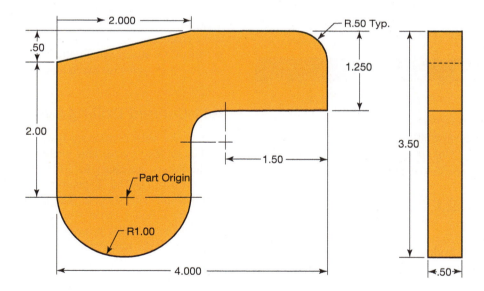

Programming Canned Cycles for Machining Centers

INTRODUCTION

This unit will cover the steps necessary to properly plan, set up, and program a machining center. This unit will focus on canned cycles. Canned cycles are sometimes called fixed cycles. Canned cycles will reduce the amount of programming necessary for repetitive machining operations such as drilling, boring, reaming, and tapping.

OBJECTIVES

After completing this unit, you should be able to:

- Demonstrate an understanding of acceptable machining center programming practices.
- Write programs using canned cycles.

CANNED CYCLES FOR MACHINING CENTERS

Canned cycles (fixed cycles) simplify the programming of repetitive machining operations such as drilling, tapping, and boring. Canned cycles are a set of preprogrammed instructions that eliminate the need for many lines of programming. Programming a simple drilled hole, without the use of a canned cycle, can take four or five lines of programming. Think of the lines that are needed to produce a hole:

- Position the *X* and *Y* axes to the proper coordinates with a rapid traverse move (G00),
- Position the *Z* axis to a clearance plane,
- Feed the tool down to depth,
- Rapid position the tool back to the clearance point.

That is four steps for one drilled hole! By using a canned drill cycle, a hole can be done with one line of code. Standard canned cycles, or fixed cycles, are common to most CNC machines. Figure M-76 lists a few commonly used canned cycles for machining centers.

G81 Canned Drilling Cycle

One of the most commonly used canned cycle is the G81 canned drilling cycle. This cycle will automatically do all of the things necessary to drill a hole in one line of code (Figure M-77). The Z position is very important when you call the canned cycle. The present Z position will become the Z initial plane. The machine will normally rapid back to the Z initial position before a rapid move to the next hole. If the tool is 12 inches above the work when the canned cycle is called, that will be the Z initial plane, and the machine will rapid up to 12 inches above the work between each hole. This would be a very inefficient program.

As you can see from Figure M-78, the canned drilling cycle consists of four moves. Those four moves are controlled by one line of programming.

Figure M-76 A few canned cycles for machining centers.

G-Code	Function	Z Axis	At Depth	Z Axis Return
G81	Drill	Feed	-	Rapid Traverse
G83	Peck Drill	Feed with Peck	-	Rapid Traverse
G84	Tapping	Feed	Reverse Spindle	Feed
G85	Bore/Ream	Feed	Stop Spindle	Cutting Feed
G86	Bore	Feed	Stop Spindle	Rapid Traverse

Figure M-77 A G81 canned-drill cycle.

Figure M-78 Example of a G81 canned drill cycle.

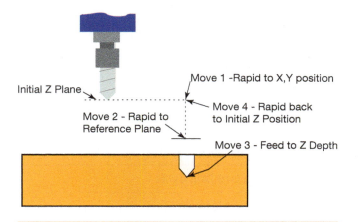

The G81 specifies a drill canned cycle, *X* and *Y* are the coordinates of the hole center, Z is the depth of the hole, R is the height for the drill to rapid down to from the initial Z position. F is the feed rate in inches per minute.

Figure M-79 shows two examples of the use of a G81. The part on the left has a raised section that needs to be avoided on moves so the tool will not be broken. The part on the right has no obstruction. If we want to use the G81 drill cycle and retract up the initial Z we would use a G98 with it. A G98 means retract to the initial Z position after drilling. The part on the right has no obstruction on the top of the part. The drill cycle will be faster if the drill only retracts to the R plane before the rapid move to the next hole location. A G99 would be used with the G81 to make the drill retract to the R plane instead of the initial Z position.

Figure M-80 shows the parameters for a G81 drilling cycle.

Figure M-80 Parameters for a G81 canned drilling cycle.

Parameters for a G81 Drill Cycle

X*	Rapid *X*-axis location
Y*	Rapid *Y*-axis location
Z	Depth of the hole
R	R Plane
F	Feed rate

*Optional

The G81 canned cycle is modal, which means that it will stay active until it is canceled by a G80. If we were drilling a series of holes, we would only need to specify the coordinates for the next hole. Figure M-81 and the program that follows are a drilling example using a G81 canned drill cycle using a G98 code that makes the tool retract to the R plane after each drill cycle.

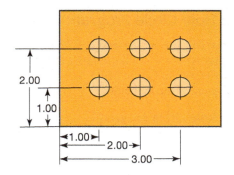

Figure M-81 Canned drilling cycle example.

Figure M-79 G81 canned cycle examples.

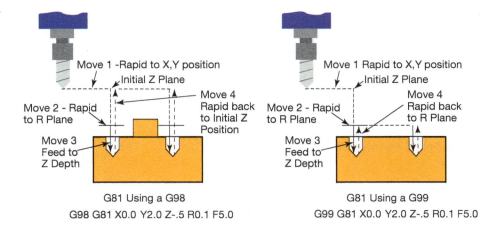

G81 Using a G98

G98 G81 X0.0 Y2.0 Z-.5 R0.1 F5.0

G81 Using a G99

G99 G81 X0.0 Y2.0 Z-.5 R0.1 F5.0

O00079 (Program name O00079)

N20 G00 G17 G40 G49 G80 G90 Rapid mode, G17 - XY plane, G40 - Cancel diameter compensation offset, G49 - Cancel tool length compensation, G90 - Absolute mode)

N30 T1 M06 (Tool Change to tool 1, 1/2 drill)

N40 G54 (Workpiece zero setting)

N50 S1000 M03 (1000 RPM, Spindle start clockwise)

N60 G00 X1. Y1. (Rapid to hole position #1)

N70 G43 H01 Z1. (Tool height offset #1, Rapid to initial level)

N80 G98 G81 Z-0.275 R0.1 F3. (Return to initial R-plane, drill hole #1 .275 deep 3.0 inches per minute feed)

N90 X2. (Drill hole #2)

N100 X3. (Drill hole #3)

N110 Y2. (Drill hole #4)

N120 X2. (Drill hole #5)

N130 X1. (Drill hole #6)

N140 G80 (Cancel drill cycle)

N150 M5 (Spindle stop)

N160 G28 (Return X and Y axis to home position)

N170 M30 (End program, reset the control)

%

G70 Bolt Hole Circle Program Example

This example will use a G81 and a G70 bolt hole canned cycle. The part is shown in Figure M-82 and the parameters for a G70 are shown in Figure M-83.

The program is shown below.

O720 (Bolt hole circle program)

N01 G90 G80 G20 G40

N02 M6 T1 (1/2 drill)

N03 G54

N04 S1200 M3

N5 G00 X2.5 Y2.5 (center position of the bolt hole circle)

N6 G43 H1 Z1.0 M8

N7 G81 G99 Z-.5 R0.1 F3.0 L0 (L0 on line 7 will cause the machine to not do this command until reading the next line, so as to not drill a hole in the center of the bolt hole circle))

N8 G70 I1.5 J60.0 L6 (G70 bolt hole circle command, I=radius of the bolt hole circle, J=starting angle for the first bolt hole from the three o' clock position, L=number of evenly spaced holes around the bolt hole circle)

N9 G90 G00 Z1.0

N10 G28

N11 M30

%

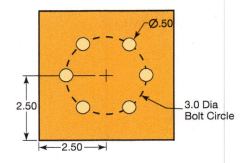

Figure M-82 Part with six holes.

Figure M-83 Parameters for a G70 canned cycle.

Parameters for a G70 Drill Cycle

X*	Rapid X-axis location
Y*	Rapid Y-axis location
Z	Z—Depth of the hole
R	Retract Plane
F	Feed rate
L	An L0 with a G81 is used mainly for positioning. Defining an L0 is used with a G81 cycle to position to the center of the bolt hole and not drill a hole. An L with a G70 is the number of evenly spaced holes in the bolt circle.
I	Radius of the bolt circle
J	Starting position of the first hole from the three o'clock position

*Optional

Peck (Deep Hole) Drilling Cycle (G83)

When deep holes are to be drilled (holes that are three to four times deeper than the diameter of the drill), a peck drilling cycle is often used. The peck drilling cycle (G83) is very similar to the G81 drilling cycle, but it uses an extra word address (Q) to specify the depth of each peck. After the drill reaches the depth of the peck, the Z axis rapids out of the hole, clearing the hole of chips (Figure M-84), and then pecks again and again until the drill depth is reached.

The canned peck drilling cycle will then rapid position the tool to the Z position specified by the R plane value and then back to the initial Z position. Note that the retract position is controlled by a G98 or a G99 code just like in the G81 drill cycle.

Figure M-85 shows the parameters for a G83 peck drill cycle.

The G83 canned peck drilling cycle is modal. The G83 cycle will stay active until it is canceled by a G80. If we were drilling a series of deep holes, we would only need to specify the coordinates for the next hole, just like we did with the standard G81 cycle. Figure M-86 is a drilling example using a G83 canned peck drilling cycle.

It is important to rapid the Z axis to an appropriate position above the workpiece before any canned cycle is called to establish the Z initial plane. An inappropriate Z initial plane could waste time or cause the tool to run into clamps or other obstacles. For example, if the spindle is 8 inches

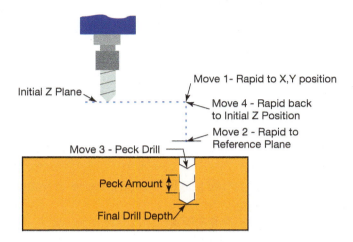

Figure M-84 Canned peck drilling cycle.

Figure M-85 Parameters for a G83 peck drilling cycle.

G83 Peck Drill Canned Cycle

X*	Rapid X-axis location
Y*	Rapid Y-axis location
Z	Z-depth
Q*	Pecking equal incremental depth amount (if I, J and K are not used)
I*	Size of the first peck depth if Q is not used.
J*	Amount reducing each peck after the first peck depth is Q is not used.
K*	Minimum peck depth if Q is not used.
P	Dwell time at Z-depth
R	R-plane (rapid point to start feeding)
F	Feed rate in inches per minute

*Indicates that it is Optional

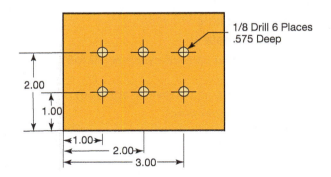

Figure M-86 Canned peck drilling cycle example.

above the work when a drill canned cycle is called, the spindle will have to rapid down to the R plane before it begins to feed to drill the hole. The correct method is to position the Z axis to a safe, close distance above the work before the canned cycle is called so that the Initial Z will be close to the work and efficient.

A G99 can be used when there are no obstacles involved. A G99 tells the control to retract the Z axis to the R plane after drilling. Then it will rapid to the next hole. This shorter retract is faster. To use this code, simply put it in the line before or in the line where the canned cycle is called, as follows:

O713 (canned Peck drilling)

N10 G20 (Inch mode)

N20 G00 G17 G40 G49 G80 G90

N30 T2 M6 (Tool Change to tool 2, 1/8 drill) N40 G54 (Workpiece zero setting)

N50 S1800 M3 (1800 RPM, Spindle start clockwise)

N60 G00 X1. Y1. (Rapid to hole position #1)

N70 G43 H2 Z1.0 (Tool height offset #1, Rapid to initial level)

N80 G98 G83 Z-.575 R.1 Q.3 F3.0

N90 X2.0 (Drill hole #2)

N100 X3.0 (Drill hole #3)

N110 Y2.0 (Drill hole #4)

N120 X2.0 (Drill hole #5)

N130 X1.0 (Drill hole #6)

N140 G80 (Cancel drill cycle)

N150 M5 (Spindle stop)

N160 G28 (Return X and Y axis to home position)

N170 M30 (End program, reset the control)

%

G73 High-Speed Peck Drill Cycle

Many machines also have a G73 high speed peck drill cycle that can be used. A G73 is faster because the spindle does not retract fully out of the hole with every peck. A G73 high speed peck drill cycle can be programmed to retract a programmed amount to clear chips or dwell to break the chips at each peck depth. Figure M-87 shows the parameters for a G73 high-speed peck cycle.

Canned Tapping Cycle (G84)

When tapped holes are needed, a G84 tapping cycle can be used. The tapping cycle is commanded much like the previous canned cycles. The difference occurs when the tap reaches the programmed depth. The spindle stops and reverses itself and automatically feeds the tap out of the hole.

The feed rate of the tapping canned cycle must be coordinated with the spindle. To do this, multiply the lead of

Figure M-87 Parameters for a G73 canned cycle.

G73 High-Speed Peck Drill Cycle

X*	Rapid X-axis location
Y*	Rapid Y-axis location
Z	Z-depth
Q*	Pecking equal incremental depth amount (if I, J and K are not used)
I*	Size of first peck depth (if Q is not used)
J*	Amount reducing each peck after first peck depth (if Q is not used)
K*	Minimum peck depth (if Q is not used)
P	Dwell time at Z-depth
R	R-plane (rapid point to start feeding)
F	Feed rate in inches per minute

*Indicates that it is Optional

the tap (lead equals 1 inch divided by the number of threads per inch) times the spindle RPM. The easier method is to divide the RPM by the threads per inch (TPI). For example 500 RPM divided by 20 TPI would equal 25 IPM feed rate.

When the G84 tapping cycle is commanded, the tap rapid positions to the specified X and Y coordinates and to the Z initial position. The tap then feeds down to the specified depth, cutting the threads. At the programmed depth, the spindle automatically reverses, and the tap is fed back to the R plane. On some older CNC machines, that are not capable of synchronous tapping, a floating tap holder is used to reduce the possibility of tap breakage.

Figure M-88 Parameters for a G84 tapping cycle.

G84 Tapping Canned Cycle

X*	Rapid X-axis location
Y*	Rapid Y-axis location
Z	Z-depth
J*	Tapping retract speed
R	R-plane (rapid point to start feeding)
F	Feed rate in inches per minute

*Optional

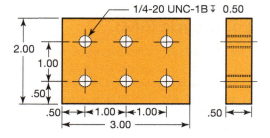

Figure M-89 Sample part with six tapped holes.

The G84 canned tapping cycle is modal and will stay active until it is canceled by a G80. If we were tapping a series of holes, we would need only to specify the coordinates for the next tapped hole. Figure M-88 shows the parameters for a G84 tapping cycle.

Figure M-89 and the program that follows are an example of tapping using a G84 canned tapping cycle.

O716 (Tapping)

N10 G20 (Inch mode)

N20 G00 G17 G40 G49 G80 G90

N30 T1 M6 (Spot drill)

N40 G54 (Workpiece zero setting)

N50 S3500 M3 (3500 RPM, Spindle start clockwise)

N60 G00 X.5 Y.5

N70 G43 H1 Z1. (Tool height offset 1, Rapid to initial level)

N80 G98 G81 Z-.09 R.1 F7.0

N90 X1.5 (Spot drill hole 2)

N100 X2.5 (Spot drill hole 3)

N110 Y1.5 (Spot drill hole 4)

N120 X1.5 (Spot drill hole 5)

N130 X.5 (Spot drill hole 6)

N140 G80 (Cancel canned cycle)

N150 M5 (Stop spindle)

N160 G28 (Return to reference position)

N170 M01

N190 T2 M6 (Tap drill)

N200 S4500 M3 (4500 RPM, Spindle start clockwise)

N210 G00 G90 X.5 Y.5

N220 G43 H2 Z1. (Tool height offset 2, Rapid to initial level)

N230 G98 G81 Z-.6 R.1 F8.50 (Return to initial R-plane, Drill hole 1 - .6 deep 8.5 inches per minute feed)

N240 X1.5 (Drill hole 2)

N250 X2.5 (Drill hole 3)

N260 Y1.5 (Drill hole 4)

N270 X1.5 (Drill hole 5)

N280 X.5 (Drill hole 6)

N290 G80 (Cancel canned cycle) N300 M5 (Stop spindle)

N320 G28 (Return to reference position)

N330 M01 (Optional stop)

N340 T3 M6 (TAP 1/4 -20)

N350 G00 G90 X.5 Y.5 S500 M3

N360 G43 H3 Z1.

N370 G98 G84 Z-.5 R.1 F25.0

N380 X1.5 (Tap hole 2)

N390 X2.5 (Tap hole 3)

N400 Y1.5 (Tap hole 4)

N410 X1.5 (Tap hole 5)

N420 X.5 (Tap hole 6)

N430 G80

N460 G28 (Return to reference position)

N470 M30 (End program, reset the control)

Bore Canned Cycle (G85)

A G85 canned cycle is used to bore holes. A G85 cycle will command the machine to make a rapid move from the Z initial position to the R plane. It will then bore the hole to a Z depth at the specified feed rate. It will then reverse and feed the spindle back to either the R plane or the Z initial position. Figure M-90 shows the parameters for a G85 boring cycle.

Figure M-91 shows an illustration of a G85 boring cycle. In the first step, the spindle rapids to the hole position specified by the X and Y parameters. The spindle then rapids down to the R plane (specified by the R parameter). In step 2, the spindle feeds down to the depth specified by the Z parameter at the feed rate specified by F. In step 3, the spindle feeds back up to the R plane at the feed rate specified by the F parameter.

The general format is

N100 G85 Xn Yn Zn Rn Fn

G85 specifies a bore cycle.

Xn and Yn are the *X* and *Y* coordinates of the hole to be bored. The X and Y can be in absolute mode (G90) or in incremental mode (G91).

Z represents the depth of the hole in absolute mode (G90) or the distance below the R plane to the hole bottom in incremental mode (G91).

R specifies the distance to the R plane in absolute mode (G90) or the distance below the initial tool position to the R plane in incremental mode (G91). If you do not specify an R plane value, the last active R plane value will be used. If there is not an active R plane, the tool will return to the initial Z position.

F specifies the feed rate.

Note that canned cycles are modal so all that would be needed to bore another hole would be the new *X Y* position if everything else remained the same.

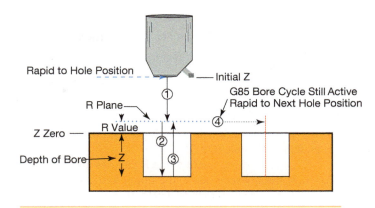

Figure M-91 Example of a bore canned cycle.

Counterbore or Spotdrill Canned Cycle (G82)

A G82 canned cycle can be used to make counterbored or spotdrilled holes (Figure M-92). A G82 cycle will command the machine to make a rapid move from the Z initial position to the R plane. It will then bore the hole to a Z depth at the feed rate specified. It will dwell for the number of seconds specified in the P value. It will then rapid the spindle back to either the R plane or the Z initial position. The parameters for a G82 are shown in Figure M-93.

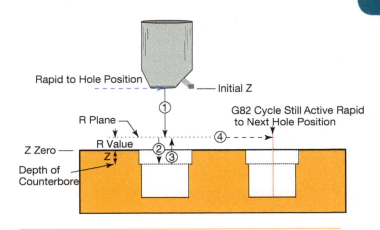

Figure M-92 Example of a counterbore cycle.

Figure M-93 Parameters for a G82 spot drill/counterbore cycle.

Figure M-90 Parameters for a G85 bore cycle.

G85 Bore in/Bore out Canned Cycle	
X*	Rapid *X*-axis location
Y*	Rapid *Y*-axis location
Z	Z-depth
R	R-plane (rapid point to start feeding)
F	Feed rate in inches per minute

*Optional

G82 Spot Drill Counterbore Canned Cycle	
X*	Rapid *X*-axis location
Y*	Rapid *Y*-axis location
Z	Z-depth
P	Dwell time at z depth
R	R-plane (rapid point to start feeding)
F	Feed rate in inches per minute

*Optional

The general format of a G82 is

N100 G82 Xn Yn Zn Rn Fn Pn

G82 specifies a counterbore cycle.

Xn and Yn are the *X* and *Y* coordinates of the hole to be counterbored. The X and Y can be in absolute mode (G90) or in incremental mode (G91).

Z represents the depth of the hole in absolute mode (G90) or the distance below the R plane to the hole bottom in incremental mode (G91).

R specifies the distance to the R plane in absolute mode (G90) or the distance below the initial tool position to the R plane in incremental mode (G91). If R plane value is not specified, the last active R plane value will be used. If there is not an active R plane, the tool will return to the initial Z position.

F specifies the feed rate.

P is the dwell time in seconds (0.01 to 99.99) at the bottom of the hole.

Note that canned cycles are modal so all that would be needed to counterbore another hole would be the new X Y position if everything else remained the same.

SUBPROGRAMS

Subprograms can be used to reduce redundant programming and shorten programs. For example, if you are machining a part with several holes that require multiple operations at each hole location, you would normally have to list those hole positions multiple times in the same program.

You could use the subprogram to hold the hole positions and just call them for each operation. Figure M-94 shows a part that requires the same hole pattern at four different locations. If we use incremental positioning, the holes in each pattern are the same.

The main program is used to prepare the control, load the tool, and get ready for machining. In line number N106 we move the tool to the hole position that is located at the 2

o'clock position on the lower-left hole pattern. The first hole is drilled in line number N110. Line number N112 is used to call the subprogram. M98 is the M code that is used to call a subprogram. The subprogram's name is 1001. At this point, the control would go to program 1001 and drill the rest of the locations of the first bolt circle. Note that the locations in the subprogram are incremental. In line N70 of the subprogram, the M99 returns the control to the main program line that follows the subprogram call (line number 128). The program then continues by drilling the first hole in the next bolt pattern, calling the subroutine to finish the other holes. The process is then repeated two more times to complete all four hole patterns.

O0000 (Main Program)

N100 G20

N102 G0 G17 G40 G49 G80 G90 (Rapid, XY Plane, Cancel diameter offsets, Cancel tool length compensation, Cancel canned cycles, Absolute mode)

N104 T2 M6 (Tool change to tool 2)

N106 G54 (Workpiece coordinates are found in G54 register)

N108 S2000 M3 (Turn on spindle clockwise at a speed of 2000 RPM)

N106 G0 X3.299 Y2.75

N108 G43 H2 Z.1

N110 G99 G81 Z-.58 R.1 F4.00

N112 M98 P1001

N128 G80

N130 G90 Y7.25

N132 G99 G81 Z-.58 R.1 F4.00

N134 M98 P1001

N150 G80

N152 G90 X7.799Y7.25

Figure M-94 Part with same hole pattern in four places.

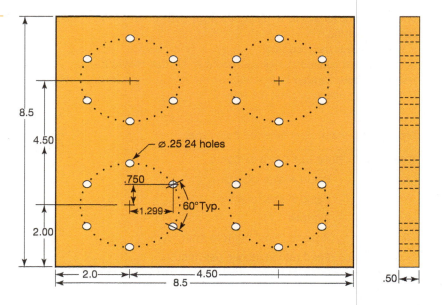

N154 G99 G81 Z-.58 R.1 F4.00

N156 M98 P1001

N172 G80

N174 G90 Y2.75

N176 G99 G81 Z-.58 R.1 F4.00

N178 M98 P1001

N194 G80

N196 M5

N198 G91 G28 Z0.

N200 G28 X0. Y0.0

N202 M30 (Program end, Memory reset)

O1001 (Subprogram)

N10 G91 (Incremental mode)

N20 X-1.299Y.75

N30 X-1.299Y-.75

N40 Y-1.5

N50 X1.299Y-.75

N60 X1.299Y.75

N70 M99 (Return from subroutine to the main program)

%

SELF-TEST

1. What type of information does the programmer get from the part drawing?
2. What must be taken into consideration when deciding on the machine to be used?
3. What are the advantages of using a canned cycle?
4. List two of the more commonly used canned cycles.
5. Calculate the feed rate for a 1/2-13 tap running at 600 RPM.
6. Using the tool data tables shown in Figure M-95, write a program to execute the profile milling and the hole operations for the part drawing shown in Figure M-97. Develop a process operation sheet as shown in Figure M-96. Use absolute programming and cutter compensation. Mill the part contour .50 deep.

Figure M-95 Tools available to program the part shown in Figure M-97.

Tool #	Operation	Tool	Speed (RPM)	Feed (IPM)
1	Mill Profile	.750 End Mill	150	6.00
2	Tap Drill	5/16 Drill	1000	3.00
3	¼ Drill	¼ Drill	1200	2.5
4	Tap	3/8-16 Tap	500	31.25

Figure M-96 Process planning sheet for question 8.

Acme Machining Inc.- Process Plan			Part #	
Operation	Tool #	Tool Description	RPM	Feed Rate

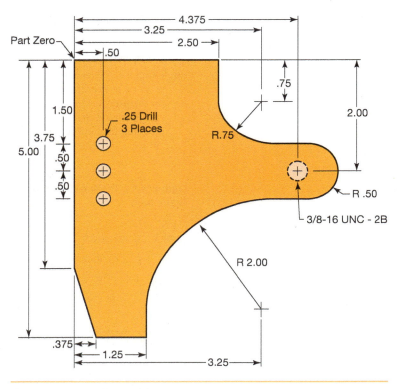

Figure M-97 Part for question 8.

7. Write a program to machine the part shown in Figure M-98.

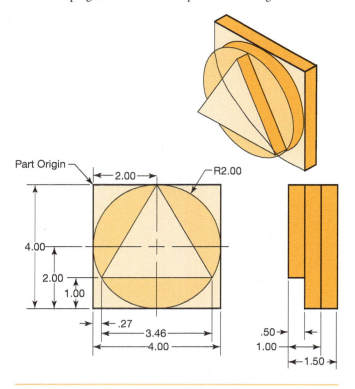

Figure M-98 Part for question 7.

8. Write a program to machine the part shown in Figure M-99. Mill the contour .5 deep.

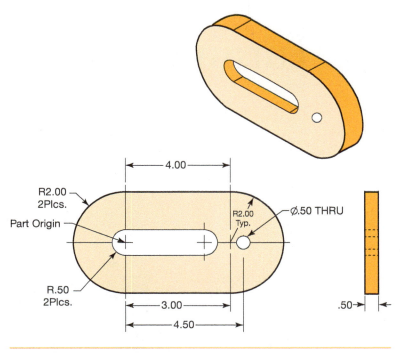

Figure M-99 Part for question 8.

CNC Turning Machines

INTRODUCTION

A CNC lathe is commonly called a turning center. Its main function is to create high-quality cylindrical parts efficiently. Turning centers can machine internal and external surfaces. Other machine operations, such as drilling, tapping, boring, and threading, are also done on turning centers.

OBJECTIVES

After completing this unit, you should be able to:

- Name the major components of the turning center.
- Correctly identify the major axes on turning centers.
- Describe three work-holding devices used on CNC lathes.
- Describe the common machining operations and the tools associated with them.
- Explain tool-wear offsets.
- Describe how geometry offsets or workpiece coordinates are set.
- Correctly identify safe working habits associated with CNC lathes.
- Identify and explain turning center controls.

TYPES OF CNC TURNING MACHINES

The standard flat-bed configuration is still evident on some CNC lathes; however, most turning machines today have a slant-bed configuration (Figure M-100). Slanting the bed on a CNC lathe allows the chips to fall away from the slideways and allows the operator easy access to load and unload parts.

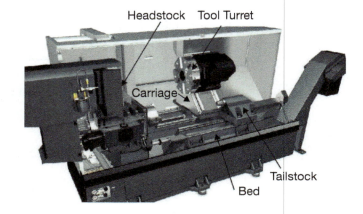

Figure M-100 Slant-bed lathe components (*Courtesy of Haas Automation Inc.*).

COMPONENTS OF CNC LATHES

The main components of a CNC lathe or turning center are the headstock, tailstock, turret, bed, and carriage (Figure M–100).

Headstock

The headstock contains the spindle and transmission gearing, which rotates the workpiece. The headstock spindle is driven by a variable speed motor. The spindle motor delivers the required horsepower and torque through a drive belt or series of drive belts. Figure M-101 shows a cutaway view of a typical spindle.

Tailstock

The tailstock is used to support one end of the workpiece. The tailstock slides along its own set of slideways on some turning machines and on the same set of slideways as the carriage on conventional-style CNC lathes. The tailstock has a sliding

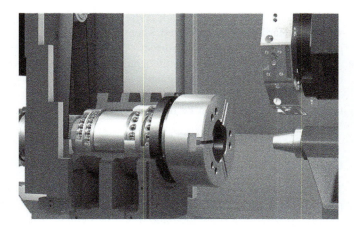

Figure M-101 Cutaway view of a headstock and spindle (*Courtesy of Haas Automation Inc*.).

Figure M-102 Programmable tailstock (*Courtesy of Haas Automation Inc*.).

Figure M-103 Twelve-station tool turret. Each turret position is numbered for identification purposes (*Courtesy of Haas Automation Inc*.).

Figure M-104 The slant bed is designed for quick chip removal and easy operator access (*Courtesy of Haas Automation Inc*.).

spindle much like that of the tailstock on a manual lathe. Two types of tailstocks are available: manual and programmable. The manual tailstock is moved into position by the use of a switch or hand wheel. The programmable tailstock can be moved manually or can be programmed like the tool turret (Figure M-102).

Tool Turrets

Tool turrets on turning machines come in all styles and sizes. The basic function of the turret is to hold and quickly index cutting tools. Each tool position is numbered for identification (Figure M-103). When the tool needs to be changed, the turret is moved to a clearance position and indexes, bringing the new tool into the cutting position. Most tool turrets can move bi-directionally to assure the fastest tool indexing time. Tool turrets can also be indexed manually, using a button or switch located on the control panel.

Bed

The bed of the turning center supports and aligns the axis and cutting tool components of the machine. The bed is made of high-quality cast iron and will absorb the shock and vibration associated with metal cutting. The bed of

the turning center lies either flat or at a slant to accommodate chip removal. The slant of the bed is usually 30 to 45 degrees (Figure M-104).

Carriage

The carriage slides along the bed and controls the movement of the tool. The basic CNC lathe has two major axes, the *X* and *Z* axes (Figure M-105).

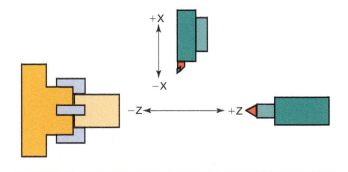

Figure M-105 Lathe axes of motion. Note that the *Z* axis is always in line with the spindle.

The Z axis is in the same plane as the spindle, just as on machining centers. The X or cross-slide axis runs perpendicular to the Z axis. Negative Z axis (–Z) motion moves the tool turret closer to the headstock. Positive Z (+Z) motion moves the turret or tool away from the headstock or toward the tailstock. Negative X (–X) motion moves the tool or turret toward the centerline of the spindle, and positive X (+X) motion moves the tool or turret away from the centerline.

WORK HOLDING

Work-holding devices are an integral part of a CNC lathe. As the demand for higher production increases, so does the need for an understanding of the different types of work-holding devices. The most common work-holding method used on turning machines is the chuck.

Chucks

There are many different types of chucks, but the most common type is the three-jaw, self-centering, hydraulic chuck (Figure M-106).

This type of chuck has three jaws, all of which move in unison under hydraulic power. The chucks are activated by a foot switch that opens and closes the chuck jaws.

Figure M-106 Three-jaw hydraulic chuck.

There are two types of chuck jaws: hardened jaws and soft jaws. Hardened jaws are used where maximum holding power is needed on unfinished surfaces.

Soft jaws are used on parts that cannot have much runout or on finished surfaces that cannot be marred. Soft jaws are typically made from soft steel and are turned to fit each type of part. Soft jaws can be machined with ordinary carbide tooling.

Collet Chucks

The collet chuck is an ideal work-holding device for small parts where high accuracy is required. The collet chuck assembly consists of a draw tube and a hollow cylinder with collet pads. Collets are available for holding hexagonal, square, and round stock. They are typically used with bar feed systems.

MATERIAL HANDLING

Material-handling devices increase production rates and reduce labor costs. The types of devices used are determined by part size, shape, and production levels.

Parts Catchers

Parts catchers are used on small-diameter workpieces. Parts catchers consist of a tray that, prior to the tool cutting off the part, tips forward and catches the completed part and delivers the part to the outside of the machine.

Chip Conveyers

Chip conveyers automatically remove chips from the bed of the machine. The chips produced by machining operations fall onto the conveyer track and are transported to scrap or recycling containers (Figure M-107).

Figure M-107 Chip conveyers provide automated chip disposal to provide a chip-free work environment.

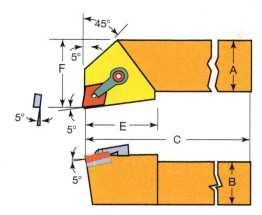

Style L Tool Holder with an 80° Insert

Figure M-108 Tool style L can be used for turning and facing using an 80-degree diamond insert.

CUTTING TOOLS

Turning centers use tool holders with indexable inserts. The tool holders on CNC machines come in a variety of styles, each suited for a particular type of cutting operation. The machining operations discussed in this chapter include facing, turning, grooving, parting, boring, and threading. Section F, unit 4 covered tool holders and carbide inserts.

Facing

Facing operations involve squaring the face or end of the stock. The tool needs to be fed into the stock in a direction that will push the insert toward the pocket of the holder (Figure M-108).

Turning

Turning operations remove material from the outside diameter of the rotating stock. Rough turning removes the maximum amount of material from the workpiece and should be done with an insert with a large included angle.

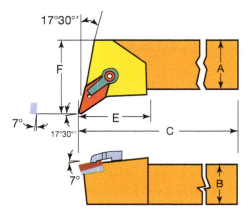

Style Q Tool Holder with an 55° Insert

Figure M-109 Tool style Q is used for profile turning using a 55-degree diamond insert.

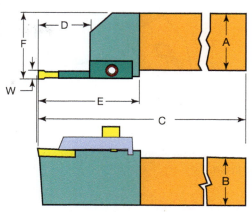

Style NG Grooving Tool Holder

Figure M-110 Tool style NG for grooving.

The large included angle will insure that the tool has the proper strength to withstand the cutting forces being exerted. Profile turning uses an insert with a smaller included angle. If the finish profile requires the use of a small, sharp-angled insert, a series of semi-finish passes are necessary to insure against tool breakage (Figure M-109).

Grooving

For internal and external grooving, the tool is fed straight into the workpiece at a right angle to its centerline. The cutting insert is located at the end of the tool (Figure M-110). Grooving operations include thread relieving, shoulder relieving, snap-ring grooving, O-ring grooving, and oil reservoir grooving.

Parting

Parting is a machine operation that cuts the finished part off from the rough stock. This operation is similar to grooving. The tool is fed into the part at a right angle to the centerline of the workpiece and is fed down past the centerline of the work, thus separating it from the rough stock. The parting tool has a carbide insert located at the end of the tool and has a slight back taper along the insert for clearance (Figure M-111).

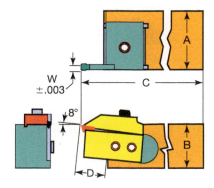

Style KGSP Parting Tool Holder

Figure M-111 Tool style KGSP for parting.

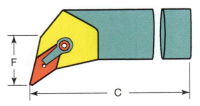

Style LP Boring Bar with 55° Diamond Insert

Figure M-112 LP-style boring bar using a standard 55-degree diamond insert.

Boring

Boring is an internal turning operation that enlarges, trues, and contours previously drilled or existing holes. Boring is done with a boring bar (Figure M-112).

Threading

Threading is the process of forming a helical groove on the outside or inside surface of a cylindrical part. Threads can be cut in multiple ways, but for this tooling section we will concentrate on single-point threading tools (Figure M-113).

Single-point threading tools are typically 60-degree carbide inserts clamped in a tool holder. The threading tool is fed into the work and along the part at a feed rate equal to the lead of the thread. (The lead of a thread is the distance it advances in one revolution.) The lead can be calculated by dividing 1 inch by the number of threads. For example, if you had eight threads per inch, the calculation would be 1/8 or .125.

MACHINE REFERENCE POSITION, TOOL CHANGE POSITION, AND PART ORIGIN

CNC turning centers have three zero points. Figure M-114 shows the typical zero points for a CNC turning center.

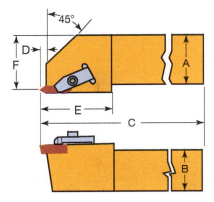

Style NS Threading Tool

Figure M-113 NS-style threading tool. This style holder can be equipped with different angled threading inserts for different types of threads.

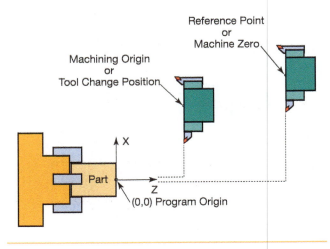

Figure M-114 Zero points for a CNC turning center.

Machine Zero

The machine zero point is set by the machine manufacturer. It is the point at which all of the axes are zeroed out.

Tool Change Position

The tool change position is also called the machining origin. The tool change position is a safe location that the machine returns to when indexing to a new tool.

The tool change position is determined during program setup. The location is input at the beginning of a program by use of the zero-offset command (G50). During operation the machine will execute all movements relative to this origin. It should be noted that on small turning centers the machine zero is often used as the tool change position. It may take too much time to return to the machine zero to change a tool so a tool change position that is closer to the chuck is usually used.

Part Origin

The part origin can also be called a program origin. The part origin is the zero point from which all of the part program dimensions were created. When a CNC turning center is setup, the operator uses tool offsets to locate the program origin from the machining origin.

OFFSETS

There are two kinds of offsets used on CNC turning machines: tool offsets and geometry offsets.

Tool Offsets

Tool offsets, also called tool-wear offsets, are an electronic feature for adjusting the length and diameter of machined surfaces. CNC turning machines have offset tables in which the operator can input or change numbers to adjust part sizes without changing the program. Adjusting the offsets is a responsibility of the CNC machine operator. If the part sizes don't meet

Figure M-115 Tool offset table. Notice that this is a wear offset table.

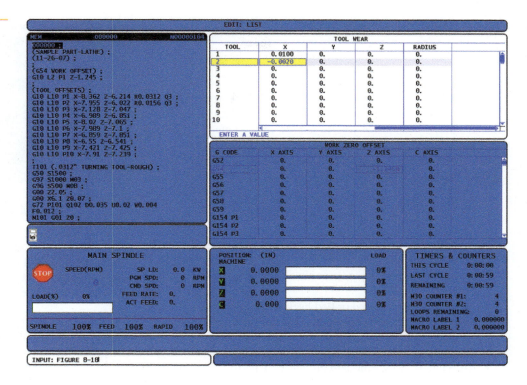

the part print requirements, the operator alters the offset, for the tool in the tool-wear offset table (Figure M-115).

Geometry Offsets

Geometry offsets, or workpiece coordinates, are used to tell the control where the workpiece is located. The workpiece coordinate is the distance from the tool tip, at the home position, to the workpiece zero point. The workpiece zero point is normally located at the end and center of the workpiece or at the chuck face and center of the machine (Figure M-116).

Geometry offsets, or workpiece coordinates, can be registered two different ways. The most common approach is to use a preparatory or G-code such as a G50 or a G92. The offset distance is determined by touching the tool tip off on the workpiece and recording the distance.

The position or distance can be determined accurately by using the position screen on the control. For example, if the distance from the home position to the face of the workpiece is 16.500 inches and the distance from the tool tip to the centerline of the workpiece is 8.500 inches, the G50 or G92 would be G92 X8.500 Z16.500.

To find the centerline of the workpiece, the operator takes a skim cut off of the outside of the workpiece. The operator then measures the turned diameter and adds that dimension to the X-axis machine position.

If a geometry offset is used, the X and Z values would be loaded into the geometry offset table under the tool offset number, and the G50 or G92 would not be needed. It is important to remember that every tool that is used in the program will need to be measured in this manner. Figure M-117 shows a tool offset table.

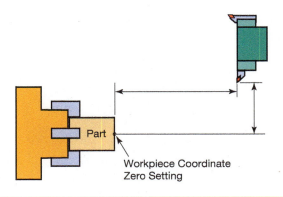

Figure M-116 Tool geometry offsets, also known as workpiece coordinate settings, are generally set as the distance from machine home to the end and center of the workpiece.

	Tool Offset (Geometry)				N0000
No.	X Axis	Z Axis	Radius	Tip	Machine Position (Relative)
01	00.000	00.000	00	00	
02	00.000	00.000	00	00	X00.0000
03	00.000	00.000	00	00	Y00.0000
04	00.000	00.000	00	00	
05	00.000	00.000	00	00	
06	00.000	00.000	00	00	
07	−12.346	−8.567	.032	3	
08	−6.5671	−6.987	.015	2	(Inch)
09	−4.5672	−3.7865	.032	3	
10	−3.789	−4.8923	.032	3	
11	00.000	00.000	00	00	
12	−10.500	−4.876	.031	1	
13	−9.5624	−5.8763	.015	3	
14	00.000	00.000	00	00	
15	00.000	00.000	00	00	
16	−8.5390	−7.9845	.015	3	

Figure M-117 Tool offset table.

MACHINE CONTROL OPERATION

Safety

Before you operate any machine, remember that no one has ever thought they were going to be injured. But it happens! It can happen in a split second when you are least expecting it, and an injury can affect you for the rest of your life.

You must be safety minded at all times. Please get to know your machine before operating any part of the machine control and please keep these safety precautions in mind.

- Wear safety glasses and side shields at all times.
- Do not wear rings or jewelry that could get caught in a machine.
- Do not wear long sleeves, ties, loose fitting clothes, or gloves when operating a machine. These can easily get caught in a moving spindle or chuck and cause severe injuries.
- Keep long hair covered or tied back while operating a machine. Many severe accidents have occurred when long hair became entangled in moving tooling and machinery.
- Keep hands away from moving machine parts.
- Use caution when changing tools. Many cuts occur when a wrench slips.
- Stop the spindle completely before doing any setup or piece loading and unloading.
- Do not operate a machine unless all safety guards are in place.
- Metal cutting produces very hot, rapidly moving chips that are very dangerous. Long chips are especially dangerous. You should be protected from chips by guards or shields. You must always wear safety glasses with side shields to prevent chips from flying into your eyes. Shorts should not be worn because hot chips can easily burn your legs. Hot chips that land on the floor can easily burn though thin-soled shoes.
- Many injuries occur during chip handling. Never remove chips from a moving tool. Never handle chips with your hands. Do not use air to remove chips. They are dangerous when blown around and can also be blown into areas of the machine where they can damage the machine.
- Securely clamp all parts. Make sure your setup is adequate for the job.
- Use proper methods to lift heavy materials. A back injury can ruin your career. It does not take an extremely heavy load to ruin your back; bad lifting methods are enough.
- Safety shoes with steel toes and oil-resistant soles should be worn to protect your feet from dropped objects.
- Watch out for burrs on machined parts. They are very sharp.
- Keep tools off of the machine and its moving parts.
- Keep your area clean. Sweep up chips and clean up any oil or coolant that people could slip on.

Figure M-118 Typical CNC Lathe control configuration.

- Use proper speeds and feeds. Reduce feed and speed if you notice unusual vibration or noise.
- Dull or damaged tools break easily and unexpectedly. Use sharp tools and keep tool overhang short.

MACHINE CONTROLS

As you look at a turning center control, you see what seems to be an endless number of buttons, keys, and switches (Figure M-118). Although every manufacturer has its own style of control, they all have basically the same features. If you have a good understanding of one machine, the next control will be much easier to learn.

Manual Control

Manual control features are buttons or switches that control machine movement (Figure M-119).

Emergency Stop Button

The emergency stop button is the most important component of the machine control. This button has saved more than one operator from disaster. The emergency stop button, which shuts down all machine movement, is a large red button. Emergency stop buttons should be used when it is evident that a collision or tool breakage is going to occur. Emergency stop buttons are often located in more than one area on the machine tool and should be located prior to operating a machine.

Figure M-119 Manual machine control features.

Moving the Axes of the Machine

Manual movement of the machine axes is done a number of different ways. Most controls are equipped with a pulse-generating hand wheel (Figure M-120).

The hand wheel gives the operator a great deal of control over the machine axes.

The hand wheel has an axis selection switch that allows the operator to choose which axis to move. The handle sends a signal or electronic pulse to the motors, which move the carriage and the cross slide.

Machines are also equipped with jog buttons (Figure M-121). When the jog button for an axis is pressed, the axis moves. The distance or speed at which the machine moves is selected by the operator prior to the move.

Some machines have a joystick that is used to move the machine axes in the direction that the joystick is moved. For example, if the joystick is moved to the down position, the cross slide moves down; if the joystick is moved to the left, the carriage moves toward the headstock, and so on. The distance or speed at which the machine moves is selected in much the same way as the jog buttons. The selection options include rapid traverse, selected feed rate, or incremental distance.

Figure M-120 This hand wheel could be used to move an axis.

Figure M-121 Axes jog buttons.

Cycle Start/Feed Hold Buttons

The two most commonly used buttons on the control are the cycle start and feed hold buttons (Figure M-122). The cycle start button is used to start execution of the program. The feed hold will stop execution of the program without stopping the spindle or any other miscellaneous functions. By pushing cycle start, the operator can restart the execution of the program.

Spindle Speed and Feed Rate Override Switches

Spindle speed and feed rate overrides are used to speed up or slow down the feed and speed of the machine during cutting operations (Figure M-123). The override controls are typically used by the operator to adjust to changes in cutting conditions, such as hard spots in the material. Feed rates can generally be adjusted from 0 to 150 percent of the programmed feed rate. Spindle speeds can often be adjusted from 0 to 200 percent of the programmed spindle speed.

Single Block Operation

The single block option on the control is used to advance through the program one block at a time. When the single block switch is on, the operator presses the button each time they want to execute a program block. When the operator wants the program to run automatically, he or she can turn the single block off and press cycle start, and the program will run through without stopping. The single block switch allows the operator to watch each operation of the program carefully. It is often used to test a new program.

Figure M-122 Cycle start and feed hold buttons.

Figure M-123 Feed rate and spindle speed override controls.

Manual Data Input

Manual data input or MDI is an input method that can be used for making changes to a previously loaded program or as a means of inputting data for the machine to act on manually, especially for setup purposes. MDI is done through the alphanumeric keyboard located on the control (Figure M-124). The keyboard is made up of letters, numbers, and symbols. These keys allow the operator to input a series of commands or a whole program. MDI is often used by the operator to input and execute a single line of code: such as M03 S800 (turn the spindle on clockwise at 800 RPM).

Figure M-124 An alphanumeric keyboard, which contains a wide variety of keys, allows the operator to input information.

SELF-TEST

1. Name four of the main components that make up the CNC turning center.
2. What is the purpose of the tool turret?
3. State one advantage of a slant-bed-style CNC turning center over the flat-bed-style CNC lathe.
4. Which of the two major axes associated with the turning center always lies in the same plane as the spindle?
5. What is the most common type of work-holding device used on the turning center?
6. Describe threading.
7. Which machining operation cuts the finished part off the rough stock?
8. When are tool-wear offsets used?
9. What is a geometry offset?
10. How can an operator accurately judge the position of the tool?
11. From what location is the workpiece zero or geometry offset typically calculated?
12. How can chips be automatically removed from the bed of the turning center?
13. Name two manual control devices that allow the operator to move the axes of the machine.
14. What tasks do override switches perform?

Programming CNC Turning Centers

INTRODUCTION

This unit covers the steps necessary to properly plan, program, and set up a part for turning. Turning centers have the capabilities to reduce part programming time and increase part quality through the use of canned cycles and tool-nose radius compensation.

OBJECTIVES

After completing this unit, you should be able to:

- Identify the two main axes of movement associated with a turning center.
- Describe the recommended sequence of operations for a turning center.
- Program parts using linear and circular moves.
- Define the term "tool-nose radius compensation."
- Explain the use of tool-nose direction vectors.

TURNING CENTERS

The turning center uses two basic axes of motion and the machining center uses three. The two basic axes of movement on the turning center are the X axis, which controls the diameter of the part, and the Z axis, which controls the length.

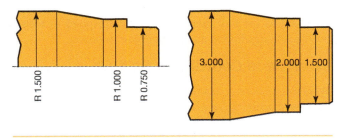

Figure M-125 CNC lathes are programmed in diameter values.

The X axis is normally programmed with diameter values rather than radius values, so the actual position of the tool would be the radial distance from the centerline (Figure M-125). The spindle centerline would be an X0 position. When positioning the tool in the X axis of travel, we will seldom position to a $-X$ position, except in the case of a facing operation. The Z axis part origin or zero position can be either at the right end of the part or a position located near the spindle of the machine.

PLANNING THE PROGRAM

When planning a program for the turning center, we need to be aware of the tooling, the work-holding device, and the part print. The part drawing or part print gives the programmer detailed information about requirements. The shape of the part, part tolerances, material requirements, surface finishes, and the quantity of parts all have an impact on the program, as well as where the part zero or datum will be located.

WORD ADDRESS PROGRAMMING FOR TURNING CENTERS

The word address programming format is a system of characters arranged into blocks of information. G-codes (preparatory functions) and M-codes (miscellaneous functions) are the main control codes in word address programming. Figures M-126 and M-127 list the more commonly used M- and G-codes found in industry. Use these or the codes for your particular turning center as a reference when completing the exercises in this chapter.

Review of Programming Procedures

Whether you are programming a machining center or a turning center, all CNC programs follow a general order.

The general order is as follows:

- Start-up procedures
- Tool call
- Workpiece location block

Figure M-126 Commonly used turning center M (miscellaneous) functions.

M00	Program stop
M01	Optional stop
M02	End of program
M03	Spindle start clockwise
M04	Spindle start counterclockwise
M05	Spindle stop
M07	Mist coolant on
M08	Flood coolant on
M09	Coolant off
M30	End of program & reset to the top of program
M98	Subprogram call
M99	End subprogram & return to main program

- Spindle speed control
- Tool motion blocks
- Home return
- Program end procedures

Start-Up (Preliminary) Procedures

Start-up procedures are those commands and functions that are necessary at the beginning of the program. A standard start-up procedure usually involves cancelling tool compensation, absolute or incremental programming, standard or metric, and the setting of the work-plane axis. Every manufacturer of machine tools has its own suggested format for start-up. Below is an example of typical turning center start-up procedure blocks.

N0001 G90; Absolute programming.

N0002 G20 G40; Inch units, tool-nose radius compensation cancel.

Tool Change and Tool Call Block

An M06 tool change code is not used on the turning center for changing or selecting tools. An M06 is usually used for clamping or unclamping the work-holding device. The T-code or tool code is sufficient to tell the control which tool turret position to change to. The tool call is also accompanied by the offset call. The offset number is the last two digits in the tool call. For example, a T0101 would load tool 1 with an offset 1. These offsets are used for offsetting the tool path to accommodate for tool wear or exact sizing of the part. The programmer must be careful to return the axis to home or to a safe position with a G28 or G29 preparatory function before calling a new tool. If the tool turret is not in a safe position to index, it could crash into the part or chuck and cause severe damage. Here is a look at a tool change block.

N0001 T0101;

The first two numbers call for tool #1, the second two numbers call for offset number 1. Normally, the tool offset

Figure M-127 Commonly used turning center G-codes or preparatory functions.

G00	Rapid traverse (rapid move)
G01	Linear positioning at a feed rate
G02	Circular interpolation clockwise
G03	Circular interpolation counter-clockwise
G04	Dwell
G09	Exact stop
G10	Programmable data input
G20	Input in inches
G21	Input in mm
G22	Stored stroke check function on
G23	Stored stroke function off
G27	Reference position return check
G28	Return to reference position
G32	Thread cutting
G40	Tool nose radius (TNR) compensation cancel
G41	Tool nose radius (TNR) compensation-left
G42	Tool nose radius (TNR) compensation-right
G50	Max. spindle speed clamp
G70	Finish machining cycle
G71	Turning cycle
G72	Facing cycle
G73	Pattern repeating cycle
G74	Peck drilling cycle
G75	Grooving cycle
G76	Threading cycle
G80	Canned cycle cancel
G81	Drill caned cycle
G82	Spot drill canned cycle
G83	Peck drilling caned cycle
G84	Tapping canned cycle
G85	Boring canned cycle
G86	Bore and stop canned cycle
G95	Feed per revolution
G96	Constant surface speed control
G97	Constant surface speed control cancel

number and the tool number should correspond to reduce mistakes in machining and damage to equipment.

Workpiece Coordinate Setting

Workpiece coordinate setting is easier on turning machines than on machining centers. The spindle or workpiece center is the location of the X0 position, and the Z0 position is typically the right end of the workpiece or the chuck face (Figure M-128).

Figure M-128 Possible workpiece coordinate location settings.

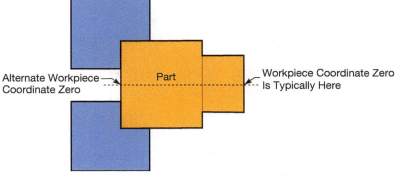

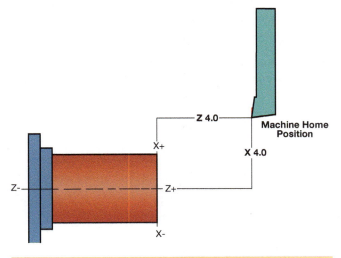

Figure M-129 G50 tool offset. In this example, the code would be G50 X4.00 Z4.00.

The workpiece location is set using an offset. Just as there are many different types of CNC controls, there are many different ways of using offsets to locate the workpiece.

G50 Method for Workpiece Offsets

When using the G50, the part datum location will accompany the code. For example, G50 X4.000 Z4.000; this would be the distance from the tool tip at the machine home position to the right end and center of the workpiece (see Figure M-129).

Tool Geometry Method for Workpiece Offsets

The most common method of workpiece location used on turning centers is a tool geometry offset. The tool geometry offset is called or activated when the tool is called. When a line in the program calls T0101, tool number 1 indexes to the cutting position and the control looks in the offset page and loads offset 01.

Tool Offset *X*-Axis Setting

A piece of material must be put in the chuck before any tool setting or zero setting operations can be performed. Index

Figure M-130 Making a cleanup cut on the diameter of the part.

the turret to the tool that is to be measured. Start the spindle using the manual mode. Make sure you are using the proper RPM. Use the turning tool to make a small cut on the diameter of the part (Figure M-130).

Approach the part carefully and feed slowly during the cut. After the small cut is done, jog away from the part using only the Z-Axis. Move far enough away from the part so that you can take a measurement with your micrometer. Stop the spindle and open the door. Take a measurement on the turned diameter of the part. On newer controls there will be a button that records the machine position and automatically records this position in the offset table. On the HAAS control, this is called the X DIAMETER MEASURE button (Figure M-131).

On the Okuma control, this button is called the *CAL* or calculate button. If you have an older control, you will have to locate the machine position screen and write down the *X*-axis location (Figure M-132). Next, you will have to add the measurement of the turned diameter to the *X*-axis machine location.

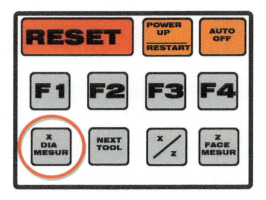

Figure M-131 HAAS control-X diameter measure button.

On some controls this is done with an add button or the Enter button. If using a G54 offset, add the machine position to the measured diameter and enter this number in the G54 X position offset line of the program. The G54 X-axis offset is the distance from the machine home position to the tool tip when it is at the center of the spindle. This procedure will need to be repeated for every tool used in your program.

Review:

- Take a cleanup cut on the diameter of the workpiece.
- Write down the X-axis machine position or press the appropriate X-axis measure button.
- Measure the diameter of the part.
- Add the diameter to the X-axis machine position.
- Put the offset (machine position + measured diameter) for this tool in the G50 command code line of your program.
- Repeat the procedure for every tool used in the program.

Tool Offset Z-Axis Setting

Index the turret to the tool that is to be measured. Make a small cut on the face of the material clamped in the spindle (Figure M-133). Approach the part carefully and feed slowly during the cut. After the small cut is done, jog away from the part using X-axis. Move far enough away from the part so that you can take a measurement with your measuring tool.

On newer controls there will be a button that records the machine position and automatically records this position in the offset table. On the HAAS control, this button is called the *Z FACE MEASURE button* (Figure M-134).

On the Okuma control, this button is called the CAL or calculate button. If you have an older control you may have to locate the machine position screen and write down the Z axis location. If using a G54 offset, enter this number in the G54 Z position offset. The G54 Z-axis offset is the distance from the machine home position to the tool tip when it is at the Z0.0 location (face) of the part. Use a piece of paper or feeler gage to measure additional tools off the face of the part and add the feeler gage thickness to the calculation (Figure M-135).

Review:

- Take a face cut.
- Write down the Z-axis machine position.
- Put this machine position in the G50 command code line of your program for this tool under the Z-axis.

Spindle Start Block

On a turning center, three codes control the spindle. A large majority of turning centers have the capability to increase or decrease the RPM as the part diameter changes. This is

Figure M-132 Machine position page.

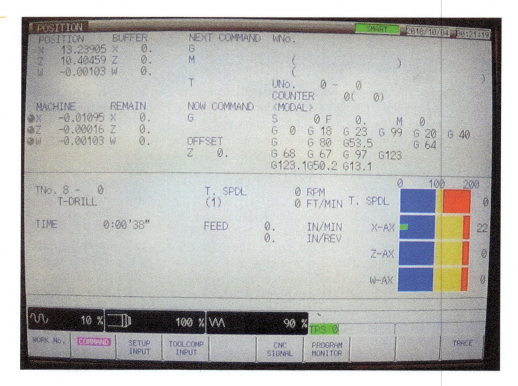

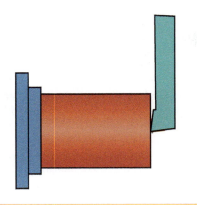

Figure M-133 Taking a facing cut.

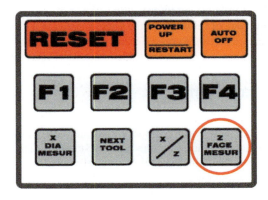

Figure M-134 HAAS control buttons for setting tool offsets.

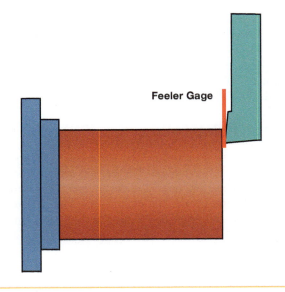

Feeler Gage

Figure M-135 Using feeler gage stock to touch off tools.

called *constant surface speed control*. Constant surface speed control is important for efficient use of cutting tools, tool life, and proper surface finish. As cutting takes place and the diameter being cut decreases, the spindle speed increases to ensure optimal cutting speeds.

Constant surface speed is controlled with a G96 preparatory code. The G96 code is followed by the proper surface footage per minute (SFPM) setting for the cutting tool material and part material. The value is set with an S: such as G96 S400.

When not using constant surface speed control, a G97 RPM input code is used. For thread cutting or drilling, a G97 code is used. The G97 will be followed by the properly calculated RPM: such as G97 S450. An M03 or M04 tells the spindle to start in a clockwise or counterclockwise direction. A typical spindle start block using constant surface speed control follows:

N0010 G96 S450 M03;

Block number 10 would set constant surface speed control to 450 SFPM and start the spindle in a clockwise direction.

Tool Motion Blocks

The tool motion blocks are the body of the program. The tool is positioned and the cutting takes place in these blocks.

Home Return

Tools need to be returned to home or a fixed position before a tool change (index) takes place. Most machine controls use a G28 command to rapid position the tool to home. A G28 is a two-step command. Two things happen whenever a G28 is commanded. First the machine will move (at rapid) the axis or axes included in the G28 command to an intermediate position. In the case of Figure M-136 the intermediate point would be X2.0 and Z0.0 absolute coordinates. Then the machine will rapid the axis or axes to the home or zero return position. If we had commanded only the X axis in the G28 command, then only the X axis would have returned to the home position. In most cases programmers like to use incremental axes commands, such as a U for X and a W for Z with the G28 command. This assures the direction of tool movement. In the case of a G28 U0.0 W0.0 command, the tool would take a direct route back to the home position in both the X and Z axes. This is covered in greater detail in practice programs later in this unit.

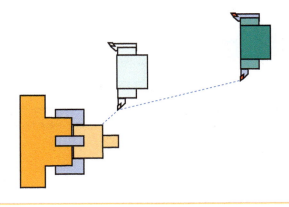

Figure M-136 G28 return to home position command.

Program End Blocks

There are a number of ways to end a program. Some controls require that you turn off the coolant and the spindle with individual miscellaneous function codes. Other controls will end the program, reset the program, and turn off miscellaneous functions, with an M30 code. No matter what type of control you have, it is always a good idea to cancel any offsets that may be active when ending your program.

Next, we examine a typical turning program that incorporates many of the elements that have been covered. The part is shown in Figure M-137.

The program for the part shown in Figure M-137 is shown below.

O0915 (Program name O0915)

N10 G90 G20 G40 (G90 - Absolute mode, G20 - Inch mode, G40 - Cancel tool nose radius compensation)

N15 G28 U0.0 W0.0 (Return to home using the straightest path, machine home in this example)

N20 G00 T0101 (Tool 1, offset 1)

N30 G50 X5.800 Z10.250 (Work offset for tool 1 at X5.8 Z10.25)

N40 G96 S400 M03 (Turn on spindle clockwise at 400 RPM constant surface speed)

N50 G50 S3600 (Set maximum spindle speed at 3600 RPM)

N60 G00 Z.1 (Rapid to Z.1)

N70 G00 X1.625 (Rapid to X1.625)

N80 G01 Z-1.25 F0.01 (Linear feed to Z-1.25 at a feed rate of 0.01 per revolution)

N90 G01 X1.75 (Linear feed to X1.75)

N100 G28 X2.00 Z0.0 M05 (G28 - Return to reference point through X2.00, Z0.0 M05 - Spindle stop)

N110 T0100 (Cancel the offset for tool 1)

N120 M30 (Program end/memory reset)

%

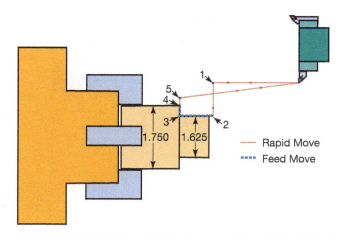

Figure M-137 The center of the workpiece is X0.0 zero and the right end of the stock is Z0.0.

CIRCULAR INTERPOLATION

Only straight-line moves were used in the program to machine the part in Figure M-138. One of the most important capabilities of a CNC machine is its ability to do circular cutting motions. CNC turning centers are capable of cutting an arc of a specified radius value. Arc, or radius, cutting is known as circular interpolation.

Circular interpolation is done in the same manner as on machining centers, with the use of G02 or G03 preparatory codes. To cut an arc, the programmer needs to follow a specific procedure. To cut an arc, the tool needs to be positioned to the start point of the arc in the line before the circular interpolation will be done. We need to tell the control the direction of the arc in the circular interpolation line: clockwise or counterclockwise. The third piece of information is the end point of the arc. The last piece of information the control needs is the position of the arc center or, if you are using the radius method of circular interpolation, the radius value of the arc (Figure M-138).

Arc Start Point

The arc start point is the coordinate location of the start point of the arc. The tool is moved to the arc start point in the line prior to the arc generation line. Simply stated, the start point of the arc is the point the tool is currently at when you want to begin cutting the arc.

Arc Direction (G02, G03)

Circular interpolation can be carried out in two directions: clockwise or counterclockwise. Two G-codes specify arc direction. The G02 code is used for circular interpolation in a clockwise direction, and the G03 code is used for circular interpolation in a counterclockwise direction. Both codes are modal. G02 and G03 codes are controlled by a feed rate (F) code, just like a G01.

Arc End Point

The tool must be positioned at the start point of the arc prior to a G02 or G03 command. The current tool position becomes the arc starting point and the arc end point is the coordinate position for the end point of the arc. The arc start point and arc end point determine the tool path, which is generated according to the arc center position.

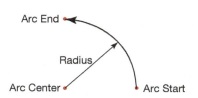

Figure M-138 The critical pieces of information needed to cut an arc are the arc start point, arc direction, arc end point, and arc centerpoint location.

Arc Centerpoints

To generate an arc path, the controller has to know where the center of the arc is. There are two methods of specifying arc centerpoints: the coordinate arc center-point method and the radius method.

When using the coordinate arc center method, a particular problem arises: how do we describe the position of the arc center? If we use the traditional *X, Y, Z* coordinate position words to describe the end point of the arc, how will the controller discriminate between the end points coordinates and the arc center coordinates? It is done by using different letters to describe the same axes. Secondary axes addresses are used to designate arc centerpoints. The secondary axes-addresses for the primary axes are:

$I = X$ *axis coordinate of an arc centerpoint*

$K = Z$ *axis coordinate of an arc centerpoint*

When we cut an arc on the turning center, the *X/Z* axes are the primary axes, and the *I/K* letter addresses are used to describe the arc centerpoint. The type of controller that you use determines how these secondary axes are located.

With some controllers, such as the HAAS or Fanuc controller, the arc centerpoint position is described as the incremental distance from the arc start point to the arc center. This is the most common method of specifying the arc center.

Some CNC controllers can calculate the centerpoint of the arc by merely stating the arc size (radius) and the end point of the arc. We will concentrate on the incremental method of defining the arc center. Keep in mind that we are locating the arc center using the incremental method.

If the arc centerpoint is located down or to the left of the start point, a negative sign (–) must precede the coordinate dimension.

Most CNC turning center controllers are programmed using diameter values, so if we move the tool out 1/4 of an inch, we change the diameter by 1/2 of an inch. If we have an arc of a .25-inch radius, the diameter of the part can change by .50 of an inch.

Study the example in Figure M-139. The line of code is G03 X.500 Z-.250 I0.0 K-.250. Note that the X value end point was a diameter value. Note also that the I and K

represented the distance from the start to the center of the arc in the X and Z directions.

The next programming example (Figure M-140) uses circular interpolation. Pay particular attention to the locations of the start, end, and centerpoint locations.

O0918 (Program name O0918)

N10 G90 G20 G40 (G90 - Absolute, G20 - Inch, G40 - Cancel tool nose-radius compensation)

N15 G28 U0.0 W0.0 (Return to reference position - machine home in this example, this makes sure that the machine is in the correct position before cutting)

N20 G00 T0101 (Tool 1, offset 1)

N30 G50 X5.800 Z10.250 (G50 - Work offset for tool 1 X5.800 Z10.25)

N40 G96 S400 M03 (Spindle on clockwise at 400 surface feet per minute constant surface speed)

N50 G50 S3600 (Set maximum spindle speed at 3600 RPM)

N60 G00 Z.1 (Rapid to Z.1)

N70 G00 X0.0 (Rapid to X0.0)

N80 G01 Z0.0 F0.008 (Linear feed to Z0.0 at a feed rate of .008 inches per revolution)

N90 G01 X.50 (Linear feed to X.50)

N100 G03 X1.00 Z-.25 I0.0 K-.25 (Counterclockwise circular interpolation)

N110 G01 Z-.50 (Linear feed to Z-.50)

N120 G02 X1.60 Z-.80 I.30 K0.0 (Clockwise circular interpolation - IJ method)

N130 G01 X2.25 Z-1.125 (Linear feed to X2.25 Z-1.125)

N140 G28 U0.0 W0.0 M05(Return to home using the straightest path, M05 - spindle stop)

N150 T0100 (Cancel tool offset values in tool 1)

N160 M30 (Program end/memory reset)

%

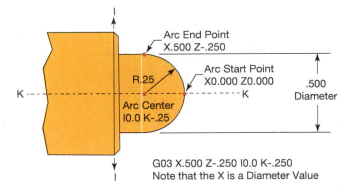

G03 X.500 Z-.250 I0.0 K-.250
Note that the X is a Diameter Value

Figure M-139 G03 arc generation.

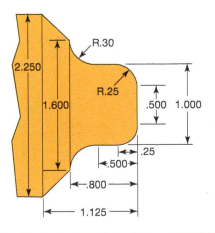

Figure M-140 Arc interpolation program example using I and K values.

Radius Programming Method for Circular Interpolation

The direct arc radius programming for arc interpolation is easier to use than the traditional I and K method. To use direct radius method, follow these steps:

- Position the tool to the start point of the arc.
- Define the direction of the arc. Arc direction can be clockwise (G02) or counterclockwise (G03).
- Define the end point of the arc in the X and Z axes.
- Define the radius value of the arc.

Study the examples shown in Figure M-141. The radius on the left was cut with G03 (counterclockwise circular interpolation). The one on the right was cut with G02 circular interpolation. Both utilized the radius method to specify the center of the arc.

The code for the one on the left was G03 X2.000 Z-.500 R.500. The R.500 in the code specified that the radius was .500. This information plus the endpoint values is enough for the machine to cut this arc.

Figure M-142 shows a sample turned part. The program for it follows. The program uses radius values to turn the arcs. Study the program.

O009 (Program name – O009)

N10 G90 G20 G40 (G90 - Absolute mode, G20 - Inch mode, G40 - Cancel tool nose-radius compensation)

N15 G28 U0.0 W0.0 (Return to reference position - machine home in this example, this makes sure that the machine is in the correct position before cutting)

N20 G00 T0101 (Tool 1, offset 1)

N30 G50 X5.800 Z10.250 (G50 - Work offset for tool 1 - X5.800 Z10.25)

N40 G96 S400 M03 (Spindle on clockwise at 400 surface feet per minute constant surface speed)

N50 G50 S3600 (Set maximum spindle speed at 3600 RPM)

N60 G00 Z.1 (Rapid to Z.1)

N70 G00 X0.0 (Rapid to X0.0)

N65 G01 Z0.0 F.008 (Linear feed to Z0.0 at a feed rate of .008 inches per revolution)

N80 G01 X.50 F0.008 (Linear feed to X.50)

N90 G03 X1.00 Z-.25 R.25 (Counterclockwise circular interpolation using the radius method)

N91 G01 Z-.50 (Linear feed to Z-.50)

N92 G02 X1.60 Z-.80 R.30 (Clockwise circular interpolation - radius method)

N93 G01 X2.25 Z-1.125 (Linear feed to X2.25 Z-1.125)

N100 G28 U0.0 W0.0 M05 (G28 - Return to home using the straightest path, M05 - Spindle stop)

N110 T0100 (Cancel tool offset values in tool 1)

N120 M30 (Program end/memory reset)

%

Tool-Nose Radius Compensation

Tool-nose radius (TNS) compensation is a type of offset used to control the shape of machined features. The turning center control can offset the path of the tool so that the part can be programmed just as it appears on the part print. There is an error that becomes apparent when we use the tool edges to set our workpiece coordinate position on a turning center. When we set the X and Z axes of the tool, we specify a single sharp point. Most of the tools we use for

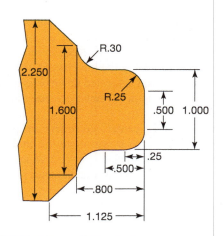

Figure M-142 Circular interpolation program example using radius programming.

Figure M-141 Arc programming.

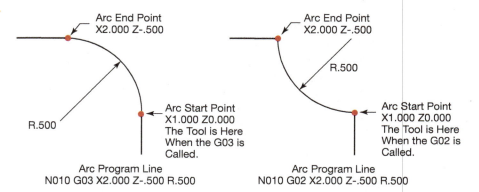

turning have radii. These radii on the tool nose can create inaccuracies in the cutter path (Figure M-143). The dotted lines represent the path the tool would take. The solid line represents the desired part shape.

You can see that the part shape would be incorrect if tool-nose radius compensation is not used. The cutting point on the tool changes as its orientation to the part path changes, leaving inaccuracies. To compensate for the radii of the cutting tool, tool-nose radius compensation must be used, which saves us from having to mathematically calculate a cutter path that would compensate for the radius of the tool.

TNR compensation also lets us use the same program for a variety of tool types. With TNR compensation capabilities, the insert radius size can be ignored and the part profile can be programmed. The exact size of the cutting tool to be used is entered into the offset file, and when the offset is called, the tool path will automatically be offset by the tool radius.

Tool-nose radius compensation can be to the right or left of the part profile. To determine which offset you need, imagine yourself walking on the programmed path in the direction of the cut. Do you want the tool to the left of the programmed path or to the right (Figure M-144)?

Compensation direction is controlled by a G-code. When compensation to the left is desired, a G41 is used.

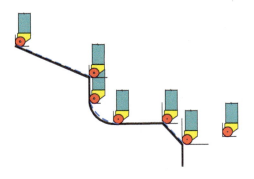

Figure M-143 The dotted lines above show the errors that could occur due to the radius of the cutting tool not being a sharp point. Tool-nose radius compensation allows us to program the part, not the tool path.

Figure M-144 Tool-nose compensation.

When compensation to the right is needed, a G42 is used. When using these cutter compensation codes, you need to specify how large the offset needs to be.

The size of the radius is placed in the nose radius offset table, which is typically located in the tool file under the tool number being used. Tool-nose radius information can be determined from catalogs or the insert package.

The other information needed to insure proper compensation is the tool-nose direction vector. The tool tip or imaginary tool tip of turning tools has a specific location or direction from the center of the tool-nose radius.

The tool-nose vector tells the control which direction it must compensate for individual types of tools. Standard tool-nose direction vectors are shown in Figure M-145. The direction vector number is usually placed in the same tool offset table as the radius value.

Lead-In and Lead-Out Moves Associated with Tool-Nose Radius Compensation

The first linear (G00 or G01) move in a line that contains a G41 or G42 is called a lead in or approach move. This first move is not compensated, but by the end of the move the tool nose radius should be fully compensated for. The distance of the lead in move should be more than the tool nose radius compensation. Lead-in moves should start away from the part (Figure M-146).

A G40 line of code cancels the tool nose radius compensation. The linear line move that is associated with the G40 is called a lead-out or departure move (Figure M-147). The tool nose radius at the start of the departure move is compensated, but by the time the tool moves to the destination the tool nose radius is no longer being compensated. The distance of the lead-out move should be more than the tool nose radius compensation. A G40 should be called when the tool is clear of the part.

Figure M-148 illustrates a typical part that uses tool-nose radius compensation. The program for the part follows the figure.

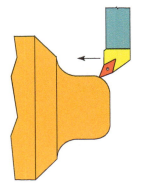
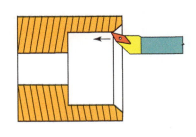

G42 TNR Compensation Right   G41 TNR Compensation Left

Figure M-145 Tool-nose radius direction vectors.

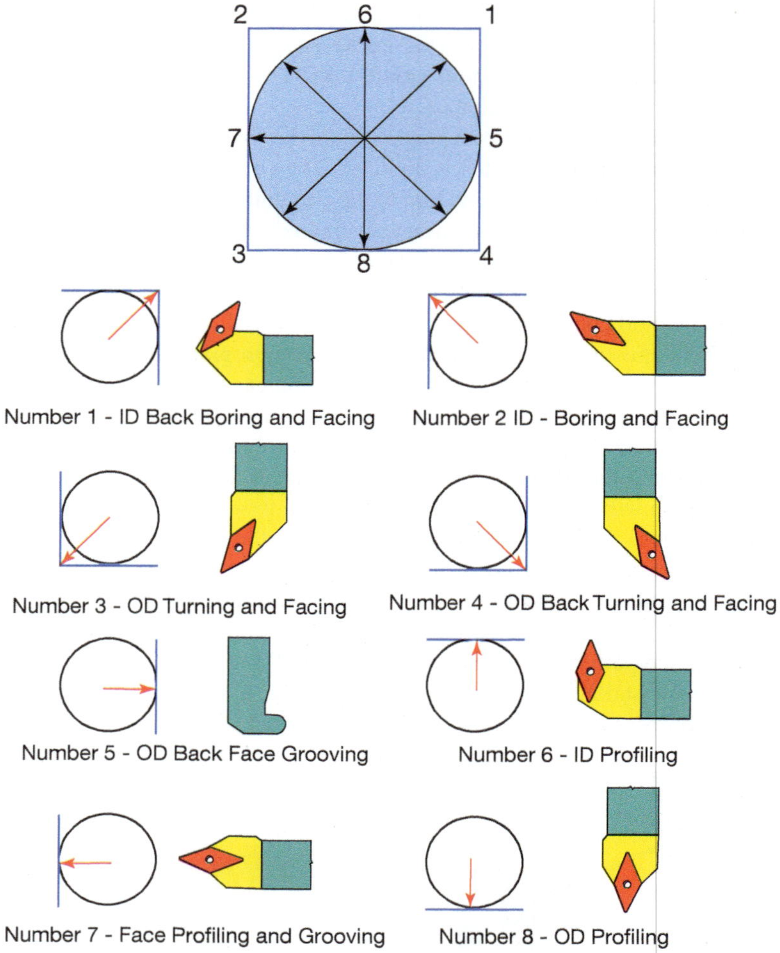

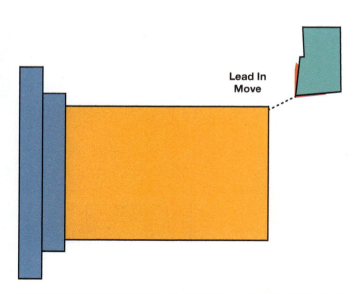

Number 1 - ID Back Boring and Facing

Number 2 ID - Boring and Facing

Number 3 - OD Turning and Facing

Number 4 - OD Back Turning and Facing

Number 5 - OD Back Face Grooving

Number 6 - ID Profiling

Number 7 - Face Profiling and Grooving

Number 8 - OD Profiling

Lead In Move

Figure M-146 Lead-in move.

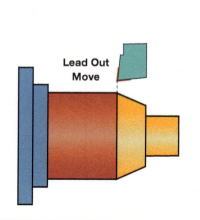

Lead Out Move

Figure M-147 Lead-out move.

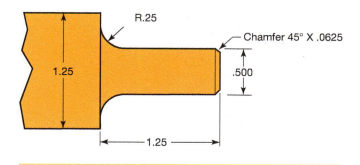

Figure M-148 Part that will utilize tool-nose radius compensation in the program.

O0915 (Program name O0915)

N5 G90 G20 G40 (G90 - Absolute mode, G20 - Inch mode, G40 - Cancel tool nose-radius compensation)

N10 G28 U0.0 W0.0 (Return to reference position - machine home in this example)

N15 G00 T0101 (Tool 1, offset 1)

N20 G50 X5.80 Z10.25 (G50 - Work Offset for tool 1 at X5.800 Z10.25)

N25 G0 Z.1 M8

N30 G50 S3600 (Set maximum spindle speed at 3600 RPM)

N35 G96 S400 M03 (Spindle on clockwise at 400 surface feet per minute constant surface speed)

N37 G00 X.375 (Start of the chamfer)

N40 G1 G42 Z0. F.01 (Linear move, tool-nose radius compensation on the right side, lead in move to Z0. at a feed rate of .01 inches per revolution)

N45 X.5 Z-.0625 (Linear move to X.5 Z-.0625)

N50 Z-1. (Linear move to Z-1.)

N55 G2 X1. Z-1.25 R.25 (Clockwise circular move)

N60 G1 X1.25 (Linear move to X1.25)

N65 G40 X1.45 (Tool-nose radius compensation cancel, lead out move)

N70 M9 (Coolant off)

N75 G28 U0. W0. M05 (G28 - Return to the reference point through incremental U0.0 W0.0 M05 - Spindle stop)

N80 T0100 (Cancel tool offset values in tool 1)

N85 M30 (Program end/Memory reset)

%

SELF-TEST

1. Which turning center axis controls the diameter of the part?
2. Define constant surface footage control.
3. What secondary axes addresses are used to define the center-point position when cutting radii on a turning center?
4. Explain why tool-nose radius compensation is needed when cutting tapers or radii.
5. Use the part print to fill in the point locations.

Point #	X (Diameter) Value	Z Value
Point 1		
Point 2		
Point 3		
Point 4		
Point 5		
Point 6		

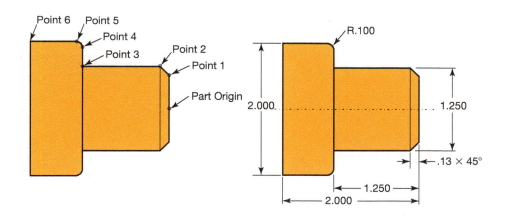

6. Use the part print to fill in the point locations. Part origin is located at the right end and center of the part.

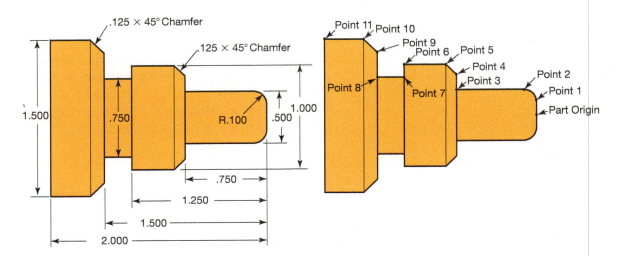

Point #	X (Diameter) Value	Z Value
Point 1		
Point 2		
Point 3		
Point 4		
Point 5		
Point 6		
Point 7		
Point 8		
Point 9		
Point 10		
Point 11		

7. Fill in the blanks to complete the program to machine the profile of the part in question 5.

O0008

N10 G90 G20 G40

N15 G28 U0.0 W0.0

N20 G00 T0101

N25 G50 X10.00 Z12.00

N30 G00 Z.1

N35 G00 X.99

N40 G50 S3600

N45 G96 S200 M____ Spindle Forward

N50 G01 G__ Z0. F.01 Tool Nose Radius Compensation Right, lead in to (POINT 1)

N55 G01 X____ Z-.13 X Coordinate at (POINT 2)

N60 G01 Z____ Z Coordinate at (POINT 3)

N65 G01 X____ X Coordinate at (POINT 4)

N70 G__ X2__ Z__ K-.1 CCW arc to X and Z Coordinate at (POINT 5)

N75 G01 Z____ Z Coordinate at (POINT 6)

N80 G01 G__ Cancel Tool Nose Radius Compensation, lead out

N85 G__ U0. W0. M05 Return to the reference point

N90 T0100

N95 M30

%

8. Number the procedures in the proper order.
_____ Tool call
_____ Start-up procedures
_____ Workpiece location block
_____ Home return
_____ Spindle speed control
_____ Tool motion blocks
_____ Program end procedures

9. Use the part print to fill in the point locations. Part origin is located at the right end and center of the part.

Point #	X (Diameter) Value	Z Value
Point 1		
Point 2		
Point 3		
Point 4		
Point 5		
Point 6		
Point 7		
Point 8		
Point 9		

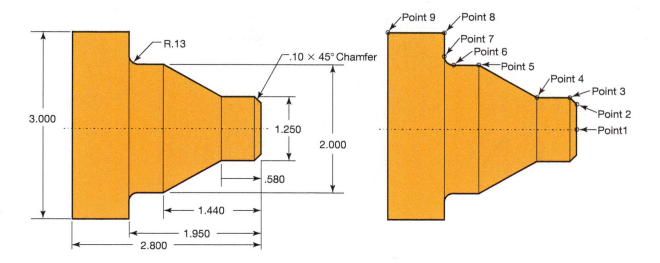

10. Fill in the blanks to complete the program to machine the profile of the part in question 9.

O0009

N10 G ____ G ____ G40 G90 Absolute mode, G20 - Inch mode

N15 G ____ U.0.0 W0.0 Return to reference position - machine home in this example

N20 G00 T0101

N25 G ____ X ____ Z ____ Work offset for tool 1 at X6.800 Z12.25

N30 G ____ S3000 Set maximum spindle speed at 3000 RPM

N35 G ____ S200 M____ Set constant surface speed, Spindle Forward

N40 G00 Z.1

N45 G00 X1.05

N50 G01 G ____ Z ____ F.01 Tool Nose Radius Compensation Right, lead in to (POINT 2)

N55 G01 X ____ Z ____ X Coordinate, Z Coordinate at (POINT 3)

N60 G01 Z ____ Z Coordinate at (POINT 4)

N65 G01 X ____ Z ____ X Coordinate, Z Coordinate at (POINT 5)

N70 G01 Z ____ Z Coordinate at (POINT 6)

N75 G ____ X ____ Z ____ R __ Arc direction, X and Z Coordinate at (POINT 7), Arc radius

N80 G ____ X ____ X Coordinate at (POINT 8)

N85 G01 Z ____ Z Coordinate at (POINT 9)

N90 G01 X3.10

N95 G ____ X3.25 Tool-nose radius compensation cancel

N100 M ____ Coolant off

N105 G ____ U____ W____ M05 Return to the reference point through incremental X0.0 Z0.0

N110 T0100

N115 M30

%

Programming Canned Cycles for CNC Turning Centers

INTRODUCTION

This unit will examine canned cycles for turning centers. There are canned cycles for many types of operations. There are canned cycles to simplify roughing, finish machining, tapping, threading, boring, drilling and deep-hole drilling, and many other common operations. Canned cycles can dramatically simplify programming. Note: you should check the manual for your particular control before using any canned cycles to be sure they are programmed in the same manner. There are differences between different machine controls.

OBJECTIVES

After completing this unit, you should be able to:

- Explain the use of various canned cycles.
- Write programs using roughing and finishing canned cycles.
- Write programs using drill canned cycles.
- Write programs using grooving canned cycles.
- Write programs using boring canned cycles.

CANNED CYCLES FOR TURNING CENTERS

Canned cycles (fixed) cycles are used to simplify the programming of repetitive turning operations, such as rough turning, threading, and grooving. Canned cycles are sets of preprogrammed instructions that can dramatically shorten the number of lines in a program. Programming a simple part without the use of a canned cycle can take up to four or five times the number of lines needed for a part programmed with canned cycles. Think of the operations that are needed to produce a thread:

1. Position the X and Z axes to the proper coordinates with a rapid traverse move (G00).
2. Position the tool for the proper lead angle.

3. Feed the tool across.
4. Rapid position the tool back to the clearance plan.
5. Feed the tool across, and repeat many times increasing the depth of cut each time.

By using a canned threading cycle, a thread can be done with one line of programming. Figure M-149 shows a general list of the most commonly used canned cycles for turning centers.

Fixed (canned cycles) may vary between CNC machines. Although the cycles work in the same basic manner, always refer to the programming manual for your specific machine when trying a new canned cycle.

Roughing Cycle (G71)

The G71 automatically takes roughing passes to turn down a workpiece to the required part profile at a specified depth of cut. The G71 roughing cycle reads a specified number of program blocks to determine the part profile, depth of cut, and feed rate. From the parameters in the program, the control then calculates all of the moves to rough the part out from the rough stock. Cutting is accomplished through parallel moves of the tool in the Z-axis direction.

Figure M-149 Common turning center canned cycles.

Common Canned Cycles for Turning Centers	
G70	Finishing Cycle
G71	Roughing Cycle
G72	Face Stock Removal
G76	Thread Cycle—Multiple Pass
G81	Drill Canned Cycle
G82	Spot Drill/Counterbore
G83	Peck-Deep Hole Drilling
G84	Tapping Canned Cycle
G85	Bore In-Bore Out Cycle
G86	Bore in-Stop-Rapid Out Cycle

See Figure M-150. This figure shows the rough stock and the finished part. Without canned cycles the programmer would need to program each of the passes to make this part. This would be quite a bit of programming. With a roughing canned cycle the operator simply defines the coordinates of the finished part in the program, the depth of cut to be used, and a feed rate. Then the operator programs a rapid move to position the cutting tool to just outside the corner of the rough stock shape before the roughing canned cycle is called. The roughing canned cycle then calculates all of the moves necessary to take several roughing cuts to machine the part. A finishing canned cycle can then be called to use the same part coordinates to calculate the moves to make a finish pass to complete the part.

Certain steps need to be followed when using canned cycles. In the first step, the tool needs to be positioned to the rough stock boundaries. This step has a dual purpose: it tells the control how large the stock is, and it creates a Z clearance position that the tool rapids back to for each pass. The G71 uses letters to give the controller information on the part profile, the amount of stock to leave for finishing, the depth of cut, and the feed rate. A G71 roughing cycle command is shown in Figure M-151.

G71 is the roughing cycle call.

P45 is the block or line number that designates the start of the part profile.

Q70 is the block or line number that designates the end of the part profile.

U.03 tells the controller that we want to leave .03 of an inch stock on the *X* axis (diameter) of the profile for finishing.

W.010 tells the controller that we want to leave .01 of an inch stock on the *Z* axis (length) of the profile for finishing.

D.060 tells the controller to take .060″ depth of cut for each pass. This has become the most common way to specify the depth of cut. Some controls are different, however. It is very important that you check the manual for your machine. One some controls the D value does not use a decimal point. On these controls the D is commanded, the controller reads from the right and decides what the depth of cut will be. Each number in each decimal position gets a value. If we wanted to take .050″ depth of cut per side, we would write it as D500. The first zero from the right has 0 tenths of a thousandths value. The next zero from the right has 0 thousandths of an inch value. The 5 in the third position from the right has 5 ten-thousandths of an inch value or 50 thousandths of an inch. Make sure you check on the correct format for your machine control.

F.010 is the feed rate of the roughing passes.

Figure M-152 shows a sample part that would be appropriate to make using a roughing canned cycle. Figure M-153 shows an example of the cutting passes that a roughing canned cycle might make.

Figure M-154 shows the coordinates of the finished part shape that need to be defined for the G71 roughing or G70 finishing caned cycle. Note that they are defined in lines

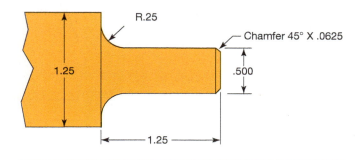

Figure M-152 Sample part.

Figure M-150 Machining passes needed to rough out a part.

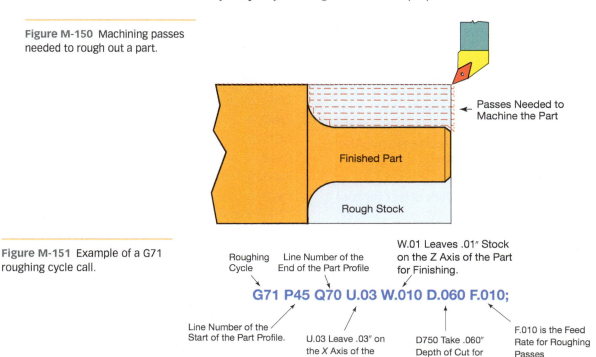

Figure M-151 Example of a G71 roughing cycle call.

Roughing Cycle

Line Number of the End of the Part Profile

W.01 Leaves .01″ Stock on the Z Axis of the Part for Finishing.

G71 P45 Q70 U.03 W.010 D.060 F.010;

Line Number of the Start of the Part Profile.

U.03 Leave .03″ on the *X* Axis of the Part for Finishing.

D750 Take .060″ Depth of Cut for Each Pass.

F.010 is the Feed Rate for Roughing Passes

Figure M-153 Roughing cycle example for the part shown in M-152.

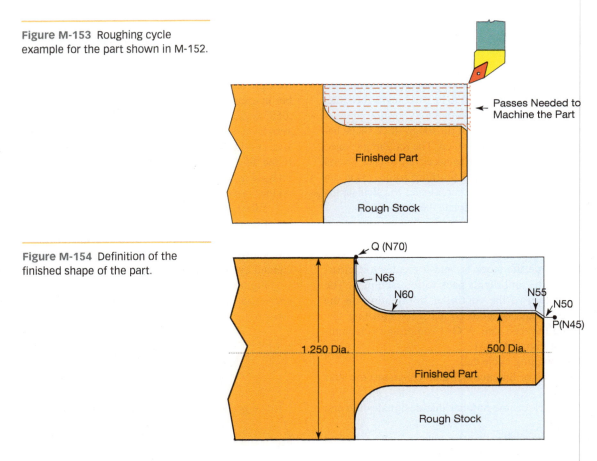

Figure M-154 Definition of the finished shape of the part.

N45 through N70 in the program that follows. In the canned cycle call P45 will point to the first line in the program that defines the start point of the part shape and Q70 will point to the line in the program that has the coordinates of the end point of the part shape.

A program that utilizes a G71 roughing cycle to make the part shown in Figure M-154 follows. TNR compensation will not be used in this example.

O0916 (Program name O0916)

N5 G90 G20 G40 (G90 - Absolute mode, G20 - Inch mode, G40 - Cancel tool- nose radius compensation)

N10 G28 U0.0 W0.0 (Return to reference position - machine home in this example)

N15 G00 T0101 (Tool 1, offset 1)

N20 G50 X5.80 Z10.25 (G50 - Work offset for tool 1 at X5.800 Z10.25)

N25 G50 S3600 (Set maximum spindle speed at 3600 RPM)

N30 G96 S400 M03 (Spindle on clockwise at 400 surface feet per minute constant surface speed)

N35 G00 X1.30 Z.100 (Rapid to X1.30 Z.100)

N40 G71 P45 Q70 U.03 W.01 D.060 F.01 (G71 - Roughing cycle, P45 - N45 first line of part profile, Q70 - last line of part profile, U -.03 Leave .03 on the X axis for finish cut, W.01- Leave .010 on the Z

axis for finish cut, D.060 –depth of cut, radius value. *(NOTE: some machines do not use a decimal point for the depth of cut value. Make sure you check your machine so that you format the depth of cut correctly, F.01 - feed of .01 inches per revolution)*

N45 G00 X.375 M8 (G00 - Rapid to X.375, M8 - coolant on)

N50 G1 Z0 F.01 (Linear move to Z0. at a feed rate of .01 inches per revolution)

N55 G1 X.5 Z-.0625 (Linear move to X.5 Z-.0625)

N60 G1 Z-1.0 (Linear move to Z-1.)

N65 G2 X1.0 Z-1.25 R.25 (Clockwise circular move)

N70 G1 X1.25 (Linear move to X1.25)

N75 M9 (Coolant off)

N80 G28 U2.0 M05 (Return to reference though U2.0 incremental)

N85 T0100 (Note that this is done to cancel the offset for tool 1. Note also that on some controls this is unnecessary. A tool call in the next line would cancel the tool 1 offset and implement the tool 2 offset.)

N90 G28 U2.0 M05 (G28 - Return to the reference point through incremental U2.0, M05 - spindle stop)

N95 M30 (Program end/Memory reset)

%

Figure M-155 Parameters for G71.

G71 Parameters

P	Block number of the start of the part shape
Q	Block number of the end of the part shape
U*	Finish stock remaining with direction (+or –), X-axis diameter value
W*	Finish stock remaining with direction (+or –), Z-axis value
I*	Last pass amount with direction (+or –), X-axis radius value
K*	Last pass amount with direction (+or –), Z-axis value
D*	Depth of cut stock removal each pass, positive radius value (HAAS Setting 72)
F	Feed rate
S**	Spindle speed in this cycle
T**	Tool and offset in this cycle

*Optional
**Rarely defined in a G71 Line

Figure M-155 shows the parameters that can be used for a G71 roughing canned cycle.

Finish Cycle (G70)

The G70 command determines the finish part profile, and then executes a finish pass on the part. The finishing cycle is called with a G70, followed by a letter address P for the start line of the finish part profile and the letter address Q for the end line of the part profile (Figure M-156). Note that the part profile would normally already have been specified by a roughing cycle. The finishing canned cycle could use the same part profile information that was already defined in the program.

A finish feed rate can also be included in the finishing canned cycle block. When the finish cycle is commanded, it reads the program blocks designated by the P and Q and formulates a finishing cycle based on the lines between P and Q that specify the part shape.

As with the G71 roughing cycle, the tool needs to be positioned to a Z clearance plane or stock boundary prior to calling the G70 finish cycle. The program below performs the roughing and finish canned cycles to machine the part shown in Figure M-158. Also note that the roughing cycle and finish canned cycles both utilize the same part shape specified in program lines N45 (P) through N70 (Q).

O0916 (Program name O0916)

N5 G90 G20 G40 (G90 - Absolute mode, G20 - Inch mode, G40 - Cancel tool- nose radius compensation)

N10 G28 U0.0 W0.0 (Return to reference position - machine home in this example)

N15 G00 T0101 (Tool 1, offset 1)

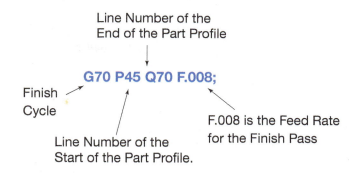

Figure M-156 A finishing canned cycle block.

N20 G50 X5.80 Z10.25 (G50 - Work Offset for tool 1 at X5.800 Z10.25)

N25 G50 S3600 (Set maximum spindle speed at 3600 RPM)

N30 G96 S400 M03 (Spindle on clockwise at 400 surface feet per minute constant surface speed)

N35 G00 X1.30 Z.100 (Rapid to X1.30 Z.100)

N40 G71 P45 Q70 U.03 W.01 D.06 F.01 (G71 - Roughing cycle, P45 - N45 first line of part profile, Q70 - last line of part profile, U.03 -leave .03 on the X axis for finish cut, D.06 -depth of cut, F.01 - feed of .01 inches per revolution)

N45 G00 X.375 M8 (G00 - Rapid to X.375, M8 - coolant on)

N50 G1 Z0. F.01 (Linear move to Z0.0 at a feed rate of .01 inches per revolution)

N55 G1 X. 5 Z-.0625 (Linear move to X.5 Z-.0625)

N60 G1 Z-1. (Linear move to Z-1.)

N65 G2 X1. Z-1.25 R.25 (Clockwise circular move)

N70 G1 X1.25 (Linear move to X1.25)

N75 M9 (Coolant off)

N80 G28 U2.0 M05 (Return to reference though U2.0 incremental)

N85 T0100 (Note that this is done to cancel the offset for tool 1. Note also that on some controls this is unnecessary. The tool call in the next line would cancel the tool 1 offset and implement the tool 2 offset.)

N90 T0202 (Tool 2, Offset 2)

N95 G50 X6.75 Z9.625 (G50 - Tool offset for tool 2 at X6.750 Z9.625)

N100 G50 S3600 (Set maximum spindle speed at 3600 RPM)

N105 G96 S600 M03 (Spindle on clockwise at 600 surface feet per minute constant surface speed)

N110 G00 X1.30 Z.100 (Rapid move to X1.30 Z.100)

N115 G70 P45 Q70 F.008

N120 G28 U2.0 M05 (G28 - Return to the reference point through incremental U2.0, M05 - spindle stop)

N125 T0200 (Cancel tool offset values in tool 2)

N130 M30 (Program end/Memory reset)

%

Internal Roughing and Finishing Canned Cycles (G71/G70)

Figures M-157 and M-158 show a part that would require a roughing and finishing cycle.

O7071 (Program name O0916)

N5 G90 G20 G40

N10 G28 U0.0 W0.0

N15 G00 T0101

N25 G50 S3600

N30 G96 S400 M03

N35 G00 X1.00 Z.100

N40 G71 P45 Q85 U-.03 W.01 D.030 F.01 (Same as external except U-.03 (negative) leaves stock for finish boring)

N45 G00 X2.00 M8

N50 G01 Z0 F.01

N55 G01 X2.0 Z-1.375

N60 G03 X1.75 Z-1.50 R.125

N65 G01 X1.50

N80 G01 Z-2.25

N85 G03 X1.0 Z-2.5 R.25

N90 M9

N95 G28 U0.0 W1.0 M05

N100 T0202

N115 G50 S3600

N120 G96 S500 M03

N125 G00 X1.00 Z.100

N130 G70 P45 Q85 F.008

N135 G28 U0.0 W1.0 M05

N140 T0200

N145 M30

%

Drill Cycle (G81)

This cycle can be used to drill holes. Study Figure M-159. The Z parameter is used to tell the machine how deep to drill. The R parameter tells the machine to rapid to the specified R Plane distance (Figure M-160). This cycle rapids to the R plane and then feeds at the commanded feed rate to the depth specified by the Z parameter. Once the drill reaches the depth, it rapids back out to the R plane.

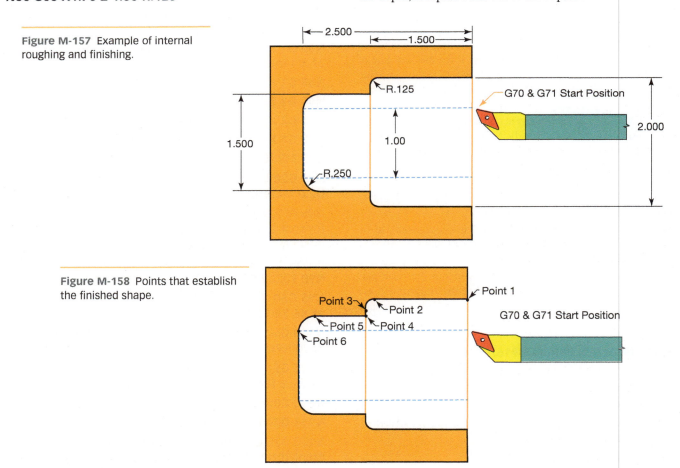

Figure M-157 Example of internal roughing and finishing.

Figure M-158 Points that establish the finished shape.

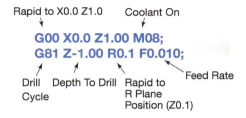

Figure M-159 A G81 drill cycle.

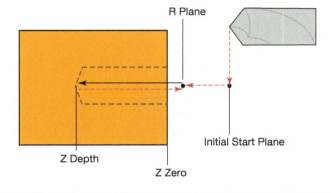

Figure M-160 A G81 drill cycle.

Deep Hole Drill Cycle (G83)

This cycle can be used to drill deep holes. Study Figure M-161. The Z parameter is used to tell the machine how deep to drill.

A program using a G83 follows. The parameters for a G83 are shown in Figure M-162.

O00120 (Program 120)

N010 G28

N020 T0101 (5/8 Spot drill) (Tool 1 - Offset 1)

N030 G97 S1450 M03

N040 G00 X0.0 Z1.0 M08 (Rapid to Initial Start Point)

N050 G83 Z-1.625 R0.1 Q0.15 F0.005 (G83 Deep Hole Drill with a full retract to R plane, Q is incremental depth of cut before full retract)

N060 G80 G00 Z1. M09

N070 G28 U0.0 W1.0

N080 M30

%

Peck Drilling Cycle (G74)

The G74 peck drilling cycle will peck drill holes with automatic retract and incremental depth of cut. The G74 command specifies the incremental depth of cut, the full depth of the hole, and the feed rate through the command variables K, Z, and F. Figure M-163 shows an example of the proper format for peck drilling.

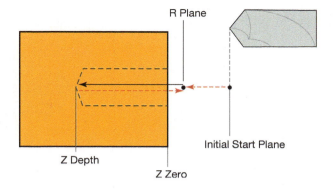

Figure M-161 Example of a G83 canned cycle.

Figure M-162 G83 canned cycle parameters.

G83 Parameters	
X	Absolute X-axis rapid location
Z	Absolute Z depth
Q	Pecking depth amount
W	Z Axis incremental pecking depth
P	Dwell time at Z-depth
R	R Plane
F	Feed rate

*Optional

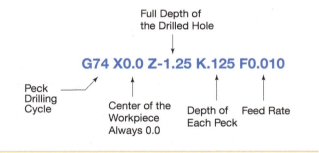

Figure M-163 Example of a G74 canned cycle call.

G74 is the peck drill canned cycle.

X0.0 is the center of the workpiece (X is always zero).

Z-1.25 is the full depth of the drilled hole.

K. 125 is the depth of each peck.

F. 01 is the drilling feed rate.

The drill must be positioned to a clearance plane in the Z axis and also to X0.0 prior to the calling the G74 peck drilling cycle. The spindle should also be reprogrammed for direct RPM using a G97 when drilling. Examine the sample peck drilling cycle in Figure M-164.

O0917 (Program name O0917)

N10 G90 G20 G40 (G90 - Absolute mode, G20 - Inch mode, G40 - Cancel tool- nose radius compensation)

Figure M-164 Peck drilling cycle.

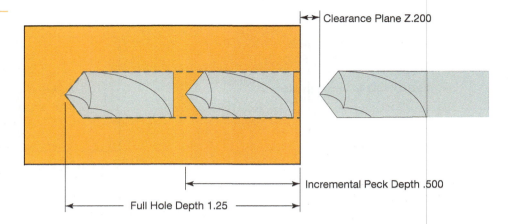

Clearance Plane Z.200

Incremental Peck Depth .500

Full Hole Depth 1.25

N20 G28 U0.0 W0.0 (Return to reference position - machine home in this example)

N30 G00 T0606 (Tool 6, offset 6)

N40 G97 S800 M03 (G97 - Spindle on clockwise at 800 RPM)

N50 G00 Z.2 (Rapid to Z.2)

N60 G00 X0.0 (Rapid to X0.0)

N70 G74 X0.0 Z-1.25 K.500 F0.01 (Peck drilling cycle)

N80 G28 U0.0 W2.0 M05 (G28 - (Backs up 2″ on Z before returning X and Z -axis home M05 - spindle stop)

N90 T0600 (Cancel tool offset values in tool 6)

N100 M30 (Program end/Memory reset)

%

A G74 canned cycle can also be used for grooving on the face of a part (Figure M-165), turning with a chip breaker or high-speed peck drilling. Figure M-166 shows the parameters for a G74 canned cycle.

Figure M-165 A G74 canned cycle being used to groove the face of a part.

Grooving Cycle (G75)

The grooving cycle is a very versatile canned cycle. To use the grooving cycle, the tool must be positioned to the start of the groove prior to calling the grooving cycle. Through a series of letter addresses, the controller can be commanded to cut a groove of varying width and depth.

Figure M-167 shows an example of a grooving cycle. This example will cut one groove. Note that the groove is wider than the tool. The grooving tool is .25″ wide. The groove is .375″ wide. Figure M-168 shows the G75 line of code that would cut this groove.

G75 is the grooving cycle call.

X.750 is the diameter of the groove at the bottom of the groove.

Z-1.375 is the end position of the groove.

F0.010 is the feed rate of the grooving tool.

I is the peck depth of cut in the *X*-axis. Note that it is a radius value.

K is the shift amount on the *Z* axis. This is used to cut a wider groove or when cutting multiple grooves. In this

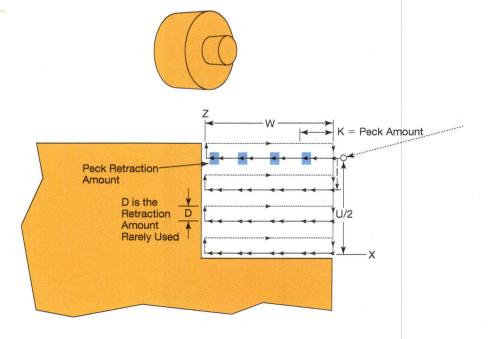

Z

W

K = Peck Amount

Peck Retraction Amount

D is the Retraction Amount Rarely Used

D

U/2

X

Figure M-166 Parameters for a G74 canned cycle.

G74 Parameters

X	Absolute X location to the furthest peck as a diameter value
Z	Absolute Z pecking depth
U	X Axis incremental distance (+ or –) to the furthest peck, diameter value
W	Z Axis incremental pecking depth
I	X-Axis shift increment between pecking cycles positive radius value
K	Z Axis pecking depth increment
D	Tool shift amount when returning to the clearance plane. Note that the D value is rarely used.
F	Feed rate

* Optional

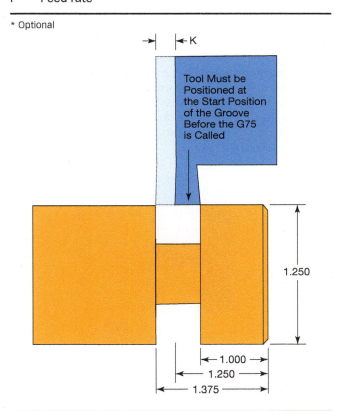

Figure M-167 This figure shows a part that would be appropriate for a G75 multiple pass grooving cycle.

example the tool is .25″ wide and the groove is .375″ wide so the K value calls for a shift of .125″.

O0918(Program name O0918)

N10 G90 G20 G40 (G90 - Absolute mode, G20 - Inch mode, G40 - Cancel tool-nose radius compensation)

N20 G28 U0.0 W0.0 (Return to reference position - machine home in this example. This is done here to make sure the machine is in the correct location before a tool is loaded. Note that it is common practice to use the U and W value here so that the control does not have to be put into incremental mode for this one line.)

N30 G00 T0505 (Tool 5, offset 5)

N40 G96 S200 M03 (Spindle on clockwise at 200 surface feet per minute constant surface speed)

N50 G00 X1.250 Z-1.250 (Rapid to X1.250 Z-1.250)

N60 G75 X.750 Z-1.375 I.125 K.125 F.010 (Grooving cycle)

N70 G28 X2.00 M05 (G28 - Return to the reference point through X2.00, M05 - spindle stop)

N80 T0500 (Cancel tool offset values in tool 5)

N90 M30 (Program end/memory reset)

%

The G75 grooving cycle is very versatile. It can be used to create one groove that is the same width as the grooving tool, multiple grooves, or grooves that are wider than the tool (Figure M-169).

Just like in our first grooving example, the tool must be positioned to the start of the groove prior to calling the grooving cycle. By the use of the cycle's parameters, the controller can be commanded to cut a groove of varying width and depth or multiple grooves. Figure M-170 shows an example of a part that has three groves that were cut with a G75 grooving cycle. Note that the first groove may not look like it is a groove as it is on the front of the workpiece.

A G75 canned cycle can be used for grooving an outside diameter. The parameters for a G75 are shown in Figure M-171. When a Z or W is used in a G75 block and Z is not the current position, then a minimum of two pecking cycles will occur. One is cut at the current Z location and another at the furthest peck location specified by the Z parameter.

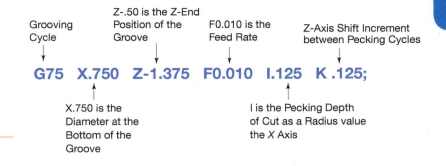

Figure M-168 Example of a G75 grooving cycle.

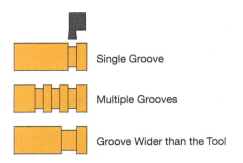

Single Groove

Multiple Grooves

Groove Wider than the Tool

Figure M-169 Examples of grooves that can be cut with a G75 grooving cycle.

The K parameter is the incremental distance between Z axis pecking cycles. The use of the K parameter will perform multiple, evenly spaced, pecking cycles between the starting position and Z. If the distance between S and Z is not evenly divisible by the value in K, the last interval along Z will be less than K.

When the I parameter is used, the peck will feed the peck amount specified in I and then retract in the opposite direction of the feed by the peck distance specified in Setting 22 in a HAAS control. *Note that for any canned cycle you should consult the manual for your particular machine as there are differences in the way canned cycles are programmed in some machines.*

Take a look at the D parameter in Figure M-171. The D parameter is the tool shift amount when returning to the clearance plane. The D code can be used in grooving and turning to provide a tool clearance shift, in the Z axis, before returning in the X axis to the clearance point. You would not want to use the D command if both sides to the groove exist during the shift, because the groove tool would break. The D parameter is rarely used.

Thread-Cutting Cycle (G76)

The G76 thread-cutting cycle can cut multi-pass threads with one block of information. By using several letter address parameters, the control will automatically calculate the correct number of cut passes, depth of cut for each pass, and the starting point for each pass.

Figure M-172 shows a part with a 1″-12 thread that would be appropriate for a G76 thread cycle.

Figure M-173 shows an example of a thread cutting canned cycle. To use the G76 thread-cutting canned cycle, the following parameters need to be programmed:

X.900 is the minor diameter of the thread. You can find these values in any reference that shows thread specifications.

Z-1.25 is the absolute Z position of the end of the thread.

I0.0 is the radial difference between the thread starting point and the thread ending point. The I is used for cutting tapered threads. For cutting straight threads, a zero should be programmed.

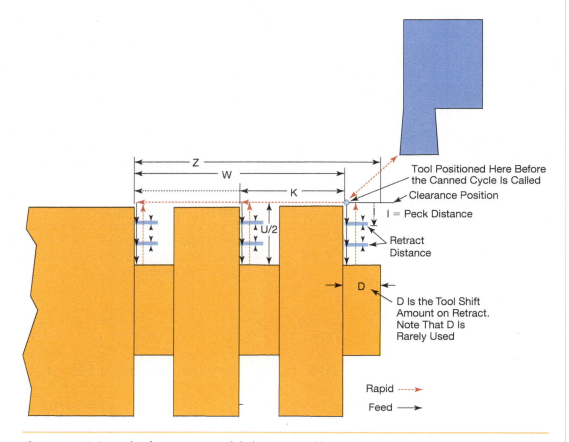

Figure M-170 Example of parameters and their use on multiple grooves.

Figure M-171 G75 grooving canned cycle parameters.

X	**X-Axis absolute pecking depth as a diameter value**
Z	Z-Axis absolute location to the furthest peck
U	X-Axis incremental pecking depth, diameter value
W	Z-Axis incremental distance and direction (+ or –) to the furthest peck
I	X-Axis pecking depth increment, radius value
K	Z-Axis shift increment between pecking cycles
D	Tool shift amount when returning to the clearance plane (note: rarely used)
F	Feed rate

* Optional

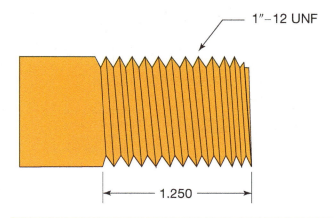

1″–12 UNF

1.250

Figure M-172 Example part for a G76 thread-cutting cycle.

K.0.05 is the thread height expressed as a radius value (i.e., [major diameter – minor diameter] divided by 2).

D0.012 is the depth of cut for the first pass (in a radius value). Note: Every pass after the first pass will be decreasing in depth. Figure M-174 shows the cutting passes for a thread. Note that the first cut depth is larger and subsequent passes get smaller. When we specify D (.012 in our example), the control uses that value and the thread height (K) to determine how many cutting passes should be made to cut the thread. So the smaller the first cut value, the more passes there are that will be made. The controller does this automatically.

If you would like to know how many passes there would be based on your D value, you could use the formula shown in Figure M-174. There are also charts available on which you select the depth of the first pass and the chart will tell you how many passes will result.

F is the thread lead (i.e., 1 divided by the number of threads per inch lead of the thread).

A is the included angle of the thread.

Prior to calling the G76 thread-cutting cycle, the tool must be positioned to the major diameter of the thread plus double the K value or thread height. The tool should also be positioned in front of the thread start position in the Z by a distance of at least double the thread lead. This insures that the proper lead will be cut throughout the length of the thread. The spindle should be running in direct RPM (G97), not constant surface footage control. Following is a program using a G76 thread cycle to cut the thread shown in Figure M-172.

O0919 (Program name O0919)

N10 G90 G20 G40 (G90 – Absolute mode, G20 - Inch mode, G40 - Cancel tool- nose radius compensation)

N15 G28 U0.0 W0.0 (Return to reference position - machine home in this example)

N20 G00 T0505 (TOOL - 5 OFFSET - 5)

N40 G97 S300 M03 (G97 - Spindle on clockwise at 800 RPM)

N60 G00 Z.200 (Rapid to Z .200 in front of the thread start)

N65 G00 X1.10 (Rapid to X1.10)

N70 G76 X.897 Z-1.25 K.05 D.0120 F.0833 A60 (Threading cycle)

N90 G28 X2.00 M05 (G28 - Return to the reference point through X2.00, M05 - spindle stop)

N100 T0500 (Cancel tool offset values in tool 5)

N110 M30 (Program end/Memory reset)

%

The parameters for a G76 threading cycle are shown in Figure M-175.

Figure M-173 Thread cutting cycle to cut the thread shown in Figure M-172.

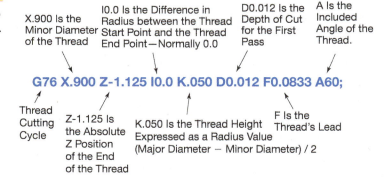

X.900 Is the Minor Diameter of the Thread

I0.0 Is the Difference in Radius between the Thread Start Point and the Thread End Point—Normally 0.0

D0.012 Is the Depth of Cut for the First Pass

A Is the Included Angle of the Thread.

G76 X.900 Z-1.125 I0.0 K.050 D0.012 F0.0833 A60;

Thread Cutting Cycle

Z-1.125 Is the Absolute Z Position of the End of the Thread

K.050 Is the Thread Height Expressed as a Radius Value (Major Diameter − Minor Diameter) / 2

F Is the Thread's Lead

Figure M-174 Diagram of multiple-pass thread cutting.

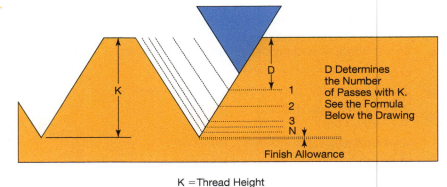

K = Thread Height
N = Number of Passes
D = Depth of First Pass

(K/Square Root of N) = D

D Determines the Number of Passes with K. See the Formula Below the Drawing

Figure M-175 Thread cutting parameters.

X	**X-Axis absolute thread finish point as a diameter value**
Z	Z-Axis absolute distance, Thread end point location
U	X-Axis incremental total distance to finish point diameter
W	Z-Axis incremental thread length finish point
K	Thread height, radius value
I	Thread taper amount, radius value
D	First pass cutting depth
P	Thread cutting method P1–P4 (HAAS control—added in version 6.05)
A	Tool nose angle, no decimal point
F	Feed rate is the thread distance per revolution (Lead of the thread)

*Optional

G84 Tapping Canned Cycle

The G84 tapping cycle is used to tap a hole on a turning center using a regular tap. Figure M-176 shows an example of a G84 tapping canned cycle block. This line would rapid to a position .3″ in front of the workpiece, control the spindle speed, and then feed the tap to thread a hole 1″ deep (Figure M-177). The spindle would then automatically reverse direction and the feed to safely feed the tap back out of the workpiece to the R plane.

Note: You don't need to start the spindle before the G84 canned cycle. The control turns it on automatically.

The parameters for a G84 tapping cycle are shown in Figure M-178.

O005 (G84 Tapping Program)
N010 G28
N020 T0202 (5-16 Tap) (Tool 2 - Offset 2)
N020 G97 S650 M05 (G84 automatically turns on the spindle)
N030 G00 X0. Z1. M08 (Rapid to Initial Start Point and Turn Coolant On)
N040 G84 Z-1.0 R0.3 F0.0625 (G84 Tapping Cycle)

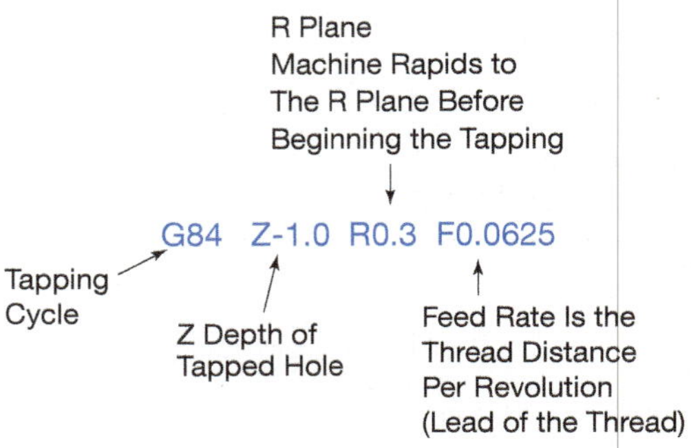

R Plane
Machine Rapids to The R Plane Before Beginning the Tapping

Tapping Cycle → G84 Z-1.0 R0.3 F0.0625

Z Depth of Tapped Hole

Feed Rate Is the Thread Distance Per Revolution (Lead of the Thread)

Figure M-176 Example of a G84 tapping canned cycle call.

N050 G80 G00 Z1. M09
N060 G28
N070 M30
%

G85 Boring Canned Cycle

This cycle is also called a bore in–bore out canned cycle. There would have to be a hole in the part before the boring cycle is called. Study Figure M-179. This boring cycle feeds in at the specified feed rate until the depth is hit. It then feeds back out to the R plane at the same feed rate.

Figure M-180 shows a G85 boring cycle call. Z is the depth of the bored hole. R is the R plane value. In this example it is .2″ in front of the part. The tool would rapid to the R plane before beginning the boring feed and then feed out of the hole to the R plane.

O008 (G85 Bore in - bore out, program 8)
N010 G28
N020 T0707 (Boring bar) (Tool 7 - Offset 7)
N030 G97 S1200 M03
N040 G00 X0.750 Z1. M08 (Rapid to initial start point)
N050 G85 Z-0.875 R0.2 F0.008 (G85 Bore in - bore out cycle)
N060 G80 G00 Z1. M09

Figure M-177 Example of a G84 tapping cycle.

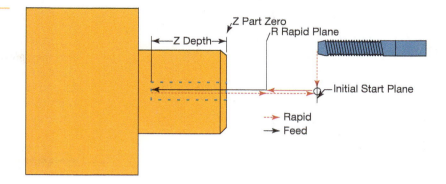

Figure M-178 Parameters for a G84 tapping cycle.

X	**X-Axis absolute rapid location**
Z	Z-Axis depth (Feeding to Z depth from the R plane)
W	Z-Axis incremental thread length finish point
R	Rapid to R Plane to start feeding
F	Feed rate is the thread distance per revolution (Lead of the thread)

*Optional

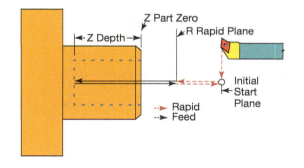

Figure M-179 A G85 boring cycle. Note that it feeds in from the R plane to the programmed depth and then feeds back out to the R plane.

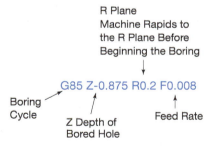

Figure M-180 A G85 boring canned cycle example.

N070 G28

N080 M30

%

Figure M-181 shows the parameters for a G85 cycle.

Figure M-181 Parameters for a G85 boring canned cycle.

X	**X-Axis absolute rapid location**
Z	Z-Axis depth (feeding to Z depth from the R plane)
U	Incremental X-axis rapid location
W	Incremental Z depth (feeding to Z depth from the R plane)
R	Rapid to R Plane to start feeding
F	Feed rate

*Optional

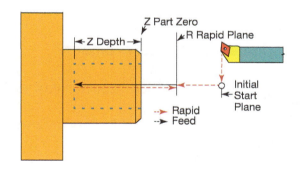

Figure M-182 A G86 boring canned cycle. Note that it feeds in from the R plane to the programmed depth, stops, and then rapids back out to the R plane.

G86 Boring Cycle

A G86 canned cycle is used to bore a hole (Figure M-182). There would have to be a hole in the part before the boring cycle is called. The boring cycle is done to machine the hole to an accurate finish size with a good surface finish. This boring cycle feeds in at the specified feed rate until the depth is hit. It then stops feeding and rapids back out to the R plane.

Figure M-183 shows a G86 canned cycle call. The depth of the bored hole is specified by the Z parameter. The R plane parameter specifies where the machine should rapid to in front of the part. The F specifies the feed rate for the boring cycle.

A program that uses a G86 boring cycle follows. Note that it feeds in from the R plane to the programmed depth, stops, and then rapids back out to the R plane.

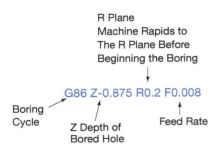

R Plane
Machine Rapids to
The R Plane Before
Beginning the Boring

G86 Z-0.875 R0.2 F0.008

Boring Cycle

Z Depth of Bored Hole

Feed Rate

Figure M-183 A G86 boring canned cycle.

Figure M-184 Parameters for a G86 boring canned cycle.

X	**X-Axis absolute rapid location**
Z	Z-Axis depth (feeding to Z depth from the R plane)
U	Incremental X-axis rapid location
W	Incremental Z depth (feeding to Z depth from the R plane)
R	Rapid to R Plane to start feeding
F	Feed rate

*Optional

O008 (G86 Bore in-stop- rapid out)

N010 G28

N020 T0707 (Boring bar) (Tool 7 - Offset 7)

N030 G97 S1200 M03

N040 G54 G00 X0.750 Z1. M08 (Rapid to initial start point)

N050 G86 Z-0.875 R0.2 F0.008 (G86 Bore in-stop- rapid out cycle)

N060 G80 G00 Z1. M09

N070 G28

N080M30

%

Figure M-184 shows the parameters for a G86 boring cycle.

Now that we have an understanding of canned cycles and how they are used, we need to put our knowledge to work. There are some part drawings after the chapter questions. Use the part drawings and the tool table shown in Figure M-185 to program these parts. The programs should include canned cycles and tool-nose radius compensation, where appropriate.

SELF-TEST

1. True or False? Canned cycles are exactly the same on all CNC machines.
2. True or False? Canned cycles vary on different machine controls so the programmer should check the manual for their particular machine before using a canned cycle.
3. If a part needs to be roughed out of bar stock a G__ would be used.
4. If a part needs to roughed out and finished a G__ and a G__ could be used.
5. Where must the tool be positioned prior to calling a roughing canned cycle?
6. To drill a deep hole in a part, a G__ would be used.
7. A G__ could be used to spot drill a hole or to counterbore a hole.
8. Which letter address controls the depth of cut when using canned cycles?
9. What is the letter address command to take .100 of an inch off the diameter of the part per pass when using a roughing cycle?
10. A G__ is used for grooving a part.
11. What letter address controls the pitch or lead of the thread when using a G76 thread-cutting cycle?
12. What must be done prior to the calling of a G76 thread-cutting cycle?

Figure M-185 Tool table.

Acme Machining Inc.- Tool Table		Part Datum			Part #	
Operation	Tool #	Tool Description	RPM	Cutting Speed	Feed Rate	
Rough Turn	T0505	80 Degree Diamond				
Finish Profile	T0303	35 Degree Diamond				
Groove	T1212	.125 Grooving Tool				
Thread	T0707	60 Degree Threading Tool				
Drill 1s Hole	T0909	1" Drill				
Bore Holes	T0808	Boring Bar				

13. Program a roughing and finish canned cycle for the part shown in Figure M-186. Use canned cycles where appropriate and the tooling shown in Figure M-185. The rough bar stock is 1.30″ in diameter.

Figure M-186

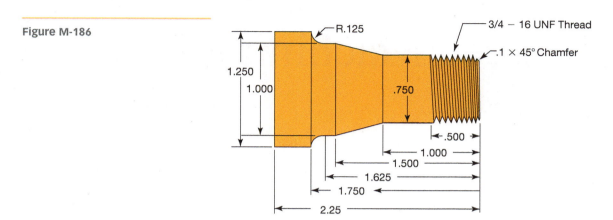

14. Program a threading canned cycle for the part shown in Figure M-186. Use the threading tool from Figure M-185.

15. Program a drill canned cycle to drill a 1″ hole through the part shown in Figure M-187. Use the 1″ drill from the tool table Figure M-185.

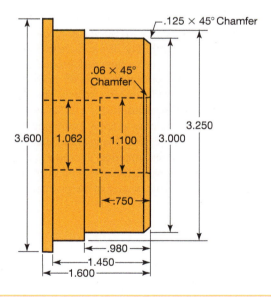

Figure M-187

16. Program boring canned cycles to bore the 1.062 and the 1.100 bores in the part shown in Figure M-187. Use G86 canned cycles. Assume a 1″ drill from the tool table Figure M-185 was used prior to boring.

Advanced Machining Processes

Most of this text covered traditional machining processes such as turning, milling, drilling, and grinding. In these processes, the cutting tool directly contacts the workpiece and removes chips of the workpiece material. Although these processes are widely used and relatively efficient, they do have limitations. For example, machining certain materials by the traditional processes may be extremely difficult or even impossible. Furthermore, the traditional processes do not always lend themselves to machining parts for certain designs necessary to the manufacturing of exotic products in aerospace and other high-technology fields. To solve these and other problems in machining manufacturing, the engineer and industrial technologist have developed other types of machining processes quite different from the classical chip-producing methods. These processes are sometimes known as *nontraditional* in that they remove workpiece material by applications of electrical energy, electrochemical processes, and ultrasound.

Included in nontraditional machining processes are the following:

1. Electrodischarge machining (EDM)
2. Electron beam machining (EBM)
3. Electrolytic grinding (ELG)
4. Electrochemical machining and deburring (ECM, ECDB)
5. Ultrasonic machining
6. Abrasive waterjet machining
7. Laser machining

OBJECTIVE

After completing this unit, you should be able to:

■ Identify common nontraditional machining processes and generally describe how these processes work and where they might be applied.

ELECTRICAL DISCHARGE MACHINING

Electrical discharge machining, commonly known as **EDM**, removes workpiece material by an electrical spark erosion process in which a large potential (voltage) difference is established between the workpiece to be machined and an electrode. A large burst (spark) of electrons travels from the electrode to the workpiece. When the electrons impinge on the workpiece, workpiece material is eroded away.

The EDM process (Figure M-188) takes place in a *dielectric* (nonconducting) oil bath. The dielectric bath concentrates the spark and also flushes away the spark-eroded workpiece material. A typical EDM system consists of a power supply, dielectric reservoir, electrode, and workpiece. The EDM machine tool has many of the same features as its conventional counterpart. These include worktable positioning mechanisms and measurement devices. Many EDM machine tools are equipped with computer numerical control systems. Thus the versatility of CNC for workpiece positioning and tool control functions can be effectively used in the process.

EDM Electrodes

In the EDM process, the shape of the electrode controls the shape of the machined feature on the workpiece (Figure M-189). EDM electrodes may be made from metal or carbon (graphite)

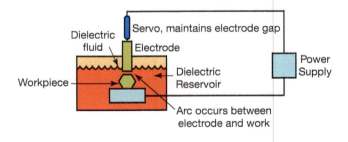

Figure M-188 EDM system.

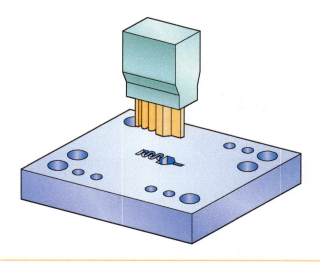

Figure M-189 The electrode on EDM cut the workpiece feature to the same shape as the electrode.

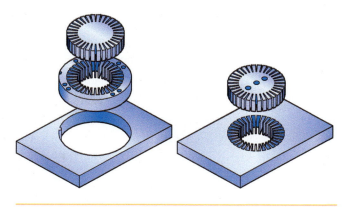

Figure M-190 Examples of complex parts that were machined on a wire EDM machine.

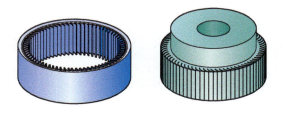

Figure M-191 Example of complex parts that were machined on a wire EDM machine.

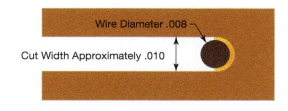

Figure M-192 Overcut produced by a wire EDM.

and shaped by molding or machining to the desired geometry. The electrode erodes as well during the EDM process. In time it becomes unusable. To circumvent this problem, a roughing electrode may be used to generally shape the workpiece and then a finish electrode may be applied to complete the process and establish final dimensions and geometry.

The grades of graphite for EDM electrodes can greatly influence tool life as well as the degree of precision obtained by the electrode tooling. The electrode material may be classified by the average graphite particle size, usually ranging from about 1 micrometer (.001 mm) up to 100 micrometers. Denser graphite has superior wear characteristics and may be more easily machined to accommodate fine details without chipping and breaking. EDM graphite electrodes may be checked by examining the microstructure of the material under a microscope.

Wire-Cut EDM

Wire EDM was once considered a nontraditional machining method. Today, EDM has become a very accepted and widely used technology. EDM can be effectively used for many complex and intricate shapes that are very difficult and time consuming to machine (Figures M-190 and M-191). They are especially valuable in making accurate, intricate parts for dies and tool and die making.

Brass wire is the most commonly used material. The most common size is .008 inch. If we add the normal .001 overcut on each side of the wire, the width of cut will be approximately .010 inch. The wire never actually touches the material it is cutting. There is a gap of approximately .001 inch (Figure M-192).

Figure M-193 shows a diagram of a wire EDM. Note the diamond guides above and below the work to accurately guide the wire. Note also that flushing takes place above and below the workpiece.

Smaller wire diameters can be used to cut smaller radii and very intricate shapes. Here, tungsten or molybdenum wires are used because their high tensile strength and high melting point allow smaller wire diameter.

The wire comes on spools that weigh between 2 and 100 pounds. A spool holds a very long continuous length of wire that permits a machine to run in excess of 500 hours on one spool of wire. The machine has a supply spool and a take-up spool (Figure M-193). The new wire is put in the supply position and the wire then travels through guides and the workpiece to the take-up spool.

Some machines have a box to catch the wire instead of a take-up spool (fewer moving parts). The wire is thus continuously new, which helps ensure accurate cutting. Most machines use take-up spools for the used wire. The rate of wire feed can be very slow (1 meter/minute) or quite fast (10 meters/minute).

EDM machines can cut with extreme accuracy; some machines are accurate to within ±.0002 inch. EDM produces very little heat and no machining forces on the material, so it doesn't distort pieces during production. Wire EDMs provide very good surface finishes, which can eliminate the

Figure M-193 A diagram of a wire EDM machine. Note the wire guide above and below the workpiece and the flushing from top and bottom.

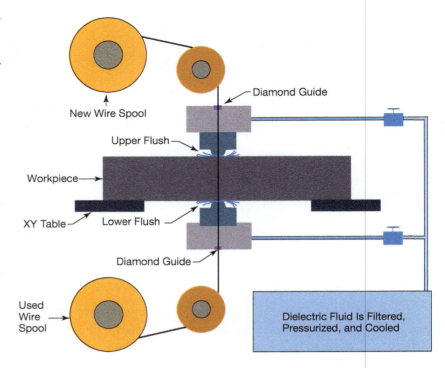

New Wire Spool

Diamond Guide

Upper Flush

Workpiece

XY Table — Lower Flush

Diamond Guide

Used Wire Spool

Dielectric Fluid Is Filtered, Pressurized, and Cooled

need for further machining. Some EDMs are capable of surface finishes of 15 RMS or less.

EDM is effective on a variety of materials, too. EDM can cut anything that is conductive, including steel, aluminum, super alloys, and even tungsten carbide. When we cut carbide, a nonconductive material, with a wire EDM, we are actually eroding the binder material. The carbide nodules are flushed away. Small finish cuts will improve the surface finish. Skim cuts also improve the surface quality. The workpiece can be hardened before EDM machining. In fact, EDM can cut hardened D2 tool steel 20 to 40 percent faster than it can cut cold rolled steel.

Cutting with EDM

Wire wear and cutting rate depend on the characteristics of the workpiece material. Speed is measured in square inches cut per hour, and manufacturer's rate equipment by cutting speed. The speed of cut is usually rated on 2.25-inch thick, hardened D2 tool steel under ideal cutting conditions. The important characteristics that determine cutting speed include the melting point of the material, the conductivity of the material, and the length and strength of the electrical pulses. Aluminum, for example, has a low melting point and is a good conductor, so it cuts much faster than steel.

EDM cuts by creating sparks between the wire and the workpiece. These rapid, high-energy sparks cause small pieces of the workpiece to melt and vaporize. Fluid is run between the wire and the workpiece. This fluid is dielectric, meaning nonconductive. This fluid serves several purposes, such as shielding the wire from the workpiece. The fluid acts like a resistor until enough voltage is applied. With sufficient voltage, the fluid ionizes and a spark melts and vaporizes a small piece of the workpiece material. These pulses occur

thousands of times each second along the length of the wire within the workpiece.

The sparks can be DC or AC current. DC current allows a very small current to flow through the dielectric fluid, which can cause electrolytic corrosion. AC power supplies do not have this problem. An AC power supply can help prevent rust formation on iron workpieces. The AC method also helps to substantially reduce minute cracking that can occur during cutting, making dies more durable. Carbide retains its hardness better with the AC method because the binder is prevented from being eluted (washed out or extracted).

After the metal has been melted, the fluid cools the vaporized metal and carries it out of the cut. Good particle removal by flushing is essential to efficient cutting. The fluid also keeps the workpiece and the wire cool.

A very important characteristic is the flushing of the vaporized material so a new energy column can form. The contaminated fluid runs through a filter to remove the vaporized particles and is then reused. The fluid also runs through a chiller to maintain the temperature, which helps maintain the machine's accuracy.

The dielectric fluid is normally deionized water. Regular water is run through an ion exchanger that removes all the impurities. Regular water is a conductor; deionized water is a good insulator. This is vital because we don't want a short circuit between the wire and the workpiece. The amount of deionization is measured by the specific resistance of the water. The lower the resistance of the water, the faster the cutting. A higher resistance is desirable for materials such as higher density graphite and carbides.

The flow rate is crucial to efficient, accurate cutting. Two valves control the fluid flow. One valve controls the top flow, and the other valve controls the fluid flow from the bottom.

The color of sparks also can be used to adjust flow rate. Blue sparks are desirable. Red sparks indicate an insufficient flow rate. Too high a flow rate is also undesirable. It can cause wire deflection, which can cause inaccuracy when cutting tapers.

Some machines now have programmable flushing that can accommodate different conditions for entering, leaving, skim cuts (finish cuts), or even changing workpiece thickness.

The electrical control on the machine maintains a gap of .001 to .002 inches between the wire electrode and the workpiece. The wire never touches the workpiece. If it did, it would short out and the machine would sense it and make a correction.

Types of Wire EDM Machines

There are three main types of wire EDM. The first is a simple two-axis EDM. This machine is like a simple XY table that can make only simple right-angle cuts. Figure M-194 shows a section view of a hole cut in a part by a wire EDM. Note the straight vertical wall.

The second type of wire EDM is a simultaneous four-axis EDM. This EDM can cut tapers through the workpiece (Figure M-195), but the top and bottom of the cut must be the same shape. For example, if the top of the cut is a square, the bottom shape must also be a square, although it can be a

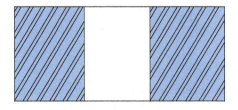

Figure M-194 A simple XY machine cut. Note that there is no taper in the hole.

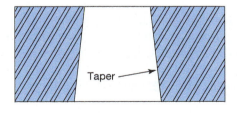

Figure M-195 A workpiece that required a hole with tapered sides.

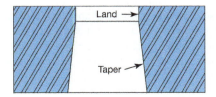

Figure M-196 A part that requires two cuts. The straight cut through the piece would be made first, and then the taper would be cut.

different size to create the tapered wall. Figure M-196 shows a part that could be cut on a four-axis wire EDM. Note that it would take two passes. The first would cut the straight wall and the second pass would cut the taper.

The third type of EDM is the independent four-axis system. In this type of system, the top shape of the cutout can be different than the bottom shape. This is especially useful for extrusion dies that are used to shape material into predefined shapes. This would require a four-axis wire EDM.

The four-axis wire EDM has a UV axis and an *XY* axis. The UV axis guides the wire above the workpiece, and the *XY* axis guides the wire below the workpiece. The UV and the *XY* axes can be controlled independently to cut tapers on workpieces.

Parts of the Wire-Feed EDM

The main parts of a wire-feed EDM are the bed, saddle, column, UV axes, XY axes, wire-feed system, dielectric fluid system, and machine control. Wire-feed machines have several servo systems to control various aspects of the machining process. The electrical current level must be accurately controlled, and the feed rate of the axes and the gap between the wire and the actual workpiece must be accurately controlled to approximately .001 to .002 inches. If the wire were to contact the workpiece, the wire could be broken or arcing could occur, which would damage the workpiece. The servo control senses the current and adjusts the drive motors to speed up or slow down to retain the proper gap between wire and workpiece. Sometimes called adaptive control, this control can also adjust the feed rate to compensate for workpiece thickness and changing cutting conditions (Figure M-197).

Important Machining Considerations

One of the problems of EDM machining is called recast. This very thin region of the metal at the cut is affected by the tremendous heat that the cutting causes. EDM removes metal by generating extreme heat on the metal, which causes it to vaporize and melt away. This extreme heat changes the molecular structure on the surface of the part that is cut. The recast layer is material that has been vaporized and has reattached itself to the remaining material and solidified before it was flushed away. These particles stick to the surface of

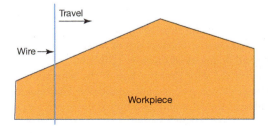

Figure M-197 How adaptive control can adjust the feed rate for varying workpiece conditions.

the cut and are what most would refer to as recast. The thickness is generally .0001 to .0002 inches. These particles can generally be easily removed by bead blasting.

The heat-affected zone below these particles is a thin region that was heated to the melting point and then cooled very rapidly. This region can be very hard and brittle, which makes it very susceptible to surface cracks because it cools more rapidly than the region below. These cracks can, in some cases, cause catastrophic failure in the part. The thin region immediately below this layer cools at a slower rate and thus is annealed.

The effects of heat can be minimized. The wire EDM process works at high current for short on-time cycles, which helps minimize the heat-affected zone. When properly run, a wire EDM can produce a part with a heat-affected zone as thin as .00004 inch. Skim cuts (finish cuts) that remove the heat-affected zone also help reduce these effects. The energy should be reduced on skim cuts, and they should remove the heat-affected zone from the previous pass.

When machining corners, you need to understand how the wire EDM cuts. The wire EDM will leave a sharp edge on the inside of the turn and a radius on the outside of the turn. The smallest radius possible will be equal to the radius of the wire plus the spark gap. If the radius of the wire was .006 and the spark gap .001, the smallest radius possible would be .007 inches: .006 inches (radius of the wire) plus .001 inches. The wire also tends to deflect in the middle because of the electromagnetic field generated. Some machines have special codes to help reduce this effect.

Machine Setup

The workpiece is clamped to the table, which must be open in the middle to allow the wire to travel through the table (Figure M-193). Many attachments are available to help clamp the workpiece to the table. Remember, the wire must be able to cut through the workpiece. You also have to make sure that the piece that is cut out will not move and short the wire. This could potentially damage the workpiece and/or the machine.

The workpiece must be properly aligned on the table. Accurate alignment is fundamental to producing a quality part. The operator must then choose and thread the appropriate wire for the job. Once the wire has been chosen, the operator mounts the spool of wire on the machine. The wire is then threaded through the machine. The wire path will vary depending on the make and model of the machine. If the wire is not threaded properly, it will cause a short to the machine and the machine will not operate. Many machines feature automatic threading.

The wire tension must then be set using a wire tension gauge (tensiometer). The amount of tension used depends on the type and diameter of the wire (Table M-1). Machine sensors monitor the wire for breakage. If the wire should break, the cutting cycle is stopped automatically. The operator then must rethread the wire and then hit the cycle start button.

The wire must then be aligned in a vertical orientation. A setup block, used to align the wire, is normally made of granite with two metal contacts on it. The setup plate is clamped to the table with the metal contacts facing the cutting area.

Table M-1 Wire Tension Table

Desired Wire Tension

Wire Diameter		Copper		Molybdenum, Brass, and Zinc Coated	
Inches	Mm	Ounces	Grams	Ounces	Grams
.002	.05	3.5	100	7	200
.003	.07	5.3	150	10.5	300
.004	.1	7	200	14	400
.005	.12	8.8	250	17.6	500
.006	.15	10.5	300	21.2	600
.007	.17	12.3	350	24.7	700
.008	.20	14	400	28.2	800
.009	.22	15.8	450	31.7	900
.010	.25	17.6	500	35.2	1000
.011	.27	19.4	550	38.8	1100
.012	.30	21.2	600	42.3	1200

The machine then does the vertical alignment. The machine control monitors the wire and moves the U and V axes until the wire is vertical. The controls that set the cutting conditions must be set very low during this step. Next, the operator must align the wire in relation to the actual workpiece. Sometimes the wire is aligned with the edges of the part, and in some cases the wire must be aligned in relation to a hole.

Edge Detection

Many machines have an automatic edge detection function. The operator locates the wire close to an edge and turns on the wire tension control switch (the wire should be running and the flushing should be on). The operator then starts the edge-finding control. The machine moves in the direction the operator has chosen until the control senses continuity (a short circuit) between the wire and the workpiece. This process is repeated several times, and the machine then averages the results to accurately determine the edge location.

Hole Location Detection

The operator threads the wire through a hole in the workpiece and aligns the table so that the wire is approximately in the center of the hole. The operator chooses the hole detection mode on the control and then turns on the wire tension control switch (the wire should be running and the flushing should be on). The operator chooses a slow feed rate and presses the desired axis button. The machine then begins the hole detection sequence and accurately centers the wire in the hole.

Slot Location Detection

The operator threads the wire through a slot in the workpiece and aligns the table so that the wire is approximately in the center of the slot. The operator chooses the slot-detection

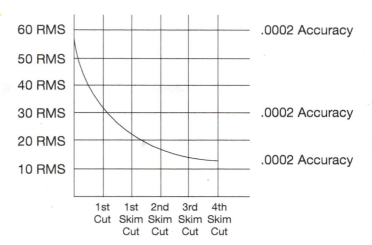

Figure M-198 Skim cuts versus surface finish and accuracy.

mode on the control and turns on the wire tension control switch (the wire should be running and the flushing should be on). The operator chooses a slow feed rate and presses the desired axis button. The machine begins the slot detection sequence and accurately centers the wire in the slot.

Test Square

The test square program, often run before cutting the actual part, cuts a .100-inch square. This permits the operator to make any offset adjustments that may be required. It also allows the operator to check the surface finish and make any changes to the cutting conditions before the actual part is cut.

Skim Cuts

Finish cuts produce better accuracy and surface finish. On a wire EDM machine, they are called skim cuts. After the part has been machined, the operator can enter a small offset value for the wire. This is often less than .001 inch. Very little material is removed in a skim cut, so the cutting speed will be faster. The same program is then run, and a very accurate part with excellent surface finish is produced (Figure M-198). Most parts, however, can be produced with one cut.

Advantages and Applications of EDM

EDM processes can accomplish machining that would be impossible by the traditional methods. Any odd electrode shape will be reproduced in the workpiece. Thus, fine detail is possible, making the process useful in tool, die, and mold work. EDM coupled to numerical control creates an excellent machining system for the tool and die maker.

Because machining is accomplished by spark erosion, the electrode does not actually touch the workpiece and is therefore not dulled by hard workpiece materials, as would be the case with conventional cutters. Thus, the EDM process can spark-erode hard metals and has found wide application in removal of broken taps without destroying an expensive workpiece in the process.

Disadvantages of EDM

The EDM process is quite slow in metal removal compared with conventional machining processes involving direct cutter contact. There is also a possibility of overcutting and local area heat-treating of the parts being machined.

However, EDM can accomplish many machining tasks that could never be done by conventional machining. Thus, the EDM process has become a well-established, versatile manufacturing process in modern industrial applications.

ELECTRON BEAM MACHINING

Electron beam machining (EBM) is in some ways related to EDM and to electron beam welding. In EDM, however, a burst of electrons (spark) impinges on the workpiece, whereas the EBM process uses electrons in a continuous beam. The workpiece material is heated and vaporized by the intense electron beam. The process, like that of electron beam welding, must be carried out in a vacuum chamber, and appropriate shielding must be employed to protect personnel from X-ray radiation.

ELECTROLYTIC GRINDING

In the process of electrolytic grinding (ELG), an abrasive wheel much like a standard grinding wheel is used. The abrasive wheel bond is metal, thus making it a conducting medium. The abrasive grains in the grinding wheel are non-conducting and aid in removing oxides from the workpiece while helping maintain the gap between wheel and work. ELG, like ECM, is a deplating process, and workpiece material is carried away by the circulating electrolyte.

ELG System

The basic ELG system (Figure M-199) consists of the appropriate power supply, the electrode (metal bonded grinding wheel), work-holding equipment, and the electrolyte supply and filtration system. Workpiece material is deplated and goes into the electrolyte solution.

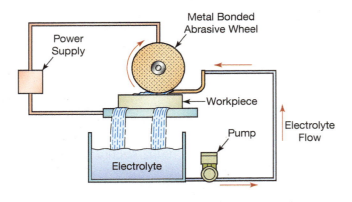

Figure M-199 ELG system.

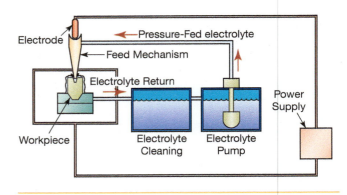

Figure M-201 ECM system.

Advantages and Applications of ELG

Because ELG is primarily electrochemical and not mechanical, as is conventional grinding, the abrasive wheel in ELG wears little in the process (Figure M-200). ELG is burr-free and will not distort or overheat the workpiece. The process is therefore useful for small precision parts and thin or fragile workpieces.

ELECTROCHEMICAL MACHINING AND ELECTROCHEMICAL DEBURRING

Electrochemical machining (ECM) is essentially a reverse metal-plating process (Figure M-201). The process takes place in a conducting fluid or electrolyte pumped under pressure between electrode and workpiece. As workpiece material is deplated, it is flushed away by the flow of electrolyte. Workpiece material is removed from the electrolyte by a filtration system.

Advantages and Applications of ECM

Like EDM, ECM can accomplish machining of intricate shapes in hard-to-machine material. The process is also burr-free and does not subject the workpiece to distortion and stress, as do conventional machining processes. This makes it useful for work on thin or fragile workpieces. ECM is also used for part deburring in the process called electrochemical deburring (ECDB). This process is useful for deburring internal workpiece features inaccessible to traditional mechanical deburring processing.

ULTRASONIC MACHINING

Ultrasonic machining is akin to abrasive processes, such as sand blasting. High-frequency sound (Figure M-202) is used as the motive force that propels abrasive particles against the workpiece. Advantages of this process include the ability to

Figure M-200 Electrolytic surface grinding.

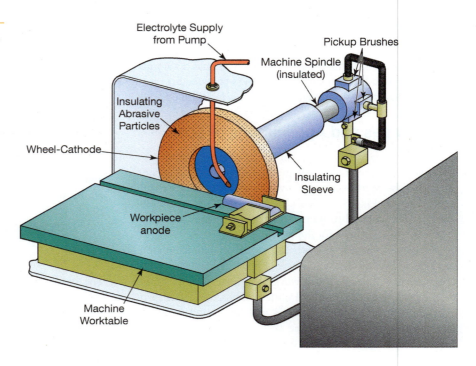

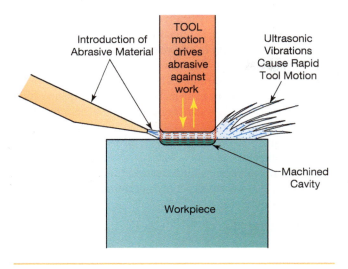

Figure M-202 Ultrasonic machining.

machine hard material with little distortion. Good surface finishes may be obtained, and part features of many different shapes can be machined.

ABRASIVE WATERJET MACHINING

Abrasive waterjet cutting can be used to dramatically reduce machining cost and dramatically increase productivity. The technology is well suited to quickly cutting complex geometry that is difficult or impossible by regular machining methods. High tolerances and high speeds are possible. Waterjets have been used for many years. Low-pressure water jets were used to mine gold in California as early as1852. In the 1960s

mining began to use high-pressure waterjets. The machining industry first began to use abrasive water jets in about 1980.

The actual cutting is often done under water to reduce splash and noise. The term waterjet is often used when referring to another system of cutting called abrasive waterjet cutting.

Waterjets are used to cut softer materials, while abrasive jets are used for harder materials. Abrasive waterjet machines are capable of cutting many industrial materials including stainless steel, aluminum, titanium, Inconel, tool steel, ceramics, granite, glass, rubber, plastic, stone, stainless steel, and many other materials. An abrasive waterjet machine is shown in Figure M-203. Abrasive waterjet cutting is fast. An abrasive jet can cut half-inch thick titanium at the rate of 7 inches per minute when a 30 HP pump is used. Materials can also be stacked so that several parts can be cut at once.

A waterjet machine uses a powerful jet of pressurized water as high as approximately 60,000 pounds per square inch (psi). Figure M-204 shows the operation of an abrasive waterjet machine and some parts that were cut on it.

Figure M-205 shows an example of the high pressure circuit for an abrasive waterjet machine. A special pump, sometimes called an intensifier, is used to create very high water pressure.

In many cases, the size, surface finish, and tolerances that can be held are good enough for the finished product and do not require further machining. Figure M-206 shows some examples of complex parts that were cut out with an abrasive waterjet.

Abrasive waterjet systems inject an abrasive, usually garnet, into the jet of water as it leaves the nozzle. Abrasive jet cutting can cut a wider variety of materials than plain

Figure M-203 An abrasive waterjet machine (*Used with permission of Omax Corporation*).

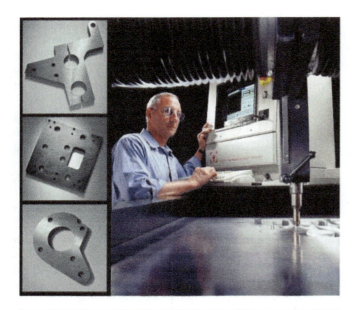

Figure M-204 On the left are shown some sample parts that were cut with an abrasive waterjet machine. On the right an operator is running the machine. Notice the nozzle (*Used with permission of Omax Corporation*).

water. The abrasive enables materials to be cut to close tolerances. Abrasive water jet cutting also can cut edges squarely and with a good finish. The material to be cut and the nozzle are usually submerged in water to reduce the cutting noise.

The principle of cutting with water is a finely focused jet of water. In a waterjet the high pressure spray is focused through a narrow nozzle. The nozzle is a jewel (Figure M-207).

The water pressure is typically between 20,000 and 60,000 pounds psi. The water is forced through a .010 to .015 inch diameter orifice (hole) in a jewel. Abrasive is added after the jewel (Figure M-207). The water–abrasive mixture leaves the nozzle at almost 1,000 mph. The latest machines can cut to within two thousandths of an inch. Abrasive waterjet machines can be used as simple waterjet or abrasive machines by changing the nozzle.

Abrasive waterjets are one of the fastest growing segments of the machine tool industry. Waterjet systems are usually less costly than laser machines and create virtually no heat-affected zone. They also work well for high-performance metals.

KERF WIDTH

Kerf width, which is the width of the cut, determines how sharp of an inside corner you can make. About the smallest practical abrasive waterjet nozzle will give you a kerf width of .030 inch diameter. Higher horsepower machines require larger nozzles, due to the amount of water and abrasive that flow through them. Some *waterjet* nozzles have very fine kerf widths (like .003).

While most cutting is done to leave a straight vertical edge on parts, it is possible to cut accurate beveled edges at angles. Figure M-208 shows a multiaxis accessory for a waterjet that makes it possible to cut angles on the sides of parts. The A-Jet is a complete software-controlled, multiaxis accessory permitting the flexibility to cut severe angles to a maximum of 60° off the vertical. The A-Jet cuts countersunk holes and jigsaw puzzle-type pieces with beveled edges. The accessory supplies additional axes of motion, allowing the operator to fabricate and shape metal edges for weld preparation.

Figure M-205 Waterjet intensifier circuit (*Used with permission of Omax Corporation*).

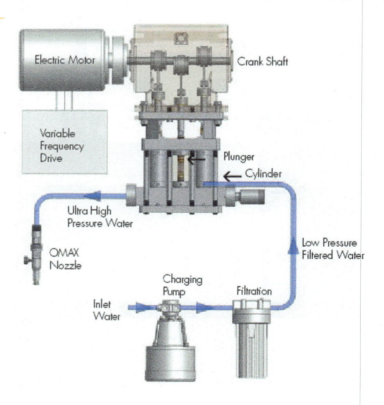

Figure M-206 Parts that were cut with abrasive waterjet. The picture on the right shows that an abrasive waterjet can also be used to mark or scribe parts (*Used with permission of Omax Corporation*).

Figure M-207 An abrasive waterjet nozzle (*Used with permission of Omax Corporation*).

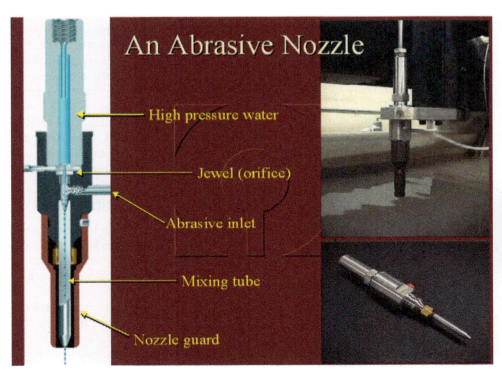

An Abrasive Nozzle

High pressure water

Jewel (orifice)

Abrasive inlet

Mixing tube

Nozzle guard

Figure M-208 Cutting a complex beveled edge part with a multiaxis attachment (*Used with permission of Omax Corporation*).

COST OF OPERATION

Abrasive jet machining is more expensive than waterjet because of abrasive cost and wear of the mixing tube. Mixing tubes have a relatively short life. Mixing tubes are one of the more significant operating expenses.

It costs approximately $15 to $50 to replace a jewel. Mixing tubes range from $100 to $200. Jewels typically last three to five times as long as mixing tubes. Jewels fail only from mineral deposits (which are often removable), or from dirt upstream in the high pressure lines. The jewel can crack, become plugged, or deposits can form in them. Proper filtration can reduce cracking and plugging.

Jewels are easily replaced and are inexpensive ($5–$50). Diamond orifices are also available but are more expensive. Diamond orifices, however, are not necessarily as accurate as sapphire ones because they are difficult to manufacture. The geometry of the jewel is critical to performance. It is difficult to achieve the correct geometry in a diamond jewel.

For precision work, a new mixing tube generally performs better than a used one. The life of a mixing tube is dependent on a number of factors. About 20 to 100 hours is fairly typical. Mixing tubes may wear faster, or last longer, depending on conditions under which they are used.

Machine providers typically say their machines hourly costs of operation are between $20 and $40.

ADVANTAGES OF ABRASIVE WATERJET MACHINING

Abrasive waterjet machines can be setup, programmed, and be producing parts very quickly. They can help increase the productivity of other machine tools also by reducing the amount of material that has to be removed. Shapes can be quickly cut on an abrasive waterjet and then if tighter tolerances are required they can be quickly finished on a traditional machine tool. In many cases parts that were made on a traditional machine tool can be made and finished on an abrasive waterjet machine much more quickly and economically. They make parts quickly out of virtually any material. Intricate shapes are easy to make.

Flat stock can be positioned by laying it on the table and putting a light weight on it. This is because there are very low side forces during the machining. Other parts may require simple positioning fixtures or hold downs to keep them stationary for cutting.

Thick material can easily be cut.

Abrasive waterjet cutting is very economical for low volume production.

Abrasive waterjet cutting creates almost no heat in the part.

No start hole is required.

PROGRAMMING AN ABRASIVE WATERJET MACHINE

Abrasive waterjet machines are very easy to program. The programmer can easily import a CAD file for the part geometry. Feed rates and cutting parameters are easily established using the programming software. Figure M-209 for a feedrate calculation screen.

Figure M-210 shows an example of a typical programming and operation screen. The part geometry can be viewed, checked, and edited on the machine control. The screen can also be used to monitor the conditions and geometry during machining.

Nesting

Nesting is the term used to maximize the number of parts that can be made from a piece of metal. Figure M-211 shows a control being used to program multiple parts. The software helps the user nest multiple parts to reduce scrap and increase productivity.

Another relatively new addition to abrasive waterjet machines has been the ability to control the angle of the nozzle. The addition of servo control to the nozzle enables the machine to cut a part with a draft angle around the part.

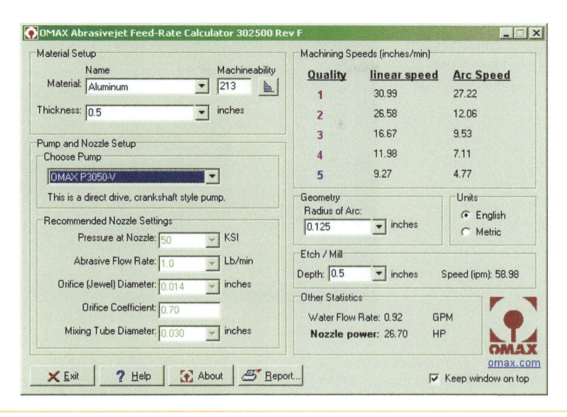

Figure M-209 Example of a programming screen. This screen assists the programmer in setting the feed rate for the cutting material and conditions (*Used with permission of Omax Corporation*).

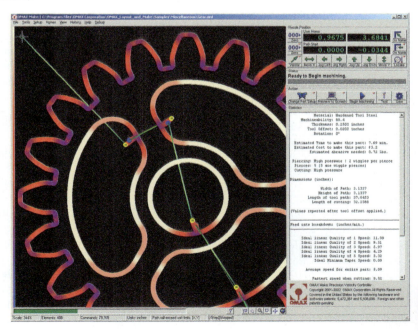

Figure M-210 Programming and operation screen (*Used with permission of Omax Corporation*).

LASER MACHINING

A laser beam is a coherent beam of high-intensity light that is focused through lenses into a small diameter. The diameter of the beam is typically about .005 inches in diameter. Laser cutting is the process of vaporizing metal or other material in a very small area. The laser itself is a single point cutting tool with a very small point. This enables very narrow cut width. The acronym **LASER** stands for **L**ight **A**mplification by **S**timulated **E**mission of **R**adiation. This is a device to produce a beam of light in which all the waves are in phase (coherent). The two main types of lasers are gas and solid-state. In gas lasers, CO_2 is mixed with helium and nitrogen to make the lasing medium. In solid-state lasers,

Figure M-211 Programming multiple parts using the nesting features of the OMAX control (*Used with permission of Omax Corporation*).

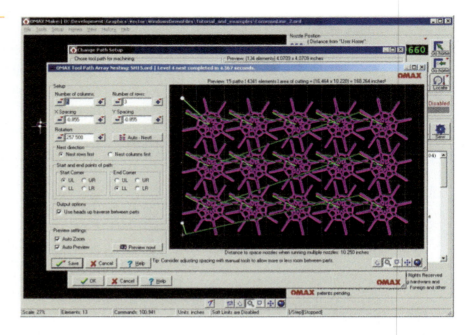

yttrium-aluminum-garnet (YAG) crystals containing neodymium ions are used as the medium. Figure M-212 shows a CO_2-based laser cutting intricate shapes in flat metal.

LASER PROCESSES

Laser manufacturing activities currently include cutting, welding, heat treating, cladding, vapor deposition, engraving, scribing, trimming, annealing, and shock hardening.

Cutting

Lasers are capable of cutting numerous materials, including steel, stainless steels, and super alloys. They can also be used to cut plastics, rubber, composites, ceramics, quartz,

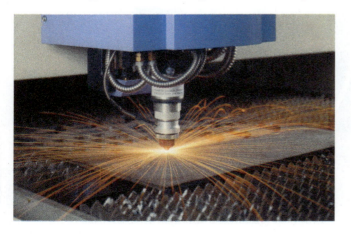

Figure M-212 Mitsubishi model 3015lxp CO_2-based laser cutting intricate shapes in flat metal substrate. The laser can cut all types of ferrous or nonferrous metals, as well as nonmetallic materials, for a wide variety of uses (*Courtesy of Mitsubishi laser*).

glass, and wood. Frequently, the process is specified for cutting extremely hard materials, such as titanium, hastelloy, and inconel. A key benefit of laser cutting is that the lack of contact between the tool (the laser cutting head) and workpiece eliminates problems of tool wear and breakage.

Marking

Laser systems have made it possible to achieve fast, permanent, noncontact marking of a wide range of materials, including metals, plastics, semiconductors, ceramics, marble, and glass.

LASER FUNDAMENTALS

Normal light moves in all directions. The standard light bulb generates light but the light does not have much power. A laser creates light that is coherent. All of the light moves in one direction. It is coherent. Coherent means that the waves of light are in phase with each other. A light bulb produces many wavelengths, making it incoherent. Regular light also consists of many colors (wavelengths). Laser light has one wavelength.

The temperature where laser light is focused is extremely hot and powerful. A 1,000 watt laser beam focused to a spot .005 inch diameter is equal to 5,000,000 watts per square inch.

The beam delivery tube directs the raw laser beam, which is about .5″–.75″ in diameter, to the focusing lens where the beam is reduced in size to a very small spot (.005″ – .009″) with intense energy.

Part cutting paths are generated from the imported CAD files and applied automatically by the CAD/CAM system. The cutting parameters for the laser can be automatically generated using E codes from technical tables, in the machine control. The most advanced machine controls have

Figure M-213 Diagram of a laser system.

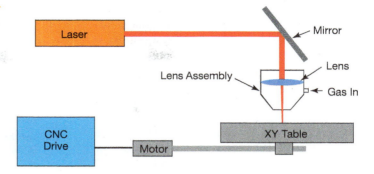

knowledge-based tables that automatically determine special processing techniques for external and internal corners.

Figure M-213 shows a block diagram of the motion system and simplified beam delivery system for one axis of a laser. In this example, the motion system is used to move the material. The cutting head is in a fixed position. Note that the coherent laser light reflects off of the mirror and then is focused in the lens assembly.

LASER CUTTING HEADS

The cutting head is the "business" end of the laser (Figure M-214). The raw laser beam passes through the lens and is focused onto the surface of the material. This heats the material to its molten state. A nozzle is used to deliver and assist gas to remove the molten metal. By moving the head or material, a path or kerf is cut. The material ejected is captured by a dust collection system to remove particles.

The purpose of the lens is to focus the laser beam. Different materials require specific focal point settings relative to the surface of the material. The focal point for carbon steel will always be on top of the material. Aluminum requires that the focal point be positioned below the bottom surface of the material.

The lens acts like a magnifying glass. The size of the lens determines where it focuses (focal point). Machines offer a choice of focal length lenses for processing materials. The focal length designates the distance between the lens and the point at which the beam is focused. Adapters are required to position the lenses at the correct height.

Automatic focus ensures that the focused point stays constant in relationship to the material surface. If the material is wavy, the focal point will adjust accordingly ensuring that the focal point stays where it is supposed to be. The main functions of the gas are to: protect the lens from contamination; remove molten material, and to cool the lens.

It is important that the beam be directed through the center of the nozzle. If it is off too much to one side it will alter the even flow of assist gas and produce inferior cuts. Since a laser beam has to travel between 12 and 20 feet and then pass through a nozzle orifice of only .070 inch to get to the material, pointing stability is critical. A misaligned nozzle will block part of the laser beam before it hits the workpiece.

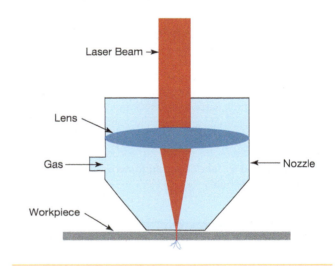

Figure M-214 Nozzle assembly.

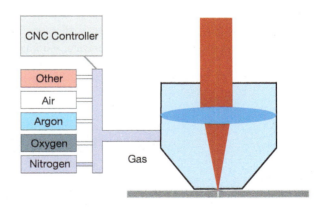

Figure M-215 Laser assist gasses connected to the manifold are controlled by the CNC control.

Assist Gasses

Gasses are used to assist in the cutting process. Gas selection is an important part of laser processing. Figure M-215 shows a block diagram of a laser with assist gasses connected to a manifold. The program in the CNC control is used to choose which gas is used for the cut.

Most steel will be processed with oxygen because it is fast. Oxygen is sometimes used with aluminum, but leaves dross that is more difficult to remove. Oxygen assist gas is the preferred gas for carbon steel cutting. It aids in the cutting process so it can cut faster and thicker than other assist gasses used with carbon steel.

Aluminum can be processed with dry, clean shop air, but it will require a compressor with sufficient capacity. Air can be an economical way of cutting thin stainless with a better result than oxygen. Nitrogen is the preferred gas for aluminum and stainless. Argon is the most frequently used gas when cutting titanium.

Advantages and Disadvantages of Laser Machining

Lasers produce the most precise cut with the smallest heat-affected zone of the thermal cutting technologies. They can cut up to 1-inch thick carbon products and ½-inch thick specialty metals. Consumables usage is less than with plasma or oxyfuel cutting. Very fragile parts can be laser cut with no support because there are no cutting forces. Laser technology is well suited to fabricating high accuracy parts, especially flexible materials.

SELF-TEST

1. Describe in general terms how the following processes work: EDM, ECM, ELG, ultrasonic, abrasive waterjet, and laser.
2. Which process would be used for die making?
3. Which process utilizes a dielectric fluid?
4. Which processes might be used for deburring applications?
5. What provides the energy in ultrasonic machining, and what does the cutting?

6. True or False? Cutting with water is a new technology.
7. True or False? Water jet machines are usually used for soft materials and abrasive waterjets are used for hard materials.
8. The pump that is used to create the high pressures for waterjet cutting is called an _____.
9. True or False? Tolerances that can be held by abrasive waterjet cutting are not very good and parts normally require finish machining on a traditional metal cutting machine.
10. List the main components found in an abrasive waterjet nozzle.
11. True or False? Mixing tubes are a significant cost in abrasive waterjet cutting.
12. True or False? Abrasive waterjet cutting is limited to relatively thin materials.
13. List at least three advantages of waterjet machining.
14. True or False? When an abrasive waterjet machine is improperly set up, particles accumulate in the bottom of the cut. This is called nesting.
15. List at least four operations that can be performed by laser machining.
16. True or False? Lasers can only be used for cutting metals.
17. Describe the main components in a typical laser machine.
18. What is the purpose of assist gasses?
19. List at least three advantages of laser cutting.

INTERNET REFERENCES

Further information on advanced machining processes:

http://www.engineersedge.com

http://www.omax.com

http://www.laserod.com

http://en.wikipedia.org/wiki/Laser_ablation

Appendices

APPENDIX 1
Answers to Self-Tests

SECTION A/CAREERS AND THE MACHINIST'S ROLE IN PROCESS PLANS

1. This question is looking for some real thought from the student about a potential career progression.
2. This question is looking for some real thought from the student about things they can do to improve the chances of reaching their career goal.
3. Attributes of a good attitude: get along with others, avoid profanity, be positive, dress appropriately, respect your co-workers and bosses, do not be a constant complainer, speak well of others, accept work willingly, and so on. Attributes of good work ethic: Be at work early, be ready to work when the shift starts, take pride in your work, look for ways to improve your productivity and the quality of your work, keep your work area clean and well organized, make sure your tools and machines are well maintained, get your work done on or ahead of time, do not miss work unnecessarily, be helpful to others, readily admit mistakes, and so on.
4. They bid on jobs. They must have a capability to do the job to meet customers' specifications and also make money on them.
5. Does the company have the capability to make the parts to specifications? Can they make the due date? Can they produce the parts at a profit? Can they get material in time?
6. A process plan details all of the things that need to be done to manufacture a part (Figure A-5). They are step-by-step instructions that are done by the machinist. The steps detail each machining operation, inspections that need to be done, deburring operations, and so on. The process plan also references the blueprint and correct revision level that needs to be used. The process plan may list the CNC program to be used, the machine, and any special setup instructions. The process plan would also detail and special handling, packaging, and or shipping requirements.
7. A routing is also called a process plan or a routing.
8. A job packet would typically include a process plan, a blueprint, inspection sheets if needed, and it may contain material certifications.

SECTION A/UNIT 2/MANUFACTURING COMPETITIVENESS AND IMPROVEMENT

1. Lean manufacturing attempts to reduce and eliminates waste that takes away from productivity and profitability.
2. The basis for improvement in the TPS was reduction or elimination of the seven wastes they identified.
3. The eight waste involved companies not utilizing the talent and brains of their workers.
4. The 5 S method has to do with organizing work. Eliminating tools and things that are not essential; organizing stations to eliminate unnecessary motion and effort; keeping things clean and maintained every day.

5. Value stream mapping is a lean manufacturing tool that can be used to analyze and design the flow of materials and information required to bring a product to a customer. It originated at Toyota. It helps eliminate nonvalue-added operations and costs in a process.
6. Kanban is not an inventory control system. It is a scheduling system that helps determine what to produce, when to produce it, and how many to produce. Kanban uses the rate of demand to control the rate of production. Demand is passed from the end customer through the chain of customer (worker) processes. Toyota first applied this technique in 1953 in their main machine shop.
7. *Poka-yoke is a Japanese word that means mistake-proofing or fail-safing. A poka-yoke is any mechanism in a lean manufacturing process that helps an equipment operator avoid mistakes. For example, the addition of a pin in a fixture to prevent the part from being mislocated would be a poka-yoke. Poka-yoke attempts to eliminate product defects by preventing, correcting, or warning the operator an error is about to occur. Shigeo Shingo first used the term Poka-yoke as part of the Toyota Production System.*

SECTION A/UNIT 3/SHOP SAFETY
Self-Test Answers

1. Eye protection equipment.
2. Wear safety goggles or a full face shield. Prescription glasses may be made as safety glasses.
3. Shoes, short sleeves, short or properly secured hair, no rings and wristwatches, shop apron or shop coat with short sleeves.
4. Use of cutting fluids and vacuum dust collectors.
5. They may cause skin rashes or infections.
6. Bend knees, squat, and lift with your legs, keeping your back straight.
7. Compressed air can propel chips through the air, implant dirt into skin, and possibly injure eardrums.
8. Good housekeeping includes cleaning up oil spills, keeping material off the floor, and keeping aisles clear of obstructions.
9. In the vertical position or with a person on each end.
10. Do I know how to operate this machine? What are the potential hazards involved?

   Are all guards in place?

   Are my procedures safe?

   Am I doing something I probably should not do?

   Have I made all proper adjustments and tightened all locking bolts and clamps?

   Is the workpiece secured properly?

   Do I have proper safety equipment?

Do I know where the stop switch is?

Do I think about safety in everything I do?

11. The standard was intended to reduce the number of deaths and injuries related to servicing and maintaining machines and equipment.

SECTION A/UNIT 4/THREADS AND FASTENERS
Self-Test Answers

1. Pitch Diameter—Sometimes called *mean diameter*, is the diameter in between the major and minor diameter. It is the theoretical diameter at which each pitch is equally divided between the mating male and female threads. Pitch diameter can be thought of as that diameter at which the male and female threads should meet. It could be thought of as the working diameter. There are various ways to inspect the size of threads, but they depend on the use of the pitch diameter. Thread micrometers measure the thread's pitch diameter. There is also a method that uses three wires that are placed in the thread and a micrometer measures the diameter over the wires. The wires rest on the pitch diameter of the thread.
2. Thread micromter or three-wire thread measurement method.
3. The minimum recommended thread engagement for a screw in an assembly is as much as the screw diameter; a better assembly will result when $1\frac{1}{2}$ times the screw diameter is used.
4. Class 2 threads are found on most screws, nuts, and bolts used in the manufacturing industry. Studs used on many types of machines are examples of class 3 fits.
5. UNC stands for Unified National Coarse thread series and UNF stands for Unified National Fine thread series. Each has its own strengths. Fine is generally stronger in hard materials and coarse is generally stronger in soft materials.
6. The formula is D = number of the machine screw $\times$ .013 inch + .060 inch; $D = 8 \times .013$ inch + .060 inch = .164 inch
7. Setscrews are used to secure gears or pulleys to shafts.
8. Stud bolts can be used instead of long bolts. Stud bolts are used to aid in the assembly of heavy parts by acting as guide pins.
9. The strength grades of bolts are identified by the markings on the heads. The grade indicates the strength of the fastener. A manufacturer's chart may be used as a guide for the proper torque of fasteners.
10. Flat washers provide a larger contact area than nuts and screw heads to distribute the clamping pressure over a larger area.
11. A helical spring lock washer prevents the unplanned loosening of nut and bolt or screw assemblies. Spring lock washers will also provide for a limited amount of take-up when expansion or contraction takes place.
12. Internal–external-tooth lock washers are used on oversized holes or to provide a large bearing surface.
13. Dowel pins are used to achieve accurate alignment between two or more parts.
14. Taper pins accurately align parts that have to be disassembled frequently.
15. Roll pins are used to align parts. Holes to receive roll pins do not have to be reamed, which is necessary for dowel pins and taper pins.
16. Retaining rings are used to hold bearings or seals in bearing housings or on shafts. Retaining rings have a spring action and are usually seated in grooves.
17. Keys transmit the driving force between a shaft and pulley.
18. Woodruff keys are used where only light loads are transmitted.
19. Gib head keys are used to transmit heavy loads. These keys are installed and removed from the same side of a hub and shaft assembly.

SECTION A/UNIT 5/BLUEPRINT READING FUNDAMENTALS
Self-Test Answers

1.

First Angle Projection   Third Angle Projection

2. First-angle projects different views. If we had the part in a glass box, in the third-angle projection the projections of the part are between the viewer and the object. In the first-angle projection, the object is between the viewer and the projections on the glass box. Third angle is the system used in the United States.

3.

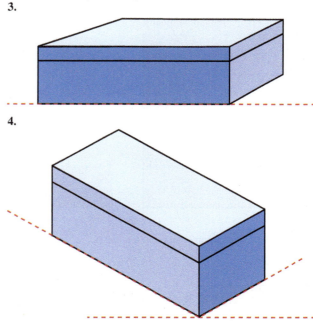

4.

5. List at least five things that are found in a title block. Part name, drawing number, revisions, notes, material specification, company name, person who drew and checked the drawing, scale, general tolerances, number required, sheet number, projection system used.
6. Revisions are shown in a revision block. They are numbered. The numbers are then shown on the print by the feature that was changed.

SECTION A/UNIT 6/VIEWS AND LINE TYPES
Self-Test Answers

1. Identify the types of lines in the drawing in Figure A-50
   1. Phantom Line
   2. Object Line
   3. Extension Line

4. Dimension Line
5. Centerline
6. Section Line
7. Hidden Line
8. Cutting plane Line
9. Leader Line

2. Assume third-angle projection. Label the views.

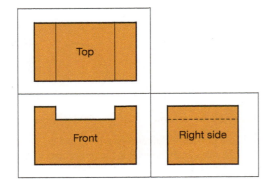

3.

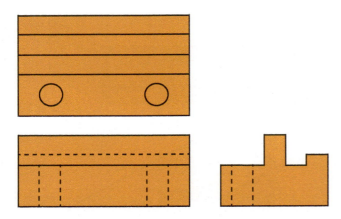

4. Draw the missing lines in the three views.

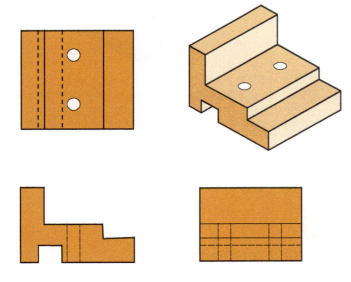

5. Fill in the numbers that represent the surface on the part in each view.

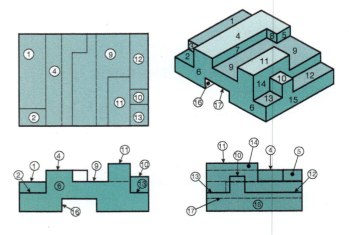

6. Fill in the numbers that represent the surface on the part in each view.

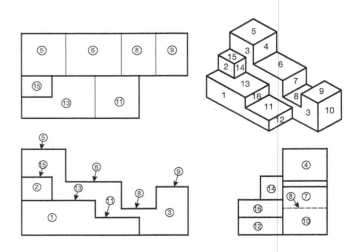

7. Draw the missing lines in the views.

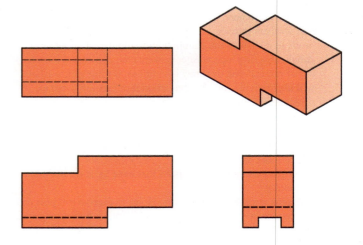

8. Draw the missing lines in the views.

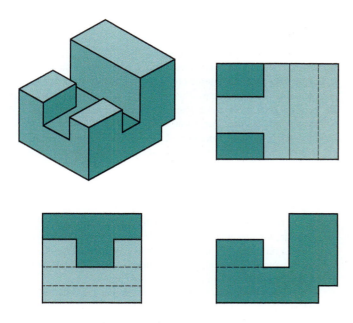

9. Draw the missing lines in the views.

10. Draw the missing lines in the views.

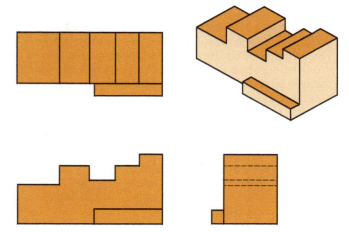

SECTION A/UNIT 7/DIMENSIONS, TOLERANCES AND FITS

Self-Test Answers

1. In the absolute dimensioning system, all dimensions are specified from the same zero point.
2. In incremental dimensioning, a dimension is specified from one feature to the next, not from a part zero.
3. What is a general tolerance and where are they found? General tolerances (also called shop tolerances) are used when no specific tolerance is specified for a feature. General tolerances are generally shown in the title block on the print.
4. 1.000 − .005 + .005
5. 1.000 + .005
6. Clearance, transitional, and interference.
7. An RC3 fit is about the closest fit that will run freely. Precision running fits are intended for precision applications at low speed with low bearing pressures. RC3 fits should not be used where noticeable temperature differences can occur.
8. Locational transition fits lie between clearance and interference. Locational interference fits are for applications where accuracy of location is important. A small amount of clearance or interference is permissible with a locational transition fit.
9. Locational interference fits are used where accuracy of location is very important. Locational interference fits are used for parts requiring rigidity and alignment. The parts can be assembled or disassembled using a press.
10. Clearance .0008 to .0021, P = 1/6 = .1666 inch, Shaft − .0008 to − .0013.

SECTION A/UNIT 8/FUNDAMENTALS OF GD&T

Self-Test Answers

1. GD&T uses a circular tolerance zone where traditional rectangular tolerancing uses a rectangle. The circular allows more variance while maintaining the same standard.
2. Datums are reference points, lines, areas, and planes that are theoretically exact for the purpose of calculations and measurements. The first machined surface on a casting, for

example, might be selected as a datum surface and used as a reference from which to locate other part features. Datums are usually not changed by subsequent machining operations and are identified by letters inside a rectangular frame

3. A datum is chosen from an important axis or feature on a part.

4. Datum A is the primary, Datum B is the secondary, and Datum C is the tertiary datum. The primary datum (Datum A) is selected to provide functional relationships, standardizations and repeatability between surfaces. Datum B is a secondary datum and is perpendicular to the primary datum so measurements can be referenced from them. Datum C, when needed, is perpendicular to both the primary and secondary datums ensuring the part's fixed position from three related datums. A datum does not have to be a surface. A datum can be a feature, such as the center of a bore. After the primary datum is located using three points, the secondary datum will typically use two points to define a line. The third (tertiary) datum is defined by a single point.

5. MMC is the maximum-material condition. For a shaft it would be the maximum allowable size.

6. Describe the term MMC as it would apply to a bore. MMC is the maximum-material condition. For a shaft it would be the minimum allowable size.

7. LMC is the least-material condition. For a shaft it would be the minimum allowable size.

8. Describe the term LMC as it would apply to a bore. LMC is the least-material condition. For a bore it would be the maximum allowable size.

RFS (regardless of feature size) means that the geometric tolerance applies no matter what the feature size is. The tolerance zone size remains the same, unlike MMC or LMC. RFS is the default condition for all geometric tolerances. Unless MMC or LMC is specified on the drawing, the tolerance zone size remains the same even though the feature size changes. Remember that the RFS symbol is very seldom used. If there is no MMC or LMC, RFS is implied.

SECTION A/UNIT 9/GEOMETRIC TOLERANCING

Quick Check 1

1. The surface indicated must be flat within .001.
2. Lines on the surface must be straight within .001. The whole surface does not have to be checked, just one or more lines. Each line is independent.
3. In a straightness tolerance, lines on the surface must be straight within the tolerance. The whole surface does not have to be checked, just one or more lines. Each line is independent. In a flatness tolerance, all points on the surface would need to be in tolerance.
4. Lines in the surface of the part must be straight within .005 at MMC. If the part diameter is 1.000, the tolerance of straightness expands to .010.
5. This is the circularity or roundness form tolerance. In this example, the part would need to be circular within .001. Not that this only refers to individual circles on the outside of the part and that each circle is independent. Usually one or a couple of circles would be checked with an indicator.
6. This is the cylindricity form tolerance. In this example, the part surface would need to be cylindrical within .001. This refers to the entire surface. All points would need to be cylindrical within .001.

Section A/Unit 9/Quick Check 2

1. Explain the profile tolerance. This is the profile of a line tolerance. In this example, the tolerance around the line profile would be a total of .005 in relation to datum A. One or more lines would be checked on the surface of the part. All points on each line would need to be in tolerance. But each line is independent.
2. Explain the profile tolerance. This is the profile of a surface tolerance. In this example, the tolerance around the surface profile would be a total of .005 in relation to datum A. All points on the surface profile would have to be in tolerance.

Section A/Unit 9/Quick Check 3

1. Angularity, perpendicularity, and parallelism.
2. Explain the orientation tolerance and complete the tolerance table.

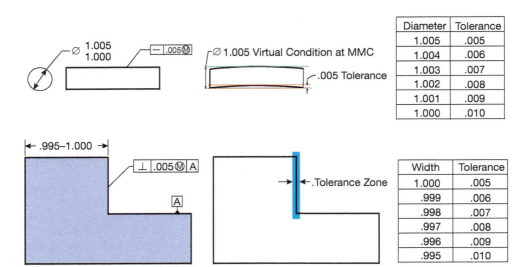

Diameter	Tolerance
1.005	.005
1.004	.006
1.003	.007
1.002	.008
1.001	.009
1.000	.010

Width	Tolerance
1.000	.005
.999	.006
.998	.007
.997	.008
.996	.009
.995	.010

Section A/Unit 9/Quick Check 4

1. Explain the location tolerance and complete the tolerance table.

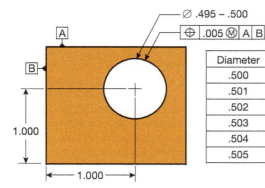

Diameter	Tolerance
.500	.005
.501	.006
.502	.007
.503	.008
.504	.009
.505	.010

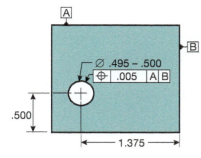

Diameter	Tolerance
.500	.010
.499	.009
.498	.008
.497	.007
.496	.006
.495	.005

2. A concentricity tolerance specifies a cylindrical tolerance zone whose axis coincides with the datum axis. A concentricity tolerance indicates that a cylinder, cone, hex, square or surface of revolution shares a common axis with a datum feature. It controls the location for the axis of the indicated feature within a cylindrical tolerance zone whose axis coincides with the datum axis.

SECTION A/UNIT 9/GEOMETRIC TOLERANCING

Self-Test Answers

1. Circular runout is used to provide control of the circular elements of a surface. The tolerance is applied independently at any circular measuring position as the part is rotated 360 degrees.

2. One way circular runout can be inspected is to put the datum diameter in a V-block and an indicator at one point on the controlled diameter. The part would then be rotated and the smallest and largest readings on the indicator would be used to determine the runout. More than one circle can be checked, but the checks are independent.

3. To inspect total runout the operator would need to check several circles along the whole length of the controlled diameter. All points of the checks would need to have .005 or less runout.

4. Number 1 is referring to the surface defined as Datum B.

Number 2 is referring to the surface defined as datum E.

Number 3 is a specification calling for the slot to be parallel with the surface represented by datum B within .003.

Number 4 is referring to the surface represented by datum E. The feature control frame says the surface must be flat within .003.

Number 5 is referring to a feature control frame that specifies that that surface must be perpendicular to datum E within .002.

5. Number 1 is referring to the surface represented by Datum C. The feature control frame specifies that this surface must be flat within .002.

Number 2 is referring to a feature control frame that controls a diameter feature. The diameter must be concentric with the diameter represented by datum A within .002.

Number 3 is referring to a feature control frame that controls the runout on the diameter represented by datum B. The runout must be less than .003 in relation to datum A.

Number 4 is referring to the .835 diameter. The .835 diameter must be concentric in relation to datum A within .002.

Number 5 is referring to the feature control frame controlling the two holes. This specifies that they have to be on location within .003 at MMC in relation to datum A.

Size	Tolerance
.203	.009
.202	.008
.201	.007
.200	.006
.199	.005
.198	.004
.197	.003

6. Explain the following based on the blueprint in Figure A-114.
   1. The distance from the right edge of the par to the centerline of the hole is .325.
   2. The four holes are to be .250 diameter ± .003 though the part, The are to be countersunk to a diameter of .32 with a 82° countersink. Their location tolerance is .0005 in relation to datums A, C, and B.
   3. This surface must be parallel to datum B within .005.
   4. The distance between the back edge of the part and the dole center is 2.20.
   5. This surface must be perpendicular to datum A within .003.
   6. This surface (datum A) must be flat within .003.
   7. This is datum B.
   8. This is datum C.

SECTION B/UNIT 1/ARBOR AND SHOP PRESSES

Self-Test Answers

1. The arbor press is used for installing and removing mandrels, bushings, and ball bearings. The hydraulic shop press is also used for straightening and bending.

2. The shaft has seized or welded in the bore because it was not lubricated with pressure lubricant.

3. A loose and rounded ram could cause a bushing to tilt or twist sideways while pressing and thus be ruined. In any case, the operator should always check to see whether a bushing is going in straight. A pressing plug with a pilot would be helpful here.

4. Just enough pressure should be applied to press in the bushing. When it stops moving, more pressure will be sensed. At that point it is time to stop.

5. The bearing should be supported on the inner race.

6. Ordinary shafts with press fits are not tapered but have the same dimension along the pressing length. Mandrels taper

.060 inch/ft, which causes them to tighten in the bore somewhere along their length.

7. The two most important steps, assuming that the dimensions are all correct, are (1) to make sure that the bore has a good chamfer (the bushing should also have a chamfer or "start") and (2) to apply high-pressure lubricant to the bore and the bushing.

8. Five ways to avoid tool breakage and other problems when broaching keyseats in the arbor press:

   Make sure that the press ram is not loose, and check to see that the proper hole in the press plate is under the work so that the broach has clearance to go through the work.

   Clean and lubricate the broach, especially the back edge between each cut.

   Do not use a broach on hard materials (over RC 35).

   Use the right-size bushing for the bore and broach.

   Make sure that at least two teeth are continuously engaged in the work.

SECTION B/UNIT 2/WORK-HOLDING AND HAND TOOLS
Self-Test Answers

1. The solid base and the swivel base types.
2. By the width of the jaws.
3. Brass, soft metal, or wood may be used to protect finishes from insert jaw serrations.
4. "Cheater" bars should never be used on the handle. The movable jaw slide bar should never be hammered on, and excessive heat should never be applied to the jaws.
5. The vise should be taken apart, cleaned, and the screw and nut cleaned in solvent. A heavy grease should be packed on the screw and thrust collars before reassembly.
6. False. C-clamps are used for clamping work. Some heavy-duty types can hold many hundreds of pounds.
7. The principal advantage of the lever-jawed wrench is its great holding power. Most types have hard, serrated jaws and so should not be used on nuts and bolt heads.
8. Soft hammers and mallets are made for this purpose. When setting down work in a drill press or milling machine vise, for instance, a lead hammer is best because it has no rebound.
9. The hard, serrated jaws will damage machine parts. Pipe wrenches should be used on pipe and pipe fittings only.
10. Standard screwdrivers should have the right-width blade to fit the screw head. They should also be shaped correctly to fit the slot.

SECTION B/UNIT 3/HACKSAWS
Self-Test Answers

1. The kerf is the groove produced in the work by a saw blade.
2. The set on a saw blade is the width of the teeth bent out from the blade back.
3. The pitch of a hacksaw blade refers to the number of teeth per inch on a saw blade.
4. The first consideration in the selection of a saw blade is the kind of material being cut. For soft materials, use a coarse-tooth blade and for harder materials use a fine-pitch blade. The second point to watch is that at least three teeth should be cutting at the same time.

5. The two basic kinds of saw blades are the all-hard blade and the flexible blade.
6. Excessive dulling of saw blades is caused by pressure on the saw blade on the return stroke, sawing too fast, letting the saw slide over the workpiece without any cutting pressure, or applying too much pressure.
7. Saw blades break if too much pressure is used or if the blade is not sufficiently tightened in the saw frame.
8. When the saw blade is used, the set wears and makes the kerf cut narrower with a used blade than the kerf cut with a new blade. If a cut started with a used blade can't be finished with that blade but has to be completed with a new blade, the workpiece should be turned over and a new cut started from the opposite end of the original cut. A new blade used in a kerf started with a used blade would lose its set immediately and start binding in the groove.

SECTION B/UNIT 4/FILES
Self-Test Answers

1. Single cut, double cut, curved cut, and rasp.
2. Four out of these: rough, coarse, bastard, second cut, smooth, and dead smooth.
3. The double-cut file.
4. bastard.
5. Too much pressure will break teeth off a file. It also will cause pinning and scratching of the work surface.
6. Files in contact with each other, files rubbing over the work without any pressure being applied, filing too fast, or filing on hardened materials causes files to dull.
7. As a safety precaution; an unprotected tang can cause serious injury.
8. A soft workpiece requires a file with coarser teeth, because there is less resistance to tooth penetration. A fine-toothed file would clog up on soft materials. For harder materials, use a fine-toothed file to have more teeth making smaller chips.
9. Pressure is only applied on the forward stroke, which is the cutting stroke.
10. Rotating a round file clockwise while filing makes the file cut better and improves the surface finish.

SECTION B/UNIT 5/HAND REAMERS
Self-Test Answers

1. Hand reamers have a square on the shank and a long starting taper on the fluted end.
2. A reamer does its cutting on the tapered portion. A long taper will help in keeping a reamer aligned with the hole.
3. Spiral-fluted reamers cut with a shearing action. They will also bridge over keyseats and grooves without chattering.
4. The shank diameter is usually a few thousandths of an inch smaller than the nominal size of the reamer. This allows the reamer to pass through the hole without marring it.
5. Expansion reamers are useful for increasing hole sizes by a small amount.
6. Expansion reamers can be adjusted only a small amount by moving a tapered internal plug. Adjustable reamers have a larger range of adjustment, from $\frac{1}{32}$ inch on small diameters to $\frac{5}{16}$ inch on large reamers. Adjustable reamers have removable blades. Size changes are made by moving these blades with nuts in external tapered slots.

7. Cutting fluids are used to dissipate the heat generated by the reaming process, but in reaming, cutting fluids are more important in obtaining a high-quality hole surface finish.
8. Reamers dull rapidly if they are rotated backward.
9. The hand reaming allowance is rather small, between only .001 and .005 inches.
10. Use a Morse taper reamer.

SECTION B/UNIT 6/IDENTIFICATION AND USES OF TAPS

Self-Test Answers

1. A bottoming tap.
2. Gun taps are used on through holes or blind holes with sufficient chip space at the bottom of a hole.
3. Fluteless spiral-pointed taps are especially useful for tapping holes in sheet metal or on soft, stringy materials where the thickness is no greater than one tap diameter.
4. Spiral-fluted taps draw the chips out of the hole and are useful when tapping a hole that has a keyseat in it bridged by the helical flutes.
5. Thread-forming or fluteless taps do not produce chips because they don't cut threads. Their action can be compared to thread rolling in that material is being displaced in grooves to form ridges shaped in the precise form of a thread.
6. Taper pipe taps are identified by the taper of the body of the tap, which is $\frac{3}{4}$ inch per foot of length, and also by the size marked on the shank.
7. When an Acme thread is cut, the tap is required to cut too much material in one pass. To obtain a quality thread, a roughing pass and then a finishing pass are needed.
8. Rake angles vary on tools depending on the kind of work machined. In general, softer, more ductile materials require larger rake angles than do harder, less ductile materials.

SECTION B/UNIT 7/TAPPING PROCEDURES

Self-Test Answers

1. Taps are driven with tap wrenches or T-handle tap wrenches.
2. A hand tapper is a fixture used to hold a tap in precise alignment while hand tapping holes.
3. A tapping attachment is used when tapping holes in a machine or for production tapping.
4. The strength of a tapped hole is determined by the kind of material being tapped, the percentage of thread used, and the length or depth of thread engagement.
5. Holes should be tapped deep enough to provide 1 to $1\frac{1}{2}$ times the tap diameter of usable thread.
6. Taps break because holes are drilled too shallow, chips are packed tight in the flutes, hard materials or hard spots are encountered, inadequate or the wrong kind of lubricant is used, or the cutting speed used is too great.
7. Tapped holes that are rough and torn are often caused by dull taps, chips clogging in flutes, insufficient lubrication, the wrong kind of lubrication, or tapping of already rough holes.
8. Broken taps can be removed by the use of a tap extractor, by drilling them out after annealing the tap, or by using an electrical discharge machine to erode the tap.

SECTION B/UNIT 8/THREAD-CUTTING DIES AND THEIR USES

Self-Test Answers

1. A die is a tool that is used to cut external threads.
2. A diestock is used to hold the die when hand threading. A special die holder is used when machine threading.
3. The size of thread cut can be changed by only a small amount on round adjustable dies. Too much expansion or contraction may break the die.
4. When assembling a two-piece die collet, be sure both die halves are marked with the same serial number and that the starting chamfers on the dies are toward the guide.
5. The chamfer on the cutting end of a die distributes the cutting force over multiple threads and aids in starting the thread-cutting operation.
6. Cutting fluids are important in threading to achieve threads with a good surface finish and close tolerance, and to give long tool life.
7. Before a rod is threaded, it should be measured to confirm that its size is no larger than its nominal size. Preferably, it is .002 to .005 inches undersized.
8. The chamfer on a rod before threading makes it easy to start a die. It also protects the starting thread on a finished bolt.

SECTION B/UNIT 9/OFF-HAND GRINDING

Self-Test Answers

1. The sharpening of tool bits and drills.
2. The wheels tend to become misshapen and out of round.
3. A Desmond dresser.
4. Overheating, which ruins the tool.
5. Safety factors are these: wear eye protection; do not grind nonferrous metals on an aluminum oxide wheel; keep workrest close to wheel; keep wheel guards in place; let newly installed wheel run idle for 1 minute; keep wheel dressed.
6. Not more than $\frac{1}{16}$-inch.
7. Running the new wheel a bit will further assure that the wheel is sound. If the wheel should have a defect, then it might show up as a wheel failure during this run-in period.
8. The blotter pads the flange pressure on the installed wheel.
9. Using a wheel that is rated for the proper rpm.
10. Ring testing may reveal a cracked wheel before it is installed on the grinder.

SECTION C/UNIT 1/MEASUREMENT AND COMMON MEASURING TOOLS

Self-Test Answers

1. Accuracy can mean a couple of different things. First, accuracy can refer to whether or not a specific measurement actually is its stated size. For example, a certain drill has its size stamped on its shank. A doubtful machinist decides to verify the drill size using a micrometer. The size is found to be correct. Therefore, the size stamped on the drill is accurate. Second, accuracy can refer to whether or not the chosen measuring tool is appropriate for the measurement. If the machinist had used a steel rule to measure the drill

to see if it was a number 7 drill (.201) or a number 6 drill (.204), it would be an inappropriate tool for the job. In this example, the act of measurement is not accurate because the wrong measuring tool was selected. **User accuracy** is also an important consideration. If the machinist is not very good at using or reading a micrometer the measurement will be inaccurate. Or if the machinist used a micrometer that was not calibrated for accuracy, the reading may be inaccurate.

2. Discrimination refers to how finely the basic unit of length of a measuring instrument is divided. The mile on an automobile odometer is divided into 10 parts; therefore, it discriminates to the nearest tenth of a mile. An inch on a micrometer is divided into 1000 or, in some cases, 10,000 parts. Therefore, the micrometer discriminates to .001 or .0001 of an inch. If a measuring instrument is used beyond its discrimination, it will not be reliable.

3. A rule of thumb for choosing which measurement device to use is that the measuring instrument should discriminate 10 times finer than the smallest unit that it will be used to measure. If the total tolerance for the length of a part is .010, we would choose a measuring instrument that could at least measure to 001 (.010/10 = .001) Note that this is not always possible.

4. In a manufacturing facility all measuring tools have their accuracy checked by the quality department. The term calibration means that the tools are calibrated against very accurate standards. If the tools are found to be out of calibration, they are adjusted or replaced. The tools are also logged into a log book and the results of the calibration are recorded along with the next required calibration date. Stickers are applied to each tool to indicate when the tools are required to be calibrated next. Calibration helps assure that a company's measurement equipment is accurate so that they can produce accurate measurements and parts that meet specifications. Workers cannot use any measurement equipment that is beyond the calibration date. Machinists are also responsible for taking good care of measuring equipment and reporting any possible damage to a device so that its accuracy can be checked.

5. A thread go/nogo gage.

6. A go/no go gage could be used to check a hole size. The go end of the gage should go into the hole and the no/go should not. If the go end will not go into the hole the hole is too small. If the nogo end of the gage goes into the hole, the hole is too large.

7. Bores can be measured with dial bore gages, air gages, or with a go/nogo gage.

8. Surface finish can be compared to a surface finish chart or an electronic surface gage can be used to determine the surface finish.

9. A Romer arm is a three-dimensional, portable, coordinate measuring device. The device duplicates and enhances the movement and reach of the human arm. The device can duplicate several degrees of freedom. This means that the Romer arm has six axes.

10. A coordinate measuring machine (CMM) is an extremely accurate instrument that can measure the workpiece in three dimensions. Coordinate measuring machines are useful for determining the location of a part feature relative to a reference plane, line, or point. They are almost a necessity for inspecting complex tolerances.

SECTION C/UNIT 2/SYSTEMS OF MEASUREMENT
Self-Test Answers

1. To find inches knowing mm, multiply mm by .03937:
$$35 \times .03937 = 1.377 \text{ inch.}$$

2. To find mm knowing inches, multiply inches by 25.4:
$$.125 \times 25.4 = 3.17 \text{mm.}$$

3. To find cm knowing inches, multiply inches by 2.54:
$$6.273 \times 2.54 = 15.933 \text{cm.}$$

4. To find mm knowing inches, multiply inches by 25.4:
$$\pm.050 \times 25.4 = \pm1.27 \text{mm.}$$

5. 10 mm = 1cm; therefore, to find cm knowing mm, divide mm by 10.

6. To find inches knowing mm, multiply mm by .03937: .02 × .03937 = .0008 inch. The tolerance would be + or − .0008 in.

7. SI refers to the International System of Units.

8. Conversions between metric and inch systems can be accomplished by mathematical procedures, conversion charts, and direct-converting calculators.

9. 1 yard = 3600/3927 m.

10. Yes, by the use of appropriate conversion dials or digital readouts.

SECTION C/UNIT 3/USING STEEL RULES
Self-Test Answers

Fractional Inch Rules (Figure C-52)

$A = 1\frac{1}{4}$ inches.

$B = 2\frac{1}{8}$ inches.

$C = \frac{15}{16}$ inch.

$D = 2\frac{5}{16}$ inches.

$E = \frac{15}{32}$ inch.

$F = 2\frac{25}{32}$ inches.

$G = \frac{63}{64}$ inch.

$H = 1\frac{59}{64}$ inches.

Decimal Inch Rules (Figure C-53)

$A = .300$ inch.

$B = .510$ inch.

$C = 1.020$ inches.

$D = 1.450$ inches.

$E = 1.260$ inches.

Metric Rules (Figure C-54)

$A = 11$ mm or 1.1 cm

$B = 27$ mm or 2.7 cm

$C = 52$ mm or 5.2 cm

$D = 7.5$ mm or .75 cm

$E = 20.5$ mm or 2.05 cm

$F = 45.5$ mm or 4.55 cm

SECTION C/UNIT 4/USING VERNIER, DIAL, AND DIGITAL INSTRUMENTS FOR DIRECT MEASUREMENTS

Self-Test Answers

Reading Inch Vernier Calipers

Figure C-66a: 1.304 inches

Figure C-66b: .724 inch

Reading Metric Vernier Calipers

Figure C-67a: 29.84 mm

Figure C-67b: 35.62 mm

Reading Inch Vernier Depth Gages

Figure C-68a: .943 inch

Figure C-68b: 1.326 inches

SECTION C/UNIT 5/USING MICROMETERS

Self-Test Answers

1. Anyone taking pride in his/her tools usually takes pride in his/her workmanship. The quality of a product produced depends to a large extent on the accuracy of the measuring tools used. Skilled workers protect their tools, because they guarantee their products.
2. Moisture between the contact faces can cause corrosion. Temperature changes can also cause pressure between the faces.
3. Even small dust particles will change a dimension. Oil or grease attracts small chips and dirt. All these can cause incorrect readings.
4. A measuring tool is no more discriminatory than the smallest division marked on it. This means that a standard micrometer can discriminate to the nearest thousandth of an inch. A Vernier scale on a micrometer will make it possible to discriminate a reading to one ten-thousandth of an inch under controlled conditions.
5. The accuracy of a micrometer depends on the inherent qualities built into it by its maker. Accuracy also depends on the skill of the user and the care the tool receives.
6. The sleeve is stationary in relation to the frame and is engraved with the main scale, which is divided into 40 equal spaces, each equal to .025 inch. The thimble is attached to the spindle and rotates with it. The thimble circumference is graduated with 25 equal divisions, each representing a value of .001 inch.
7. There is less chance of accidentally moving the thimble when reading a micrometer while it is still in contact with the workpiece.
8. Measurements should be made at least twice. On critical measurements, checking the dimensions additional times will ensure that the size measurement is correct.
9. As the temperature of a part is increased, the size of the part will increase. When a part is heated by the machining process, it should be permitted to cool down to room temperature before being measured. Holding a micrometer by the frame for an extended period of time will transfer body heat through the hand and affect the accuracy of the measurement taken.
10. The purpose of the ratchet stop or friction thimble is to enable equal pressure to be repeatedly applied between the measuring faces and the object being measured. Use of the ratchet stop or friction thimble will minimize individual differences in measuring pressure applied by different persons using the same micrometer.

Outside Micrometer Readings

Figure C-108a: .669 inch

Figure C-108b: .787 inch.

Figure C-108c: .237 inch

Figure C-108d: .994 inch

Figure C-108e: .072 inch

Depth Micrometer Readings

Figure C-121a: .535 inch

Figure C-121b: .815 inch

Figure C-121c: .732 inch

Figure C-121d: .938 inch

Figure C-121e: .647 inch

Vernier Micrometer Readings

Figure C-128a: .3749 inch

Figure C-128b: .5377 inch

Figure C-128c: .3123 inch

Figure C-128d: .1255 inch

Figure C-128e: .2498 inch

Inside Micrometer Readings

Figure C-115a: 1.617

Figure C-115b: 2.000

Figure C-115c: 2.254

Figure C-115d: 2.784

Figure C-115e: 2.562

Metric Micrometer Readings

Figure C-126a: 21.21 mm

Figure C-126b: 13.27 mm

Figure C-126c: 9.94 mm

Figure C-126d: .559 mm

Figure C-126e: 4.08 mm

SECTION C/UNIT 6/USING COMPARISON MEASURING INSTRUMENTS

Self-Test Answers

1. In comparison measurement, the measurement of an unknown dimension is compared with a known dimension. This often involves a transfer device that represents the unknown and is then transferred to the known, where the reading can be determined.
2. Cosine error is error incurred when misalignment exists between the axis of measurement and the axis of the measuring instrument.
3. Cosine error can be reduced by making sure that the axis of the measuring instrument is exactly in line with the axis of measurement.
4. One way to measure the slot would be with an adjustable parallel and a micrometer.
5. One way to do it would be with a dial test indicator in conjunction with a height gage. It could also be done with a test indicator and planer gage.
6. An optical comparator could be used.
7. A combination square can be checked against a precision square, or if the actual amount of deviation is required, the cylindrical square or micrometer square could be used.
8. A telescoping gage and outside micrometer could be used.

SECTION C/UNIT 7/USING GAGE BLOCKS
Self-Test Answers

1. The wringing interval is the space or interface between wrung gage blocks.
2. Wear blocks are made from hard material such as tungsten carbide. Wear blocks are used in applications where direct contact with gage blocks might damage them.
3. If gage blocks should become heated or cooled above or below room temperature, normalizing is the process of returning them to room temperature.
4. Grade 1: + or − .000002 inch

    Grade 2: + .000004 inch

    − .000002 inch

    Grade 3: + .000008 inch

    − .000002 inch

5. The conditioning stone is a highly finished piece of granite or ceramic material and is used to remove burrs from the wringing surface of a gage block.
6. A microinch is 1 millionth of an inch. On surface finish, it refers to the deviation of a surface from a uniform plane.
7. Gage block accuracy depends on the following factors:

    extreme cleanliness,

    no burrs,

    minimum use of the conditioning stone,

    leaving stacks assembled only for minimum amounts of time,

    cleaning before storage,

    application of gage block preservative.

8. 3.0213
    <u>.1003</u>
    2.9210
    <u>.121 </u>
    2.800
    <u>.800 </u>
    2.000
    <u>2.000 </u>
    0.0000

9. 1.9643
    <u>.100 </u>   (wear blocks 2 × .050)
    1.8643
    <u>.1003</u>
    1.7640
    <u>.114 </u>
    1.6500
    <u>.650 </u>
    1.0000
    <u>1.000 </u>
    0.000

10. Gage blocks can be used to check other measuring instruments (calibration), to set sine bars for angles, as precision height gages for layout, in direct gaging applications, and for setting machine and cutting tool positions.

SECTION C/UNIT 8/USING ANGULAR MEASURING INSTRUMENTS
Self-Test Answers

1. Plate protractor and machinist's combination set bevel protractor.
2. Five minutes of arc.
3. The sine bar becomes the hypotenuse of a right triangle. Angles are measured or established by elevating the bar a specified amount or calculating the amount of bar elevation, knowing the angle.
4. Figure C-210a: 56°20′

    Figure C-210b: 50°

    Figure C-210c: 34°30′

    Figure C-210d: 96°15′

    Figure C-210e: 61°45′

5. Bar elevation = bar length × sine of angle desired

    = 5 inches × sin 37°

    = 5 × .6018

    = 3.0090 inches

6. Sine of the angle desired = elevation/bar length

    = 2.750/5

    Sine of angle = .550

    Angle = 33°22′

7. The 10-inch bar requires stacks twice as high as the 5-inch bar for equivalent angles.
8. 2.6124 inches.
9. 1.8040 inches.
10. 30 degrees.

SECTION C/UNIT 9/QUALITY IN MANUFACTURING
Self-Test Answers

1. ISO 9001 is an international standard that specifies requirements for a quality management system (QMS). Organizations can use the ISO standard to demonstrate the ability to consistently provide products and/or services that meet customer and regulatory requirements.
2. False
3. False
4. True
5. True
6. False
7. True
8. False
9. True
10. False
11. Attribute is the simplest type of data. The product either has the characteristic (attribute) or it doesn't. If blue is

the desired attribute and the product is chairs, we would have two piles of products after we inspected them. One pile would have the desired attribute (blue), and the other pile would have chairs of any other color. Attribute data can also be go/no-go type data. Go/no-go gages are often used to check hole sizes. If the go end of the gage fits in the hole and the no-go end doesn't, we know the hole is within tolerance.

12. Variable data can be much more valuable. It not only tells us *whether* parts are good or bad, it tells us how good or how bad they are. Variable data are generated using measuring instruments such as micrometers, verniers, and indicators. If we are measuring with these types of instruments, we generate a range of sizes, not just two (good or bad) as

with attribute data. We could end up with many piles of part sizes. Variable data can also be called *analog data* because it can assume a range of sizes, not just good or bad.

13. Make sure that the data is accurate. Consider the gages, methods, and personnel. Clarify the purpose of collecting the data. Everyone involved should realize that the purpose is quality improvement. We are not collecting data to make people work harder or get them in trouble. Take action based on the data. When we have learned statistical methods, we will only make changes or adjustments to processes based on the data.

14. Numbers can become difficult to work with when they have many digits. Coding is a method of representing a size by typically a single digit.

15. A histogram is a graphical representation of data.

Blueprint Specification $= 1.126$				
$1.124 = -2$	$1.128 = 2$	$1.123 = -3$	$1.127 = 1$	$1.121 = -5$
$1.119 = -6$	$1.127 = 1$	$1.125 = -1$	$1.119 = -7$	$1.127 = 1$
$1.129 = 3$	$1.126 = 0$	$1.118 = -8$	$1.121 = -5$	$1.116 = -10$

Blueprint Specification $= 1.2755$			
$1.2752 = -3$	$1.2749 = -6$	$1.2752 = -3$	$1.2754 = -1$
$1.2759 = 4$	$1.2750 = -5$	$1.2761 = 6$	$1.2752 = -3$
$1.2748 = -7$	$1.2756 = 1$	$1.2752 = -3$	$1.2756 = 1$
$1.2753 = -2$	$1.2755 = 0$	$1.2747 = -8$	$1.2749 = -6$

Blueprint Specification $= 2.105$			
$2.109 = 4$	$2.103 = -2$	$2.108 = 3$	$2.113 = 8$
$2.102 = -3$	$2.101 = -4$	$2.096 = -9$	$2.104 = -1$
$2.101 = -4$	$2.100 = -5$	$2.100 = -5$	$2.111 = 6$
$2.098 = -7$	$2.100 = -5$	$2.099 = -6$	$2.109 = 4$

Blueprint Specification $= 5.00$				
$5.03 = 3$	$5.06 = 6$	$5.02 = 2$	$5.02 = 2$	$5.07 = 7$
$5.02 = 2$	$5.05 = 5$	$5.01 = 1$	$5.04 = 4$	$5.03 = 3$
$4.98 = -2$	$4.97 = -3$	$5.00 = 0$	$5.01 = 1$	$4.93 = -7$

16.

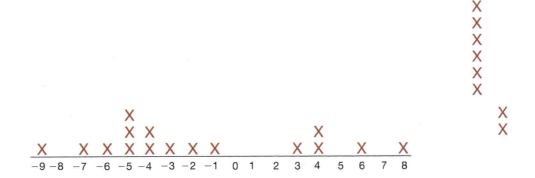

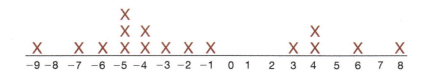

17. The first rule of variation is straightforward: *No two things are exactly alike.* Rule number two, *variation can be measured*, is also straightforward. No matter what our product or process, no two will be alike (rule 1), and we will be able to measure and find differences in every part (rule 2). The third rule of variation is that *individual outcomes are not predictable.* No, because we cannot predict individual outcomes. Rule number four states that *groups form patterns with definite characteristics.*

18. The amount of salt dropped also varied. Could we expect a person to drop exactly 3 ounces of salt every time? The amount of salt is also an assignable cause. How could we remove this cause of variation? Statistical methods can be used to identify assignable causes of variation. Chance causes of variation always exist. Chance causes of variation are those minor reasons which make processes vary. We cannot quantify or even identify all of the chance causes.

19. a. mean—Mean is another word for average.

   b. histogram—Graphical representation of data.

   c. normal distribution—A normal distribution looks like a bell curve.

   d. bell curve—If a distribution of part sizes is normal, a histogram of the data would look like a bell shape (bell curve).

   e. range—Difference between the largest and smallest value in a sample.

   a. Code the data. (Hint: 1 should equal 2.251.)

   b. 2.48

a. Code the data. (Hint: 2.0005 = 5 .)

b. 2.26

c. 3.61

20.

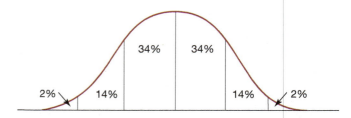

21. a. Calculate the mean. .14

   b. Calculate the sample standard deviation. 2.03

22. a. 1.77

   b. 3.65

23. a. Calculate the mean. .05

   b. Calculate the sample standard deviation. 2.24

24. a. Draw a bell curve.

   b. Draw lines where the six standard deviations would be.

   c. Label them with actual sizes from this process. (Hint: 99.7 percent of all parts should lie between −1 and +11.)

   d. Label the percentages.

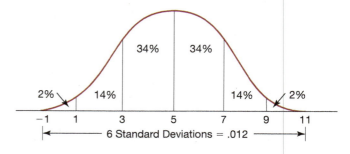

Subgroup 1	Subgroup 2	Subgroup 3
2.0005 = 5	2.0005 = 5	2.0002 = 2
1.9997 = −3	2.0004 = 4	2.0004 = 4
2.0009 = 9	2.0001 = 1	2.0000 = 0
2.0006 = 6	2.0004 = 4	1.9995 = −5
1.9999 = −1	2.0002 = 2	2.0001 = 1

1	2	3	4	5	6
2.252 = 2	2.249 = −1	2.248 = −2	2.252 = 2	2.246 = −4	2.250 = 0
2.252 = 2	2.249 = −1	2.254 = 4	2.253 = 3	2.248 = −2	2.249 = −1
2.253 = 3	2.243 = −7	2.252 = 2	2.251 = 1	2.247 = −3	2.252 = 2
2.250 = 0	2.248 = −2	2.251 = 1	2.250 = 0	2.246 = −4	2.248 = −2
2.252 = 2	2.249 = −1	2.247 = −3	2.250 = 0	2.249 = −1	2.251 = 1

25. a.

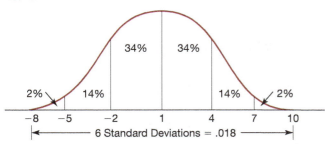

b. 140

c. 680

d. 180

26. a.

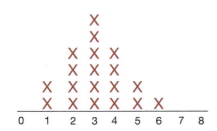

b. Yes, it does look like a normal distribution.

27. a.

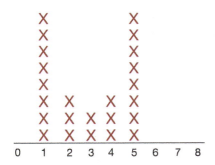

b. This one does not look like a normal distribution. This means that something was wrong with our process. Data from a normal process should be normally distributed.

SECTION C/UNIT 10/STATISTICS IN MANUFACTURING

Self-Test Answers

1. There is a very small chance that seven in a row would be on one side of the centerline. A process can be adjusted if seven in a row fall on one side of a centerline. Assume seven part sizes in a row fall below center. The odds are so low that this could happen that we could assume that something has changed in the process. This is the second rule: if seven fall on one side of the centerline, the process has changed and an adjustment or change is necessary. If the average of the seven sizes was calculated, it would give the exact adjustment needed. The chart not only tells when to adjust, but also how much to adjust.

In other words, if seven in a row are above or below center, it means that the mean (average) has shifted up or down.

The only other rule is that a process has changed if seven in a row increase or decrease. This is called a *trend*. If each of the seven in a row gets larger (or smaller), this trend means that the process has changed.

2.
   - Reduces unnessacary adjustmens.
   - Reduce variation in parts
   - Can predict needed maintenance
   - Higher quality at lower cost
   - Helps a company understand the capability of their machines
   - Helps bid on jobs and machine assignment

3. 99.7%

4. No adjustments should be made during the study. Because we are trying to study the process, it is imperative that we not change the process through adjustments during the study. We need accurate data on how good the process is, not the operator.

5.
   a. Calculate the CP.

   Six standard deviations on this machine would be .018. The tolerance for the job is a total of .012. The capability would only be 66%. The would be a poor job and would result in scrap being run.

   b. Will there be scrap? If so, how much? (Hint: Draw a bell curve to help find the answer.)

   There will be approximately 4% scrap.

6.
   a. Calculate the CP.

   b. CP = .020/.018 = 1.11

   c. Will there be scrap? If so, how much?

   d. There should not be scrap.

7. CP = 1.11 B/C

8. CP = .66 Poor

9. No this job will produce at a minimum about 4% scrap.

   They could bid high knowing they would have to inspect and reject parts. They could change the process to make it more capable. They could contact the customer and see if the tolerance can be larger. They could run the par on a more capable machine.

10.

Subgroup 1	Subgroup 2	Subgroup 3	Subgroup 4	Subgroup 5
2.252 = 2	2.249 = −1	2.248 = −2	2.252 = 2	2.246 = −4
2.252 = 2	2.249 = −1	2.254 = 4	2.253 = 3	2.248 = −2
2.253 = 3	2.243 = −7	2.252 = 2	2.251 = 1	2.247 = −3
2.250 = 0	2.248 = −2	2.251 = 1	2.250 = 0	2.246 = −4
2.252 = 2	2.249 = −1	2.247 = −3	2.250 = 0	2.249 = −1

11. **See the chart for the answers.**
   a. Standard deviation 2.64
   b. Process mean
   c. Average range
   d. Is the data normally distributed? (Check with a histogram.)

   The histogram shows that the process is basically normal. It does show some evidence of a bimodal distribution. It would be wise to run some additional parts and check again.

   e. Upper control limit averages
   f. Lower control limit averages
   g. Upper control limit range

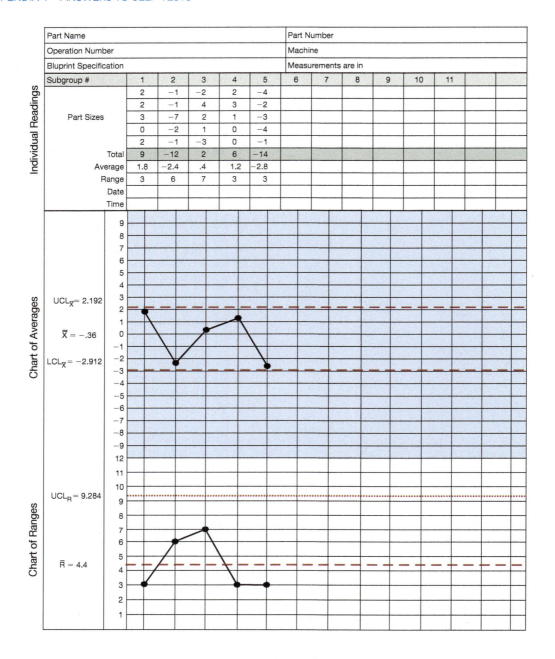

Chart form details:

Part Name						Part Number						
Operation Number						Machine						
Bluprint Specification						Measurements are in						

Individual Readings

Subgroup #	1	2	3	4	5	6	7	8	9	10	11		
Part Sizes	2	−1	−2	2	−4								
	2	−1	4	3	−2								
	3	−7	2	1	−3								
	0	−2	1	0	−4								
	2	−1	−3	0	−1								
Total	9	−12	2	6	−14								
Average	1.8	−2.4	.4	1.2	−2.8								
Range	3	6	7	3	3								
Date													
Time													

Chart of Averages

UCL$_{\bar{X}}$ = 2.192
$\bar{\bar{X}}$ = −.36
LCL$_{\bar{X}}$ = −2.912

Chart of Ranges

UCL$_R$ = 9.284
$\bar{R}$ = 4.4

12.

Subgroup 1	Subgroup 2	Subgroup 3	Subgroup 4	Subgroup 5
1.253 = 3	1.254 = 4	1.251 = 1	1.249 = −1	1.247 = −3
1.250 = 0	1.249 = −1	1.245 = −5	1.250 = 0	1.251 = 1
1.247 = −3	1.250 = 0	1.251 = 1	1.256 = 6	1.252 = 2
1.251 = 1	1.253 = 3	1.248 = −2	1.252 = 2	1.251 = 1
1.248 = −2	1.249 = −1	1.252 = 2	1.248 = −2	1.245 = −5

Use with question 12.

See the chart for answers.
a. Code the data.
b. Is the process normal?
   A histogram shows that it is relatively normal.
c. Mean.
d. Average range.
e. Complete an XR chart (enter the data, calculate control limits, and plot the data).
f. Compute the capability.
   Standard Deviation = 2.66
   Total tolerance = .006
   .006/15.96 = .38% —Very poor job.

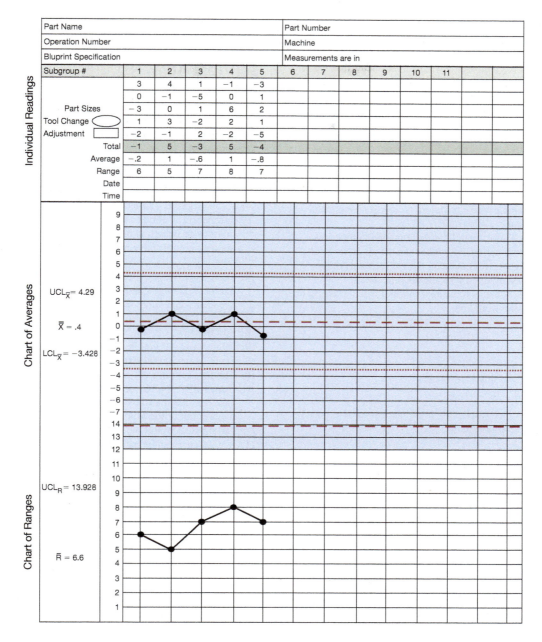

		Subgroup #	1	2	3	4	5	6	7	8	9	10	11		
			3	4	1	−1	−3								
			0	−1	−5	0	1								
		Part Sizes	−3	0	1	6	2								
		Tool Change ⬭	1	3	−2	2	1								
		Adjustment ▢	−2	−1	2	−2	−5								
		Total	−1	5	−3	5	−4								
		Average	−.2	1	−.6	1	−.8								
		Range	6	5	7	8	7								
		Date													
		Time													

Part Name / Part Number
Operation Number / Machine
Bluprint Specification / Measurements are in

Chart of Averages:
UCL$_{\bar{X}}$ = 4.29
$\bar{\bar{X}}$ = .4
LCL$_{\bar{X}}$ = −3.428

Chart of Ranges:
UCL$_R$ = 13.928
$\bar{R}$ = 6.6

SECTION D/UNIT 1/SELECTION AND IDENTIFICATION OF STEELS

Self-Test Answers

1. Carbon and alloy steels are designated by the numerical SAE or AISI system. The UNS system is also used.
2. The three basic types of stainless steels are martensitic (hardenable) and ferritic (nonhardenable), both magnetic and of the 400 series, and austenitic (nonmagnetic and nonhardenable, except by work hardening) of the 300 series.
3. Each piece would be identified as follows:
   a. AISI C1020 CF is a soft, low-carbon steel with a dull metallic luster surface finish. Use the observation test, spark test, and file test for hardness.
   b. AISI B1140 (G and P) is a medium-carbon, resulfurized, free-machining steel with a shiny finish. Use the observation test, spark test, and machinability test.
   c. AISI C4140 (G and P) is a chromium–molybdenum alloy of medium carbon content with a polished, shiny finish. Because an alloy steel is harder than a similar carbon- or low-carbon-content steel, a hardness test such as the file or scratch test should be used to compare with known samples. The machinability test would be useful as a comparison test.
   d. AISI 8620 HR is a tough low-carbon steel used for carburizing purposes. A hardness test and a machinability test would immediately show the difference from low-carbon hot-rolled steel.
   e. AISI B1140 (ebony) is the same as the resulfurized steel in part b, but the finish is different. The test would be the same as for part b.
   f. AISI D1040 is a medium-carbon steel. The spark test would be useful here, as well as the hardness and machinability tests.
4. A magnetic test can quickly determine whether it is a ferrous metal or perhaps nickel. If the metal is white in color, a spark test will be needed to determine whether it is a nickel casting or one

of white cast iron, because they are similar in appearance. If a small piece can be broken off, the fracture will show whether it is white or gray cast iron. Gray cast iron will leave a black smudge on the finger. If it is cast steel, it will be more ductile than cast iron, and a spark test should reveal a smaller carbon content.

5. O1 refers to an alloy-type oil-hardening (oil-quench) tool steel. W1 refers to a water-hardening (water-quench) tool steel.

6. The 40-inch-long, $2\frac{7}{16}$-inch-diameter shaft weighs 1.322 lb/ inches. The cost is \$.95/lb; $1.322 \times 40 \times .95 = 50.236$ cost of the shaft.

7. **a.** No.
   b. Hardened tool steel or case-hardened steel.

8. Austenitic (having a face-centered cubic unit cell in its lattice structure). Examples are chromium, nickel, stainless steel, and high-manganese alloy steel.

9. Nickel is a nonferrous metal that has magnetic properties. Some alloy combinations of nonferrous metals make strong permanent magnets, for example, the well-known Alnico magnet, an alloy of aluminum, nickel, and cobalt.

10. Some properties of steel to be kept in mind when ordering or planning for a job would be

    strength,

    machinability,

    hardenability,

    weldability (if welding is involved),

    fatigue resistance,

    corrosion resistance (especially if the piece is to be exposed to a corrosive atmosphere).

SECTION D/UNIT 2/SELECTION AND IDENTIFICATION OF NONFERROUS METALS

Self-Test Answers

1. Because aluminum is about one-third less dense than steel, it is used extensively in aircraft. It also forms an oxide on the surface that resists further corrosion. The initial cost is much greater. Some higher-strength aluminum alloys cannot be welded.

2. The letter *H* following the four-digit number always designates strain or work hardening. The letter *T* refers to heat treatment.

3. Magnesium is approximately one-third less dense than aluminum and is approximately one-fourth as dense as steel. Magnesium will burn in air when finely divided.

4. Copper is most extensively used in the electrical industries because of its low resistance to the passage of current when it is unalloyed with other metals. Copper can be strain hardened or work hardened, and certain alloys may be hardened by a solution heat-treat and aging process.

5. Bronze is basically copper and tin. Brass is basically copper and zinc.

6. Nickel is used to electroplate surfaces of metals for corrosion resistance, and as an alloying element with steels and nonferrous metals.

7. All three resist deterioration from corrosion.

8. 2024-T6 is an aluminum–copper alloy, solution heat treated and artificially aged. Cast aluminum alloys generally have lower tensile strength than wrought alloys.

9. 5056-H18 is an aluminum–magnesium alloy, strain hardened to a full hard temper.

10. Die-cast metals.

SECTION D/UNIT 3/HARDENING, CASE HARDENING, AND TEMPERING

Self-Test Answers

1. No hardening would result, as 1200°F (649°C) is less than the lower critical point, and no dissolving of carbon has taken place.

2. There would be almost no change. For all practical purposes in the shop, these low-carbon steels are not considered hardenable.

3. Air- and oil-hardening steels are not so subject to distortion and cracking as W1 steels, and they are deep hardening.

4. 1,450°F (788°C), 50°F (10°C) above the upper critical limit.

5. Tempering is done to remove the internal stresses in martensite, which is brittle. The temperature used gives the best compromise between hardness and toughness or ductility.

6. Tempering temperature should be specified according to the hardness, strength, and ductility desired. Mechanical properties charts give these data.

7. 525°F (274°C). Purple.

8. 600°F (315°C). It would be too soft for any cutting tool.

9. Immediately. If you let it set for any length of time, it may crack from internal stresses.

10. The low-carbon steel core does not harden when quenched from 1,650° (899°C), so it remains soft and tough, but the case becomes hard. No tempering is therefore required, as the piece is not brittle all the way through, as a fully hardened carbon steel would be.

11. A deep case can be made by pack carburizing or by a liquid bath carburizing. A relatively deep case is often applied by nitriding or by similar procedures.

12. Three methods of introducing carbon into heated steel are roll, pack, and liquid carburizing.

13. The surface decarburizes or loses surface carbon to the atmosphere as it combines with oxygen to form carbon dioxide.

14. Dispersion of carbon atoms in the solid solution of austenite may be incomplete, so little or no hardening in the quench takes place as a result. Also, the center of a thick section takes more time to come to the austenitizing temperature.

15. Circulation or agitation breaks down the vapor barrier. This action allows the quench to proceed at a more rapid rate.

16. The furnace control display.

17. They run from the surface toward the center of the piece. The fractured surfaces usually appear blackened. The surfaces have a fine crystalline structure.

18. Overheating.

    Wrong quench media.

    Wrong selection of steel.

    Poor part design.

    Time delays between quench and tempering.

    Wrong angle into the quench.

    Not enough material to grind off decarburization.

19. Controlled atmosphere furnace.

    Wrapping the piece in stainless steel foil.

    Covering with cast iron chips.

20. An air-hardening tool steel should be used when distortion must be kept to a minimum.

SECTION D/UNIT 4/ANNEALING, NORMALIZING, AND STRESS RELIEVING

Self-Test Answers

1. Medium-carbon steels that are not uniform and have hardened areas from welding or prior heat treating need to be normalized so they can be machined. Forgings, castings, and tool steel in the as-rolled condition are normalized before any further heat treatments or machining is done.
2. 1,550°F (843°C), 50°F (10°C) above the upper critical limit.
3. The spheroidization temperature is quite close to the lower critical temperature line, about 1,300°F (704°C).
4. The full anneal brings carbon steel to its softest condition as all the grains are re-formed (recrystallized), and any hard carbide structures become soft pearlite as it slowly cools. Stress relieving will recrystallize only distorted ferrite grains and not the hard carbide structures or pearlite grains.
5. Stress relieving should be used on severely cold-worked steels or for weldments.
6. High-carbon steels (.8 to 1.7 percent C).
7. Process annealing is used by the sheet and wire industry and is essentially the same as stress relieving.
8. In still air.
9. Slowly. Packed in insulating material or cooled in a furnace.
10. Low-carbon steels tend to become gummy when spheroidized, so the machinability is worse than in the as-rolled condition. Spheroidization sometimes is desirable when stress relieving weldments on low-carbon steels.

SECTION D/UNIT 5/ROCKWELL AND BRINELL HARDNESS TESTERS

Self-Test Answers

1. Resistance to penetration is the one category utilized by the Rockwell and Brinell testers. The depth of penetration is measured when the major load is removed on the Rockwell tester, and the diameter of the impression is measured to determine a Brinell hardness number.
2. As the hardness of a metal increases, the strength increases.
3. The A scale and a Brale marked "A" with a major load of 60 kgf should be used to test a tungsten carbide block.
4. It would become deformed or flattened and give an incorrect reading.
5. No. The Brale used with the Rockwell superficial tester is always marked or prefixed with the letter N.
6. False. The ball penetrator is the same for all the scales using the same diameter ball.
7. The diamond spot anvil is used for superficial testing on the Rockwell tester. When used, it does not become indented, as is the case when using the spot anvil.
8. Roughness will give less-accurate results than would a smooth surface.
9. The surface "skin" would be softer than the interior of the decarburized part.
10. A curved surface will give inaccurate readings.
11. A 3,000-kg weight would be used to test specimens on the Brinell tester.
12. A 10-mm steel ball is usually used on the Brinell tester.

SECTION E/UNIT 2/BASIC SEMIPRECISION LAYOUT PRACTICE

Self-Test Answers

1. The workpiece should have all sharp edges removed by grinding or filing. A thin, even coat of layout dye should be applied.
2. The towel will prevent spilling layout dye on the layout plate or layout table. The layout die can stain a layout table.
3. The punch should be tilted so that it is easier to see when the point is located on the scribe mark. It should then be moved to the upright position before it is tapped with the layout hammer.
4. The combination square can be positioned on the rule for measurements. The square head acts as a positive reference point for measurements. The rule may be removed and used as a straight edge for scribing.
5. The divider should be adjusted until you feel the tip drop into the rule engraving.
6. Pocket, engineer's, machinist's.
7. A very sharp edge may be ground on the tool bit scriber, which allows it to be placed right next to a rule for maximum scribing accuracy.
8. A magnified crosshair may be located with great accuracy.
9. The tips of the divider will fit into the rule engravings.
10. The prick punch has a much sharper tip (30 degrees) as opposed to the center punch tip (90 degrees).

SECTION E/UNIT 3/BASIC PRECISION LAYOUT PRACTICE

Self-Test Answers

1. Zero reference is checked by bringing the scriber to rest on the reference surface and then checking the alignment of the beam and Vernier zero lines.
2. The position of the Vernier scale may be adjusted on height gages with this feature.
3. By turning the workpiece 90 degrees.
4. 18 inches.
5. The sine bar.

SECTION F/UNIT 1/MACHINABILITY AND CHIP FORMATION

Self-Test Answers

1. At lower speeds, negative-rake tools usually produce a poorer surface finish than do those with positive rakes.
2. It is called a built-up edge and causes a rough, ragged cutting action that produces a poor surface finish.
3. Thin uniform chips indicate the least surface disruption.
4. No. It slides ahead of the tool on a shear plane, elongating and altering the grain structure of the metal.
5. Negative rake.
6. Higher cutting speeds produce better surface finishes. There also is less disturbance and disruption of the grain structure at higher speeds.
7. Surface irregularities such as scratches, tool marks, microcracks, and poor radii can shorten the working life of the part considerably by causing stress concentration that can develop into a metal fatigue failure.

8. The property of hardness is related to machinability. Machinists sometimes use a file to determine hardness.
9. The 9-shaped chip.
10. Machinability is the relative difficulty of a machining operation with regard to tool life, surface finish, and power consumption.

SECTION F/UNIT 2/SPEEDS AND FEEDS FOR MACHINE TOOLS

Self-Test Answers

1. The rpm for the $\frac{1}{8}$-inch twist drill should be 2,880, and for the $\frac{3}{4}$-inch drill, 480.
2. They would be the same.
3. Use the next-lower speed setting or the highest setting on the machine if the calculated speed is higher than the top speed of the machine.
4. 100 rpm.
5. The rpm should be 3,000.
6. Inches per revolution (ipr).
7. .010 ipr.
8. Milling machines, both vertical and horizontal.

SECTION F/UNIT 3/CUTTING FLUIDS

Self-Test Answers

1. Cooling and lubrication.
2. Synthetic and semisynthetic fluids, emulsions, and cutting oils.
3. Remove the cutting fluid, clean the tank, and replace with new cutting fluid.
4. Because the tramp oil will contaminate the soluble oil–water mix.
5. They tend to remove skin oils. The "oilier" types do not cause this problem.
6. Using a pump oiler or brush with the appropriate cutting fluid.
7. Because the rapidly spinning grinding wheel tends to blow the coolant away.
8. Because only a small amount of liquid that contains the fatty particles ever reaches the cutting area.

SECTION F/UNIT 4/CARBIDE TOOLING SPECIFICATION AND SELECTION

Self-Test Answers

1. Hard and brittle.
2. Carbide tools come in a variety of grades. The grade is based on the carbide's wear resistance and toughness. As an insert becomes harder or more wear resistant, it becomes brittle (less tough).
3. Carbide is a very hard, durable cutting tool, but it still wears. The wear resistance of cemented carbide can be greatly increased by using coated carbide inserts. Wear-resistant coatings can be applied to the carbide substrate (base material) through the use of plasma coating or vapor deposition. The coating is very thin but very hard. The most common types of coatings include titanium carbide (TiC), titanium nitride (TiN), and aluminum oxide (AlO). Aluminum oxide is a very wear-resistant coating used in high-speed finish cuts and light roughing cuts on most steels and all cast irons. Titanium nitride coatings are very hard and have the strength characteristics to perform well under heavy rough-cutting conditions. All three coatings will perform well on most steels, as well as on cast iron.

4. The most common types of coatings include titanium carbide (TiC), titanium nitride (TiN), and aluminum oxide (AlO).
5.

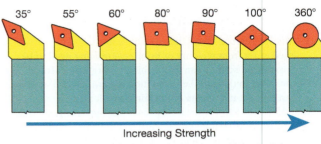

35°   55°   60°   80°   90°   100°   360°

Increasing Strength

6. Cobalt acts as the binder that holds the carbide particles together.
7. The nose radius of the tool directly affects tool strength and surface finish, as well as cutting speeds and feeds. The larger the nose radius, the stronger the tool. If the tool radius is too small, the sharp point will make the surface finish unacceptable, and the life of the tool will be shortened. Larger nose radii will give a better finish and longer tool life and will allow for higher feed rates. If the tool nose is too large, it can cause chatter. It is usually best to select an insert with a tool-nose radius as large as the machining operation will allow.
8. Shape of the part and strength of the insert.
9. The size of the insert is based on the inscribed circle (IC), which is the largest circle that will fit inside the insert.
10. The side relief angle, also known as the side rake angle, is formed by the top face of the cutting tool and side cutting edge. The angle is measured in the amount of relief under the cutting edge

Top View

Side View

Side Relief Angle

Neutral Rake   Positive Rake   Negative Rake

11. Increasing the lead angle will reduce tool breakage when roughing or cutting interrupted surfaces.

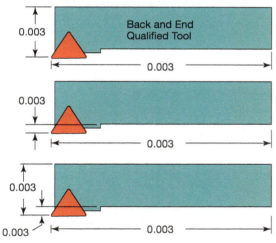

Strongest Point   Weakest Point

Positive   Negative

12. Tools that are used in CNC machines are machined to a high level of accuracy. Qualified tools are typically guaranteed to be within .003 of an inch.

The accuracy of the cutting tip is referenced to specific points or datums located on the holders. The higher level of accuracy enables the operator to change inserts without having to remeasure the tools. The figure below shows an example of where the measurements are qualified from.

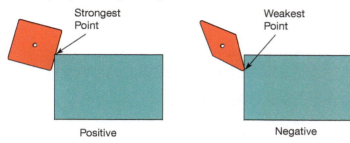

0.003

Back and End Qualified Tool

0.003

0.003

0.003

0.003

0.003

0.003

0.003

13. Triangular shape, positive relief, hole and chip breaker, 1/2″ IC, 3/16 thick, 1/32 corner radius.

14. 80-degree diamond shape, negative relief, M specifies the tolerances, hole and chip breaker, 3/8 IC, 3/8″ thick, 1/32 corner radius.

15. Circular shape, negative relief, M specifies the tolerances, hole and chip breaker, 1/2″ IC, 3/16″ thick, 1/32 corner radius.

16. Square shape, positive relief, hole and chip breaker, 1/2″ IC, 3/16″ thick, 1/32 corner radius.

17. Interrupted cuts can be a problem and should be avoided; use a softer or tougher grade of insert. Increasing the cutting speed and lowering the feed rate will sometimes eliminate chipping.

18. I believe that you are experiencing excessive edge wear, the probable cause is friction. Excessive friction causes heat to build along the cutting edge, which causes the binders to fail. One possible

cause is that the lead angle is too great. Choose a holder that reduces the lead angle. Check the tool height. A crash or bump of the tool turret may be causing the tool to be too high. Another possible cause may be that the feed rate is too low. Increasing the feed rate will cause the chips to concentrate away from the cutting edge. Finally, it could be a grade selection problem.

19. The major causes of insert breakage are a lack of rigidity, too hard of an insert grade, and low operating conditions. When making roughing cuts through hard spots or sand inclusions, use a tougher, not harder, grade of carbide.

20. The type of tool failure that is occurring is called depth-of-cut notching. Depth of cut notching is an unnatural chipping away of the insert right at the depth of cut line. Depth-of-cut notching is usually a grade selection problem. If you are using an uncoated insert, consider changing to a coated insert. Depth-of-cut notching may also be solved by lowering the feed rate and/or by reducing the lead angle.

SECTION G/UNIT 1/TYPES OF CUTOFF MACHINES AND SAFETY

Self-Test Answers

1. Horizontal cutoff saw.
2. The ability to cut stock on an angle.
3. A cold saw uses a circular blade to cut metal. They are fast and accurate.
4. Round stock should not be cut on a vertical saw unless the stock is held securely in a vise or other holding device.
5. Waterjets and lasers.

SECTION G/UNIT 2/USING HORIZONTAL CUTOFF MACHINES

Self-Test Answers

1. Raker, wave, and straight.
2. Workpiece material, cross-sectional shape, and thickness.
3. On the back stroke.
4. The tooth offset on either side of the blade. Set provides clearance for the back of the blade.
5. Standard, skip, and hook.
6. Cutoff material can bind the blade and destroy the set.
7. The horizontal band saw.
8. Cooling, lubrication, and chip removal.
9. Scoring and possible blade breakage.
10. The workpiece must be turned over and a new cut started.

SECTION G/UNIT 3/PREPARING A VERTICAL BAND SAW FOR USE

Self-Test Answers

1. The ends of the blade should be ground with the teeth opposed. This will ensure that the ends of the blade will match and be in line.
2. The blade ends are placed in the welder with the teeth pointed in. The ends must contact squarely in the gap between the jaws. The welder must be adjusted for the band width to be welded. You should wear eye protection and stand to one side during the welding operation. The weld will occur when the weld lever is depressed.

3. The weld is ground on the grinding wheel attached to the welder. Grind the weld on both sides of the band until the band fits the thickness gage. Be careful not to grind the saw teeth.
4. The guides support the band. This is essential to straight cutting.
5. Band guides must fully support the band except for the teeth. A wide guide used on a narrow band will destroy the saw set as soon as the machine is started.
6. Annealing is the process of softening the band weld to improve strength qualities.
7. The band should be clamped in the annealing jaws with the teeth pointed out. A small amount of compression should be placed on the movable welder jaw prior to clamping the band. The correct annealing color is dull red. As soon as this color is reached, the anneal switch should be released and then operated briefly several times to slow the cooling rate of the weld.
8. Band tracking is the position of the band as it runs on the wheels.
9. Band tracking is adjusted by tilting the idler wheel until the band just touches the backup bearing.

SECTION G/UNIT 4/USING A VERTICAL BAND SAW
Self-Test Answers

1. The three sets are straight, wave, and raker. Straight set may be used for thin material, wave for material with a variable cross section, and raker for general-purpose sawing.
2. Scalloped and wavy edged bands might be used on nonmetallic material where blade teeth would tear the material being cut.
3. Band velocity is measured in feet-per-minute.
4. The upper guidepost must be adjusted so that it is as close to the workpiece as possible.
5. Band pitch must be correct for the thickness of material to be cut. Generally, a fine pitch will be used on thin material. Cutting a thick workpiece with a fine-pitch band will clog saw teeth and reduce cutting efficiency.

SECTION H/UNIT 1/DRILL PRESS FUNDAMENTALS
Self-Test Answers

1. The upright and radial-arm drill presses are the two basic types. The upright is a similar but heavy-duty drill press equipped with power feed. The radial-arm drill press allows the operator to position the drill over the work where it is needed, rather than to position the work under the drill, as with other drill presses.

    The upright and radial-arm drill presses all perform much the same functions of drilling, reaming, counterboring, countersinking, spotfacing, and tapping, but the upright and radial machines do heavier and larger jobs. The radial-arm drill can support large, heavy castings and work can be done on them without the workpiece being moved.
2. Sensitive drill press:

Spindle, A	Base, F
Quill lock handle, G	Power feed, C
Column, E	Motor, J
Switch, L	Variable-speed control, D
Depth stop, B	Table lift crank, K
Head, N	Quill return spring, I
Table, O	Guard, M
Table lock, H	

3. Radial drill press:

Column, B	Base, D
Radial arm, C	Drill head, E
Spindle, A	

SECTION H/UNIT 2/DRILLING TOOLS
Self-Test Answers

1.

Web, T	Chisel edge angle, B
Margin, U	Body clearance, J
Drill point angle, D	Helix angle, I
Cutting lip, P	Axis of drill, Y
Flute, K	Shank length, N
Body, O	Tang, C
Lip relief angle, E	Taper shank, X
LAND, G	Straight shank, W

2.

Decimal Diameter	Fractional Size	Number Size	Letter Size	Metric Size
(a) .07811	$\frac{5}{62}$			
(b) .1495		25		
(c) .272			1	
(d) .159		21		
(e) .1969				5
(f) .323			P	
(g) .3125	$\frac{5}{16}$			
(h) .4375	$\frac{7}{16}$			
(i) .201		7		
(j) .1875	$\frac{3}{16}$			

SECTION H/UNIT 4/OPERATING DRILLING MACHINES
Self-Test Answers

1. The rpm rates of the drills would be as follows:
    $\frac{1}{4}$-inch diameter: 1,440 rpm
    2-inch diameter: 180 rpm
    $\frac{3}{4}$-inch diameter: 480 rpm
    $\frac{3}{8}$-inch diameter: 960 rpm
    $1\frac{1}{2}$-inch diameter: 240 rpm
2. Worn margins and outer corners broken down. The drill can be ground back to its full size and resharpened.
3. The feed is about right when the chip rolls into a close helix. Long, stringy chips can indicate too much feed.

4. Feeds are designated by a small, measured advance movement of the drill for each revolution. A .001-inch feed for a $\frac{1}{8}$-inch-diameter drill, for example, would move the drill .001 inch into the work for every turn of the drill.

5. Besides having the correct cutting speed, a sulfurized oil-based cutting fluid helps reduce friction and cool the cutting edge.

6. Drill jamming can be avoided by a "pecking" procedure. The operator drills a small amount and pulls out the drill to remove the chips. This is repeated until the hole is finished.

7. The depth stop is used to limit the travel of the drill so that it will not go into the table or vise. The depth of blind holes is preset and drilled. Countersink and counterbore depths are set so that several can be easily made the same.

8. The purpose of using work-holding devices is to keep the workpiece rigid, to prevent it from turning with the drill, and for operator safety.

9. Included in a list of work-holding devices would be strap clamps, T-bolts, and step blocks. Also used are C-clamps, V-blocks, vises, jigs, fixtures, and angle plates.

10. Parallels are mostly used to raise workpieces off the drill press table or to lift a workpiece higher in a vise, thus providing a space for the drill breakthrough. They are made of hardened steel, so care should be exercised in their use.

11. Thin materials tend to spring downward from the drilling force until drill break-through begins. The drill then "grabs" as the material springs upward, and a broken drill is often the result. This can be avoided by placing the support or parallels as near the drill as possible.

12. Angle drilling is done by tilting a drill press table (not all types tilt) or by using an angular vise.

13. The wiggler is used for locating a center punch mark under the center axis of a drill spindle.

14. Some odd-shaped workpieces, such as gears with extending hubs that need holes drilled for setscrews, might be difficult to set up without an angle plate.

15. One of the difficulties with hand tapping is the tendency for taps to start crooked or misaligned with the tap-drilled hole. Starting a tap by hand in a drill press with the same setup as used for the tap drilling ensures a perfect alignment.

SECTION H/UNIT 5/COUNTERSINKING AND COUNTERBORING

Self-Test Answers

1. Countersinks are used to chamfer holes and to provide tapered holes for flat head fasteners such as screws and rivets.

2. Countersink angles vary to match the angles of different flat head fasteners or different taper hole requirements.

3. A center drill is used to make a 60-degree countersink hole in workpieces for lathes and grinders.

4. A counterbore makes a cylindrical recess concentric with a smaller hole so that a hex head bolt or socket head cap screw can be flush mounted with the surface of a workpiece.

5. The pilot diameter should always be a few thousandths of an inch smaller than the hole, but not more than .005 inch.

6. Lubrication of the pilot prevents metal-to-metal contact between it and the hole. It will also prevent the scoring of the hole surface.

7. A general rule is to use approximately one-third of the cutting speed when counterboring, as when using a twist drill with the same diameter.

8. Feeds and speeds when counterboring are controlled to a large extent by the condition of the equipment, the available power, and the material being counterbored.

9. Spotfacing is performed with a counterbore. It makes a flat bearing surface, square with a hole to set a nut, washer, or bolt head.

10. Counterboring requires a rigid setup with the workpiece securely fastened and provisions made to allow the pilot to protrude below the bottom surface of the workpiece.

SECTION H/UNIT 6/REAMING IN THE DRILL PRESS

Self-Test Answers

1. Machine reamers are identified by the design of the shank— either a straight or tapered shank and usually a 45-degree chamfer on the cutting end.

2. A jobber's reamer is a finishing reamer like a chucking reamer, but it has a longer fluted body.

3. An accurate hole size cannot be obtained without a high-quality surface finish.

4. As a general rule, the cutting speed used to ream a hole is about one-third to half of the speed used to drill a hole of the same size in the same material.

5. The feed rate, when reaming as compared with drilling the same material, is approximately two to three times as great. As an example, for a 1-inch drill the feed rate is about .010 to .015 inches per revolution. A 1-inch reamer would have a feed rate of between .020 and .030 inches.

6. The reaming allowance for a $\frac{1}{2}$-inch diameter hole would be $\frac{1}{64}$ inch.

7. Cutting fluid is required to ream a hole with a good surface finish. Cutting fluid will cool the workpiece and the tool and will also act as a lubricant between the chip and the tool to reduce friction and heat buildup.

8. Chatter may be eliminated by reducing the speed, increasing the feed, or using a piloted reamer.

9. Bellmouthed holes are usually caused by a misaligned reamer and workpiece setup. Piloted reamers, bushings, or a floating holder may correct this problem.

10. Surface finish can be improved by decreasing the feed and checking the reaming allowance. Too much or not enough material will cause poor finish. Use a large volume of coolant.

11. Carbide-tipped reamers are recommended for long production runs where highly abrasive materials are reamed.

12. Cemented carbides are hard but also brittle. The slightest amount of chatter or vibration may chip the cutting edges.

SECTION I/UNIT 1/ENGINE LATHE FUNDAMENTALS

Self-Test Answers

Part A

For this part a lathe in your shop should be used and the student should be able to point out and name the parts of the lathe.

Part B

1. Fine chips, filings, and grindings form an abrasive sludge that wears and scores the sliding surfaces. Frequent cleaning will help prevent damage to the machine.
2. Heavier chips should be removed with a brush; never use an air jet. The ways should then be wiped clean with a cloth and lightly oiled.
3. Because most nicks come from dropping chucks and heavy workpieces on the lathe, a lathe board used every time a chuck is changed or heavy work is installed will prevent much of this damage. A tool board will help keep tools such as files from being laid across the ways.
4. Once daily.
5. No. The oil on the ways may have collected dirt or grit from the air to form an abrasive mixture. The ways should first be cleaned and oiled.
6. The chips should be cleaned from the lathe and swept up on the surrounding floor area. The lathe ways and slides should be wiped and oiled.
7. Straight gibs and tapered gibs.
8. The gib on the cross slide should be adjusted so it will have a slight drag, but the compound should be set up fairly tight when it is not being used.

SECTION I/UNIT 2/TOOLHOLDERS AND TOOLHOLDING

Self-Test Answers

1. The T-bolt holding the tool post to the compound rest was not sufficiently tightened.
2. Quick-change toolholders are adjusted for height with a micrometer collar.
3. Tool height on rocker-arm toolholders is adjusted by swiveling the rocker in the tool post ring.
4. Toolholder overhang affects the rigidity of a setup; too much overhang may cause chatter.
5. Drilling machine tools are used in a lathe tailstock.
6. The lathe tailstock is bored with a Morse taper hole to hold Morse taper shank tools.

SECTION I/UNIT 3/CUTTING TOOLS FOR THE LATHE

Self-Test Answers

1. High-speed steel is easily shaped into the desired shape of cutting tool. It produces better finishes on low-speed machines and on soft metals.
2. Its geometric form: the side and back rake, front and side relief angles, and chip breakers.
3. Unlike single-point tools, form tools produce their shape by plunging directly into the work.
4. When a "chip trap" is formed by improper grinding on a tool, the chip is not able to clear the tool; this prevents a smooth flow across the face of the tool. The result is tearing of the surface on the workpiece and possibly a broken tool.
5. Some toolholders provide a built-in back rake of about 16 degrees; to this is added any back rake on the tool to make a total back rake that is excessive.

6. A zero rake should be used for threading tools. A zero to slightly negative back rake should be used for plastics and brass, because they tend to "dig in."
7. The side relief allows the tool to feed into the work material. The end relief angle keeps the tool end from rubbing on the work.
8. The side rake directs the chip flow away from the cut and also provides for a keen cutting edge. The back rake promotes smooth chip flow and good finishes.
9. The angles can be checked with a tool grinding gage, a protractor, or an optical comparator.
10. Long, stringy chips or those that become snarled on workpieces, tool post, chuck, or lathe dog are hazardous to the operator. Chip breakers and correct feeds can produce an ideal chip that does not fly off but will simply drop to the chip pan and is easily handled.
11. Chips can be broken up by using coarse feeds and maximum depth of cuts for roughing cuts and by using tools with chip breakers on them.
12. Overheating a tool causes small cracks to form on the edge. When a stress is applied, as in a roughing cut, the tool end may break off.

SECTION I/UNIT 4/LATHE SPINDLE TOOLING

Self-Test Answers

1. The lathe spindle is a hollow shaft that can have one of three mounting devices machined on the spindle nose. It has an internal Morse taper that will accommodate centers or collets.
2. The spindle nose types are the threaded, long taper key drive, and the camlock.
3. The independent chuck is a four-jaw chuck in which each jaw can move separately and independently of the others. It is used to hold odd-shaped workpieces.
4. The universal chuck is most often a three-jaw chuck, although these chucks are made with more or fewer jaws. Each jaw moves in or out by the same amount when the chuck wrench is turned. They are used to hold and quickly center round stock.
5. A drive plate.
6. The live center is made of soft steel, so it can be turned to true it up if necessary. It is made with a Morse taper to fit the spindle taper or special sleeve if needed.
7. Workpieces and fixtures are mounted on faceplates. These are identified by their heavy construction and the T-slots. Drive plates have only slots.
8. Collet chucks are accurate work-holding devices. Spring collets are limited to smaller material and to specific sizes.

SECTION I/UNIT 5/OPERATING LATHE CONTROLS

Self-Test Answers

1. Levers located on the headstock can be shifted in various arrangements to select speeds.
2. The feed reverse lever.
3. These levers are used for selecting feeds or threads per inch for threading.

4. Because the crossfeed is geared differently (about one third of the longitudinal feed), the outside diameter would have a coarser finish than the face.
5. The half-nut lever is only used for threading.
6. They are graduated in English units. Some metric conversion collars are being made and used that read in both English and metric units at the same time.
7. You can test with a rule and a given slide movement such as .125 or .250 inch. If the slide moves half that distance, the lathe is calibrated for double depth and reads the same amount as that taken off the diameter.

SECTION I/UNIT 6/FACING AND CENTER DRILLING
Self-Test Answers

1. A lathe board is placed on the ways under the chuck, and the chuck is removed, because it is the wrong chuck to hold rectangular work. The mating parts of an independent chuck and the lathe spindle are cleaned and the chuck is mounted. The part is roughly centered in the jaws and adjusted to center by using the back of a tool holder or a dial indicator.
2. The tool should be on the center of the lathe axis.
3. The compound must be swung to either 30 or 90 degrees so that the tool can be fed into the face of the work by a measured amount. A depth micrometer or a micrometer caliper can be used to check the trial finish cut.
4. $\text{rpm} = \dfrac{300 \times 4}{4} = 300.$
5. Center drilling is done to prepare work for turning between centers and for spotting workpieces for drilling in the lathe.
6. Center drills are broken as a result of feeding the drill too fast and having the lathe speed too slow. Breakage also can result from having the tailstock off center, the work off center in a steady rest, or a lack of cutting oil.
7. The sharp edge provides a poor bearing surface and soon wears out of round, causing machining problems such as chatter.
8. $\frac{3}{16}$ inch.

SECTION I/UNIT 7/TURNING BETWEEN CENTERS
Self-Test Answers

1. The other method is that the workpiece is held in a chuck on one end and in the tailstock center on the other end.
2. Coarser feeds, deeper cuts, and smaller rake angles all tend to increase chip curl. Chip breakers also make the chip curl.
3. Dead centers are hardened 60-degree centers that do not rotate with the work but require high-pressure lubricant. Ball bearing centers turn with the work and do not require lubricant. Pipe centers turn with the work and are used to support tubular material.
4. With no end play in the workpiece and the bent tail of the lathe dog free to move in the slot.
5. Because of expansion of the workpiece from the heat of machining, it tightens on the center, thus causing more friction and more heat. This could ruin the center.
6. Excess overhang promotes lack of rigidity. This causes chatter and tool breakage.

7. $\text{rpm} = \dfrac{90 \times 4}{1\frac{1}{2}} = 360 \div \dfrac{3}{2} = 360 \times \dfrac{2}{3} = 240$, or

$$\dfrac{360}{1.5} = 240.$$

8. For most purposes where liberal tolerances are allowed, .015 to .030 inches can be left for finishing. When closer tolerances are required, two finish cuts are taken, with .005 to .010 inches left for the last finish cut.

SECTION I/UNIT 8/ALIGNMENT OF LATHE CENTERS
Self-Test Answers

1. The workpiece becomes tapered.
2. The workpiece is tapered with the small end at the tailstock.
3. By the witness mark on the tailstock, by using a test bar, and by taking a light cut on a workpiece and measuring.
4. The dial indicator.
5. With a micrometer. The tailstock is set over with a dial indicator.

SECTION I/UNIT 9/OTHER LATHE OPERATIONS
Self-Test Answers

1. Drilled holes are not sufficiently accurate for bores in machine parts, as they would be loose on the shaft and would not run true.
2. The chief advantage of boring in the lathe is that the bore runs true with the centerline of the lathe and the outside of the workpiece, if the workpiece has been set up to run true (with no runout). This is not always possible when reaming bores that have been drilled, because the reamer follows the eccentricity or runout of the bore.
3. Ways to eliminate chatter in a boring bar are as follows:
   Shorten the bar overhang, if possible.
   Reduce the spindle speed.
   Make sure the tool is on center.
   Use as large a diameter bar as possible without binding in the bore.
   Reduce the nose radius on the tool.
   Apply cutting fluid to the bore.
   Use tuned or solid carbide boring bars.
4. Grooves and thread relief are made in bores by means of specially shaped or ground tools in a boring bar.
5. Cutting speeds for reaming are half that used for drilling; feeds used for reaming are twice that used for drilling.
6. A spiral-point tap works best for power tapping.
7. You can avoid chatter when cutting off with a parting tool by maintaining a rigid setup and keeping enough feed to produce a continuous chip, if possible.
8. Knurling is used to improve the appearance of a part, to provide a good gripping surface, and to increase the diameter of a part for press fits.
9. Ordinary knurls make a straight or diamond pattern impression by displacing the metal with high pressures.
10. When knurls produce a double impression, they can be readjusted up or down and moved to a new position. Angling the toolholder 5 degrees may help.

SECTION I/UNIT 10/SIXTY-DEGREE THREAD INFORMATION AND CALCULATIONS

Self-Test Answers

1. The sharp V-thread can be easily damaged during handling if it is dropped or allowed to strike against a hard surface.
2. The pitch is the distance between a point on a screw thread to a corresponding point on the next thread measured parallel to the axis. "Threads-per-inch" is the number of threads in one inch.
3. American National Standard and Unified Standard threads both have the 60-degree included angle and are both based on the inch measure with similar pitch series. The depth of the thread and the classes of thread fits are different in the two systems.
4. To allow for tolerancing of external and internal threads to promote standardization and interchangeability of parts.
5. This describes a diameter of $\frac{1}{2}$ inch, 20 threads per inch, and Unified coarse series external thread with a class 2 thread tolerance.
6. The flat on the end of the tool for 20 threads per inch should be P $=$ .050 $\times$ .125 $=$ .006 inch for American National threads and for Unified threads.
7. The fit of the thread refers to classes of fits and tolerances, while percent of thread refers to the actual minor diameter of an internal thread, a 100-percent thread being full-depth internal threads.
8. Both are 60-degree forms.

SECTION I/UNIT 11/CUTTING UNIFIED EXTERNAL THREADS

Self-Test Answers

1. A series of cuts are made in the same groove with a single-point tool by keeping the same ratio and relative position of the tool on each pass. The quick-change gearbox allows choices of various pitches or leads.
2. The chips are less likely to bind and tear off when feeding in with the compound set at 29 degrees, and the tool is less likely to break.
3. The 60-degree angle on the tool is checked with a center gage or optical comparator.
4. The number of threads per inch can be checked with a screw pitch gage or by using a rule and counting the threads in one inch.
5. A center gage is used to align the tool to the work.
6. No. The carriage is moved by the thread on the leadscrew when the half-nuts are engaging it.
7. Even-numbered threads may be engaged on the half-nuts at any line, and odd-numbered threads at any numbered line. It would be best to use the same line every time for fractional-numbered threads.
8. The spindle should be turning slowly enough for the operator to maintain control of the threading operation, usually about one-fourth of turning speeds.
9. The lead screw rotation is reversed, which causes the cut to be made from the left to the right. The compound is set at 29 degrees to the left. The threading tool and lathe settings are set up in the same way as for cutting right-hand threads.
10. Picking up the thread or resetting the tool is a procedure used to position a tool to existing threads.

SECTION I/UNIT 12/CUTTING UNIFIED INTERNAL THREADS

Self-Test Answers

1. The minor diameter of the thread.
2. 75 percent.
3. Large internal threads of various forms can be made, and the threads are concentric to the axis of the work.
4. To the left of the operator.
5. A screw pitch gage should be used.
6. Boring bar and tool deflection cause the threads to be undersize from the calculations and settings on the micrometer collars.
7. The minor diameter equals $D(P \times .541 \times 2)$.

   $P = \frac{1}{8}$ in. $= .125$ in.

   $D = 1$ inch $- (.125 \times .541 \times 2) = .8648,$ or $.865$ inch
8. A thread plug gage, a shop-made plug gage, or the mating part.

SECTION I/UNIT 13/CUTTING TAPERS

Self-Test Answers

1. Steep tapers are quick-release tapers, and slight tapers are self-holding tapers.
2. Tapers are expressed in taper-per-foot, taper-per-inch, and by their angle.
3. Tapers are turned by hand feeding the compound slide, by offsetting the tailstock and turning between centers, or by using a taper attachment. A fourth method is to use a tool that is set to the desired angle and form cut the taper.
4. No. The angle on the workpiece would be the included angle, which is twice that on the compound setting. The angle on the compound swivel base is the angle with the work centerline.
5. The reading at the lathe centerline index would be 55 degrees, which is the complementary angle.
6. Offset $= \dfrac{10 \times (1.125 - .75)}{2 \times 3} = \dfrac{3.75}{6} = .625$ inch.
7. Four methods of measuring tapers are using the plug and ring gages, using a micrometer on layout lines, using a micrometer with precision parallels and drill rod on a surface plate, and using a sine bar, gage block, and dial indicator.
8. The taper plug gage and the taper gage are the simplest and most practical means to check a taper.

SECTION I/UNIT 14/USING STEADY AND FOLLOWER RESTS

Self-Test Answers

1. When workpieces extend from the chuck more than four or five workpiece diameters and are unsupported by a dead center; when workpieces are long and slender.
2. The steady rest is placed near the tailstock end of the shaft, which is supported in a dead center. The steady rest is clamped to the lathe bed, and the lower jaws are adjusted to the shaft finger tight. The upper half of the frame is closed, and the top jaw is adjusted with some clearance. The jaws are locked and lubricant is applied.
3. The jaws should be readjusted when the shaft heats up from friction, to avoid scoring. Also, soft materials are sometimes used on the jaws to protect finishes.

4. A center punch mark is placed in the center of the end of the shaft. The lower two jaws on the steady rest are adjusted until the center punch mark aligns with the point of the dead center.
5. No.
6. No. When the surface is rough, a bearing spot must be turned for the steady rest jaws.
7. By using a cat head.
8. A follower rest.

SECTION I/UNIT 15/ADDITIONAL THREAD FORMS

Self-Test Answers

1. Translating-type screws are mostly used for imparting motion or power and to position mechanical parts.
2. Square, modified square, Acme, stub Acme, and Buttress are five basic translating thread forms.
3. Because the pitch for 4 TRI would be .250 in., the depth of thread would be $P/2 = .250/2 = .125$ inch.
4. Because the pitch for 4 TPI would be .250 in., the depth of the Acme thread would be $.5P + .010$ inch, or $.5 \times .250$ inch $+ .010$ inch $= .135$ inch.
5. 29 degrees.
6. The Acme thread form.
7. Buttress threads are used where great forces or pressures are exerted in one direction.
8. .375 inch.
9. The distance from a point on one thread to a corresponding point on the next.
10. The distance the nut travels in one revolution.
11. Single lead.
12. Accurately slotted faceplate.

   Indexing a gear on the drive train.

   Using the thread-chasing dial.

   The compound rest method.

13. They provide rapid traverse, are more efficient, have a larger minor diameter, are stronger, and furnish more bearing surface area than a single thread.
14. You can determine the number of leads by counting the number of starting grooves at the end of a bolt or screw.
15. Roughing of coarse threads may move the tool slightly. If one thread has already been finished, there is no more allowance for adjustment. Finishing both threads consecutively, however, gives a much greater assurance that the setup will not move.

SECTION I/UNIT 16/CUTTING ACME THREADS ON THE LATHE

Self-Test Answers

1. The thread angle and the thread form (shape).
2. The included angle, the relief angle, and the flat on the end of the tool.
3. $P = \frac{1}{6} = .1666$ in.; depth $= .5P + .010$ in. $= .93$ in.
4. The tool is aligned by using the Acme tool gage.
5. P $= .1666$ inch; minor diameter .750 inch $- .1666$ inch $= .583$ inch.
6. With an Acme tap set.
7. An Acme thread plug gage.
8. Light finishing cuts with a honed tool and a good grade of sulfurized cutting oil will help make good thread finishes. A rigid setup and low speeds will also help.

SECTION J/UNIT 1/VERTICAL MILLING MACHINES

Self-Test Answers

1. The column, knee, saddle, table, ram, and toolhead.
2. By using the table traverse handwheel and the table power feed.
3. By using the cross traverse handwheel.
4. By using the quill feed hand lever and handwheel.
5. The table clamp locks the table rigidly and keeps it from moving while other table axes are in movement.
6. The spindle brake locks the spindle while tool changes are being made.
7. The spindle has to stop before speed changes from high to low are made.
8. The ram movement increases the working capacity of the toolhead.
9. Loose machine movements are adjusted with the slide gibs.
10. The quill clamp is tightened to lock the quill rigidly while milling.

SECTION J/UNIT 2/CUTTING TOOLS AND CUTTING TOOL HOLDERS FOR THE VERTICAL MILLING MACHINE

Self-Test Answers

1. When viewed from the cutting end, a right-hand cut end mill will rotate counterclockwise.
2. An end mill has to have center cutting teeth to be used for plunge cutting.
3. End mills for aluminum usually have a fast helix angle and highly polished flutes and cutting edges.
4. Carbide end mills are effective when milling abrasive or hard materials. Carbide end mills are often used because they can dramatically improve productivity.
5. Roughing mills are used to remove large amounts of material.
6. Tapered end mills are mostly used in mold or die making to obtain precisely tapered sides on workpieces.
7. Carbide insert tools are used, because new cutting edges are easily exposed. They are available in grades to cut most materials and are efficient cutting tools.
8. Straight shank mills are held in collets or solid holders.
9. A four-flute end mill is stronger than a two- or three-flute end mill.

SECTION J/UNIT 3/SETUPS ON THE VERTICAL MILLING MACHINE

Self-Test Answers

1. Workpieces can be aligned on a machine table by measuring their distance from the edge of the table, by locating against stops in the T-slots, or by indicating the workpiece side.
2. To align a vise on a machine table, the solid vise jaw needs to be indicated.
3. Toolhead alignment is checked when it is important that the spindle is perpendicular to the machine table for precision machining.
4. When the knee clamping bolts are loose, the weight of the knee makes it sag; but when the knee clamps are tightened, the knee is pulled into its normal position in relation to the column.

5. Tightening the toolhead clamping bolts usually produces a small change in the toolhead position.
6. A machine spindle can be located over the edge of a workpiece with an edge finder.
7. The spindle axis is half the tip diameter away from the workpiece edge when the tip suddenly moves sideways.
8. An offset edge finder works best at 600 to 800 rpm.
9. To eliminate the effect of backlash, always position from the same direction.
10. The center of a hole is located with a dial indicator mounted in the machine spindle.

SECTION J/UNIT 4/VERTICAL MILLING MACHINE OPERATIONS

Self-Test Answers

1. Lower cutting speeds are used to machine hard materials, tough materials, and abrasive material and on heavy cuts.
2. Higher cutting speeds are used to machine softer materials, to obtain good surface finishes, with small-diameter cutters, for light cuts, on frail workpieces, and on frail setups.
3. The calculated rpm is a starting point and may change depending on conditions illustrated in the answers to problems 1 and 2.
4. The thickness of the chips affects the tool life of the cutter. Thin chips dull a cutting edge quickly. Chips that are too thick cause tool breakage or the chipping of the cutting edge.
5. The rule of thumb for depth of cut for a HSS end mill is that the depth should not exceed half the diameter of the cutter in mild steel.
6. The feed rate for a two-flute, $\frac{1}{4}$-inch-diameter carbide end mill in low-carbon steel is

$$\text{Feed rate} = f \times \text{rpm} \times n$$
$$f = .0005$$
$$\text{rpm} = \frac{\text{CS} \times 4 \times}{D} = \frac{300 \times 4}{\frac{1}{4}} = 4{,}800$$
$$n = 2$$

$$\text{feed rate} = .0005 \times 4{,}800 \times 2 = 4.8\,\text{ipm}$$

7. A cutter is accurately centered over a shaft by using an edge finder or by using a piece of paper and the end mill with the spindle off.
8. The feed direction against the cutter rotation ensures positive dimensional movement. It also prevents the workpiece from being pulled into the cutter because of any backlash in the machine.
9. End mills can work themselves out of a split collet if the cut is too heavy or when the cutter dulls.
10. Angular cuts can be made by tilting the workpiece or by tilting the workhead.
11. Circular slots can be milled by using a rotary table or an index head.
12. Squares, hexagons, or other shapes that require surfaces at precise angles to each other can be made by using a dividing head.
13. Exchange one regular insert with a wiper insert to improve the surface finish.
14. A T-slot or dovetail cutter only enlarges the bottom part of a groove. The groove has to be made before a T-slot or dovetail cutter can be used.

SECTION J/UNIT 5/USING AN OFFSET BORING HEAD

Self-Test Answers

1. An offset boring head is used to produce standard and non-standard size holes at precisely controllable hole locations.
2. Parallels raise the workpiece off the table or other work-holding device to allow through holes to be bored.
3. Unless the locking screw is snugged after tool slide adjustments are made, the toolslide might move during the cutting operation, resulting in a tapered or odd-sized hole.
4. The toolslide has multiple holes so that the boring tool can be held in different positions for different size bores.
5. It is important that you know whether one graduation is one thousandth of an inch or two thousandths of an inch in hole size change.
6. The best boring tool to use is the one with the largest diameter that can be used and with the shortest shank.
7. It is important that the cutting edge of the boring tool be on the centerline of the axis of the tool slide. This ensures that the rake and clearance angles are correctly positioned.
8. The hole size obtained for a given amount of depth of cut can change depending on the sharpness of the tool, the amount of tool overhang (boring bar length), and the amount of feed per revolution.
9. Boring tool deflection changes when the tool dulls, the depth of cut increases or decreases, or the feed is changed.
10. The cutting speed is determined by the kind of tool material and the kind of work material, but boring vibrations set up through an unbalanced cutting tool or a long boring bar may require a smaller-than-calculated rpm.

SECTION K/UNIT 1/HORIZONTAL SPINDLE MILLING MACHINES

Self-Test Answers

1. The student should identify the common parts of a horizontal milling machine.
2. The student should lubricate a horizontal milling machine.

SECTION K/UNIT 2/TYPES OF SPINDLES, ARBORS, AND ADAPTERS

Self-Test Answers

1. Face mills with a diameter of over 6 inches.
2. A taper with a larger included angle will align two items accurately but will not stick (self-release). The opposite is a self-holding taper such as is found on taper shank drills with a taper of only $\frac{5}{8}$ inch per foot.
3. Milling machine tapers are 3.5 inches per foot.
4. When small-diameter cutters are used, on light cuts, and where little clearance is available.
5. A style B arbor gives more rigidity when using larger cutters on heavier cuts.
6. Most cutters should be driven by the two keys in the arbor. These keys should extend into the spacing collars on either side of the cutter.
7. Spacing collars are used to take up the space between the cutter and the end of the arbor. They are also used to space

straddle milling cutters. Bearing collars ride in the arbor support bearing. They provide support for the outer end of the arbor. Spacing and bearing collars are precision accessories and should be protected against nicks and burrs.

8. Dirt or burrs between collars can cause cutter runout and inaccurate machining.
9. It is important that the clearance between the bearing collar and the arbor support be adjusted correctly. High spindle rpm requires more clearance then machining at a low rpm. The second point, just as important, is regular lubrication.
10. The arbor support reservoir holds oil that lubricates the arbor bushing.

SECTION K/UNIT 3/ARBOR-DRIVEN MILLING CUTTERS

Self-Test Answers

1. Profile-sharpened cutters and form-relieved cutters.
2. Light-duty plain milling cutters have many teeth. They are used for finishing operations. Heavy-duty plain mills have few but coarse teeth, designed for heavy cuts.
3. Plain milling cutters do not have side cutting teeth. This would cause extreme rubbing if used to mill steps or grooves. Plain milling cutters should be wider than the flat surface they are machining.
4. Side milling cutters, having side cutting teeth, are used when grooves are machined.
5. Straight-tooth side mills are used only to mill shallow grooves because of the limited chip space between the teeth and their tendency to chatter. Staggered-tooth mills have a smoother cutting action because of the alternate helical teeth; more chip clearance allows deeper cuts.
6. Half side milling cutters are efficiently used when straddle milling.
7. Metal slitting saws are used in slotting or cutoff operations.
8. Gear tooth cutters and corner rounding cutters.
9. To mill V-notches, dovetails, or chamfers.
10. A right-hand cutter rotates counterclockwise when viewed from the outside end.

SECTION K/UNIT 4/WORK-HOLDING METHODS AND STANDARD SETUPS

Self-Test Answers

1. A clamping bolt should be close to the workpiece and a greater distance away from the support block.
2. Finished surfaces should be protected with shims from being marked by clamps or rough vise jaws.
3. Screwjacks are used to support workpieces or to support the end of a clamp.
4. A stopblock counters the cutting forces to keep the workpiece from slipping.
5. Toe clamps have a low profile and allow the workpiece to be clamped directly to the table. The top surface of the workpiece is totally accessible for machining. They exert pressure both horizontally and downward.
6. A swivel vise allows machining of angles on workpieces.
7. The closer a workpiece is clamped to the machine table, the more rigid the machining setup will be.

SECTION K/UNIT 5/MACHINE SETUP AND PLAIN MILLING

Self-Test Answers

1. When milling, the direction that the workpiece is being fed can either be in the same direction as the cutter rotation or opposed to the direction of cutter rotation. When the direction of feed is opposed to the rotation direction of the cutter, this is said to be *up* or *conventional milling*. When the direction of feed is the same as the rotation direction, this is said to be *down* or *climb milling* because the cutter is attempting to climb onto the workpiece. If excessive backlash exists in the table or saddle, the workpiece may be pulled into the cutter during climb milling. This can result in a bent arbor, broken cutter, damaged workpiece, and possible injury to the operator. Climb milling should be avoided unless the mill is equipped with adequate backlash control. Remember that any cutter can be operated in an up-milling or down-milling mode depending only on which side of the workpiece the cut is started.
2. When selecting a mill arbor, use one that has minimum overhang beyond the outer arbor support.
3. To secure and arbor in the spindle, insert the tapered shank into the spindle socket. Be sure that the socket is clean and free from burrs or nicks. Large mill arbors are heavy, and you may need help holding them in place until the drawbolt is engaged. Do not let the arbor fall out onto the machine table. Thread the drawbolt into the arbor shank all the way and then draw the arbor into the spindle taper by turning the drawbolt locknut. Tighten the locknut with a wrench.
4. The will reduce the stress on the arbor.
5. The helps make the setup and axes more rigid and reduce the effects of backlash and looseness in the ways.

SECTION K/UNIT 6/USING SIDE MILLING CUTTERS

Self-Test Answers

1. Full side milling cutters are used to cut slots and grooves and where contact on both sides of the cutter is made.
2. Half side milling cutters make contact on one side only, as in straddle milling, where a left-hand and a right-hand cutter are combined to cut a workpiece to length.
3. The best cutter is the one with the smallest diameter that will work, considering the clearance needed under the arbor support.
4. Usually a groove is wider than the cutter.
5. A layout shows the machinist where the machining is to take place. It helps in preventing errors.
6. Accurate positioning is done with the help of a paper feeler strip. Adequate accuracy often is achieved by using a steel rule or by aligning by sight with the layout lines.
7. If a workpiece is measured while it is clamped in the machine, additional cuts can be made without making additional setups.
8. Shims and spacers control workpiece width in straddle milling operations.
9. After the cutter is aligned with either a steel rule or the machine dials, start the spindle. Then, advance the table by hand slowly until the cutter makes a small cut on the edge of the

workpiece. Make only enough of a cut to take a measurement. Stop the machine, measure, and if the cut is not exactly correct, make an adjustment. Because the cut is only on the edge of the workpiece, it will probably disappear when a chamfer is put on that edge.

10. Interlocking side mills are used to cut slots over 1 inch wide and also when precise slot width is to be produced.

SECTION L/UNIT 1/TYPES OF GRINDERS

Self-Test Answers

1. Horizontal spindle with reciprocating table, horizontal spindle with rotary table, vertical spindle with either reciprocating or rotary table.
2. Less than + or − .0001.
3. The universal cylindrical grinder is a very versatile machine. A universal cylindrical grinder can be used to grind the outside or inside cylindrical surfaces of parts. Cylindrical tapers can also be ground internally or externally on parts. Work pieces can be held in a chuck on the headstock or between centers.
4. A centerless grinder is used to grind cylindrical workpieces without the use of centers. It can be used for short or long cylindrical workpieces, such as lengths of drill rod.
5. Universal cylindrical grinders often have an internal grinding attachment. This is a spindle that can be swiveled down to be in line with the headstock centerline. Internal grinding uses mounted abrasive wheels. These are grinding wheels attached to a shank for mounting into the machine's grinding spindle. Internal grinding can be done on concentric workpieces parallel to the machine spindle axis. Internal tapers may also be ground.

SECTION L/UNIT 2/SELECTION AND USE OF GRINDING WHEELS

Self-Test Answers

1. Grinding wheels have two major components—the abrasive grains and the bond. The abrasive grains do the cutting. The bond holds the grains together to form the wheel.
2. The abrasive type used in a wheel is selected based on the work material. The proper abrasive will remain sharp during the grinding operation. If the correct abrasive and bond is chosen, the wheel will self-sharpen during cutting. Dull grains are released by the bond and new ones are exposed.
3. The ability for the bond to crumble and expose new grains is called friability.
4. The grade of the grinding wheel designates the strength of the bond in the grinding wheel. The bond is a hard grade if the spans between each abrasive grain are very strong and securely holds the grains during grinding. A soft wheel releases the grains under small grinding forces. The relative amount of bond in the wheel determines the wheel's grade or hardness. Hard grade wheels have longer wheel life.
5. Aluminum oxide.
6. After a grain type has been chosen, grit size must be selected. Every grinding wheel has a number that designates its grit size. Grit size is the size of the individual abrasive grains in the wheel. When grit is manufactured, it is run through a series of ever finer screens to establish various sizes of grains. Grit size

is established by the number of openings per linear inch in the final screen used to size the grain. The higher the number, the smaller the openings in the screen. Coarse grains would have lower numbers such as 10, 16 or 24. Coarse grains are used for rapid stock removal where a fine finish is not required. Fine grit wheels have higher numbers such as 70, 100 and 180. They are used for fine finishes and for hard, brittle materials.

7. Straight, cylinder, straight cup, flaring cup, shallow dish.
8. Type of abrasive, grit size, grade or hardness, structure, and bond.
9. Choose a wheel with a maximum RPM rating to match or exceed the grinder's spindle speed.
10. A vitrified bond grinding wheel can be "ring tested" for possible cracks (Figure L-18). Hold the wheel on your finger or on a small pin. Tap the wheel lightly with a wooden mallet or screwdriver handle. A good wheel will give off a clear ringing sound, whereas a cracked wheel will sound dull.
11. When a new wheel is installed on a grinder, it must be trued before it is used. The cutting surface of a new wheel will run out slightly due to the clearance between the wheel bore and machine spindle. Truing a wheel will make every point on its outer cutting surface concentric with the machine spindle. This concentricity is important for achieving smooth and accurate grinding. Dressing is the process of sharpening a grinding wheel. Truing and dressing are performed in the same manner.
12. Precision grinders are trued and dressed with single- (Figure L-20) or multiple-point diamond dressers. The diamond dresser is mounted on the grinder chuck so that it can be traversed across the cutting surface of the wheel. The dresser must be positioned off center on the wheel on the outgoing rotation side (Figure L-21) to prevent the dresser from getting caught and being pulled under the wheel. Truing and dressing should always be done with coolant to cool the diamond dresser.
13. The bore of a diamond wheel is machined so that it will closely fit the grinder spindle. When a resinoid-bonded diamond or CBN wheel is mounted, it may be adjusted to run true by using a dial indicator. The wheel is tapped lightly using a block of wood. These wheels are not dressed normally.
14. Balancing is usually required on large wheels (over 14-inch diameter) but may not be required for smaller wheels. An out-of-balance wheel can cause chatter marks in the workpiece finish.
15. Wheels are balanced on a disk balancing tool (Figure L-23) or on parallel ways (Figure L-24). The parallel ways balancing tool must be level. The grinding wheel is mounted on a balancing arbor and placed on the ways. The heavy point rotates to the lowest position. By adjusting weights in the flanges (Figure L-25), balance can be achieved. Balancing should be carefully done, to the point where the weight of a postage stamp applied to the grinding surface in a horizontal position will cause the wheel to move. It should be borne in mind that balancing on ways or with the overlapping disk tool results simply in a static balancing. Sometimes even a carefully static-balanced wheel can be quite unbalanced in operation. Sometimes it is necessary to measure the needed weight for static balance and use two smaller weights arrayed about 60 degrees from the light spot.
16. The wheel will become glazed with dull abrasive.
17. Structure 8 (open); bond J; grit size 80; abrasive, silicon carbide; bonding material, resinoid.

SECTION L/UNIT 3/SETUP OF SURFACE GRINDERS

Self-Test Answers

1. In general, large crossfeed movements combined with small downfeeding movements are preferred. This combination keeps more of the wheel working and tends to keep the wheel surface flat longer between dressings. Narrow crossfeed movements tend to round the edges of the wheel excessively.
2. Magnetic table.
3. Laminated parallels and V blocks are specially designed for grinder work. These laminated accessories are made with nonmagnetic and soft steel inserts so that the lines of magnetic force will be conducted through to the workpiece. Laminated parallels must be treated with great care because the magnetically permeable materials are soft. They should be checked carefully for burrs each time they are used.
4. Nonmagnetic materials may be held on a magnetic chuck by blocking the parts with steel (magnetic) spring-tooth clamps. Double-sided tape is also used for certain applications.
5. A thin-warped part must be shimmed so that the part is not pulled flat on the magnetic chuck. The top is then ground flat. The part is flipped over and ground. The result will be a flat, parallel part.
6. They cool a workpiece for size control, lubricate to prevent chip adherence to the wheel, flush swarf, and control grinding dust.
7. As close as possible to the grinding contact area.
8. Mist coolant is sprayed between the wheel and workpiece with air pressure.
9. They are usually transparent, which helps with visibility. Some have additives for rust control. They are not prone to bacterial growth. They provide a high level of cooling capacity but have less lubricating ability than either the soluble or straight oils.
10. It is important to apply grinding fluid where the grinding is taking place. This is the area where the wheel contacts the workpiece. One of the greatest difficulties in both surface and cylindrical grinding is "grinding dry with fluid." On some heat-sensitive steels, inadequate grinding fluid application can result in the development of grinding cracks from having the high-temperature workpiece contact area with the grinding wheel quenched an instant later with the grinding fluid.

SECTION L/UNIT 4/USING THE SURFACE GRINDER

Self-Test Answers

1. Mark the entire surface of the chuck with bluing or pencil lines. As you are grinding, the removal of the lastmarks will show you when the whole chuck surface has been trued.
2. The grade of the wheel is either too hard or too soft or there is vibration.
3. Improve the coolant flow on the work. Dress the wheel more coarsely. Change to a softer grade wheel.
4. Causes could be insufficient coolant, dirt or burrs between the work and chuck, poor setup of workpiece, stroke of table movement is too long, the chuck is out of line.

SECTION L/UNIT 5/CYLINDRICAL GRINDING

Self-Test Answers

1. On the plain cylindrical grinder the wheelhead does not swivel. The workpiece is typically traversed past the grinding wheel in plain cylindrical grinders. On the universal cylindrical grinder, all the major components are able to swivel, and the grinding head can be equipped with a chuck for internal grinding. The plain cylindrical grinder is usually a rigidly constructed production machine, whereas the universal grinder is much less rigidly constructed because of all the motions that are designed into the machine. The universal cylindrical grinder is used mainly in toolroom applications, where versatility is critical.
2. This prevents the duplication of headstock bearing irregularities into the workpiece.
3. This prevents the force of the grinding toward the footstock end from deflecting the part away from the grinding wheel.
4. Yes, the table that carries the centers is usually capable of being swiveled for this purpose.
5. A center-hole grinding machine can be used to prepare workpieces for cylindrical grinding.
6. This is grinding in which the rotating workpiece is moved across the face of the grinding wheel, which generates the surface.
7. This is cylindrical grinding, in which the rotating workpiece is moved directly into the grinding wheel. This action imparts a "mirror image" of the form of the periphery of the grinding wheel into the part.
8. The workpiece is supported on two conical work centers that project into matching conical holes in the ends of the workpiece.
9. The interruption should be narrower than the width of the grinding wheel face.
10. A chuck or other work-holding device, like a faceplate, to hold and rotate the part relative to the grinding spindle. A "high-speed" attachment that mounts to the wheelhead assembly and carries a mounted internal grinding wheel.

SECTION L/UNIT 6/USING A CYLINDRICAL GRINDER

Self-Test Answers

1. The combined curvature of the wheel and the workpiece creates a narrow line of contact from which the grinding swarf can easily escape.
2. It does not have to be mounted for each use, hence it is convenient for the operator.
3. It is useful for aligning the swivel table to parallel or for offsetting the table for accurate conical (tapered) surfaces.
4. None. The part should be adjusted with nil end play but be able to rotate freely on centers lubricated with a special high-pressure (lead-free) center lubricant.
5. The "tail" of the driving dog must not touch the bottom of the driving slot, or the workpiece could be forced from its conical seat. This would result in poor accuracy.
6. At least one-third of the wheel width, where possible. Half the wheel width would be optimal to permit full grinding to size.
7. The tarry control permits a dwell at the end of the traverse to stop "table bounce." (Without a tarry control, it is difficult to obtain the highest possible dimensional accuracy.)
8. About one-fourth of the wheel width.

9. About one-eighth of the wheel width.
10. .0002 inch (or less).

SECTION L/UNIT 7/UNIVERSAL TOOL AND CUTTER GRINDER

Self-Test Answers

1. Vertical height control of the wheelhead and, in some designs, the ability to tilt the wheelhead.
2. The cutter edges break down quickly for lack of proper support.
3. The larger clearances that are optimal for aluminum would lead to early failure in steels.
4. The amount necessary to take off the cutter to restore the cutting edge can easily be seen, so less of the cutter material is lost in regrinding.
5. Helical cutters must rotate while being sharpened to follow the tooth form. Mounting the stop on the table works only for straight-tooth cutters.
6. It permits checking of the setup for parallel and correcting if necessary. The measured difference should not exceed .001 inch.
7. The distance from the cutting edge to the axis of the cutter must be uniform, or the slot made by the cutter will be stepped.
8. Form-relieved cutters are sharpened only on the face. To preserve the form of the cutter, the grind should be radial to the cutter axis.
9. A free-turning accessory spindle. (An air-bearing type is preferred because of its uniformly low friction.)
10. The workhead. It is capable of being both rotated and tilted for these clearances. A tilting spindle design of tool and cutter grinder is also desirable.

SECTION M/UNIT 1/CNC MACHINES AND POSITION DIMENSIONING SYSTEMS

Self-Test Answers

1. A milling machine typically has three axes of motion. The X axis is the table movement right and left as you face the machine. The Y axis is the table movement toward and away from you. The Z axis is the spindle movement up and down. A move toward the work is a negative $Z(-Z)$ move. A move up in this axis would be a positive $Z(+Z)$ move.
2. The Z axis always denotes movement parallel to the spindle axis, the up and down movement. Toward the work is a negative Z move.
3. Make a sketch of the four Cartesian quadrants and identify the signs of each quadrant.

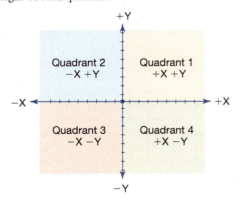

4. Incremental programming specifies the movement or distances from the point where you are currently located. A move to the right or up from this position is always a positive move ($+$); a move to the left or down is always a negative move ($-$). With an incremental move, we are specifying how far and in what direction we want the machine to move.
5. Absolute positioning systems have a major advantage over incremental positioning. If the programmer makes a mistake when using absolute positioning, the mistake is isolated to the one location.
6.

Point	X	Y
A	3	1
B	3	−4
C	1	5
D	−3	2
E	−5	6
Point	X	Y
Origin to Point A	3	1
Point A to Point B	0	−5
Point B to Point C	−2	9
Point C to Point D	−4	−3
Point D to Point E	−2	4

SECTION M/UNIT 2 FUNDAMENTALS OF MACHINING CENTERS

Self-Test Answers

1. Tool changers are an automatic storage and retrieval system for the cutting tools. They can very quickly change tools automatically when commanded to do so.
2. X, Y, and Z.
3. Z.
4. A rotary table.
5. By using a hand wheel or by using jog buttons.
6. The workpiece coordinate or program zero is the point or position from which all of the programmed coordinates are established. For example, when the programmer looks at the part print and notices that all of the dimensions come from the center of the part, these datums are then used to establish the program zero or workpiece coordinate.

   The workpiece coordinate system can be set from any datum feature. Pick the feature that would allow the programmer to do the least number of calculations.

   The part origin is the X0, Y0, Z0 location of the part in the rectangular or Cartesian coordinate system. In absolute programming, all of the tool movements would be programmed with respect to this point. If all of the dimensions were located from the center of the bored hole, then that point would become the program zero. During part setup, the X and Y zero position of the part has to be located. Using the hand wheels or other manual positioning devices and an edge finder or probe, the setup person locates the point at which the center of the spindle and the part origin are the same. The "home zero" is then entered as a G-code in the appropriate area of the program or in an offset table.

SECTION M/UNIT 3/FUNDAMENTALS OF PROGRAMMING MACHINING CENTERS

Self-Test Answers

1. Preparatory codes are used to set conditions or cancel conditions. When a parent with a child approaches a corner and wants to cross a street, the parent may give the child several instructions. The parent tells the child to stop running and stand still before they go into the street. The parent tells the child to look both ways before stepping into the street. The parent tells the child to hold their hand and walk across the street. The instructions given to the child are to ensure that the child crosses the street safely and successfully.

   Preparatory functions for a CNC have some of the same purposes. We tell the machine how we want it to operate in order to produce a quality part in a safe manner. In effect, we tell the machine how we want it to operate. Some preparatory G-codes are used to cancel modes. For example, if we just finished running a program that used offsets, we would like to cancel any offsets that may be in memory. A G40 would cancel diameter offsets and a G49 would cancel height offsets. We would like to cancel any canned cycles that may have been active. So those are two examples of canceling modes that may be active for safety and proper operation. Other codes are used to set the mode we want to operate in. For example, if our program is written in inch mode (not metric), we would use a G20. If our program is developed in the XY plane, we would tell the control to use the XY plane by using a G17.

2.

M00	Program stop	Non-Modal
M01	Optional stop	Non-Modal
M02	End of program	Non-Modal
M03	Spindle start clockwise	Modal
M04	Spindle start counterclockwise	Modal
M05	Spindle stop	Modal
M06	Tool change	Non-Modal
M07	Mist coolant on	Modal
M08	Flood coolant on	Modal
M09	Coolant off	Modal
M30	End of program & reset to the top of program	Non-Modal
M40	Spindle low range	Modal
M41	Spindle high range	Modal
M98	Subprogram call	Modal
M99	End subprogram & return to main program	Modal

3. To program a part, you need to determine where the workpiece zero (or part datum) should be located. The part datum is a feature of the part from which the majority of the dimensions of the part are located. Because all of the dimensions of the part in Figure E-7 come from the lower left-hand corner, this was the logical choice for the workpiece zero point. It is good practice to choose a part feature that is easy to access with an edge finder and one that will involve the fewest number of calculations. This approach will help avoid errors.

4. G-codes or preparatory functions fall into two categories: modal or nonmodal. Nonmodal or "one-shot" G-codes are those command codes that are only active in the block in which they are specified. Modal G-codes are those command codes that will remain active until another code of the same type overrides it. For example, if you had five lines that were all linear feed moves, you would only have to put a G01 in the first line. The other four lines would be controlled by the previous G01 code. The feed rate was modal in the first example. The feed rate does not change unless a different feed rate is commanded.

5.

Code	Function
G00	Rapid move
G01	Linear move at a programmed feed rate
M03	Spindle on—clockwise
G54	Workpiece coordinate preset
G92	Workpiece coordinate preset
G02	Clockwise circular interpolation
G03	Counterclockwise circular interpolation
G70	Inch programming
M08	Flood coolant on

6. G91 G01 X10.0 F10.0.
7. G91 G01 Y-5 F10.0.
8. M03 S800.
9. G90 G70.
10. M06 T4.
11. G00 Z5.0.
12. The G54-G59 workpiece coordinate is the absolute coordinate position of the part zero. These are not available on all machines. Six are available, and all serve the same function. This allows the programmer to have six different workpiece coordinates established on a machine. This would be very beneficial for repetitive jobs that could be located at the same position on the machine table. For example, a job that is run once each week might use the G59. The G59 would be used to establish the location for that particular job.

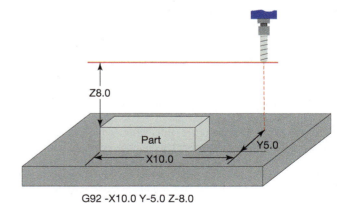

G92 -X10.0 Y-5.0 Z-8.0

G92 workpiece coordinates.

The G92 workpiece coordinate is the incremental distance from the workpiece datum (X, Y, and Z zero) to the center of the spindle. In effect, it tells the machine where the spindle is in relation to the workpiece. The spindle must then be in that position when you start to run the program.

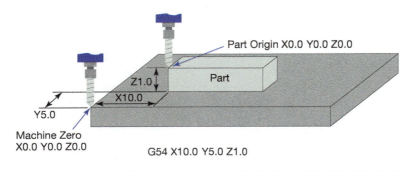

Part Origin X0.0 Y0.0 Z0.0

Part

Z1.0

X10.0

Y5.0

Machine Zero
X0.0 Y0.0 Z0.0

G54 X10.0 Y5.0 Z1.0

Workpiece coordinates.

13. G90
    G01 X8.000 Y5.250 F10.0
14. G91
    G01 X5.000 F10.0
15. G90
    G01 X-7.500 Y-3.250
16. G02 X3.000 Y0.000 I1.0 J0.0 F5.0
17. G03 X1.000 Y0.000 I-1.0 F8.0
18. G03 X4.000 Y3.000 I0.0 J-1.0 F5.0
19. G03 X3.0 Y4.0 I0.0 J-1.0 F5.0
20. G02 X5.000 Y4.000 I0.0 J-1.0 F7.0
21. G02 X-8.25 Y4.250 R-10.0
22. G02 X4.125 Y 2.125 R5.0 F5.0
23. G03 X-4.125 Y -2.125 R5.0 F5.0
24. G03 X-4.125 Y 2.125 R5.0 F5.0
25. Length offsets make it possible for a CNC machine to adjust to different tool lengths. Every tool is going to be a different length, but CNC machines can deal with this quite easily. CNC controllers have a special area within the control to store tool length offsets. The tool length offset is the distance from the tool tip at home position to the workpiece Z zero position. This distance is stored in a table that the programmer can access using a G-code or tool code. On a machining center a G43 code is typically used. The letter address G43 code is accompanied by an "H" auxiliary letter and a two-digit number. The G43 tells the control to compensate the Z axis, while the H and the number tell the control which offset to call out of the tool offset table. The tool length offset typically needs to be accompanied by a Z axis move to activate it.
26. Cutter compensation can be to the right or to the left of the part profile. To determine which offset you need, imagine yourself walking behind the cutting tool. When compensation to the left is desired, a G41 is used. When compensation to the right is desired, a G42 is used. When using the cutter compensation codes, you need to tell the controller which offset to use from the offset table. The offset identification is a number that is placed after the direction code. A typical cutter compensation line would look like this: G41 D12;.

    To initialize cutter compensation, the programmer has to make a move (ramp on). This additional move must occur before cutting begins. This move allows the control to evaluate its present position and make the necessary adjustment from centerline positioning to cutter periphery positioning. This move must be larger than the radius value of the tool. The machine corrects for the offset of the tool during the ramp move. In the figure below the machine compensates for the offset in the move between point 1 and point 2.

To cancel the cutter compensation and return to cutter centerline programming, the programmer must make a linear move (ramp off) to invoke a cutter compensation cancellation (G40). This is an additional move after the cut is complete.

27. N10 G20 G90 GG40 G80 G17; (Inch programming, absolute programming, cancel diameter compensation, cancel canned cycles, XY plane)

N20 M06 T02; (Tool change, tool #2)

N30 G54; (Workpiece zero setting)

N40 M03 S800; (Spindle start clockwise, 800 RPM)

N50 G00 X-1.00 Y-1.00; (Rapid to position #1)

N60 G00 Z.1; (Rapid down to .100 clearance above the part)

N70 G01 Z-.25 F5.0; (Feed down to depth at 5 inches per minute)

N80 G01 X-.25 Y-.25; (Feed to position #2, offsetting for the tool radius)

N90 G01 X-.25 Y2.5; (Feed to position #3)

N100 G01 X2.25 Y2.5; (Feed to position #4)

N110 G01 X2.25 Y2.25; (Feed to position #5)

N120 G01 X3.25 Y2.25; (Feed to position #6)

N130 G01 X3.25 Y.500; (Feed to position #7)

N140 G01 X1.750 Y.500; (Feed to position #8)

N150 G01 X1.750 Y-.25; (Feed to position #9)

N160 G01 X-1.0 Y-.25; (Feed to position #10, 1 inch to the left of the part)

N170 G28; (Return all axes to home position)

N180 M06 T0; (Tool change, tool 0)

N190 M30; (End program, rewind program to beginning)

28. O061

N10 G17 G40 G49 G80 G90

N15 T1 M6

N20 G54

N25 S1500 M3

N30 G0 X-.5 Y-.5

N35 G43 H1 Z1.

N40 Z.1

N45 G1 Z-.35 F10.

N50 G41 D1 X0. F6.

N55 Y2.5

N60 X2.5

N65 G2 X3. Y2. I0. J-.5

N70 G1 Y1.5

N75 X1.75

N80 G3 X1.5 Y1.25 I0. J-.25

N85 G1 Y.75

N90 X.735 Y0.

N95 X-.5

N100 G40 Y-.375

N105 Z.1 F10.

N110 G28

N115 M30

N120 %

SECTION M/UNIT 4/PROGRAMMING EXAMPLES

Self-Test Answers

1. O061

   N10 G17 G40 G49 G80 G90

   N15 T1 M6

   N20 G54

   N25 S1500 M3

   N30 G0 X-.5 Y-.5

   N35 G43 H1 Z1.

   N40 Z.1

   N45 G1 Z-.35 F10.

   N50 G41 D1 X0. F6.

   N55 Y2.5

   N60 X2.5

   N65 G2 X3. Y2. I0. J-.5

   N70 G1 Y1.5

   N75 X1.75

   N80 G3 X1.5 Y1.25 I0. J-.25

   N85 G1 Y.75

   N90 X.735 Y0.

   N95 X-.5

   N100 G40 Y-.375

   N105 Z.1 F10.

   N110 G28

   N115 M30

   N120 %

2. O062

   N10 G17 G40 G49 G80 G90

   N15 T5 M6

   N20 G54

   N25 S1000 M3

   N30 G0 X-.5 Y-3.5

   N35 G43 H5 Z1.

   N40 Z.1

   N45 G1 Z-.5 F10.

   N50 G41 D5 X0.

   N55 Y-.375

   N60 G2 X.375 Y0. I.375 J0.

   N65 G1 X2.5

   N70 Y-.75

   N75 G3 X3.5 Y-.75 I.5 J0.

   N80 G1 Y0.

   N85 X5.625

   N90 G2 X6. Y-.375 I0. J-.375

   N95 G1 Y-1.5

   N100 X4. Y-3.

   N105 X2.

   N110 X0. Y-1.5

   N115 Y.5

   N120 G40 X-.5

   N125 Z.1

   N130 G0 Z1.

   N135 M5

   N140 G28

   N145 M30

   %

3. O63

   N10 G17 G40 G49 G80 G90

   N15 T1 M6

   N20 G54

   N25 S1000 M3

   N30 G0 X-1.5 Y1.

   N35 G43 H1 Z1.

   N40 Z.1

   N45 G1 Z-.5 F10.

   N50 G41 D1 X-1.

   N55 Y2.

   N60 X1. Y2.5

   N65 X2.5

   N70 G2 X3. Y2. I0. J-.5

   N75 G1 Y1.25

   N80 X1.5

   N85 G3 X1. Y.75 I0. J-.5

   N90 G1 Y0.

   N95 G2 X-1. Y0. I-1. J0.

   N100 G1 Y2.5

   N105 G40 X-1.5

   N110 Z.1

   N115 G28

   N120 M30

   %

SECTION M/UNIT 5/PROGRAMMING CANNED CYCLES FOR MACHINING CENTERS

Self-Test Answers

1. The part drawing provides detailed information about the part. The shape of the part, the tolerances, material requirements, surface finishes, and the quantity required all have an impact on the program. From the part drawing, the programmer will determine what type of machine is required, work-holding considerations, and part datum (workpiece zero) location. Once these questions have been answered, the programmer can develop a process plan.

2. The machine will be selected based on its size, horsepower, accuracy, tooling capacity, and the number of axes of travel required. The size of the machine is based on the amount of travel of the axes. Does the part safely fit on the machine? Machine tools have a weight capacity. It can be unsafe to overload the capacity of the machine. What are the machining power requirements? The size of tools, depth of cut, and part material have a direct effect on the horsepower requirements of the machine. Is the machine rigid enough to withstand rough-machining operations? The rigidity of the machine affects the depth of cut, feed and speed rates, and surface finishes. To maximize production rates, choose a machine that is rigid enough to handle the task. What are the part tolerances? Is our machine capable of the accuracy required? How many different tools will be needed to manufacture this part? Will the machine's tool carousel accept that many tools? Will the machine be available when you are ready to make these parts?

3. Canned cycles can dramatically simplify programming and make shorter, more efficient programs.

4. G81, G73, G82, G84

5. 46.15 IPM

6. O78

   N10 G17 G40 G49 G80 G90

   N15 T1 M6

   N20 G54

   N25 G0 X-1.5 Y.75

   N30 S150 M3

   N35 G43 H1 Z1.

   N40 Z.1

   N45 G1 Z-.5 F6.

   N50 G41 D1 Y0.

   N55 X2.5

   N60 Y-.75

   N65 G3 X3.25 Y-1.5 I.75 J0.

   N70 G1 X4.375

   N75 G2 X4.375 Y-2.5 I0. J-.50

   N80 G1 X3.25

   N85 G3 X1.25 Y-4.5 I0. J-2.

   N90 G1 Y-5.

   N95 X.375

   N100 X0. Y-3.750

   N105 Y1.5

   N110 G40 X-.75

   N115 Z.1 F6.0

   N120 G0 Z1.

   N125 G28

   N130 M01

   N135 T2 M6

   N140 S1000 M3

   N145 G0 X4.375 Y-2.

   N150 G43 H2 Z1.0

   N155 G98 G81 Z-.6 R.1 F3.

   N160 G80

   N165 G28

   N170 M01

   N175 T3 M6

   N180 S1200 M3

   N185 G0 X.5 Y-1.5

   N190 G0 G43 H3 Z1.0

   N195 G98 G81 Z-.575 R.1 F2.5

   N200 Y-2.

   N205 Y-2.5

   N210 G80

   N215 M5

   N220 G28

   N225 M01

   N230 T4 M6

   N235 S500 M3

   N240 X4.375 Y-2.

   N245 G43 H4 Z1.0

   N250 G98 G84 Z-.6 R.1 F31.25

   N255 G80

   N260 M5

   N265 G28

   N270 M30

   N275 %

7. O79

   N10 G17 G40 G49 G80 G90

   N15 T1 M6

   N20 G54

   N25 S1000 M3

   N30 G0 X-.75 Y-5.

   N35 G43 H1 Z1.

   N40 G0 Z.1

   N45 G1 Z-1. F10.

   N50 G41 D1 X0. Y-5.

   N55 X.0 Y-2.

   N60 G2 X0. Y-2. I2. J0.

   N65 G1 X0. Y.75

   N70 G40 X-1.0 Y.75

   N75 Z.1

N80 G0 Z1.

N85 X-2. Y-4.0

N90 Z.1

N95 G1 Z-.5 F10.

N100 G41 D1 Y-3.0

N105 X.27 Y-3.

N110 X2. Y0.

N115 X3.73 Y-3.

N120 X-.75

N125 G40 Y-4.

N130 Z.1

N135 G0 Z1.

N140 M5

N145 G28

N150 M30

%

8. O710

N10 G17 G40 G49 G80 G90

N15 T1 M6

N20 G54

N25 S1000 M3

N30 G0 X1.5 Y.125

N35 G43 H1 Z1.

N40 Z.1

N45 G1 Z-.5 F3.

N50 G41 D1 Y.5 F6.

N55 X0.

N60 G3 X0. Y-.5 I0. J-.5

N65 G1 X3.

N70 G3 X3.0 Y.5 I0. J.5

N75 G1 X1.5

N80 G40 Y.125

N85 Z.1 F10.

N90 G0 Z1.

N95 X-2.75 Y-2.75

N100 Z.1

N105 G1 Z-.5 F3.

N110 G41 D1 X-2.0 F6.

N115 G1 Y0.

N120 G2 X0. Y2. I2.0 J0.

N125 G1 X4.

N130 G2 X4. Y-2. I0. J-2.

N135 G1 X0.

N140 G2 X-2. Y0. I0. J2.

N145 G01 Y2.75

N150 G1 G40 X-2.75 Y3.0

N155 Z.1 F10.

N160 G0 Z1.

N165 G28

N170 M01

N175 T2 M6

N180 S500 M3

N185 G0 X4.5 Y0.

N190 G43 H2 Z1.

N195 G98 G81 Z-.65 R.1 F4.

N200 G80

N205 M5

N210 G28

N215 M30

%

SECTION M/UNIT 6/ CNC TURNING MACHINES

Self-Test Answers

1. Headstock, tailstock, tool turret, bed, carriage.
2. The basic function of the turret is to hold and quickly index cutting tools. Each tool position is numbered for identification. When the tool needs to be changed, the turret is moved to a clearance position and indexes, bringing the new tool into the cutting position. Most tool turrets can move bi-directionally to assure the fastest tool indexing time. Tool turrets can also be indexed manually, using a button or switch located on the control panel.
3. The bed of most turning centers lies at a slant to accommodate chip removal.
4. X
5. Chuck
6. Threading is the process of forming a helical groove on the outside or inside surface of a cylindrical part.
7. Parting
8. If the part sizes don't meet the part print requirements, the operator alters the offset, for the tool in the tool-wear offset table.
9. Geometry offsets, or workpiece coordinates, are used to tell the control where the workpiece is located.
10. The position or distance can be determined accurately by using the position screen on the control.
11. The workpiece coordinate is the distance from the tool tip, at the home position, to the workpiece zero point. The workpiece zero point is normally located at the end and center of the workpiece or at the chuck face and center of the machine.
12. Bar feeders automatically load rough stock into the work-holding device. The raw stock is fed into the machine by pneumatic or hydraulic pressure. The stock is fed the same distance each time through the use of stock stops. When the stock reaches the stock stop, the work-holding device closes and clamps the workpiece in place. Bar feeders eliminate the need for the operator to manually load individual part blanks.
13. Chip conveyers automatically remove chips from the bed of the machine. The chips produced by machining operations fall onto the conveyer track and are transported to scrap or recycling containers.
14. Most controls are equipped with a pulse-generating hand wheel and jog feed buttons.
15. Spindle speed and feed rate overrides are used to speed up or slow down the feed and speed of the machine during cutting operations.

SECTION M/UNIT 7/PROGRAMMING CNC TURNING CENTERS

Self-Test Answers

1. X

2. Constant Surface Speed enables the machinist to specify spindle speed in terms of surface speed instead of RPM. The machine will automatically maintain the rpm's needed to make sure cutting happens at the desired surface speed.

3. I, K

4. Tool-nose radius compensation is a type of offset used to control the shape of machined features. The turning center control can offset the path of the tool so that the part can be programmed just as it appears on the part print. There is an error that becomes apparent when we use the tool edges to set our workpiece coordinate position on a turning center. When we set the X and Z axes of the tool, we specify a single sharp point. Most of the tools we use for turning have radii. These radii on the tool nose can create inaccuracies in the cutter path. TNR compensation takes care of this.

5.

Point #	X (Diameter) Value	Z Value
Point 1	1.066	0
Point 2	1.250	−.092
Point 3	1.250	−1.250
Point 4	1.800	−1.250
Point 5	2.000	−1.350
Point 6	2.000	−2.000

6.

Point #	X (Diameter) Value	Z Value
Point 1	.300	0
Point 2	.500	−.100
Point 3	.500	−.750
Point 4	.824	−.750
Point 5	1.000	−.838
Point 6	1.000	−1.250
Point 7	.750	−1.250
Point 8	.750	−1.500
Point 9	1.324	−1.500
Point 10	1.500	−1.588
Point 11	1.500	−2.000

7. O0008

N10 G90 G20 G40

N15 G28 U0.0 W0.0

N20 G00 T0101

N25 G50 X10.00 Z12.00

N30 G00 Z.1

N35 G00 X.99

N40 G50 S3600

N45 G96 S200 M03 Spindle Forward

N50 G01 G42 Z0. F.01 Tool Nose Radius Compensation Right, lead in to (POINT 1)

N55 G01 X1.25 Z-.13 X Coordinate at (POINT 2)

N60 G01 Z-1.25 Z Coordinate at (POINT 3)

N65 G01 X1.80 X Coordinate at (POINT 4)

N70 G03 X2.0 Z-1.35 K-.1 CCW arc to X and Z Coordinate at (POINT 5)

N75 G01 Z-2.0 Z Coordinate at (POINT 6)

N80 G01 G40 Cancel Tool Nose Radius Compensation, lead out

N85 G28 U0. W0. M05 Return to the reference point

N90 T0100

N95 M30

%

8. • Start-up procedures
   • Tool call
   • Workpiece location block
   • Spindle speed control
   • Tool motion blocks
   • Home return

   Program end procedures

9.

Point#	X (Diameter) Value	Z Value
Point 1	0	0
Point 2	1.15	−.10
Point 3	1.25	−.58
Point 4	1.25	−.58
Point 5	2.00	−1.44
Point 6	2.00	−1.82
Point 7	2.26	−1.95
Point 8	3.00	−1.95
Point 9	3.00	−2.80

10. O0009

N10 G90 G20 G40 - Absolute mode, G20 - Inch mode

N15 G28 U.0.0 W0.0 Return to reference position - machine home in this example

N20 G00 T0101

N25 G50 X6.80 Z12.25 Work offset for tool 1 at X6.800 Z12.25

N30 G50 S3000 Set maximum spindle speed at 3000 RPM

N35 G96 S200 M03 Set constant surface speed, Spindle Forward

N40 G00 Z.1

N45 G00 X.80

N50 G01 G42 Z0 F.01 Tool Nose Radius Compensation Right, lead in to (POINT 2)

N55 G01 X1.25 Z-.10 X Coordinate, Z Coordinate at (POINT 3)

N60 G01 Z-.58 Z Coordinate at (POINT 4)

N65 G01 X2.00 Z-1.4 X Coordinate, Z Coordinate at (POINT 5)

N70 G01 Z-1.82 Z Coordinate at (POINT 6)

N75 G02 X2.26 Z-1.95 R.13 Arc direction, X and Z Coordinate at (POINT 7), Arc radius

N80 G01 X3.00 X Coordinate at (POINT 8)

N85 G01 Z-2.80 Z Coordinate at (POINT 9)

N90 G01 X3.10

N95 G40 X3.25 Tool-nose radius compensation cancel

N100 M09 Coolant off

N105 G28 U0.0 W0.0 M05 Return to the reference point through incremental X0.0 Z0.0

N110 T0100

N115 M30

%

SECTION M/UNIT 8/PROGRAMMING CANNED CYCLES FOR CNC TURNING CENTERS

Self-Test Answers

1. **False**
2. **True**
3. **G71**
4. A **G71** and a **G71** could be used.
5. In the first step, the tool needs to be positioned to the rough stock boundaries.
6. A **G83** would be used.
7. A **G82** could be used.
8. **D**
9. **D.100**
10. **G75**
11. **F**
12. Prior to calling the G76 thread-cutting cycle, the tool must be positioned to the major diameter of the thread plus double the K value or thread height. The tool should also be positioned in front of the thread start position in the Z by a distance of at least double the thread lead. This insures that the proper lead will be cut throughout the length of the thread. The spindle should be running in direct RPM (G97), not constant surface footage control.

    O1002

    N10 G20 G90 G40

    N15 G28 U0.0 W0.0

    N20 T0505

    N25 G96 S200 M03

    N30 G50 X8.25 Z12.50

    N35 G50 S3600

    N40 G0 Z.10

    N45 G0 X1.35

    N50 G71 P55 Q90 U.02 W.01 D.05 F.01 (Roughing Cycle)

    N55 G0 X.550

    N60 G1 Z.0

    N65 X.75 Z-.10

    N70 G1 Z-1.0

    N75 G1 X1.0 Z-1.5

    N80 G1 Z-1.625

N85 G2 X1.25 Z-1.75 R.125

N90 G1 Z-2.25

N95 G28 U0.0 W0.0

N100 T0303

N105 G96 S300 M03

N110 G50 X8.35 Z12.55

N115 G50 S3600

N120 G0 Z.10

N125 G0 X1.35

N130 G70 P55 Q90 F.008 (Finishing Cycle)

N135 G28 U0.0 W0.0

N140 T0300

N145 M30

%

13. N10 G90 G20 G40

    N15 G28 U0.0 W0.0

    N20 G00 T0707

    N40 G97 S300 M03

    N60 G00 Z.200

    N65 G00 X.825

    N70 G76 X.675 Z-.50 K.036 D.01 F.0625 A60

    N90 G28 U0.0 W0.0 M05

    N100 T0500

    N110 M30

    %

    O01003

    N10 G90 G20 G40

    N15 G28 U0.0 W0.0

    N20 G00 T0707

    N25 G97 S200 M03

    N30 G00 Z.200

    N35 G00 X.83

    N40 G76 X.675 Z-.50 K.0375 D.0120 F.0625 A60

    N45 G28 X2.00 M05

    N50 T0500

    N55 M30

    %

    O1004

    N010 G28 U0.0 W0.0

    N020 T0909

    N030 G97 S225 M03

    N040 G00 Z1.0 M08

    N050 G00 X0.0

    N060 G81 Z-1.75 R.5 F0.01

    N070 G80 G00 Z1. M09

    N080 G28 W1.0

    N090 T0900

    N090 M30

    %

SECTION M/UNIT 8/UNIT 9/ ADVANCED MACHINING PROCESSES

Self-Test Answers

1. EDM is electrodischarge machining, where workpiece material is removed by spark erosion.

   ECM is electrochemical machining, and ELG is electrolytic grinding. The processes take place in a conducting medium, and workpiece material is deplated by electrochemical action.

   Ultrasonic machining uses ultrasound to induce sympathetic vibrations in an anvil. The anvil's vibrations drive an abrasive material against the workpiece, providing the cutting action.

   Abrasive waterjet machining uses a high-pressure stream of water containing an abrasive material as a cutting tool.

   Laser machining uses the energy from an intense coherent beam of light energy as a cutting tool.

2. Wire-Feed EDM.
3. EDM.
4. ECM and ECDB.
5. Ultrasound vibrations provide the energy. Abrasive materials do the cutting.
6. False
7. True
8. Intensifier
9. False
10. Jewel (orifice), abrasive inlet, mixing tube, nozzle guard
11. True
12. False
13. Accurate, fast, cost effective, intricate shapes can be easily cut.
14. False
15. Cutting, marking, welding, heat treating, cladding, vapor deposition, engraving, scribing, trimming, and shock hardening.
16. False
17. Laser, Mirror, lens, XY table, CNC drive.
18. Gasses are used to assist in the cutting process. Gas selection is an important part of laser processing.

    Most steel will be processed with oxygen because it is fast. Oxygen is sometimes used with aluminum, but leaves dross that is more difficult to remove. Oxygen assist gas is the preferred gas for carbon steel cutting. It aids in the cutting process so it can cut faster and thicker than other assist gasses used with carbon steel.

    Aluminum can be processed with dry, clean shop air, but it will require a compressor with sufficient capacity. Air can be an economical way of cutting thin stainless with a better result than oxygen. Nitrogen is the preferred gas for aluminum and stainless. Argon is the most frequently used gas when cutting titanium.

19. Lasers produce the most precise cut with the smallest heat-affected zone of the thermal cutting technologies. They can cut up to one-inch thick carbon products and half-inch thick specialty metals. Consumables usage is less than with plasma or oxyfuel cutting. Very fragile parts can be laser cut with no support because there are no cutting forces. Laser technology is well suited to fabricating high accuracy parts, especially flexible materials.

APPENDIX **2**
General Tables

BASIC DESIGNATIONS

ISO metric threads are designated by the letter M followed by the *nominal size* in millimeters, and the *pitch* in millimeters, separated by the multiplication sign ×.

EXAMPLE

M16 × 1.5

Numbers in the table marked with an asterisk are the commercially available sizes in the United States.

TOLERANCE SYMBOLS

3 4 5 |6|
7 8 9

Numbers are used to define the amount of product tolerance permitted on either internal or external threads. Smaller-grade numbers carry smaller tolerances; that is, grade-4 tolerances are smaller than grade-6 tolerances, and grade-8 tolerances are larger than grade-6 tolerances.

e |H| |G| g

Letters are used to designate the "position" of the product thread tolerances relative to basic diameters. Lowercase letters are used for external threads, and capital letters for internal threads.

In some cases the position of the tolerance establishes an allowance (a definite clearance) between external and internal threads.

By combining the tolerance amount number and the tolerance position letter, the *tolerance symbol* is established that identifies the actual maximum and minimum product limits for external or internal threads. Generally, the first number and letter refer to the pitch diameter symbol. The second number and letter refer to the crest diameter symbol (minor diameter of internal threads or major diameter of external threads).

EXAMPLE

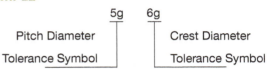

Where the pitch diameter and crest diameter tolerance symbols are the same, the symbol need be given only once.

EXAMPLE

6g
⊤
Pitch Diameter and Crest

Diameter Tolerance Symbol

It is recommended that the *coarse series* be selected whenever possible, and that *general-purpose* grade 6 be used for both internal and external threads.

Tolerance positions g for external threads and H for internal threads are preferred.

Other product information may also be conveyed by the ISO metric thread designations. Complete specifications and product limits may be found in the ISO Recommendations or in the B1 report "ISO Metric Screw Threads."

Some examples of ISO metric thread designations are as follows:

M10
M18 × 1.5
M6—6H
M4—6g
M12 × 1.25—6H
M20 × 2—6H/6g
M6 × 0.75—7g 6g

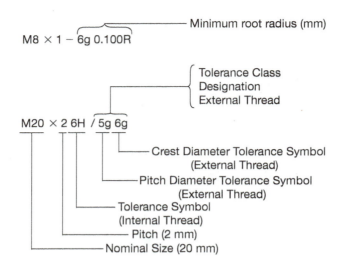

Table 1 Decimal Equivalents of Fractional Inches

Fraction Inch		Decimal Inch	Decimal Millimeters	Fraction Inch		Decimal Inch	Decimal Millimeters
	$\frac{1}{64}$	.015625	0.39688		$\frac{33}{64}$	.515625	13.09690
$\frac{1}{32}$		.03125	0.79375	$\frac{17}{32}$		.53125	13.49378
	$\frac{3}{64}$	.046875	1.19063		$\frac{35}{64}$	.546875	13.89065
$\frac{1}{16}$		.0625	1.58750	$\frac{9}{16}$		.5625	14.28753
	$\frac{5}{64}$	.078125	1.98438		$\frac{37}{64}$	.578125	14.68440
$\frac{3}{32}$		.09375	2.38125	$\frac{19}{32}$		.59375	15.08128
	$\frac{7}{64}$	.109375	2.77813		$\frac{39}{64}$	.609375	15.47816
$\frac{1}{8}$		.1250	3.17501	$\frac{5}{8}$		.6250	15.87503
	$\frac{9}{64}$	.140625	3.57188		$\frac{41}{64}$	.640625	16.27191
$\frac{5}{32}$		.15625	3.96876	$\frac{21}{32}$		.65625	16.66878
	$\frac{11}{64}$	.171875	4.36563		$\frac{43}{64}$	.671875	17.06566
$\frac{3}{16}$		.1875	4.76251	$\frac{11}{16}$		.6875	17.46253
	$\frac{13}{64}$	.203125	5.15939		$\frac{45}{64}$	.703125	17.85941
$\frac{7}{32}$		.21875	5.55626	$\frac{23}{32}$		.71875	18.25629
	$\frac{15}{64}$	.234375	5.95314		$\frac{47}{64}$	.734375	18.65316
$\frac{1}{4}$		.2500	6.35001	$\frac{3}{4}$		.7500	19.05004
	$\frac{17}{64}$	.265625	6.74689		$\frac{49}{64}$	.765625	19.44691
$\frac{9}{32}$		.28125	7.14376	$\frac{25}{32}$		.78125	19.84379
	$\frac{19}{64}$	.296875	7.54064		$\frac{51}{64}$	.796875	20.24067
$\frac{5}{16}$		.3125	7.93752	$\frac{13}{16}$		.8125	20.63754
	$\frac{21}{64}$	.328125	8.33439		$\frac{53}{64}$	.828125	21.03442
$\frac{11}{32}$		.34375	8.73127	$\frac{27}{32}$		.84375	21.43129
	$\frac{23}{64}$	.359375	9.12814		$\frac{55}{64}$	.859375	21.82817
$\frac{3}{8}$		.3750	9.52502	$\frac{7}{8}$		.8750	22.22504
	$\frac{25}{64}$	.390625	9.92189		$\frac{57}{64}$	.890625	22.62192
$\frac{13}{32}$		.40625	10.31877	$\frac{29}{32}$		.90625	23.01880
	$\frac{27}{64}$	.421875	10.71565		$\frac{59}{64}$	.921875	23.41567
$\frac{7}{16}$		.4375	11.11252	$\frac{15}{16}$		.9375	23.81255
	$\frac{29}{64}$	.453125	11.50940		$\frac{61}{64}$	.953125	24.20942
$\frac{15}{32}$		.46875	11.90627	$\frac{31}{32}$		.96875	24.60630
	$\frac{31}{64}$	.484375	12.30315		$\frac{63}{64}$	.984375	25.00318
$\frac{1}{2}$		.5000	12.70003	1		1.0000	25.40005

Table 2 Inch/Metric Conversion Table

The table is printed in five side-by-side column groups (Drill No. or Letter | Inch | mm), read left to right, top to bottom.

Drill No. or Letter	Inch	mm
	.001	0,0254
	.002	0,0508
	.003	0,0762
	.004	0,1016
	.005	0,1270
	.006	0,1524
	.007	0,1778
	.008	0,2032
	.009	0,2286
	.010	0,2540
	.011	0,2794
	.012	0,3048
80 .0135	.013	0,3302
79 .0145	.014	0,3556
	.015	0,3810
1/64	.0156	0,3969
78	.016	0,4064
	.017	0,4318
77	.018	0,4572
	.019	0,4826
76	.020	0,5080
75	.021	0,5334
	.022	0,5588
74 .0225	.023	0,5842
73	.024	0,6096
72	.025	0,6350
71	.026	0,6604
	.027	0,6858
70	.028	0,7112
69 .0292	.029	0,7366
	.030	0,7620
68	.031	0,7874
1/32	.0312	0,7937
67	.032	0,8128
66	.033	0,8382
	.034	0,8636
65	.035	0,8890
64	.036	0,9144
63	.037	0,9398
62	.038	0,9652
61	.039	0,9906
	.0394	1,0000
60	.040	1,0160
59	.041	1,0414
58	.042	1,0668
57	.043	1,0922
	.044	1,1176
	.045	1,1430
56 .0465	.046	1,1684
3/64	.0469	1,1906
	.047	1,1938
	.048	1,2192
	.049	1,2446
55	.050	1,2700
	.051	1,2954
	.052	1,3208
	.053	1,3462
	.054	1,3716
54	.055	1,3970
	.056	1,4224
	.057	1,4478
	.058	1,4732
	.059	1,4986
53 .0595	.060	1,5240
	.061	1,5494
	.062	1,5748
1/16	.0625	1,5875
	.063	1,6002
52 .0635	.064	1,6256
	.065	1,6510
	.066	1,6764
51	.067	1,7018
	.068	1,7272
	.069	1,7526
50	.070	1,7780
	.071	1,8034
	.072	1,8288
49	.073	1,8542
	.074	1,8796
	.075	1,9050
48	.076	1,9304
	.077	1,9558
47 .0785	.078	1,9812
5/64	.0781	1,9844
	.0787	2,0000
	.079	2,0066
	.080	2,0320
46	.081	2,0574
45	.082	2,0828
	.083	2,1082
	.084	2,1336
	.085	2,1590
44	.086	2,1844
	.087	2,2098
	.088	2,2352
43	.089	2,2606
	.090	2,2860
	.091	2,3114
	.092	2,3368
42 .0935 3/32	.093	2,3622
	.0937	2,3812
	.094	2,3876
	.095	2,4130
41	.096	2,4384
	.097	2,4638
40	.098	2,4892
	.099	2,5146
39 .0995	.100	2,5400
38 .1015	.101	2,5654
	.102	2,5908
37	.103	2,6162
	.104	2,6416
36 .1065	.105	2,6670
	.106	2,6924
	.107	2,7178
	.108	2,7432
	.109	2,7686
7/64	.1094	2,7781
35	.110	2,7940
34	.111	2,8194
	.112	2,8448
33	.113	2,8702
	.114	2,8956
	.115	2,9210
32	.116	2,9464
	.117	2,9718
	.118	2,9972
	.1181	3,0000
	.119	3,0226
31	.120	3,0480
	.121	3,0734
	.122	3,0988
	.123	3,1242
	.124	3,1496
1/8	.125	3,1750
	.126	3,2004
	.127	3,2258
	.128	3,2512
	.129	3,2766
30 .1285	.130	3,3020
	.131	3,3274
	.132	3,3528
	.133	3,3782
	.134	3,4036
	.135	3,4290
29	.136	3,4544
	.137	3,4798
	.138	3,5052
	.139	3,5306
28 .1405	.140	3,5560
9/64	.1406	3,5719
	.141	3,5814
	.142	3,6068
	.143	3,6322
27	.144	3,6576
	.145	3,6830
	.146	3,7084
26	.147	3,7338
	.148	3,7592
	.149	3,7846
25 .1495	.150	3,8100
	.151	3,8354
24	.152	3,8608
	.153	3,8862
23	.154	3,9116
	.155	3,9370
	.156	3,9624
5/32	.1562	3,9687
22	.157	3,9878
	.1575	4,0000
	.158	4,0132
	.159	4,0386
21	.159	4,0386
	.160	4,0640
20	.161	4,0894
	.162	4,1148
	.163	4,1402
	.164	4,1656
	.165	4,1910
19	.166	4,2164
	.167	4,2418
	.168	4,2672
	.169	4,2926
18 .1695	.170	4,3180
	.171	4,3434
11/64	.1719	4,3656
	.172	4,3688
17	.173	4,3942
	.174	4,4196
	.175	4,4450
	.176	4,4704
16	.177	4,4958
	.178	4,5212
	.179	4,5466
15	.180	4,5720
	.181	4,5974
14	.182	4,6228
	.183	4,6482
	.184	4,6736
13	.185	4,6990
	.186	4,7244
	.187	4,7498
3/16	.1875	4,7625
	.188	4,7752
12	.189	4,8006
	.190	4,8260
11	.191	4,8514
	.192	4,8768
	.193	4,9022
10 .1935	.194	4,9276
	.195	4,9530
9	.196	4,9784
	.1969	5,0000
	.197	5,0038
	.198	5,0292
8	.199	5,0546
	.200	5,0800
7	.201	5,1054
	.202	5,1308
	.203	5,1562
13/64	.2031	5,1594
6	.204	5,1816
5 .2055	.205	5,2070
	.206	5,2324
	.207	5,2578
	.208	5,2832
4	.209	5,3086
	.210	5,3340
	.211	5,3594
	.212	5,3848
3	.213	5,4102
	.214	5,4356
	.215	5,4610
	.216	5,4864
	.217	5,5118
	.218	5,5372
7/32	.2187	5,5562
	.219	5,5626
	.220	5,5880
2	.221	5,6134
	.222	5,6388
	.223	5,6642
	.224	5,6896
	.225	5,7150
	.226	5,7404
	.227	5,7658
1	.228	5,7912
	.229	5,8166
	.230	5,8420
	.231	5,8674
	.232	5,8928
	.233	5,9182
A	.234	5,9436
15/64	.2344	5,9531
	.235	5,9690
	.236	5,9944
	.2362	6,0000
	.237	6,0198
B	.238	6,0452
	.239	6,0706
	.240	6,0960
	.241	6,1214
C	.242	6,1468
	.243	6,1722
	.244	6,1976
	.245	6,2230
D	.246	6,2484
	.247	6,2738
	.248	6,2992
	.249	6,3246
E 1/4	.250	6,3500
	.251	6,3754
	.252	6,4008
	.253	6,4262
	.254	6,4516
	.255	6,4770
	.256	6,5024
F	.257	6,5278
	.258	6,5532
	.259	6,5786
	.260	6,6040
G	.261	6,6294
	.262	6,6548
	.263	6,6802
	.264	6,7056
	.265	6,7310
17/64	.2656	6,7469
H	.266	6,7564
	.267	6,7818
	.268	6,8072
	.269	6,8326
	.270	6,8580
	.271	6,8834
I	.272	6,9088
	.273	6,9342
	.274	6,9596
	.275	6,9850
	.2756	7,0000
	.276	7,0104
J	.277	7,0358
	.278	7,0612
	.279	7,0866
	.280	7,1120
K	.281	7,1374
9/32	.2812	7,1437
	.282	7,1628
	.283	7,1882
	.284	7,2136
	.285	7,2390
	.286	7,2644
	.287	7,2898
	.288	7,3152
	.289	7,3406
L	.290	7,3660
	.291	7,3914
	.292	7,4168
	.293	7,4422
	.294	7,4676
M	.295	7,4930
	.296	7,5184
19/64	.2969	7,5406
	.297	7,5438
	.298	7,5692
	.299	7,5946
	.300	7,6200
	.301	7,6454
N	.302	7,6708
	.303	7,6962
	.304	7,7216
	.305	7,7470
	.306	7,7724
	.307	7,7978
	.308	7,8232
	.309	7,8486
	.310	7,8740
	.311	7,8994
	.312	7,9248
5/16	.3125	7,9375
	.313	7,9502
	.314	7,9756
	.3150	8,0000
	.315	8,0010
O	.316	8,0264
	.317	8,0518
	.318	8,0772
	.319	8,1026
	.320	8,1280
	.321	8,1534
	.322	8,1788
P	.323	8,2042
	.324	8,2296
	.325	8,2550
	.326	8,2804
	.327	8,3058
	.328	8,3312
21/64	.3281	8,3344
	.329	8,3566
	.330	8,3820
	.331	8,4074
Q	.332	8,4328
	.333	8,4582
	.334	8,4836
	.335	8,5090
	.336	8,5344
	.337	8,5598
	.338	8,5852
R	.339	8,6106
	.340	8,6360
	.341	8,6614
	.342	8,6868
	.343	8,7122
11/32	.3437	8,7312
	.344	8,7376
	.345	8,7630
	.346	8,7884
	.347	8,8138
S	.348	8,8392
	.349	8,8646
	.350	8,8900
	.351	8,9154
	.352	8,9408
	.353	8,9662
	.354	8,9916
	.3543	9,0000
	.355	9,0170
	.356	9,0424
	.357	9,0678
T	.358	9,0932
	.359	9,1186
23/64	.3594	9,1281
	.360	9,1440
	.361	9,1694
	.362	9,1948
	.363	9,2202
	.364	9,2456
	.365	9,2710
	.366	9,2964
	.367	9,3218
U	.368	9,3472
	.369	9,3726
	.370	9,3980
	.371	9,4234
	.372	9,4488
	.373	9,4742
	.374	9,4996
3/8	.375	9,5250
	.376	9,5504
V	.377	9,5758
	.378	9,6012
	.379	9,6266
	.380	9,6520
	.381	9,6774
	.382	9,7028
	.383	9,7282
	.384	9,7536
	.385	9,7790
W	.386	9,8044
	.387	9,8298
	.388	9,8552
	.389	9,8806
	.390	9,9060
25/64	.3906	9,9219
	.391	9,9314
	.392	9,9568
	.393	9,9822
	.3937	10,0000
	.394	10,0076
	.395	10,0330
	.396	10,0584
X	.397	10,0838
	.398	10,1092
	.399	10,1346
	.400	10,1600
	.401	10,1854
	.402	10,2108
	.403	10,2362
Y	.404	10,2616
	.405	10,2870
	.406	10,3124
13/32	.4062	10,3187
	.407	10,3378
	.408	10,3632
	.409	10,3886
	.410	10,4140
	.411	10,4394
	.412	10,4648
Z	.413	10,4902
	.414	10,5156
	.415	10,5410
	.416	10,5664
	.417	10,5918
	.418	10,6172
	.419	10,6426
	.420	10,6680
	.421	10,6934
27/64	.4219	10,7156
	.422	10,7188
	.423	10,7442
	.424	10,7696
	.425	10,7950
	.426	10,8204
	.427	10,8458
	.428	10,8712
	.429	10,8966
	.430	10,9220
	.431	10,9474
	.432	10,9728
	.433	10,9982
	.4331	11,0000
	.434	11,0236
	.435	11,0490
	.436	11,0744
	.437	11,0998
7/16	.4375	11,1125
	.438	11,1252
	.439	11,1506
	.440	11,1760
	.441	11,2014
	.442	11,2268
	.443	11,2522
	.444	11,2776
	.445	11,3030
	.446	11,3284
	.447	11,3538
	.448	11,3792
	.449	11,4046
	.450	11,4300
	.451	11,4554
	.452	11,4808
	.453	11,5062
29/64	.4531	11,5094
	.454	11,5316
	.455	11,5570
	.456	11,5824
	.457	11,6078
	.458	11,6332
	.459	11,6586
	.460	11,6840
	.461	11,7094
	.462	11,7348
	.463	11,7602
	.464	11,7856
	.465	11,8110
	.466	11,8364
	.467	11,8618
	.468	11,8872
15/32	.4687	11,9062
	.469	11,9126
	.470	11,9380
	.471	11,9634
	.472	11,9888
	.4724	12,0000
	.473	12,0142
	.474	12,0396
	.475	12,0650
	.476	12,0904
	.477	12,1158
	.478	12,1412
	.479	12,1666
	.480	12,1920
	.481	12,2174
	.482	12,2428
	.483	12,2682
	.484	12,2936
31/64	.4844	12,3031
	.485	12,3190
	.486	12,3444
	.487	12,3698
	.488	12,3952
	.489	12,4206
	.490	12,4460
	.491	12,4714
	.492	12,4968
	.493	12,5222
	.494	12,5476
	.495	12,5730
	.496	12,5984
	.497	12,6238
	.498	12,6492
	.499	12,6746
1/2	.500	12,7000

Table 2 Inch/Metric Conversion Table (*continued*)

Drill No. or Letter	Inch	mm	Drill No. or Letter	Inch	mm	Drill No. or Letter	Inch	mm	Drill No. or Letter	Inch	mm	Drill No. or Letter	Inch	mm
				.600	15,2400					.800	20,3200			
	.501	12,7254		.601	15,2654		.701	17,8054		.801	20,3454		.901	22,8854
	.502	12,7508		.602	15,2908		.702	17,8308		.802	20,3708		.902	22,9108
	.503	12,7762		.603	15,3162		.703	17,8562		.803	20,3962		.903	22,9362
						45/64	.7031	17,8594						
	.504	12,8016		.604	15,3416		.704	17,8816		.804	20,4216		.904	22,9616
	.505	12,8270		.605	15,3670		.705	17,9070		.805	20,4470		.905	22,9870
	.506	12,8524		.606	15,3924		.706	17,9324		.806	20,4724		.9055	23,0000
													.906	23,0124
	.507	12,8778		.607	15,4178		.707	17,9578		.807	20,4978	29/32	.9062	23,0188
	.508	12,9032		.608	15,4432		.708	17,9832		.808	20,5232		.907	23,0378
							.7087	18,0000						
	.509	12,9286		.609	15,4686		.709	18,0086		.809	20,5486		.908	23,0632
			39/64	.6094	15,4781									
	.510	12,9540		.610	15,4940		.710	18,0340		.810	20,5740		.909	23,0886
	.511	12,9794		.611	15,5194		.711	18,0594		.811	20,5994		.910	23,1140
	.5118	13,0000												
	.512	13,0048		.612	15,5448		.712	18,0848		.812	20,6248		.911	23,1394
									13/16	.8125	20,6375			
	.513	13,0302		.613	15,5702		.713	18,1102		.813	20,6502		.912	23,1648
	.514	13,0556		.614	15,5956		.714	18,1356		.814	20,6756		.913	23,1902
	.515	13,0810		.615	15,6210		.715	18,1610		.815	20,7010		.914	23,2156
33/64	.5156	13,0968												
	.516	13,1064		.616	15,6464		.716	18,1864		.816	20,7264		.915	23,2410
	.517	13,1318		.617	15,6718		.717	18,2118		.817	20,7518		.916	23,2664
	.518	13,1572		.618	15,6972		.718	18,2372		.818	20,7772		.917	23,2918
						23/32	.7187	18,2562						
	.519	13,1826		.619	15,7226		.719	18,2626		.819	20,8026		.918	23,3172
	.520	13,2080		.620	15,7480		.720	18,2880		.820	20,8280		.919	23,3426
	.521	13,2334		.621	15,7734		.721	18,3134		.821	20,8534		.920	23,3680
	.522	13,2588		.622	15,7988		.722	18,3388		.822	20,8788		.921	23,3934
												59/64	.9219	23,4156
	.523	13,2842		.623	15,8242		.723	18,3642		.823	20,9042		.922	23,4188
	.524	13,3096		.624	15,8496		.724	18,3896		.824	20,9296		.923	23,4442
	.525	13,3350	5/8	.625	15,8750		.725	18,4150		.825	20,9550		.924	23,4696
	.526	13,3604		.626	15,9004		.726	18,4404		.826	20,9804		.925	23,4950
										.8268	21,0000		.926	23,5204
	.527	13,3858		.627	15,9258		.727	18,4658		.827	21,0058		.927	23,5458
	.528	13,4112		.628	15,9512		.728	18,4912		.828	21,0312		.928	23,5712
									53/64	.8281	21,0344			
	.529	13,4366		.629	15,9766		.729	18,5166		.829	21,0566		.929	23,5966
				.6299	16,0000									
	.530	13,4620		.630	16,0020		.730	18,5420		.830	21,0820		.930	23,6220
17/32	.531	13,4874		.631	16,0274		.731	18,5674		.831	21,1074		.931	23,6474
	.5312	13,4937												
	.532	13,5128		.632	16,0528		.732	18,5928		.832	21,1328		.932	23,6728
	.533	13,5382		.633	16,0782		.733	18,6182		.833	21,1582		.933	23,6982
	.534	13,5636		.634	16,1036		.734	18,6436		.834	21,1836		.934	23,7236
						47/64	.7344	18,6532						
	.535	13,5890		.635	16,1290		.735	18,6690		.835	21,2090		.935	23,7490
	.536	13,6144		.636	16,1544		.736	18,6944		.836	21,2344		.936	23,7744
	.537	13,6398		.637	16,1798		.737	18,7198		.837	21,2598		.937	23,7998
												15/16	.9375	23,8125
	.538	13,6652		.638	16,2052		.738	18,7452		.838	21,2852		.938	23,8252
	.539	13,6906		.639	16,2306		.739	18,7706		.839	21,3106		.939	23,8506
	.540	13,7160		.640	16,2560		.740	18,7960		.840	21,3360		.940	23,8760
			41/64	.6406	16,2719									
	.541	13,7414		.641	16,2814		.741	18,8214		.841	21,3614		.941	23,9014
	.542	13,7668		.642	16,3068		.742	18,8468		.842	21,3868		.942	23,9268
	.543	13,7922		.643	16,3322		.743	18,8722		.843	21,4122		.943	23,9522
									27/32	.8437	21,4312			
	.544	13,8176		.644	16,3576		.744	18,8976		.844	21,4376		.944	23,9776
													.9449	24,0000
	.545	13,8430		.645	16,3830		.745	18,9230		.845	21,4630		.945	24,0030
	.546	13,8684		.646	16,4084		.746	18,9484		.846	21,4884		.946	24,0284
35/64	.5469	13,8906												
	.547	13,8938		.647	16,4338		.747	18,9738		.847	21,5138		.947	24,0538
	.548	13,9192		.648	16,4592		.748	18,9992		.848	21,5392		.948	24,0792
							.7480	19,0000						
	.549	13,9446		.649	16,4846		.749	19,0246		.849	21,5646		.949	24,1046
	.550	13,9700		.650	16,5100	3/4	.750	19,0500		.850	21,5900		.950	24,1300
	.551	13,9954		.651	16,5354		.751	19,0754		.851	21,6154		.951	24,1554
	.5512	14,0000												
	.552	14,0208		.652	16,5608		.752	19,1008		.852	21,6408		.952	24,1808
	.553	14,0462		.653	16,5862		.753	19,1262		.853	21,6662		.953	24,2062
												61/64	.9531	24,2094
	.554	14,0716		.654	16,6116		.754	19,1516		.854	21,6916		.954	24,2316
	.555	14,0970		.655	16,6370		.755	19,1770		.855	21,7170		.955	24,2570
	.556	14,1224		.656	16,6524		.756	19,2024		.856	21,7424		.956	24,2824
			21/32	.6562	16,6687									
	.557	14,1478		.657	16,6878		.757	19,2278		.857	21,7678		.957	24,3078
	.558	14,1732		.658	16,7132		.758	19,2532		.858	21,7932		.958	24,3332
	.559	14,1986		.659	16,7386		.759	19,2786		.859	21,8186		.959	24,3586
									55/64	.8594	21,8281			
	.560	14,2240		.660	16,7640		.760	19,3040		.860	21,8440		.960	24,3840
	.561	14,2494		.661	16,7894		.761	19,3294		.861	21,8694		.961	24,4094
	.562	14,2748		.662	16,8148		.762	19,3548		.862	21,8948		.962	24,4348
9/16	.5625	14,2875												
	.563	14,3002		.663	16,8402		.763	19,3802		.863	21,9202		.963	24,4602
	.564	14,3256		.664	16,8656		.764	19,4056		.864	21,9456		.964	24,4856
	.565	14,3510		.665	16,8910		.765	19,4310		.865	21,9710		.965	24,5110
	.566	14,3764		.666	16,9164	49/64	.7656	19,4469		.866	21,9964		.966	24,5364
	.567	14,4018		.667	16,9418		.766	19,4564		.8661	22,0000		.967	24,5618
	.568	14,4272		.668	16,9672		.767	19,4818		.867	22,0218		.968	24,5872
	.569	14,4526		.669	16,9926		.768	19,5072		.868	22,0472	31/32	.9687	24,6062
				.6693	17,0000									
	.570	14,4780		.670	17,0180		.769	19,5326		.869	22,0726		.969	24,6126
	.571	14,5034		.671	17,0434		.770	19,5580		.870	22,0980		.970	24,6380
			43/64	.6719	17,0656									
	.572	14,5288		.672	17,0688		.771	19,5834		.871	22,1234		.971	24,6634
	.573	14,5542		.673	17,0942		.772	19,6088		.872	22,1488		.972	24,6888
	.574	14,5796		.674	17,1196		.773	19,6342		.873	22,1742		.973	24,7142
	.575	14,6050		.675	17,1450		.774	19,6596		.874	22,1996		.974	24,7396
	.576	14,6304		.676	17,1704		.775	19,6850	7/8	.875	22,2250		.975	24,7650
	.577	14,6558		.677	17,1958		.776	19,7104		.876	22,2504		.976	24,7904
	.578	14,6812		.678	17,2212		.777	19,7358		.877	22,2758		.977	24,8158
37/64	.5781	14,6844												
	.579	14,7066		.679	17,2466		.778	19,7612		.878	22,3012		.978	24,8412
	.580	14,7320		.680	17,2720		.779	19,7866		.879	22,3266		.979	24,8666
	.581	14,7574		.681	17,2974		.780	19,8120		.880	22,3520		.980	24,8920
	.582	14,7828		.682	17,3228		.781	19,8374		.881	22,3774		.981	24,9174
						25/32	.7812	19,8433						
	.583	14,8082		.683	17,3482		.782	19,8628		.882	22,4028		.982	24,9428
	.584	14,8336		.684	17,3736		.783	19,8882		.883	22,4282		.983	24,9682
	.585	14,8590		.685	17,3990		.784	19,9136		.884	22,4536		.984	24,9936
													.9843	25,0000
	.586	14,8844		.686	17,4244		.785	19,9390		.885	22,4790	63/64	.9844	25,0031
	.587	14,9098		.687	17,4498		.786	19,9644		.886	22,5044		.985	25,0190
			11/16	.6875	17,4625		.7874	20,0000						
	.588	14,9352		.688	17,4752		.787	19,9898		.887	22,5298		.986	25,0444
	.589	14,9606		.689	17,5006		.788	20,0152		.888	22,5552		.987	25,0698
	.590	14,9860		.690	17,5260		.789	20,0406		.889	22,5806		.988	25,0952
	.5906	15,0000							57/64	.8906	22,6219			
	.591	15,0114		.691	17,5514		.790	20,0660		.890	22,6060		.989	25,1206
	.592	15,0368		.692	17,5768		.791	20,0914		.891	22,6314		.990	25,1460
	.593	15,0622		.693	17,6022		.792	20,1168		.892	22,6568		.991	25,1714
19/32	.5937	15,0812												
	.594	15,0876		.694	17,6276		.793	20,1422		.893	22,6822		.992	25,1968
	.595	15,1130		.695	17,6530		.794	20,1676		.894	22,7076		.993	25,2222
	.596	15,1384		.696	17,6784		.795	20,1930		.895	22,7330		.994	25,2476
	.597	15,1638		.697	17,7038		.796	20,2184		.896	22,7584		.995	25,2730
						51/64	.7969	20,2402						
	.598	15,1892		.698	17,7292		.797	20,2438		.897	22,7838		.996	25,2984
	.599	15,2146		.699	17,7546		.798	20,2692		.898	22,8092		.997	25,3238
				.700	17,7800		.799	20,2946		.899	22,8346		.998	25,3492
										.900	22,8600		.999	25,3746
													1.000	25,4000

Table 3 Tap Drill Sizes

Tap	Tap Drill	Decimal Equivalent of Tap Drill	Tap	Tap Drill	Decimal Equivalent of Tap Drill	Tap	Tap Drill	Decimal Equivalent of Tap Drill
0–80	56	.0465		28	.1405		Q	.3320
	$\frac{3}{64}$	.0469		$\frac{9}{64}$	.1406		R	.3390
1–64	54	.0550	10–24	27	.1440	$\frac{7}{16}$–14	T	.3580
	53	.0595		26	.1470		$\frac{23}{64}$	.3594
1–72	53	.0595		25	.1495		U	.3680
	$\frac{1}{16}$	.0625		24	.1520		$\frac{3}{8}$	.3750
2–56	51	.0670		23	.1540		V	.3770
	50	.0700		$\frac{5}{32}$	.1563	$\frac{7}{16}$–20	W	.3860
	49	.0730		22	.1570		$\frac{25}{64}$	.3906
2–64	50	.0700	10–32	$\frac{5}{32}$	.1563		X	.3970
	49	.0730		22	.1570	$\frac{1}{2}$–13	$\frac{27}{64}$	.4219
3–48	48	.0760		21	.1590		$\frac{7}{16}$	.4375
	$\frac{5}{64}$	.0781		20	.1610	$\frac{1}{2}$–20	$\frac{29}{64}$	.4531
	47	.0785		19	.1660	$\frac{9}{16}$–12	$\frac{15}{32}$	.4688
	46	.0810	12–24	$\frac{11}{64}$	.1719		$\frac{31}{64}$	.4844
	45	.0820		17	.1730	$\frac{9}{16}$–18	$\frac{1}{2}$	.500
3–56	46	.0810		16	.1770		$\frac{33}{64}$	.5156
	45	.0820		15	.1800	$\frac{5}{8}$–11	$\frac{17}{32}$	.5313
	44	.0860		14	.1820		$\frac{35}{64}$	.5469
4–40	44	.0860	12–28	16	.1770	$\frac{5}{3}$–18	$\frac{9}{16}$	.5625
	43	.0890		15	.1800		$\frac{37}{64}$	.5781
	42	.0935		14	.1820	$\frac{3}{4}$–10	$\frac{41}{64}$	.6406
	$\frac{3}{32}$	.0938		13	.1850		$\frac{21}{32}$	.6563
4–48	42	.0935		$\frac{3}{16}$	.1875	$\frac{3}{4}$–16	$\frac{11}{16}$	.6875
	$\frac{3}{32}$	.0938	$\frac{1}{4}$–20	9	.1960	$\frac{7}{8}$–9	$\frac{49}{64}$	.7656
	41	.0960		8	.1990		$\frac{25}{32}$	.7812
5–40	40	.0980		7	.2010	$\frac{7}{8}$–14	$\frac{51}{64}$	.7969
	39	.0995		$\frac{13}{64}$	.2031		$\frac{13}{16}$	.8125
	38	.1015		6	.2040	1–8	$\frac{55}{64}$	.8594
	37	.1040		5	.2055		$\frac{7}{8}$	.875
5–44	38	.1015		4	.2090		$\frac{57}{64}$	.8906
	37	.1040	$\frac{1}{4}$–28	3	.2130		$\frac{29}{32}$	.9063
	36	.1065		$\frac{7}{32}$	.2188	1–12	$\frac{29}{32}$	.9063
6–32	37	.1040		2	.2210		$\frac{59}{64}$	.9219
	36	.1065	$\frac{5}{16}$–18	F	.2570		$\frac{15}{16}$	.9375
	$\frac{7}{64}$	.1094		G	.2610	1–14	$\frac{59}{64}$	.9219
	35	.1100		$\frac{17}{64}$	.2656		$\frac{15}{16}$	.9375
	34	.1110		H	.2660	$1\frac{1}{8}$–7	$\frac{31}{32}$	.9688
	33	.1130	$\frac{5}{16}$–24	H	.2660		$\frac{63}{64}$	.9844
6–40	34	.1110		I	.2720		1	1.0000
	33	.1130		J	.2770			
	32	.1160	$\frac{3}{8}$–16	$\frac{5}{16}$	.3125			
8–32	29	.1360		O	.3160			
	28	.1405		P	.3230			
8–36	29	.1360	$\frac{3}{8}$–24	$\frac{21}{64}$	.3281			

Table 4 Metric Tap Drill Sizes[a]

Metric Tap Size	Recommended Metric Drill				Closest Recommended Inch Drill			
	Drill Size (mm)	Inch Equivalent	Probable Hole Size (in.)	Probable Percent of Thread	Drill Size	Inch Equivalent	Probable Hole Size (in.)	Probable Percent of Thread
M1.6 × .35	1.25	.0492	.0507	69	—	—	—	—
M1.8 × .35	1.45	.0571	.0586	69	—	—	—	—
M2 × .4	1.60	.0630	.0647	69	#52	.0635	.0652	66
M2.2 × .45	1.75	.0689	.0706	70	—	—	—	—
M2.5 × .45	2.05	.0807	.0826	69	#46	.0810	.0829	67
*M3 × .5	2.50	.0984	.1007	68	#40	.0980	.1003	70
M3.5 × .6	2.90	.1142	.1168	68	#33	.1130	.1156	72
*M4 × .7	3.30	.1299	.1328	69	#30	.1285	.1314	73
M4.5 × .75	3.70	.1457	.1489	74	#26	.1470	.1502	70
*M5 × .8	4.20	.1654	.1686	69	#19	.1660	.1692	68
*M6 × 1	5.00	.1968	.2006	70	#9	.1960	.1998	71
M7 × 1	6.00	.2362	.2400	70	$\frac{15}{64}$	.2344	.2382	73
*M8 × 1.25	6.70	.2638	.2679	74	$\frac{17}{64}$	.2656	.2697	71
M8 × 1	7.00	.2756	.2797	69	J	.2770	.2811	66
*M10 × 1.5	8.50	.3346	.3390	71	Q	.3320	.3364	75
M10 × 1.25	8.70	.3425	.3471	73	$\frac{11}{32}$	.3438	.3483	71
*M12 × 1.75	10.20	.4016	.4063	74	Y	.4040	.4087	71
M12 × 1.25	10.80	.4252	.4299	67	$\frac{27}{64}$	.4219	.4266	72
M14 × 2	12.00	.4724	.4772	72	$\frac{15}{32}$	.4688	.4736	76
M14 × 1.5	12.50	.4921	.4969	71	—	—	—	—
*M16 × 2	14.00	.5512	.5561	72	$\frac{35}{64}$	.5469	.5518	76
M16 × 1.5	14.50	.5709	.5758	71	—	—	—	—
M18 × 2.5	15.50	.6102	.6152	73	$\frac{39}{64}$	.6094	.6144	74
M18 × 1.5	16.50	.6496	.6546	70	—	—	—	—
*M20 × 2.5	17.50	.6890	.6942	73	$\frac{11}{16}$	.6875	.6925	74
M20 × 1.5	18.50	.7283	.7335	70	—	—	—	—
M22 × 2.5	19.50	.7677	.7729	73	$\frac{49}{64}$	.7656	.7708	75
M22 × 1.5	20.50	.8071	.8123	70	—	—	—	—
*M24 × 3	21.00	.8268	.8327	73	$\frac{53}{64}$	.8281	.8340	72
M24 × 2	22.00	.8661	.8720	71	—	—	—	—
M27 × 3	24.00	.9449	.9511	73	$\frac{15}{16}$	.9375	.9435	78
M27 × 2	25.00	.9843 ⎫	.9913	70	$\frac{63}{64}$	.9844	.9914	70
*M30 × 3.5	26.50	1.0433 ⎪						
M30 × 2	28.00	1.1024 ⎪						
M33 × 3.5	29.50	1.1614 ⎪						
M33 × 2	31.00	1.2205 ⎬ Reaming recommended to the drill size shown						
M36 × 4	32.00	1.2598 ⎪						
M36 × 3	33.00	1.2992 ⎪						
M39 × 4	35.00	1.3780 ⎪						
M39 × 3	36.00	1.4173 ⎭						

[a]Formula for metric tap drill size: Basic major diameter (mm) $- \dfrac{\% \text{ thread} \times \text{pitch (mm)}}{76.980} = $ drilled hole size (mm)

Formula for percent of thread: $\dfrac{76.980}{\text{Pitch (mm)}} \times [\text{basic major diameter (mm)} - \text{drilled hole size (mm)}] = $ percent of thread

Table 5A Tapers

Taper per Foot	Amount of Taper Length Tapered Portion (in.)													
	$\frac{1}{32}$	$\frac{1}{16}$	$\frac{1}{8}$	$\frac{3}{16}$	$\frac{1}{4}$	$\frac{5}{16}$	$\frac{3}{8}$	$\frac{7}{16}$	$\frac{1}{2}$	$\frac{9}{16}$	$\frac{5}{8}$	$\frac{11}{16}$	$\frac{3}{4}$	$\frac{13}{16}$
$\frac{1}{16}$	.0002	.0003	.0007	.0010	.0013	.0016	.0020	.0023	.0026	.0029	.0033	.0036	.0039	.0042
$\frac{3}{32}$	.0002	.0005	.0010	.0015	.0020	.0024	.0029	.0034	.0039	.0044	.0049	.0054	.0059	.0063
$\frac{1}{8}$	.0003	.0007	.0013	.0020	.0026	.0033	.0039	.0046	.0052	.0059	.0065	.0072	.0078	.0085
$\frac{1}{4}$	.0007	.0013	.0026	.0039	.0052	.0065	.0078	.0091	.0104	.0117	.0130	.0143	.0156	.0169
$\frac{3}{8}$	.0010	.0020	.0039	.0059	.0078	.0098	.0117	.0137	.0156	.0176	.0195	.0215	.0234	.0254
$\frac{1}{2}$	.0013	.0026	.0052	.0078	.0104	.0130	.0156	.0182	.0208	.0234	.0260	.0286	.0312	.0339
$\frac{5}{8}$	.0016	.0033	.0065	.0098	.0130	.0163	.0195	.0228	.0260	.0293	.0326	.0358	.0391	.0423
$\frac{3}{4}$	.0020	.0039	.0078	.0117	.0156	.0195	.0234	.0273	.0312	.0352	.0391	.0430	.0469	.0508
1	.0026	.0052	.0104	.0156	.0208	.0260	.0312	.0365	.0417	.0469	.0521	.0573	.0625	.0677
$1\frac{1}{4}$	.0063	.0065	.0130	.0195	.0260	.0326	.0391	.0456	.0521	.0586	.0651	.0716	.0781	.0846

Taper per Foot	Amount of Taper Length Tapered Portion (in.)													
	$\frac{7}{8}$	$\frac{15}{16}$	1	2	3	4	5	6	7	8	9	10	11	12
$\frac{1}{16}$	.0046	.0049	.0052	.0104	.0156	.0208	.0260	.0312	.0365	.0417	.0469	.0521	.0573	.0625
$\frac{3}{32}$	.0068	.0073	.0078	.0156	.0234	.0312	.0391	.0469	.0547	.0625	.0703	.0781	.0859	.0937
$\frac{1}{8}$	.0091	.0098	.0104	.0208	.0312	.0417	.0521	.0625	.0729	.0833	.0937	.1042	.1146	.1250
$\frac{1}{4}$	.0182	.0195	.0208	.0417	.0625	.0833	.1042	.1250	.1458	.1667	.1875	.2083	.2292	.2500
$\frac{3}{8}$	.0273	.0293	.0312	.0625	.0937	.1250	.1562	.1875	.2187	.2500	.2812	.3125	.3437	.3750
$\frac{1}{2}$	.0365	.0391	.0417	.0833	.1250	.1667	.2083	.2500	.2917	.3333	.3750	.4167	.4583	.5000
$\frac{5}{8}$	.0456	.0488	.0521	.1042	.1562	.2083	.2604	.3125	.3646	.4167	.4687	.5208	.5729	.6250
$\frac{3}{4}$	.0547	.0586	.0625	.1250	.1875	.2500	.3125	.3750	.4375	.5000	.5625	.6250	.6875	.7500
1	.0729	.0781	.0833	.1667	.2500	.3333	.4167	.5000	.5833	.6667	.7500	.8333	.9167	1.0000
$1\frac{1}{4}$	.0911	.0977	.1042	.2083	.3125	.4167	.5208	.6250	.7292	.8333	.9375	1.0417	1.1458	1.2500

Taper per Foot	Amount of Taper Length Tapered Portion (in.)													
	13	14	15	16	17	18	19	20	21	22	23	24	...	...
$\frac{1}{16}$	.0677	.0729	.0781	.0833	.0885	.0937	.0990	.1042	.1094	.1146	.1198	.1250	...	...
$\frac{3}{32}$	.1016	.1094	.1172	.1250	.1328	.1406	.1484	.1562	.1641	.1719	.1797	.1875	...	...
$\frac{1}{8}$	.1354	.1458	.1562	.1667	.1771	.1875	.1979	.2083	.2187	.2292	.2396	.2500	...	...
$\frac{1}{4}$	.2708	.2917	.3125	.3333	.3542	.3750	.3958	.4167	.4375	.4583	.4792	.5000	...	...
$\frac{3}{8}$	.4062	.4375	.4687	.5000	.5312	.5625	.5937	.6250	.6562	.6875	.7187	.7500	...	...
$\frac{1}{2}$	.5417	.5833	.6250	.6667	.7083	.7500	.7917	.8333	.8750	.9167	.9583	1.0000	...	...
$\frac{5}{8}$	.6771	.7292	.7812	.8333	.8854	.9375	.9896	1.0417	1.0937	1.1458	1.1979	1.2500	...	...
$\frac{3}{4}$	.8125	.8750	.9375	1.0000	1.0625	1.1250	1.1875	1.2500	1.3125	1.3750	1.4375	1.5000	...	...
1	1.0833	1.1667	1.2500	1.3333	1.4167	1.5000	1.5833	1.6667	1.7500	1.8333	1.9167	2.0000	...	...
$1\frac{1}{4}$	1.3542	1.4583	1.5625	1.6667	1.7708	1.8750	1.9792	2.0833	2.1875	2.2917	2.3958	2.5000	...	...

Table 5B Tapers and Angles

Taper per Foot	Included Angle			Angle with Center Line			Taper per Inch	Taper per Inch from Center Line
	Deg.	Min.	Sec.	Deg.	Min.	Sec.		
$\frac{1}{8}$	0	35	47	0	17	54	.010416	.005208
$\frac{3}{16}$	0	53	44	0	26	52	.015625	.007812
$\frac{1}{4}$	1	11	38	0	35	49	.020833	.010416
$\frac{5}{16}$	1	29	31	0	44	46	.026042	.013021
$\frac{3}{8}$	1	47	25	0	53	42	.031250	.015625
$\frac{7}{16}$	2	5	18	1	2	39	.036458	.018229
$\frac{1}{2}$	2	23	12	1	11	36	.041667	.020833
$\frac{9}{16}$	2	41	7	1	20	34	.046875	.023438
$\frac{5}{8}$	2	59	3	1	29	31	.052084	.026042
$\frac{11}{16}$	3	16	56	1	38	28	.057292	.028646
$\frac{3}{4}$	3	34	48	1	47	24	.062500	.031250
$\frac{13}{16}$	3	52	42	1	56	21	.067708	.033854
$\frac{7}{8}$	4	10	32	2	5	16	.072917	.036456
$\frac{15}{16}$	4	28	26	2	14	13	.078125	.039063
1	4	46	19	2	23	10	.083330	.041667
$1\frac{1}{4}$	5	57	45	2	58	53	.104166	.052084
$1\frac{1}{2}$	7	9	10	3	34	35	.125000	.062500
$1\frac{3}{4}$	8	20	28	4	10	14	.145833	.072917
2	9	31	37	4	45	49	.166666	.083332
$2\frac{1}{2}$	11	53	38	5	56	49	.208333	.104166
3	14	2	0	7	1	0	.250000	.125000
$3\frac{1}{2}$	16	35	39	8	17	49	.291666	.145833
4	18	55	31	9	27	44	.333333	.166666
$4\frac{1}{2}$	21	14	20	10	37	10	.375000	.187500
5	23	32	12	11	46	6	.416666	.208333
6	28	4	20	14	2	10	.500000	.250000

Table 6 General Measurements

Measurement Rules
Length

Side of square of equal periphery as circle = diameter × 0.7854.
Diameter of circle of equal periphery as square = side × 1.2732.
Length of arc = number of degrees × diameter × 0.008727.

Area

Triangle = base × half perpendicular height.
Parallelogram = base × perpendicular height.
Trapezoid = half the sum of the parallel sides × perpendicular height.
Trapezium, divide two triangles and find area of the triangles.
Parabola = base × $\frac{2}{3}$ height.
Ellipse = long diameter × short diameter × 0.7854.
Regular polygon = sum of sides × half perpendicular distance from center to sides.
Surface of cylinder = circumference × length + area of two ends.
Surface of pyramid or cone = circumference of base × $\frac{1}{2}$ of the slant height + area of the base.
Surface of a frustrum of a regular right pyramid or cone = sum of peripheries or circumferences of the two ends × half slant height + area of both ends.
Area of rectangle = length × breadth.

General Information

To find the circumference of a circle, multiply diameter by 3.1416.
To find diameter of a circle, multiply circumference by .31831.
To find area of a circle, multiply square of radius by 3.1416.
Area of rectangle: Length multiplied by breadth. Doubling the diameter of a circle increases its area four times.
To find area of a triangle, multiply base by $\frac{1}{2}$ perpendicular height.
To find side of square inscribed in a circle, multiply diameter by 0.7071, or multiply circumference by 0.2251, or divide circumference by 4.4428.
To find diameter of circle circumscribing a square, multiply one side by 1.4142.
A side multiplied by 4.4428 equals circumference of its circumscribing circle.
A side multiplied by 1.128 equals diameter of a circle of equal area.
A side multiplied by 3.547 equals circumference of a circle of equal area.

Equivalent Measures
Measures of Length

1 Meter =			**1 Inch =**	
39.37	inches		1000.	mils (thousandths)
3.28083	feet		0.0833	foot
1.09361	yards		0.02777	yard
1000.	millimeters		25.40	millimeters
100.	centimeters		2.540	centimeters
10.	decimeters		**1 Foot =**	
0.001	kilometer		12.	inches
1 Centimeter =			1.33333	yards
0.3937	inch		0.0001893	mile
0.0328083	foot		0.30480	meter
10.	millimeters		30.480	centimeters
0.01	meter		**1 Yard =**	
1 Millimeter =			36.	inches
39.370	mils		3.	feet
0.03937	inch (or $\frac{1}{25}$ inch nearly)		0.0005681	mile
0.001	meter		0.914402	meter
1 Kilometer =			**1 Mile =**	
3280.83	feet		63360.	inches
1093.61	yards		5280.	feet
0.62137	mile		1760.	yards
1000.	meters		320.	rods
1 Mil =			8.	furlongs
0.001	inch		1609.35	meters
0.02540	millimeter		1.60935	kilometers
0.00254	centimeter			

Table 6 General Measurements (*continued*)

Measures of Volume and Capacity

1 Cubic Meter =		**1 Liter =**	
61023.4	cubic inches	1.	cubic decimeter
35.3145	cubic feet	61.0234	cubic inches
1.30794	cubic yards	0.353145	cubic foot
1000.	liters	1000.	cubic centimeters or centiliters
264.170	gallons U.S. liquid = 231 cubic inches	0.001	cubic meter
1 Cubic Decimeter =		0.26417	U.S. gallon liquid
61.0234	cubic inches	1.0567	U.S. quarts
0.0353145	cubic foot	2.202	pounds of water at 62 degrees Fahrenheit
0.26417	U.S. liquid gallon	**1 Cubic Yard =**	
1000.	cubic centimeters	46656.	cubic inches
0.001	cubic meter	27.	cubic feet
1 Cubic Centimeter =		0.76456	cubic meter
0.0000353	cubic foot	**1 Cubic Foot =**	
0.0610234	cubic inch	1728.	cubic inches
1000.	cubic millimeters	0.03703703	cubic yard
0.001	liter	28.317	cubic decimeters or liters
1 Cubic Millimeter =		0.028317	cubic meter
0.000061023	cubic inch	7.4805	gallons
0.0000000353	cubic foot	**1 Cubic Inch =**	
0.001	cubic centimeter	16.3872	cubic centimeters
		1 Gallon (British) =	
		4.54374	liters
		1 Gallon (U.S.) =	
		3.78543	liters

Measures of Weight

1 Gram =		**1 Grain =**	
15.432	grains	0.064799	gram
0.0022046	lb (avoir.)	**1 Ounce =**	
0.03527	oz (avoir.)	437.5	grains
1 Kilogram =		0.0625	pound
1000.	grams	28.3496	grams
2.20462	lb (avoir.)	**1 Pound =**	
35.2739	oz (avoir.)	7000.	grains
1 Metric Ton =		16.	ounces
2204.62	pounds	453.593	grams
0.984206	ton of 2240 pounds	0.453593	kilogram
22.0462	cwt	**1 Ton (2240 pounds) =**	
1.10231	tons of 2000 pounds	1.01605	metric tons
1000.	kilograms	1016.05	kilograms

Table 7A Density or Specific Gravity of Metals and Alloys

Material	Specific Gravity	Weight in Lb.		Cu. In. in 1 Lb.
		Cu. Ft.	Cu. In.	
Aluminum —cast	2.569	160.	.093	10.80
Aluminum —bronze	7.787	485.	.281	3.56
Aluminum —wrought	2.681	167.	.097	10.35
Antimony	6.712	418.	.242	4.13
Arsenic	5.748	358.	.207	4.83
Benedict nickel	8.691	542.6	.3140	3.19
Bismuth	9.827	612.	.354	2.82
Gold (pure)	19.316	1203.	.696	1.44
Standard 22 carat fine	17.502	1090.	.631	1.59
Iron — cast	6.904	430.	.249	4.02
	7.386	499.	.266	3.76
	7.209	464.	.260	3.85
Iron —wrought	7.547	470.	.272	3.56
	7.803	486.	.281	3.68
	7.707	480.	.278	3.60
Lead —cast	11.368	708.	.410	2.44
Lead —sheet	11.432	712.	.412	2.43
Manganese	8.012	499.	.289	2.46
Nickel —cast	8.285	516.	.299	3.35
Nickel —rolled	8.687	541.	.313	3.19
Platinum	21.516	1340.	.775	1.29
Silver	10.517	655.	.379	2.64
Steel	7.820	487.	.282	3.55
	7.916	493.	.285	3.51
	7.868	490.	.284	3.53
Tin	7.418	462.	.267	3.74
White metal (Babbitt's)	7.322	456.	.264	3.79
Zinc —cast	6.872	428.	.248	4.05
Zinc —sheet	7.209	449.	.260	3.84

Table 7B Approximate Melting Points of Metals and Various Substances

Solid	Degrees Celsius	Degrees Fahrenheit
Alloy —3 lead, 2 tin, 5 bismuth	100	212
Alloy —1 lead, $1\frac{1}{2}$ tin	200	392
Alloy —1 lead, 1 tin	215	419
Aluminum	657.3	1215
Antimony	430 to 630	806 to 1166
Bismuth	269.2	517
Brass	1030	1886
Bronze	920	1688
Cadmium	320	608
Calcium	760	1400
Chromium	1487 to 1515	2709 to 2749
Cobalt	1463 to 1500	2665 to 2732
Copper	1054 to 1084	1929 to 1893
Gold	1045 to 1064	1913 to 1947
Iridium	1950 to 2500	3542 to 4532
Iron — cast gray	1220 to 1530	2228 to 2786
Iron — cast white	1050 to 1135	1922 to 2075
Iron — wrought	1500 to 1600	1912 to 2732
Lead	327	620
Magnesium	750	1382
Manganese	1207 to 1245	2205 to 2273
Mercury	−39.7	−39.5
Nickel	1435	2615
Osmium	2500	4532
Palladium	1546 to 1900	2815 to 3452
Platinum	1753 to 1780	3187 to 3276
Potassium	62	144
Rhodium	2000	3632
Ruthenium	2000+	3632
Silver	960	1760
Sodium	79 to 95	174.2 to 203
Steel	1300 to 1378	2372 to 2532
Steel — hard	1410	2570
Steel — mild	1475	2687
Tin	232	449
Titanium	1700	3092
Tungsten	3000	5432
Vanadium	1775	3227
Zinc	419	786
Phosphorus	44.4	112

Table 8 Right-Triangle Solution Formulas

$\angle A$	$\angle B$	Side a	Side b	Side c
$\sin A = \dfrac{a}{c}$		$a = c \times \sin A$		$c = \dfrac{a}{\sin A}$
$\cos A = \dfrac{b}{c}$			$b = c \times \cos A$	$c = \dfrac{b}{\cos A}$
$\tan A = \dfrac{a}{b}$		$a = b \times \tan A$	$b = \dfrac{a}{\tan A}$	
$\cot A = \dfrac{b}{a}$		$a = \dfrac{b}{\cot A}$	$b = a \times \cot A$	
$\sec A = \dfrac{c}{b}$			$b = \dfrac{c}{\sec A}$	$c = b \times \sec A$
$\csc A = \dfrac{c}{a}$		$a = \dfrac{c}{\csc A}$		$c = a \times \csc A$
	$\sin B = \dfrac{b}{c}$		$b = c \times \sin B$	$c = \dfrac{b}{\sin B}$
	$\cos B = \dfrac{a}{c}$	$a = c \times \cos B$		$c = \dfrac{a}{\cos B}$
	$\tan B = \dfrac{b}{a}$	$a = \dfrac{b}{\tan B}$	$b = a \times \tan B$	
	$\cot B = \dfrac{a}{b}$	$a = b \times \cot B$	$b = \dfrac{a}{\cot B}$	
	$\sec B = \dfrac{c}{a}$	$a = \dfrac{c}{\sec B}$		$c = a \times \sec B$
	$\csc B = \dfrac{c}{b}$		$b = \dfrac{c}{\csc B}$	$c = b \csc B$

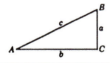

Table 9 Wire Gages and Metric Equivalents

Gage No.	American or Brown & Sharpe 's		Gage No.	American or Brown & Sharpe 's	
	in.	mm		in.	mm
000000	.5800	14.732	21	.02846	.723
00000	.5165	13.119			
0000	.4600	11.684	22	.02535	.644
000	.4096	10.404	23	.02257	.573
00	.3648	9.266	24	.02010	.511
0	.3249	8.252	25	.01790	.455
			26	.01594	.405
1	.2893	7.348	27	.01420	.361
2	.2576	6.543	28	.01264	.321
3	.2294	5.827			
4	.2043	5.189	29	.01126	.286
5	.1819	4.620	30	.01003	.255
6	.1620	4.115	31	.008928	.227
7	.1443	3.665	32	.007950	.202
			33	.007080	.180
8	.1285	3.264	34	.006305	.160
9	.1144	2.906	35	.005615	.143
10	.1019	2.588			
11	.09074	2.305	36	.005000	.127
12	.08081	2.053	37	.004453	.113
13	.07196	1.828	38	.003965	.101
14	.06408	1.628	39	.003531	.090
			40	.003145	.080
15	.05707	1.450	41	.002800	.071
16	.05082	1.291	42	.002494	.063
17	.04526	1.150			
18	.04030	1.024	43	.002221	.056
19	.03589	.912	44	.001978	.050
20	.03196	.812			

Table 10 Cutting Speeds for Commonly Used Materials

	Tool Material						
Work Material	**High-Speed Steel**	**Uncoated Carbide**	**Coated Carbide**	**Cermet**	**Ceramic**	**CBN**	**Diamond**
Aluminum							
Low silicon	300–800	700–1400					1000–5000
High silicon							500–2500
Bronze	65–130	500–700					1000–3000
Gray cast iron	50–80	250–450	350–500	400–1000	700–2000	700–1500	
Chilled cast iron					250–600	250–500	
Low-carbon steel	60–100	250–350	500–900	500–1300	1000–2500		
Alloy steel	40–70		350–600	300–1000	500–1500	250–600	
Tool steel	40–70		250–500			500–1200	150–300
Stainless steel							
200 and 300 series	30–80	100–250	400–650		300–1100		
400 and 500 series			250–350		400–1200		
Nonmetallics		400–600					400–2000
Superalloys		70–100	90–150		500–1000	300–800	

Table 10A Feeds for High-Speed Steel End Mills (Feed per Tooth in Inches)

Cutter Diameter	Aluminum	Brass	Bronze	Cast Iron	Low-Carbon Steel	Low-Carbon Steel	Medium-Alloy Steel	Stainless Steel
$\frac{1}{8}$	.002	.001	.0005	.0005	.0005	.0005	.0005	.0005
$\frac{1}{4}$	.002	.002	.001	.001	.001	.001	.0005	.001
$\frac{3}{8}$	.003	.003	.002	.002	.002	.002	.001	.002
$\frac{1}{2}$	.005	.002	.003	.0025	.002	.002	.001	.002
$\frac{3}{4}$	.006	.004	.003	.003	.004	.003	.002	.003
1	.007	.005	.004	.0035	.005	.003	.003	.004
$1\frac{1}{2}$	.008	.005	.005	.004	.006	.004	.003	.004
2	.009	.006	.005	.005	.007	.004	.003	.005

Table 10B Coolants and Cutting Oils Used for General Machining

Material	Dry	Water-Soluble Oil	Synthetic Coolants	Kerosene	Sulfurized Oil	Mineral Oil
Aluminum		X	X	X		
Brass	X	X	X			
Bronze	X	X	X			X
Cast iron	X					
Steel						
Low carbon		X	X		X	
Alloy		X	X		X	
Stainless		X	X		X	

Precision Vise Project Drawings

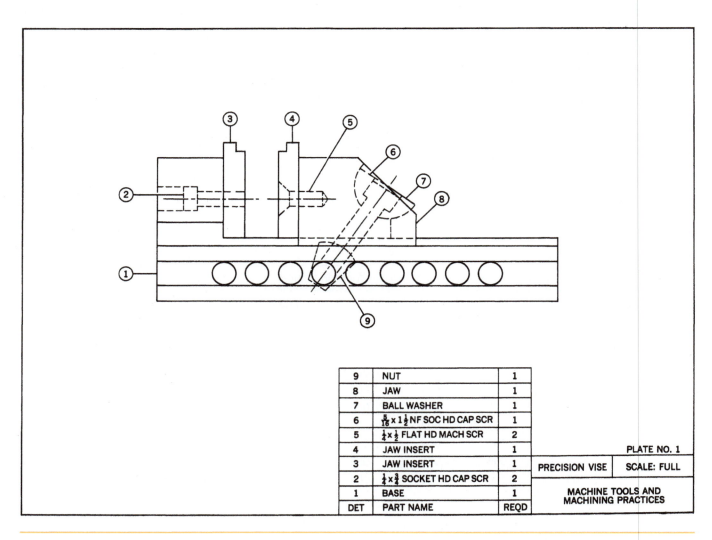

9	NUT	1
8	JAW	1
7	BALL WASHER	1
6	$\frac{5}{16} \times 1\frac{1}{2}$ NF SOC HD CAP SCR	1
5	$\frac{1}{4} \times \frac{1}{2}$ FLAT HD MACH SCR	2
4	JAW INSERT	1
3	JAW INSERT	1
2	$\frac{1}{4} \times \frac{3}{4}$ SOCKET HD CAP SCR	2
1	BASE	1
DET	PART NAME	REQD

PLATE NO. 1

PRECISION VISE	SCALE: FULL
MACHINE TOOLS AND MACHINING PRACTICES	

Drawing I

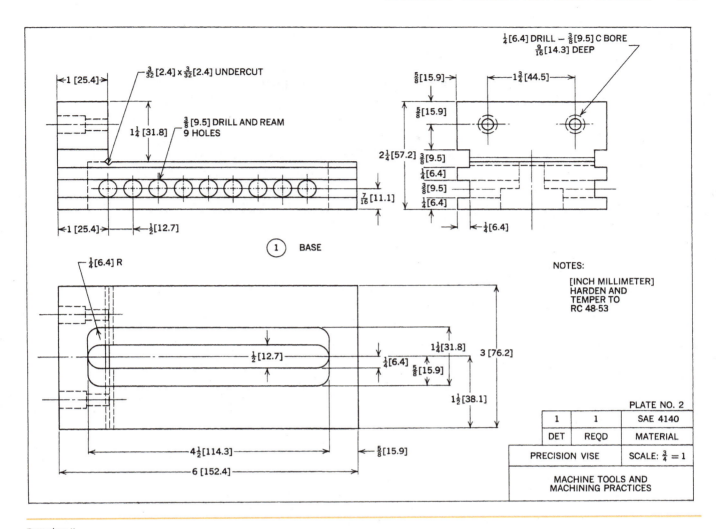

Drawing II

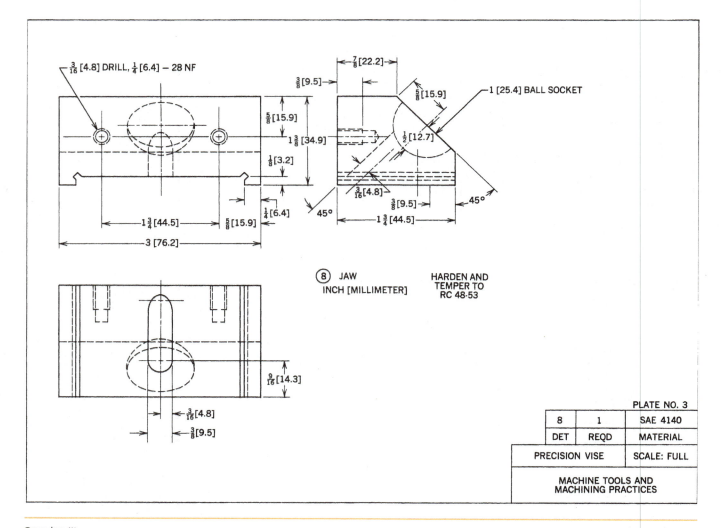

Drawing III

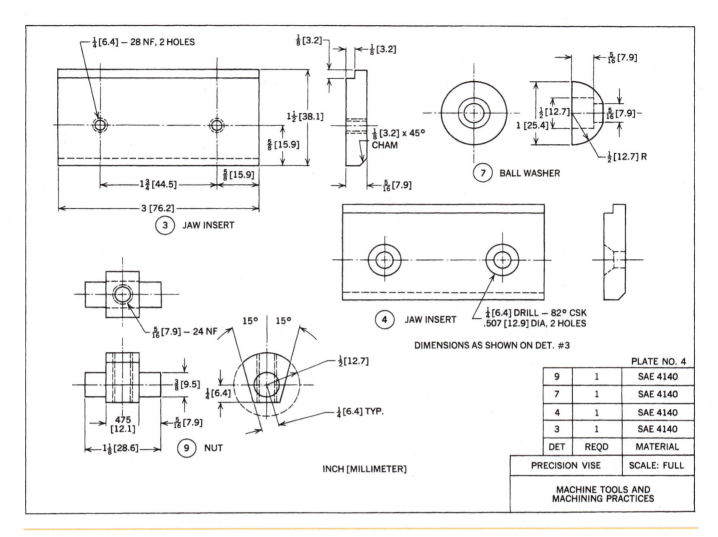

$\frac{1}{4}$[6.4] — 28 NF, 2 HOLES

$\frac{1}{8}$[3.2]

$\frac{1}{8}$[3.2]

$1\frac{1}{2}$[38.1]

$\frac{5}{8}$[15.9]

$\frac{1}{8}$[3.2] x 45°
CHAM

$\frac{5}{16}$[7.9]

$1\frac{3}{4}$[44.5]

$\frac{5}{8}$[15.9]

3 [76.2]

(3) JAW INSERT

$\frac{5}{16}$[7.9] — 24 NF

$\frac{3}{8}$[9.5]

$\frac{1}{4}$[6.4]

475
[12.1]

$\frac{5}{16}$[7.9]

$1\frac{1}{8}$[28.6]

(9) NUT

15°   15°

$\frac{1}{2}$[12.7]

$\frac{1}{4}$[6.4] TYP.

$\frac{5}{16}$[7.9]

$\frac{1}{2}$[12.7]

$\frac{5}{16}$[7.9]

1 [25.4]

$\frac{5}{16}$[7.9]

$\frac{1}{2}$[12.7] R

(7) BALL WASHER

(4) JAW INSERT

$\frac{1}{4}$[6.4] DRILL — 82° CSK
.507 [12.9] DIA, 2 HOLES

DIMENSIONS AS SHOWN ON DET. #3

INCH [MILLIMETER]

PLATE NO. 4

9	1	SAE 4140
7	1	SAE 4140
4	1	SAE 4140
3	1	SAE 4140
DET	REQD	MATERIAL

PRECISION VISE	SCALE: FULL

MACHINE TOOLS AND
MACHINING PRACTICES

Drawing IV

Glossary

***a* axis** In CNC, the rotational axis about the *X* axis.

Abrasive A substance such as finely divided aluminum oxide or silicon carbide used for grinding (abrading), smoothing, or polishing.

Absolute dimensions On drawings, a style in which all dimensions start at a zero or absolute point.

Absolute positioning A CNC programming mode called by the G90 code, in which all tool and workpiece positions are measured from a point of absolute zero.

Acicular Needlelike; resembling needles or straws dropped at random.

Acute angle An angle less than 90 degrees.

Alignment The proper positioning or state adjustment of parts in relation to one another, especially in line, as in axial alignment.

Allotropic Materials that can exist in several different crystalline forms are said to be allotropic.

Alloy A combination of two or more substances, specifically metals such as alloy steels or aluminum alloys.

Aluminum oxide Also alumina (Al_2O_3). Occurs in nature as corundum and is used extensively as an abrasive. Today, most aluminum oxide abrasives are manufactured.

Ammonia A pungent colorless gaseous alkaline compound of nitrogen and hydrogen (NH_3). It is soluble in water.

Amorphous Having no definite form or outline; materials such as glass that have no definite crystalline structure.

Angular Having one or more angles; measured by an angle; forming an angle.

Angular measure The means by which an arc of a circle is divided and measured. This can be in degrees (360 degrees in a full circle), minutes (60 minutes in one degree), and seconds (60 seconds in one minute), or in radians. See *Radian*.

Angularity The quality or characteristic of being angular.

Anhydrous Free from water.

Annealing A heat treatment in which metals are heated and then cooled slowly for the purpose of decreasing hardness. Annealing is used to improve machinability and to remove stress from weldments, forgings, and castings. It is also used to remove stresses resulting from cold work and to refine and make uniform the microscopic internal structures of metals.

Anvil The fixed measuring face of an outside micrometer.

Arbor A rotating shaft on which a cutting tool is fastened. Often used as a term for *mandrel*.

Arc Part of the circumference of a circle or a dimension defined by an angle.

As rolled When metals bars are hot-rolled and allowed to cool in air, they are said to be in the "as rolled" or natural condition.

Assembly drawing A working-type engineering drawing depicting a complete unit, usually included with detail drawings of all parts in a set of working drawings.

ATC Used in CNC machining centers, the automatic tool changer.

Austenite A solid solution of iron and carbon or iron carbide in which gamma iron, characterized by a face-centered cubic crystal, is the solvent.

Auxiliary view On orthographic drawings, a view projected off the standard orthographic views to reveal the true shape and size of a part feature otherwise distorted in the standard views.

Axial (1) Having the characteristics of an axis (that is, centerline or center of rotation); situated around and in relation to an axis as in axial alignment. (2) In CNC, related to the fundamental programmable axes *X*, *Y*, *Z*, *a*, *b*, and *c*.

Axial rake An angular cutting surface rotated about the axial centerline of a cutting tool such as a drill or reamer.

Axis (1) Centerline or center of rotation of an object or part; the rotational axis of a machine spindle, which extends beyond the spindle and through the workpiece. Machining of the object imparts the machine axis to that area of metal cutting. (2) On CNC machine tools, the line along which a major machine tool component such as a mill table, saddle, lathe cross slide, or turret travels.

***b* axis** In CNC, the rotational axis around the *Y* axis.

Backlash A condition created owing to clearance between a thread and nut. The amount of thread turn before a component begins to move.

Beam The scale on a vernier caliper or height gage graduated in true or full-sized units.

Bellmouth A condition in a machined hole in which the end is flared out in a bell shape to a dimension larger than the nominal size of the hole.

Bezel A rim that holds a transparent face of a dial indicator that can be rotated to bring the index mark to zero.

Bimetallic Made from two different metals.

Bill of materials A bill of materials is a list of the parts or components that are required to build a product.

Blind hole A hole that does not go completely through an object.

Blotter A paper disk placed between a grinding wheel and the retaining flange, often marked with wheel type and speed rating.

Bolster plate A structural part of a press designed to support or reinforce the platen (base surface) on which the workpiece is placed for press work.

Bond On a grinding wheel the material and its relative strength that holds the abrasive in place.

Bore (1) A machined hole. (2) The process of enlarging a drilled hole to a larger size.

Boring The process of removing metal from a hole by using a single-point tool. The workpiece can rotate with a stationary bar, or the bar can rotate on a stationary workpiece to bore a hole.

Brinell hardness The hardness of a metal or alloy, measured by hydraulically pressing a hard ball (usually 10 mm in diameter) with a standard load into the specimen. A value is derived by measuring the indentation with a special microscope.

Brittleness The property of a material that causes it to suddenly break at a given stress without bending or distortion of the edges of the broken surface. Glass, ceramics, and cast iron are examples of brittle materials.

Broaching The process of removing unwanted metal by pulling or pushing a tool on which cutting teeth project through or along the surface of a workpiece. The cutting teeth are progressively longer by a few thousandths of an inch to give each tooth a chip load. One of the most frequent uses of broaching is for producing internal shapes such as keyseats and splines.

Buffing wheel A disk made up of layers of cloth sewn together. Fine abrasive is applied to the periphery of the cloth wheel to provide a polishing surface as the wheel is rotated at a high speed.

Burnish To make shiny by rubbing. No surface material is removed by this finishing process. External and internal surfaces are often smoothed with high-pressure rolling. Hardened plugs are sometimes forced through bores to finish and size them by burnishing.

Burr (1) A small rotary file. (2) A thin edge of metal, usually sharp, left from a machining operation. See *Deburr*.

Bushing A hollow cylinder used as a spacer, reducer for a bore size, or bearing. Bushings can be made of metals or nonmetals such as plastics or formica.

Button die A thread-cutting die that is round and usually slightly adjustable. It is held in a diestock or holder by means of a cone-point setscrew that fits into a detent on the periphery of the die.

c **axis** In CNC, the rotational axis around the *Z* axis.

CAD Computer-aided design.

Calibration The comparison and adjustment of a measuring instrument such as a micrometer or dial indicator with a known measurement standard so that the tool will measure accurately.

CAM Computer-aided manufacturing.

Cam A rotating or sliding part with a projection or projecting geometry that imparts motion to another part as it slides or rotates past.

Carburizing compound A carbonaceous material that introduces carbon into a heated solid ferrous alloy by the process of diffusion.

Cavity (1) A machine feature, such as a hole, groove, or slot, enclosed on all sides in two dimensions. (2) The space in a casting mold where molten metal will flow to form a cast part.

Celsius A temperature scale used in the SI metric systems of measurement on which the freezing point of water is 0° and the boiling point is 100°. Formerly called *centigrade*.

Cementite Iron carbide, a compound of iron and carbon (Fe_3C) found in steel and cast iron.

Centerline A reference line on a drawing or part layout from which all dimensions are located.

Chamfer A bevel cut on a sharp edge of a part to improve resistance to damage and as a safety measure to prevent cuts.

Chasing a thread In machining terminology, making successive cuts in the same groove with a single-point tool. Also done when cleaning or repairing a damaged thread.

Chatter Vibration of workpiece, machine, tool, or a combination of all three due to looseness or weakness in one or more of these areas. Chatter may be found in either grinding or machine operations and is usually noted as vibratory sound and seen on the workpiece as wave marks.

Checked A term used mostly in grinding operations indicating a surface having many small cracks (checks). The terms *heat checked* or *crazed* are used in reference to friction clutch surfaces.

Chip trap A deformed end of a lathe cutting tool that prevents the chip from flowing across and away from the tool.

Chips The particles removed when materials are cut; also called *filings*.

Circularity The extent to which an object has the form of a circle; the measured accuracy or roundness of a circular or cylindrical object such as a shaft. A lack of circularity is referred to in shops as out of round, egg-shaped, or having a flat spot.

Circumference The periphery or outer edge of a circle. Its length is calculated by multiplying π (3.1416) times the diameter of the circle.

Closed architecture Proprietary control systems in which access to the machine control hardware and software is limited or "closed" to the end user. In a proprietary CNC, only the CNC builder provides replacement parts or modifies and reconfigures the components.

Clutch A component usually found in a mechanical drive that permits a driven component and driving component to be mechanically disconnected and reconnected at will.

CNC Computer numerical control. Control of machine tools and other manufacturing equipment using computer programs.

Coarseness (1) A measure of grit size in grinding. (2) Spacing of teeth on files and other cutting tools.

Coincident The alignment of two graduations on separate graduated scales with each other, such as the coincident line of vernier and true-scale graduations.

Cold finish The surface finish obtained on metal by any of several means of cold working, such as rolling or drawing.

Cold working Any process such as rolling, forging, or forming a cold metal in which the metal is stressed beyond its yield point. Grains are deformed and elongated in the process, causing the metal to have a higher hardness and lower ductility.

Complementary angles Two angles whose sum is 90 degrees. Often referred to in machine shop work, because most angular machining is done within one quadrant, or 90 degrees.

Concave An internal arc or curve; a dent.

Concentricity The extent to which an object has a common center or axis. Specifically, in machine work, the extent to which two or more surfaces of a shaft rotate in relation to each other; the amount of runout on a rotating member.

Contour Machining an uneven but continuous path on a workpiece in two or three dimensions.

Convex An external arc or curve; a bulge.

Coolant Any of several products using oil and water mixtures to cool and lubricate cutting tools and grinding wheels during machining operations.

Coordinate dimensions A method of specifying point locations in a two-dimensional plane system defined by two perpendicular axes.

Cosine error A condition in which the axis of a measuring instrument is out of line with the axis of the measurement to be taken, resulting in an error equal to the measuring instrument reading multiplied by the cosine of the misalignment angle.

Counterboring The process of enlarging a drilled hole for a portion of its length, as to permit sinking a screw head.

Countersinking Enlargement of the upper part of a hole by chamfering, to receive the cone-shaped head of a screw, bolt, etc.

Crest of thread Outer edge (point or flat) of a thread form.

Critical temperatures The upper and lower transformation points of iron that delineate the range in which ferrite changes to austenite as the temperature rises.

Cutting fluid Any of several materials used in cutting metals: cutting oils, synthetics, soluble or emulsified oils (water based), and sulfurized oils.

Cyanogen (CN)2 A colorless flammable poisonous gas with characteristic odor. It forms cyanic and hydrocyanic acids when in contact with water. Cyanogen compounds are often used for case hardening.

Deburr To remove a sharp edge or corner caused by a machining process.

Decarburization The loss of surface carbon from ferrous metals when heated to high temperatures in an atmosphere containing oxygen.

Decibel A unit for expressing the relative intensity of sounds on a scale from zero (least perceptible sound) to about 130 (the average pain level).

Degrees (1) A circle is divided into 360 degrees, four 90-degree quadrants. Each degree is divided into 60 minutes and each minute into 60 seconds. Degrees are measured with protractors, optical comparators, and sine bars, to name a few methods. (2) Divisions of temperature scales.

Dendrite A formation that resembles a pine tree in the microstructure of solidifying metals. Each dendrite usually forms a single grain or crystal.

Detail drawing A detail drawing is a drawing giving extra detail to a portion of a larger drawing.

Diagonal A straight line from corner to corner on a square, rectangle, or any parallelogram.

Diameter Twice the radius; the length of any straight line going through the center of a figure or body, specifically a circle, in drafting the layout.

Diametral pitch The ratio of the number of teeth on gears to the number of inches of pitch diameter.

Die (1) Cutting tool for producing external threads. (2) A device mounted in a press for cutting and forming sheet metal.

Die-cast metal Metal alloys, often called *pot metals*, forced into a die in a molten state by hydraulic pressure. Thousands of identical parts can be produced from a single die or mold by this process of die casting.

Dimension A measurement in one direction; one of three coordinates—length, width, and depth. Thickness, radius, and diameter are given as dimensions on drawings.

Discrimination The degree to which a measuring instrument divides the units in which it measures. A .001-in. micrometer can discriminate to one thousandth of an inch. With a vernier, it can discriminate to one ten-thousandth of an inch (.0001 in.).

Distortion The alteration of the shape of an object that would normally affect its usefulness. Bending, twisting, and elongation are common forms of distortion in metals.

Dovetail An angular shape used on many types of interlocking slide components, especially on machine tools.

Dowel pins are used as precise locating devices in machinery. Dowel pins are machined to tight tolerances, as are the corresponding holes, which are typically reamed. A dowel pin may have a smaller diameter than its hole so that it freely slips in, or a larger diameter so that it must be pressed into the hole.

Ductility The property of a metal to be deformed permanently without rupture while under tension. A metal that can be drawn into a wire is ductile.

EB machining Electron beam machining. A process whereby a gas is passed through an electric arc, creating a plasma or high-temperature gas, which is then used as the cutting tool. EB processes are effective for welding materials.

Ebonized Certain cold-drawn or cold-rolled bars that have black-stained surfaces are said to be ebonized. This is not the same as the black, scaly surface of hot-rolled steel products.

Eccentricity A rotating member whose axis of rotation is different or offset from the primary axis of the part or mechanism. Thus, when one turned section of a shaft centers on a different axis than the shaft, it is said to be eccentric or to have "runout." For example, the throws or cranks on an engine crankshaft are eccentric to the main bearing axis.

ECM Electrochemical machining or electrochemical deburring (ECDB), a process whereby a conducting fluid (electrolyte) is pumped between work and electrode to remove the workpiece material.

Edgefinder A tool fastened in a machine spindle that locates the position of the workpiece edge in relation to the spindle axis.

EDM Electrodischarge machining. With this process, a graphite or metal electrode is slowly fed into the workpiece, which is immersed in oil. A pulsed electrical charge causes sparks to jump to the workpiece, each tearing out a small particle. In this way, the electrode gradually erodes its way through the workpiece that can be a soft or an extremely hard material such as tungsten carbide.

Elasticity The property of a material to return to its original shape when stretched or compressed.

Electrode A tool or other device used to make an electrical contact such as the cutting tool in EDM machining equipment.

ELG Electrolytic grinding. A machining process in which an abrasive with a conducting bond is used to deplate the workpiece material.

Emulsifying oils An oil containing an emulsifying agent such as detergent so that it will mix with water. Oil emulsions are used extensively as coolants in machining operations.

EOB In CNC programming, the code denoting end of block or end of sequence.

Expansion The enlargement of an object, usually caused by an increase in temperature. Metals expand when heated and contract when cooled in varying amounts, depending on the coefficient of expansion of the particular metal.

Expansion fit The fitting of mating parts by heating one so as to expand its dimensions slightly. On cooling, the parts are locked together.

Extruding A form of metal working in which a metal bar, either cold or heated, is forced through a die that forms a special cross-sectional shape such as an angle or channel. Extrusions of soft metals such as aluminum and copper are common.

F code word In CNC programming, the code used to call a feed rate for milling or drilling.

Face (1) The side of a metal disk or end of a shaft when turning in the lathe. A facing operation is usually at 90 degrees to the spindle axis of the lathe. (2) The periphery or outer cylindrical surface of a straight grinding wheel.

Fahrenheit A temperature scale calibrated with the freezing point of water at 32 degrees and the boiling point at 212 degrees. The Fahrenheit scale is gradually being replaced with the Celsius scale used with the metric system of measurement.

Ferrite The microstructure of iron or steel that is mostly pure iron and appears light gray or white when etched and viewed with a microscope.

Ferromagnetic Metals or other substances that have unusually high magnetic permeability, a saturation point with some residual magnetism, and high hysteresis. Iron and nickel are both ferromagnetic.

Ferrous From the Latin word *ferrum,* meaning "iron." An alloy containing a significant amount of iron.

Fillet (1) A concave junction of two surfaces. (2) An inside corner radius of a shoulder on a shaft. (3) An inside corner weld.

Finishing (surface) The control of roughness by turning, grinding, milling, lapping, superfinishing, or a combination of any of these processes. Surface texture is designated in terms of roughness profile in microinches, waviness, and lay (direction of roughness).

Finish mark A finish mark is a symbol used to indicate surface finishes by drawing a modified check mark and placing the degree of finish in the angle of the check mark.

Fixture A device that holds workpieces and aligns them with the tool or machine axis with repeatable accuracy.

Flammable Any material that will readily burn or explode when brought into contact with a spark or flame.

Flash (1) Excess material extruded between die halves in die castings or forging dies. (2) The upset material formed when welding band saws.

Floating Free to move about over a given area, for example, a floating edge finder tip, floating die holder, or floating reamer holder.

Flute The groove in a drill, tap, reamer, or milling cutter.

Forging A method of metal working in which the metal is hammered into the desired shape or is forced into a mold by pressure or hammering, usually after being heated to a more plastic state. Hot forging requires less force to form a given point than does cold forging, which is usually done at room temperature.

Formica A trademark used to designate several plastic laminated products; especially, a laminate used to make gears.

Forming A method of working sheet metal into useful shapes by pressing or bending.

Fractional inch Fractional inch describes the system in which Fractions of an inch on a ruler are broken down in way that emphasizes halves, then quarters, then eighths, then sixteenths.

Friable Related to the brittleness of a grinding wheel abrasive.

Friction (1) Rubbing of one part against another. (2) Resistance to relative motion between two parts in contact, usually generating heat.

G code word In CNC programming, a preparatory function program code that calls a particular mode of operation such as rapid traverse, linear interpolation, or circular interpolation.

Galling Cold welding of two metal surfaces in intimate contact under pressure. Also called *seizing*, it is more severe and more likely to happen between two similar soft metals, especially when they are clean and dry.

Gantry A type of CNC machining center in which the spindle is supported on both sides of the table by vertical supports or gantries.

Gib A part of a slide mechanism used to adjust the clearance between two sliding parts.

Glazing (1) A work-hardened surface on metals resulting from using a dull tool or too rapid a cutting speed. (2) A dull grinding wheel whose surface grains have worn flat, causing the workpiece to be overheated and "burned" (discolored).

Grade (1) Hardness of a grinding wheel. (2) Level of precision of gage block sets.

Graduations Division marks on a rule, measuring instrument, or machine dial.

Grain (1) In metals, a single crystal consisting of parallel rows of atoms called a *space lattice*. (2) Abrasive particles in a grinding wheel.

Grain boundary The outer perimeter of a single grain where it contacts adjacent grains.

Grain growth Called *recrystallization*. Metal grains begin to re-form to larger and more regular size and shape at certain temperatures, depending to some extent on the amount of prior cold working.

Graphite Carbon used as the material for EDM electrodes.

Grit (1) Any small, hard particles such as sand or grinding compound. Dust from grinding operations settles on machine surfaces as grit, which can damage sliding surfaces. (2) Diamond dust, aluminum oxide, or silicon carbide particles used for grinding wheels.

Ground and polished (G and P) A finishing process for some steel alloy shafts during their manufacture. The rolled, drawn, or turned shafting is placed on a centerless grinder and precision ground, after which a polishing operation produces a fine finish.

Gullet The bottom of the space between teeth on saws and circular milling cutters.

H code word In CNC programming, the H code refers to tool offset file numbers.

Hardenability The property that determines the depth and distribution of hardness in a ferrous alloy induced by heating and quenching.

Hardening Metals are hardened by cold working or heat-treating. Hardening causes metals to have a higher resistance to penetration and abrasion.

Harmonic chatter A harmonic frequency is a multiple of the fundamental frequency of sound. Any machine part, such as a boring bar, has a fundamental frequency and will vibrate at that frequency and also at several harmonic or multiple frequencies. Thus, chatter or vibration of a tool may be noted at several different spindle speeds.

Hazard (1) A situation dangerous to any person in the vicinity. (2) A danger to property, such as a fire hazard.

Heat treated Metal whose structure has been altered or modified by the application of heat.

Helical The geometry of a helix where a point both rotates and moves parallel to the axis of a cylinder. Examples include threads, springs, and drill flutes.

Helix The path described by a point rotating about a cylinder while being moved along the cylinder. The distance of movement compared with each revolution is the lead of the helix.

High-pressure lube A petroleum-grease or oil-containing graphite or molybdenum disulfide that continues to lubricate even after the grease has been wiped off.

Hog To remove large amounts of material from a workpiece with deep heavy cuts.

Horizontal Parallel to the horizon or baseline level.

Hot rolled Metal flattened and shaped by rolls while at a red heat.

Hub A larger diameter at the center of a wheel, gear, pulley, sprocket, or other shaft-driven member that provides a bore in its center to receive a shaft. The hub also provides extra strength to transfer power to or from the shaft by means of a key and keyseat.

Hydrojet machining A machining process in which high-pressure water containing an abrasive material is directed toward the workpiece.

I code word In CNC programming, the I and J words define arc centers.

Increment A single step of a number of steps; a succession of regular additions; a minute increase.

Incremental positioning In CNC, programming a mode called by the G91 preparatory function code, in which each positioning move is measured from the point where the tool is presently located.

Inert gas A gas, such as argon or helium, that will not readily combine with other elements.

Infeed The depth a tool is moved into the workpiece.

Interface (1) The point or area of contact between tool and workpiece. (2) The contact point or area of two mating parts in an assembly.

Interference fit Force fit of a shaft and bore, bearings, and housings or shafts. Negative clearance in which the fitted part is slightly larger than the bore.

Internal stress Also called *residual stress*. Stress in metals built in by heat treatment or by cold working.

Interpolation In CNC programming, a "best fit" tool path along an angular, circular, or helical programmed path.

Intruding A surface-hardening treatment for ferrous alloys obtained by heating an alloy in the presence of disassociated ammonia gas, which releases nitrogen to the steel. The formation of iron nitride causes the hardened surface.

Involute Geometry found in modern gears that permits mating gear teeth to engage each other with rolling rather than sliding friction.

ipm rate In machining, a feed rate measured in inches per minute.

ipr rate In machining, a feed rate specified in inches per revolution of the machine spindle.

Iron carbide Also called cementite (Fe_3O_4); a compound of iron and carbon, which is quite hard.

J code word In CNC programming, the I and J words specify arc center locations.

Jig A device that guides a cutting tool and aligns it to the workpiece.

Jig boring is the use of a jig boring machine to make quick- yet-very-precise location of hole centers. It can be viewed as a specialized type of milling machine that provided a higher degree of positioning precision (repeatability) and accuracy. Although capable of milling, a jig borer is more suited to highly accurate drilling, boring, and reaming, where the quill or headstock does not see the significant side loading. A jig bore is a machine designed for locational accuracy.

Journal The part of a rotating shaft or axle that turns in a bearing.

Just-in-time Manufacturing system that produces a smaller number of parts at one time to reduce inventory for both the producer and the consumer.

Kerf The width of a cut produced by a saw.

Key A removable metal part that when assembled into keyseats provides a positive drive for transmitting torque between shaft and hub.

Keyseat An axially located rectangular groove in a shaft or hub.

Keystock Square or rectangular cold-rolled steel bars used for making and fitting keys in keyseats.

Keyway Same as keyseat (British terminology).

Knurl Diamond or straight impressions on a metal surface produced by rolling with pressure. The rolls used are called knurls.

Laminated Composed of multiple layers of the same or different materials.

Laser An intense source of coherent light energy often used as a cutting tool.

Lattice The regular rows of atoms in a metal crystal.

Lead The distance a thread or nut advances along a threaded rod in one revolution.

Loading A grinding wheel whose voids are being filled with metals, causing the cutting action of the wheel to be diminished.

Lobe The offset or projection on a cam that contacts the part to which motion is to be imparted.

Lockout/Tagout is a safety procedure which is used in business and industry to ensure that dangerous machines are properly shut off and not started up again prior to the completion of maintenance or servicing work.

Longitudinal Lengthwise, as the longitudinal axis of the spindle or machine.

M code word In CNC programming, M codes are miscellaneous function codes such as M06 for tool change or M03 for spindle stop.

Machinability The relative ease of machining, which is related to the hardness of the material to be cut.

Magnetic Having the property of magnetic attraction and permeability.

Major diameter is the largest diameter of a screw thread, measured at the crest for an external thread and at the root for an internal thread.

Malleability The ability of a metal to deform permanently without rupture when loaded in compression.

Mandrel A cylindrical bar on which the workpiece is affixed and subsequently machined between centers. Mandrels, often erroneously called *arbors*, are used in metal turning and cylindrical grinding operations.

Mar To scratch or otherwise damage a machined surface.

Martensite The hardest constituent of steel, formed by quenching carbon steel from the austenitized state to room temperature. The microstructure can be seen as acicular or needlelike.

Mechanical properties (of metals) Some mechanical properties of metals are tensile strength, ductility, malleability, elasticity, and plasticity. Mechanical properties can be measured by mechanical testing.

Metal cementation Introducing a metal or material into the surface of another by heat treatment. Carburizing is one example of metal cementation.

Metal spinning A process in which a thin disk of metal is rapidly turned in a lathe and forced over a wooden form or mandrel to form various conical or cylindrical shapes.

Metallizing Applying a coating of metal on a surface by spraying molten metal on it. Also called *spray weld* and *metal spray*.

Metrology The science of weights and measures or measurement.

Microstructure Structure visible only at high magnification.

Modal commands In CNC programming, modal commands once invoked remain in force until canceled or changed.

Mode A particular way in which something is done or a machine is operated, such as manual or automatic mode on machines.

Mushroom head (1) An oversized head on a fastener or tool that allows it to be easily pushed with the hand. (2) A deformed striking end of a chisel or punch that should be removed by grinding.

Neutral In machine work, neither positive nor negative rake is a neutral or zero rake; a neutral fit is neither a clearance nor an interference fit.

Nitrogenous gas Ammonia (NHA) used in nitriding.

Nomenclature Pertaining to the names of individual parts of machines or tools; a list of machine parts indicating their names.

Nominal Usually refers to a standard size or quantity as named in standard references.

Nonferrous Metals other than iron or iron alloys, for example, aluminum, copper, and nickel.

Normalizing A treatment consisting of heating to a temperature above the critical range of steel followed by cooling in air. Normalizing produces in steel a "normal structure" consisting of free ferrite and cementite or free pearlite and cementite, depending on the carbon content.

Nose radius The rounding of the point of a lathe cutting tool. A large radius produces a better finish and is stronger than a small one.

Obtuse angle An angle greater than 90 degrees but less than 180 degrees.

Open architecture A control system in which the PC platform allows a CNC machine manufacturer to use hardware and software from different vendors, various motors, and memory and storage upgrades. Users can run their own choice of programming systems, SPC software, and shop management software, all from a single PC.

Origin The intersection of the X, Y, and Z axes.

Orthographic drawing Projections of a single view of an object in which the view is projected along lines perpendicular to both the view and the drawing surface.

Oxide scale At a red heat, oxygen readily combines with iron to form a black oxide scale (Fe_2O_3), also called *mill scale*. At lower temperatures, 400°F to 650°F (204°C to 343°C), various oxide scale colors (straw, yellow, gold, violet, blue, and gray) are produced, each color within a narrow temperature range. These colors are used by some heat treaters to determine temperatures for tempering.

Oxidize (1) To combine with oxygen. (2) To burn or corrode by oxidation.

Oxyacetylene Mixture of oxygen and acetylene gases to produce an extremely hot flame used for heating, welding, and flame cutting.

Parallax error An error in measurement caused by reading a measuring device, such as a rule, at an improper angle.

Parallel The condition in which lines or planes are equidistant from one another.

Parting Also called *cutting off*; a lathe operation in which a thin blade tool is fed into a turning workpiece to make a groove that is continued to the center to sever the material.

Pascal A unit of pressure equal to one newton per square meter.

Pearlite Alternating layers of cementite and ferrite in carbon steel. Under a microscope, the microstructure of pearlite sometimes appears like mother-of-pearl, hence the name. It is found in carbon steels that have been slowly cooled.

Pecking A process used in drilling deep holes to remove chips before they can seize and jam the drill. The drill is fed into the hole a short distance to accumulate chips in the flutes and then drawn out of the hole, allowing the chips to fly off. This process is repeated until the correct depth of the hole is reached.

Pedestal A base or floor stand under a machine tool.

Penetrant A thin liquid that can enter small cracks and crevices. Penetrant oils are used to loosen rusted threads; dye penetrants are used to find hidden cracks.

Periphery The perimeter or external boundary of a surface or body.

Perpendicular At right angles to any geometry or part feature, that is, at a right angle to or forms a 90-degree angle with another part feature, line, or surface.

Pin Straight, tapered, or cotter pins are used as fasteners of machine parts or for light drives.

Pinion The smaller gear of a gear set, especially in bevel gears.

Pinning A condition in which chips of workpiece material jam in the teeth of a file.

Pitch (1) In saw teeth, the number per inch. (2) In threads, one divided by the number per inch. (3) The diameter or radius to the centerline of a feature or features located on the circumference of a circle, such as pitch circle or pitch diameter.

Pitch diameter For threads, the pitch diameter is an imaginary circle, which on a perfect thread occurs at the point where the widths of the thread and groove are equal. On gears, it is the diameter of the pitch circle.

Plasma beam machining A machining process in which a high-temperature gas (plasma) is used as the cutting tool. Plasma arcs are effective for cutting materials in sheet form.

Point to point In CNC machining, operations such as drilling and tapping that occur at single locations or points.

Pot metals Die-casting alloys, which can be zinc, lead, or aluminum based, among others.

Potassium cyanide A poisonous crystalline salt (KCN) used in electroplating and for case-hardening steel.

Precipitation hardening A process of hardening an alloy by heat treatment in which a constituent or phase precipitates from a solid solution at room temperature or at a slightly elevated temperature.

Precision A relative but higher level of accuracy within certain tolerance limits. Precision gage blocks are accurate within a few millionths of an inch, yet in some shops precision lathe work may be within a few thousandths of an inch tolerance.

Pressure Generally expressed in units as pounds per square inch (psi) and called *unit pressure*, whereas force is the total load.

Profile An outline view; also, a side or elevation view.

Proportion An equality of two ratios.

Prototype A full-scale original model on which something is patterned.

Pulley A flat-faced wheel used to transmit power by means of a flat belt. Grooved pulleys are called *sheaves*.

Quadrant The plane areas bounded by the perpendicular X and Y axes.

Quench A rapid cooling of heated metal for the purpose of imparting certain properties, especially hardness. Quenchants are water, oil, fused salts, air, and molten lead.

Quench cracking Cracking of heated metal during the quenching operation caused by internal stresses.

Safety Checklist A safety checklist is a list of precautions that should be followed for a particular machine or process to ensure the operator's safety.

Scale Blueprints are usually "scale drawings," meaning that the plans are drawn at a specific ratio relative to the actual size of the object.

Seize A condition in which two metal parts are pressed together without the aid of lubrication, resulting in frictional forces that tear metal from each part and cause a mechanical welding (seizing) of the two.

Semiprecision Layout done to tolerances of $\pm\frac{1}{64}$ in.

Serrated Small grooves, often in a diamond pattern, used mostly for a gripping surface.

Set The width of a saw tooth. The set of saw teeth is wider than the blade width.

Setup The arrangement by which the machinist fastens the workpiece to a machine table or work-holding device and aligns the cutting tool for metal removal. A poor setup is one in which the workpiece could move from the pressure of the cutting tool, thus damaging the workpiece or tool, or when chatter results from lack of rigidity.

sfpm Surface feet per minute on a moving workpiece or tool.

Shallow hardening Some steels such as plain carbon steel (depending on their mass), when heated and quenched, harden to a depth of less than $\frac{1}{8}$ in. These are shallow-hardening steels.

Shank The part of a tool held in a work-holding device or in the hand.

Shearing action A concentration of forces in which the bending moment is virtually zero and the metal tends to tear or be cut along a transverse axis at the point of applied pressure.

Sheaves Grooved pulleys such as those used for V-belts or cables.

Sherardized Zinc-inoculated steel, a process in which the surface of steel is given a protective coating of zinc. It is not the same as galvanized or zinc-dipped steel. Zinc powder is packed around the steel while it is heated to a relatively low temperature in the sherardizing process.

Shim (1) A thin piece of material used to take up space between workpiece and work-holding device. (2) A piece used to fill space between machinery and foundations in assemblies.

SI (Système International) The international metric system of weights and measures.

Silicon carbide A manufactured abrasive. Silicon carbide wheels are used for grinding nonferrous metals, cast iron, and tungsten carbide, but are not normally used for grinding steel.

Sine bar A small precision bar with a given length (5 or 10 in.) that remains constant at any angle. It is used with precision gage blocks to set up or to determine angles within a few seconds of a degree.

Sintering Holding a compressed metal powder briquette at a temperature just below its melting point until it fuses into a solid mass of metal.

Slot Groove or depression as in a keyseat slot.

Snagging Rough grinding to remove unwanted metal from castings and other products.

Soluble oils Oils that have been emulsified and will combine with water.

Solution heating treating See *Precipitation hardening.*

Solvent A material, usually liquid, that dissolves another. The dissolved material is the solute.

Spark testing A means of determining the relative carbon content of plain carbon steels and identifying some other metals by observing the sparks given off while grinding the metals.

SPC Statistical Process Control The methods and tools used to determine the results of a manufacturing process by recording and graphing part dimensions and then generating various statistical information about the results.

Specifications Requirements and limits for a particular job.

Speeds Machine speeds are expressed in revolutions per minute; cutting speeds are expressed in surface feet per minute.

Sphericity (1) A condition of circularity in all possible axes. (2) The quality of being in the shape of a ball. (3) The extent to which a true sphere can be produced with a given process.

Spheroidize anneal A heat treatment for carbon steels that forms the cementite into spheres, making it softer and usually more machinable than by other forms of annealing.

Spiral A path of a point in a rotating plane that is continuously receding from the center is called a *flat spiral.* The term *spiral* is often used, though incorrectly, to describe a helix.

Spline A shaft on which teeth have been machined parallel to the shaft axis that will engage similar internal teeth in a mating part to prevent turning.

Spotfacing A machined feature in which a certain region of the workpiece is faced, providing a flat, accurately located surface.

Sprockets Toothed wheels used with a chain for drive or conveyor systems.

Squareness The extent of accuracy that can be maintained when making a workpiece with a right angle.

Statistics Mathematical data generated by measuring a few samples of parts produced by a manufacturing process and then making inferences about the suitability of a larger production run. Statistics may also be used to generate graphs showing how close parts are being manufactured to design specifications.

Stepped shaft A shaft having more than one diameter.

Stick-slip A tendency of some machine parts that slide on ways to bind slightly when pressure to move them is applied, followed by a sudden release that often causes the movement to be greater than desired.

Straightedge A comparison measuring device used to determine flatness. A precision straightedge usually has an accuracy of about .0002 in. in a 24-in. length.

Strength The ability of a metal to resist external forces. This can be tensile, compressive, or shear strength.

Stress An external force applied to an object.

Stress relief anneal A heat treatment, usually under the critical range, for the purpose of relieving stresses caused by welding or cold working.

Stroke A single movement of many movements, as in a forward stroke with a hacksaw.

Structure The density of abrasive in a grinding wheel, open to dense.

Surface plate A cast iron or granite surface having a precision flatness for precision layout, measurement, and setup.

Swarf The chips produced by grinding operations.

Symmetrical Usually bilateral in machinery, in which two sides of an object are alike, but usually as a mirror image.

Synthetic oils Artificially produced oils given special properties such as resistance to high temperatures. Synthetic water-soluble oils or emulsions are replacing water-soluble petroleum oils for cutting fluids and coolants.

T-nut A threaded nut in a T shape designed to fit into the T-slot on a machine tool table.

T-slot The slot in a machine tool table shaped like a T and used to hold T-nuts and studs for various clamping setups or hold-down requirements.

Tang The part of a file on which a handle is affixed.

Tap extractor A tool sometimes effective in removing broken taps.

Tapered thread A thread made on a taper, such as a pipe thread.

Tapping A method of cutting internal threads by means of rotating a tap into a hole sufficiently under the nominal tap size to make a full thread.

Telescoping gage A transfer type of tool that assumes the size of the part to be measured by expanding or telescoping. It is then measured with a micrometer.

Temper (1) The cold-worked condition of some nonferrous metals. (2) Also called draw, a method of toughening hardened carbon steel by reheating it.

Temperature The level of heat energy in a material as measured by a thermometer or thermostat and recorded with any of several temperature scales: Celsius, Fahrenheit, or Kelvin.

Template A metal, cardboard, or wooden form used to transfer a shape or layout when it must be repeated many times.

Tensile strength The maximum unit load that can be applied to a material before ultimate failure occurs.

Tension A stretching or pulling force.

Terminating threads Methods of ending the thread, such as undercutting, drilled holes, or tool removal.

Test bar A precision ground bar that is placed between centers on a lathe to test for center alignment using a dial indicator.

Thermal cracking Checking or cracking caused by heat.

Thread axis The centerline of the cylinder on which the thread is made.

Thread chaser A tool used to restore damaged threads.

Thread crest The top of the thread.

Thread die A device used to cut external threads.

Thread engagement The distance a nut or mating part is turned onto the thread.

Thread fit classes Systems of thread fits for various thread forms range from interference fits to loose fits; extensive references on thread fits may be found in machinists' handbooks.

Thread lead The distance a nut travels in one revolution. The pitch and lead are the same on single-lead threads but not on multiple-lead threads.

Thread pitch The distance from a point on one thread to a corresponding point on the next thread.

Thread relief Usually, an internal groove that provides a terminating point for the threading tool.

Tolerance The allowance of acceptable error within which the mechanism will still fit together and be totally functional.

Tool geometry The proper shape of a cutting tool that makes it work effectively for a particular application.

Tool offset In CNC programming, the distance from the tool end to the workpiece.

Tooling Generally, any machine tool accessory separate from the machine itself. Tooling includes cutting tools, holders, work-holding accessories, jigs, and fixtures.

Toolmaker An experienced general machinist, often involved with high-precision work making other tools, dies, jigs, and fixtures used to support regular machining and manufacturing.

Torque A force that tends to produce rotation or torsion. Torque is measured by multiplying the applied force by the distance at which it is acting to the axis of the rotating part.

Toxic fumes Gases resulting from heating certain materials are toxic, sometimes causing illness (as metal fume fever from zinc fumes) or permanent damage (as from lead or mercury fumes).

Transfer measurement A step in measurement in which a transfer measuring tool such as a telescoping gage is set to the unknown dimension and subsequently measured with a direct measuring tool such as a micrometer.

Transformation temperature Same as critical temperature; the point at which ferrite begins to transform to astatine.

Traverse To move a machine table or part from one point to another.

Truing (1) In machine work, the use of a dial indicator to set up work accurately. (2) In grinding operations, to make a wheel concentric to the spindle with a diamond.

Truncation To remove the point of a triangle (as of a thread), cone, or pyramid.

Tungsten carbide An extremely hard compound formed with cobalt and tungsten carbide powders by briquetting and sintering into tool shapes.

Turning Operations in which the work is rotated against a single-point tool.

Turning center A CNC lathe.

Ultrasonic machining A machining process whereby high-intensity sound (above 20 kHz) is used to propel abrasive as a material removal tool.

Vernier A means of increasing the discrimination of a graduated measuring instrument by adding a shorter scale alongside the main scale so as to employ the mathematical principle of the vernier.

Vertical Turret Lathe Some lathes have a vertical spindle instead of a horizontal one, with a large rotating table on which the work is clamped. These machines are called vertical turret lathes (VTLs).

Vibration An oscillating movement caused by loose bearings or machine supports, off-center weighting on rotating elements, bent shafts, or nonrigid machining setups.

Vise A work-holding device. Some types are bench, drill press, and machine vises.

Wedge angle Angle of keenness; cutting edge.

Wheel dressing Sharpening the grinding surface of an abrasive wheel by means of a dressing tool such as a diamond or Desmond dresser.

Wrought Hot or cold worked; forged.

X axis On CNC machining centers, the table axis; on CNC turning centers, the cross slide axis.

X–Y plane The plane formed by the X and Y axes.

X–Z plane The plane formed by the X and Z axes.

Y axis On a CNC machining center, the saddle axis.

Y–Z plane The plane formed by the Y and Z axes.

Z axis On CNC machine tools in general, the spindle axis.

Zero back rake Also neutral rake; neither positive nor negative level.

Zero index Also zero point. The point at which micrometer dials on a machine are set to zero, and the cutting tool is located to a given reference, such as a workpiece edge.

Index

Note: Page numbers in *italics* refer to figures and illustrations.

Decimal Equivalents of Fractional Inches

Fraction Inch	Decimal Inch	Decimal Millimeters	Fraction Inch	Decimal Inch	Decimal Millimeters
1/64	.015625	0.39688	33/64	.515625	13.09690
1/32	.03125	0.79375	17/32	.53125	13.49378
3/64	.046875	1.19063	35/64	.546875	13.89065
1/16	.0625	1.58750	9/16	.5625	14.28753
5/64	.078125	1.98438	37/64	.578125	14.68440
3/32	.09375	2.38125	19/32	.59375	15.08128
7/64	.109375	2.77813	39/64	.609375	15.47816
1/8	.1250	3.17501	5/8	.6250	15.87503
9/64	.140625	3.57188	41/64	.640625	16.27191
5/32	.15625	3.96876	21/32	.65625	16.66878
11/64	.171875	4.36563	43/64	.671875	17.06566
3/16	.1875	4.76251	11/16	.6875	17.46253
13/64	.203125	5.15939	45/64	.703125	17.85941
7/32	.21875	5.55626	23/32	.71875	18.25629
15/64	.234375	5.95314	47/64	.734375	18.65316
1/4	.2500	6.35001	3/4	.7500	19.05004
17/64	.265625	6.74689	49/64	.765625	19.44691
9/32	.28125	7.14376	25/32	.78125	19.84379
19/64	.296875	7.54064	51/64	.796875	20.24067
5/16	.3125	7.93752	13/16	.8125	20.63754
21/64	.328125	8.33439	53/64	.828125	21.03442
11/32	.34375	8.73127	27/32	.84375	21.43129
23/64	.359375	9.12814	55/64	.859375	21.82817
3/8	.3750	9.52502	7/8	.8750	22.22504
25/64	.390625	9.92189	57/64	.890625	22.62192
13/32	.40625	10.31877	29/32	.90625	23.01880
27/64	.421875	10.71565	59/64	.921875	23.41567
7/16	.4375	11.11252	15/16	.9375	23.81255
29/64	.453125	11.50940	61/64	.953125	24.20942
15/32	.46875	11.90627	31/32	.96875	24.60630
31/64	.484375	12.30315	63/64	.984375	25.00318
1/2	.5000	12.70003	1	1.0000	25.40005